The Triumphant ORFFYREAN Perpetual Motion
Finally Explained!

KENNETH W. BEHRENDT

authorHOUSE

AuthorHouse™
1663 Liberty Drive
Bloomington, IN 47403
www.authorhouse.com
Phone: 1 (800) 839-8640

© 2019 Kenneth W. Behrendt. All rights reserved.

No part of this book may be reproduced, stored in a retrieval system, or transmitted by any means without the written permission of the author.

Published by AuthorHouse 02/06/2019

ISBN: 978-1-5462-7646-3 (sc)
ISBN: 978-1-5462-7645-6 (hc)
ISBN: 978-1-5462-7675-3 (e)

Library of Congress Control Number: 2019900572

Print information available on the last page.

Any people depicted in stock imagery provided by Getty Images are models, and such images are being used for illustrative purposes only.
Certain stock imagery © Getty Images.

This book is printed on acid-free paper.

Because of the dynamic nature of the Internet, any web addresses or links contained in this book may have changed since publication and may no longer be valid. The views expressed in this work are solely those of the author and do not necessarily reflect the views of the publisher, and the publisher hereby disclaims any responsibility for them.

This volume is dedicated to craftsman extraordinaire Johann Ernst Elias Bessler.

His self-sacrifice and tireless efforts to finally achieve the "impossible" have been an inspiration to generations of inventors who followed him seeking to do the same…

CONTENTS

Introduction xiii

Chapter 1 Johann Bessler's Story in a Not So Small Nutshell 1

        Figure 1(a) - The Bidirectional Merseburg Wheel Constructed in Early 1715 159

        Figure 1(b) - Johann Borlach Shows How He Thought Bessler's Merseburg Wheel Really Worked 160

        Figure 1(c) - Christian Wagner's Fake Perpetual Motion Wheel Mechanics 161

        Figure 2(a) – The Bidirectional Kassel Wheel Constructed in 1717 162

        Figure 2(b) - The Kassel Wheel Powering Various Devices. 163

        Figure 2(c) - Important Actors in the Bessler Story ... 164

        Figure 2(d) - The "Toys Page" from MT. 165

Chapter 2 I Meet Herr Doktor Orffyreus 166

Chapter 3 Is Perpetual Motion Really Possible? 211

Chapter 4 A Few Notes Before We Begin 249

Chapter 5 Details of His One-Directional, 3 Feet Diameter,
Gera Prototype Wheel............................ 261

    Figure 3(a) - Left Ascending Side of 3 Feet
              Diameter Gera Prototype Wheel in
              Clockwise Rotation................. 350

    Figure 3(b) - Right Descending Side of 3 Feet
              Diameter Gera Prototype Wheel in
              Clockwise Rotation................. 351

    Figure 4 - Details of the 3 Feet Diameter Gera
              Prototype Wheel's Weighted Lever..... 352

    Figure 5(a) - Layer 1 - Left Ascending Side -
              Spring Cords..................... 356

    Figure 5(b) - Layer 1 - Right Descending Side -
              Spring Cords..................... 357

    Figure 6(a) - Layer 2 - Left Ascending Side - Long
              Lifter Cords 358

    Figure 6(b) - Layer 2 - Right Descending Side -
              Long Lifter Cords................ 359

    Figure 7(a) - Layer 3 - Left Ascending Side - Main
              and Stop Cords................... 360

    Figure 7(b) - Layer 3 - Right Descending Side -
              Main and Stop Cords 361

    Figure 8(a) - Layer 4 - Left Ascending Side - Long
              Lifter Cords 362

    Figure 8(b) - Layer 4 - Right Descending Side -
              Long Lifter Cords................ 363

    Figure 9(a) - Layer 5 - Left Ascending Side -
              Spring Cords..................... 364

    Figure 9(b) - Layer 5 - Right Descending Side -
              Spring Cords..................... 365

    Figure 10 - More Details of the 3 Feet Diameter
              Gera Prototype Wheel's Construction .. 366

Chapter 6	The Bidirectional 12 Feet Diameter Merseburg and Kassel Wheels	368
	Figure 11(a) - 12 Feet Diameter Drum with Its Cloth Covers Removed	471
	Figure 11(b) - Details of the 12 Feet Diameter Merseburg Wheel's Lever	472
	Figure 12(a) - Merseburg Wheel Axle Pivot Pin and Bearing	474
	Figure 12(b) - Pendulum Details	475
	Figure 12(c) - More Details of the Merseburg Wheel's Construction	476
	Figure 12(d) - Method Used to Prevent Drum Tympanic Oscillation	477
	Figure 12(e) - More Details of the Kassel Wheel's Construction	478
Chapter 7	How Bessler Made the Merseburg and Kassel Wheels Bidirectional	479
	Figure 13(a) - Effect of Latching Mechanisms on Each Internal Wheel's CoG	537
	Figure 13(b) - Details of the "Cat's Claw" Gravity Latches	538
	Figure 13(c) - The Gravity Latches Installed on Merseburg Wheel Weighted Lever	539
	Figure 14 - The Gravity Latches in Operation	540
Chapter 8	Introduction to the DT Portraits	541
	Figure 15 - The Two DT Portrait Frontispieces	564
	Figure 16 - The First DT Portrait	565
	Figure 17 - The Second DT Portrait	566

Chapter 9   Clues Found in the Second DT Portrait 567

    Figure 18 - Second DT Portrait Table Items 656

    Figure 19 - Second DT Portrait Clues Give Angle
               of 9:00 Weighted Lever 657

    Figure 20 - Second DT Portrait Circular Sector
               Paper Piece Shows Cord Attachment
               Points. 658

    Figure 21(a) - Second DT Portrait Clues That
               Verify the Circular Sector Paper
               Piece's Cord Attachment Points 659

    Figure 21(b) - Some Additional Cord Clues 660

    Figure 22(a) - Second DT Portrait Clues for
               Spring Constants, Weight and Lever
               Masses, and Main Arm Lengths. 661

    Figure 22(b) - Second DT Portrait "L" Square
               Gives Gravity Latch Parameters 662

    Figure 23 - Second DT Portrait Clues about Cord
               Layers. 663

    Figure 24 - Second DT Portrait Clues Give Lever
               Shape 664

Chapter 10   Clues Found in the First DT Portrait 665

    Figure 25 - First DT Portrait Table Items. 728

    Figure 26 - First DT Portrait Clues Give Angle of
               9:00 Lever 729

    Figure 27 - First DT Portrait Hand Gives Cord
               Attachments 730

    Figure 28 - First DT Portrait Wig Gives Spring Clues. 731

    Figure 29 - First DT Portrait Clues about Weight
               and Lever Masses, Axle Center to
               Lever Pivot Center Distances, and
               Lever Lengths. 732

Figure 30 - First DT Portrait Clues about Lever
           Shape and Cord Layers 733

Chapter 11   What About the Second Gera, Draschwitz, and
             Karlshafen Wheels? 735

             Table 1 - External Parameter Values for All of
                       Bessler's Wheels 764
             Table 2 - Internal Parameter Values for All of
                       Bessler's Wheels 765

Epilogue. 767

             Figure 31 - A Permanent Magnet Motor Design
                         Using Distorted Magnetic Fields 781

About the Author. 783

INTRODUCTION

In the Introduction to my previous book, *Essays from the Edge of Science*, published in April of 2015, I lamented that the volume could not have been devoted to an early 18th century inventor named Johann Ernst Elias Bessler. As I mentioned there, he had invented and publicly exhibited a total of four rather incredible inventions during his lifetime. These consisted of wheels formed from hollow drums mounted on horizontal wooden axes. According to Bessler, their drums contained special mechanisms he had discovered that allowed them to remain continuously out of balance so they could "turn of themselves" and, while doing so, actually perform useful outside work. Having been fascinated by the possibility of such an invention since my adolescence, I very much wanted to devote an entire volume to this unique inventor and his inventions. However, there was a major obstacle preventing me from producing such a book on him at that time.

The problem was that I did not want to just produce another biography of the man and the external features of his several wheels because that had already been done adequately by other authors and, originally, even by Bessler himself in several works he published during his lifetime. Rather, I wanted to produce a work wherein I actually gave the specifics of *how* he constructed his marvelous wheels with enough detail so that they could again be duplicated! Considering that the secret perpetual motion mechanics Bessler discovered and used was never clearly revealed by him during his lifetime and that over the last three centuries literally thousands, perhaps tens of thousands of inventors had tried without any hint of success to reproduce his wheels, this, obviously, was not going to be an easy goal for me to achieve. Adding to the disparagement I faced was the fact that today's scientific orthodoxy still, almost in its entirety, considers the construction of any type of self-motive machinery as a physical impossibility.

Yet, despite these daunting obstacles and only through the most incredible and improbable occurrence of good fortune, I finally did manage to solve the mystery of Bessler's wheels with a solution that I found to be both plausible and, most importantly, in agreement with the overall properties of his wheels as described in his published works. But, that miracle could not have taken place without a lot of help.

That help took the form of a set of alphanumerical, geometrical, and mathematical clues that Bessler had carefully hidden in two frontispiece portraits that appear in his last published book which specified the physical construction parameters of the secret "perpetual motion structures" contained within the drums of his wheels. Those clues were designed to appear completely random and meaningless to the curious who merely glanced over them while reading his published work. However, to the serious reverse engineer of his inventions who was actually making a consistent and ongoing effort to reconstruct his wheels, those clues would, by the most careful of analysis, take on a totally new meaning. For such a person, the clues were like the markings on stones and trees that can enable an observant person wandering through a dark and foreboding forest to eventually make his way out of it and back to safety and sunshine again.

Another critically important factor in my eventual success was that, unlike past hopeful reverse engineers of Bessler's wheels, I had long ago given up the laborious task of handcrafting physical models in favor of using the latest computer simulation software to "construct" virtual models of his first prototype wheel that only existed as stored data on my computer's hard drive and could only be seen and run on its monitor's screen. With such assistance, I was able to compress, literally, decades of work into only a few years and to do so with a minimum of expense and the risk of injury that always exists whenever one is working with tools in a shop environment. Such an approach, however, has other risks.

Even the best of computer models is only an approximation of physical reality and all such virtual models can only be shown to be valid when they are "reduced to practice" or used to construct actual physical models. However, with practice, I quickly learned to recognize the various kinds of glitches that could occur in my virtual models that might make them *seem* to display what I call the "perpetual motion effect" or the "pm effect" when, in fact, they were not doing so. I did not want to waste my and others precious time chasing an erroneous model. I used every conceivable safeguard to assure that the virtual wheel models I was working with would have a *very* high probability of eventually leading to working physical prototypes if the day came when I finally found one that was actually displaying the pm effect.

And that day did come for me as will be described in the second chapter of this book. As had Bessler on a fateful day late in the year 1711, I experienced great excitement and pleasure when I first beheld a virtual wheel model, which was an exact duplicate of the one he first found success with, undeniably displaying the pm effect on my computer's monitor screen. That was followed by a prolonged state of calm and contentment as I realized my quest was finally over and I could now rest my mind and then move on to other mysteries that had fascinated me during my life and which I wanted to address before my allotted time on Earth had come to an end.

Apparently destiny, unpredictable as it can be, had selected me to be the first to unravel the Gordian knot of the Bessler wheel mystery after three long centuries of repeated failures by others to do so had passed. That is now sufficient enough success for me even though I must leave the final verifying physical duplication of one of his amazing wheels to someone else better equipped for the task than I currently am. My role in the Bessler saga is, apparently, merely to serve as a middleman between the long dead inventor and the craftsman of today that is in search of some guidance as to which direction he should proceed in an effort to duplicate one of Bessler's wheels. That is a task, especially for one of Bessler's 12 feet diameter wheels, which will require a considerable investment in effort, money, and time. However, for the builder of average ability and means who might want to attempt a duplication of the smallest and easiest to construct of Bessler's wheels, I have included detailed illustrations for his 3 feet diameter, one-directional prototype wheel which he constructed and first found success with in Gera, Saxony in late 1711.

Just so there will be no misunderstandings about the material I am providing in this volume, let me emphasize that it is, basically, theoretical in nature. I do not have a physical wheel model currently spinning away in my basement workshop although I certainly wish I did. I can also not guarantee that anyone attempting a duplication of the various types of

Bessler wheels to be described will find success. A failure might be due to his lack of competence in constructing handmade models, modifications that he makes to the various part parameters I've specified in the name of expediency, or, possibly, due to various construction obstacles that may arise which were not anticipated by my virtual models. There are, obviously, many risks involved all for which neither I nor the publisher will accept any liability. However, for those few individuals determined to produce a working replica of a Bessler wheel and secure a place in the history of the subject, I can say without hesitation that I have made every effort I can to provide the reader with technically accurate information and that if I was to personally attempt a physical duplication of a Bessler wheel myself, then I would be using exactly the information provided in this volume as my guide in doing so.

I was recently asked why, if I've found the secret of Johann Bessler's perpetual motion wheels, I do not immediately patent the design and try to make a fortune off of such a remarkable invention. That would then serve as my just compensation for the thousands of hours of effort it took me to find the secret method he used to do the alleged "impossible".

My answer to that is, firstly, the US Patent Office does not accept patent applications for perpetual motion machines unless the inventor has a working *physical* model that he can let them examine to determine to *their* satisfaction that it is, indeed, outputting energy in excess of what it consumes to keep itself in operation and, thereby, appears to be violating the First Law of Thermodynamics which is also known as the Law of Energy Conservation. Since I do not currently have such a physical model, any patent application in the United States I might make would immediately be rejected and simply be a waste of my time and filing fees.

But, secondly and most importantly, I would never try to obtain a patent on someone *else's* invention! The basic design I am providing in this volume is not mine, but, rather, that of Johann Bessler. Additionally, although the design's details were carefully encrypted for almost three centuries, they were, technically, "in the public domain" and freely available to anybody who could properly decode them during that time. That reality would, no doubt, be a major obstacle to anybody attempting to patent and profit off of his invention by presenting it as his own "original" design.

Indeed, Bessler several times in his writings hints that, should his invention not sell during his lifetime, which it unfortunately did not, that others would eventually find his secret and use it to replicate his wheels. Such duplication, of course, would finally allow the world to know that the wheels were genuine after all and that he was not a swindler and a charlatan as his many detractors over the centuries have tried to portray him. Such a suggestion by the inventor, in my opinion, was his way of bequeathing, upon his death, his invention to humanity. Thus, Bessler's wheels now actually belong to all of humanity and it is my sincere hope that this volume will be the beginning of the means by which that legacy will be delivered.

I'm sure that if an afterlife actually exists and Bessler's immortal soul is there now and aware of current affairs back on planet Earth, then it would give him much pleasure to know that his name will finally be cleared of the suspicion of fraud attached to it and that he will thereafter finally be granted his rightful place in the history of science. Not as a thief extracting money from gullible people with a bogus wheel propelled by trickery, but, rather, as a gifted craftsman who, having been obsessed with the vision of achieving the age old dream of perpetual motion, used all of his skills to do exactly that.

But, he wanted more than to just revolutionize the world with self-motive wheels. He also wanted to revolutionize the ethical and moral standards of humanity and produce a world where there would be more peace and prosperity for all. I view the successful reverse engineering of Bessler's wheels as just the beginning of a soon coming revolution in self-moving machinery which, along with other renewable sources of energy, will help provide for the future energy needs of humanity without adding further to the damage done to our planet by the Industrial Age's almost exclusive reliance on fossil fuels whose atmospheric carbon emissions are, even now, a growing source of global concern.

In closing this introduction, let me briefly describe the layout of this volume. I will begin by stating what this book is *not*.

Firstly, it is not an attempt to provide a very detailed biography of Johann Bessler although it is impossible to discuss the inventor and his wheels without delving into the major events of his life. This I have done in the first chapter.

Secondly, this is not a book of translations. The particular early 18th century German dialect Bessler used contains many now obsolete words and idiomatic expressions that only someone who specializes in that era's language can hope to fully comprehend. While simple sentences can often be adequately translated even by using today's online translator sites and modern German, the translation of more complex sentences can become somewhat subjective in nature because the translator is subtly influenced by assumptions he makes about the intended meanings of certain words and phrases based on his preconceived notions of how Bessler's wheels may have worked. This can be a source of much error when one begins to try to translate Bessler's few descriptions of the interior mechanics of his wheels. Only after I finally found success in unraveling the secret mechanics of the inventor's wheels, did I fully appreciate this source of error.

In most cases, when I do use an English translation of a passage from one of Bessler's writings, I will use one that British author John Collins commissioned to be done. However, on occasion, if I have found what I consider a better translation or interpretation I will use that and mention that I have done so.

Now it's time to discuss what the reader will find in this volume.

I start with a "nutshell" summary of the high points of the inventor's life. That nutshell, however, turned out to be a bit larger than I expected because the Bessler story, even when based on only a portion of the fragmentary information we have about him, still contains too many twists and turns in it to be adequately treated with only the brief chronology which I had originally intended to provide. Then I provide a brief autobiographical chapter that documents how my fascination with the Bessler story came about and what that led me to do about it. That will be followed by a discussion of the subject of perpetual motion machines, in general, and how, with certain qualifications, such a device is, indeed, possible. Next, I will devote much space to a precise description of the first 3 feet diameter, one-directional, self-starting prototype wheel with which Bessler first found success. The information provided should allow anyone of modest crafting ability and means to duplicate that wheel. Following that will be a chapter that deals with his largest 12 feet diameter, bidirectional wheels.

Finally, and perhaps most importantly, I will give a detailed accounting of the many secret clues, previously completely unsuspected, that he left in his last published book about his wheels and how, with much effort and incredibly good fortune, they were finally properly understood as they slowly led me to rediscovering the secret perpetual motion mechanics he used. That material alone almost deserves an entire volume devoted to it, but I've tried to compress it all into a few chapters so that the present book on Bessler and his marvelous self-moving wheels will be as complete as possible for a single volume treatment.

They say that many authors inevitably write the book they wished they had available when they originally began reading about a subject in an effort to learn as much about it as possible. Perhaps that is why the present volume exists. It represents the book I wish had existed many decades ago when I first learned about Bessler and his wheels. Now, a half century later it finally exists and I can only hope that, like the published material I first encountered about this unique inventor and historical figure, it will help to stimulate the imaginations of future "seekers after perpetual motion" and motivate some of them to try to physically duplicate Bessler's achievements. With such duplication, the field of "over unity" devices will gain new respect and, perhaps, eventually lead to machines capable of outputting serious amounts of energy without the need for conventional fuels. Considering the adverse environmental impact of the escalating use of fossil fuel generated power in the last few decades alone, the time for such a breakthrough could never be better.

CHAPTER 1

Johann Bessler's Story in a Not So Small Nutshell

Unfortunately, it is not possible to provide a full and very detailed story of Johann Bessler's life today. We really only have bits and pieces of it derived from his own often vague writings, various newspaper articles of his time, and some of his personal correspondence as well as the private correspondence of others in which he is mentioned. From these, however, it becomes obvious that he was a rather complex and unique individual who was influenced during his lifetime by his interactions with, literally, thousands of people from all of the social strata of the late 17th and early to mid-18th centuries (the reader will find images of a few of the more important people in his life in Figure 2(c) at the end of this chapter).

In trying to reconstruct his life story, one is, therefore, somewhat in the same position as one trying to reconstruct a movie based on a few feet of film from the original complete film. However, despite these limitations, it is possible to get a fairly accurate idea of the man and his motives from what surviving information we do possess. What follows is my attempt to provide the reader new to this remarkable inventor with a condensed, but accurate account derived from various sources. On occasion, I have filled in gaps in his story with what I consider to be plausible conjectures based on my own research in order to make the whole seem as complete and logical as possible. I remain confident that this version will prove to be a valid one as additional information about Bessler and his invention comes to light in the future.

Of his physical appearance not much can be said because there is only one known image of his face which appeared in the frontispiece of a book he published in 1719. However, from that printed drawing one can conclude that, as an adult, he was of average height and weight for his time and somewhat handsome. His refined appearance made him look more like he belonged to the educated upper classes of his day rather than to the less educated or even illiterate lower laboring class into which he was born. Possibly, this discrepancy between his physical appearance and his social status was a powerful motivating force in his life that made him seek to greatly improve his financial and social status whenever possible.

The most important aspect of the Bessler story, however, is the fact that his true legacy to humanity has, finally, been re-discovered. That legacy consists in a previously unsuspected collection of mathematical and textual clues he left that give the *precise* parameters of the parts used inside of his wheels which allowed them to produce the amazing and long sought "perpetual motion effect" or, as I often simply write it, the "pm effect". But, the full revelation of those clues must await future chapters. In this chapter, the main focus is on the inventor himself and the many interesting people he interacted with as he pursued the secret of mechanical perpetual motion and actually managed to finally obtain it. However, this beginning chapter will also provide some technical details of his wheels along the way in order to prepare the serious student of his inventions for the more technical chapters to come.

Bessler's name at birth was Elias Bessler, but later as an adult he added the names of Johann and Ernest to it for reasons which are unknown but may have been connected with his religious beliefs. He was born in the region around the town of Zittau in the hilly Kingdom of Saxony (which is now in the modern nation of Germany near its eastern border with Poland and about 130 miles southeast of Berlin) in the year 1680 although his baptism was not recorded in Zittau until Tuesday, May 6th, 1681. This was a bit unusual because, at that time, a baptism rarely took place more than three days after an infant's birth and was immediately recorded. The exact date of Bessler's birth is not known, but it could have been as early as November of 1680.

His father, a common day laborer, was named Andreas Bessler and he married Bessler's mother, Maria Maucke on February 11th of 1680 in the town of Zittau. Bessler had a younger brother, Gottfried, who was born on May 26 of 1688 in Zittau. There were also three other children that either died young or had been stillborn.

By today's standards, Bessler came from what might be considered a lower middle class working family that earned its living by performing various manual labor jobs that required skills in carpentry, construction, plumbing, etc. Normally, the children of such families received little if any formal education back then and, often, a male child would be apprenticed to a certain tradesman and then practice that trade after he had gained sufficient experience and knowledge of it or he might become a professional soldier and hope to move up through the military ranks as far as his abilities would allow. Girls usually received no formal education other than the domestic skills that they learned from their mothers. Their primary function was to bear children and take care of them and the home.

However, Bessler's father may have done work for someone who was impressed by the assistance given by the young Bessler on the job and, as a result, managed to get him enrolled in a "gymnasium" or high school in Zittau so he could complete what today would be the equivalent of a rather comprehensive secondary school education. That school was run by a headmaster named Christian Weise who specialized in preparing young males for civil service type jobs in the local towns. Weise was so impressed by young Bessler's potential that he took personal charge of his education and pressured him to learn as much as he could during his formal training at the school. Bessler had no problems complying with that demand. Weise was also a prolific playwright and many of his plays criticized the rich while being sympathetic to the common people. Weise's influence had an impact on Bessler that would stay with him for the rest of his life.

After he completed his education around the age of 15 years or so, Bessler wandered extensively throughout the Europe of his day (even visiting universities as far west as Ireland) in search of adventure, knowledge, and wealth as he added many skills to his abilities along the way. He even tried unsuccessfully to do treasure hunting in an effort to quickly improve his financial situation and still somewhat lowly social status. He had an insatiable thirst for knowledge and eventually did studies in fields as diverse as astronomy, calendar making, clockmaking, copperplate engraving, carpentry, glass blowing, gunsmithing, mathematics, mechanics, music, organ making, painting, sculpture, smelting, surveying and theology. For hobbies, he enjoyed such activities as dancing, dueling, horseback riding, and playing different types of musical instruments. He could also read and write Latin which was unheard of for someone who came from his level of the social strata.

During his travels he would, upon arriving in a new region, immediately inquire where he could find some temporary employment. He would take whatever was available and offered and, as his repertoire of skills grew, he found it easier to obtain work and improved wages. He presented himself as a diligent, jack of all trades and perhaps accumulated letters of recommendation which he carried with him and which served as a resume for a future employer to read. He always made sure he made his employers happy so that he could get a good recommendation from them.

In one of his youthful adventures, he accompanied a rich older gentleman to Italy and while there he visited a monastery and somehow found himself in the kitchen (perhaps he parted company with his traveling companion and visited the monastery because they were known for offering meals and temporary lodging to strangers). While in the kitchen Bessler noticed they had a most unusual roasting spit in their hearth. It used a clockwork mechanism that would slowly rotate a piece of meat on the spit rod as either a weight on the side of the device slowly descended or, possibly, as a large mainspring slowly unwound itself. In either case, the device would have to occasionally be powered up manually by using of a detachable metal crank.

Bessler had never seen such a device and it fascinated the young man so much that he could not get the image of it out of his mind. As he thought about its construction, he wondered if it might be possible to design a device that could rotate the spit *without* needing to have its weight raised or spring rewound repeatedly. In other words, he wondered if the device's parts could somehow be arranged so that they would self-rotate without needing occasional outside assistance. He reasoned that if that was possible, then the device should continue to operate forever.

With experiences like this, it wasn't long before Bessler found himself deeply drawn into the science of mechanics and he even briefly became an apprentice watch and clockmaker and used these skills whenever he could. From his study of the then available literature on the subject of mechanics, he eventually became aware of the history of humanity's long and futile search for a working perpetual motion wheel design and began to fill a notebook up with various possible designs for such a device some of which he would actually find the time to construct models of and test.

After years of effort and many attempts with devices of various types, however, he found himself no closer to success. But he continued to pursue a working design because he was convinced that achieving it would be the ultimate key to finding the fame and fortune he so desperately wanted and believed he deserved. He knew that a working perpetual motion wheel was something that would attract the attention of everyone, rich and poor alike, and he hoped to be able to sell the invention for a great sum of money to the richest who were interested in it. Until then, all he could do was continue his search for such a machine and, most importantly, earn sufficient income while doing so to keep himself fed and sheltered.

One day after he had deserted some army involved in a local conflict that he had been forced to serve by attending to their wounded as a medic, he saved the life of an alchemist that had fallen into a pond and was in danger of drowning. That man then rewarded Bessler by teaching him how to prepare and prescribe various types of balms, elixirs, lotions, ointments, pills, and potions. Unlike the medicinal treatments Bessler had previously tried to prepare and sell, the ones the alchemist taught him to make were far more effective. Such remedies were in constant demand by the common people for the many ailments and injuries they suffered due to the hard work they had to perform daily and even the upper classes had health issues and could benefit from quality medical compounds.

One of his best sellers for upper class gentlemen who had engaged in activities that violated the moral standards of the day was an expensive lotion that could be used to treat the skin lesions that would appear when one contracted a syphilis infection. Of course that treatment would not work to eliminate the infection deeper inside of the body which would still make the victim infectious to others. Eliminating that would require some medicine be taken internally and, perhaps, Bessler even had some herbal remedy he could supply that was effective in the pre-antibiotic 18th century.

As he began to derive some reliable income from selling his newly formulated medications, Bessler decided to take the professional Latin name of "Orffyreus" which he derived by arranging the 26 letters of the German alphabet in a circle and then selecting the letters that were diametrically opposite to the letters in his surname "Bessler". This yielded the word "Orffyre" which he then Latinized by adding the ending "-us". Thereafter he referred to himself as "Doctor Orffyreus" and was, finally, beginning to attain some of the social standing he so much desired. Apparently, in the early 18th century it was far easier to become a medical doctor than it is nowadays and it was possible then for a person with no formal medical training to give himself the title of "Doctor" after he began prescribing his own homemade remedies to people or performing minor surgeries. No doubt many people were harmed as a result of such lax standards, but Bessler's patients generally seemed to benefit from his medications thanks to the grateful alchemist who had trained him in their preparation.

In one early newspaper account of Bessler, his professional name was given as "Orpheus" because, possibly, the reporter misspelled it due to it sounding similar to the name Bessler came up with. Yet, I don't think that similarity was coincidental. I believe that Bessler purposely chose "Orffyreus" because it did sound like "Orpheus" who was a legendary musician, poet, and prophet in ancient Greek religion and myth.

There are various stories about the original Orpheus which all tend to involve his ability to produce music of a divine nature with a magic lyre he possessed. That music had the power to mesmerize birds, fish, wild beasts, and even make trees and inanimate rocks dance while also changing the course of rivers! One story described how he used his magic lyre in the underworld of Tartarus to temporarily stop an *ever* spinning, fiery wheel to which a king named Ixion had been bound by Zeus as punishment for having tried to seduce the latter's wife, Hera, during the king's visit to Olympus, the home of the Greek gods and goddesses. Another legend of Orpheus credits him with giving humanity the gift of medicine while another tells of how he journeyed down into the underworld to bring back his dead wife Eurydice's soul so she could live again.

These were all details that would have resonated with Bessler who was also very interested in music and realized that some of his medicinal compounds were actually saving the lives of people and, therefore, helping to keep their souls in their bodies and out of a premature arrival in the afterlife. Bessler also desperately wanted to produce a genuine working perpetual motion machine and, in his mind, that was equivalent to bestowing life to inanimate matter just as Orpheus did when his divine music made stones dance. Indeed, in one of Bessler's books, he refers to Orpheus and the skill he used in playing his magical lyre which indicates Bessler's admiration for the real person in ancient times whose abilities inspired the myths surrounding Orpheus. It also indicates Bessler's admiration for those who got results by highly developing and using their manual skills as would be required to play a musical instrument like a lyre.

In the region around Prague in the year 1703 and at the age of 23 Bessler had become so religious that he actually earned a living as an itinerant preacher who was espousing his own interpretations of the Christian scriptures. He may have even worn clerical garb that he had purchased even though, technically, he had never been ordained as a priest by any recognized school of theology. Most likely he was hoping to recruit followers into a new religion that he would create and probably used the appearance of being an ordained priest to get people to listen to his preaching. Unfortunately, very few of them seemed interested. This activity, however, did bring him into contact with a priest and rabbi and he absorbed much of their religious philosophies and incorporated them into his own. From the priest he learned the significance of various numbers used in the Bible and from the rabbi he learned of the numerology and codes used in the Cabbala which is the ancient Jewish mystical interpretation of the Bible. Bessler thereafter became fascinated with such things as secret knowledge and the various coding systems used to protect it from being revealed to outsiders. He intended to use what he learned from them to conceal the secrets of any perpetual motion wheel he managed to invent.

However, his progress with producing such a machine was not that impressive. He had already spent several years pursuing designs based on various principles that he hoped would work and still had found no success. Yet, despite his repeated failures he was convinced it could be done and that, with enough effort, he would eventually find a design that worked.

He soon realized that he disliked the crowds and even the smell of larger cities and preferred the quieter life found in small towns. Indeed, he considered city living to be unhealthy due to the congestion of the people living there. He only spent time in larger cities when it was absolutely necessary for his medical career or his pursuit of the "Holy Grail of Mechanics" which was a working perpetual motion machine.

Wherever he stayed, it was not long before he was passing what little idle time he had working on some perpetual motion machine design or another that had occurred to him. Although he preferred simple imbalanced wheel designs, on occasion he would construct small models of devices that utilized hydrostatic and pneumatic principles to keep a wheel or axle in rotation. Nothing worked, however. So obsessed he became with his various models that he would often spend what money he had earned on parts for a machine he was building rather than on food!

But, Bessler always seemed to stay out of debt and even when he did not privately treat patients, he could make money by selling his homemade remedies to local doctors. These were mostly standard herbal remedies and ointments that these doctors all used to treat minor cuts and prevent them from becoming infected and Bessler would manufacture supplies of them for the doctors at an attractive price. When word got around that the young man produced quality products, he found a steady demand for them. In those days there were no corner drugstores and patients were usually supplied with their medications directly from their doctors along with instructions on how to properly use them. Also, most doctors would visit the patients in their homes and usually had no offices for patients to visit as they do today.

Bessler would not hesitate to prescribe a medicinal compound for a patient if consulted on his case by a local physician and even privately for a patient who could not afford the services of a local doctor, but still needed a treatment. In many cases, he probably supplied remedies to the poor for free and then passed the cost on to wealthier patients without them even being aware of it. As a Christian, he felt it his moral duty to help those less fortunate than himself whenever possible. One of his most eagerly sought remedies was a treatment for the "French pox" which was first recorded in Europe after French troops invaded Italy during the Italian War of 1494 to 1498. There's no way of knowing if he could actually cure the infection or if his compounded medicinal just eased the initial acute symptoms. This treatment was probably one of his more expensive ones which he sold to members of the upper classes who had contracted the disease while abroad or by visiting local prostitutes. Its effectiveness and the steady demand for it assured that Bessler was never completely without some money.

To further his knowledge of the human body and his income, he studied human anatomy and may even have developed a side business of preparing skeletons for display. The doctors of his time, many self educated, would pay well for such a specimen which they then used to aid their understanding of anatomy and how to set broken bones and such. Bessler would pay for the corpse of some old vagrant which had been found, dismember it, and then boil the pieces in water to remove the tissues from the bones. The clean white bones were then dried and reassembled into a complete skeleton with the joints held together with wire. This was grisly work at best, especially as a corpse started to deteriorate, but to Bessler the body was really irrelevant after a person's soul had left it upon death.

As his youthful need to wander waned, he found himself returning to the region near where he was born and for a while during the year 1709 he and his younger brother, Gottfried, worked as apprentices for a relative who was an organ maker. They would install and repair church organs in various towns and Bessler was fascinated by the mechanical complexity of the organs of his day. In later life, we would credit the skills he learned in this temporary occupation with allowing him to finally be able to construct genuine working perpetual motion wheels. Today's organs are mostly electronic, but the ones Bessler worked on were mechanical marvels that used cords, levers, springs, and valves in intricate arrays to produce their unique tonal music.

Also, coincidentally, in 1709 an Italian inventor named Bartolomeo Cristofori introduced a new type of musical instrument that was the forerunner of the modern piano. Previously, there was a keyboard instrument known as a "harpsichord" which, when one of its keys was pressed, would use the tip of a quill to pluck a single cord of a horizontally mounted harp. Cristofori came up with the idea of using the keys to cause a carefully counterbalanced hammer to fly up, hit a cord, and then drop back down into its resting place beneath the cords. Bessler learned of the new mechanism and was fascinated by how counterbalancing allowed a small force applied to the end of a key to cause a much more massive hammer to so quickly rise inside of the instrument.

In the year 1711 at the age of about 30, Bessler was called to the town of Annaberg-Bucholz to treat the youngest daughter, Barbara Elisabeth Schuhmann, of the mayor, Dr. Christian Schuhmann. The young woman had become the victim of a strange malady that caused her to scream and thrash about uncontrollably either in bed or on the floor of her home and to suffer from strange hallucinations. Besides being the mayor, Schuhmann was also the town's head physician, but his medical skills were unable to help his daughter. The girl's mother (who, incidentally, was also named Barbara) feared that if something was not done soon to help her daughter, then she might actually die.

Bessler arrived and, after spending an hour with Barbara, all of her symptoms suddenly vanished! Most likely, the young woman had consumed bread made with rye flour that had been contaminated with a fungus called ergot that produces alkaloids capable of inducing hallucinations and spasms. Bessler would have been familiar with this ailment and cured the woman by administering an elixir containing a tranquilizing plant extract and then simply making sure she did not have any additional servings of the bread made from the contaminated rye. Her parents would then have been overjoyed to see her make such a rapid recovery and have made sure that the contaminated bread was discarded.

In Bessler's mind, the young woman would have seemed like another Eurydice who he, like the Orpheus of Greek mythology, had led out of a dismal underworld and back into the world of the fully living again. Needless to say, as Bessler continued to make follow up visits to see how his patient was recovering, he found himself falling deeper and deeper in love with her and they were soon engaged. Eventually, Bessler would marry Barbara Elisabeth Schuhmann and collect a nice dowry which would do much to improve his financial and social standing. He would also have four children with her, but only one daughter would survive and eventually marry a widowed preacher in 1735.

Word of his miraculous curing of Barbara's bizarre condition got around and, as a result, his fame as a healer began to get him more and more patients eager for one of his cures whether it was an elixir to treat the body, an ointment for a wound, or his prayers to treat a troubled mind or soul.

During the fall of 1711 in a cramped workshop in which he had been living in the town of Gera (which is about 120 miles west of Zittau and

52 miles west-northwest of the town of Annaberg-Bucholz where Barbara Elisabeth Schuhmann lived), the inventor began working on a particular design for an imbalanced wheel that he seemed to have been led to by the sum of all of his previous failed attempts. He instinctively knew that it had to work, but it seemed no matter how many modifications he made to it, he just could not get it to run continuously. Something, probably just a minor detail, was wrong, but no matter how much time he spent staring at his latest model wheel and no matter how many calculations he made as to where the center of gravity of its collection of little lead weight carrying levers was located and should remain during the rotation of its drum, the wheel just stubbornly refused to keep its center of gravity on the descending side of the axle as its drum rotated. As soon as he released the wheel during a test, its center of gravity would always rotate down and around with the drum until it was located directly under the center axis of the axle and then, after oscillating a little, the wheel would come to a complete stop.

These repeated failures were beginning to create a state of deep despair in Bessler. He even contemplated declaring his whole pursuit a total waste of time and money, quitting the pursuit, and then just concentrating fully on his career in medicine. He had already labored on the project for about a decade during which time he had found no success whatsoever despite having constructed a hundred or more machines of various types and spending a considerable amount of money for parts in the process. He was mentally and physically exhausted from his failed pursuit of the "Holy Grail of Mechanics" and ready for a long and, possibly, permanent vacation from it all.

That would, of course, have made his new fiancé very happy because, although she loved Bessler very much, Barbara was concerned about his obsession with perpetual motion. Whenever they talked, it seemed that it was not more than a minute or so before the topic changed to the problems he was having with his latest model wheel and how he would proceed once he finally got it working. She had heard it all many times over and really just wanted Bessler to marry her and then "settle down" and apply himself to the field of medicine where he was finding success.

To be sure, during any of the discussions he had with Barbara about his plans for much larger and more powerful wheels, he would have downplayed the expenses involved with his "hobby". While a small table top sized model was relatively inexpensive to construct and modify, a larger edition would require parts that needed to be specially fabricated and would be expensive to obtain. Although Bessler was a knowledgeable craftsman, he was, mainly, a carpenter and a mechanic and not a blacksmith even though he may have had some minor experience with their difficult craft. He could obtain stock pieces of white oak wood rather inexpensively and then cut and assemble them into a larger wheel's axle and drum. But, such things as the metal screws, bolts, pins, springs, and cast lead weights needed for the wheel's internal mechanics would have to be purchased from others who were better equipped and more experienced to manufacture them. And, any parts that had to be custom made would raise construction costs even further.

However, rather than just quitting, Bessler decided to make one last effort to get his latest model wheel working. He redid all of his calculations for it and, to add a measure of good luck, he actually incorporated various numerologically significant numbers into its various parameters. Then, being very religious, he turned to God for divine help. He started to believe that he would only find success if he actually received a miracle in response to his prayers. And pray he did several times a day as he worked on what, if it failed, would be his last attempt.

His last effort to achieve success began to take its toll on the young inventor. He was not eating properly and, often, was tired most of the day because he had been up too late working on his model wheel by lamp light. Then, finally, the miracle he sought was delivered to him by a means occasionally used in the Bible!

Sometime during the late fall of 1711, after a particularly tiring day, Bessler fell into a deep sleep one night and had an unusual dream that inspired him to make a small modification to the one-directional, 3 feet diameter wooden prototype model wheel he had been working on which he had not considered making previously. He may have been starting to suffer from his usual seasonal depression around this time of year, but it quickly lifted as a result of this dream which he described as invigorating. This dream was so vivid that Bessler believed that it was actually a vision from God and a message to him commanding him to continue working on his latest wheel design. He dared not ignore what this dream was telling him to do.

Because of this unusual dream, Bessler immediately set to work making the modifications he thought the dream was directing him to make. Then a month or so later in late 1711, the time came to test the effectiveness of the modifications that had been suggested to him in his usually vivid dream. He released the drum of his table top model wheel and it immediately began turning on its own and continued to do so until he grasped its axle and caused it to stop! When he released the axle again, it then immediately began turning again. He repeated this several times and then just let the wheel run for several minutes until he finally stopped it for fear that the wear and tear on its delicate parts might cause it to experience a critical failure that would require repair.

In order to stop his model wheel from running continuously, Bessler had placed bolts into its upright axle supports which could be tightened so that they pressed downward and against the axle pivot pins and acted as two brakes on the ends of the axle. Whenever the bolts were both loosened, the model wheel would immediately begin rotating and then, after manually stopping the wheel's rotation by grasping its axle, he would again tighten these bolts so as to lock the axle in place.

He could not believe his eyes when he first beheld the sight of a genuine working imbalanced type perpetual motion wheel spinning rapidly before him. But, there it was…success at long last! After a decade of effort and over one hundred attempts, he had finally achieved what most in the scientific community of his day had declared a physical impossibility: a wheel which could continuously rotate *without* the need to be wound up like a clock movement or which was dependent upon the external motion of the wind or of a flow of water. Through incredible personal sacrifice almost to the point of endangering his physical and mental health and with what he considered divine help, he had finally uncovered the secret of mechanical perpetual motion, a secret that had eluded thousands of inventors and craftsmen over the centuries including some of the leading scientists and intellectuals of their days.

At the sight of this, the young inventor fell to his knees in his small workshop and immediately began offering prayers of gratitude to God for the miracle that He delivered to his humble servant. For the rest of his life, Bessler would say that if it was not for a literal divine intervention, he would never have found success. This miracle also convinced Bessler that God approved of his future plans regarding religion and those will be explained a bit later on in this retelling of the fascinating Bessler story.

He intended this first prototype wheel to be used to give his friends and certain important and wealthy visitors a quick demonstration in the hope that the latter would supply him with the funds for the larger and far more powerful wheels he wanted to make. Also, this small prototype incorporated all of the important geometry in its various parts that would always be used no matter how large he made a wheel.

However, so fearful was he that someone might break into his little Gera workshop in his absence and steal this first working wheel, he made a practice of dissembling it by removing all of the components within its small drum and hiding them in different places around his workshop. When he knew he needed it for a demonstration, he could reassemble it in less than a half an hour for the occasion. Then, once that was over with, it was quickly dissembled again and its parts scattered in different locations that only Bessler knew about. If a thief broke into his workshop when the inventor was not there, then the largest intact piece he would have found was the prototype's empty drum which, by itself, would be of little value in determining the wheel's secret imbalanced pm wheel mechanics.

Bessler's concerns proved to be well founded because, only a few weeks after he had found the design for his working Gera prototype wheel that worked, his workshop was actually burglarized, but, aside from stealing some of his tools, the thief removed no parts of his wheel.

Also, shortly after his discovery, Bessler married Barbara Schuhmann and, needing a larger place to live, he, his new wife, and her young maid, Anne Rosine Mauersbergerin (who had formerly been the maid of Barbara's mother) moved in as guests with Bessler's cousin Herr Detter Langen and his wife who lived in another part of the town of Gera. But, their quarters there were still too cramped and did not provide Bessler with a secure place to set up his workshop. After only a month or so as guests of his cousin, Bessler rented a house from a Herr Richter that was located on Niclaus Hill (where today stands St. Saviour's Church which is a Lutheran church) and moved into it. Finally, he had enough of a steady income from his medical practice and time to raise a family while continuing to work with his finally successful design for a working perpetual motion wheel.

Bessler's first working wheel, his 3 feet diameter, one-directional Gera prototype wheel, was really just a well made toy that could do little more than keep itself in motion. Its ability to do "external" work was limited almost entirely to overcoming its thin rotating drum's aerodynamic drag as it spun in excess of 60 revolutions per minute.

Based on various carefully hidden clues he left us in one of his books, I eventually determined that his Gera prototype wheel used eight small wooden weighted levers each of which had an ingot shaped lead weighted attached to it.

The reader will learn in the following chapters that each of the weighted levers of the Gera prototype wheel (and all of Bessler's later wheels as well) consisted of a parallel pair of wooden pieces *each* of which had three arms. Thus, each weighted lever actually consisted of three parallel pairs of wooden arms that were arranged in precise angular relationships to each other. One of those parallel pairs of wooden arms, which I refer to as the "main arms", held between its ends by two small screws the center axis of a small lead ingot end weight that, together with its two little attachment screws, weighted exactly 6 ounces or 0.375 pound. *Without* its attached 6 ounce, lead ingot end weight and two screws, the remainder of a little wooden lever in the Gera prototype wheel weighed exactly 1.25 ounces or 0.078125 pound. Consequently, after a lead ingot end weight and its two screws were attached between its odd shaped wooden lever's main arms in the Gera prototype wheel, the assembled lever had a total mass of 7.25 ounces or 0.453125 pound.

Hereafter, I shall refer to a weighted lever from which its lead ingot end weight and its attachment screws have been removed as an *unweighted* lever and the reader should always keep this nomenclature in mind in order to prevent confusion as he proceeds through the remainder of this volume. This term shall also be applied to a lever Bessler used in any of the larger diameter wheels that he constructed *after* his Gera prototype wheel whenever that lever does not have its lead ingot end weight and its attachment screws or its three cylindrical lead end weights and their attachment bolts secured between the ends of the parallel pair of main arms of its lever. Thus, one sees that the mass of a *weighted* lever is always the *sum* of the masses of its lead end weight(s) and the attachment screws or bolts which attached it to an unweighted lever *and* the mass of the remaining unweighted lever's three parallel pairs of wooden arms and anything else attached between those pairs of arms (the reader will learn about those extra attachments between an unweighted lever's parallel pairs of arms later in the chapter devoted exclusively to the construction details of the one-direction, 3 feet diameter, Gera prototype wheel).

In each of the Gera prototype wheel's weighted levers, the lead ingot end weight's center axis, which also contained the center of gravity of the weight, was located at an exact distance of 3.5 inches away from the center axis of the steel pivot pin about which the weighted lever could rotate. The two ends of this pivot pin were tightly press fitted into two very slightly smaller diameter holes which had been drilled into a parallel pair of wooden radial frame pieces of the prototype wheel's drum and that made the lever's inserted pin a stationary part of the drum. The drum contained a total of eight of these fixed pivot pins and each pin had a single weighted lever that could rotate, within limits, about the pin. Each steel pivot pin was only 0.0625 inch = 1/16 inch in diameter.

The center axes of the eight steel pivot pins of the 3 feet diameter Gera prototype wheel's drum were, in turn, each located exactly 14 inches from the center axis of the wheel's small wooden axle which was 1.5 inches in diameter x 12 inches = 1 foot in length. The drum's eight pivot pins were arranged so that their center axes were located at the vertices of an imaginary octagon that surrounded the axle. The pivot pin for each weighted lever had to be pressed through a wooden radial frame piece on one side of the drum, was then more easily pushed through the two *very* slightly larger diameter pivot holes of the weighted lever's two wooden parallel 3 arm side pieces, and was finally pressed into the slightly smaller diameter hole in the other parallel wooden radial frame piece on the other side of the drum.

Amazingly, because of Bessler's work with small clock movement parts, the drum of the one-directional Gera prototype wheel was only 4 inches thick and, of course, all of the wheel's eight steel pivot pins were also each 4 inches in length since each of them extended from the outside facing surface of one of the drum's wooden radial frame pieces to the outside facing surface of that radial frame piece's parallel partner on the opposite face of the drum.

Each of the Gera prototype wheel's eight weighted levers was also attached by two cords to two small helical steel extension springs whose other ends were attached by small metal hooks to a 0.0625 inch = 1/16 inch diameter to a steel spring hook attachment anchor pin that passed through two of the drum's parallel pair of *outer* octagonal frame pieces that also served as braces between two adjacent parallel pairs of radial frame pieces. The two ends of each anchor pin were, like the two ends of each weighted lever's pivot pin, tightly press fitted into the midpoint of the length of each of the two parallel outer octagonal frame pieces. This then placed the center axes of the drum's eight spring hook attachment anchor pins at an exact distance of 16.5 inches from the center axis of the prototype drum's axle.

If the reader has had difficulty in following the description of the 3 feet diameter Gera prototype wheel's weighted levers just given and how they were placed inside of the drum, then I can assure him that he will have a complete understanding of them by the time he has completed studying the material in Chapter 5. For the purpose of this chapter, however, one need only know that each of the eight weighted levers found in one of Bessler's one-directional, self starting wheels had a single lead ingot end weight attached to it that was tightly held between the ends of a parallel pair of the lever's wooden main arms by two small steel attachment screws so that the center axes of the screws, which also contained the center of gravity of the lead ingot weight, was parallel to the center axis of the lever's steel pivot pin and located at a precise distance from it. As a consequence of this, the central axes of the all of the lead ingot end weight steel attachment screws, of all of the weighted lever steel pivot pins, and of the single wooden axle of the drum itself were *all* parallel to each other!

Using this basic preliminary understanding of the weighted levers used in Bessler's 3 feet diameter, one-directional Gera prototype wheel, we can now very briefly describe how they moved during drum rotation so that their collective center of gravity would always remain on the descending side of the wheel (or more accurately, on the descending side of the axle since the center of gravity of the eight weighted levers was located within a fraction of an inch of the center axis of the axle and, thus, actually located *within* the body of the 1.5 inch diameter axle) so that a chronic state of imbalance would be maintained as the wheel's drum and its attached axle rotated.

As a weighted lever's pivot pin's center axis, during clockwise drum rotation, moved around the axle from the drum's 6:00 position to its 9:00 position on its ascending side, the lever's pair of parallel main arms and their attached lead ingot end weight would, due to the pull of gravity, slowly rotate counterclockwise away from the lever's wooden radial stop piece that was located on and physically attached to the pair of parallel radial drum frame pieces that held the lever's steel pivot pin and would thereby try to assume a more vertical orientation.

At the 9:00 position of the drum, the two little steel extension springs attached to another of the three parallel pairs of arms of the weighted lever would achieve their maximum stretch which was only 0.5 inch. Then, as the weighted lever's pivot pin's center axis rotated past the drum's 9:00 position and eventually onward to its 4:30 position located on the descending side of the drum, the weighted lever would change the direction it was rotating about its pivot pin. For this part of the lever's journey around the drum's axle, the pair of parallel main arms holding the 6 ounce lead ingot end weight would begin rotating clockwise and back toward its radial stop again until final contact with the stop piece was made shortly after the weighted lever's pivot pin's center axis had passed the 3:00 position of the drum. At this location, the two springs attached to the weighted lever would also have returned to their normal unstretched lengths which were 1 inch at a minimum.

The net effect, as all of the one-directional Gera prototype wheel's eight weighted levers rotated in their own directions and rates about their particular pivot pins during drum rotation, was that their collective center of gravity would always remain on one side of the center axis of the axle which was on its descending side. This imbalance then produced a constant torque that accelerated the prototype wheel. The axle torque was always greatest when the prototype wheel was stationary and its constant power output, just a mere 0.2 watts, was greatest immediately after its self-starting had commenced.

Unlike the perpetual motion wheels of other inventors, the imbalance in Bessler's prototype model wheel was maintained as the wheel began to rotate and accelerate and it experienced a constant driving torque as a result. However, the wheel's rotation rate would not continue to increase until the drum and its internal parts flew apart due to unsustainable centrifugal forces acting on them.

As the wheel sped up, its weights and levers would naturally experience increasing centrifugal forces acting on them. These forces always pushed a lever's center of gravity away from the axle and thereby tended to delay the counterclockwise rotations of the weighted levers around their pivot pins as the center axes of these pivot pins moved between the drum's 6:00 and 9:00 positions on its ascending side of the drum and, as wheel speed increased, the resulting increasing time delay then caused the center of gravity of all of the eight weighted levers to slowly rotate clockwise about the axle's center axis. If this process continued long enough, the center of gravity of the eight weighted levers would eventually be located directly below the center axis of the axle at a point Bessler referred to as the "*punctum quietus*" or, in English, "stopping point" where the center of gravity could produce no torque at all to continue accelerating the wheel.

However, in practice, the center of gravity of the eight weighted levers never actually reached the *punctum quietus*. As it tried to do that, the driving torque applied to the wheel's axle would slowly decrease until a point was reached at which there was just enough driving torque remaining to exactly counter balance the impeding torques acting on the axle which were due to the aerodynamic drags acting on its swiftly moving drum and swinging internal levers and also by the various bearing drags created as its various parts rotated so that the wheel would then maintain a constant high rate of rotation. That terminal rotation rate for the Gera prototype wheel was in excess of 60 rotations per minute and could have been as high as 90 rotations per minute!

Contrary to what has been generally assumed about Bessler's one-directional imbalanced wheel designs, their levers' lead end weights never actually made physical contacts with wooden stops that were attached directly to the inner curving surface of their drums' outer rim walls. In reality, it was only two of the leading surfaces of a parallel pair of main arms of a weighted lever that made contact with a wooden stop piece that was attached to the two parallel radial frame pieces of the drum which held that weighted lever's pivot pin. I refer to these wooden stops as "radial stop pieces".

Also, contrary to what many believe, the diminutive weighted levers inside of Bessler's one-directional, 3 feet diameter prototype wheel were not physically isolated from each other. Rather, their swinging motions about their respective pivot pins as the center axes of these orbited the drum's axle were carefully coordinated with each other by a collection of cords that interconnected adjacent levers in a *very* precise way. Indeed, it has been the failure of *all* later inventors to properly coordinate the motions of the levers within their various imbalanced wheel designs which is mainly responsible for them failing to maintain their imbalances during rotation. This was the same problem that Bessler encountered and it was only his unrelenting effort over a decade to solve it which allowed him to finally achieve the success denied many thousands of others.

Bessler had come up with the idea of placing his prototype design's various sets of coordinating cords into five well defined "layers" that subdivided the cylindrical volume of space inside of the drum between the two planes defined by the *inward* facing surfaces of the Gera prototype wheel's eight weighed levers' sixteen 3 arm side pieces. This arrangement was critically necessary in order to prevent the cords from rubbing together during drum rotation and eventually fraying and breaking as a result.

The ends of the cords of each set had to be attached to particular coordinating cord hook attachment pins inside of a 3 arm weighted lever whose center axes were at specific distances from the center axis of a lever's pivot pin. To attach these cords, Bessler affixed tiny metal hooks to the ends of the cords which could then be quickly and easily slipped over the thin metal hook attachment pins which had been inserted into a weighted lever's parallel pairs of arms by simply being pressed through a slightly smaller hole drilled into the wood of one arm until the end of the pin reached another predrilled hole in the other corresponding parallel arm and penetrated it. Once attached to its assigned hook attachment pin in a lever's pair of arms, the hook would stay securely in place even though it and the cord tied to it were free to rotate a little around the hook attachment pin and might be under varying tension during wheel rotation. To reduce friction from the metal to metal contact between the steel coordinating cord hooks and their steel hook attachment pins, he simply lubricated them with droplets of olive oil.

If a coordinating cord broke while the Gera prototype wheel was running and caused it to stop, Bessler would simply open one of the cloth coverings that he had attached to both sides of the drum to conceal its interior mechanics and then replace the two broken pieces of cord and their metal attachment hooks with a new single whole cord with its two attached hooks of the correct type that he had previously made for repair purposes. An occasional repair could be made in a few minutes and he kept a supply of spare coordinating cords, each with their two end hooks already attached, on hand for just such emergencies. While there were five cord layers inside of the Gera prototype wheel's one-direction turning drum, they contained only four different types of coordinating cords with their attached hooks and the reason for this will become clear as the reader later studies the material of Chapter 5 which delves into the construction of Bessler's 3 feet diameter, one-directional Gera prototype wheel in far greater detail than is given here.

It had taken Bessler many hours of mathematical calculating and even a divine inspiration in the form of a vivid dream coupled with all of his learned skills in clock and organ making to finally be able to produce this wonderful little toy wheel. Indeed, it also required an enormous amount of sheer luck because countless others over the centuries and even millennia had tried to achieve what Bessler had and failed even after a lifetime of effort (however, most likely, a handful of them did manage to find other designs that also maintained imbalance during rotation, but those working designs have been "lost to history"). In his writings, Bessler said that, aside from the help he was convinced he had received from God, he believed another reason that he finally found success was because he had simply made more attempts to construct such a device that others had.

Obviously, the extraordinary effort he made, aside from his phenomenal luck, played a critical role in his success. What he had learned as an assistant organ maker was also invaluable in helping him come up with the method of arranging his prototype wheel's forty coordinating cords into the five distinct layers that kept them neatly separated from each other during wheel rotation.

The 3 feet diameter Gera prototype wheel's axle and its attached drum with its collection of eight wooden weighed levers (each holding a single six ounce lead ingot end weight with two steel screws), 40 coordinating cords, 16 steel extension springs, and 96 little metal attachment hooks (80 for the cords and another 16 to attach one end of each spring to its place on a spring hook attachment anchor pin in the drum) only weighed a little over six pounds. The wooden base that supported the drum's axle end pivots also weighted about six pounds and like the drum frame pieces and axle was most likely made from European white oak which is a very strong wood. When placed on a tabletop and allowed to self start and slowly begin rotating until it reached its full terminal speed, it was a marvel to behold and had a sort of mesmerizing effect on those few to whom Bessler demonstrated it.

To conceal its internal mechanics, Bessler covered the open sides of his first working model wheel with tightly stretched sheets of linen that had been dyed a dark color so that there would be no revealing shadows of the internal levers cast on the cloth covering of one side of the drum when a bright light source such as an oil lamp or sunny window was illuminating the other side of the drum. These linen sheets had to be tacked to the drum's thin wooden radial frame pieces and, if some part failed inside of the 3 feet diameter wheel's drum, he could remove the tacks, peel a portion of the linen sheet away from one side of the drum, and then locate and replace the part. In almost all cases, the part that needed to be replaced would have been either a broken coordinating cord or spring.

Despite its very low start up power output of about 0.2 watt, his Gera prototype wheel's ability to mysteriously self-start when released and to continuously accelerate until it reached a maximum terminal rotation rate of more than one full rotation per second was rather amazing and Bessler would use this prototype wheel for the next fifteen years of his life to give his personal friends and private visitors a quick demonstration of what he had achieved. But, this model wheel's power output would certainly not impress a businessman looking for a new source of energy to, for example, quickly pump water out of a flooded mine or turn the huge stone grinding wheels of a grain mill. Bessler needed to construct a larger edition of his wheel that would begin to perform some impressive feats during public and private demonstrations. Once he possessed the correct and precise shapes of the parts used in his successful 3 feet diameter prototype wheel, he immediately set about constructing a larger and more power wheel which he intended to publicly exhibit in the town of Gera. (Although this next larger wheel would, technically, be the *second* working wheel he constructed in the town of Gera, in his writings Bessler often refers to it as his first wheel and gives the date of his discovery of a working imbalanced pm wheel design as 1712. It seems that he preferred not to count his earlier prototype as being part of the history of the evolution of his wheels. Perhaps he was a bit embarrassed by its low power output and wanted his discovery to only be associated with the next and more power wheel he constructed in Gera. I, however, in order to be historically accurate, will consider his 3 feet diameter, one-directional Gera prototype wheel as being the first working wheel in the series of wheels that he constructed during his lifetime.)

This next, larger, and more powerful version of his original one-directional, 3 feet diameter prototype wheel was ready for public exhibition on June 6th of 1712. Its drum was 4.5 feet in diameter and also only 4 inches thick and, upon being released, could reach a top rotation rate of about 60 rotations per minute. Being much heavier than his prototype wheel, it required a 4 inch diameter axle that was probably about 2.5 feet in length. The total weight of the wheel's axle and drum with its eight internal weighted levers installed was about 30 pounds.

The wheel itself was mounted inside of a larger frame made of oak pieces that was portable and weighted about 50 pounds and, to enhance portability, would have had four small metal wheels attached to its base that added about another 20 pounds. The two steel end pivots of the wheel's wooden axle both rested on brass bearing plates located in the vertical members of the oak frame. Since the wheel's drum, axle, and end pivots weighed about 30 pounds, the total weight of the wheel and its support frame with caster wheels attached was, therefore, about 100 pounds. This weight was low enough to have allowed two strong men working together to lift and move the wheel when it was mounted in its frame or one strong man to lift and move the wheel and wooden frame separately and then mount the wheel on its bearing plates in the frame.

This frame with its attached small attached metal wheels made the second Gera wheel it carried portable so that Bessler could easily move it around to different places on the ground floor of his rented home in

Gera and show visitors that it was not physically connected to any cleverly concealed mechanisms under the floor that were driving it. Its portability also meant it could be moved into another room and securely locked up while Bessler was not home so there would be less chance of someone breaking in and getting access to the wheel in order to examine the interior of its drum and thereby steal the secret of its perpetual motion mechanics. This portability factor also meant that his second Gera wheel, intended for public exhibition, could be taken out of doors and placed on a wagon so that it could be shown to a large crowd of people.

He decided that the first public demonstration of his second Gera wheel would take place in the town square at noon on June 6th, 1712 and placed an announcement about this earlier in the Gera town newspaper. At the appointed time a large crowd assembled and surrounded Bessler who had placed his wheel and its support frame on a small flat bed cart whose bed was raised several feet off of the ground. The audience could clearly see under the cart and around each side of the one-directional wheel. Unlike his previous prototype wheel's drum, however, Bessler concealed the inner mechanics of the second Gera wheel by completely covering both circular faces of its drum with thin slats of wood that were closely fitted together and attached to the outside facing surfaces of its sixteen radial frame pieces with small nails or screws.

After loudly announcing to those who had gathered in the square what the nature of his invention was, Bessler proceeded to release its drum by removing a small tether cord he attached to the rim of the drum which prevented its rotation. Immediately, the 4.5 feet diameter, one-directional wheel began to turn. It did so slowly at first, but then picked up speed over the next few minutes until it was turning at about 60 rotations per minute. Murmurs of surprise rose from the crowd. Those closest to the cart, however, could make out scraping or scratching sounds coming from inside of the wheel's drum. Perhaps, upon hearing these sounds, one of those closest to the wheel yelled out "Hey, Herr Bessler, let that trained rat you have trapped inside of there out before he suffocates!" at which point the people in the crowd began to laugh.

Such heckling most certainly would have annoyed Bessler, but he would have dutifully continued this demonstration by attaching a weight of a few pounds to the wheel's axle by means of a short length of cord which had been tied to the weight at one of its ends. This cord would have also had a small loop at its other end that was quickly slipped over a metal pin projecting from near one end of the wheel's rapidly moving axle. As soon as the loop at the end of the cord caught onto the pin, the cord would have began wrapping around the axle as the weight began to rise and, by the time it reached almost to the axle itself, the weight would have stopped rising because it would have slowed the rapidly turning wheel to a complete standstill. At that point, the inventor would have grabbed the weight, detached it from the cord, put it back down on the bed of the cart, slowly unwound the cord from the axle, and, perhaps, repeated the demonstration several times again, *after* its drum had again reached its full terminal rotation rate, for those who arrived late to the second Gera wheel's first public exhibition.

To Bessler's great surprise, however, this demonstration, other than getting some mention a few days later in the local newspaper, did not result in anyone coming forth and expressing interest in purchasing his invention. People found it to be an interesting curiosity, but what good would lifting a small weight a few feet be when they needed something that could reliably provide the same power output as a wind or water mill?

Such outdoor demonstrations made Bessler nervous. Aside from the possibility of being physically attacked by rowdy hecklers in a crowd who might try to tear the wood slat covering off the sides of his second Gera wheel to expose it as a trained animal powered fraud, he also worried that, while being moved about outside, the wheel and its wooden support base might fall off of the flat bed cart and then the drum could be damaged or even open up to reveal his secret lever design to those standing nearest to the cart.

As far as Bessler was concerned, the less people getting close to his newest wheel the better. So, his outside demonstrations days would have been short lived after which he then decided to only give the demonstrations inside of his rented home. This allowed him to better control the crowds viewing the wheel and also eliminated the need to have it dangerously elevated on a cart so that those at the back of a large crowd could also see it. Naturally, Frau Bessler would not have been too enthusiastic about this change, but any objections his wife Barbara had to this plan were quickly overcome by Bessler telling her that they were on the verge of becoming fabulously wealthy and that the inconvenience of having regular crowds of strangers passing through the ground floor of their new home to achieve that status was a rather small sacrifice to make.

To achieve increased power, Bessler used more massive weights in his second one-directional, 4.5 feet diameter Gera wheel than he did in his first one-directional, 3 feet diameter Gera prototype wheel. In his second Gera wheel, the lead ingot weight attached to the ends of each weighted lever's parallel pair of main arms had a mass of 1.125 pounds and the mass of the unweighted lever it was attached to was 0.234 pound. This resulted in a weighted lever with a mass of 1.359 pounds. The wooden levers he used placed the center axes of the larger lead ingot weight's two attachment screws at a distance of exactly 5.25 inches away from the center axis of the lever's steel pivot pin whose ends, again, were firmly embedded in the drum's wooden radial frame pieces. The center axes of the wheel's eight weighted levers' pivot pins, which formed the vertices of an imaginary octagon inside of the drum, were all located exactly 21 inches from the center axis of the wheel's 2 inch diameter axle. Because of the greater masses of its eight weighed levers and higher tensions in their coordinating cords, the inventor now had to also switch from using the thread-like cords found in his 3 feet diameter Gera prototype wheel to using stronger string-like cords to coordinate the various rotational motions of the weighted levers of the second 4.5 feet diameter Gera wheel during the rotation of its drum.

These string-like cords were made of either cotton, flax, hemp, or linen fibers. It was important for Bessler's imbalanced wheel mechanics that any cords he used be flexible yet as inelastic as possible and would also be able to hang in a plane and not tend to form curls when it was loose because it had been previously wrapped around a supply spool. If a coordinating cord stretched too much when it was under tension, that small change in its length could result in the center of gravity of a wheel's eight weighted levers not being kept displaced as far, horizontally, onto the axle's descending side as possible and the wheel's torque at any rotation rate would be reduced. It's also possible that he may have used thin strips of leather instead of cords or even catgut strings (which are not made from the intestines of cat's, but, rather, from those of cattle) whose use he would have been familiar with from his study of music since these are strongly resistant to the stretching of their lengths that occurs when a stringed instrument has the tension on its strings adjusted in order to tune each of them to produce a certain fundamental sound frequency.

With its increased weighted lever mass and drum diameter, the second Gera wheel had a maximum start up power that was about 4.5 times as much as provided by his 3 feet diameter Gera prototype wheel or about 0.9 watts. That was only about enough to continuously power a small modern incandescent light bulb! But, with every public demonstration of the wheel, Bessler would promise onlookers that even greater amounts of work could be performed with a wheel of a larger diameter containing more massive weights.

Bessler remained hopeful that, with the continuing in-home demonstrations of his second Gera wheel, it would not be too long before a rich businessman came forth to buy it outright for a fabulous sum. The new owner could then, with some occasional continuing technical advice from Bessler, further develop the invention and use the profits it would produce to recover the money he had to invest to obtain it.

What made a transaction particularly difficult, however, was that Bessler demanded that the payment for his invention be made in a single, upfront lump sum amount. He described the deal he was looking for in very simple terms. He said that he would draw a line on the floor. Then, when the buyer pushed the lump sum amount to his side of the line, he would respond by pushing his wheel to the other side. Obviously, only the richest of individuals, most likely a member of the aristocracy or a consortium of wealthy businessmen, would be able to make such a payment. Bessler, however, was not deterred by this difficulty. He was confident that, as word of his miraculous invention spread throughout Europe, it was only a matter of time before a buyer would come forth willing to pay a large sum for it. It was this faith in the inevitability of the sale and the wealth it would instantly bring him and his family that helped keep the inventor motivated and demonstrating his wheel on a nearly daily basis.

Bessler even managed to get an official test done of his second Gera wheel shortly after he began exhibiting it publicly in June of 1712. He had approached the Count of Reuss and his mother the Dowager Countess and asked them to set up a test. Apparently, Bessler was a friend of the Reuss family and had sold several of its members some of his various medicinal concoctions which had cured their particular maladies.

We know little of exactly how the test was conducted from Bessler's mentioning of it in his authored works about his wheels, but the test was probably done at the residence of the countess and all present seemed to have been thoroughly satisfied that the inventor had, indeed, managed to construct a genuine perpetual motion wheel. Four months after the official test on October 9th of 1712, Bessler finally received a written certificate from the Count of Reuss announcing that the former had demonstrated a true perpetual motion to those who signed the certificate. The document bore the signatures of fourteen people of high reputation who affirmed that Bessler had something that actually seemed to be real. But, of course, none of them were allowed to look inside of the second Gera wheel's drum to view its internal mechanics so most of Bessler's critics simply dismissed this test as worthless because of that restriction.

Shortly after this successful first official test, the news of Bessler's marvelous self-moving wheel began to spread beyond Gera to the neighboring towns. People high and low born, rich and poor, educated and uneducated soon flocked to his home to see what all the talk was about. Quite to Bessler's surprise, however, while many praised his technical achievement, most seemed to think that it was either a hoax or that he was actually in league with the Devil and practicing sorcery!

As previously mentioned, the second Gera wheel, as it began to accelerate from a standstill, produced some noticeably loud and rhythmic scraping or rubbing sounds. These were most likely caused by the protruding rounded heads of the lead ingot weights' attachment screws momentarily rubbing up against the inwardly facing surfaces of the drum's wooden radial frame pieces as the levers swung clockwise about their pivot pins and their parallel pairs of main arms finally came to land on their radial stops pieces as a weighted lever's pivot pin's center axis just passed the 3:00 position of the drum during its clockwise rotation. More rubbing and scrapping sounds might have issued from the drum's 6:00 position as weighted levers whose pivot pins' center axes passed that drum position there had their lead ingot weights' attachment screw heads again rubbing against their nearest radial frame pieces' inward facing surfaces as a lever's parallel pair of mains swung away from its radial stop piece and began rotating counterclockwise about its lever's pivot pin. These rubbing contacts then caused vibrations in the radial frame pieces that produced sounds which seemed to issue from the drum's lower descending side.

This would have been one of the unintended consequences of Bessler trying to make the drum as thin as possible and, in the process, not allowing enough of a gap between its weighted levers' lead ingot weights' attachment screw heads and the inner surfaces of its drum's wooden radial frame pieces. As he continued to demonstrate the wheel, however, grooves would have been carved out of the inner facing surfaces of the radial frame pieces by the screw heads and the sounds would have lessened in volume. Fortunately, these grooves were not deep enough to seriously weaken the structural integrity of the drum which, if they had been, might have resulted in the outer portion of the drum detaching from its 16 radial frame pieces when the drum was at its maximum terminal rotation rate.

Many visitors noticing this undesirable effect concluded that Bessler probably had a trained animal like a small rat or squirrel hidden inside of the wheel's 4 inches thick drum which, upon a subtle signal from the inventor, would begin climbing up the inside surface of the lower descending half of the drum in order to make it turn and it was, therefore, the animal's claws contacting and trying to gain a hold on the interior curving surface of the thin wooden outer rim wall of the drum that then accounted for the scratching sounds. Some even believed that the inventor had, like a former historical person named Johann Georg Faust (1480 to 1540) was believed to have done for his success, sold his soul to the Devil for the power to make his wheel turn by imprisoning a small demon inside of it! Such accusations were particularly hurtful to a religious man like Bessler who had made as much effort as possible, short of actually opening its drum and revealing its interior, to convince visitors to his home demonstrations that his second Gera wheel was genuine and not an animal powered hoax.

Worst of all, occasional hecklers, envious of the attention Bessler was getting, would even show up and stand outside of his home shouting that he was a fraud as they tried to talk others out of entering to view the wheel. These incidents began to happen on Sundays and Bessler, finally annoyed that the Sabbath was being disrespected in such a manner, stopped any public viewings on that day (note that in Judaism the Sabbath is observed on Saturday and not on Sunday as is done by most Christians). When these disrespectful annoyances continued on other days of the week, however, he finally resolved to move to another town.

He, his wife Barbara, their first infant daughter, and his wife's maid finally moved to the town of Draschwitz in 1713. This town is about 20 miles north of Gera and the same distance southwest of Leipzig. On New Year's Eve of 1713 Bessler dismantled the second Gera wheel which he had brought with him and immediately got to work planning the construction of what would become his third wheel. It was finally finished in Draschwitz sometime in late September of 1713 and, as was done with the previous second Gera wheel, again put on public display on the ground floor room of a house he had rented.

This third wheel was also a self-starting, one-directional wheel, but it had a diameter fully double that of the second Gera wheel or 9 feet with a thicker drum that was 6 inches thick. Its drum's thickness was, therefore, only 1/18th of its diameter giving it an almost wafer-like appearance and this, obviously, was again intended to further convince skeptics that its speed and power could not possibly be provided by a trained animal running along the curving interior rim wall at the bottom of its drum. Skeptics, however, would still claim a trained animal, such as a heavier cat in this case, was the secret of the Draschwitz wheel's propulsion.

This third wheel Bessler built would be the last to use a single lead ingot end weight. It had a mass of 9 pounds and was attached to an unweighted lever with a mass of 1.875 pounds so that the mass of each of the eight wooden weighted levers inside the large diameter drum was 10.875 pounds. He began to notice that the ingots he was casting for his wheels were starting to become excessively large. At this point he was beginning to think about using three smaller separate lead end weights that would be attached in close proximity to each other at the end of a lever's parallel pair of main arms.

The center axis containing the center of gravity of the single lead ingot weight that the Draschwitz wheel used on its weighted levers was again attached by two steel screws and held at an exact distance of 10.5 inches from the center axis of the lever's pivot pin.

As with his two previous wheels, the pivot pins for the one-directional Draschwitz wheel's eight weighted levers were embedded in the drum's wooden radial frame pieces so the center axes formed the vertices of an imaginary octagon, but the center axes of the pivot pins were now located at a larger distance of 42 inches from the center axis of the drum's 6 inch diameter oak axle which was 2.5 feet in length. The total weight of this much larger wheel's axle and attached drum was about 120 pounds.

With an eightfold increase in the mass of its weighted levers as well as a doubling of its diameter compared to those of the publicly demonstrated second Gera wheel, the Draschwitz wheel had a maximum start up power output of about 14 watts which was about 15.6 times as powerful as second Gera wheel and about 70 times as powerful as his first working 3 feet diameter Gera prototype wheel.

Again, Bessler concealed the open framework of the Draschwitz wheel's drum with thin slats or strips of wood that were nailed to the drum's wooden radial frame pieces. There was a hand sized gap in one of these strips located about halfway between the axle and the drum's outer rim wall which Bessler could use to inspect the internal mechanics of the wheel. On occasion, when some part inside of the drum broke and needed to be replaced, Bessler could peer in through this small gap to locate the part and would then know where on the full round face of the drum to begin removing strips of wood near the problem so that he could reach in there and replace the failed part.

One wonders how, by peering through a palm sized hole and thus blocking any outside light from entering it with his head, Bessler would have had enough light to inspect the wheel's internal mechanics in order to determine where a failure had occurred. Apparently, he had not fitted the thin wooden strips that covered the drum's side together too tightly and, as a consequence, some light from the exterior could enter the drum. Most likely, the Draschwitz wheel's drum was well illuminated during the day by sunlight entering the windows of the room in which it was located and enough of this light entered the drum through crevices between the wooden slats so that a broken part could be located.

It was previously mentioned that Bessler did not allow the lead ingot end weights held between the ends of a weighted lever's parallel pair of main arms to make actual physical contact with a wooden radial stop piece inside of the drum. Rather, it was only a *portion* of the leading lengthwise surfaces of a wooden lever's two parallel main arms that made the actual contact. As the pivot pin center axes of the weighted levers inside of a clockwise rotating, one-directional wheel passed the drum's 3 o'clock position, this feature allowed the mass of the lead end weight(s) to be distributed over a larger surface area of wood than would have happened if only one surface of an end weight was to come to rest on a flat wooden stop piece. Bessler knew that there were many rival inventors lurking in the crowds visiting his home to see the wheel who would be carefully listening to any sounds his wheels emitted in an effort to discern the nature of their secret imbalanced pm wheel mechanics. To deaden the Draschwitz wheel's revealing weighted lever main arms to radial stop pieces impact sounds taking place during drum rotation, the inventor simply glued thin sheets of felt to the portions of the wooden radial stops that the weighted levers' main arms made contact with.

Despite his fear of another inventor stealing the secret of his wheels, there is the possibility that, for a small fee, Bessler would actually let some carefully selected people take a quick peek inside the Draschwitz wheel through its small inspection hole in the drum's wood slat cladding. However, to do so they probably had to first swear an oath never to reveal what they had seen other than mentioning to others that they had neither seen nor heard any small trained animals inside of the drum. He occasionally allowed this since he knew the interior of the wheel's drum was only dimly illuminated and that the person would only be able to barely see the weighted levers and even so would not understand exactly how the various springs and cords they might have caught a glimpse of would function when the wheel was turning. At best, they would only know that the drum was filled with oddly shaped wooden levers near its outer rim and that there was a network of cords interconnecting them. Aside from earning Bessler some extra thalers, these people, being good Christians and obeying the oath they had been required to take, would then go off and only verify to others that they had seen no hidden animals inside the drum which could have made it turn. With such doubt eliminated, even more people could be expected to come to his home to see his third and largest working one-directional, 9 feet diameter wheel.

I'm quite sure that Bessler would have had some subtle method of denying such a viewing privilege to any individuals who struck him as knowing too much about mechanics and who might be inventors interested in stealing his secret design so that they could claim that they and not Bessler had really discovered it. He would have wanted only average people to get a glimpse of the drum's internal mechanics who he would not have to fear would try to replicate it.

Having a somewhat lower rotation speed than his second Gera wheel, his Draschwitz wheel would only reach a terminal rotation rate in excess of 50, but less than 60 revolutions per minute when allowed to run freely without any external devices being attached to and worked by its axle. If, however, a rope was suddenly attached to a projecting pin on its axle so that the rope wound around the axle, the wheel could, using a rope and an overhead pulley, raise a 40 pound weight up several yards high off of the floor before the wheel slowed to a stop during a braking test of the machine.

A 1 foot section of axle projected from each side of the one-directional Draschwitz wheel's 9 feet diameter drum and, this time, the pivots at the ends of the axle rested on brass bearing plates that were embedded in heavy wooden vertical support planks that spanned the distance from the floor of the wheel's room to its ceiling. This, of course, meant that the wheel was not portable, but, for the protection of its secret mechanics, Bessler would have made sure it was located in a room whose doors and windows could be securely locked.

The axle section on one side of the drum had three metal pins in it that were arranged around its circumference in such a way that they could raise and drop three long solid wooden stamps weighing a few pounds apiece that were a miniature version of those used in pounding mills. These stamps were arranged in a row and held in line with each other by a wooden frame that could be moved around.

When the portable stamping mill's frame was slid into position near the side of the Draschwitz wheel's rotating axle carrying the metal pins, the pins would, sequentially, catch a notch cut into the outward facing surface of a stamp, raise it a few inches, and then suddenly drop it as the pin was pulled out of the notch by the continuing rotation of the axle. Thus, the first stamp would be lifted and then dropped by the axle's first metal pin. After that had happened and almost an additional 120° of axle rotation had occurred, the second metal pin on the axle would then have rotated into position and begun lifting the second stamp which, after rising a few inches, would likewise be dropped. Finally, after the second stamp had dropped and almost an additional 120° of axle rotation had occurred, the third metal pin on the axle would come into position and cause the third and last wooden stamp to be raised and dropped whereupon the sequence would begin again with the first stamp after an almost additional 120° of axle rotation had taken place.

The resulting impacts of the bottoms of the wooden stamps against an empty box located at the bottom of the stamping mill's frame would produce a tremendous amount of noise that was nearly deafening. Aside from showing a useful application of the wheel that might appeal to a businessman looking for an alternative to wind or water power for his early Industrial Age factory, this noise maker was a great way to prevent the reverse engineers in the crowds viewing his wheel from detecting other sounds that might betray the way its secret perpetual motion mechanics worked.

The axle bearings on this impressive 9 feet diameter, one-directional wheel were described as "open" which meant that Bessler had secured the brass plates that supported the axle's end pivots into holes that he had carefully cut into the vertical wooden planks. These holes allowed the bearings to be visually inspected during wheel rotation so as to allay any suspicions that there might be hidden mechanisms inside of the planks themselves which were actually turning the wheel through its pivots.

Originally, Bessler may have used two steel bolts with oversized heads that could be hand screwed into the wheel's two vertical support planks that they penetrated so that their ends would provide enough friction against the axle pivot pins to keep the constantly imbalanced Draschwitz wheel from rotating. Most likely, since these bolts could not be placed directly above the axle pivot pins as they were in the Gera prototype wheel, each of their threaded shafts penetrated the wood of a vertical support plank at a 45° angle from vertical line passing through the center axes of the pivots. But, their use over time would have damaged the steel pivots by cutting grooves into their outer surfaces when the bolts were repeatedly tightened by witnesses to slow and stop the axle and its attached drum from turning. To prevent such damage, which would eventually require two axle pivot pin replacements to correct, Bessler simply began manually slowing the wheel by, perhaps, using thick leather workman's gloves to apply drag to the outer rim wall of its ascending side and then, once fully stopped, securing the drum to the floor below it by using a rope tether that had one end bolted to the floor and the other end attached by a small metal hook to a small eyelet that was screwed directly into the end of one of the drum's radial frame pieces which had been exposed by cutting away the piece of the wooden slat that covered it.

In the town of Draschwitz Bessler received a somewhat better reception for his third wheel than he had initially gotten in Gera for his second wheel. The people there were a bit more refined and he was visited by a constant stream of members from all levels of the social strata. There were academics, clerics, occasional nobility, and, of course, mostly the average people who had heard of his latest and greatest wheel and wanted to see for themselves if what they were hearing was true. However, there were still many who doubted that he actually had the "real thing"; that is, the long sought perpetual motion which most of the highly educated of his day were beginning to declare to be a physical impossibility. Particularly annoying to Bessler was their claim that, since his wheels only turned in one direction, their internal mechanics must, like those of clock movements, be powered by low profile, tightly wound up spiral mainsprings or by carefully aligned, slowly descending weights inside of the drums that did not touch their interior rotating surfaces. Such speculation made sense to Bessler's critics because it rationalized how his drums could be so thin compared to their diameters which would tend to eliminate the possibility of trained animals being used to power them.

Bessler, when questioned by his visitors about the mechanical system his Draschwitz wheel used, would, in order to protect its secret, avoid giving them any detailed answers. He would just be vague or describe its internal actions to them by using confusing Latin terminology for parts and actions which he had made up on the spot. However, he made no secret of the fact that his perpetual motion wheel's mechanics did use weights and springs, but he would also quickly point out that the springs were not employed in the same way that they were in the then available spring powered clock movements. In his later writings, though, Bessler tended to avoid discussing springs as much as possible because, I suspect, he knew that any discussion of them would only have served to strengthen the claims of his detractors that his wheels were really just spring powered fakes.

While he was in Draschwitz, Bessler began to receive letters from one Andreas Gartner who was a Dresden-based master model maker to the King of Poland at the time (Dresden is located about halfway between Draschwitz and Zittau). He had heard of Bessler's amazing self-moving wheels and wanted to know more about them.

Gartner, an old man by that time, was famous throughout Europe for the many complex clock movements he had created for royalty. Of course, believing perpetual motion to be impossible, he naturally assumed that Bessler, also being a clockmaker, must have created a very clever windup clockwork powered wheel. Bessler found his insinuation rather insulting and, after exchanging several letters, all communication ceased between them. Gartner, not taking up Bessler's challenge to come and personally inspect one of his wheels at that time, continued to believe and state that they had to be clever fakes and this emboldened Bessler's other detractors to make similar claims.

Sometime during early September of 1714 Bessler was visited by the famous German historian, mathematician, philosopher, and scientist Gottfried Wilhelm Leibniz as he was traveling around the area on business. Like others, he had heard of Bessler's much larger and second publicly exhibited wheel in Draschwitz and wanted to personally view it. Bessler was thrilled to have someone of such eminence interested in his invention and Leibniz actually spent several hours conducting various tests on the wheel in Bessler's presence (Bessler and his brother or another trusted assistant, being younger men, would have done the actual testing while Leibniz directed them).

Leibniz was very impressed by what he saw, but he could not accept that the wheel was running purely because of its mechanics. He thought that, somehow, Bessler had managed to tap a previously unsuspected force of nature and that was the real source of the wheel's power. To Leibniz, Bessler's wheels were like water wheels being driven by invisible water.

Bessler told him of all of the problems he was having finding a buyer and at that time Leibniz suggested that the real problem was that potential buyers were not seeing the wheel run long enough so that the possibility of it being powered by hidden wound up mainsprings could be completely eliminated. Leibniz then suggested a duration test that would last several weeks and not just the brief tests so far provided that were really proving nothing to the satisfaction of the skeptics. It was also Leibniz who suggested that the best way to prove to the skeptics that his wheels were not powered by spring driven clockwork type mechanisms was to build a wheel that could run in *either* of its two possible directions of rotation. Since clock movements only make a clock's hands turn in one direction, such a modification would surely prove that the mechanics of Bessler's wheels were different from those of timepieces. Bessler intended to heed the older scientist's advice and probably immediately began considering how he could produce a two-directional or bidirectional wheel. Exactly how he did that will be treated in a later chapter.

Around this time Bessler announced that he was willing to sell his invention outright to anyone who could afford the price he decided would be sufficient compensation for the amount of time and effort he had invested during a decade in obtaining the working design he had found. The figure he came up with was 100,000 thalers which was probably a figure he derived after learning that the English government in 1714 had promised to pay a prize of an equivalent amount to anyone who could provide a device or method for determining the longitude of a ship at sea to within 30 nautical miles. The "thalers" that Bessler wanted were silver coins weighing about an ounce each and were the major denomination used in Saxony at the time. Interestingly, our English word "dollar" actually derives from the name of this coin.

The amount Bessler demanded was rather steep considering the average person's income then was only a few hundred thalers a year. The amount Bessler wanted was actually equal to the value of a *ton* of gold at the time and today would be worth about $37.9 million US dollars at a price of $1,300 US dollars per troy ounce of pure gold! However, to convince a potential buyer of his sincerity, Bessler stated that if someone purchased his one of his wheels and then, upon inspection, it proved *not* be a genuine perpetual motion wheel, not only would the buyer receive a complete refund of the purchase price, but Bessler would allow his own head to be severed from his body with an axe as punishment for having deceived the buyer!

As the year of 1714 began to draw to an end, Bessler found his situation in Draschwitz becoming progressively less to his liking. Despite his many public demonstrations and successful tests, skepticism was increasing and he was again beginning to be harassed by those that thought he was a charlatan. No doubt he was even beginning to fear for his own and his family's safety. Perhaps he imagined what would happen if a mob of hooligans, emboldened by a night of drinking, decided to break into his rented home and try to tear open the drum of his wheel to satisfy themselves that it was a hoax. They might show up armed with guns and someone might get shot and killed. Most likely his wife had the same fear and was pressuring him to move on to another town and start anew there. That would have been enough to help Bessler decide to, again, destroy his creation, the product of months of work after salvaging some of its more valuable components such as its precisely cast lead ingot weights and its expensive extension springs, and then move on to a different town to begin another and, hopefully, more impressive wheel's construction. He, his wife Barbara, an infant daughter, and his wife's ever present maid finally left Draschwitz in November and were forced to spend the winter of 1714 / 15 in a rented house in the small village of Obergreisslau.

This action on his part would be repeated several times again before Bessler's death. Before he relocated to a new town, he would first destroy whatever wheel he had previously constructed in the town that he was leaving after removing its more expensive components. He made sure he left nothing behind that might reveal the secret of his imbalanced perpetual motion wheels' mechanics. Indeed, if he had left only a single intact lever behind, it might have been sufficient to allow the entire interior mechanics of his wheels to be reverse engineered by a skilled craftsman.

Bessler had hoped to begin building an even more powerful version of the Draschwitz wheel in this small town, but it soon became apparent to him that the house he rented there was just too small for what he had in mind which was a wheel that would be a full 12 feet in diameter! Quite to his surprise, he found the relative quiet of Obergreisslau to be such a change of pace from all of the attention he and his wondrous invention were receiving in Draschwitz that he fell into a deep depression and could get little done. It's possible that Bessler may have suffered from what is today referred to as "Seasonal Affective Disorder" or "SAD". This is a form of depression that in some people is triggered by the shortening of the days and consequent reduction in exposure to ultraviolet light that happens during the winter months. It tends to be mild and, generally, passes as the spring months and longer days arrive. However, in some individuals the depression can become deep enough to be disabling.

Finally, as his depression lifted, Bessler and his family left Obergreisslau and arrived in the town of Merseburg shortly after Easter of 1715. There he found rooms to rent at the Green Manor by the sixth gate of the town that had ceilings high enough to accommodate a 12 feet diameter wheel (which meant the ceiling of its ground floor was about 14 feet high) and he even had a new innovation to add to this wheel which he believed would finally silence any skeptic that dared to suggest that his wheels used spring wound clockwork mechanisms to obtain their long running motions. He would silence such detractors by constructing a single wheel that was capable of being started so that its drum could rotate in either of its two possible directions of rotation!

After settling into their new residence, Bessler's days were roughly divided between selling his medicinal preparations to those in need of them, working on the construction of what would be his third and most impressive publicly exhibited wheel, writing a small book to sell to the public during his wheel demonstrations to bring in some additional income, and raising his family.

That first booklet he wrote would be titled *Grundlicher Bericht* or, in English, *Thorough Report* and was intended to impress the reader with the details of the official testing done on his previously destroyed Draschwitz wheel. It would also set the pattern for all of his later written works. They would describe the external structural features of his wheels and various "official" tests confirming that they were not moved by any external sources of power, but, when it came to treating their internal mechanics, Bessler would only offer tantalizing, but vague, descriptions involving weights being moved nearer to the center of the axle on the ascending side of a wheel and then farther from the axle on the descending side. How this was done was not covered in detail. I shall, hereafter, if and when the need arises, refer to this first written work by Bessler as simply "GB" and will also use abbreviations, to be defined, for the other of his written works as they are mentioned later in this and other chapters.

For much of the month of April to the beginning of June of 1715,

Bessler must have worked almost daily on his newest wheel. That is a time period of, probably, less than just two months and in that time, he managed to construct his largest and most powerful wheel yet and give it the remarkable ability to be bidirectional.

This fourth wheel was a full 12 feet in diameter and had a drum that was double the thickness of that of the Draschwitz wheel or 12 inches. Its axle was, like that of the Draschwitz wheel, 6 inches in diameter, but now the axle was more than doubled to 6 feet in length. At each end of the axle, Bessler had inserted a steel pivot pin that was 0.75 inch in diameter and which projected out to a distance of about 6 inches from the cutoff end of the axle. My best estimate for the combined masses of the Merseburg wheel's axle with its two steel end pivots, the attached drum, and its internal mechanics is about 550 pounds.

This wheel was also the first to use a closely spaced trio of cast cylindrical lead end weights that were each 1.78 inches in diameter and 4 inches in length so that they each weighed very close to 4 pounds when a steel attachment bolt was inserted into small diameter shaft bored out along the center axis of the weight from one of its circular ends to the other. Three of these 4 pound lead end weights and their three attachment bolts, with a total mass of 12 pounds, were then held between the ends of a parallel pair of main arms of a wooden unweighted lever that, before the addition of its three cylindrical lead end weights and their attachment bolts, had a mass of 2.5 pounds. Thus, the mass of a weighted lever after its three end weights were bolted into place was 14.5 pounds.

Each weighted lever of the Merseburg wheel kept the center axis of its middle cylindrical lead end weight's attachment bolt, on which the center of gravity of the three weights as located, at a distance of exactly 14 inches from the center axis of the weighted lever's pivot pin. Now, however, because this first bidirectional wheel actually contained *two* one-directional wheels that were placed side by side and back to back inside of its enlarged drum, eight 12 inch long steel pivot pins were used each of which was shared by two weighted levers! An extended pivot pin's ends and middle section were in turn embedded in the wood of *three* parallel radial frame pieces of the drum so that the center axis of the pivot pin was located at an exact distance of 56 inches from the center axis of the wheel's

6 inch diameter wooden axle. The center axes of these eight lengthier pivot pins were located at the vertices of an imaginary octagon.

The Merseburg wheel's larger diameter and greater total end weight mass gave it a start up power output that was about 1.8 times that of the Draschwitz wheel or about 25 watts upon start up. Bessler now had a wheel that was about 125 times as powerful as the small 3 feet diameter, one-directional prototype wheel with which he first found success back in the town of Gera in late 1711. Finally, Bessler felt that he had a wheel powerful enough to impress a businessman looking for a new source of power for a small factory. It could do a lot more than just keep itself in motion and was able to perform a degree of continuous external work that amazed everybody to whom it was demonstrated.

Apparently, Bessler had even constructed some sort of simple water lifting device that he could quickly attach to the Merseburg wheel's axle and which would allow the wheel to raise water from a supply trough to a higher elevation from where it would pour back down again into the trough and make a constant splashing sound as it did so. Such a demonstration would appeal to a farmer who might want to use several of the Bessler's wheels to pump ground water out of a well in order to irrigate his fields and improve their crop yield. A more abundant crop would then produce additional income for the farmer at a market and, in time, the wheels would pay for themselves!

Unlike his first two publicly demonstrated wheels, this fourth wheel of his, after being untied, was stationary and then had to be given a gentle push to start it turning. Once pushed, it was capable of running in either of two directions with a maximum terminal rotation rate of about 40 rotations per minute when it was freely turning and its axle was not attached to and operating any outside machinery.

In order to bestow bidirectionality to his latest creation, Bessler simply built two one-directional wheels and enclosed them in a larger common drum which accounted for the sudden doubling of the Merseburg wheel's drum thickness compared to that of the previous Draschwitz wheel. However, the Merseburg wheel's two internal one-directional wheels were arranged in opposition to each other so that their two separate sets of eight weighted levers produced two separated centers of gravity which were on opposite sides of the axle's center axis. The torques produced by these two centers of gravity were exactly equal in magnitude, but opposed to each other in direction so that there was no net torque acting on the axle and, thus, the wheel initially remained motionless.

When the drum was given a slight push in either direction, however, something really amazing would happen. Using a delicate system of gravity powered latches installed near the end weights of each of its eight weighed levers, the one-directional wheel forced to rotate *counter* to the direction of motion it was designed to normally spontaneously self-start in would, after a half of a complete rotation, find *all* eight of its weighted levers' main arms securely held in physical contact against their wooden radial stop pieces by small metal "cat's claw" latches. These latches had arms with notches cut into them that would engage a coordinating cord hook attachment pin that passed through the parallel pair of wooden main arms of the weighted lever and was located between the three cylindrical end weights and the weighted lever's pivot pin.

This latching action, when completed for all of the retrograde rotating internal one-directional wheel's eight weighted levers, then produced a radially symmetrical configuration of the weighted levers that immediately caused their center of gravity to be withdrawn up into the exact center axis of the axle where it could contribute no torque to the axle. At that time it was only the remaining "active" internal one-directional wheel inside of the bidirectional wheel's drum which then provided all of the torque that drove the axle, its attached drum, and the other inactive, latched up and retrograde rotating internal one-directional wheel inside of the drum. This was because the active internal one-directional wheel was rotating in the direction it was designed to normally spontaneously self-start in and its eight *un*latched weighted levers' center of gravity remained horizontally offset onto the axle's descending side,

When the axle and its attached drum of the bidirectional Merseburg wheel were slowed to a stop, which an assistant could accomplish by placing a thick sheet of leather on the axle and leaning on it, it was then necessary to back rotate the drum a half of a rotation in order to *unlatch* half of the weighted levers of the previously inactive internal one-directional wheel while simultaneously latching up half of the levers of the previously active internal one-directional wheel. Once that was done, the centers of gravity of the two *separate* internal one-directional wheels inside of the drum would again be located on opposite sides of the axle's center axis. But, at that time the *composite* center of gravity of the *two* one-directional internal wheels would actually be located in the *same* plane which would be perpendicular to the center axis of the axle and located *between* the drum's two internal one-directional wheels. Additionally, that composite center of gravity would be located directly under the center axis of the axle at its *punctum quietus* so that there would be no net torque acting on the axle. As a result, the huge drum would again remain stationary when released.

It was a simple, yet ingenious system for achieving bidirectionality, but it was probably one of the most difficult improvements Bessler had ever made to one of his wheels. As the reader will learn in a later chapter, each of a bidirectional wheel's 16 weighted levers had to have its own set of two independently operating latches. That's a total of 32 separate latches and they all had to work flawlessly as the wheel began rotating and picking up speed in either of its two possible directions of rotation. Only Bessler's skill as a clockmaker allowed him to conceive of such a system and to make it work. He complained in his writings that the careful adjustment of this system of latches gave him many a headache and he probably, at one point, may have considered eliminating them altogether. But, he had an obsessive need to prove to his skeptics that his earlier wheels were *not* one-directional because they employed clockwork mechanisms that could only turn in one direction as their previously wound up mainsprings slowly unwound in only one direction. He probably also wanted to show Leibniz that he respected his advice about making his wheels bidirectional and how skillful he could be in making the suggested performance capability modification.

It was with this third publicly exhibited and first bidirectional wheel in Merseburg that Bessler also made a major change in the way he concealed the internal mechanisms of his larger wheels. Now, instead of slats or strips of wood that had to be nailed to the outer surfaces of a drum's wooden radial frame pieces, he decided, as he had done with his original 3 feet diameter Gera prototype wheel, to tack on linen cloth that had been dyed a dark greenish color in order to make it as opaque as possible so that anyone viewing one of the full round faces of the cloth covered drum would not be able to see any shadows of its internal mechanics cast onto the cloth by a bright sunlit window located on the other side of the drum. Oiling the cloth also eventually became necessary because some of the olive oil lubricant Bessler used on the wheel's many weighted lever pivot pins had started oozing out onto the dry linen cloth he originally used and gave a clear indication of the number and arrangement of these pivots as well as their distance from the axle's center axis. This was information Bessler dared not let any of the reverse engineers viewing his Merseburg wheel at public demonstrations have. This change in drum interior cloaking material also helped reduce the total weight of the wheel so that its axle's brass bearing plates would last longer.

    The attached pieces of linen cloth on the backside of the Merseburg wheel's drum which were not readily visible to the public then had strategically placed openings cut into them that made it much easier to access the wheel's internal weighted levers from that side should one of their coordinating cords or springs fail in operation and require a quick repair so that the wheel could continue to turn. These openings in the linen were made by cutting eight trapezoidal flaps that could be folded back to access the interior of the drum and then closed up again and held in place against the radial and middle octagonal frame pieces under the cloth by short pins with large area heads on them which may have looked like somewhat oversized pushpins used to tack notes to a bulletin board. Bessler placed these inspection and service openings on the backside of the drum in order to assure that if, while in rapid rotation, one of the flaps came loose and resulted in one of the drum's shifting weighed levers being exposed, there would be little risk of anyone getting a good look at it.

    One witness, a Herr Christian Wolff, who was present during the official test of the Merseburg wheel claimed that, through a gap that formed where a cloth flap covering one of the drum's inspection openings had come loose, he could see one of the wheel's wooden weighted lever's main arms aligning with its wooden radial frame pieces and then making contact with a small wooden piece that was attached to the radial frame pieces near the drum's outer rim. This independent observation agrees with Bessler's claim that there was no weight hanging from the hidden section of axle inside of the drums of his wheels which then applied a torque that turned the entire wheel. All of the important "perpetual motion structures" inside of one of Bessler's wheels were located as close to its drum's outer periphery as possible. This assured that the offset center of gravity of the *active* one-directional internal wheel's eight weighted levers that drove the bidirectional Merseburg wheel would then be displaced as far as possible, horizontally, onto the descending side of the axle so as to maximize the amount of torque that the axle could deliver at any particular rotation rate in order to perform outside work.

The reader will find an illustration of Bessler's Merseburg wheel depicted in Figure 1(a) at the end of this chapter. This printed drawing was taken from the third book Bessler authored in 1719 and shows both a profile view of the wheel on the left side and an axial view on the right side. The service flaps in its drum's linen cloth coverings are not shown because they are on the hidden backside of the drum and just knowing how many there were and where they were located might have given away important information about the wheel's internal mechanics to a reader of the book.

Like his first two wheels, each end of the much thicker Merseburg wheel's oak axle had a securely implanted steel pin that served as a pivot pin projecting from it. Each pivot pin rested in a half moon shaped groove or pocket in a brass bearing plate that was itself held inside of a cutout section of one of the axle's two wooden vertical support planks. The axle was 6 inches in diameter and its steel end pivots were probably 0.75 inch in diameter. These pivots were then lubricated with some greasy substance like lard to minimize the wearing of the softer brass plates upon which they rested and to which each pin applied a force of about 225 pounds. This arrangement provided a secure support for the wheel, yet also allowed witnesses to easily inspect the wheel's axle bearings so that they could see with their own eyes that the axle pivot pins were not being turned by any kind of mechanism hidden inside of either of the vertical support planks that was in contact with a pivot.

On the left side of the Merseburg wheel illustration of Figure 1(a) one can see various devices that Bessler could attach to the wheel and which it was capable of continuously operating.

There was a set of four wooden stamps that were similar to those used in the stamping mill that Bessler had constructed for the second Gera wheel. They impacted the top of an empty box and were capable of producing a loud rhythmic pounding sound that was almost deafening. Unlike the three stamps used with the Draschwitz wheel that would each be lifted once during a complete axle rotation to produce a total of three impact sounds per axle rotation, the Merseburg wheel's axle lifted each of its mill's four stamps twice during a complete rotation for a total of eight sounds per axle rotation.

As had been the case with the previous Draschwitz wheel's stamping mill, this add on device for the Merseburg wheel was really intended to mask the eight impact sounds that its active internal one-directional wheel's eight weighted levers' main arms made as they made contact with their radial stop pieces during each drum rotation. Masking these internal sounds with louder external ones was an effective way of preventing witnesses of the Merseburg wheel from determining from what location inside the drum those sounds issued. This was even more necessary with the bidirectional Merseburg wheel because, unlike the one-directional, 9 feet diameter Draschwitz wheel, the 12 feet diameter Merseburg wheel did not have felt pieces glued to its wooden radial stop pieces in order to muffle the easily counted sounds of its weighted levers' main arms as they contacted their respective wooden radial stop pieces shortly after their lever's pivot pin center axis passed the 3:00 position of the drum when it was viewed rotating in a clockwise direction.

Why did Bessler eliminate the felt pieces from the larger bidirectional Merseburg wheel? Possibly, he learned from the Draschwitz wheel that the felt cushions were quickly worn away and, consequently, rendered useless by the constant impacts they received on the descending side of the drum. That then resulted in a quantity of the material used to make the felt, wool fibers, being released inside of the turning drum, tumbling around as it rotated, and eventually adhering to places he had lubricated to reduce metal to metal wear.

As the felt material stuck to oiled surfaces and accumulated in these many places, it tended, via the wick effect, to draw any lubricating oil away from the contacting metal parts' surfaces there and, thus, could increase the rate at which these parts wore away. The Merseburg wheel, being bidirectional, had twice as many places that required lubrication as did the prior one-directional Draschwitz wheel and the extra maintenance needed to clean away the debris from the larger felt cushions the Merseburg wheel and periodically replace all 32 of them would have required twice as much time to perform. In addition, the presence of this undesirable waste from its worn felt cushions might have interfered with the delicate functioning of the Merseburg wheel's many newly added gravity activated latches whose flawless operation allowed the wheel to achieve bidirectionality. Clearly, the sound deadening felt cushions needed to be eliminated from the Merseburg wheel and Bessler just compensated for that by attaching very noisy external machinery like the stamping mill to the wheel's axle.

It's interesting that Bessler chose to have the Merseburg wheel's add on stamping mill produce exactly eight pounding sounds per axle rotation. This construction detail of the 12 feet diameter, bidirectional wheel is yet further evidence that its drum's active internal one-directional wheel, when in motion, actually relied upon the use of eight weighted levers that, on the enlarged drum's descending side, were each in their turn making contact with a wooden radial stop piece as their pivot pins passed the 3:00 position of the drum. If any of the reverse engineers in the crowds visiting Bessler's home to view public demonstrations of the large wheel managed to hear and accurately count eight sounds issuing from the drum's descending side as it rotated, he might just think that he was actually only hearing the pounding sounds of the external wooden stamps and that their number of impacts per wheel rotation really had nothing to do with the wheel's internal mechanics when, to the contrary, their number did.

The impact sounds of a weighted lever's parallel pair of wooden main arms against their wooden radial stop piece were probably the least loudest when the drum was first started and only noticeable by a witness standing within a few feet of the front circular side of the drum, but then grew progressively louder as the drum accelerated until they became noticeable to spectators watching a demonstration and located at a distance of ten or more feet from the wheel. Perhaps Bessler only slid the noise making stamping mill machine into position against the rear section of the Merseburg wheel's axle containing the eight metal pins when he noticed that the internal sounds from the drum were getting louder with increasing speed or when he suspected that a rival inventor trying to steal the secret of its mechanics might be among his visitors during a particular demonstration. That would then have been more than sufficient to mask the main arms impact sounds issuing from the cloth covered front side of the large 12 feet diameter drum.

However, the Merseburg wheel produced other sounds due to the automatic operation of the pairs of gravity activated latches assigned to each of the wheel's sixteen weighted levers that were necessary to achieve bidirectionality. When the stamping mill was not attached to the axle and producing its near deafening pounding sounds, these extra internal sounds produced by the gravity latches were described as "clatters" or "rattles". How these additional sounds were produced will become obvious in a later chapter where the addition of the gravity latched required by Bessler's bidirectional wheels are described in detail.

On the front end of the Merseburg wheel's axle there was a projecting screw with a wing shaped head on it to which one could attach a small noose at the end of a rope when the wheel had reached its maximum free running rotation rate. The noose would then tighten as the rope quickly wound about the wheel's rapidly turning axle. A pair of wooden deflector pieces attached to the front end of the axle near the drum would keep the rope windings from bunching up there and would also force them to begin creating a second layer of windings that would move back toward the projecting screw again. Via two strategically placed pulleys, the rope from the wheel's axle was then able to lift a sturdy wooden box containing six bricks. The box and bricks together weighed about 100 pounds and would rise through a distance of about 15 feet from the ground up to one of the pulleys which was located outside of the house and near the top of an open window of the room on the ground floor in which the wheel was located. During this braking rest, the wheel would maintain a fairly constant speed during the ascent of the load of bricks and suddenly slow down and stop just before the load reached the top pulley outside of the window.

Once the box of bricks had reached an elevation just below the top pulley, the wheel and its axle would suddenly rapidly slow down and stop turning and the load of bricks outside of the window would momentarily stop rising and then begin descending. As it did this, the wheel would begin rotating in the opposite direction and load of bricks would, quite unexpectedly, descend at about the same constant speed that it had ascended. Eventually, the descending load of bricks reached and contacted the soft ground which cushioned its impact so that the box would not break and could come to rest. At that instant the rope attached to the wheel's axle would then be complete unwound, but still securely attached to the axle by the wing head screw. The wheel's drum and axle, however, would continue turning in the *same* direction and then begin winding the rope around the axle again, but this time its winds would be formed in the opposite direction on the axle.

This action then caused the load of bricks to begin rising again and its original, nearly constant vertical velocity ascent would be repeated only this second time by the bidirectional Merseburg wheel's axle turning in the opposite direction it had turned during the load's first ascent. As the box of bricks reached the same elevation just below the top pulley as it had on its previous ascent, the wheel's drum and its axle would quickly slow down and then stop turning again. At that instant they would immediately reverse their direction of rotation again and begin lowering the box back down to the ground at a fairly constant velocity only this time by the wheel turning in the opposite direction that it had turned during the load's first descent.

Every time the load of bricks came to a stop near the outside top pulley, the wheel's drum and its axle would thereafter reverse their direction of rotation. And, every time the load of bricks came to a stop on the ground, the wheel's drum and its axle would continue their same direction of motion, but the rope attached to the axle would reverse the direction that its windings were being wrapped around the axle. Each repetitive cycle in this demonstration actually required a complete ascent *and* descent of the box of bricks for *each* of the *two* possible directions in which the Merseburg wheel was capable of rotating. Thus, a *single* "cycle" consisted of four vertical motions of the load of bricks and their containing box as they first ascended and then descended while the axle turned in the one direction and they then ascended and descended again while the axle turned in the opposite direction. This cyclic action could be allowed to continue nonstop for as long as Bessler allowed it to and the wooden box containing the bricks, its lifting rope, or either of the two pulleys did not fail. Such a dramatic demonstration, easily visible to everyone in a small crowd gathered outside of the inventor's residence, must have been amazing to all who witnessed it.

One of the most fascinating aspects of this dramatic demonstration of Bessler's 12 feet diameter, bidirectional Merseburg wheel that many of the engineers in the visiting groups noticed was that the load of bricks that were hoisted and then lowered would maintain a nearly constant velocity throughout their vertical motions!

This indicated that the eight weighted levers of the bidirectional wheel's active internal one-directional wheel were able to supply extra torque to the axle as the wheel started to slow down that would then resist the slowing of down of the axle's rotation rate and more evenly supply the rising load of bricks (which included the mass of their containing box) with the energy they needed to rise in the Earth's gravity field. This was because, as wheel rotation started to decrease with an increase in the elevation of the load of bricks, the offset center of gravity of the active internal one-directional wheel's eight weighted levers began moving horizontally farther away from the center axis of the drum's axle on its descending side which then provided more lifting torque to the axle. When the load of bricks had almost reached the top pulley, the axle and drum of the bidirectional wheel and their 16 weighted levers, having finally lost all of their rotational kinetic energy to the load of bricks, would rapidly stop turning in one direction, and begin turning in the opposite direction as the load of bricks then descended.

That action, after a half rotation of the drum, then resulted in all of the weighted levers of the *previously* active internal one-directional wheel having their parallel pairs of main arms locked down against their radial stop pieces by their respective pairs of gravity latches while all of the locked down weighted levers of the *previously* inactive internal one-directional wheel were freed from their gravity latches so they could then rotate about their respective pivot pins again as need to place their center of gravity on the axle's descending side. However, because this other internal wheel was then being accelerated at a rate greater than its normal acceleration rate by the descending load of bricks (a condition we can call "over speeding"), the center of gravity of its eight weighted levers would actually swing from the axle's descending side, though its *punctum quietus*, and over to the axle's *ascending* side with a horizontal displacement from the center axis of the axle that increased with increasing drum and axle's rotation rate. This created a counter torque acting on the axle that then produced a braking effect that increased with increasing axle and drum rotation rate and tended to resist any increase in that rotation rate. Because of this automatic and self-regulating braking effect, the descending load of bricks was forced to give up its supply of gravitational potential energy to the axle, its drum, and the drum's 16 weighted levers at a fairly constant rate as the load of bricks also descended at a fairly constant rate.

This kind of demonstration could also have been done using a simple flywheel which had been brought up to speed by an external energy source. However, in the case of such a flywheel, the energy losses, which would mainly come from the friction of the rope traveling around the pulleys and the aerodynamic drag acting on the box of bricks passing through the air, would eventually cause the action to stop as the load of bricks reached a lower and lower elevation with each successive ascent until, finally, it just rested motionless on the ground as the flywheel and its axle also came to a stop. With Bessler's bidirectional Merseburg wheel, however, the action was continuously repetitive because the energy losses in the system were constantly replaced by the secret perpetual motion mechanics of the wheel itself! Also, in the case of a simple flywheel, the vertical ascent velocity of a load of bricks being hoisted would have its maximum rate as it lifted off of the ground which would then steadily decrease with the elevation of the rising load of bricks as it moved closer to the top pulley. During the descent of the load, its vertical velocity would also steadily increase as its elevation decreased and it would more forcefully hit the ground each time before it briefly came to rest there. This would be due to the lack of an automatic and self-regulated resistance to change in rotation rate by the flywheel.

Also, it is interesting to note the location of the first pulley that the rope from the axle ran around on its way to the load of bricks. That pulley was securely attached to the floor just below the point at which the noose of the rope would be attached to the wing head screw near the end of the axle. This location for the pulley was necessitated by the sudden jerking force that the axle would experience when a rope attached to a heavy external load suddenly tightened as it was wrapped around the rapidly rotating axle by the projecting screw head. That jerking force would then, because the pulley was located under the axle, tend to pull the axle's end pivot pin near the pulley down tighter against the brass half moon bearing plate that it rested on and thereby help keep the axle pivot pin more securely held by the plate.

If Bessler had instead located the pulley on the ceiling above the axle, then that sudden jerking force on the axle might have caused the axle's end pivot pin to lift up and out of an exposed or open bearing plate and that could have allowed that end of the axle to come completely away from its vertical support plank. This would probably have resulted in the pivot pin at the other end of the axle also riding up off of its bearing plate if open and the entire axle and drum might then have escaped from their vertical support planks as the rapidly rotating drum fell to and contacted the floor. If that happened, then the drum would have been severely damaged or even completely destroyed. Obviously, this was something that Bessler would have wanted to prevent at any cost even though it required him to pass the rope from the floor pulley through a small opening he had cut in the bottom of the front vertical support plank for the rope to pass through. That cutout reduced the amount of weight that the plank could support, but, still allowed it to support more than its share of the Merseburg wheel's total axle and attached drum mass of about 550 pounds.

Attached to the ends of the curving crank arms whose other ends were attached to the very ends of the Merseburg wheel's two axle pivot pins were rigid metal drive rods that would "pump" the cross beam rods of two large compound pendulums with each rotation of the bidirectional wheel's axle. The exact purpose of these two pendulums has been the subject of much speculation.

Although it is not clear from studying the illustration shown in Figure 1(a), the curving cranks attached to the ends of the pivot pins of the Merseburg wheel's 6 feet long axle were *not* in opposition to each other which would mean that the ends of the two cranks that attached to the pendulum drive rods would appear to be spaced angularly 180° apart from each other in an axial view which is an imaginary x-ray vision "see through" view along the length of the axle that would appear to superimpose its front and back circular ends and the curving crank arms attached to their axle pivot pins. If the ends of the curving arms of the two cranks had been 180° apart from each other, then that would have caused the individual centers of gravity of both compound pendulums to rise and fall *together* as both pendulums always swung in the *same* direction. Nor were the ends of the two cranks in alignment with each other so as to appear to be 0° apart in an axial view. If that had been the case, then the centers of gravity of both compound pendulums would again rise and fall *together* as the two pendulums always swung in *opposite* directions. My research convinces me that Bessler actually used a compromise between these two extremes to attach the ends of the curving arms of the two cranks used on his bidirectional wheels to their pendulums' drive rods.

In this compromise approach, the two curving arms of the cranks would have appeared to be 90° apart from each other in an axial view and the result would be that the two pendulums only swung in the same directions during *alternating* 90° segments of a complete axle rotation while they swung in opposite directions during the other alternating 90° segments of a complete axle rotation. Thus, their two swing directions during a full 360° of axle rotation were: same > opposite > same > opposite. This would then result in one pendulum's center of gravity always falling while the other pendulum's center of gravity was always rising and *vice versa*. As the axle rotated, there was, therefore, no change in the elevation of the *combined* center of gravity of the two pendulums which only oscillated through a short horizontal distance from one side of an imaginary vertical line passing through the center axis of the wooden axle to the other side of the vertical line. Because of this unique curving crank arm axial alignment, a bidirectional wheel, regardless of which direction its drum rotated, only had to expend a small and negligible amount of energy to keep the combined center of gravity and its two pendulums swaying from side to side horizontally while undergoing no net vertical motions.

Bessler does mention in his writings that the pendulums served the purpose of "modifying" the motion of his wheels. Due to their low natural resonance frequency of oscillation, the compound pendulums were intended to act as speed governors and reduce the maximum terminal rotation rate of a free running wheel in rotation. This was done to reduce the wear on the wheel's axle and internal weighted lever pivot pins during any long duration test and would certainly have been preferable to using some sort of external friction brake to limit the wheel's speed. During the beginning of a public demonstration, the two pendulums were probably only temporarily used to amaze the public because they gave the impression that the wheel was capable of outputting a lot of power. Most of the visitors viewing their actions would be unaware of how false that impression actually was.

Why none of the witnesses mentions them in the known Bessler literature is somewhat obvious. They were always removed during a demonstration of the wheel's ability to rapidly lift, via the flywheel effect, a heavy load because at that time it was necessary for the wheel to reach its maximum free running terminal rotation rate and angular momentum before the external load's lifting rope was suddenly attached to its axle's wing head screw. Only by disconnecting the two pendulums' drive rods and their curving arm cranks from the ends of the axle's pivot pins could that maximum speed to be reached. Also, when the agents of potential buyers were examining the wheel or it was being given an official test and the open bearings were being observed while the wheel was running, it was desirable to not have the pendulums swinging about and, possibly, hitting and injuring one of the examiners! I will have much more to say about these interesting pendulums in Chapter 6.

During the month of July in 1715, Bessler had an accident that nearly killed him. He had been walking down one of the streets of Merseburg and somehow slipped. Perhaps the street was paved with cobble stones and, having just rained, they were still wet and slippery (I once lived on a street paved with these stones and can personally attest to how dangerous they can be when wet). Upon landing, Bessler smacked his head and was so dazed that he could not get up without assistance. He had obviously sustained a serious concussion. As a result, he would have been immediately taken to a local doctor to be examined and then home to recover. Bessler's injury was so severe that it incapacitated him for nearly a month and kept him in bed and being tended by his wife and her maid. During that time his brother and perhaps another hired and trusted assistant would have been in charge of providing demonstrations of the Merseburg wheel to the crowds of people who would begin to gather in front of the house starting in the morning.

On the early morning of July 22nd, while Bessler was still lying in bed half asleep and hadn't even had breakfast yet, something quite unexpected happened. Bessler's home was visited by three individuals that Bessler had been previously communicating with by letters and whom he was very wary of because of their suggestions that his wheels were actually just clever fakes powered by conventional sources of power.

There was Andreas Gartner mentioned above and he was accompanied by a mining engineer named Johann Borlach and a mathematics professor from Leipzig University named Christian Wagner. (Wagner had actually visited Bessler a year earlier to view his Draschwitz wheel and even had breakfast with him. Wagner, apparently, was not that impressed with the performance of that wheel and Bessler heard nothing more from him after his brief viewing of it.). The three would have arrived bearing various letters of invitation they had collected prior to their relationships with Bessler turning sour and perhaps hoped, if Bessler was not at home at the time, that they could talk they way into the room where the wheel was kept in order to be able to minutely examine it and even try to see inside of its drum when no one was watching.

They were, however, met at the door by Bessler's brother who, realizing who they were let them enter and explained how ill Bessler was. He told them they could only see the wheel if Bessler authorized it and allowed Gottfried to use the only key to the room which Bessler would have kept on a strong cord hung around his neck. At that point Gottfried would have entered Bessler's sickroom and quietly told him of their arrival.

Perhaps, upon recognizing their names, Bessler hoped their personal viewing of the Merseburg wheel would help convert them into believers in its authenticity and so he gave Gottfried the key to use and permission to give them a private demonstration of the inventor's first bidirectional wheel. However, as a safeguard, he would have told Gottfried that only he could actually conduct the demonstration and the three visitors would have to keep their distance from it.

Standing behind the guard rails that had been installed around the wheel, the three unexpected visitors would have only been allowed to view it from a distance of about four to five feet and only then from the front side of its drum that did not have the cloth flaps covering the inspection and service holes that had been cut into the dyed linen cloth covering most of the drum's large back circular face.

Gottfried would have then chatted politely with them about the wheel, in general, and, after giving it a gentle push to start it turning, allowed it to accelerate up to its maximum terminal rotation rate of about 40 rotations per minute and then let it run at that speed for, perhaps, ten minutes or so. The dual oscillating pendulums would not have been attached. Possibly, to demonstrate that the wheel was outputting energy, the stamping mill machine was moved into position against the rear axle so that the eight projecting steel pins on it could repeatedly raise and drop its four solid wooden stamps. Bessler might have been able to hear the pounding sounds of the falling wooden stamps and knew that Gottfried was in charge of the demonstration as he wanted him to be.

During this brief at a distant examination of the Merseburg wheel as it ran at its maximum terminal speed, the three visitors may have noticed a few details that raised their suspicions. One was that the rear vertical support plank of the axle (on the left side of Figure 1(a)) was "hopping" or rapidly jumping up and down a fraction of an inch in synchronization with the wheel's rotation. To them, this would have indicated that support plank was probably hollowed out and there was something inside of it that was applying torque to the portion of the axle's rear pivot pin that passed through the plank and was hidden from view by it. Next, they might have noticed that there were several cuts made into the side of the rear vertical support plank near the axle pivot pin it supported. These cuts would have reinforced their belief that Bessler needed to expose the section of the plank there above the axle pivot pin so he could attach a long metal drive rod to a hidden crank that was part of the pivot pin.

Convinced that Bessler's 12 feet diameter, bidirectional Merseburg wheel had to be a clever fake and the inventor a charlatan, the three left and it was shortly after their unannounced visit that they intensified their efforts to refute and ridicule Bessler's claim of having constructed a genuine perpetual motion wheel. Borlach even published an illustration in which he showed how a hidden drive rod inside of the wheel's rear vertical support plank was used to power the wheel. As can be seen in Figure 1(b) at the end of this chapter that contains Borlach's illustration, that drive rod would have had to have passed through a carefully hidden hole drilled through the ceiling and then been connected to one end of a pivoted rocking beam. The other end of the beam would have had the end of a rope tied to it so that an accomplice in an adjoining room could periodically pull down on the rope as the drive rod to axle pivot pin crank connection passed its bottom dead center position. That action would then have caused the drive rod to pull up on the crank and thereby provide a momentary torque to the axle and the wheel to accelerate them. An enlarged view of the crank and its attached drive rod is shown above the upper right corner of Borlach's illustration in Figure 1(b).

This hypothesis, however, requires one to believe that the person doing the necessary yanking of the rope was somehow warned that there were three unexpected visitors viewing the wheel on the particular morning that Borlach, Gartner, and Wagner showed up and that he began working the rope just as soon as Gottfried gave the initially stationary bidirectional wheel its starting push. Thus, one must believe that an accomplice was kept in the hidden room throughout the daylight hours as visitors arrived to ensure that he would be ready at a moment's notice to manually accelerate the wheel during brief demonstrations.

This scenario seems highly unlikely to me although, admittedly, such a method using a hidden accomplice was successfully used during hoaxed public demonstrations of another claimed perpetual motion wheel in the city of Philadelphia, Pennsylvania, USA in 1812 and a year later in New York City, New York, USA constructed by an inventor named Charles Redheffer. His fraudulent wheel, however, was powered by a carefully concealed pulley system that used catgut cords (the axle of his far smaller wheel, unlike those of Bessler's wheels, was vertically oriented). The hoax was finally exposed during a public demonstration when the famous American mechanical engineer Robert Fulton (who developed the first commercially successful steamboat) removed some boards on a wall near the device to expose the catgut cords that propelled it. These cords were followed upstairs to a room containing an old man who was eating a piece of bread with one hand while trying to turn a crank handle with the other that moved the cords. The group of people at the public demonstration, realizing that they had been viewing a fake wheel after paying good money to see a genuine one, became enraged and tore the device apart while Redheffer ran for his life!

It is far more likely that the three visitors to Bessler's Merseburg wheel were merely noticing the jumping of the rear vertical support plank which was caused by a combination of two factors.

One factor was a slight misalignment of the center of gravity of the drum, weighted levers, and axle with the center axes of the axle's two end pivot pins that created a wobble in them at the wheel's maximum terminal rotation rate. That misalignment could have happened later *after* the wheel and its internal components had been perfectly balanced and might have been due the weight of the wheel causing a slight shift in the position of the axle's back pivot pin when axle was mounted on its two lower bearing plates. This wobble could, in turn, then apply a periodic lifting force to the rear wooden vertical support plank through the contact between the axle's rear pivot pin and the *upper* of the two brass half moon bearing plates installed in the plank that embraced the pivot pin.

The second factor was that the vertical support plank had been cut a bit shorter than 14 feet in length so that it could be fitted into position between the room's floor and ceiling surfaces which were exactly 14 feet apart. That created a slight gap of a fraction of an inch to exist between the top of the rear vertical support plank and the ceiling of the room when the bottom of the plank was in contact with the floor. That gap then allowed the rear vertical support plank to slide up and down a fraction of an inch inside of the boxes of wood pieces attached to the floor and ceiling that kept the plank's ends in position when the wheel achieved its maximum speed of about 40 rotations per minute. Thus, the rear vertical support plank would have "hopped" about once every 1.5 seconds and the motion would have been very noticeable to anyone who was positioned to get a good view of the rear vertical support plank as the wheel rotated at its maximum rate.

A similar "hop" was probably not noticed in the front vertical support plank because its bottom end that rested against the floor may have been better secured with nails to its box of wooden pieces that were attached to the floor to restrain its horizontal motion. Perhaps there was an oversight during the installation of the rear vertical support plank and the necessary nails to secure it to the floor's wooden box pieces were not used or, if they had been, then maybe the constant periodic upward jerking force on the rear support plank caused the nails to shear or somehow work themselves loose.

Finally, the cuts the three visitors noted in the rear vertical support plank itself would have been the seams between the plank's larger surfaces and the smaller sections of the plank that could be removed to expose the axle's end pivot and the half moon brass bearing plate upon which it rested.

Christian Wagner eventually constructed his own smaller bidirectional wheel and published its schematic in a pamphlet of his own in which he challenged Bessler's claim to having invented genuine perpetual motion wheels. Wagner's bogus wheel, which he openly admitted was not a perpetual motion wheel, used dual clock mainsprings that, after being manually wound up, would allow the wheel to run continuously for several hours at a fairly rapid pace. Once it had reached its maximum running speed of about 50 rotations per minute, it was even capable of lifting a small weight several feet into the air during a braking test. In Figure 1(c) at the end of this chapter, the reader will find the schematic for the mechanics used in Wagner's version of Bessler's wheels. Wagner's design is rather ingenious and he was quite convinced Bessler's used a similar mechanism in his various wheels.

In Wagner's wheel, two wound up mainsprings housed in barrel gears near the bottom of a simple linear train of gears eventually supplied pushing and pulling force to a connecting rod near the top which was attached to a crank pin formed from a bent metal rod whose two elongated ends held together two four sections of the wheel's segmented wooden axle. As the connecting rod attempted to move the crank pin, the entire clockwork mechanism would swing a little to one side of a vertical line passing through the center axis of the axle since the mechanism was, literally, hung by two long metal arms from two other exposed sections of the crank's elongated ends whose center axes coincided with the center axis of the axle. Note how compact the design is with there only being a fraction of an inch clearance between the mechanism's thin metal suspension arms and the wooden drum of the wheel. Wagner wanted to show people that he could replicate Bessler's wheels' performances with mechanics packaged into a similarly thin drum.

As the angular displacement of the clockwork movement increased and it swung farther onto what would be the drum's descending side, the torque applied to the split axle's crank pin would increase and, at some point, the axle and the drum would begin to rotate. This motion did not, however, cause the clockwork mechanism to rotate along with the drum and its axle. As drum rotation rate increased, the mechanism would continue to hang under the axle, but the initial angle that it made with the vertical would slowly decrease and so would the torque applied to the wheel's axle. At around 50 revolutions per minute, the reduction in the torque from the mechanism, which was then almost vertically oriented, would exactly equal the aerodynamic and various bearing drags present in the wheel and the rate of its rotation would then became constant. The design shown in Figure 1(c) was not self-starting, but was bidirectional if given an initial push start to begin its rotation. It drum probably had a diameter of about 6 feet.

Wagner never made any pretense that his wheel was "perpetual" and said that it would stop after only a few hours of continuous running. His calculations had also convinced him that Bessler's larger bidirectional wheels could not run for more than a few days even if their entire drums were filled with wound up mainsprings and he challenged Bessler to demonstrate one of his wheels running for longer than that duration. There would be a future test of the largest of Bessler's 12 diameter bidirectional wheels which would amply demonstrate how very wrong Wagner was. Also, note the extensive use of gears by Wagner's fake perpetual motion wheel mechanics. In his writings, Bessler repeatedly mentioned that his wheels did not use the same type of gear containing mechanics found in clocks. Wagner did not believe that and could conceive of no other way that Bessler's self-motive wheels might work.

For over a year Borlach, Gartner, and Wagner continued their efforts to refute Bessler's claims and said that the official tests of his wheels were just too short in duration to prove that they were genuine perpetual motion machines. Perhaps the real problem was that people like Gartner, a highly respected builder of mechanisms, considered it impossible for someone from Bessler's rather lowly social class to have found the secret of perpetual motion when they, with all of their years of experience in mechanics, had failed to do so. As for Wagner, he, being a mathematician, had already proved to himself, using mathematical analysis, that mechanical perpetual motion was a physical impossibility.

Nothing short of actually personally inspecting and testing Bessler's secret imbalanced wheel mechanics would have convinced these men that the former's self-moving wheels were genuinely perpetual and, of course, Bessler had no intention of giving them such invasive access to his invention unless paid a grand sum to do so.

Interestingly, all three of Bessler's major detractors seemed unaware of his first exhibited wheel in Gera. It was mounted in an open wooden framework which could be rolled on small metal wheels around a room to different locations. There was no possibility of any metal drive rod being hidden in one of its somewhat small vertical support boards which could then be powered from an adjoining room. However, if the three had known about that wheel, then they probably would have then just dismissed it as mainspring driven.

Again, considering Figure 1(b), the reader will note the guard rail around the Merseburg wheel which runs from the wall behind the rear portion of the wheel's axle, then makes a turn to its right and runs past the front end of the axle. That section of the rail has a table behind it. We only see part of that front section of rail, however, and it would probably have extended farther until it met a wall or actually made another turn to its right and then extended along the other side of the wheel until it again met the wall behind the rear portion of the wheel's axle again.

In previous public demonstrations, visitors had been allowed to freely approach and touch the Draschwitz wheel, but Bessler noticed that some of them were doing things when his back was turned that he could not tolerate. Some would place their ears near the wheel's rotating axle in an attempt to hear what sounds were being produced by its internal mechanics. Others, when the wheel's 9 feet diameter drum was still tethered to the floor, were even trying to secretly peel back some of the wooden slats covering its drum's sides so they could take a peek inside! With the Merseburg wheel, because of its linen cloth coverings and inspection holes and flaps on the back circular face of the drum, Bessler decided to use methods to better control the access of the visitors to the 12 feet diameter bidirectional wheel.

Now, upon entering the large room containing his first bidirectional wheel, they were kept about five feet away from it by a sturdy guard rail that surrounded the wheel. However, this rail would have had a locked down center section or gate in it so that only certain visitors Bessler trusted would, after the section was unlocked and raised, be allowed to enter in order to get a better view of his workmanship or to actually perform tests with the wheel under Bessler's close supervision. That central gate also served an extra and very profitable purpose.

Every once in a while there would be a heckler in the group of people inside the room viewing the Merseburg wheel. That person would loudly proclaim to all present that the wheel had to be a fake and there had to be weights hanging off of the hidden section of axle inside the drum that were somehow being used to propel the wheel. At this time, I believe Bessler would challenge the person to put his money where his mouth was. Bessler would ask the person if he was willing to bet a thaler (which was about a day's wages back then) that there was something attached to the axle and, if he lost his bet because there was nothing attached to the drum's axle, he would have to admit it to the crowd. If the person did feel a rope on the axle, however, then Bessler may have said he would pay the person a hundred thalers and admit he was a hoaxer! Bessler, of course, not wanting to profit off of such wagering would then promise to give the thaler to some local church to help the poorest in their congregation.

Many thalers would have gone to help the needy with these bets when the skeptics were allowed to use the central gate to approach the front side of the wheel. There a visitor could insert a hand into a small slit in the cloth covering the front side of the drum near its front projecting axle (Bessler, of course, would have always been present to make sure that the skeptic did not go around to the drum's backside and try to insert his hand under one of the eight large cloth flaps that were covering the drum's eight inspection and service holes there which had been cut into the linen covering the drum's back face). All that the disbeliever would then be able to feel would be the smooth surface of the hidden portion of the axle, some of the wooden radial frame pieces extending away from it, and the small wedge shaped wood pieces that helped secure the radial frame pieces to the axle. There were no chains, leather belts, or ropes that could be felt. No weights, gears, or pulleys. In short, the person's hand felt nothing hanging from the axle that could be providing a propelling torque to the axle. The lost wager and a former heckler's public confession that he had been wrong would then prove to the others present that Bessler's wheels were not the fakes that they may have heard they had to be.

The official test of the Merseburg wheel was done on October 31st of 1715 by a committee of "learned men" from the area. The details of that test were documented in a book that Bessler authored and published in 1716 called *Apologia Poetica Perpetuum Mobile* in Latin which translates to the English *Poetic Defense (of his) Perpetual Motion* which I will hereafter simply refer to as "AP".

We learn from it that the official testing was simply an external examination of all parts of the wheel with special attention paid to its axle's two steel end pivot pins to make sure that nothing was concealed in the vertical support planks that might be providing propelling torque to the axle through those pivot pins.

Each of the axle's two end pivot pins' bearings could be "opened" or exposed for visual examination while a pivot pin rested on the *bottom* "half moon" brass bearing plate of the two plates that were embedded in a vertical support plank. The term "half moon" here refers to the cross sectional shape of the semi-cylindrical recess that had been formed in the brass plate upon which half of a cylindrical steel pivot pin rested. The two half moon brass bearing plates themselves would have been formed by simply drilling a single cylindrical hole with a diameter slightly larger than the diameter of an axle's pivot pin into the center of a solid brass cube and then carefully sawing the cube in half along the center axis of the hole. No doubt, Bessler used this simple type of bearing on all of his perpetual motion wheels with the possible exception of his 3 feet diameter, one-directional Gera prototype wheel whose design will be given in detail in a later chapter.

When an axle's pivot pin's bearing was "closed", the bearing's two half moon brass plates would then completely surrounded their steel pivot pin yet were just loose enough to allow the pivot pin to freely rotate inside of them as long as it was well lubricated. Once an axle's pivot pin bearing was closed, it was physically impossible for the steel pivot pin to slip laterally out from between its two brass bearing plates which, should that have happened, could then have allowed that end of the axle to roll itself completely out of the vertical support plank and fall toward the floor.

This simple type of bearing construction allowed an examiner of the wheel to completely view an axle's pivot pin as it rotated and was supported only by its lower brass bearing plate while the wheel was turning and to assure himself that there was nothing inside of the vertical support plank that was moving the pivot pin and the axle in which its other end was embedded. However, to do so he would have to stand on a small wooden box or platform in order to see the rotating pivot pin up close.

During the official examination of the Merseburg wheel, it was made to run in both directions repeatedly and also used to demonstrate its quick lifting of a heavy load via the flywheel effect,. As part of the examination, Bessler was even required to move the entire drum and its attached axle to another nearby set of wooden vertical support planks to absolutely eliminate the possibility that the original supports were supplying it with power. (Perhaps to accommodate this extra set of vertical support planks, the railing on one side of the room was removed. If so, then, most likely, it would have been the railing shown in Figure 1(b) that was located between the right side outer rim wall of the drum, as viewed from the front of the axle pointing toward the cross legged table, and the wall in the background with three windows in it, one of which is in the hidden room behind the rear vertical support plank, which would then allow the same window shown on the right side of Figure 1(a) to be used for the rapid hoisting of weight loads with the pulley mounted outside of that figure's window.) We can be certain that, once the axle's end pivot pins were off of their original set of lower brass bearing plates whose half moon shaped recesses they normally hid from view, those plates and their recesses were again carefully scrutinized to make sure that they were solid brass and did not contain anything that could possibly contact and provide torque to the axle's pivot pins.

The 12 feet diameter Merseburg wheel's drum containing all of its mechanics and its attached axle with its two steel end pivot pins had a gross mass of about 550 pounds.

After removing its 48 cylindrical lead end weights and their 48 attachment bolts, which had a total mass of 192 pounds, from the ends of bidirectional drum's 16 weighted levers' parallel pairs of main arms (remember that there were three end weights and attachment bolts per lever), Bessler would have reduced the drum and its attached axle's remaining mass down to about 358 pounds. That was just light enough so that he and an assistant could, by reaching over their heads and grasping opposite ends of the axle, lift the axle's two end pivot pins off of their opened up lower brass half moon bearing plates and then, after moving the axle a foot or so horizontally away from the support planks, gently lower the bottom middle surfaces of the projecting sections of the axle so that they came to rest in two padded cradles attached to the tops of temporary vertical support planks that were themselves attached to a reinforced cart.

These cradles would have been shaped like circular pockets with a depth of 2 inches and a radius of curvature of 3 inches so as to prevent the Merseburg wheel's 6 inch diameter axle from rolling off of them. The bottommost inside surface of each cradle's circular pocket would have been exactly 6 feet and 9 inches above the floor and that would have put it *less* than a foot away from the axle's center axis when the axle was mounted normally on its usual vertical support planks.

Since the center axes of the axle end pivot pins, when normally mounted in their usual vertical support planks, were exactly 7 feet above the level of the floor, they would have had to have been raised about 3 inches upward to a height of 7 feet and 3 inches off of the floor before the 6 inch diameter axle was moved over to and then lowered onto the cart's two 2 inch deep cradle pockets that would then hold the axle high enough off of the floor so that the center axes of its axle pivot pins were still 7 inches above the floor. This means that each man would have had to have lifted the bottommost curving surface of his end of the axle from a height of 6 feet and 9 inches above the floor to a height of exactly 7 feet above the floor or through a vertical distance of only 3 inches in order to "seat" his side of the axle into the cart's cradle pocket nearest him.

Assuming that both men were at least 5 feet and 8 inches tall, then, with sufficient strength, they both should have been able to simultaneously lift the bottom surface of his end of the axle to a height of exactly 7 feet above the floor, which was 1 inch more than the minimum necessary for the bottommost curving surface to his end of the axle to clear the highest surface of the cart's cradle nearest him before his end was then lowered into the pocket of the cradle.

If the elbows of their raised arms were 2 inches above the tops of their heads as they both fully extended their arms directly over their heads and both also had elbow to web of thumb distances of 14 inches, then we get the maximum distance of the webs of their thumbs above the floor by simply adding their heights, or 5 feet and 8 inches, to the distances of their elbows above the tops of their heads, or 2 inches, to the distances from their elbows to the webs of their thumbs, or 14 inches which is 1 foot and 2 inches, to finally get a distance that places the webs of their thumbs at a height of exactly 7 feet above the floor. This detailed analysis demonstrates that it was physically possible for only two sufficiently strong men to have moved the Merseburg wheel to and from a specially designed cart during a translocation of the wheel.

Obviously, performing this type of maneuver as quickly and safely as possible twice during an official test of the Merseburg wheel (once to remove the axle and attached drum from its original set of vertical support planks and place them on the cart and again to move them from the cart to the second set of vertical support planks) would have required the two to have practiced the feat several times in advance. They would have positioned the cart's two cradles as close to the axle as possible so that should one of the men weaken during the two man lift, then, as his end of the axle fell, it would have landed either back on the bottom half moon brass bearing plate in the axle's wooden vertical support plank or in one of the cart's two cradles and not fallen toward the floor between the axle's vertical support plank and the cart's cradle. Should that have ever happened, the impact would have seriously damaged the wheel's drum.

This cart would have had four small metal wheels, two of which were caster type wheels, attached to its base which then would then have allowed the two men to easily push the entire 12 feet diameter, bidirectional Merseburg wheel's axle and attached drum, once emptied of its 48 end weights, to its second set of permanent vertical support planks in another part of the wheel's room. The two men would then have had to have been able to twice press upward through a distance of only 3 inches with a force greater than about 179 pounds each in order to accomplish this which, with some effort, is certainly possible for two young men in prime physical condition. If two additional volunteers with a height each of at least 5 feet and 8 inches joined in, then each of the four men would only have had to press upward with a force greater than about 89.5 pounds making the task much easier. With six men doing the lifting, the load each would have to lift would be reduced to only 59.6 pounds.

Once the men could feel the axle's two end steel pivot pins resting on the flat horizontal surfaces of the cutout sections of the wooden vertical support planks that contained each pivot pin's lower brass bearing plate, they would have given a horizontal push to the axle ends until their pivot pins either slid or even rolled over the flat lubricated surfaces of the planks' cutouts and finally seated themselves into their brass bearing plates' half moon shaped recesses. Only after so secured, would the translocation cart have been moved away.

As an alternative to applying upward pressing forces to directly lift the axle off of its two lower half moon brass bearing plates, two separated overhead block and tackle pulley systems attached to the ceiling beams over the external axle sections could have been used. Their lower moving blocks would have been attached to straps placed around the ends of the drum's axle and, if each pulley system increased the lifting force applied to the moving block by a factor of four, then Bessler and his assistance could have easily lifted the axle off of its bearings by pulling on each pulley system's rope with a downward force greater than about 32.25 pounds.

Study of the left side profile view of the Merseburg wheel in Figure 1(a) reveals that there was some sort of a narrow band of material attached on each side of its outer rim wall near its two round drum faces which I've concluded was most likely just a series of thick strips of leather that had been tacked to the wood of the rim wall. However, these strips of leather were *not* intended to allow the drum, once lightened, to be placed on a smooth floor and then rolled over to a new set of vertical support planks without damaging the outer rim wall. This was because, despite the rigidity of the Merseburg wheel's drum, the sections of wood which formed its outer rim wall were not strong enough in most places to even support the reduced weight of the drum, about 258 pounds, after all 48 of its cylindrical lead end weights and their steel attachment bolts had been removed. If the drum was placed directly on the floor so that its axle's center axis was parallel to the plane of the floor in an attempt to roll the drum on its rim wall from one location of a room to another, then the wooden frame pieces to which the drum's outer wooden rim wall was attached would have quickly broken and splintered and the wheel would then have required extensive reconstruction to make it operational again.

Prior to the official test, all of the rooms surrounding the one that contained the wheel were inspected to make sure that there was nothing suspicious going on in them. Bessler's Merseburg wheel easily passed its various performance tests and on December 4th of 1715 the examiners signed and presented to Bessler a certificate that stated that they considered his invention to be a genuine "perpetual motion...having the property to move right and left, easily moved, but requiring great effort to stay its movement; with the power of raising...a box of stones weighing 70 pounds 8 ells high perpendicularly..." which would have been to a height of about 15 feet. (Note than an "ell" at the time was a unit of length used in the city of Leipzig and was equivalent to about 22.3 inches, but not all towns defined "their" ells by the Leipzig standard.)

Unfortunately, although the committee of experts was allowed to externally examine the Merseburg wheel and test its performance capabilities, as with the second Gera and Draschwitz wheels that were tested by others, they were not allowed to view its internal mechanics which were always kept carefully concealed from view by the opaque sheets of darkly dyed, oiled, and tightly stretched linen that Bessler had covered the drum's sides with in order to prevent its secret from being disclosed. They also only tested the wheel's running duration in both of its two possible directions of rotation for about a half an hour prior to and after its translocation to the new set of vertical support planks. That's a total running time of only about an hour.

Critics of Bessler's wheels immediately claimed, however, that this "official" test had not been nearly long enough to establish that Bessler had, indeed, constructed a genuine working perpetual motion machine. They said that the time it ran for could just as easily have been performed by a wheel concealing a hidden spring powered clockwork mechanism. They also found it troubling that, after its translocation, Bessler was allowed to install the weights back in the wheel's large drum out of view of the examiners behind its rear side. They were suspicious that, while out of sight behind the drum when the cloth inspection and service flaps located there had been opened for the reinstallation of the bidirectional wheel's 48 cylindrical lead end weights, Bessler might have been secretly winding up a set of mainsprings inside of the drum whose tension actually provided the torque to the axle that made the giant wheel rotate during its brief running time. But, despite these valid criticisms, the fact remains that none of the examiners of his invention could find any obvious indication that he was perpetrating a fraud even though they made every reasonable effort to do so by examining the wheel's exterior and its environment.

An interesting incident happened during the official testing of the Merseburg wheel that gives a very valuable clue as to how its internal mechanics worked.

After the bidirectional wheel had been moved to its new set of vertical support planks, but before its 48 cylindrical lead end weights and attachment bolts were reinstalled on the ends of the drum's 16 unweighted levers to ready the wheel for additional testing, Bessler actually let some of the assembled witnesses handle a few of the end weights which he had carefully wrapped in some handkerchiefs. According to a letter written later to Leibniz by one of the people to handle one of the end weights named Christian Wolff, it weighed about 4 pounds and was, very obviously, cylindrical in shape. As previously mentioned, my research indicates that it was 1.78 inches in diameter by 4 inches in length or about the size and shape of a small juice can.

Of particular note is the fact that, while the examiners were allowed to handle these weights, they were not permitted to touch the flat circular ends of their cylindrical shapes. By this prohibition Bessler was hiding the method he used to attach these lead weights to their levers within the wheel. Each cylindrical end weight had a long narrow shaft bored out of it along its center axis through which a slightly longer, snuggly fitting bolt was passed in order to attach the weight between the parallel pair of wooden main arms of the unweighted lever that would carry the end weight.

When the translocation of the Merseburg wheel was completed, Bessler began re-installing the 48 cylindrical lead end weights back into the bidirectional wheel's drum through the inspection and servicing openings in the linen cloth covering on the backside of the drum which was facing away from those gathered in the wheel's room. At that time, all of the examiners would have been required to stand patiently near a location in the room from which they could only see the front circular face of the relocated 12 feet diameter drum.

At one point, those assembled noticed that Bessler seemed to be struggling to attach one of the cylindrical lead end weights to its lever when suddenly there was a sharp cracking sound as though he had lost his grip on the lever which, being under spring tension, then slapped loudly against another structure inside of the drum. This sound was immediately followed by a brief twanging sound.

These sudden, unexpected sounds were so loud that they actually startled those present and it was obvious to them from Bessler's location and arm movements on the hidden side of the drum that the end weights were being attached to some sort of arms near the periphery or outer rim of the drum. What was not obvious to them was that each of the drum's unweighted levers to which three of these end weights were to be attached was itself attached by a pair of cords to two steel extension springs that were then both anchored to a pin at a certain nearby location in the drum whose center axis was a critical distance from the center axis of the axle.

As the inventor struggled to pull on the lever's parallel pair of main arms so that he could attach the first of their three end weights back on it, the two main arms just slipped out of one of his oil stained hands and then slapped against another structure inside of the drum due to the force applied to it by the two stretched extension springs that were attached to it. In this case, since there was no end weight yet attached between the parallel pair of the unweighted lever's main arms, it was only those two wooden arms which had slapped violently against that lever's wooden radial stop and had produced the loud cracking sound. The twanging sound that followed would have been due to the sudden contraction and vibration of the unweighted lever's two attached steel springs. Because of the mass of the unweighted lever, the springs could not completely contract and hold the lever's main arms in firm contact with the wooden radial stop piece they had struck. They would have been remained slightly stretched so that they could continue vibrating momentarily.

I have often wondered what Bessler would have said to those gathered for the official test when this unexpected event occurred. It would have been obvious to them that the inventor's design employed springs and he might have then immediately reassured everyone that they had, indeed, heard a spring, but that the wheel was not actually powered by these springs as was the case in a mainspring driven clock movement. Once that was out of the way, he would have hurried to install all 48 weights back in the 12 feet diameter drum. If it took him 60 seconds, on average, to install each weight and tighten its attachment bolt, then the task would have required 48 x 60 seconds = 2,880 seconds = 48 minutes. Most likely Bessler remained behind the wheel's drum and took the weights and attachment pins from wooden trays containing them that were placed on a table close to him.

I think this incident also provides us with a general idea of how the inventor would have initially installed the 48 cylindrical lead weights back into the Merseburg wheel's front and back internal one-directional wheels.

As viewed from the front side of the drum, Bessler would have initially been seen standing just behind the left side outer rim wall of the drum with the undone linen flap hanging down from the rim of the drum. His assistance would have been near him, but remained on the left side of the outer rim wall so that he could not see through the opened inspection hole and into the drum's interior. The Bessler would have had a table next to him, perhaps the cross legged one, upon which rested a wooden tray containing some of the cylindrical lead weights and bolts which were covered by a sheet of darkly dyed linen.

The tray probably held no more than a dozen weights and their attachment bolts at a time (enough for 4 levers) whose mass was 48 pounds which, along with the mass of the tray, might have had a total mass of 60 pounds. As the tray emptied, Bessler would have removed it and replaced it with another covered tray while his assistant continued to brace the giant drum. The assistant acted as a human brake by leaning against the outer rim wall of the drum so as to prevent the drum's rotation while Bessler, one by one, reattached the lead weights to the ends of the main arms of the levers.

Assuming that the loud sound that startled everyone was produced as the unweighted lever slipped out of Bessler's grip as he was trying to reinstalled the first weight into the drum from its back side, then he was probably reaching deep into the drum, past its back internal one-directional wheel, to grasp the unweighted lever of the *front* internal one-directional wheel whose pivot pin's center axis was located at the drum's 9:00 position as viewed from the front side of the drum by the assembled witnesses during the official testing. The proximity of that lever to the cloth covering on the front circular face of the drum is why the sound of the lever's main arms striking their radial stop piece was so loud.

Possibly, instead of having grease slicked hands, Bessler was using some installation method that required him to both pull an unweighted 9:00 lever down with his left hand while also holding the cylindrical lead weight and trying to align its central bolt shaft with the holes at the ends of the lever's parallel pair of main arms so that he could then slip the attachment bolt through the hole facing him, through the shaft drilled through the center axis of the weight, and, finally, by using his right hand, begin finger tightening the threaded end of the bolt into a square nut that was actually embedded in the outside facing surface of one of the main arms. Once the bolt was partially threaded into the nut, the inventor, while still holding the lever and its newly attached end weight in position with his left hand, would then finally tighten the slotted head of the bolt with a short shank screwdriver using his right hand. This is a somewhat clumsy way to attach the weights, but Bessler would have been forced to use it since he had to do the job completely by himself and, for security purposes, to also work through a single servicing hole in the linen covering on the backside of the drum. He had probably gotten much practice with the method from his initial installation of the weights after the linen had originally been attached to the drum, but, even with much practice, one can still have an accident, especially when he is trying to rush a task to keep witnesses from becoming impatient.

Bessler would have, working from the backside of the drum, installed all 24 of the front internal one-directional wheel's cylindrical lead weights first, and then, after moving over to the other side of the back of the drum, which would then make him appear just behind the right side of the drum's outer rim wall as viewed from the front of the wheel, continued to install all 24 of the *back* internal one-directional wheel's cylindrical lead weights to that internal wheel's eight levers. This time, however, he would have been pulling down on unweighted levers whose pivot pin center axes were located at the 3:00 position of the drum as viewed from the front of the drum.

It is a good thing that the official test of the Merseburg wheel had not been conducted later and shortly before Christmas of 1715 because a rather embarrassing thing happened as a visitor during a public demonstration was witnessing the wheel in rotation for several minutes. The wheel suddenly slowed by itself to a complete stop! The witness was a bit surprised and asked Bessler what had just happened to which the Bessler said there was some sort of unexpected drag on its mechanical parts that had caused the stoppage.

Investigation eventually showed that the problem was due to the cold air in the wheel's room causing the lubricants he had used, lard for the axle pivot pins and olive oil for the various lever pivot pins, to congeal. The consequence of this was to then a cumulative increase in the viscous drags acting on the wheel's moving components that exceeded the maximum start up torque produced the imbalanced center of gravity of its one active internal one-directional wheel and the wheel came to a stop. To prevent such embarrassing failures in the future, especially if a potential buyer might be viewing the wheel, Bessler would have made sure that he did not allow the room temperature to get too low on cold days. The addition of a simple wood burning stove would have sufficed and also kept his visitors more comfortable. The fact that this large 12 feet diameter, bidirectional wheel, weighing around 550 pounds, could be gradually slowed to a stop by the viscous drag of cold lubricants attests to how low its axle's maximum start up torque really was.

After the Merseburg wheel's successful official examination and testing, the numbers of people visiting Bessler's home increased dramatically. It then became necessary for Bessler to hire assistants in order to handle the crowds and all of these employees would have been required to take a sacred vow not to attempt to learn the secret of his wheel. To pay for them, Bessler began to charge a small admission fee to each visitor. But, still the crowds swelled. Bessler then added a "poor box" for visitors to see before entering the line of people waiting to view the famous wheel. Any money deposited in it would then be donated to help the poor of Merseburg. Being almost entirely Christians, the arriving visitors were morally obligated to make a donation and, quite to Bessler's surprise, this box tended to suppress the size of his crowds!

Then, in late 1715, Bessler received a letter from Gartner who had found out about the admiring crowds at the former's home, the admission fee, and the newly added poor box. In the letter Gartner suggested, aside from being a charlatan, that Bessler was also a crook who was gouging visitors for extra money by taking advantage of their charitable Christian generosity. Needless to say, Bessler's contempt and disgust for Gartner escalated dramatically after this and vice versa.

The beginning of 1716 had another unpleasant surprise for Bessler. In January, a tax law was passed by authorities in Dresden requiring all businesses to put aside a small amount of money from a day's income which was needed by the government for various projects. Bessler was not opposed to that, but it required him to also remove some of the money that accumulated in his poor box and he felt that was a violation of his interpretation of the Bible concerning the treatment of the poor. Just coincidentally, all of Bessler's major detractors happened to reside in Dresden at the time and it was only natural that Bessler assumed they were behind this new tax that he was obliged to pay. Once again, he was seriously considering destroying his latest wheel and moving on to another town that would be exempt from such an onerous tax. There he would build yet another wheel and it would be the grandest yet.

Sometime in June of 1716 Bessler received two very important letters. They were from Count Karl von Hesse-Kassel, the reigning landgrave of the small and independent state of Hesse-Kassel which was located in a part of Saxony that is now in central Germany. He had heard of Bessler's Merseburg wheel and the successful testing it had undergone and was fascinated by the possibilities of such an invention *if* it was real. He wanted Bessler to visit him so they could discuss the matter of his wheels in detail and what the count might do to help Bessler find a buyer for them. Bessler probably could not believe his eyes as he read these letters. The count wanted him to come to his castle in Hesse-Kassel and there was the possibility that he might fund Bessler's further development of his wheels if he could be convinced they were real. Karl would also pay all of the expenses for Bessler's trip if he agreed to make it.

Karl was an enthusiast of the newly developing sciences, well versed in mathematics, and had a huge private collection of scientific instruments, clocks, steam engines, and many different kinds of machines. He also collected various art objects, gems, coins, and maps and even built a museum for them that his citizens could visit and enjoy viewing. He was a big believer in trying to raise the educational level of the people in Hesse-Kassel, especially of the children. Fascinated by the abundance of new information coming forth about our solar system by the invention of the telescope a century earlier, Karl even established an astronomical observatory. (Karl, by the way, was the German equivalent of the English name Charles, a name which became popular in continental Europe after the reign of Charles the Great or "Charlemagne", born in 742 AD and died 814 AD, who was a Frankish king that eventually came to rule most of Europe.)

The count had previously funded other inventors such as Denis Papin who the French to this day claim was the original inventor of the steam engine. One of the projects the two worked on was the development of a novel steam powered piston pump that the count hoped could pump water up to a supply tank on top of his castle after which it would flow down through pipes to operate various small fountains in the gardens surrounding the building. Unfortunately, the pump Papin came up with worked very inefficiently because of excessive leakage of the high pressure steam from its various joints and valves and the project was abandoned.

Ever since Papin had left his wife behind in Saxony and retuned to London in 1707 hoping to get employment from the Royal Society and also to try to develop and sell his steam engine technology there (sadly, he was unsuccessful and finally died in poverty there in 1712), the count had been looking around for some new and promising inventor to fund and when he had heard of Bessler and his marvelous self-moving wheels, he believed he had finally found him.

Karl had come to power as a landgrave or count in 1677 after he had married, at his mother's insistence, the widow, Marie Amalie, of his deceased older brother in 1673. During their 38 year long marriage Amalie would eventually bear him 17 children, but only nine of them lived to be twenty years of age or older.

As the hereditary count of Hesse-Kassel, Karl had, via taxation, access to the wealth of an entire state. But, he was not a greedy nobleman. Most of the money he took in went out to benefit his small state with certain types of public works projects, infrastructure improvements, education, etc. He was also in demand by the aristocracy of various nations to act as an impartial arbitrator of any disputes they might have with each other. As a result he was involved in peace treaties that prevented wars and was known and, more importantly, liked by the heads of nations like England, Sweden, Prussian, and even Russia. Counterbalancing his efforts to achieve peaceful settlements between states, he had no qualms about hiring out his own soldiers to serve as mercenaries in the wars of other states and is probably the first to practice this. The income from that further enriched his small principality. His political influence was enormous considering the size of the territory he ruled and he was on a first name basis with some of the richest and most powerful people in Europe. He was, in short, "the" person that Bessler needed to aid him in his efforts to find a rich buyer for his revolutionary invention. The fact that Karl was a Calvinist and Bessler a rather pious Lutheran was not an insurmountable obstacle to an alliance between the two.

However, before Bessler could receive Karl's patronage, the count insisted that the inventor would have to prove that he actually did have a genuine perpetual motion wheel. That, of course, immediately presented a problem. Bessler had sworn he would not reveal the secret to anyone until *after* he had sold the invention and had physical possession of the money, yet Karl wanted to know the secret before he would fund the construction of a future improved version of the Merseburg wheel and help Bessler find a buyer for it. The count could not risk the damage to his reputation that would occur if he was taken in by a clever hoaxer.

Finally, a compromise was reached after the exchange of several letters. Bessler would reveal the secret to Karl and, if he was convinced that Bessler had actually done what most of the intelligentsia of the day considered impossible, the count would pay Bessler a fee of 4,000 thalers for the privilege of being shown the secret and give his solemn word never to reveal the secret or try to use it himself before the invention was sold. Karl, eager to learn the secret and also to put his mind to rest about Bessler's invention possibly being a clever hoax even signed a brief oath to this effect which would serve as proof that he had made such a promise in the unlikely event that he ever broke it. To sweeten the deal further, if the wheel was genuine, then the count would also provide Bessler with employment in Hesse-Kassel, lodging at his splendid Weissenstein Castle residence for the inventor and his extended family, and, of course, whatever funds he needed to improve the wheel so that the chances of it being sold to one of the many rich and influential visitors to the castle would be maximized.

It was the deal of a lifetime and, obviously, something that Bessler could not afford to let slip through his fingers. The dark side of this, however, was that if Bessler could not convince the count that he had the real thing, then he would forfeit the 4,000 thaler payment and the count would immediately denounce him as a liar and a swindler. His reputation would then be so tarnished that he could forget about ever selling his wheel and would have to go back to peddling medicinal potions and salves for the rest of his life. If Karl had been a tyrant, he might even have had Bessler executed in order to make an example out of him as to what can happen when a charlatan shows up at his court and tries to swindle him out of the money that, ultimately, belonged the citizens of Hesse-Kassel and needed to be spent as wisely as possible so as to derive the most benefit from it for as many people as possible.

I have often wondered how Bessler managed to reveal the secret of his wheels to Count Karl. The Merseburg wheel was massive and not portable and the count, a man getting on in years, would certainly not have wanted to travel over a hundred miles to Merseburg to view the wheel and do so in an environment which might help Bessler deceive the count in the event that his wheel was a fake. I can think of only one likely scenario.

Sometime in August of 1717 Bessler would have hired a horse drawn coach in Merseburg to deliver him to Weissenstein Castle in Hesse-Kassel for a private demonstration of his wheel. For this demonstration, he would have taken his small and portable 3 feet diameter Gera prototype wheel with him (the same one whose construction details will be given later in this book). Bessler, an accomplished equestrian, could have easily made the trip on horseback by himself, but it would have been cumbersome to carry the 3 feet diameter wheel's drum while on horseback and he would not have wanted to risk having it damaged by the continuous jarring motions to which it would have been subjected during the journey.

The town of Hesse-Kassel is located about 120 miles to the west of Merseburg and would have required Bessler and the coachman to make an overnight stop at a midpoint way station that provided a bed for travelers on a trip exceeding the limit of their horses to complete in a single trotting run. The travelers would rest there, dine, and then sleep while their horses did much the same. The next day, they would been refreshed, washed up, had breakfast, and been ready to complete the second half of the trip. That night would have been one in which Bessler's mind was filled with the possibilities awaiting him. Of course, he would have kept the wheel and its portable base right in the private bedroom with him and the door securely locked all night. That wheel's drum and base, wrapped separately in two large cushioning pieces of cloth, would at no time have been out of Bessler's possession.

The next day the coach would have arrived at the magnificent Weissenstein Castle. Bessler had never been there before and had only heard of its grandeur. But, now he was right in the middle of the huge estate and about to meet one of the most powerful men in the region!

Bessler, tightly clutching his invention would have been conducted into a grand hall where the count greeted visiting dignitaries and held occasional parties for his fellow noblemen. There was no regal throne, only a large room with an ornate fireplace and a massive table on one side surrounded by comfortable chairs where Karl could greet people and discuss business with them while servants kept them supplied with food and beverages. Karl was a man more interested in promoting mutually beneficial trade deals than one concerned with putting on displays of wealth and power.

Upon the announcement of Herr Doctor Orffyreus' arrival, I can imagine the count, then 62 years of age, quickly excusing himself from any company he had and rushing over to shake the inventor's hand while telling him how impressed he had been with the stories he had been hearing about the fabulous Merseburg wheel. Bessler would have been nearly in shock to be suddenly in the company of someone he had heard about for years, but never previously met. All he would be able to say was that he was greatly honored to be meeting Karl and that he had brought a small model of his wheel to show him since the terms of the revelation of its secret mechanics had been previously arranged and agreed to by both parties.

The wheel? Oh yes, there would be plenty of time for that later in the evening, but, since it was probably late afternoon by the time Bessler arrived at the castle, dinner would have to come first. I can imagine Bessler being seated at a long dining table with Karl at the head and surrounded by those who lived at court with the count or, like Bessler, were just visiting him while the count's many servants scurried about making sure everyone got their meal courses on time.

Bessler would have sat there slowly eating the best of cuisine, drinking the finest of local wines, smiling at those around him, and, of course, keeping an arm in reassuring contact with the drum of his wheel at the his side just in case someone would try to slip away with it. Those questioning the strange circular bundle the inventor was so protective of would have been told that it was a special surprise gift for the count that Bessler intended to give him in private later and that would have ended any further inquiry.

His invention and all of the hard work required to achieve it had finally landed him right at the heart of the economic and political power of a small, but influential state and now he sat at the same table with the leader of that state! To Bessler, this would have seemed like a miracle and the result of divine forces working in his favor. He had had some contact with aristocracy in the past, but nothing could compare with what he was experiencing at Weissenstein Castle.

After dinner and a bit of socializing, the count would have approached Bessler and asked him to join him in his private study so that he could finally find out if the young inventor actually had something real or was mistaken or, worst of all, was trying to pass off a fake to the count and defraud him and his state out of a not insignificant sum of money by doing so. Karl would have had guards at the door of the study who, at his command, would have grabbed Bessler and ejected him and his bogus wheel from the castle if it turned out he was wasting the landgrave's valuable time.

With the door to the study locked, Bessler would have carefully unwrapped his model wheel and its base and, after sliding the drum's axle pivot pins into the bearings of the base's vertical wooden support pieces, placed them on a nearby table that was well lit by several oil lamps. At this point the toy wheel would have been stationary with its axle's end pivots clamped tightly in place by two wing head bolts located at the top of each of the vertical axle supports. The sides of the drum, as mentioned previously, had been covered with sheets of dark colored linen that were secured to the drum with small tacks at strategic locations. Bessler would have then stood aside and beckoned the count to approach and personally loosen both of the brake bolts that prevented axle rotation.

Karl would have done this and noticed that, once free to turn, the wheel slowly began to accelerate and, after perhaps a minute or so, had reached a dizzying speed. It was incredible, but the count, quite familiar with mechanics, knew this could easily be achieved with a hidden clockwork mechanism running off of a wound up mainspring inside of the drum. He needed to actually *see* the drum's inner mechanics in operation to be finally convinced. It was time to open the drum up and reveal the secret it had taken Bessler a decade to discover.

When the pieces of stretched linen were finally removed from both sides of the 3 feet diameter drum of the Gera prototype, Karl would have beheld a set of eight small wooden 3 arm levers whose thin steel pivot pins, firmly set in the drum's radial frame pieces, formed the vertices of a perfect octagon. However, each lever did not consist of just a single 3 arm piece, but, rather, two such identical pieces that were parallel with each other and held apart by additional glued in slips of wood and pressed in metal pins. Between *one* pair of each lever's three parallel pairs of wooden arms, its main arms, a small, 6 ounce, lead ingot weight was securely held by two small screws.

Each of the toy wheel's eight weighted levers also had a set of strong thread-like cords attached to various pins inside of it that were located at certain precise distances from the center axis of the lever's fixed pivot pin. Furthermore, each weighted lever had two cords that attached it to two small stretched springs whose other ends were attached to another pin that pierced and was held securely in place by another parallel pair of the drum's frame pieces. The entire device was very carefully made and quite delicate from its appearance. Karl had seen and examined many intricate mechanisms in his time, but none of them were capable of achieving sustained imbalance during the rotation of the device as did the arrangement of interconnected weighted levers he now beheld.

Bessler would have explained to Karl that it had taken him many months of calculations and laborious trial and error experiments to find the precise shape of the levers to use, the ratio of a lever's mass to that of its attached end weight's mass, the various points within a lever to attach its various coordinating cords, and the cord lengths and spring constants necessary so that the center of gravity of all of the weights and levers was, when the drum was stationary, located on the its descending side and, more importantly, would remain there as the drum began to turn and accelerate as a result of this persistent imbalance.

Perhaps, to convince the count that there was nothing inside the model wheel's wooden support base that was responsible for its motion, Bessler even removed the axle and drum from the base's vertical support pieces and, holding the two axle pivot pins at arm's length on his two upturned pinky fingers which had been made slippery with a bit of olive oil, showed Karl how the wheel still accelerated. Perhaps the inventor even let the count do this himself!

So, there it was: a genuine working imbalanced type perpetual motion wheel.

The count would have manually stopped the drum and then released it several times and, each time, it would have immediately begun accelerating and, if allowed to run, it would finally reach a speed of more that one rotation per second. At that speed the open framework of the drum would have been creating a noticeable breeze and Karl would have marveled that the design was so simple. So simple, he would later tell people, that a mere carpenter's young apprentice could easily duplicate it after being allowed to study the parts and their motions for a while. Indeed, the count was amazed that no one else had discovered this arrangement before Bessler. Karl could see from the small wooden pieces in the prototype wheel that there was no possibility of any sort of hidden clockwork mechanisms being used. Bessler had finally done it! He had a genuine working perpetual motion wheel and Karl would be the one to help the inventor find a buyer who would then make it available as a power source to the whole world!

Karl, however, was examining a model of a one-directional wheel and he would naturally have asked Bessler how he made a bidirectional wheel work. Bessler would have explained to the count that his recently constructed, bidirectional Merseburg wheel was actually just two side by side, back to back one-directional wheels built inside of an enlarged drum and that he had figured out a way, using sets of gravity activated latches, to only allow the eight weighted levers of *one* of the drum's two contained internal one-directional wheels to have an imbalanced center of gravity and thereby provide axle torque *if* the drum was given a push so that internal wheel happened to be turning in the direction it was intended to designed to normally self-start turn in.

However, the other internal one-directional wheel inside of the drum that would be rotating *counter* to its intended normal self-starting direction after the drum was given a push would have the parallel pairs of main arms of its eight weighted levers all securely locked into place against wooden radial stop pieces attached to the ends of their radial frame pieces by that one-directional wheel's set of gravity activated latches. Bessler would have mentioned that he had made this enhancement to his one-directional wheels in an effort to silence those skeptics that claimed his wheels only turned in one direction because they were fakes secretly powered by spring driven clockwork type movements inside of their drums. The young inventor would also have told the Karl that he intended to use these same type of latches to make any wheel whose construction the count funded also bidirectional with the hope that would make it more attractive to a potential buyer.

As the two men exited the study with the again carefully wrapped up Gera prototype wheel's drum and stand safely tucked under Bessler's arms, the count would have turned to one of the guards outside the room and told him to complete his part of the prearranged deal with Bessler by loading the hired coach's luggage section up with leather bags containing the 4,000 thalers Karl had agreed to pay for the unique privilege of being allowed to know a secret that had eluded thousands of craftsmen over the centuries. The two loyal guards would then have taken turns guarding the coach throughout the night. Those silver coins probably weighed about 275 pounds and make one realize why paper money was eventually introduced! (It's also possible that Bessler only received an initial partial payment with a promise to receive the final installment at a later time. Bessler, of course, would not have liked this change of plan, especially if it was a surprise, but he would not have said anything for fear of offending the count.)

And so the deal was completed. Karl was fully convinced that the wheel was genuine and Bessler would soon be moving himself and his family to Weissenstein Castle so he could begin constructing a new wheel for the count. He would also be employed as Counselor of Commerce whose job it was to promote the economy of Hesse-Kassel by meeting with various local and outside business people while, incidentally, telling them all about what his marvelous invention might do for them if they could raise the money to purchase it and even begin manufacturing it for other businesses to purchase and use. He was officially appointed to the position of counselor upon receipt of a notice from the count that was dated August 15th, 1716.

Bessler, in order to impress his new benefactor, promised that he would construct a bidirectional wheel at Weissenstein Castle that was at least *twice* as powerful as the one at the town of Merseburg had been! The inventor confidently predicted this wheel would soon be the talk of all Europe and help to greatly expand business in Hesse-Kassel and, indeed, all of Saxony. Needless to say, this was music to the ears of Karl because more business activity meant more tax revenue and that meant more improvements in the lives of the citizens of his principality through the construction of various civil projects such as roads, bridges, and canals and the employment their construction and maintenance required.

In 1716 Karl also had a pet project of his own that he had planned for years and which was nearing completion. At an elevation of almost 360 feet above the garden grounds surrounding his beautiful walled castle and about 4,000 feet to the west of them, he had a huge octagonal shaped stone base topped by a narrow pointed pyramid constructed just below the crest of a hill. From the ground under it, this structure was about 204 feet high and at the top level of its pyramid there was a large copper statue of the Greek hero Hercules that was 27 feet in height! Karl's plan was for water to gush forth from a spout near the bottom of the octagonal base which would then run down as a rushing cascade over 200 gradually descending steps toward his castle. There were also several oval shaped pools along the way that the water would temporarily collect in before it continued on its way down the next set of steps.

These steps and their intervening oval pools stretched for a horizontal distance of about 820 feet and were made, like the octagonal base and its pyramid, from a soft volcanic ash rock called "tuff" that was quarried locally. At the bottom of the cascade, the water would form a stream that would then flow toward and, finally, drain into a large artificial lake behind the castle and some of it would even spray out of a fountain in the middle of the lake to a height of tens of feet. To keep the lake from overflowing, excess water would then go over a spillway and be conducted to a stream below the lake. The plan looked great on paper, but it had one serious problem which defied solution.

The water gushing out of the statue's base had to be supplied by several reservoir ponds at a slightly higher elevation that were, in turn, filled by rain water, but the volume of water they could supply was variable and not reliable. Using them meant that, on occasion, if the rainfall was insufficient, then the cascade would have to be turned off until it rained enough to again fill up the reservoirs again. This required having to manually close the reservoir locks that supplied water to a system of underground pipes that connected the reservoirs to the octagonal base's water spout.

From early spring through late fall when the air temperature was well above the freezing point of water, Karl wanted the air around his castle to be filled by the moist mist and soothing sounds from the rushing water of this cascade which he believed would be both refreshing and healthy for those visiting his castle. Unfortunately, that plan would be frustrated unless the cascade could be kept running continuously during those months. The obvious solution was to find some means of recirculating the water from the lake near the castle back up again and directly to the spout located in the octagonal structure supporting the figure of Hercules. The higher reservoir lakes need then only be occasionally used to compensate for the loss of water in the castle's lake due to its evaporation in the heat of the summer sun.

One gallon of water weighs about 8.34 pounds and Karl's cascade would have required a minimum of 1,000 gallons of water to be pumped up through a height of about 360 feet every minute around the clock. That would have required a constant expenditure of energy of about 67.9 kilowatts and, unfortunately, none of the then available power sources, whether animal, steam, or wind, seemed up to the task. On the bright side, though, pump technology had progressed to the point where pumps were becoming available that were efficient due to the manufacture of reliable check valves and piston seals. But those pumps were useless unless they could be provided with energy at a rate sufficient for them to perform their intended tasks. Desperate for an adequate source of power to run his cascade, Karl even sent his court engineer over to England to learn about improvements made in the steam engine invented by Thomas Newcomen in 1712 and see if they looked like they would enable one or more of these engines to drive the water pumps needed for the cascade.

To Bessler, however, powering the pumps for Karl's cascade presented no serious obstacle. He would have told the count that the job could easily be done by just using several of his perpetual motion wheels driving pumps at various elevations along the 360 feet climb up the hill from the lake behind Karl's castle to the water spout in the octagonal base. Such a novel approach would actually be fitting tribute to Karl whose willingness to give Bessler and his invention a fair hearing had made it all possible. The very idea of this would certainly have fascinated Karl considering his interest in all things scientific. Now all the count had to do was help Bessler find a buyer for his invention and then Karl could provide the buyer with his first sale by commissioning him to construct the wheels necessary for the cascade project. The whole plan certainly sounded feasible when being discussed over glasses of the finest wines from the vineyards of Saxony.

However, despite his optimism, there was a problem with Bessler's scheme.

In order to move the water at the necessary minimum, yet impressive rate of 1,000 gallons per minute required the construction of several oversized "super wheels" that would have been 40 feet in diameter! The iron girder open drum of each such wheel would then have contained eight iron unweighted levers (again, these are the levers *before* their lead end weight(s) and their attachment bolt(s) were finally attached to them) each of which had a mass of 200 pounds and held a cylindrical lead end weight and attachment bolt with a mass of 960 pounds at the end of its parallel pair of main arms (these end weights could have been cast as a single piece, but, more likely, would have been a collection of lead discs with holes in their centers which were mounted on a heavy steel bolt). The distance between the center axis of a weighted lever's pivot pin and the center axis of its cylindrical lead end weight's attachment bolt would have been 3.89 feet. The center axis of the weighted lever's pivot pin, which was fixed to the drum, would have been located at a distance of 15.56 feet from the center of the huge drum's axle.

Strong ropes or greased kink resistant chains would have coordinated the shifting of the massive weighted levers, each weighing 1,160 pounds, during the drum's slow rotation and each lever would have been attached to the huge drum's frame by two 14 inch long extension springs each of which would require 240 pounds of force to stretch its length through a distance of a single inch! Most likely, however, if the wheel had ever been built, then *each* of its two large required springs would have been replaced by a single flexible metal bar that had one of its ends bolted to an iron outer octagonal frame piece and its other end near the frame piece's middle attached to a chain whose other end was attached to a weighted lever's A1 spring cord hook attachment pin. When in motion, one such super wheel would be able to output about 6.667 kilowatts of power continuously!

If one constructed twelve of these huge wheels and connected the water pumps they powered in series, then their combined power output would have been 80 kilowatts which would have been more than enough to keep the water in the count's cascade gushing along at a rate of about 1,000 gallons per minute, 24 hours per day, 7 days per week from the early spring through to the late fall! That task would actually only have required 67.9 kilowatts, but the extra 12.1 kilowatts of power produced by a dozen super wheels would have been needed to compensate for the percentage of power that would have been wasted through conversion to thermal energy by the friction in the mechanics of the pumps and wheels.

The next step would be to "gang" or attach three such super wheels to a single, solid 28 inch diameter wooden axle to form a triplet of wheels on a common axle and, in order to power the count's magnificent cascade, one would then need to construct *four* such tripled up super wheels. One would then place one such triplet of wheels at the level of the lake near the castle, a second triplet at an elevation of 90 feet above the lake, a third one at an elevation of 180 feet above the lake, and the final fourth triplet at an elevation of 270 feet above the lake which would be the one that would finally send the water the last 90 feet of height to the octagon's spout. And, of course, one would need four pumps each of which would be powered by a single triplet of super wheels and a line of joined pipes that would interconnect the four super wheel powered pumps between the lake and the discharge spout in the base of the Hercules Octagon.

By using even more massive weighted levers and stiffer springs, one could thereby increase the power output of a single super wheel so that less than 12 of them would be required to recirculate the water to the cascade. Conceivably, by doubling the masses of the super wheels' cylindrical lead end weights and their iron unweighted levers, only six super wheels would have been needed to do the job, but they would have required metal drum frames strong enough to carry the more massive weighted levers. With only six super wheels to use, they could have been put in pairs onto a common axle to provide *three* doubled up wheels. Starting at the level of the supply lake near the castle again, the first doublet of super wheels would send the recirculated lake water to a height of 120 feet above the lake, the second doublet from there would send the water through another 120 feet to a height of 240 feet above the lake, and the third and last doublet of super wheels would send the water the remaining vertical distance of 120 feet to the octagon's water spout. With this approach, one need only use three, though higher flow rate pumps which could lower the cost of the project somewhat.

One might think that the construction of such huge wheels would be a daunting task at best, but Bessler would have assured the count that this could easily be accomplished if only enough workmen could be assigned to the project.

Once completed, these doubled and tripled up super wheels could be concealed by placing their axles at ground level with the bottom half of their drums turning in a dug out trench lined with masonry. The top

20 feet high half of the wheels could then be hidden behind the foliage on one side of the cascade and further concealed by a small roofed over containment structure to keep rain water from flooding out the trenches. If done carefully, none of the visitors enjoying the soothing white noise of the cascade's splashing waters would even be aware of their presence. Any rattling sounds produced by their constantly moving coordination chains would be masked by the gushing water sounds of the cascade itself and by the fact that the lower halves of the huge drums were located below ground level. Also, since all of the interconnecting pipes between the lake and octagonal base would be buried, that would add to the concealment even though having them above ground and hidden by shrubbery would make servicing easier and less expensive in the event that a joint between two lengths of pipe started leaking. The necessary installation, however, would have to await the sale of the invention and it would be the new owner who would be doing the work although, of course, Bessler would make himself available to provide some technical support if any problems were encountered during construction.

A few days later after his meeting with Karl, Bessler and his toy prototype wheel were back again in the town of Merseburg and his wife Barbara must have been exuberant when she saw the heavy bags of silver coins being unloaded from the coach. Her husband had managed to finally get the attention of the people most likely to quickly find a buyer for his invention and they would all soon be dining with nobility in an actual castle! It was truly a dream come true for a family from the lower strata of 18th century German society.

Once back in Merseburg, Bessler openly and proudly described his experience at Weissenstein Castle to everyone he met and the townspeople were in awe of him. This was truly one of the happiest periods in his life. Finally, all of his years of hard work looked like they were beginning to pay off. The crowds to see the Merseburg wheel continued without diminishing even with the presence of the prominent poor box. But, their opportunity to view the wheel in that town was short lived. Not long after his return, Bessler destroyed his first bidirectional wheel and only salvaged its most costly parts such as its 48 cylindrical lead end weights and their attachment bolts, its 32 steel extension springs, and the delicate gravity activated latching mechanisms that allowed it to achieve bidirectionality.

By August of 1716 Bessler, his wife Barbara and their daughter, and his wife's maid Anne Rosine Mauersbergerin were relocated in Weissenstein Castle on the outskirts of the town of Kassel in the principality of Hesse-Kassel within the independent Germanic state of Saxony and getting used to a standard of opulent living they had never known before. It wasn't too long after they settled in that they were joined by Bessler's parents, Andreas and Maria Bessler, his brother, Gottfried, and some of Bessler's wife's relatives including her mother and her new maid Rosina Kuntzmann. It is a testament to Count Karl's fascination with Bessler's wheels and their potential that he would allow an invasion of this magnitude, but he was rich and had enough extra room in his spacious castle to accommodate them all. The nobility there were not used to interacting socially with members of the lower classes and, perhaps, Karl felt it would teach them some humility and make them more appreciative of their fortunate and leisurely status in life.

Bessler soon began a routine of organizing the various legal affairs of commerce for the count early in the day and then later in the day, while the sun was still up and even at night by lamp light after dinner, he worked with the assistance of his brother, Gottfried, to construct his most powerful wheel yet in a small building attached to the castle in its surrounding garden which was referred to as the "machine room". This room at ground level was probably used to store various gardening tools used in the maintenance of the castle's surrounding gardens which were mostly contained within its high perimeter walls. Those implements would have been moved to another storage area to make way for the wheel Bessler and his brother would soon construct.

The machine room probably had few windows and these could be covered with heavy cloth so that, during the day, only the top portions were open to let air and sunlight in. During the night, while Bessler and his brother were sleeping, the windows could be locked and completely covered. This large room also had very secure doors to deter any thieves and two guards loyal to the count were on duty at all times to make sure no one other than Bessler, his brother, or the count would be able to enter the work area. Bessler, fearful, that these guards might try to steal his secret or sleep on the job, even hired two additional ones of his own so they could watch the two guards posted by the count!

Other than Karl, Gottfried was probably the only person on Earth Bessler would trust around his wheel. His brother, who was about eight years younger than him, was very loyal to Bessler and greatly impressed by his knowledge of almost everything. Bessler and Gottfried had even worked together in the past assembling, installing, and repairing church organs and they worked effectively together.

Gottfried was the type of person that was willing to help without asking too many detailed questions about a wheel's precise specifications. He, like Bessler and Karl, knew that the wheels were genuine and that was sufficient for him. He, also like Bessler, wanted the invention to sell because he knew the money would benefit the entire family in the long run and not just Bessler. Meanwhile, Bessler's wife began to feel a little neglected and is said to have complained that his wheel project was all that her husband ever talked about with her.

It wasn't long after his arrival at Weissenstein Castle when Bessler found himself the center of attention at all of the castle parties as he energetically told the guests of his plans for his wheels and the progress being made on the one he was constructing under the count's patronage. Everyone wanted to know when it would be completed and what it could do when it was. They were all told that the construction of his most powerful wheel to date was well underway and, most likely, well before the end of the year of 1717, it would finally be ready for viewing by everyone at court and Bessler himself would be conducting the demonstrations.

It was also probably shortly after Bessler's arrival at Weissenstein Castle and he and his brother Gottfried had begun constructing the Kassel wheel that the inventor finished one of his most important literary works. Due to the influence his former tutor Christian Weise (who died in 1708) had on him, Bessler completed an almost 7,500 line long poem in German titled *"Apologia Poetica"* or, as mentioned before, "Poetic Defense (of the method of perpetual motion he discovered)". AP, as I abbreviate the title, consists almost entirely of rhymed couplets written in German in which Bessler recounts his experiences from his youth up through the completion of his first bidirectional Merseburg wheel and its successful testing. It was another "tell everything, but the actual secret" literary work that the inventor intended to sell to those who had heard about or actually viewed a demonstration of the Merseburg wheel and wanted to have some more information about its mechanics. After they read it, however, they would have been sadly disappointed.

Although this book gives some important clues as to the processes going on inside of the drums of his wheels, their exact mechanics were not described in detail other than to inform the curious reader that the end weights inside of the drum of a clockwise rotating one-directional wheel's ascending side moved closer to the axle and then, following this action, they climbed back away from the axle toward the rim again as they reached the other descending side of the wheel. That continuous action during the rotation of the drum was what kept the center of gravity of the drum's actively shifting weighted levers on the descending side of the axle despite the rotation of the drum.

Most certainly, Karl would have been given one of the first copies of this book by Bessler himself and would, as an educated man, have enjoyed pouring over its many verses and realizing how they perfectly described, although rather vaguely, what he had seen the 3 feet diameter Gera prototype wheel's diminutive mechanisms doing months earlier. Now that Bessler was residing at Weissenstein Castle itself, he made copies of this book available for free to any of the important visitors to the castle that expressed interest in the greatest of wheels he was constructing, but which had not yet been completed. The cost of producing the publication would have been fully paid by Karl.

There was also some sad news for Bessler before the close of the year 1716 at Weissenstein Castle when he learned that his friend Gottfried Leibniz died on the 14th of November, 1716. He was probably the most eminent of scientists who had backed the reality of Bessler's wheels and helped spread the news of them throughout Europe and even as far east as to the court of the Tsar of Russia, Peter "The Great". Leibniz did this, mainly through the sheer volume of correspondence he maintained with six hundred other scientists, noblemen, and aristocrats. In one of his many letters, Leibniz simply stated that "Bessler is my friend." Considering how suspicious Bessler could become about anybody asking too many questions about his wheels, that single sentence shows the great respect that Bessler must have had for Leibniz.

Once Bessler moved to Weissenstein Castle, all of the frequently published harassments he was the target of from his many skeptics suddenly ceased. The word got around that Bessler was then under the protection of Karl and anyone spreading false and demeaning rumors about the inventor or his wheels would find Karl making things "difficult" for him. So, detractors like Borlach, Gartner, and Wagner had to cease their libel and slander against the inventor until such time as he was again vulnerable or they could figure out a way to change their method of attack.

It was during early 1717, before the bidirectional Kassel wheel was completed, that Bessler's three major detractors again renewed their attacks on him, only this time, fearing the wrath of Karl if they resorted to outright libel and slander, their attacks took on a new and more subtle form. Their plan was concocted after learning that the inventor was working on his biggest and grandest wheel yet and doing so right in Karl's castle. They would simply publish challenges to Bessler to prove he had a genuine perpetual motion wheel by submitting it to a duration test in which it would be required to run for four weeks continuously without being touched accept for one permitted repair of its mechanism if needed. To further get him to agree to this test, they even offered to pay him 1,000 thalers if his wheel could pass the test. While they did not actually explicitly insult him with these challenges, the implication was clear. If he did not take up such a challenge, then it meant that he could not because he did not have a genuine perpetual motion wheel. And, if that was the case, then, obviously, he had been defrauding the public for years and, even worse, the much beloved count himself. I'm sure that Bessler and Karl quickly became aware of the implication of these challenges, but were unconcerned by them since they knew the wheels Bessler had previously constructed were genuine as would be the bidirectional wheel soon to be completed at Weissenstein Castle.

Finally, after many distractions and delays, by August of 1717, almost a year after starting, Bessler and his brother finished the construction of the inventor's fourth, heaviest, and most powerful wheel ever and, when it was ready for viewing, practically everyone in the castle visited the "machine room" to see Bessler demonstrate its abilities. All were tremendously impressed by its size and apparent power.

The drum of the Kassel bidirectional wheel was, like the Merseburg wheel before it, a full 12 feet in diameter, but now the drum was 18 inches thick which made it 50% thicker than the drum of the Merseburg wheel.

Each of the two internal one-directional wheels housed in the Kassel wheel's expanded drum, as in the case of the previous Merseburg wheel's drum, contained eight weighted levers that had been constructed from parallel pairs of wooden 3 arm side pieces. Each of the Kassel wheel drum's weighted lever's parallel pair of main arms held at its ends three cylindrical lead weights as had been the case in the previous Merseburg wheel and the center axis of the steel attachment bolt of its middle cylindrical lead end weight was located at an exact distance of 14 inches from the center axis of a weighted lever's pivot pin as was done in the Merseburg wheel. Also, as had been the case in the previous Merseburg wheel, the weighted levers' eight shared pivot pins had lengths equal to the drum's thickness, but were now 18 inches long instead of 12 inches long. The center axes of the eight weighted lever pivot pins for the Kassel wheel's drum, as in the case of the eight of the Merseburg wheel's drum, were embedded in the drum's wooden radial frame pieces at a distance of exactly 56 inches from the center axis of its larger 8 inch diameter wooden axle.

However, in the case of the Kassel wheel, each cylindrical lead end weight at the end of the main arms of a weighted lever was 2.04 inches in diameter x 6 inches in length which then gave the weight a mass of 8 pounds or 128 ounces instead of the 4 pounds or 64 ounces mass of the end weights Bessler used in the Merseburg wheel and that he allowed its examiners to handle during the official testing of that wheel. For the Kassel wheel Bessler had changed the diameter of the cylindrical lead end weights he had previously used in the Merseburg wheel from 1.78 inches to 2.04 inches. When these cylindrical end weights had holes bored through their center axes and their steel attachment bolts were added, they had a mass of almost exactly 1.333 pounds per inch of cylinder length which was 33.333% more than the mass of 1 pound per inch of cylinder length that he had previously used for the cylindrical lead end weights in his Merseberg wheel.

The wooden 3 arm side pieces used to construct each of the Kassel wheel's 16 weighted levers were also 1.333 times the thickness of those in the Merseburg wheel, 0.5 inch thick instead of 0.375 inch thick, and all of the weighted levers' pivot pins of the Kassel wheel were also larger in diameter to withstand the greater centrifugal forces that would be applied to them during drum rotation.

The new thicker *un*weighted wooden levers of the Kassel wheel each had a mass of 5 pounds without their attached end weights and their steel attachment bolts in place. (To be more accurate, however, the mass a wheel's unweighted has to also include the masses of the various metal coordinating cord attachment hooks and half of the masses of any coordinating cords and unfixed springs attached to the various hook attachment pins held between the parallel 3 arm side pieces of the unweighted lever. In the case of the Gera prototype wheel, these extra masses are so small that they do not have to be taken into consideration, but as wheel diameter increases and the metal hooks, cords, and springs become more massive, they should be taken into account otherwise the center of gravity of the wheel's eight weighted levers will not be displaced as far as possible onto the axle's descending side and, consequently, the wheel will not run with the maximum torque that it can have at any drum rotation rate.)

With its three 8 pound cylindrical lead end weights and their attachment bolts in place, each Kassel wheel weighted lever then had a mass of 29 pounds. The centrifugal forces that were applied to the Kassel wheel's weighted levers at any drum rotation rate were always double those for the Merseburg wheel turning at the same rate, but by doubling the thickness of the oak radial frame pieces used in the Kassel wheel's drum, from 0.5 inch to 1.0 inch, the drum could easily withstand the additional stresses applied to it by its 16 weighted levers' eight 18 inch long steel pivot pins even when the rotation rate was at its maximum terminal value of 26 revolutions per minute.

The Kassel wheel's huge hollow drum was attached to and supported by a solid oak axle that was 8 inches in diameter and from whose ends steel pivots a full 1.0 inch in diameter projected out about 6 inches. Having double the mass of the Merseburg wheel, this wheel including its axle and its two embedded steel end pivots, probably weighted around 1100 pounds.

On October 31st of 1717, as part of the preparation for the official testing of his invention intended to finally silence all of its skeptics, Count Karl requested that Bessler disassemble and move the 12 diameter, bidirectional wheel from its "machine room" at ground level to another room on the second floor of the castle which he did (note that the castle, shown in the painting reproduced in Figure 2(c) actually had four floors or stories above a ground level story, so this next room would have been on the second story above the ground story and, thus, had two stories above it). This room was in one of the most secure parts of the castle and had walls of solid stone almost 2 feet thick! The translocation of the wheel was intended to provide additional proof to those that examined it that it was genuine and not dependent in any way upon its surroundings for its motive power.

Two illustrations of the bidirectional Kassel wheel are shown in Figure 2(a) and 2(b). These drawings are from a book with the long winded German title of *Das Triumphirende Perpetuum Mobile Orffyreanum an Alle Potentaten, Hohe Häupter, Regenten und Stände der Welt* or, in English, *The Triumphant Perpetual Motion Machine of Orffyreus, to All the Potentates, High Leaders, Regents and Ranks of the World* which Bessler, at the request of Count Karl, wrote in 1719. This was Bessler's most professionally produced book and had the same text written in both German on the left column of each page in Gothic type for the average reader and Latin on the right column of each page in humanist type for the highly educated since, at the time, Latin was the official language being used by scientists and other scholars around the world. Lectures given in universities at the time were in Latin and one had to able to understand this difficult language whether it was read, spoken, or written if one wanted to get a higher education. This book, like the inventor's earlier ones, gives much detail about the official tests performed on his wheels, including, of course, the Kassel wheel at Weissenstein Castle, yet only gives vague hints concerning their internal mechanics. Hereafter, I will simply refer to this book as "DT".

On the left side of Figure 2(a) we see the bidirectional Kassel wheel in the process of almost completing the lifting of a sturdy wooden box full of bricks or stones vertically upward through a height of *three* full stories from the ground level of the courtyard of the castle to a pulley located outside of the window at the top of the second floor room (which, again, was located on the second story above the castle's first story which was above its ground story so that the wheel's room was three full stories above ground level) that the wheel had been moved to at the count's request. The arm to which the pulley was attached was supported with a diagonal member, represented by two closely spaced lines, that had one of its ends attached to the side of the exterior wall surface of the castle (another close by pair of diagonal lines above it probably represents the slanting roof of the castle). Just below and to the right of the window there is something small on the floor which may be some items used to lock the closed window panes to prevent anyone from entering through one of them when there was no one in the room. The Kassel wheel, like the Merseburg wheel, was capable of causing a heavy load to rise and then lower repeatedly through a distance of tens of feet which, as mentioned previously, can only be done if the wheel was supplying enough energy to replace that lost by various bearing frictions and aerodynamic drags as this attention getting demonstration was performed.

The left side of Figure 2(a) shows that Bessler had again placed the first pulley that the rope went around on the floor below the axle, but he had turned the pulley's plane through 90° because, unlike as in the case of the Merseburg wheel's room, the window in the second floor room of Weissenstein Castle was located at a right angle to the axle of the 12 feet diameter, bidirectional Kassel wheel.

The placement of the pulley directly below the wheel's axle was again necessary so that the sudden jerking force applied by the rope to one end of the axle would not lift the pivot at that end of the axle off of its bearing plate as the load of bricks at ground level was initially lifted from the ground or when the load later descended to the ground and then nearly instantly reversed direction again. Such an undesirable sudden lifting force applied to the end of the axle with the rope attached to it could possibly, if the axle's two pivot pin bearings were exposed for inspection purposes at the time, have resulted in the rapidly spinning axle rolling up and out of its end pivot pins' two lower brass half moon bearing plates so that the 1100 pound axle and drum would then crash to the floor and be severely damaged. Placing the pulley under the axle caused the jerking force on one end of the axle to pull its pivot pin downward and more tightly into its lower bearing plate and thus keep the axle properly positioned. Because of their large cross sectional area, the strength of the oak vertical support planks that contained the bearing plates was more than sufficient to withstand such a transitory force being applied to them.

There are a few other interesting features on the left side of Figure 2(a) such as the triangular lock at the bottom of the drum which was used, not to prevent the wheel from spontaneously self-starting because this required a push start, but, rather, to prevent unauthorized testing of the wheel in Bessler's absence. Such testing by someone not knowing the proper method for stopping a bidirectional wheel so that it would remain stationary could result in the wheel being left to run continuously. If Bessler was not present for days after the unauthorized tester had simply left the room, then the wheel might have just run for all of that time during which a lot of unnecessary wear would occur to its various bearings and coordinating cords.

One also notes on the left side of Figure 2(a) that the two vertical support planks have been eliminated which makes the wheel appear to be floating in midair with a cross sectional axial view of the circular front end of the axle showing the circular cross section of the steel pivot embedded in it at its center. This, aside from making the axle visible, was intended to show the reader that the axle was solid wood and, therefore, contained no hidden spring that could be used to power the wheel. Also seen is a single wind of a heavy rope about an inch in diameter that is wrapped around the axle. As in practically all of Bessler's illustrations, the important components are labeled with numbers. I will not be listing these numbers along with the part names assigned to them because, in most cases, the function of the parts is obvious.

On the right side of Figure 2(a) one sees the front side of the Kassel wheel's drum with front end of its now hidden axle supported by the front vertical support plank. Also seen is the front one of the wheel's two swinging compound pendulums whose metal drive rod is being moved up and down by an oddly curved crank attached to the end of the axle pivot pin that projects forward and out of the plane of the front side of the vertical support plank. I will have much more to say about the use of these pendulums in a later chapter and it must suffice at this time to just mention that they were primarily intended to reduce a wheel's maximum terminal rotation rate and thereby prolong the amount of time it could run before some internal part failure brought it to a stop. The pendulums actually functioned as a novel form of inertial braking on the axle and how they did this is yet another testament to Bessler's inventive genius.

The two spherical weights at the ends of the pendulum's horizontal cross beam appear to be iron cannon balls and probably weighted about 30 pounds each. The football shaped lead bob weight at the bottom of its wooden support piece would have had a mass of about 60 pounds. The cross beam two which the two spherical weights were attached was probably solid iron and weighed about 60 pounds and would have been forge welded to an equally long iron "movable pivot rod" that served as a fulcrum for the entire pendulum and probably weighed another 70 pounds (that rod is not visible in the axial view of Figure 2(a), but we do see that it had a square cross section which would have helped it resist bending from the weight of the compound pendulum attached to it). The wooden support piece connecting the cross beam to the football shaped lead bob weighted only about 30 pounds. The mass of *each* of the two iron diagonal braces connecting the ends of the cross beam to the lead bob weight was probably about 60 pounds. Thus, the total mass of *each* of the Kassel wheel's two compound pendulums was about 400 pounds which was just about 150 less than the weight of the entire Merseburg wheel!

The lead bob weight and its wooden support piece were obviously shaped so as to minimize the aerodynamic drag acting on the lower end of the support piece and the bob weight. Aside from not wasting energy by stirring up the air surrounding these moving objects, this was necessary in order to allow them to move as fast as possible so that the aerodynamic drag they experienced as they were accelerated in each direction by the drive rod activated cross beam would not apply forces to the wooden support piece which might cause it to flex along its length and then crack and, possibly result in the lead bob weight separating from the pendulum and taking flight inside of the wheel's room. The use of the two diagonal brace rods attached to the lead bob weight also helped reduce this possibility.

One very important detail shown on the right side of Figure 2(a) is the two heavy iron bridge pieces between which the Kassel wheel's vertical support planks were held. Those bridge pieces show an air *gap* of several inches between the ends of the planks and the surface of the wooden floor or ceiling to which the ends of the bridge pieces are attached with oversized screws. This was Bessler's way of demonstrating that there was nothing passing down through hidden channels in the vertical support planks in order to supply some torque to accelerate the wheel after it was given a push. He probably added this expensive feature as a way of refuting the hypothesis advanced by Borlach in the illustration he published that showed how he thought Bessler was faking the demonstrations of his Merseburg wheel as was shown in Figure 1(b). I shall have more details to offer about these iron bridge pieces later in this chapter.

Like the DT illustration for the Merseburg wheel shown in the right side of Figure 1(a), we only see the front side of the bidirectional Kassel wheel's 12 feet diameter drum in both sides of Figure 2(a).

Figure 2(b) also provides us with two views of the Kassel wheel and we see some other interesting ways it could be used.

On the left side of Figure 2(b) we see an almost profile view of the wheel's drum and a little of the *front* circular face of the drum is visible and appears as a darkened ellipse. In the foreground of the figure there is a large vat containing a cylindrical Archimedean water screw about 12 feet in length and 18 inches in diameter. The cross section of this water lifting machine's cylindrical drum is actually a polygon and the length of the drum itself consists of two cylindrical sections each of which is made from several dozen 6 feet long pieces of wood that surround and are attached to a central wooden screw shaped piece. It is interesting to note that the length and diameter of the water screw's drum just happens to equal the diameter and thickness of the Kassel wheel's drum. Bessler probably handcrafted this ancient device using the same 6 feet long pieces of wood he used to construct the drums of both the Merseburg and Kassel wheels. These pieces had a rectangular cross section that was 0.50 inch thick x 2 inches wide in the Merseburg wheel's drum and had to be doubled up to make pieces with a rectangular cross section that was 1 inch thick x 2 inches wide in the Kassel wheel's drum.

What is unusual about the Archimedean water screw in Figure 2(b) is that it is rotated about its center axis using a square pulley system. There is a large square pulley surrounding the midsection of the water screw's polygonal cylinder that is, via a twisted loop of rope, attached to the wheel's *back* axle section. But this attachment is not directly to the axle's surface itself, but, rather, to four "Y"-shaped wooden forks that project from the axle at angular intervals of 90° from each other. Why this particular arrangement? Obviously, it was intended to prevent slippage of the wet rope, part of which was always submerged beneath the water's surface in the vat.

From the 2:1 ratio of the distances of the rope from the center axis of the Archimedean water screw and from the center axis of the wheel's back axle as well as the noted decrease in the maximum free running rotation rate of the wheel from 26 down to 20 rotations per minute when its axle was driving the water screw, we can say that the polygonal cylindrical body of the screw would have been rotating at only 10 rotations per minute. In one second, therefore, the water screw's polygonal cylindrical body only rotated through 1/6th of its average circumference and effectively lifted a volume of water about 3 feet from the surface of the water in the vat up to the raised end of the screw whereupon it spilled out and then landed in a tilted gutter that carried it back into the vat so that it could be continuously recirculated as the water screw rotated.

I won't burden the reader with all of the calculations involved in my estimate here, but will simply state that the water was probably being ejected from the top of the screw's polygonal cylinder at a rate of about 1.5 quarts per second. This would have required a constant energy expenditure at a rate of about 12.72 watts. If we assume another 12.72 watts was wasted in overcoming any friction in the polygonal cylinder's end pivot pin bearings, hydrodynamic drag acting on the various external and internal surfaces of the water screw's wooden components, and drag in the square pulley system as it rotated the water screw's polygonal cylinder, we can probably safely assume that the total power output of the Kassel wheel, when turning at 20 rotations per minute, was about 25.44 watts.

This huge wheel's maximum power output, however, would have been at start up when the center of gravity of the eight weighted levers of the single active internal one-directional wheel that drove its drum and axle was at its maximum horizontal displacement onto the axle's descending side because centrifugal forces had not yet begun to move that center of gravity horizontally closer to the *punctum quietus* located directly vertically below the center axis of its 8 inch diameter oak axle. At that time the maximum power output was probably about double what the wheel was limited to expending while operating the attached Archimedean water screw or about 50 watts. At a maximum power output of 50 watts, the Kassel wheel was then twice as powerful as the Merseburg wheel had been and 250 times more powerful than Bessler's little 3 feet diameter Gera prototype wheel. We see from this maximum start up power estimate for the Kassel wheel that, with its construction, Bessler kept his promise to Count Karl to make a wheel for him that was twice as powerful as the Merseburg wheel had been at start up.

In the event that the vat in which the lower end of the Archimedean water screw sat began leaking water or actually spilled its entire contents on the floor should one of its reinforced corners pull apart, we see in the right side drawing of Figure 2(b) that Bessler installed a drain in the floor of the room containing the wheel that would allow a long and narrow open trough suspended under the room's floor and attached to the ceiling of the room below the wheel's room to conduct the water out of the castle through an open upper pane of a window in the wall of the room containing the drainage trough.

What the reader of DT is usually not aware of is that the right side drawing of Figure 2(b) actually shows the *backside* of the Kassel wheel's drum. Bessler made sure that this view did not show the eight pinned down flaps of cloth that covered the eight inspection and servicing holes there that had been cut into the sheets of dyed and oiled linen that covered an otherwise open backside face of the drum. The existence and positions of those flaps was information he did not want the many reverse engineers who might obtain a copy of DT to know about.

The vat's water drainage trough under the floor of the wheel's room had to be about 14 feet in length so that it could reach all the way from under the floor drain near the vat to the outside of the opened window of the room below the wheel's room. Bessler could have easily fabricated this shallow open trough by using a single 1 inch thick x 12 inches wide x 14 feet long plank of wood to which he had glued and nailed two 1 inch thick x 4 inches wide x 14 feet long pieces of wood to form the two side walls of the trough and to also prevent its bottom piece from bending under the weight of any water it was carrying out of the window. The large 14 feet long planks needed to make this hidden drainage trough were probably standard pieces of lumber used for the inner walls of rooms with 14 feet high ceilings.

Once all of the trough pieces were attached to each other and the glue dried, the inside surfaces of the trough that carried the drainage water would have been coated with oil olive or wax in order to render them water repellant and protect the wood from damage by any water that clung to them after the system was used. The outside wall that the vat's drained water is shown exiting from on the right hand side of Figure 2(b) is the same wall containing the window through which a rope is shown hoisting a load of bricks in the left side drawing of Figure 2(a) even though he does not show the window in Figure 2(b) because it is now located behind the plane of the Kassel wheel's drum and our view of it is blocked by another axle powered stamping mill.

Why did Bessler put the vat and its Archimedean water screw on the side of the Kassel wheel farthest from the wall containing the window, one floor down, through which any water spilling from the vat would exit? Would it not have been much easier to just place the trough between the wheel and the wall so that he could dispense with having to use such a long trough under the floor? There are actually several good reasons why he used the arrangement shown in the right side of Figure 2(b).

The obvious reason is that there wasn't sufficient space between the wall containing the window, which was one floor down, that supported the end of the trough from which the drain water exited and the nearest side of the Kassel wheel's drum so as to be able to locate the vat and its long Archimedean screw there. There was far more space available on the other side of the drum even though it required attaching a 14 feet long suspended trough to the ceiling of the room beneath the wheel's room.

By using the arrangement shown in Figure 2(b), a visitor to the second floor room in Weissenstein Castle would enter and immediately see the front side of the 12 feet diameter, bidirectional wheel's drum. To the right of it he would see the raised end of the Archimedean water screw with the constant stream of water pouring out of it and dropping down into its return trough to the vat. The viewer would also not be distracted by the mechanics that made this possible which consisted of the twisted pulley rope, an odd shaped square pulley at the mid-section of the water screw, and the even more oddly shaped "Y" pieces embedded in the surface of the wheel's rear axle section that pulled on the twisted rope.

One can only imagine what went through Karl's mind when he first saw the water screw being operated by the Kassel wheel. He, no doubt, would have imagined that the short return trough to the water supply vat represented his awe inspiring cascade less than a mile from his residence and the vat itself the lake behind his castle. The ancient water lifting device in continuous rotation would have symbolized to him the series of super wheel driven pumps that would lift the recirculating water from the lake back up through 360 feet until it came gushing out of the spout in the huge stone base of the Hercules Octagon at the top of the cascade. I suspect that this is exactly the impression Bessler wanted to create in the Karl's mind and it would all become possible soon after the inventor, with the count's help, found a buyer for the imbalanced pm wheel design that could meet the very high price demanded for it.

Several other very important features are visible in the left hand portion of Figure 2(b). Let us begin with the first and see to what conclusions it leads.

There was a bucket attached by a rope directly to the front axle section of the Kassel wheel into which precise masses of water could be placed and then hoisted up from the floor only to an elevation at which the bucket's handle would come into contact with the side of the wheel's axle (my best estimate is that the bucket, made of wood, weighed about 5 pounds and could hold, without risk of spilling, 2 gallons of water which meant that, by adding water to the initially empty bucket, the combined weight of the bucket and its water could be accurately varied from 5 pounds up to 21.72 pounds).

The bucket shown in Figure 2(b), however, was primarily used for the most accurate testing of both the Merseburg and Kassel wheels by serious examiners interested in measuring the *direct* lifting force of a wheel's axle at start up which was the best way to determine its maximum power output. This maximum power output, however, would slowly decrease as a wheel's drum rotation rate increased and, at some maximum terminal rotation rate, would have decreased until it just exactly equaled the small amount of power being dissipated by the aerodynamic and bearing drag forces constantly present and acting to slow a wheel.

To perform this test, either the bidirectional Merseburg or Kassel wheel was slowly manually rotated at least a half of a turn in either direction. This was to assure that all of the eight weighted levers in the *one* of its two internal one-directional wheels would have their eight parallel pairs of main arms locked down against their wooden radial stop pieces by the gravity activated latches assigned to each lever. Once that state was achieved, the center of gravity of those eight weighted levers would be located exactly at the center axis of the axle and that internal one-directional wheel inside of the bidirectional wheel's drum could then provide no torque to the axle. However, after that half turn of the bidirectional wheel's drum had been completed, any locked down weighted levers in the *other* internal one-directional wheel which had made a half turn in the direction that it was designed to spontaneously self-start in would be freed so that all eight of that internal one-directional wheel's weighted levers could function normally again and provide their full start up torque to the axle if the drum of the bidirectional wheel was allowed to continue rotating by releasing it.

At this point, the bidirectional wheel was considered ready for this special test. If the drum was then released, it would immediately begin to accelerate and so, since its axle was now experiencing the maximum start up torque of *one* of its two internal one-directional wheels, it was just the start up torque of that single internal wheel that was being tested. The handle of the empty bucket would then be attached with a short length of rope to a protruding wing head bolt screwed into the bidirectional wheel's axle and the drum released to see if the axle torque was sufficient to lift the bucket. When the bucket began to rise, some water would be added to the bucket to increase their combined weight and, again, the drum would be released to see if the bucket and its added water could be raised by the axle's torque. If so, then more water was added to the water already in the bucket to further increase their combined weight and the test continued. At some point the combined weight of the bucket and its added water would to too heavy for the bidirectional wheel's single fully enabled internal one-directional wheel's maximum start up torque to lift. The combined weight of the bucket and its added water could then be used to determine the maximum start up torque of a bidirectional wheel (I am not sure if this method was used for Bessler's second Gera or Draschwitz one-directional wheels, but it could have been by using, perhaps, a smaller lighter bucket to hold a lesser volume and mass of water. If used with a one-directional wheel that was always spontaneously self-starting, the test could be done immediately without the need to first prepare the wheel by rotating its drum in order to place the center of gravity of its eight weighted levers as far as possible, horizontally, onto its axle's descending side. This was probably the simplest and most direct way of measuring the actual maximum constant power of one of Bessler's wheels.

This special test with the water filled bucket was much different from those earlier mentioned demonstrations using a wheel's maximum accumulated rotational kinetic energy when it was running freely at its terminal rotation rate to suddenly and rapidly lift a box of bricks from the outside ground level to a pulley located outside of a wheel room's window.

Those spectacular braking tests, especially when a load was allowed to continuously rise and fall like a hand toy known as a yo-yo, were intended to amaze the masses attending the inventor's public demonstrations and gave a deceptive impression of his wheels' actual constant power output.

Let us use a term from the sport of weight lifting to refer to this special type of test with the water carrying bucket as a "dead lift test" of one of Bessler's wheels because it did not allow a much heavier mass to be suddenly lifted using a wheel's axle *after* the wheel had reached some maximum terminal rotation rate or the rope from the axle to the bucket of water had been passed through a system of hoisting pulleys such as a block and tackle in order to increase the lifting force applied to the mass of the bucket and its contents. In other words, for a dead lift test of one of Bessler's wheels to be performed, it had to be performed starting with a stationary wheel and have the rope from the load represented by the mass of the bucket and any water it contained *directly* connected to a wheel's axle so that the load hung below the axle and no other torque modifying machines could be employed.

Even though Bessler used a floor mounted pulley and another pulley located outside of an open window for his audience amazing rapid lift demonstrations, these pulleys did not actually increase the lifting force on a load located outside of the window. As such, one could have chosen to perform a dead lift test on either the Merseburg or Kassel wheels with one of the pulleys and a load outside of the wheel's room if he so desired. The problem with doing that was that the drag of the rope traveling through two additional pulleys would have somewhat diminished the start up lifting power of the wheel. Doing a dead lift test with the load attached directly to the axle was the most accurate type of test and obviously more convenient when it came time to add or subtract some of the water from the bucket to determine the exact maximum mass of the load that the axle could not lift at start up. That amount of mass would always be a bit less if the rope to a load located outside of the wheel's room had to pass through another two pulleys before the lifting occurred.

Using any sort of hoisting pulley system would, however, because of the gain in mechanical advantage it created, increase the amount of mass that could be lifted at start up by the axle of one of the inventor's wheels, but that gain was then *always* paid for by a reduction in the vertical lifting *speed* of the load.

Thus, if one used a block and tackle type hoist to, say, double the amount of mass that could be lifted by a rope winding around a wheel's axle at start up as compared to the mass of the load that could be lifted when the hoist was not used and the rope was directly attached from the axle to the load being lifted, then the rate at which the load vertically ascended would be halved when the hoist was used. This was because, in order for the block and tackle to double the mass of the load that could be lifted for a given force being applied to the rope by a wheel's axle, double the amount of rope would have to be drawn through the pulleys of the block and tackle and it would take the axle twice as long to draw that extra length of rope through the pulleys of the block and tackle.

If the block and tackle type hoist was then modified (by adding pulleys to it or using a different block and tackle) to lift, say, four times the amount of mass that could be lifted by a wheel's axle as compared to when the block and tackle was not employed and a rope from the axle was connected directly to the load, then the rate of vertical ascent of the load would be reduced to only a quarter of what it was when the rope from the load was simply attached directly to a wheel's axle. This is because, in order to quadruple the lifting force applied to a load by the block and tackle, four times as much rope would have to be drawn through its pulleys and wound about the wheel's axle.

While testing Bessler's wheels using hoists that increased mechanical advantage so that heavier loads could be lifted, the consequent reductions in lifting speed meant that there really was no gain in the rate at which the axle was doing external work and, thus, no gain in a wheel's constant power output. The engineers and scientists of Bessler's day would have been well aware of this detail while most of the common people watching such a hoist assisted lift of a heavy load would not have been.

For the Merseburg wheel, my calculations based on eyewitness accounts indicate that, using the dead lift test, its 6 inch diameter oak axle would not have had enough torque to lift the bucket and its added water if their combined weight exceeded about 10 pounds. In the case of the Kassel wheel with double the starting torque of the Merseburg wheel, the same dead lift test would show that its 8 inch diameter oak axle would not have had enough torque to lift the bucket and its added water if their combined weight exceeded about the 15 pounds. Thus, the maximum start up torque of the Merseburg wheel was only about 2.5 foot-pounds and for the twice as massive Kassel wheel it was about 5 foot-pounds. Needless to say, the hired engineers dispatched to inspect these wheels by businessmen interested in using them to power their early Industrial Age factories and shops would not have been too impressed by such low maximum start up torques when they considered the much higher torques that could be delivered by the axles of conventional water and wind mills despite their intermittent unreliability.

Of course, Bessler would have immediately dismissed their concerns by merely stating that the wise buyer of his imbalanced pm wheel design could easily obtain any degree of start up load lifting ability he required by simply using wheels containing more massive weighted levers in drums of greater diameter and then, if necessary, placing several of these on a common axle. The inventor probably also mentioned that Count Karl himself, a man well versed in mathematics and a student of all things scientific and technological, was interested in using a collection of Bessler's planned 40 feet diameter super wheels to power his awe inspiring cascade just as soon as a buyer for the invention was found. Bessler's assurances of virtually unlimited power plus the added mentioning of the use of the design as an attraction in the most important city in the most influential principality in all of Saxony was information he wanted to make sure any agent investigating his wheels for a potential buyer would take back to him.

It's possible, using the dead lift test values above, to calculate how far, *horizontally*, onto their axle's descending side the centers of gravity of the Merseburg and Kassel wheels' *active* internal one-directional wheel's eight weighted levers were located.

Since the maximum weight of a load directly attached by a rope to an axle that is being lifted during a dead lift test is equal in magnitude to the torque resulting from the horizontal displacement of the center of gravity or "CoG" of the active one-directional wheel's eight weighted levers, we can use the following simple formula to express the equality of these two opposed forces acting on a wheel's axle:

(axle radius in inches) x (maximum dead lift load in pounds) = (horizontal CoG displacement from axle center axis in inches) x (total weight of active internal one-directional wheel's eight weighted levers in pounds)

This is now rearranged to yield:

(horizontal CoG displacement from axle center axis in inches) = [(axle radius in inches) x (maximum dead lift load in pounds)] / (total weight of active internal one-directional wheel's eight weighted levers in pounds)

For the Merseburg wheel we use an axle radius of 3 inches, a maximum dead lift load of about 10 pounds, and a total weight of its active internal one-directional wheel's eight weighted levers of 116 pounds. Substituting these values into the equation above then gives us:

(horizontal CoG displacement from axle center axis in inches) = [(3 inches) x (10 pounds)] / (116 pounds) = 0.259 inch

For the Kassel wheel we must use a larger axle radius of 4 inches, a maximum dead lift load of about 15 pounds, and a total weight of its active internal one-directional wheel's eight weighted levers of 232 pounds. Substituting these values into our mechanics equation then gives us:

(horizontal CoG displacement from axle center axis in inches) = [(4 inches) x (15 pounds)] / (232 pounds) = 0.259 inch

So, we see from these calculations that the center of gravity of the eight weighted levers of the *active* internal one-directional wheel inside of the bidirectional Merseburg wheel's 12 feet diameter drum was *horizontally* displaced onto the descending side of the axle by a distance of only about 0.259 inch from an imaginary vertical line passing through the center axis of the axle and for the doubly massive bidirectional Kassel wheel's 12 feet diameter drum by an equal distance of 0.259 inch from an imaginary vertical line passing through the center axis of its larger axle. It should be emphasized here that the actual centers of gravity of the two wheels' eight active weighted levers were *not* spatially located on imaginary *horizontal* lines passing through the center axes of their axles, but, rather, were at points in space that were *vertically* located at distances of about 2.6 inches *below* those imaginary horizontal lines (this value was not derived from mathematical calculations, but, rather, from extrapolating the vertical displacement distance of the center of gravity in computer models of the 3.5 feet diameter, one-directional Gera prototype wheel). Thus, the vertical displacement of the center of gravity of a 12 feet diameter, bidirectional wheel's active weighted levers *below* an imaginary horizontal line passing through its axle's center axis was ten times greater than its horizontal displacement onto the axle's descending side from an imaginary vertical line passing through its axle's center axis.)

It seems that Bessler may have purposely designed his 12 feet diameter, bidirectional wheels so that the centers of gravity of their active internal one-directional wheels eight weighted levers would have horizontal displacements from the center axes of their axles of only about a quarter of an inch. This should dispel any notion students of Bessler's wheels have that their mechanics produced large horizontal displacements of their centers of gravity from the center axes of their axles. What little torque they produced was mainly due to the large total mass of their eight active weighted levers rather than to the location of the center of gravity of that total mass. Even in Bessler's one-directional, 40 feet diameter super wheel which was never constructed, the center of gravity of its eight weighted levers would have had a total mass of 9280 pounds, but was only located about 0.867 inch from the center axis of its 28 inches diameter axle!

At the other lowest end of the spectrum of drum diameters that Bessler used or contemplated, in the case of his 3 feet diameter Gera prototype wheel, the center of gravity of its eight little wooden weighted levers would only have been displaced horizontally by about 0.0625 inch = 1/16 inch from the center axis of its 1.5 inch = 1-8/16 inch diameter wooden axle. This is why the reverse engineer who begins his journey to duplicate one of Bessler's larger wheels and decides to start, as I suggested previously, by first replicating the 3 feet diameter Gera prototype wheel needs to be as precise in its construction as possible.

Its drum, once constructed and attached to the axle, but before its eight weighted levers are installed, must be as carefully balanced as possible so as to assure that its center of gravity is located exactly on the center axis of the axle even if additional balancing weights must be attached to the drum's outer octagonal frame pieces to bring the whole into such precise balance. Its eight wooden weighted levers must each have a mass as close to 7.25 ounces or 0.453125 pound as possible and the center axes of their pivot pins must all be located as close to exactly 14 inches from the center axis of the axle as possible. Those unaccustomed to such precision work may have to make several attempts before they finally get the construction correct so that the start up center of gravity of their Gera prototype wheel duplicate's eight weighted levers will be displaced onto the axle's descending side and as close as possible to a horizontal distance of 0.0625 inch from an imaginary vertical line passing through the center axis of its axle.

And, we must remember at all times that these were maximum horizontal displacements that only existed when the wheel was stationary and about to begin rotation. As soon as wheel's rotation began and accelerated and, consequently, centrifugal forces also began to increase and interfere with the timely shifting of a one-directional wheel's eight weighted levers, the horizontal displacement of the center of gravity of that wheel's weighted levers would begin to decrease as it slowly swung around the axle's center axis in the direction of wheel rotation. This would cause the center of gravity to move toward the *punctum quietus* or equilibrium point that was located *on* an imaginary vertical line passing through the center axis of the axle. As this was happening, the torque that the offset center of gravity applied to the axle also slowly decreased. But, of course, in the case of one of Bessler's self-moving wheels, the center of gravity of its active eight weighted levers would never quite reach its *punctum quietus*.

Now, to finally continue with the analysis of the remaining two items in the background of the left side drawing of Figure 2(b), we see that Bessler had replaced the two compound pendulums with something else. They appear to be long wooden planks with a semi-circular flanges extending at right angles from their lower ends (only one of which, the one powered by the front end of the axle, is visible). Their purpose is clear. When forced to swing from the left to right side of the wheel's axle and back again, their flanges presented a large cross sectional area to the direction of their motion and this then created a lot of aerodynamic drag on the flanges which was then used to provide drag on the wheel's axle and slow it down so as to reduce the wheel's maximum terminal rotation rate.

The illustration does not show exactly how the flanged planks were driven by the wheel, but we can safely assume that this was done using small, but strong metal drive rods between the ends of the curved arm metal cranks attached to the axle's steel pivot pins and offset metal arms at the tops of the planks which themselves oscillated in vertical planes perpendicular to the center axis of the wheel's axle. This arrangement transferred a large amount of drag induced counter torque to the wheel's rotating axle as it forced the two large semi-circular flanges to sweep through the air. This method was really just a simple form of aerodynamic braking that dissipated the build up of rotational kinetic energy in the Kassel wheel by turning it into the motion of the air surrounding the wheel and, ultimately, into an increase in the thermal energy of that air and any objects it came into contact with.

The left side of Figure 2(b) is also important because it gives some of the details of how Bessler placed the much more massive compound pendulums into place when he wanted to attach them, via metal drive rods, to the curved metal cranks at the ends of the Kassel wheel's axle. A drive rod was, through a small bolt that served as a pivot, permanently attached to one side of a pendulum's top horizontal metal cross beam and each pendulum also had an additional movable pivot rod with a square cross section (to prevent rod flexing) affixed perpendicularly to the center of the horizontal cross beam as can be seen as a cross section having a square shape with a dot at its center in the middle of the cross beam on the right side of Figure 2(a). The center axis of this additional movable pivot rod was parallel to the center axis of the wheel's axle and provided the fulcrum around which a compound pendulum swung.

While Bessler only attached iron movable pivot rods, because of their greater strength and resistance to flexing, to the heavy compound pendulums he used with both his Merseburg and Kassel wheels, inspection of the left side of Figure 2(b) convinces me that the movable pivot rods attached to the planks with the air drag enhancing semi-circular flanges near their lower ends were probably made from long pieces of wood having a circular cross section that then had steel pins inserted into their ends to serve as end pivots. They were actually smaller versions of the axles used in his wheels. This change from a metal to wooden pivot rod was justified by the lower masses of the flanged planks as compared to the compound pendulums.

However, the ends of each iron pivot rod attached to a compound pendulum of the Kassel wheel were simply filed down to give them a circular cross section with a diameter smaller than the width of the rod's square cross section so that they could serve as actual pivots when they were dropped down into vertical slots cut into two heavy vertical plank mounted metal brackets in order to support the pendulum (a slightly different method, however, was used to mount the compound pendulums in the Merseburg wheel as shown in Figure 1(a) which involved inserting the outermost end pivot pin of each metal pivot rod into a lubricated hole that had been drilled into a short iron bar that was, in turn, mounted horizontally between two parallel wooden support planks). When the air drag flanged planks were used, the steel pins at the ends of their wooden pivot rods that served as pivots were also just dropped down into the slots of the same heavy metal brackets. Periodically, Bessler would apply a few drops of olive oil to the pivot pins at the ends of the pivot rods to reduce their friction and wear as they twisted around inside of their metal support brackets when either the compound pendulums or flanged planks were being used.

For each of the compound pendulums in the Kassel wheel, one of its supporting metal brackets was attached by a single large screw to an outward facing surface of one of the drum axle's vertical support planks and the other metal bracket was attached to a separate and somewhat smaller in cross sectional area vertical support plank. Near the front side of the Kassel wheel's axle, the smaller separate vertical support plank for the pendulum's movable pivot rod was about four feet away from the front axle's main vertical support plank. Near the backside of the Kassel wheel's axle, however, the smaller separate vertical support plank for the pendulum's movable pivot rod was about six feet away from the back axle's main vertical support plank. All four of the movable pivot rod end pin metal brackets' vertical slots and the center axis of the wheel's axle were in the same vertical plane.

Most importantly, the metal brackets' four vertical slots and the movable pivot rod end pins they supported were open and easily visually inspected by examiners of the Kassel wheel. That eliminated the possibility of one of the wheel's curved axle cranks actually being driven by its attached drive rod that connected it to a movable pivot rod whose end pin, in turn, would have been shaped into a crank and driven by a concealed metal connecting rod hidden inside one of the vertical planks to which the metal bracket was attached.

Once a compound pendulum's iron or semi-circular flanged plank's wooden movable pivot rod's metal end pins were in place in the vertical slots of their two open plank mounted metal support brackets and the vertical piece of the pendulum with a football shaped lead bob weight attached to its end or the plank with a semi-circular flange attached to its end was pointing straight down toward the floor, the wheel's drum would be slowly rotated until the end of its curved arm axle crank could be easily connected with a small bolt to the middle of the drive rod whose other upper end was attached to the compound pendulum's metal cross beam or to a perpendicular wooden arm attached to the top of the flanged plank. In the case of the Merseburg and Kassel wheels, attaching each drive rod to the end of its curved arm axle crank probably only required a minute or so of effort.

However, in the case of the Kassel wheel, because of the heavy mass of *each* of the compound pendulums Bessler used, which was about 400 pounds, the two 30 pound iron cannon balls, 60 pound football shaped lead bob weight, and two 60 pound diagonal brace rods were only attached to complete a pendulum *after* its metal cross beam and attached square cross section movable pivot rod, wooden vertical bob weight piece, and its two lateral metal diagonal brace rods, *as a single unit*, were first mounted into the vertical slots of their pair of metal end pin support brackets. Each of these incomplete pendulums would have had a mass of about 160 pounds and could have been easily lifted up and then lowered into place by three of Karl's stronger servants or by a single person using a block and tackle until the two end pins of its movable pivot rod were maneuvered into and finally securely held in the vertical slots of the two heavy metal brackets attached to the vertical support planks. The two vertical slots would have been separated from each other by a distance that was fraction of an inch more than the length of the movable pivot rod which had *not* been filed down to form the rod's end pins. This then allowed for a small amount of play between the movable pivot rods and the support brackets which helped reduce the rubbing friction between them when a pendulum was being actively driven for a long period of time.

If three servants manually placed an incomplete pendulum into place between its two vertical support plank mounted brackets, then each servant would have had to have been able to vertically lift at least 53.3 pounds which is certainly possible for a young man in his prime. They would then have used ladders to reach the ends of the metal cross beam and attach the two iron cannon balls to them. These cannon balls would have had holes drilled through them so they could be slid onto the ends of the cross beam and then they would have been locked into place by a single washer and bolt placed through holes drilled into the very ends of the cross beam. The same attachment method would have been used to attach the football shaped lead bob weight to the ends of the two diagonal brace rods and the wooden vertical piece. The bottom ends of all three would then have passed through an appropriately shaped slot bored out of the lead bob weight and then, after emerging from the bottom of the bob weight, passed through an oval shaped metal washer plate with three holes cut into it. The very ends of all three pieces would then have had small bolts inserted through holes drilled into them so that the weight of the slightly loose bob weight on the metal washer plate was transferred to all of them.

The last thing Bessler would have wanted was for a potential buyer or his agent to be injured or even killed should one of the cannon balls come loose from the end of a cross beam and then fall toward the floor. This is probably another reason why these compound pendulums are not mentioned in the various letters written by those who privately viewed the Kassel wheel. Bessler did not use them for brief private demonstrations in order to reduce the risk to those examining the wheel. They, however, were very useful in reducing a wheel's maximum terminal rotation rate so that it could run for a far longer time interval between part failures that would prevent it from running and require immediate repair.

Next, for the left hand side of Figure 2(b), we note the way that Bessler secures the ends of the Kassel wheel's vertical support planks to the room's floor and ceiling. We see that he is using those same heavy iron bridge pieces as are shown in the right hand side of Figure 2(a), but something does not quite look right about them. We see that the bridge pieces look *identical* in both figures even though our angle of view changes by 90° as we go from Figure 2(a) to 2(b)! How can that be possible?

At first, I thought Bessler simply artistically twisted the angle of the brackets in Figure 2(b) by 90° in order to emphasize to any skeptical readers of DT that there was a sizable air gap between the end of a plank and the surface of the floor or ceiling which was big enough to insert one's hand into and thereby prove to the reader that there was no drive rod extending up or down into a hidden channel in a vertical support plank that could then propel the huge wheel by "pumping" a hidden crank formed into the section of the axle pivot pin concealed inside of the vertical support plank. However, there is a much simpler explanation for this apparent anomaly in the two DT illustrations of the Kassel wheel.

The thick iron bridge pieces Bessler used were not just rectangular pieces, much longer than they were wide, whose shorter opposite ends were, via a blacksmith's efforts, heated and then hammered down so that, after cooling and having holes drilled into them, they could be attached by using only two large screws to the floor or ceiling, but, rather, they were actually much wider in shape and had had their four *corners* deformed so that, after having holes drilled in them, they could be attached by *four* large screws to the floor or ceiling. This resulted in a thick metal rectangular bridge piece which was like a low table with its four "legs" formed from its deformed corners.

Assuming that these two rectangular bridge plates, before having their corners heated to soften them and then hammered down, were 2 inches thick x 12 inches wide x 24 inches long and were made of iron, then one can calculate that, after the holes were drilled in their four corners, their weight was about 160 pounds each. Of course, the four corners of each rectangular bridge plate on the floor would together have needed to have been able to support the weight of the portion of the plate not touching the floor which was probably about 120 pounds, the weight of the 160 inch long = 13.333 feet long wooden vertical support plank attached to it, about 288 pounds, half of the weight of the Kassel wheel, about 550 pounds, and half of the weight of one of the compound pendulums, about 200 pounds (the other half of the weight of a compound pendulum was supported by the slotted metal bracket attached to a separate smaller vertical support plank). That means that the four corners of *each* of the floor mounted rectangular bridge plates needed, *together*, to be able to support a weight of about 1158 pounds or 289.5 pounds of weight per corner without being flattened by the load placed on them. Although not visible in Figure 2(b), each of the ends of a vertical support plank, 12 inches wide by 6 inches thick and really more of a beam than a plank, was attached to the center of a rectangular bridge piece by at least two large screws that passed through holes drilled in the bridge piece and then penetrated the wood of the support plank to a depth of perhaps 6 inches.

These two rectangular bridge plates could, however, support much more than just 1158 pounds each because, occasionally, someone would attempt to quickly stop the Kassel wheel's maximum speed, free running drum rotation by firmly grabbing onto its rim which would then result in him being lifted several feet off of the floor as the drum came to a rapid stop. Such a feat would momentarily add about an additional 75 to 100 pounds to the load being supported by each of the floor's rectangular bridge plates.

If specially made, then the fabrication of these four special bridge plates must have cost a small fortune and I'm sure Bessler was very glad he managed to obtain Karl's financial backing for the construction of this, his most powerful wheel ever. However, it's also possible that the two rectangular bridge pieces were stock pieces that had already being made for other applications and Bessler just obtained them and used them for his wheel's vertical support planks. If the gap they formed between the surface of the floor or ceiling and the plate was 2 inches, then it would have been possible for an examiner of the Kassel wheel to insert his hand or a metal or wooden ruler into the gap far enough to reach the center of the plate and thereby verify that there were no thin metal rods or wires passing between the floor or ceiling and the end of a vertical support plank.

Finally, we must consider another machine attached to the Kassel wheel's axle which is illustrated on the right side of Figure 2(b). It consisted of two solid wooden stamps weighing about 25 pounds each that were, as the axle rotated, repeatedly raised and then dropped onto a short wooden beam intended to convert their impacts with it into pounding sounds that would fill the room. Unlike the device used with the Merseburg wheel, the Kassel wheel's stamps were not directly lifted by projecting metal pins on the wooden axle. Rather, the stamps were alternately lifted and then dropped by the shorter end of a long wooden lever whose other longer end was pushed down by a metal piece projecting perpendicularly from the surface of the axle. The short ends of the levers did not disengage from the holes for them in the wooden stamps.

There were four such projecting metal pieces at 90° angular intervals from each other around the circumference of the axle for each of the two wooden stamps in the mill and these two sets of four metal pieces were staggered so that, in an axial view, their outermost ends formed the vertices of an imaginary octagon. This arrangement always allowed one of the stamps to be lifted while the other was dropped so as to produce eight impact sounds per axle rotation rather than causing both stamps to be lifted and dropped together which would only have produced four somewhat louder impact sounds per axle rotation.

So, the Kassel wheel's stamping mill produced a total of eight impact sounds per axle rotation which was equal to the number of impact sounds produced by the Merseburg wheel's stamping mill. However, in the case of the Kassel wheel's stamping mill, Bessler had figured out a way to reduce the number of working stamps from four to only two. Those examining the Kassel wheel were always impressed when Bessler removed one of the wooden stamps from the mill's frame and let them handle it. Because of its 25 pound weight, they quickly realized that it was not hollow and, thus, intended to give a false impression of its true mass.

There was also a mechanical complication with the back axle of the Kassel wheel which required Bessler to place its stamping mill at a distance from the axle and then have its wooden stamps raised and dropped by the ends of two long wooden levers that had to have their common pivot pin held by yet another pair of 14 feet long, floor to ceiling wooden vertical support planks. Bessler could not just use the metal pins projecting from the surface of the axle, as had been done with the Draschwitz and Merseburg wheels, to directly operate the wooden stamps because, in order to operate the Archimedean water screw with the wheel's rear axle section, he had to attach four of those somewhat clumsy "Y" shaped wooden pieces to the axle to act as a no slip pulley for the drive rope of the water screw. Those wooden pieces, which were permanently attached to the axle, prevented the stamping mill from being brought up close to the axle.

The somewhat cluttered drawing shown in Figure 2(b) might have given the reader of DT the impression that the Kassel wheel was capable of simultaneously operating all of the devices shown. That, of course, was a false impression and one Bessler did not make much of an effort to correct. He wanted potential buyers to view his wheels as having no limit as to how much power they could produce if only enough of them were employed and they were big enough. The reality, however, was that the Kassel wheel could only run *one* of the various devices illustrate in Figure 2(b) at a time and needed to have its rotation impeding compound pendulums or semi-circular flange planks removed to do so. This is why the latter are shown disconnected from the curved arm cranks as the wheel's axle operates the water screw that raises water on the left side of the figure and we don't see the stamps also being operated. On the right side of the figure, however, we see the stamps being operated, but the water screw is not being operated. Its supply vat is empty and we note that all of the water has been drained out of it via the drain in the floor and the extended gutter attached to the ceiling of the room below the wheel's room.

Bessler's critics, upon hearing of the completion and beginning exhibition of the Kassel wheel were sure that it had to be a fake and probably pitied the count for having been so gullible as to have been taken in by a swindler like Bessler. That charlatan, they lamented, had not only managed to move himself and his many dependents into the luxurious Weissenstein Castle, but was probably also pocketing a significant share of the funds Karl was providing for the construction of the fake wheel. It was their moral duty, they felt, to finally expose Bessler, make sure he received just punishment for his fraud, and try to salvage the count's reputation if that was still possible. To achieve these aims, they publicly suggested that, if the Kassel wheel was genuine, then it should be able to run continuously for a period of four weeks if left untouched. They were quite confident that it could not and that, in fact, no such test would ever be conducted.

To finally silence all of these criticisms, Bessler and Karl decided on performing a duration test of the Kassel wheel whose result would undeniably prove to all that Bessler had, indeed, actually invented a genuine perpetual motion wheel!

On November 12th, 1717, this fourth exhibited 12 feet diameter, bidirectional wheel (which, including his 3 feet diameter Gera prototype wheel, was actually the fifth he had constructed) was started up in the new room on the second floor above the ground floor of Weissenstein Castle into which it had been installed after a very intensive examination of the bearings that its drum's horizontal axle pivot pins rested on and also of all of the inside surfaces of the room to eliminate the slightest possibility that any hidden power sources or secret passages might somehow be used to maintain the huge wheel's motion. All of these precautions were done under the personal supervision of Count Karl and with the assistance of a small group of technically competent witnesses he had gathered which included such notable people as Doctor Dietrich of Bohsen, another famous doctor named Friedrich Hoffman who was an authority on mechanics, Christian Wolff who was the Chancellor of the University of Halle, and John Rowley who was a famed maker of mathematical instruments.

All windows in the wheel's room were closed and boarded up from the inside and the only door leading into the room was then closed, locked, and had hot wax placed across the crevices it formed with the surrounding wall. The various seals of the invention's investigators were then impressed into the warm wax before it hardened. This was to absolutely guarantee that no one would be able to enter the room without detection during the test period that ensued. To further reduce the possibility of anyone gaining unauthorized entry to the room during the test, two armed guards where stationed at the door around the clock. Aside from making sure no unauthorized persons attempted to enter the room, they each made sure the other could not enter and tamper with the wheel.

Bessler, however, was not content with that level of security. What if the two guards had been secretly bribed by the inventor's detractors to pick the door's lock, gain entry to the room, and then either sabotage the test or even open and fold back one of the linen flaps that covered an inspection and servicing opening in the drum's cloth covering to make a few sketches of the specially shaped levers it used along with the system of springs and coordinating cords that allowed it to achieve perpetual rotation! To prevent this possibility and sleep more soundly at night, Bessler again hired two additional guards to watch the two provided by the count as well as each other! It's really amazing what paranoia can motivate a person to do.

During this test, several people arrived at the castle wishing to view the wheel they had heard so much about, but could not do so because it was in the sealed room and one of them was Dutch hydraulic engineer Georg Michael Meetsma, who worked in the Kassel area. Such visitors would have been politely told that the official test was in progress, but allowed, if they wished, to place an ear to the heavy oak and metal door to the room containing the wheel so they could hear it producing noises as its drum rotated. Meetsma in a poem he later wrote described what he heard by placing his ear against the sealed door as a thumping sound. That sound was made as the 29 pound weighted levers of the Kassel wheel's active internal one-direction wheel made contact with their wooden radial stop pieces as their pivot pin axes passed the 3:00 position of the huge drum's descending side.

In order to guarantee his wheel would run for the required time period during its test, Bessler probably connected the two long pendulums with the semi-circular flanges on them shown on the left hand side of Figure 2(b) to the axle's curved arm end cranks. Thus, the wheel would not have been running freely at its maximum terminal rotation rate of 26 rotations per minute, but at a slower rate, probably between 15 and 20 revolutions per minute, in an effort to reduce the wear on its various bearings and the coordinating cords interconnecting its weighted levers.

One may also wonder why they did not drill a spy hole in the wooden door so that they could visually observe the wheel to see if it was running. That was not done, I suspect, because it might have then allowed someone who visited the castle during the test to later lie and claim he had looked through the opening in the door and did not see the wheel running. Suddenly opening the door and seeing the huge wheel either running or not running would also have been a lot more dramatic, especially if the test was a success and it was still in motion.

Bessler, despite his Lutheran piety, had a flare for the dramatic and was a bit of a showman. He knew what the members from each of the various social strata visiting his wheels wanted to see and he delivered that for them.

The common people used to a life of toil would have seen his wheels making pendulums swing to and fro and also dazzling them with the staccato sounds of a constantly running stamping machine. (Although the Kassel wheel could not run both the Archimedean water screw *and* stamping mill simultaneously, it would have been able to operate either while also swinging its two compound pendulums because the latter dissipated virtually none of the wheel's constant energy output when in operation. But, because the semi-circular flanges did dissipate the wheel's constant energy output, they could not be used with either the water screw or the stamping mill.) After that he would give them spectacular demonstrations of the rapid lifting of heavy loads using a wheel's accumulated flywheel rotational kinetic energy. For the better educated and curious intelligentsia he delivered elaborate mechanical descriptions involving Latin names they were unfamiliar with but would often pretend to understand so as not to appear ignorant. For the rich and powerful he portrayed his invention as the ultimate portable power source that could greatly enhance their income. It could pump out a flooded mine, run a factory, or continuously irrigate elevated land if one just constructed wheels big enough and used enough of them to supply the power needed for a particular application and it could do all of this without any need for animal, steam, water, or wind power. He, literally, had something for everyone, but it would only be theirs *after* he was *fully* paid an enormous sum in advance for its secret.

Two weeks later, at the midpoint of an originally planned four week duration test, on November 26th, 1717 the seals which had not been violated were broken and the room was entered by the original investigators who had sealed it. They found the large wheel to be rotating at exactly the same rate as when they had initially sealed the room. Without Bessler's assistance, the investigators, including the count himself, briefly stopped the wheel to inspect its axle's pivot bearings and check them for excessive wear. At this time they may have been relubricated. They then restarted the wheel, exited the room, and its door was locked and more hot wax was applied to the crevices as the examiners again impressed their seals into the wax before it cooled completely.

The room was finally entered six weeks later on January 4th, 1718 to reveal that the wheel was still in motion. At this point Bessler wished to continue with the demonstration for several more weeks, but Karl decided that the duration test, having lasted 54 days or almost eight weeks and twice the originally planned four weeks, had been more than sufficient to prove that Bessler had, indeed, invented a genuine perpetual motion machine. Karl was concerned that, if the wheel was allowed to continue to run further, it might experience some internal part failure that would stop it since it was only a model and then critics would seize upon such a mechanical failure as a justification for claiming that the wheel really wasn't perpetual after all.

The news of this successful test at Weissenstein Castle, verified by many reputable witnesses, quickly spread throughout Hesse-Kassel and the surrounding principalities. Bessler's critics were in shock and could not figure out how someone they considered a crook had managed, once again, to deceive everyone with one of his fake wheels. Gartner eventually came up with a method he believed Bessler might have used.

Since the wheel was only seen running when the door to its room in the castle was opened, that obviously meant that closing the door somehow resulted in a braking mechanism being activated that then stopped the great wheel's rotation. That stoppage would conserve the limited supply of energy Gartner assumed was contained by the drum's hidden clockwork mechanism. Later, when the door was opened again, the brake was released and the wheel would then immediately start turning.

He reasoned that the door to the room had been cleverly modified so that, when it was closed, it depressed a hidden button in the door frame which, through an elaborate mechanical linkage system, then caused a pin inside one of the wheel's vertical support planks to press down against one of the axle's pivot pins so that the resulting drag would eventually stop the wheel completely. This system would operate automatically even if Bessler was not present. It sounded plausible to him because he had heard that the walls and adjoining rooms of the room containing the great wheel had been meticulously examined, but, most likely, he thought no one had examined the door frame itself!

The problem with this scenario is that, obviously, Gartner did not know that the Kassel wheel's heavy wooden vertical support planks were not in direct contact with either the room's floor or ceiling. As mentioned earlier, those planks were attached to thick steel rectangular bridge plates that maintained a gap of several inches between the end of a plank and a floor or ceiling surface. Those gaps would have been carefully examined as would the axle pivot pin bearings of the wheel. Anything, no matter how slight, that seemed suspicious would have been caught by those who examined that wheel and its room on that day of November 12th, 1717. Another problem with this method was that, once stopped, a bidirectional wheel needed to be given a push to start it turning again in either direction. If opening the door allowed a hidden brake on the Kassel wheel's axle pivot pin to be released, the wheel would still have remained stationary because there was no one around to give the drum a push start before the examiners entered the room. One gets the impression that Gartner did not even know that the Kassel wheel was bidirectional!

Count Karl, because he was busy with affairs of state, was not able to give Bessler an official certificate verifying the successful completion of the test until the next year on May 27th, 1718. That document was another dream come true for Bessler. Finally, one of the most respected aristocrats in all of Europe had officially verified that the inventor's wheel was genuine. The certificate stated that the results of all of the recent testing of his machine had completely eliminated any "hint or suspicion" that his invention was not a genuine perpetual motion machine. Surely, at this point Bessler must have been convinced that the sale of his marvelous invention would probably take place before the end of 1718.

But 1718 passed and still Bessler could not find a buyer that could meet his strict demand for a single, complete lump sum payment for the invention in advance. Various deals were considered such as the count acting as a middleman who would hold the buyer's money until he was satisfied that the wheel was genuine and, only then, would the 100,000 thalers be released to Bessler. That plan sounded plausible, but still no one seemed interested enough to accept it. During 1718 the room at Weissenstein Castle devoted to the wheel was visited by many representatives of potential buyers, but, despite his official certificate from Karl, Bessler could not convince them that the wheel was not a hoax or that it could be made much more powerful.

In one newspaper report concerning his past Merseburg wheel, Bessler was even quoted as saying that the secret of his wheels could be found in the words of Jesus! This might be a reference to Matthew 12:25 where one reads "And Jesus knew their thoughts, and said unto them, Every kingdom divided against itself is brought to desolation; and every city or house divided against itself shall not stand:"

Bessler may have interpreted this verse as some sort of message from Jesus stating that an imbalanced wheel could keep in constant motion if only one could find a design that would maintain the imbalance as the wheel rotated. In Bessler's mind, the drum of one of his one-directional imbalanced pm wheels may have represented a city or house "divided against itself" or experiencing some sort of internal conflict or asymmetry because the horizontal distance of the center of gravity of the four weighted levers in the descending side of the drum from the center axis of the axle was always slightly greater than the horizontal distance of the center of gravity of the four weighted levers in the ascending side of the drum from the center axis of the axle. Because of this chronic state of asymmetry, the "house" or drum "could not stand" or remain stationary. Bessler was a student of Christian scripture and probably considered this verse to be some sort of divine prophecy about the future development, by him, of a working imbalanced pm wheel.

Finally, the year 1719 arrived and still no buyer was found for the Kassel wheel despite the many interested in it to which Bessler had spent so much time giving private demonstrations. This delay was starting to depress Bessler and his patron and friend Karl, out of concern for him, tried to distract him by getting him to spend his idle time working on the definitive literary treatment of his wheels. It was during this year that Bessler authored DT which, as mentioned above, was his most professionally written text. Once completed, copies of it would be freely given to the visiting agents of those interested in buying the invention so that they could see what the wheel looked like even though they were too distant or busy to travel to Kassel to personally view it.

It is a long work written in both German and Latin. Approximately one third of the text was used by Bessler to praise God for having allowed the inventor to discover the secret of perpetual motion, another third of the text is filled with praise for Count Karl who had the wisdom to see the value of Bessler's work and to financially support it, and the remaining third of the book deals with the secret of his self-moving wheels. Of course, he never really quite reveals this secret in sufficient detail for his wheels to be replicated and it contains no illustrations of the wheel's internal mechanisms. Indeed, it is this frustrating quality of all of Bessler's writings which is one of the main reasons I eventually felt the need to write the book you now hold. It is actually an attempt to supply what was missing from his works and which would only have eventually become publicly available if Bessler had sold his invention which, unfortunately, he was not destined to do.

However, despite the limited value of DT to those attempting to reverse engineer his wheels, we are given a few tantalizing hints about the internal mechanics of them that he eventually found and used. For example, one passage, in English, states:

> "The internal structure of the machine is of a nature according to the laws of mechanical perpetual motion, so arranged that certain disposed weights, once in rotation, gain force from their own swinging, and must continue this movement as long as their structure does not lose its position."

> "...these weights, on the contrary, are the essential parts, and constitute the perpetual motion itself; since from them is received the universal movement which they must exercise so long as they remain out of the center of gravity; and when they come to be placed together, and so arranged one against another that they can never obtain equilibrium, or the *punctum quietus* which they unceasingly seek in their wonderfully speedy flight, one or the other of them must apply its weight at right angles to the axis, which in its turn must also move."

An alternative and, I feel, more accurate interpretation would read:

> "The internal mechanics of a wheel's drum consists of a particular arrangement of weighted levers which rotate around their own pivot pins as these pivot pins, in turn, rotate around the center axis of the axle of the drum. These combined motions then cause the center of gravity of these weighted levers, despite the drum's rotation, to remain on the descending side of the axle and, thus, they are able to provide a continuous torque to rotate the drum and will continue to do this so long as they remain interconnected and precisely coordinated with each other (by a system of coordinating cords)."

"...these weighted levers are critically necessary to producing the drum's perpetual motion from the constant torque they produce and must continue to produce it as long as their center of gravity is horizontally offset from an imaginary vertical line passing through the center of rotation of the drum which is at the center axis of its axle. And, when all of a drum's weighted levers are installed and interconnected and coordinated with each other (by a system of coordinating cords) so that their various motions during drum rotation never allow their center of gravity to reach their equilibrium point located directly vertically below the center axis of the axle which they continuously try to do as they rapidly revolve around the axle, these weighted levers will continue to supply the torque that will cause the drum to rotate perpetually."

Granted that my interpretation / translation is wordier than the usual ones given for this passage, but I believe it more accurately expresses the general description of the secret perpetual motion wheel mechanics that Bessler used. However, the inventor could not state what is in my interpretation because it would have, in his opinion, given the reverse engineers reading DT too much of a hint of what direction to proceed in as they attempted to duplicate his wheels' mechanics.

We see from his description that the inventor was using some sort of an imbalanced wheel design. Many attempts had been made down through the centuries prior to Bessler's effort to make self-moving wheels that used this principle and none that we know of, until that of Bessler, had ever been successful.

These devices usually consisted of small metal weights, even liquid mercury in glass vials, arranged around the circumference of a wheel with a horizontal axis of rotation. Their basic principle of operation then involved the weights either rolling, sliding, or otherwise shifting about during the wheel's rotation in such a way that the center of gravity of the weights, regardless of what position the wheel was in, would always be located on the wheel's descending side. Such a chronically imbalanced condition should then make the wheel continuously rotate and perform useful external work in addition to keeping itself in motion.

In actual practice, however, what always happened, with the single exception of Bessler's design, was that, as such a wheel began to rotate, its center of gravity would tend to also rotate along with the wheel, swing below its axle's center axis to the *punctum quietus* location as the torque it provided momentarily dropped to zero, and then begin oscillating back and forth below the center axis of the axle and through the *punctum quietus* until it finally came to rest at the *punctum quietus* itself. When this occurred, the wheel would, of course, have no torque available and would remain motionless thereafter.

Those examining Bessler's wheels were well aware of this dismal history of humanity's attempts to construct a working imbalanced perpetual motion wheel and they would have had mixed emotions upon viewing the Kassel wheel. On the one hand, they knew it had never been done, yet on the other hand there was precisely such a wheel right in front of their eyes that appeared to be breaking that long history of failures. It's hard to deny the evidence of one's eyes, yet, in their minds, they must have been asking themselves what the chance would be that someone like Bessler, who had only a limited education and was mostly just a self-taught craftsman, might be the first to achieve success while many other far more profound thinkers over the centuries had tried and failed. It just did not seem too probable to them, yet the Kassel wheel had passed a very impressive test. I suspect that the majority of those viewing Bessler's wheels tended to think it was more likely a trick rather than the real thing. Despite that opinion, however, the minutest scrutiny of his wheel's external details failed to indicate even the slightest hint of fraud.

Occasionally, when Bessler was out of earshot, those examining the Kassel wheel would try to obtain some confidential information about it from Count Karl since he was the only one, other than Bessler and his brother Gottfried, who had ever seen the internal mechanics that were used. But Karl, a man who had vowed never to reveal the inventor's secret, would only make a few vague statements concerning it in order to satisfy their curiosity. He said that the wheel's secret mechanics were "very simple" and, in fact, so simple that a "carpenter's boy" could understand and duplicate them after having seen the inside of a wheel's drum. He even said that he was amazed that, due to its simplicity, no one else had come up with the same design prior to Bessler!

The year 1720 was a long and frustrating one for Bessler. He and his growing family continued to reside at Weissenstein Castle and he continued to serve as the Commerce Counselor of Hesse-Kassel while showing his most powerful 12 feet diameter, bidirectional wheel to all of the important visitors who requested to see it. But, still no one came forward with the money he demanded for the secret of his wheels despite all of the private demonstrations he gave and letters he wrote. The purchase would not just be for a set of plans on paper, but would also include the Kassel wheel itself which the new owner could then disassemble so its parts could be copied and used to make many more wheels. After the sale, Bessler wanted to be free to begin founding his "Fortress of Wisdom" which would actually be a sort of engineering school that taught the many useful technical crafts in an environment that instilled the students with the various religious and moral principles that Bessler considered essential for everyone. While his attempt at an "Orffyrean Church" had recently failed, he was confident that his school would eventually serve to convert the entire world into believing and following his particular and correct interpretations of Biblical principles. This would be yet another attempt by him to unify the religions of the world and help eliminate warfare and the many miseries it inflicted on humanity.

With the sale of the invention, it would then become the new owner's problem to protect its secret from those that would simply purchase one and then start making and selling their own unauthorized copies. This was a very real danger in most of Europe at the time because they were only beginning to enact patent laws to protect inventors and their inventions. If Bessler had lived in England at that time, he would have had that protection, but it did not really exist for him in the Kingdom of Saxony.

While at the castle, Bessler acquired a few enemies who found his piety and high moral standards somewhat troublesome. One of his most powerful enemies took the form of a beautiful woman who just happened to be his benefactor's mistress!

After the death of his wife in 1711, Count Karl found himself coming under the influence of a lady named the Marquise de Langallerie who had arrived in Kassel in 1713 with her husband, a French adventurer, who left her there and took off, as adventurers tend to do, in search of adventure. It was not long before this attractive woman caught the attention of the lonely widower Karl. They formed a rather close relationship. How close? The rumors at the castle were that the two children she bore in 1714 and 1716 were actually the count's illegitimate progeny! Years later in 1722 another younger woman also arrived named Christine von Bernhold. Both women naturally immediately hated each other and became bitter rivals for Karl's affections and only by showering them both with gifts and excessive annual stipends did he manage to keep some peace between the two.

Needless to say, Bessler found all of these amorous antics morally unacceptable, yet they involved the one person on the planet who had offered him and his family refuge, him employment, and the financial backing for his largest and most powerful wheel yet. Obviously, Bessler could not just come out and tell Karl that his behavior was immoral and would result in his soul being cast into eternal hellfire upon his death when the man was using all of his influence trying to help Bessler achieve a very grand dream, indeed: a religious school that would serve to transform the entire world to the high ethical and moral principles that Bessler's research of the Scriptures revealed to him.

Not being able to openly express his feelings to Karl was agony for Bessler, and he had to settle for just trying to suggest to the count as politely as possible what the most moral thing to do with these women would be which, obviously, would be to marry one of them if that could be arranged legally and would not be prohibited by Karl's Calvinist religious affiliation. Other than that, Bessler had to tolerate the situation because the greater good that the sale of his invention would enable had to take priority over his personal objections to the immoral behavior on the part of Karl.

The Marquise de Langallerie was a Catholic and became suspicious of Bessler just as soon as he and his family, all Lutherans, arrived at Weissenstein Castle. Bessler would have immediately disliked her just because she was a Catholic and her behavior with the count would have labeled her as a harlot in Bessler's mind. And, of course, the last thing Langallerie wanted was anybody at Weissenstein Castle telling Karl that he should not be spending so much on her or sharing her bed.

She so disliked Bessler that she eventually managed to talk Karl into getting him and his family to move out of Weissenstein Castle permanently. Her persuasive reasoning would have been simple. She would have claimed that Bessler and his family were actually commoners and really did not belong at Karl's court. She would have suggested that Bessler's duties as a commerce counselor could probably be better conducted by one of Karl's adult children who had more formal training in law. She would have pointed out that the inventor's wheel had not sold after years and might never sell while Bessler's family would continue to swell in size and take over more of the rooms at the castle. She would have added that he and his family really needed to have their own home away from the castle until a buyer was found for the wheel. The Kassel wheel could still remain at the castle and kept locked up in its room until potential buyers or their agents wished to view it and, at that time, Bessler could return and give a demonstration and even stay for a dinner and talk with Karl. Yes, it all made complete sense to the marquise, at least.

Karl did not like this idea, but in order to placate his mistress he went along with it. The plan was that Bessler's employment by Karl would also be ended and that would then give Bessler time to get involved in other projects in neighboring principalities and also promote his wheel to potential buyers in them. To make the transition as painless as possible, Karl would also give Bessler a house with surrounding gardens in the nearby town of Karlshafen which is about 23 miles north of Kassel along with a severance payment equal to five years of his salary or 1,750 thalers to help himself and his growing family relocate to the new residence and support themselves until other sources of income became available to them. (The count also permitted many Huguenots fleeing persecution in northern France at the time to settle in this town. The Huguenots were French Protestants who had been inspired by the writings of John Calvin, the founder of Calvinism.) And, to further help Bessler, he and his family could live in the house as long as they wanted to and would be exempt from paying any taxes on it.

Bessler so disliked Langallerie and her relationship with Karl that he agreed to this new arrangement almost immediately, but he did not like the idea of making a move and leaving one of his wheels behind intact. However, the count assured him that it would be securely locked up inside of its new location and no one would be allowed to enter unless Bessler was also present which helped to greatly relieve Bessler's fears about the security of his perpetual motion wheel's secret mechanics. That promise, however, would be broken before the end of the year 1721 with tragic results.

In August of the year 1721, Count Karl invited a Professor Willem Jacob s'Gravesande to Weissenstein Castle to view some of the latest scientific instruments that he had recently acquired for his private collection. s' Gravesande was a professor of astronomy and mathematics at the University of Leiden in the Netherlands as well as being a Fellow of the Royal Society over in London and a friend of Sir Isaac Newton. He was someone whose opinion on technical matters Karl trusted.

When s'Gravesande arrived, he found out that there was an architect already there named Joseph Emanuel Fischer von Erlach who had come to help in the construction and installation of what would be one of the most powerful steam powered water pumps in Europe at the time. Karl had been told it was really the only practical way to recirculate the water from the artificial lake in which it collected at the rear of his castle back up again through a 360 feet vertical elevation to its spout located in the Hercules octagon's base so that the count could keep his cascade running continuously throughout the warm weather months. Unfortunately, the steam engine needed for this pump was a somewhat complex device that was proving difficult to assemble. Consequently, its installation was experiencing one delay after another while the count's cascade still depended upon an unreliable system of rain water filled reservoirs to stay in operation. This delay did not please the count who had anticipated it being installed and fully operational by the spring of 1721 which had not happened.

Karl had had his mind so filled with visions by Bessler of the job being done continuously without the need for any fuel by a collection of oversized super perpetual motion wheels that he decided to ask both of these men to test and evaluate the Kassel wheel. Neither had seen the wheel previously, but, of course, had heard much about it. Since Fisher von Erlach had just tested the wheel, it was s'Gravesande's turn and the count allowed him, on August 16th, 1721, to spend some time with the wheel testing it in Karl's presence even though Bessler was not there. Of course, being a gentleman, s'Gravesande would have given the count his solemn promise not to try to look inside of the huge wheel's drum and did not do so. He only performed a thorough examination of the wheel's external features and performance that fully satisfied him that the wheel was, indeed, a genuine imbalanced perpetual motion wheel that was truly self-motive.

Like others before him, without the masking sounds of the attached stamps, s'Gravesande would have counted the eight gently thumping sounds per drum rotation that seemed to emanate from around the 3:00 position of the drum when it was rotating in a clockwise direction. While the professor would, undoubtedly, have been very impressed by Bessler's creation, it certainly would not have convinced him that it was powerful enough to pump water at the rate that would be needed to operate Karl's cascade. As mentioned previously, to achieve that feat would have required a dozen super wheels, each with a drum 40 feet in diameter, working together.

The next day Bessler found out about these tests at the castle that were conducted behind his back and which were permitted by the count who, aside from Bessler, was the only one that had a key to the room in which the 12 feet diameter wheel was kept. Bessler immediately left for Weissenstein Castle, gained entry, made his way to the second floor room above the ground floor containing the wheel, unlocked the door, and then proceeded to use an axe to chop the wheel into as many small fragments as he could. He was in a rage and felt like a fool for having let his guard down so much when it came to maintaining the security on his invention. In a matter of perhaps ten or fifteen minutes he had destroyed what had taken him and his brother Gottfried almost a year to construct! Then, before he left to return to Karlshafen, he scrawled a message of the door to the room that now held the pulverized and shredded remains of the Kassel wheel. That message blamed its destruction of the impertinence and refusal of 's Gravesande to accept the results of the elaborate testing that had been done previously on the wheel. Maybe Bessler feared that the next time someone showed up to examine his invention, the count, being busy, would just give him the key to the wheel's room and trust him to examine it without either the count or Bessler being present! That was something Bessler could not allow and only destroying the wheel could guarantee that it could never happen.

It's really too bad that Bessler chose this response because it turned out that, after his examination, 's Gravesande began writing to his colleagues about how impressed he was by the Kassel wheel and even wrote to Newton to give him a general description of the wheel and the tests he conducted on it in the count's presence. Unfortunately, there is no record of Newton responding, but I suspect, if he did, that he was probably not as skeptical about the possibility of perpetual motion as many believe he would have been. In fact, there is some evidence that Newton himself once proposed a design for a perpetual motion wheel that, like a water wheel, would use the impacts of incoming gravity particles from outer space on its upper surfaces to produce a driving torque on its axle. I have often wondered what Newton's reaction would have been if he had made the pilgrimage to Kassel to personally test the wheel there. I think if he had, then he would have given his approval to the device, formed a friendship with Bessler, and with that the chances of Bessler selling his invention would have escalated greatly. Such a meeting never took place because of Newton's advancing age and worsening health problems.

After the destruction of the Kassel wheel, the Bessler story tends to become a bit more difficult to track down.

Although Karl was initially very angry at the destruction of a wheel that had cost him a small fortune for materials, he eventually forgave Bessler and the two remained good friends. And, besides, it was Karl's fault for allowing someone access to the wheel without Bessler being present even though he promised that would not happen. Karl knew that the wheel was genuine and could appreciate Bessler's fear of the secret being stolen by someone without paying the inventor for all of the hard work he had performed to find that secret.

Once out of the castle, Bessler even had some plans to build one of the 40 feet diameter super wheels that he had mesmerized Karl with for years and was confident it would be able to power anything that a conventional water or wind mill might and to do so day and night in any sort of weather or season without the need for any sort of external power supply. The cost would, of course, be far more that for the Kassel wheel and would have to be funded privately by a consortium of investors since, after the sudden destruction of the Kassel wheel, the count would not risk any more of the tax money he was collecting on such an expensive project that might also be destroyed in the event that Bessler's paranoia once again seized control of him in the future. If his super wheel could be completed, then Bessler promised it could be tested by 's Gravesande again only this time with Bessler present!

However, before the inventor could seek investors for that grand project, he needed another smaller, less expensive wheel that he could personally show potential buyers in his Karlshafen home and which would be more powerful than his little toy 3 feet diameter Gera prototype wheel that he still retained. To provide that, he most likely began the construction of a 6 feet diameter, one-directional wheel that was mounted between floor to ceiling wooden vertical support planks and which he kept in a locked workshop in his new home in Karlshafen.

This wheel would have had eight weighted levers each of whose parallel pair of wooden main arms held a single 3 pound cylindrical lead end weight whose attachment bolt's center axis was located at an exact distance of 7 inches from the lever's pivot pin center axis. That pivot pin's center axis, in turn, was located at an exact distance of 28 inches from the center axis of the wheel's 5 inch diameter axle. For the eight cylindrical lead end weights of this Karlshafen wheel, Bessler simply reused eight of the 4 pound end weights and their attachment bolts that had been installed in the bidirectional Merseburg wheel and which he had kept after its destruction. By carefully sawing an inch of length off of their 4 inch lengths and also the lengths of their attachment bolts, he would have obtained cylindrical lead end weights and attachment bolts that weighted almost exactly 3 pounds each.

When completed, this new wheel, because of its heavier end weights, would have had a start up power output of about 3.1 watts which made it almost three and a half times more powerful than the 4.5 feet diameter second Gera wheel that, at start up, outputted about 0.9 watts. It's start up power certainly was nowhere near as impressive as those of the Merseburg and Kassel wheels had been, but it was adequate to show a potential buyer of the design that it could, at a minimum, do such practical things as run a small grindstone for sharpening tools or run a window fan to keep the air moving in a stuffy factory to keep workers more comfortable on a hot and humid summer day. And, as usual, Bessler would promise a potential buyer that by enlarging the design, there was no limit to how much free labor the device could perform for him once he had purchased it, of course.

Gone were the days of the Archimedean screw's continuous outpouring of water and the hypnotic swinging of massive dual compound pendulums that were easily powered by the 1100 pound bidirectional Kassel wheel, but Bessler hoped this little 6 feet diameter, one-directional wheel would still be impressive enough to, finally, persuade a rich buyer to give him his just compensation for all of the years of struggle he had to endure to find a design for an imbalanced perpetual motion wheel that actually worked in practice. Once again, his hope of finding a buyer began to escalate and his depression to lessen. Yet, he probably eventually realized that the episode of anger which had compelled him to destroy the Kassel wheel was a huge mistake and perhaps the worst of his entire life. With the destruction of that huge wheel, Bessler had nothing to show interested dignitaries and potential buyers or their agents visiting Weissenstein Castle other than his small table top model prototype wheel that could only output a fraction of a watt of power. It was this realization which helped motivate Bessler to begin the construction of his next far less powerful 6 feet diameter wheel which would be the sixth working wheel that he would actually construct during his lifetime.

And, again, as he began to get enthusiastic about the construction of this wheel, his fear also began to grow about the possibility that, when he was not at home in Karlshafen, someone might break in and try to steal its secret as could have happened at Weissenstein Castle on that fateful day in August of 1721. Bessler would make sure that never happened again. The door to his workshop would have had several locks on it and no one would ever see the new wheel unless Bessler was also present.

Meanwhile, Bessler's trusted brother Gottfried left for England without his family with the hope of finding wealthy investors there that might be interested in purchasing the new, but smaller wheel that Bessler was working on in Karlshafen. He was hoping to obtain employment there with the Royal Society and use that to assist his older brother in selling his invention. At the time early steam engines were starting to come into regular use in England, but, although powerful, they were inefficient and consumed large quantities of coal to keep the water in their boilers boiling. In addition, people were beginning to notice how dirty they made the air. One solution, of course, was to convert to a nonpolluting power source that would also be able to operate if no animal, water, or wind power sources were available and Bessler had such a source.

When not working on his sixth smaller and less impressive one-directional wheel in his new home in Karlshafen, Bessler still found time to try to found an Orffyrean Church in the principality and, in order to promote it, published a book in 1723 titled, in German, *Der rechtglaubige Orffyreer, oder die einige Vereinigung der uneinigen Christen in Glaubens-Sachen* or, in English, *The Proper Orffyrean Belief on How to Unify Various Christian Denominations in Matters of Faith.*

His aim was to try to resolve the differences between Catholics and Protestants which had caused so much strife as well as bloodshed in the 16th century. No doubt his church failed because Catholics saw it as a thinly disguised attempt to convert Catholics into Protestants and, especially, into Bessler's particular version of Lutheranism. General histories of Bessler and his wheels often fail to mention how religious he was. He had thoroughly studied the Bible, prayed devoutly and regularly, and, from his attempts to establish his own church, had come to his own conclusions as to what the accounts in that collection of ancient writings really meant. His independence in interpreting Scriptures no doubt brought him into conflict with many of the leaders of his local Lutheran congregations that resented anyone questioning their beliefs and practices and this was probably another reason he wanted to start a new religion based on his particular beliefs.

Just as Bessler was again beginning to have some hope that he would soon find a buyer for his next wheel something happened that really escalated his paranoia concerning the security of his perpetual motion wheels' secret.

On the Sunday after Easter in 1724 in Karlshafen, Bessler found out that the maid, Anne Rosine Mauersbergerin, who he had kept on as an employee only because of his wife's attachment to her and the assistance that she provided with the care of their children, had been secretly communicating with Count Karl's mistress, the Marquise de Langallerie, who was probably Bessler's main enemy at the castle! Apparently, the marquise wanted to know all about what Bessler was up to in his new home, especially in his workshop.

Needless to say, this type of security leak would have infuriated Bessler. Only because she was so close to Bessler's wife did Mauersbergerin avoid being immediately fired and even then because she agree to sign a scary oath vowing never to reveal anything she heard while in the inventor's employ or in any way demean him if she left his employ or else she should be cursed for all eternity. Bessler eventually had everyone in his family sign the same oath. What a shock it must have been to him to realize that the very people his inventions were supporting were secretly undermining its security and, possibly, even plotting against him!

To Bessler, the marquise was a dishonored woman that would stop at nothing to enrich herself and increase her power. Perhaps he had often wondered, when he and his family had lived at the castle, if one night the marquise might steal the only other key to the wheel room from Karl and then, after everyone was asleep, allow an accomplice to sneak into that room, open one of the linen inspection and service flaps on the backside of the wheel's drum, and then make a few detailed sketches of the secret lever shape he used and the particular arrangement of coordinating cords and springs that made mechanical perpetual motion possible. With possession of that critical information, anyone would be in a position to duplicate the wheel, claim he was really the original inventor, and sell if off for a fraction of the enormous sum that Bessler demanded!

To further add to Bessler's concerns, in 1725 it began to become apparent that Count Karl, who was then 71, was beginning to show the early signs of dementia. He started to make bad decisions regarding money and had trouble remembering things he had only recently done. His many adult children were worried about him. They were also not too happy that their father's mistress, the Marquise de Langallerie, was slowly becoming his caretaker and, thus, even more powerful at court. What if she managed to get their father to change his will and leave her the bulk of his estate and them relatively nothing? Such things were known to happen and they became even more suspicious of her motives than had been Bessler. Not surprisingly, as Karl's mental faculties began to decline, her annual pension provided by the count kept increasing. The aging count was, obviously, being manipulated by the marquise.

Peter the Great, Czar of Russia, had heard of Bessler's wheels by way of correspondence from Leibniz. The czar wanted to modernize Russia by introducing the latest sciences and technologies there and enhancing education, at least for the upper classes. When he found out about Bessler's Merseburg wheel, he, like many others, was fascinated by it. Yes, the price was steep, but Peter was in the very fortunate position of being easily able to pay that price. He had years earlier dispatched two emissaries to Saxony in an attempt to buy the Merseburg wheel in July of 1716. When they arrived, however, they discovered that, it had been destroyed and Bessler was still recovering from the after effects of the fall he had taken. At most, they could only see a quick demonstration of the inventor's 3 feet diameter, one-directional Gera prototype wheel.

However, in 1725 the opportunity again arose for Bessler to sell one of his wheels and get the huge compensation he sought. He received news that Peter the Great was coming to visit him and, if he was convinced the inventor had a real wheel, he would immediately pay Bessler the 100,000 thalers he demanded and take the wheel back to Russia with him. Bessler busied himself in an effort to complete the smaller 6 feet diameter wheel he was working on in his Karlshafen home's workshop. Now, finally, after more than a decade of struggle, he was on the verge of finally selling his invention. The money would all immediately be his and he could then proceed with the second biggest project of his life: the establishment of a religious school that would produce godly craftsmen who would go on to inspire proper religious moral principles in humanity and thereby change the world for the better. This, I believe, Bessler considered even more important than the successful quest to achieve perpetual motion. The latter was just the means he hoped to realize a far grander vision.

Then the bad news arrived. On February 8th, 1725, Peter died in St. Petersburg, Russia en route to his meeting with Bessler after he finished a lengthy inspection tour of various other projects he had initiated. In the winter of 1723 he had started to have problems with his bladder and urinary tract and in 1724 underwent surgery to try to correct the problem. Then, in early January of 1725, he developed the symptoms of uremia which is a toxic state that results when failing kidneys allow excessive levels of urea and creatinine to build up in one's blood. Finally, the czar lapsed into a coma as a result and died at the age of almost 53 years after having been emperor of all Russia for 42 years. An autopsy indicated he died of gangrene of the bladder which caused a fatal bacterial infection to spread to his kidneys and from there throughout the rest of his body. If modern antibiotics had been available then and immediately given, he might have survived. In any event, with his death, the last hope for a quick sale of his wheel ended for Bessler. (Had the meeting ever taken place, then, aside from the language difficulties, it might have been a bit awkward for another reason. Bessler was probably only of average height for a European male back then which would have been about five feet, six inches. By comparison, Peter was a giant of a man at six feet, eight inches tall and, thus, over a foot taller than Bessler!)

Then, quite sadly, Bessler's first wife, Barbara Elisabeth Bessler, died from tuberculosis on May 10th, 1726 at the age of 40 and her end was probably hastened by the death of her infant son Karl Friedrich Bessler having occurred only five days earlier. It must have been very depressing for Bessler, the "doctor", not to have been able to save either of them as he had so easily done the first time he had met his future wife. Only Bessler's deep religious beliefs allowed him to cope with such a devastating loss. But his wife's maid, Anne Rosine Mauersbergerin, then about 38 years of age, remained employed and continued to care for Bessler's still surviving children.

In 1726 Bessler was about 46 years old and so, quite possibly, his deceased wife's maid had visions of marrying Bessler and then taking the place of her former mistress in their home in Karlshafen. Unfortunately, her plans were thwarted because Bessler did not trust her and he finally decided to permanently end their relationship in order to get her out of his home sometime before November of 1727 after she had been in his employ for about 15 years. Possibly, she threatened, if dismissed, to violate the oath she had signed and use it to show everyone what a horrible employer Bessler had been. That, of course, would have quickly sent the inventor into a rage. She actually fled from his Karlshafen house because he threatened to strangle her! Obviously, they had a stormy relationship, at best, after the death of Bessler's wife.

When some of Bessler's detractors found out about the maid's firing, they saw it as a golden opportunity to, finally, expose Bessler and his wheels as fraudulent and prevent him from using them to steal money from anyone else. All they needed to do was help Bessler's ex-maid financially and get her to swear that the inventor's wheels were being manually turned from an adjoining room in Weissenstein Castle. Bessler's wife was dead and Gottfried had left for England. His surviving children were still minors and could not testify. That meant it was the maid's word against Bessler's word. The inventor's detractors and the maid could all get even with Bessler if the maid was just willing to provide false testimony which is exactly what she did. Since Gartner died on February 2nd of 1727, he was not involved in this bizarre plot.

On November 28th of 1727 Bessler' maid, Anne Rosine Mauersbergerin, signed a statement claiming that the Kassel wheel was a fake and Bessler was soon arrested on a charge of fraud. The maid claimed that she, Bessler's wife, the inventor's young daughter, and the inventor's brother had taken turns operating a hand crank mechanism in an adjoining room in order to keep the 12 feet diameter Kassel wheel in steady motion during the brief periods when it was being examined and tested. She further claimed that there was a barb at the end of a metal rod hidden in one of the wheel's vertical support planks that would move up and down in order to catch onto and rotate the hidden portion of the shaft of one of the steel pivot pins projecting from an end of the wooden axle to which the wheel's drum was attached.

The maid, of course, knew that she was not really telling the truth based on personal knowledge because she was not actually part of any such alleged fraud, but her handlers had convinced her that the crime she was charging Bessler with was how the Kassel wheel *must* have actually been turned even if she did not know it for a fact. Perhaps they added that God would certainly forgive her for breaking the Old Testament commandment that forbids lying since doing so would prevent an evil sinner like Bessler from breaking another commandment that forbids stealing. Their logic as well as financial assistance to her was sufficient to secure her cooperation.

Fortunately for Bessler, the count ordered him immediately released upon learning of his arrest, the maid's testimony was soon discredited, and the charge against the inventor was dropped. I think it was Karl's insistence that the Kassel wheel was genuine and his sterling integrity that managed to get the charge quickly dropped. It was obvious that the testimony of a disgruntled ex-employee would never stand up against the eye witness testimony of a man of the count's reputation in a court of law if the case had been allowed to go that far which it was not.

As for Bessler's fired maid, Anne Rosine Mauersbergerin, she did not become destitute when those who convinced her to commit perjury stopped supporting her, but was eventually rehired as a maid by Bessler's former mother in law, Barbara Schuhmann, and even got married to someone other than Bessler. Most likely, she eventually regretted her actions. Frau

Schuhmann, along with her other two still living daughters and their husbands as well as her rehired maid, Anne Rosine Mauersbergerin, even moved so that they could all live closer to Bessler's home in Karlshafen. They continued to be a constant annoyance to him after his dear wife's death and always seemed to be in need of money from him which, of course, they never paid back. They knew he really was a generous Christian man and used that to exploit him in whatever way they could. As his financial resources began to dwindle, however, they slowly faded out of his life.

This experience with his wife's maid had greatly unsettled Bessler and made him even more concerned about maintaining the security of his inventions. He suddenly realized that it was actually legally possible to have him removed, by force if necessary, from his home and even, with the proper paperwork, for someone to enter his locked workshop and then confiscate either his 3 feet diameter Gera prototype wheel or the larger 6 feet diameter one-directional wheel he had finally finished and intended to sell to Peter the Great. Once he lost control of these wheels, there was no telling who would get possession of them and could then steal the secrets of their mechanics and sell them to the highest bidder.

Starting as early as 1722, Bessler had been working on a large book on the subject of the mechanics of perpetual motion machines which was intended to instruct the reader in the principles needed to build working devices of this nature. This volume was to include illustrations of many of the devices he had conceived and built over the years along with descriptions of their operation and why they did *not* work and what was to be learned from their failures. At the end of the work, he would reveal, in detail, the secret of his finally successful one-directional and bidirectional wheels with appropriate illustrations. This monumental work would, however, only be published and sold to the public *after* one of his wheels was sold and that edition would not contain the secret of the inventor's imbalanced pm wheels. If one of his wheels was sold before the public publication of the work, then the inventor would offer the edition to the buyer which did include the details of the inventor's secret pm wheel mechanics and the buyer could then decide what to do with it as far as public publication was concerned. He might just decide to keep the entire text secret and then try to improve on Bessler's many failed designs, using his newly acquired knowledge of the imbalanced pm wheel mechanics that did work, to somehow improve and make them all workable and perpetual in their functioning.

As previously mentioned, Bessler intended to use most of the money from the sale of his invention to fund the founding of a school to be called the "Fortress of Wisdom" in which he would teach students all of the various crafts along with his version of Christianity. Most likely, he intended to use this book documenting the many failed pm machine designs he had tried before finding success as an introductory textbook for new students interested in the development of pm machines. Incredibly, though Bessler never published the book, we still have most of the contents of this work today. It has been given the German title of *Maschinen Tractate* or, in English, *Treatise on Machines*. It contains 140 illustrations made by Bessler himself and, hereafter, I will simply refer to it as "MT".

As soon as Bessler was released by the local authorities, he returned to his home in Karlshafen and immediately removed, burned, and buried all of the printed illustrations in his unpublished book along with the woodcut printing blocks that produced them which showed the secret of his marvelous self-moving wheels. However, he left the remainder of the printed illustrations, many of which contain his handwritten descriptions on them, along with their woodcut printing blocks intact because they did not reveal the actual secret perpetual motion mechanics he had found at Gera in 1711. About five and a half years later on May 1st, 1733 he added a brief introduction to his still unpublished work saying that anyone with a "discerning mind" who studied the surviving illustrations would, finally, be able to find the secret of perpetual motion as he had! In the drawings which have survived to this day, it can be seen that four original ones numbered 138 to 141 had been removed and replaced with a single one which is today referred to as the "Toys Page" by researchers because it illustrates five popular children's toys of his time.

Most likely, the four original drawings and their woodcut printing blocks removed from MT and destroyed by Bessler showed the arrangement of the weighted levers inside of one of his one-directional wheels along with their system of coordinating cords, the structural details of the weighted levers he used, the arrangement of the two internal back to back one-directional wheels within the enlarged single drum of a bidirectional wheel, and a drawing of the special gravity activated latches he used in a bidirectional wheel to give it its ability to rotate in either direction. The Toys Page substitution, however, does show some interesting mechanical principles and the note for the page in Bessler's handwriting suggests that if someone can understand the mechanics of the toys shown and apply them in a different way, he will be able to duplicate one of Bessler's wheels. For the curious, I have included the Toys Page from MT at the end of this chapter and it is shown in Figure 2(d).

Let us briefly analyze some of the symbolism and encoded information contained in that most out of place page before returning to the Bessler story.

As can be seen in the test printing Bessler made from its wooden MT printing block, the Toys Page illustrates five children's toys. They are, left to right, a scissor jack labeled "E" which was operated by squeezing the handles at the bottom together to make the point at the top fly rapidly upward or even horizontally outward so that a child could hit some normally out of reach object or play a prank on a friend (and, hopefully, not hit him in the eye with the point as it flew out!), two "hammermen" toys labeled with double "C's" and "D's" and operated by moving the horizontal sticks of their parallelogram shaped mechanisms upon which the opposed pairs of hammer wielding figures are mounted, a "Jacob's ladder" which is shown in two views labeled "A" and "B", and, below these four toys, a kind of top or, actually, early gyroscope that was popular in Bessler's day. Although I will be focusing extensively on the interpretations of the symbols in the two DT portraits later in this volume and showing how they were used by Bessler to preserve and hide the various parameters of the parts used in the mechanics of his imbalanced pm wheels, the Toys Page also contains some interesting clues about these parameters and I will now give the reader a brief analysis of them.

The two most prominent toys in the illustration are, of course, the two hammermen toys that occupy the center area. The child playing with these would alternately pull the little handles at the ends of the two horizontal sticks below the men apart and then push them toward each other again. This would then cause the two hammermen to take turns smacking the object between them with their hammers and making a bit of a racket in the process. It will be noted that the two hammermen of the top pair grip the handles of their hammers with both of their hands, but the two hammermen of the bottom pair have no arms and their hammers with enlarged heads appear to emerge from spring-like bodies. This symbolism was Bessler's way of telling a reader of MT that the top hammermen toy contained clues about the *un*weighted levers to which the lead end weights were attached while the bottom hammermen toy contained clues about the lead end weights themselves as well as the extension springs attached to the weighted levers inside of his wheels. But, what were these clues?

The information in the bottom hammermen toy is fairly easy to extract. Note that Bessler placed the letter "D" near the left hammer's head. "D" is the 4th letter of the alphabet with an alphanumeric value of 4 and this tells us that the mass of the end weight in the Merseburg wheel was 4 pounds. But, *two* "D" letters are present and this tells us that the Merseburg wheel's cylindrical lead end weight's mass was doubled for use in the Kassel wheel since 4 + 4 = 8 and that the end weights in the Kassel wheel were 8 pounds each. How do we know that the information pertains to the Merseburg and Kassel wheel's cylindrical lead end weights? The answer is gotten by evaluating other alphanumeric clues in the Toys Page.

Note that there is an "E" drawn inside of the scissor jack to the left of the bottom pair of hammermen. Thus, we have the letter "E" and *two* letter "D's" associated with the bottom pair of hammermen. "E" is the 5th letter of the alphabet with an alphanumeric value of 5 and "D" is the 4th letter with an alphanumeric value of 4. If we add up the alphanumeric values of these *three* letters, then we get 5 + 4 + 4 = 5 + 8 = 13 which is the alphanumeric value of the 13th letter of the alphabet or the letter "M" which stands here for the first letter in the beginning of the word "Merseburg"!

Obtaining the alphanumeric value of 11 for the letter "K" which is the 11th letter of the alphabet and stands for the first letter in the word "Kassel" is a bit more difficult. Note that the letter "E" is inside of two of the sections of the scissor jack that look like two letter "X's" stacked vertically one on top of the other. We can write these "X's" horizontally as "XX" which is also the Roman numeral equivalent of the number 20. Once we have that number, we simply subtract the alphanumeric value of the letter "E" *and* the nearest "D" from it to get 20 − (5 + 4) = 20 − 9 = 11.

Thus, the above alphanumeric analyses of the letters drawn near the lower pair of hammermen in the Toys Page assures us that the information it contains pertains to both the Merseburg and Kassel bidirectional wheels.

But, what about those spring-like bodies of the two lower pair of opposed hammermen? They tell us that *each* of the weighted levers in Bessler's wheels had two springs attached to it.

For the Merseburg wheel, each spring had a spring constant value of 10 pounds per inch and the two springs attached to each weighted lever had a combined spring constant value of 20 pounds per inch. To get the spring constant value of 10 for one of the springs, we take the alphanumeric value of one of the "X" shaped sections inside of the scissor jack on the left side of the illustration which is 10 and the alphanumeric value of two of the "X's" or 20 to get a spring constant value of 20 pounds per inch for the combined spring constant value for both of the extension springs attached to each of the 16 weighted levers in the Merseburg wheel.

For the Kassel wheel, each spring had a spring constant value of 20 pounds per inch and the two springs attached to each of that bidirectional wheel's 16 weighted lever had a combined spring constant value of 40 pounds per inch. To get the spring constant value of 20 for one of the springs, we simply multiply the value of "E" or 5 times the value of the nearest "D" or 4 and get 5 x 4 = 20 which stands for 20 pounds per inch (that we must perform a multiplication operation is indicated by a letter "X" shaped section of the scissor jack which is also the symbol for multiplication in arithmetic). The combined spring constant value for the two springs attached to each weighted lever of the Kassel wheel of 40 pounds per inch is then gotten by multiplying the alphanumeric value of the "E" or 5 times the *sum* of the alphanumeric values of the *two* letter "D's". This then gives us 5 x (4 + 4) = 5 x 8 = 40 which stands for 40 pounds per inch.

The top pair of hammermen, as previously mentioned, provides information about the masses of the *un*weighted levers used in Bessler's wheels and this was, again, for the Merseburg and Kassel wheels. In the Merseburg wheel the unweighted levers had a mass of 2.5 pounds and in the Kassel wheel this was doubled to 5 pounds. Here's how to obtain these values from the top hammermen toy.

To the immediate right of the right top hammerman one sees the side view of the top of the Jacob's ladder with its two tiles beginning to part and the letter "B" drawn between them almost as though it is forcing them to divide. The alphanumeric value of "B" is 2 and the implication of its location between the top two tiles is that we must divide something by the number 2 and thereby split it into two equal parts. If one extends a line downward along the length of the *falling* tile, which is located to the right of the letter "B", far enough and toward the left side of the illustration in the Toys Page, it will pass between the hands of the top left hammerman who is grasping the handle of his hammer whose head has just come down and struck the piece of metal resting on the anvil between the two hammermen. This tells us that we are going to be finding some information about the mass of the unweighted levers in Bessler's Merseburg and Kassel wheels.

Continuing to extend the line downward and to the left of the illustration causes it to eventually intersect one of the pivot joints in the scissor jack which is the one located immediately above the letter "E" and is horizontally level with the hat worn by the lower left hammerman. Dividing the alphanumeric value of "E" or 5 by the alphanumeric value of "B" or 2 then gives 5/2 = 2.5 which represents that mass of an unweighted lever in the Merseburg wheel. To get the mass of an unweighted lever in the Kassel wheel, we simple take the "X" shape of the section of the scissor jack that the pivot joint holds together and treat it like the Roman numeral for the number 10. Dividing that 10 by the value of "B" or 2 then gives us 10/2 = 5 which was the mass of the unweighted levers in the Kassel wheel.

How do we know that this information applies to the unweighted levers in Merseburg and Kassel wheels?

Again, Bessler confirms it with various alphanumeric and other number values in the Toys Page which can be combined to produce the alphanumeric values of 13 for the "M" which is the first letter in the word "Merseburg" and of 11 for the value of "K" which is the first letter in the word "Kassel". For example, if one counts the number of tiles visible in the side view of the Jacob's ladder that has the letter "B" near its top end, he will find that they total 11 tiles and this then stands for the alphanumeric value of the "K" in "Kassel". By adding the alphanumeric value of the letter "B" which is 2 to the number of tiles or 11, we get 2 + 11 = 13 which stands for the alphanumeric value of "M" in "Merseburg".

However, there is even more interesting symbolism used in the Toys Page.

For example, there are eight "X"-shaped sections to the scissor jack on the left side of Figure 2(d) (the arrowhead at the top is not included in the count) and they represent the eight weighted levers found in a one-directional Bessler wheel (even if it was one of the two internal one-directional wheels found in a large bidirectional wheel's drum) and the structure of the scissor jack with its connected "X" shaped sections indicates that all eight weighted levers were also connected together and affecting each others motion during drum rotation. If one multiplies the number of scissor jack sections, 8, by the alphanumeric value of the letter "E" at the bottom of them, one obtains the number 40 which just so happens to be the total number of coordinating cords found in one of Bessler's one-directional wheels!

The side view of the Jacob's ladder toy on the right side of the top hammermen toy labeled "B" shows the ladder as one of its two top tiles is about to flip down to the next lower section so as to allow a tile there to flip down to the next section below it so as to then allow a tile there to flip down to the next lower section below it and so on and so on until a final tile flips down to the lowest or 10th section and there are then two tiles in that bottom section (at this point a child could invert the toy while holding onto one the tiles in the bottom section and that would trigger a repeat this sequence of flipping and descending tiles).

This flipping action of the Jacob's ladder's tiles is symbolic of what was happening inside of one of Bessler's one-directional wheels during drum rotation. As the pivot pin center axis of one weighted lever after another *approached* the 9:00 position of a clockwise turning drum, that lever's parallel pair of main arms rotated first counterclockwise about the lever's pivot pin so as to cause the center of gravity of the end weight(s) it held to draw *closer* to the *axle's* center axis, then the arms took a brief pause in their rotation about their lever's pivot pin for about 6° of drum rotation until the lever's pivot pin center axis actually reached the drum's 9:00 position. As the lever's pivot pin center axis passed that drum position, the parallel pair of main arms of the lever would suddenly begin rotating clockwise about the lever's pivot pin as the center of gravity of the end weight(s) that the pair of main arms held moved *farther* from the center axis of the axle.

When the first tile at the top of the Jacob's ladder illustrated in Figure 2(d) flipped down to the second level below the top level, it rotated clockwise. That action then caused a tile on the second level to be released that then flipped down to third level and rotated counterclockwise as it did so. That tile then caused a tile in the third level below the top level to be released and that tile would then rotate clockwise as it flipped down to the fourth level. That tile would then release another tile in the fourth level below the top level which would rotate counterclockwise at it flipped down to the fifth level below the top level and so forth until there were two tiles located in the bottommost level and this odd chain reaction process finally stopped. Thus, the Jacob's ladder's tiles constantly reversed the direction that they swung as one followed them down from the top level to the bottom level of the toy. As one followed the movement of the weighted levers in one of Bessler's one-directional wheels as their pivot pin center axes passed a clockwise rotating drum's 9:00 position, he would notice them each first swinging counterclockwise, then clockwise, and be immediately followed by the next rising weighted lever that duplicated the action of the first lever and swung around its pivot pin first counterclockwise and then clockwise. This would then be followed by the next weighted lever until each of the wheel's eight levers had completed the same reversal of swing directions about their respective pivot pins as they passed the drum's 9:00 position.

This sudden change in direction of a weighted lever's rotation about its pivot pin as the center axis of the pivot pin approached a clockwise rotating drum's 9:00 position is also indicated by the two hammermen of the upper pair. The upper left hammerman's hammer comes closer to and eventually strikes the metal piece on the anvil while the upper right hammerman's hammer is moving farther away from it. This tells us that, at the 9:00 position of a clockwise rotating drum, the center of gravity of an approaching weighted lever's end weight(s) drew as close as possible to the center axis of the drum's axle, symbolized by the metal piece on the little anvil that is centered between the two hammermen, and, after the lever's pivot pin's center axis passed the 9:00 position of the drum, the distance of the center of gravity of the end weight(s) from the axle's center axis increased which is symbolized by the right hammerman raising his hammer. How do we know this happened at the 9:00 position of the rotating drum? At this point, the reader may realize that this clock time is gotten by simply multiplying the alphanumeric values of the two "C's" or 3 x 3 = 9. The alphanumeric value, 3, of the "C's" also tells us that each weighted lever in the Merseburg and Kassel wheels had three cylindrical lead end weights attached between the ends of its parallel pair of main arms.

Just as we can use the two "C's" of the upper hammermen to determine at what clock time position of a clockwise rotating drum that a weighted lever began rotating clockwise about its pivot pin so that its parallel pair of main arms began to draw closer to its radial stop piece, we can also use the two "D's" of the lower pair of hammermen to determine at which drum clock time position a weighted lever's two wooden main arms were finally in contact with their radial stop piece. To do this, we simply multiply the alphanumeric values of the two "D's" or 4 x 4 = 16. If one starts at midnight on a clock dial and advances the dial's hour hand through 16 hours, he will find that the hand eventually arrives at the 4 o'clock marker on the dial at which the time indicated will be 4 pm. That this applies to a weighted lever's main arms finally making contact with their radial stop piece and the main arms then being perfectly radially aligned with their lever's parallel pair of radial frame pieces is symbolized the lower right hammerman whose oversized hammer's head, which is actually an axe head, makes contact with a *wooden* log that is located between the two lower hammermen. Thus, each weighted lever's parallel pair of main arms finally came to rest on its *wooden* radial stop shortly *after* its lever's pivot pin's center axis passed the 3:00 position of a clockwise rotating drum and *before* the pivot pin's axis center arrived at the drum's 4:00 clock time position. This is also exactly what happens in the computer model using the final successful design that I found for Bessler's imbalanced pm wheel mechanics.

If we add up the alphanumeric values of the two "C's" and the two "D's" in the illustration, we get 3 + 3 + 4 + 4 = 14 which just happens to be the exact distance, in inches, between the center axis of the middle cylindrical lead end weight of a weighted lever in either the Merseburg or Kassel wheel and the center axis of the lever's pivot pin which was 14 inches. As was done above, by again multiplying the two alphanumeric values of the "D's", we get 4 x 4 = 16 which was also the total number of weighted levers found in the *two* internal wheels of one of Bessler's bidirectional wheels.

To conclude this discussion of the symbolism the items selected for the Toys Page illustration shown in Figure 2(d), we must consider the little toy located at the very bottom of the group of children's toys which, visually, was most likely placed there so it would be the last item a reader of the alternative version of MT would consider. What mechanical principle used in Bessler's wheels could it have been intended to symbolize?

That little toy was actually an egg shaped top that would have a string inserted through a hole in its outer hollow shell and then into another hole in a central rotor axle contained within the hollow shell (the axle was then attached to two metal discs which are not visible because they were inside the egg's shell and located above and below its external opening). The external portion of the metal shaft of the rotor axle with its pointed end protruding below the egg would then be twisted by hand until a length of string about 18 inches or so was fully wound around the rotor axle section inside of the hollow shell. At that point the child would quickly pull the string out of the egg and its rotor axle would spin at high speed. The egg would then be placed on the ground or a table and its rotor's center axis would remain in a vertical orientation until the rotor inside finally slowed down enough from friction whereupon the whole egg would begin to wobble and eventually fall over.

Bessler must have been fascinated with this toy whose behavior was not fully understood until the 19th century when the gyroscope was invented. I suspect he included it in the Toys Page as a way of saying to the reader that the page's illustration provided clues about building one of his self-moving wheels that could keep the center of gravity of its various weighted levers in a constant state of imbalance relative to the center axis of its drum's axle as the levers continuously oscillated about their own individual pivot pins as well as revolved around the axle's center axis during wheel rotation in a manner roughly analogous to how the little toy top could, as it precessed about on a supporting surface such as a floor or table top, keep the center of gravity of its rapidly spinning internal metal rotor discs and their common metal axle located at a fixed distance from an imaginary vertical line drawn through the rotor axle's point of contact with the surface which supported the top.

Thus, just as the center of gravity of the active weighted levers in one of Bessler's one-directional wheels could not fall below the center axis of a wheel's axle to its *punctum quietus* inside of the axle, neither could the center of gravity of the top's rotating metal axle and its attached metal discs fall until it coincided with the plane of the top's supporting surface (of course, the top would, after its central rotor and its attached discs slowed down enough, eventually fall over, but, even when the rotor axle and its attached discs had completely stopped moving and the top was finally laying motionless on its side of the supporting surface, their center of gravity would still not be able to contact the plane of the surface because of the shape of the top!). Also, as mentioned above, the particular top he drew probably contained two metal rotor discs mounted on the same axle: one below the hole for the string in the rotor axle and the other above it and this is suggested by the two encircling bands painted on the exterior of the egg-shaped hollow shell. The obvious symbolism of this was that his bidirectional wheels contained two separate one-directional wheels which my research definitely concludes was the case.

The alphanumerical analysis just given for the various clues contained in MT's Toys Page during this brief digression will help to begin to familiarize the reader with the type of coding system that Bessler used to record the numerical values of the various parts used in his imbalanced perpetual motion wheels which will be studied in far greater depth later in this volume. Now let us return to the telling of the Bessler story.

Sadly, following his release from a local jail, Bessler, realizing how easily the secrecy of his invention might be violated due to bogus charges made by anyone he dealt with, promptly destroyed the 6 feet diameter, one-directional wheel he had completed after the destruction of his previously Kassel wheel and which he had hoped to sell to Czar Peter the Great as well as the small 3 feet diameter Gera prototype wheel with which he had first found success back in late 1711. One of the last to view that small model wheel running before its destruction was one of Count Karl's sons named Maximillian who had visited Bessler earlier in the year of 1727.

The basic details of the inventor's imbalanced wheel design were, however, permanently embedded in his memory and he could always quickly build one of them again if a serious buyer came forth. Count Karl and Bessler's brother had seen the interior mechanics of Bessler's wheels, but they did not fully realize how very critical to their operation were the shape of their levers, the masses of both the variously shaped lead end weights and the levers to which they were attached, the spring constants of the steel extension springs used, and the locations of the center axes of the various cord hook attachment pins inside of the weighted levers to which the four different types of coordinating cords were attached by their end hooks. More importantly, what knowledge they did possess was not detailed enough to allow them to successfully duplicate a wheel. At this point in time, there was really only one person on Earth that knew the *precise* details of the inventor's imbalanced perpetual motion wheels. That person was Bessler and he wanted to make sure it stayed that way until and unless he received the 100,000 thalers he demanded as full payment for his marvelous invention.

On March 23rd, 1730 Count Karl died at the age of 76 after a lingering illness, possibly cancer, and this, undoubtedly, was a very sad time in Bessler's life. He and the count had become good friends over the years and his aging patron had always stood by the inventor and asserted that his inventions were genuine working perpetual motion devices while carefully maintaining their secret which Bessler had revealed to him. It was the count who had provided him with the employment, income, and housing which had greatly reduced the amount of struggle in his and his family's lives. During the count's final days, it must have again been emotionally stressful for Bessler, the "doctor", to again find himself helpless to preserve the life of someone so close to him. And, again, only the inventor's deep religious convictions helped him through this difficult time.

With the death of Karl, Bessler eventually found himself excluded from the affairs going on at Weissenstein Castle. Even though he had no wheels and no patron to finance the construction of another one, several of Karl's sons occasionally visited him at his home in Karlshafen. He and his family had been living in Karlshafen for almost nine years and most of Bessler's severance pay was gone. However, he did manage to get an occasional job overseeing various construction projects and picked up some money from that. He probably also supplemented that income by again compounding the various medicinal supplies for local physicians. But, the reality was that he was entering his fifties, competing for work with younger men, and the job offers he was getting were becoming less frequent. Despite his declining financial status, however, Bessler kept actively trying to interest potential buyers in many interesting and useful inventions that he had conceived all of which, of course, would require one or more of his perpetual motion wheels for its operation.

Interestingly, after Karl's death, it was discovered that the count had actually spent over 100,000 thalers on his two mistresses! Some have wondered why, if he knew for a fact that Bessler's wheels were actually genuine perpetual motion machines, he did not use that money to purchase the secret for himself.

The answer to this riddle is that Karl had spent that large sum of money over the course of almost 20 years so that the amount he spent on the two women per year only averaged out to about 5,000 thalers. Apparently, as rich and powerful as the count was, he simply did not have a lump sum of 100,000 thalers available to meet the payment Bessler demanded. Also, one must remember that Karl was not really a businessman. He was an aristocrat and primarily interested in promoting commerce and prosperity in his small principality from which he could derive taxes to further benefit his principality and, of course, support his noble rank. As such he would only have seen his role as being someone who might help Bessler find a buyer for the invention and then, after someone else further developed it, the count could encourage the various business owners in Hesse-Kassel to begin purchasing the wheels for use in their small factories.

In 1731, and at the age of 51, Bessler asked the court blacksmith at Weissenstein castle, Johann Adam Krone, for the hand of his then 18 year old daughter, Catharine Elisabeth Krone, in marriage. It's possible that Bessler may have impregnated the girl earlier and had to "do the honorable thing" and marry her. She eventually bore five children, but only one survived to adulthood. There is a letter in existence in which, as a condition for being allowed to marry Catharine, Bessler promised to reveal the secrets of his self-moving wheels to her father. I suspect, however, that promise was never kept.

Years later in 1738, Bessler made an announcement that he had designed three unusual inventions. They were an eternal fountain that would cause water to continuously spurt from an otherwise quiet pool of water, a large self-playing organ, and an "Orffyrean Ship" which was intended to be an emergency preservation device that could save lives and goods in the event of a shipwreck. Most likely this last invention was inspired by a youthful experience of his in which he was aboard a ship that was in danger of sinking after its wooden hull was damaged just below its water line. Apparently, the crew tried to prevent their ship from sinking by throwing some of its cargo overboard in order to lighten the ship so that the damaged portion of its hull would rise above the water line and water would then stop flooding in through it. They even threw much of the passengers' baggage overboard including some of Bessler's possessions! Unfortunately, there are no drawings of these three interesting inventions available for our analysis, but we can probably safely assume that they all would have utilized small, but powerful versions of his one-directional wheels.

Bessler had managed to maintain a fairly decent standard of living for about a decade after the death of Count Karl in 1730, but then his income began to seriously decline. During this decade his second wife had four children that he had to support. Since his imbalanced pm wheel invention still remained unsold, he was forced, despite his advancing years to travel to the towns surrounding Karlshafen and take whatever work he could find that might use his skills in the planning and construction various projects. He always had the hope that, maybe, if he was just lucky enough, he might still find a buyer for his perpetual motion wheels.

In 1743 he found himself in the region of Brunswick, which is about 75 miles northeast of Karlshafen, and overseeing the construction of several factories for the 30 year old Duke Karl I of Brunswick-Wolfenbüttel. During that job, the duke decided he wanted Bessler to install a windmill in the town of Fürstenberg, which was conveniently only about 7 miles away from the inventor's home in Karlshafen, in order to produce vegetable oils and wheat. Bessler decided to try building the duke a new type of windmill that had not been used previously in Europe although the design was commonplace in the Middle East. Rather than having its wind catching sails in a vertical plane and attached to a horizontal axle, Bessler intended to build one with its sails in a horizontal plane that were attached to a vertical axle. The advantage of this design was that the windmill could operate regardless of which direction the incoming wind struck its sails.

Originally, Bessler wanted to put his horizontal windmill on the cliff of a nearby mountain so that an updraft from a river below the cliff would power its sails. But the duke thought that made it too inaccessible. He decided to have it constructed at another site inside the town of Fürstenberg itself despite Bessler's protests that the winds at that location were not really that strong. But, since the duke was paying for the construction, Bessler was obliged to change the location.

The first two stories of the windmill had thick walls of sandstone while the upper two and half stories were made partially of stones and partially of wooden timbers. Since the blades were in a horizontal plane and their axle ran down into the interior of the mill, some means had to be provided to keep rain from entering the structure. Bessler simply constructed a large four-sided roof to cover the sails which were just large canvas sheets that had been stretched and nailed onto four wooden frames that were then attached to the central vertical axle. However, despite its design, the location the duke selected for the windmill prevented it from operating at its maximum possible power.

While Count Karl had been quite generous in funding Bessler's 12 feet diameter, bidirectional Kassel wheel at Weissenstein Castle, Duke Karl I was somewhat of a tightwad and had a habit of doling out the funds for his projects as sparingly as possible and that included the salaries of those he hired to work on them. For most of the year 1745 Bessler, then 65 years of age, had kept writing to the duke's court counselor, who was in charge of providing Bessler with the funds for the windmill project, and, literally, begging this counselor for his salary so that he and his family could eat.

His letters, of which one written at the end of April, 1745 still survives, were really pathetic. At his peak Bessler was living in a castle and he and his family were dining with nobility. As 1745 drew to an end, he was on the verge of starvation, suffering from various illnesses, severely depressed, drowning in debt, and responsible for a new family he had to support. Indeed, his status had become so lowly that people were refusing to lend him any more money because they feared he would not be able to pay it back!

However, despite all of his suffering, we learn from the surviving letter he wrote that by midnight of April 14th, 1745 he had finally completed the construction of another one-direction wheel at his home in Karlshafen that was probably identical to the 6 feet diameter one he had constructed for Peter the Great, but which was destroyed in late 1727 after his arrest due to his former maid's bogus fraud charge against him. This new wheel we learn, the seventh and *last* working imbalanced pm wheel he would ever construct, had been commissioned from London via letter after the involvement of a senior district magistrate of the duke named Baron Anton von Mannsberg who just so happened to be a Fellow of the Royal Society and also Bessler's landlord. Apparently, although Bessler had become somewhat obscure as his fortunes declined, there was still some interest in purchasing his imbalanced pm wheel invention and finally learning the secret of its operation.

Then, on November 30th of the year 1745, at the age of 65, Bessler had a tragic accident. He had been working on the windmill in Fürstenberg when he fell off of one of its upper stories. He did not die immediately upon impact with the cold ground, but sustained some very serious internal injuries. Whether he was conscious or not after his fall is not known. He lingered for a day or so and then died in that small town. His body was returned to Karlshafen and placed in a family crypt on the grounds of the property which Karl had given him permission to construct while the count was still alive. This crypt was located in a garden and would have been an underground stone construction with many niches in its walls that could hold a large number of caskets. After placement of a casket in its assigned niche, the opening might have been covered over with a stone or metal plaque that would have had the deceased's name, a short description and possibly favorite motto or saying, and also his or her life dates somehow inscribed on it. Bessler's mother and father were already placed there along with his first wife and his deceased children by her. Bessler's second wife Catharina and his brother Gottfried as well as his widow would also be placed there when they finally passed on.

Some have speculated that Bessler must have committed suicide from the depression caused by his dismal situation. I, however, feel that is unlikely. Bessler was a very religious man and he knew that suicide was considered a great sin and punishable by an eternity in Hell. Most likely, he was underfed and overworked as he tried to complete the windmill for the duke. Possibly, he had only a meager breakfast on the morning of the accident and, as a result, was hypoglycemic. That state of low blood sugar combined with excessive exertion could then have further lowered his blood sugar to the point where he fainted and fell off one of the top floors of the four and a half story high windmill's base.

If he had fallen head first and then broken his neck on impact, he might have died immediately. But, he probably tumbled as he fell and the force of the impact was spread out as he landed more or less lengthwise on the ground. Such a landing spares the limbs from being broken, but can cause the tearing of internal organs that then results in internal hemorrhaging. With the limited medical treatments available for such injuries in the early 18th century, such an accident was usually fatal within a day or so.

After his burial, the locked door to his workshop was finally forced open in the hope of finding anything of value that would then become part of his estate and which could benefit his heirs. However, what they found inside was of little monetary value. Most of it was donated to the state by his second wife, then a widow. These included his many letters and books, the woodcut printing blocks for the illustrations to his unpublished treatise, MT, on perpetual motion devices, various scientific items he had collected over the years, and, most importantly, the remains of the small 6 feet diameter wheel he had completed at the request of Baron von Mannsberg for the Royal Society over in London.

I don't believe that these remains of Bessler's last constructed imbalanced pm wheel were just the still intact components that he had merely disassembled and hid for the sake of protecting the wheel's secret in the event that someone broke into the workshop when he was absent from his home. If that had been the case, then a skilled craftsman, preferably also a pursuer of pm, hired by the duke or von Mannsberg should have been able to rebuild the wheel from these parts. I think, however, the destruction was far more thorough and included the axle, drum frame, individual levers, and the sets of coordinating cords of various lengths including all spare components. Bessler would only have resorted to such total destruction as a consequence of the rage he felt when he finally realized that he had been the victim of a cruel deception by individuals in whom he had previously placed much trust.

Most likely, at the last minute, when it came time to arrange the payment to Bessler of the 100,000 thalers for the wheel that he had struggled to complete, the money was not produced. The inventor was probably again given the same condition for payment that he had received since he was in his early 30's: *first* give us the wheel to inspect and, *if* we believe it's genuine, *then* we'll hand over the money. To Bessler any such deal was totally unacceptable because, once a buyer knew the secret, in detail, he would then have nothing obliging him to pay Bessler for it except his word as a gentleman. For Bessler, there was only one such gentleman in the world, Count Karl, whose word he could actually trust (except in one instance) and he was now dead and buried. Just as soon as he received this response again, Bessler promptly broke up the commissioned but unpaid for wheel and left its fragments in a pile on the floor of his workshop in his home in Karlshafen and that is where his widow found it after she managed to gain entry after his death. Possibly, upon entering his locked workshop, his much younger widow discovered a single mostly intact 3 arm side piece of one of the weighted levers of his last constructed 6 feet diameter, one-directional wheel and personally placed it in his niche along side of his casket as something for future researchers to find. (This crypt most likely still exists near where the original home stood, but it has since been covered over with a parking lot!)

Once Bessler died, his widow who may have still been caring for young children was left without any income, but, fortunately for her the state granted a small weekly stipend of money and firewood to her which was enough to allow them to all survive.

On several occasions during his life, Bessler had been on the verge of finding a buyer for his marvelous self-moving wheel invention, but something always seemed to prevent the final sale. His situation was made all the more complicated by the fact that there were no patent laws in his region of Europe at the time and any buyer for the wheel, after having paid an enormous sum for it, would, like Bessler, then have to worry that the design might be stolen, copied, and sold by those who had paid nothing for it.

Bessler himself was also a major obstacle to the sale of his invention. Although becoming a very religious man after overcoming the excesses of his youth, he had a very cynical personality and appears to have been subject to occasional bouts of extreme paranoia. He also stubbornly insisted right to the end of his life that, unless he was paid the *entire* huge amount he was asking for his invention in a single, up front, lump sum payment, he would not sell it so that its secret could be revealed. Such traits are, unfortunately, not exactly conducive to the marketing of a new invention, especially one for which so much money was demanded and which produced so little useable power in its operation relative to its gross weight. If he had been a more flexible person, he might have asked for a far lower initial sum of money and taken a percentage of the future profits of the wheels as they were manufactured and sold. In this way, over his lifetime, he might have actually received far more than 100,000 thalers and would not have had to suffer the indignities he did in his last few years of life.

There's also something rather symbolic in the way Bessler died. With his amazing self-moving wheels, he had managed to harness gravity and make it do something it normally would not do. Then, it was almost as though, out of resentment for being so treated, gravity sought its revenge. Once, in July of 1715 gravity slammed Bessler's head down against the ground and almost killed him. But, he managed to survive for another 30 years until in November of 1745 gravity finally won by slamming his entire aging body down against the ground. That time he did not survive, but the story of his wheels does survive and, with their soon coming successful duplication, it will continue to do so for many more centuries to come.

Figure 1(a) - The Bidirectional Merseburg Wheel Constructed in Early 1715

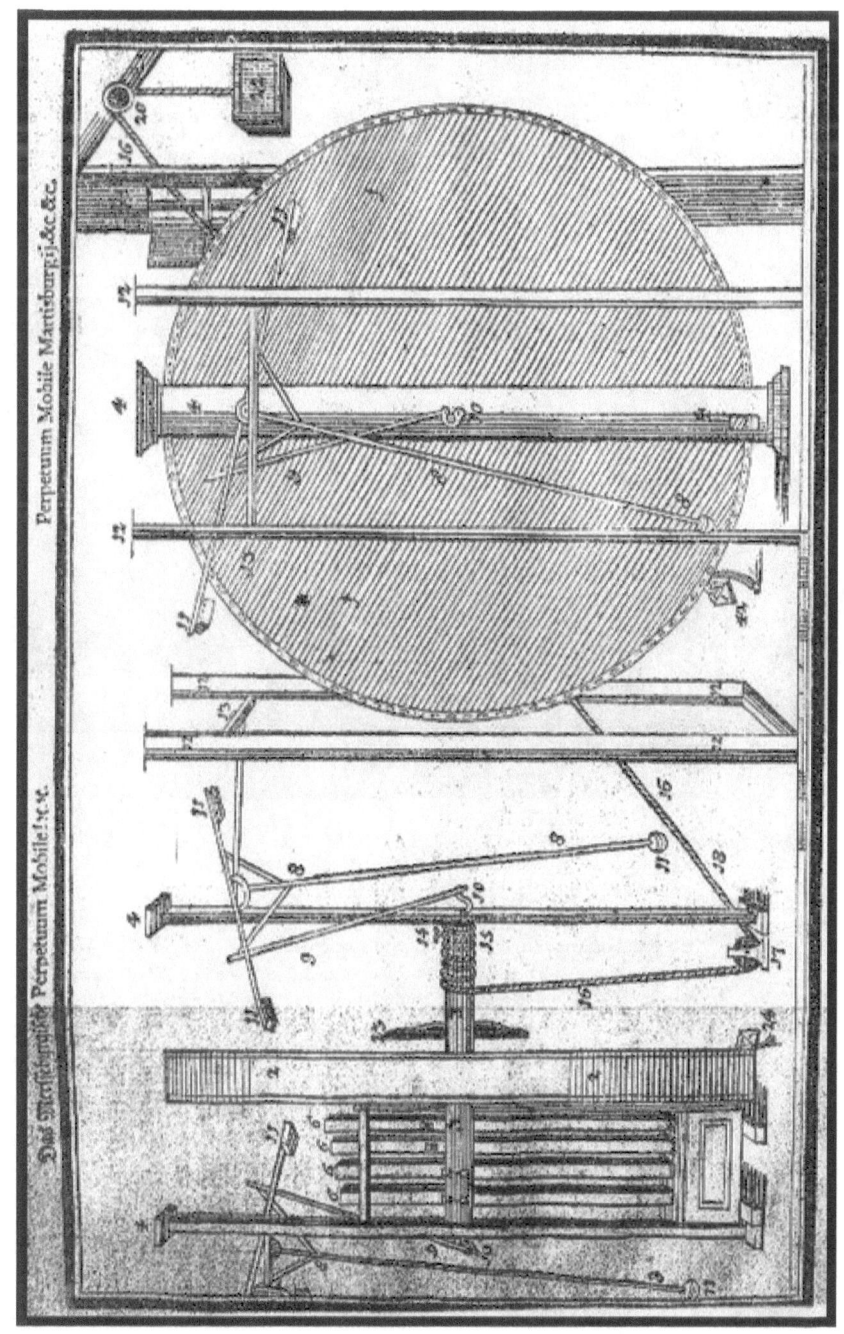

Figure 1(b) - Johann Borlach Shows How He Thought Bessler's Merseburg Wheel Really Worked

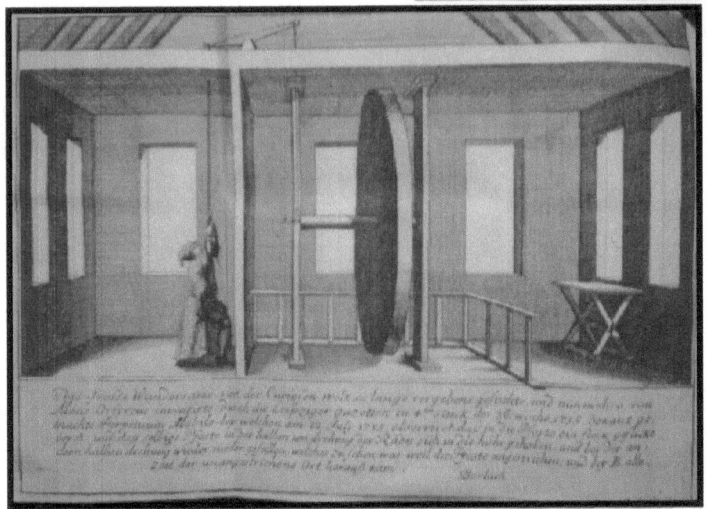

This figure published by a mining engineer named Johann Gottfried Borlach suggests that the back vertical support plank of the Merseburg wheel had a concealed channel in it through which a metal connecting rod joined a hidden crank formed in the axle's rear pivot pin. The connecting rod was then attached to a rocking beam on the floor of the attic above the wheel. The other end of the beam had a rope attached to it by which an accomplice in a small room hidden behind the wheel's room could "pump" the axle's hidden crank to make the wheel rotate. This figure gives a rare view of the room that contained the Merseburg wheel. The ceiling is 14 feet high and a railing is shown partially surrounding the wheel. The vertical support planks are not held between shims, but, rather, "boxed in" by wooden pieces nailed to the ceiling and floor that prevent lateral movement of a plank. Borlach did not accurately depict the location of the drum on the axle and neglects to show the external curved cranks attached to the ends of the axle's two pivot pins.

Figure 1(c) - Christian Wagner's Fake Perpetual Motion Wheel Mechanics

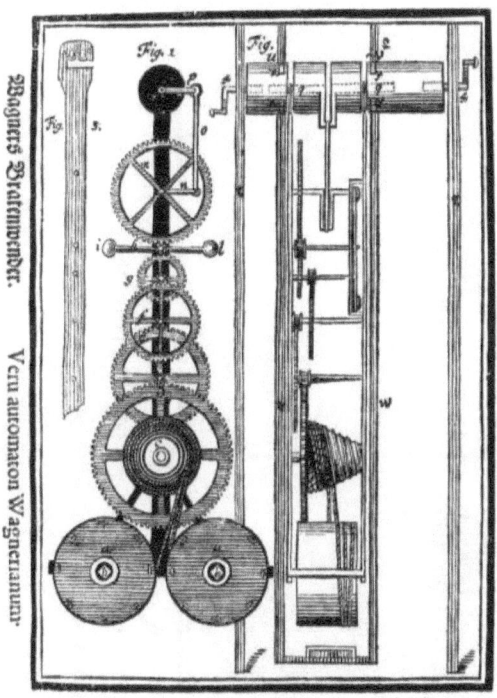

This illustration was published in one of Wagner's "Criticisms" of Bessler's genuine perpetual motion wheels and shows the mechanics that Wagner used in his fake wheel and which he believed was being used in Bessler's wheels. On the left side is an axial view of the clockwork movement which is powered by two large mainsprings that are concealed in drums. Through a train of gears, the torque of the mainsprings drive a connecting rod attached to a crank pin that joins together the two sections of a split axle. On the right a profile view shows how the entire movement hung on other pins in the split axle so it could remain relatively stationary as the drum rotated around it.

Figure 2(a) - The Bidirectional Kassel Wheel Constructed in 1717

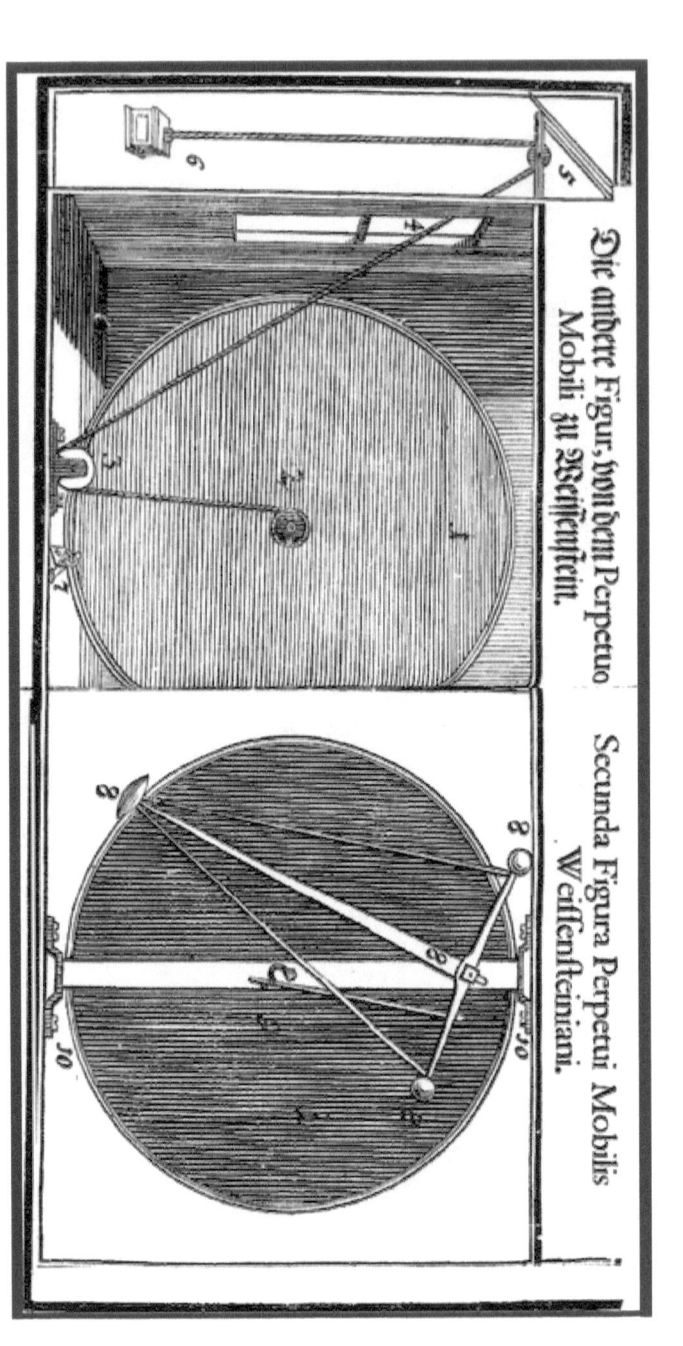

These two illustrations from DT show the Kassel wheel lifting a box of bricks from the Weissenstein Castle courtyard and swinging the front one of the two pendulums attached to its axle.

Figure 2(b) - The Kassel Wheel Powering Various Devices

These two illustrations from DT show the Kassel wheel in a room at Weissenstein Castle powering an Archimedean water screw and a stamping machine. In the left side background there are two motion dampening flanged planks that could be attached to and run by cranks at the ends of the wheel's axle pivots.

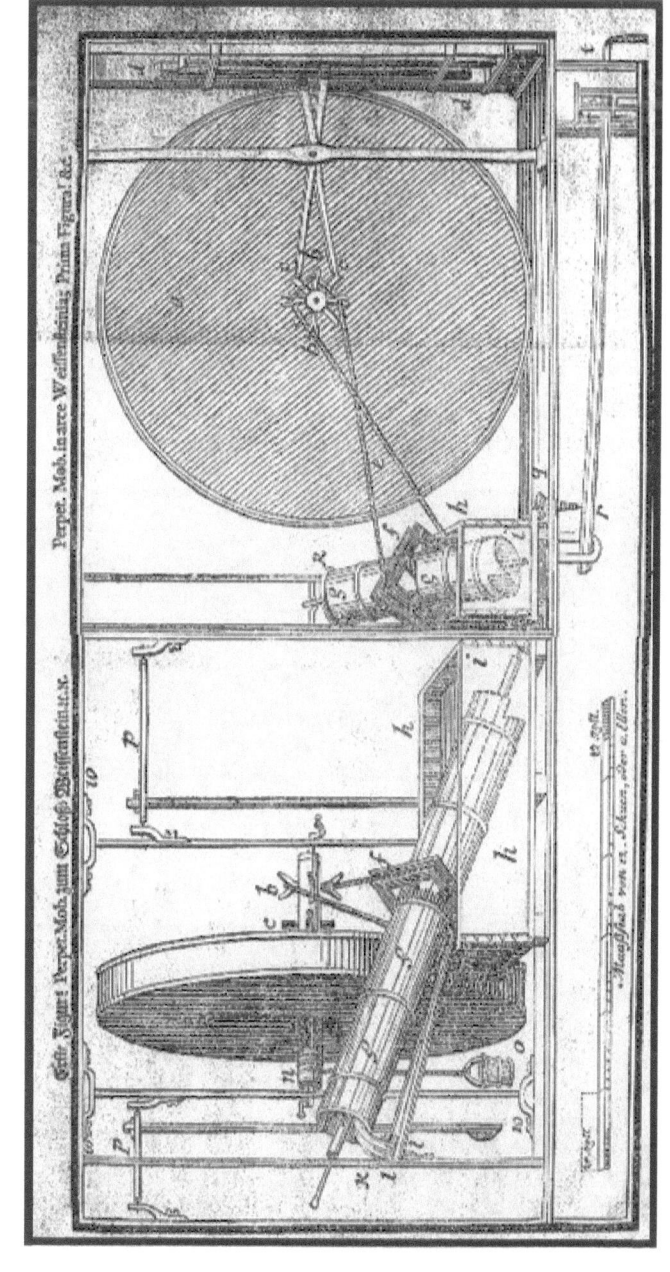

Figure 2(c) - Important Actors in the Bessler Story

Johann Ernst Elias Bessler   1680 to 1745

Count Karl von Hesse-Kassel
1654 to 1730

Gottfried Wilhelm Leibniz
1646 to 1716

View from the southeast of part of the front of Weissenstein Castle (right side) and Count Karl's Casade in the distance behind the castle (center).

Willem Jacob s'Gravesande
1688 to 1742

View from the southwest of the back of Weissenstein Castle from a painting by Johann Heinrich Tischbein done around 1778. Note the surrounding wall, enclosed courtyard, and gardens to the left and right sides of the castle.

Peter the Great
1672 to 1725

Figure 2(d) - The "Toys Page" from MT

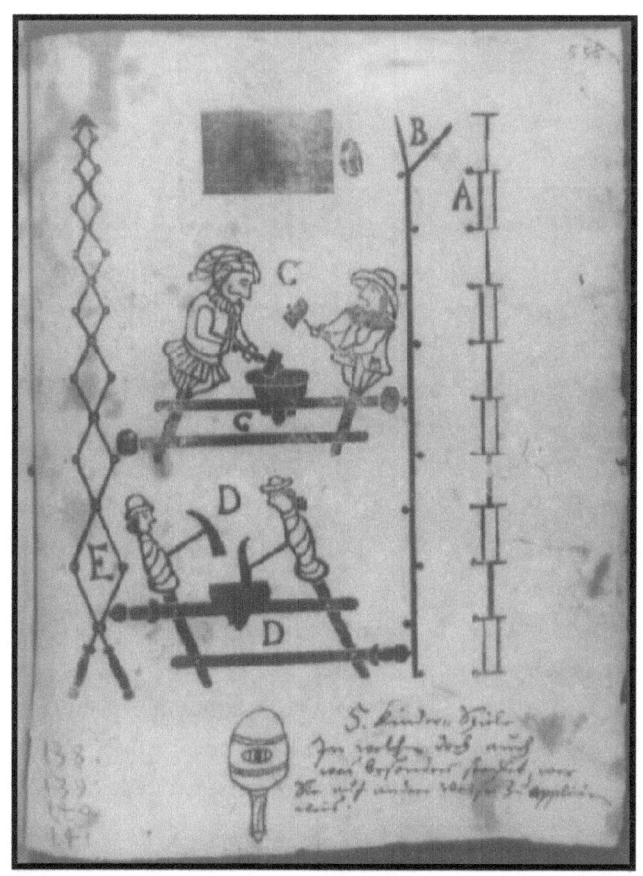

CHAPTER 2

I Meet Herr Doktor Orffyreus

In order to properly convey the effect that my "meeting" with Johann Bessler had on my mind and life, it's probably a good idea to give some of the details of my background first. Generally, I'm somewhat of a private person, but, perhaps, if the reader understands my history up to the point of my becoming aware of the Bessler story, then my reaction to it (as well as that of others) will become more understandable.

I was born on May 10th, 1951 in the city of Elizabeth, New Jersey to an American first generation Polish mother and German father and was their only child. They could both speak fluent English as well as the two native languages mostly spoken in the American homes that they grew up in during the early decades of the 20th century. Growing up, I could only speak English, however, because my parents rarely spoke their parents' native languages in our home.

I remember, as a small child, going on shopping trips to the city center with my mother who would let me walk at her side as she held my hand and then carry me when my little legs grew tired. I looked forward to such trips because she would always buy me some little toy that caught my eye and that I said I wanted. When I got home with these toys, I'd play with them for a while, grow bored, and then start to wonder what was inside of them that made them work. I remember little toy cars and boats made of stamped metal parts held together with tiny metal tabs. As a small child, I would sit for hours slowly bending the tabs up until I could take a toy apart and see what was inside. Usually, I found that the axle of a toy's rear pair of rubber wheels was geared to a small metal flywheel inside that could be revered up to a high rotation rate by rapidly and repeatedly dragging the little car's wheels over the surface of the floor and then, when the car was again placed back on the floor, the energy stored in its hidden flywheel would be used to power its rear wheels and send the little car racing across the floor. Occasionally, I would discovered some little spring wound movement that, after being wound with an externally projecting key, would make a toy car's rear wheels spin so it could race across the floor. Rarely, however, I'd find a more elaborate system of levers and rods that animated a toy and I would carefully study how they worked.

Needless to say, my mother was not too happy when she discovered my disassemblies and I was constantly trying to figure out ways to hide what I was doing from her. She, of course, would always find out and then, one day, she said, "Okay, Kenneth, now put it back together again." I dutifully tried and, at first, could never get a toy back together and working properly again. But, then the day came when I was successful and the toy worked as it should. I was exuberant and it made my mother happy to see me succeed.

I wasn't long after that that my father began to provide me with old alarm clocks that had stopped running and I began to disassemble them to see, literally what made them tick. I was fascinated by their intricate arrays of intermeshing gears and pinions. I also quickly learned how dangerous it could be to try to remove the gears from a fully wound up clock movement. There was a tremendous amount of energy stored in a big steel spring, the mainspring, which was contained in a single large drum shaped "barrel wheel". Its sudden expansion, upon being pulled out of this wheel, could easily blind a person if part of the spring struck his eye. I found, however, that by removing some of the other gears in a clock's movement, I could cause the remaining gears to run at high speed so that the tension in the mainspring that was driving them was decreased as it unwound. As this happened, the hands of the clock would race around the dial and it looked like an entire day of time was passing in only a few seconds! Only after this was done could the barrel wheel and its internal spring be safely removed. Before long, I was actually managing to occasionally successfully repair a clock and I was only five or six years of age!

By the time I was seven years of age, we had moved to a working class neighborhood in that city which was only a few blocks away from the old Singer Manufacturing Company plant. They made sewing machines there and, during the war years of the 1940's, machine gun parts for our military.

It was then 1958 and I attended the local Benjamin Franklin Public School #13 which was a short walk from our second floor flat in a rear building on our landlord's property.

There were many young children about my age on the street I lived on, but I quickly noticed that not a single one of them went to the same public grammar school that I was attending. They all were attending a Catholic grammar school down the street from my school. My parents told me that it was too expensive to send me there and my father kept telling me that he did not want anybody "forcing" religion on me. Rather, he wanted me to grow up and make up my own mind about religion. He was an atheist and, I suspect, was mad at God for the early death of his own mother when he was a teenager growing up in Brooklyn, New York. Such traumas can make one lose his religious belief.

It was great to be a kid growing up in the late 1950's and early 1960's, especially in a working class neighborhood where there were many other children of about the same age available to play with.

I remember long summer vacations from school when we'd spend most of the day out of doors starting at about 10 am in the morning. We'd play all sorts of games like kick ball, hide and seek, hopscotch, jump rope, jacks, marbles, 1, 2, 3 red light, Simon says, and then there were the board games in which we'd, literally, spend half a day playing one game after another of Monopoly, Life, checkers, or chess. We all had collections of plastic guns and sometimes would play "War" which was just hide and seek with water pistols and cap guns used to symbolically kill each other as we took turns being the "good" or the "bad" guys.

We'd all take a break to "refuel", also known as having lunch, between noon and 1 pm and then be back out on the street or on someone's porch or in their backyard having fun until dinner time which varied depending upon to what family a kid belonged. There were close friendships, rivalries, and occasional fist fights. For a dime one could get a Coca Cola and a candy bar and the sugar calories in those powered us through all sorts of physical activities. We'd race our bicycles, hop over fences, jump from garage roof to garage roof, and run as fast as our legs could carry us if, during a baseball game, one of our home run hits shattered someone's window.

I remember one incident in particular when I was about nine years old

The Triumphant Orffyrean Perpetual Motion Finally Explained!

and we were playing softball with a not so soft ball. Right behind second base was a delicatessen with a large pane glass display window on each side of its front door. I was covering second base and my friend whacked the ball very hard and it flew over my head and right through one of the store's windows. Within seconds, about a dozen kids sudden disappeared from the scene and I found myself still standing by the base I was covering. Suddenly, the front door of the store flew open and the owner, his face beet red with rage, emerged carrying a giant stick. He looked like he could easily kill someone!

Fortunately, he did not see me and I managed to duck down behind a car that was parked in front of his store and just behind second base. I slowly slid myself under the car and lay there as quietly and motionless as I could. Meanwhile, I could see his shoes and feet walking back and forth in front of his store and hear him cursing loudly about "those damn kids" and how he was going to "call the police". It was really scary and I was convinced if he found me I would be killed on the spot. But, after a few minutes his wife emerged and he started to simmer down. A while later they went back into the store and I slipped out from under the car and slowly moved away from the scene by hiding under other cars parked on the street. His store window was replaced a few days later and his insurance paid for it. We kids tended to treat the zone in front of his store like it was off limits for a long time after that. But, the next year we were back again and this time using a new invention: the "wiffle ball". It was a very light, hollow, polyethylene plastic ball with vents cut into it. When it was struck with a bat, it only flew a few tens of feet and did not have enough momentum to shatter a window. Once again, all was peaceful on our street.

As we entered our teenage years in the early 1960's we started to become builders.

I remember making rubber band guns from a foot long piece of wood, a rubber band, and a clothespin. It could fire out a metal bottle cap with considerable accuracy and we'd compete to see who could hit a tin can with his gun's cap and knock it off of a brick. Then we made racing cars from old wooden boxes, boards, and roller skate parts. Those with more generous allowances could afford to purchase kits to make balsa wood model airplanes and we'd spend minutes winding up their rubber band powered propellers and seeing how long they would fly before making a crash landing and, most likely, being damaged.

In the early 1960's there was much talk on the television about a "space race" and how American children needed to get ready for it. During the summer of 1961 I noticed that there was a guy walking around our neighborhood with a Geiger counter making measurements. I asked him what he was doing and he mumbled something about checking for radioactivity. I knew about that from the many sci-fi movies I watched on television. It was something "bad" that usually indicated a monster was about to appear and attack people! Then, in October of 1962 my mother seemed a bit worried.

I asked her why and she said that she read in the newspaper that there might be an atomic war and that a lot of people could get killed! In one of the sci-fi movies I'd seen on television they showed how houses could be disintegrated by an atomic bomb blast and I imagined that happening to the building we lived in. I did not realize it at the time, but the "Cuban Missile Crisis" was in progress and, in reality, it could have led to the end of human civilization if our President Kennedy had not kept his cool despite what his military "experts" were telling him he should do which was to begin an armed invasion of the nearby island of Cuba. What the President's advisors did not know at the time was that there were about a hundred mobile tactical nuclear missiles in that country and that the soviet "advisors" there were authorized to launch them at their discretion *without* needing any prior permission from the Kremlin over in Moscow in the event of an invasion of the island by the US. If we had invaded Cuba as Kennedy's generals suggested and those soviet advisors decided to launch some of their nuclear missiles at our invading fleet's ships, then that could easily have escalated into WWIII and this volume would never have been written!

I had graduated from my grammar school's 6th grade in 1963 and began attending the Grover Cleveland Junior High School in Elizabeth, New Jersey that taught 7th, 8th, and 9th grades. It was a lot farther from my home than the grammar school was and I had to begin taking a bus twice a day to attend. I did not like having to travel that far but had no choice. There was a Catholic high school within convenient walking distance of our home, but, again, my parents decided not to pay any tuition for me to attend. Within months of my beginning at the public junior high school, Kennedy was assassinated and the school closed for two weeks while the nation mourned the loss of our President. It was a sad time and I was beginning to realize just how dangerous our world and society could be.

As I attended junior high school, I began to take an interest in space travel and exploration. I had sent a letter to NASA asking for some pictures of their rocket ships and they sent me back a large manila envelope filled with dozens of photographs of various types of missiles and rockets. I started to build plastic model rockets and even a wheel shaped space station that I learned later was based on a design that the head of NASA, a former German rocket builder named Werner von Braun, had designed. I regularly fantasized about what life would be like when I was eventually became an astronaut and was living on the Moon or maybe even Mars. Television and movies were presenting sci-fi movies that depicted a dazzling future filled with adventure and it would all be made possible through science and technology.

Like most of the kids on my block I had a lot of fun with those red transparent plastic "water rockets" that we occasionally received as a gift. These were just hollow plastic rockets with fins in the rear that surrounded a small opening in the tail end of the rocket's egg shaped fuselage. One inverted the rocket and then inserted a small funnel into the tail opening through which he poured enough water to fill half the rocket's internal volume up. The funnel was then remove and a hand pump's outlet nozzle was inserted into the tail opening of the rocket as a slide was pushed into place that locked the pump to the tail of the rocket. Then one had to start pushing on the handle of the pump. This forced air into the space above the water inside of the rocket and with each pump the air pressure inside of the rocket increased. At some point it was no longer possible to pump any more air into the rocket by hand because the back pressure on the pump's handle was too great.

Finally, the rocket was inverted so that its rounded nose pointed straight up and the water then flowed to the bottom of its fuselage while the compressed air collected at the top of the fuselage. This, of course, was done outside in the middle of an empty street (but, preferably in a large field one of which was only a block away from my home). A kid would then do a slow 10 to 0 countdown at which time he pulled back on the slide that secured the bottom of the rocket's fuselage to the pump. As soon as the rocket was thus unlocked from the pump, the pressure of the highly compressed air inside would force the water out of the opening at the bottom of the fuselage at high velocity and the little plastic rocket would immediately streak skyward until it was almost out of sight! At the apex of its flight, the pressurized water and the thrust it provided would be gone and the rocket would simply drop back down to Earth. Since it only weighed a few ounces when empty, its terminal rate of descent was not fast enough to damage it when it landed even on an asphalt paved street.

It was in junior high school, however, that I began building model rockets that actually used solid fuel motors. I found a company out west that sold kits for making single and even multiple stage rockets out of cardboard and balsa wood. One then inserted a solid fuel rocket motor cartridge into the tail ends of their tubular cardboard fuselages. The end of a motor was prepared by wrapping it with several layers of paper tape and that then created enough friction to hold the motor in place inside of the rocket's fuselage when it was pushed into the tail end of the fuselage.

To launch the rocket vertically, one had to glue a hollow straw-like tube along the length of its fuselage and then slide that tube along with its attached rocket over a vertically oriented steel rod whose bottom end was pressed into a hole that had been drilled into a heavy wooden block. The bottom points of the rocket's balsa wood fins touched the wooden base and there was a space of a few inches between the base and the rocket's motor that one could use to insert the nickel chromium igniter wire into the motor's nozzle. To protect the wooden base from the motor's hot exhaust gases, there was a sheet of woven asbestos that was first slid over the metal rod before the rocket with its motor cartridge was slid down the length of the rod. At the time we did not fully recognize how dangerous handling asbestos could be and a simple piece of steel should have been used.

Launching these model rockets was fairly simple. One simply connected the ends of a long piece of lamp cord with alligator clips attached to them to the ends of the nickel chromium wire which had been pushed up inside of the rocket motor's ceramic exhaust nozzle. That then formed an electrical circuit with a handheld switch and a pack of batteries. After a dramatic, sci-fi movie countdown, the launch button was pressed and current began flowing through the nickel chromium wire loop. That loop then became so hot that it actually glowed orange and the heat from it would ignite the rocket motor's solid propellant chemicals. As the motor's solid fuel ignited, the wire loop would be expelled from the motor cartridge and the thrust of the fuel's rapidly produced exhaust gas would send the little model rocket soaring high into the air. Achieving an altitude in excess of 1,000 feet was possible with a single stage rocket using one of the company's more powerful rocket motor cartridges.

When these rockets reached their maximum altitude after, perhaps, less than ten seconds of thrust, the solid fuel would burn through the cartridge to its other end and then ignite a chemical there that would produce a lot of dark colored smoke. The pressure from that smoke would actually eject a small plastic parachute that had been carefully folded up and inserted into the top end of the rocket's fuselage before a loosely fitted nose cone was pressed into place there. As the parachute was pushed forward by the expanding cloud of smoke, the nose cone would pop off and, from the ground, one would see a sudden puff of dark smoke in the sky indicating that the recovery system had been activated. Then one would see his carefully build and painted rocket's parachute open and knew the rocket was again returning to Earth. The several strings of the parachute were joined together and then attached to a small steel eyelet that had been screwed into the back of the balsa wood nose cone. The nose cone was also attached to the inside surface of the leading end of the rocket's fuselage by a long thin ribbon of rubber that acted like a shock absorber when the parachute suddenly opened and yanked on the ejected nose cone.

A successful launch and recovery of one's solid fuel model rockets required having a large open field to use as a launch site. This posed a problem in a crowded city like Elizabeth, New Jersey. My father and I tried launching them from a public park, but were chased away by the park police because they considered my rockets to just be "fireworks" which were not permitted in the park without a special permit which I did not have. Fortunately, we managed to find a large undeveloped field to use and visited it regularly every time I had finished building a new rocket and wanted to take it for a test flight.

I remember one particular model rocket I made that had a somewhat tragic ending. It was about 20 inches in length and I had painted it a bright yellow color and decorated it with black stripes that encircled its body. It was absolutely beautiful of behold and the pride of my growing collection of model rockets. I must have spent a month working on it in my spare time.

On the morning of its first flight we arrived at the field and I noticed that it was a particularly windy day. Such winds are undesirable because, once a rocket's parachute deployed, they could carry it for some distance horizontally away from the place where it was launched. To compensate for this, it was necessary to launch the rocket at an angle *into* the wind so that, after its parachute opened, the wind would carry in back in the direction of the launcher. Well, since I had spent so much time building my beautiful model rocket, I made my best estimate as to the angle to adjust its launcher's guide rod (by this time I had gone beyond using the heavy wooden block type launch pad and was using a fancy launcher with tripod legs whose rod could be tilted at an angle away from the vertical). I was highly confident that my rocket would return and land almost at my feet.

Then I launched the rocket at about a 45º angle from the vertical and watched as it flew almost out of sight. But, then something went wrong. There was no puff of dark smoke and I did not see a deployed parachute. Apparently, the charge that was ignited when the solid fuel burned its way through the motor cartridge was not sufficient to push out the large plastic parachute I had stuffed into the front end of the rocket's tube and my favorite model rocket just kept going and going and going. I never saw it again. I've often wondered what would have gone through the mind of anyone who might have found it after it landed. Most likely, if it had landed on soft ground, he would have found it embedded in the soil, nose cone down with the back portion of the fuselage intact.

But failures like these did not discourage my interest in space and I eagerly viewed any movies that depicted space travel and read all of the popular science fiction stories whose plots involved the subject. I even joined the "Science Fiction Book Club" and soon had a growing library of the works of authors like Asimov, Bradbury, Clarke, Heinlein, and even Wells and Verne.

One film, in particular, really caught my attention. It was titled *Forbidden Planet* and starred a gracefully aging Walter Pidgeon, a youthful Leslie Nielsen, and a beautiful young actress named Anne Francis. This film was made in 1956 and was one of the most expensive sci-fi films produced during its decade because of its costly props such as "Robbie the Robot" and a full scale, 170 foot diameter flying saucer. The film is historically important because it was the first ever to depict human beings exploring a star system other than our own in a saucer shaped spacecraft of their own construction. Their craft certainly did not look like the sleek cylindrical rockets with fins that I thought would be required for space exploration, models of which I had been building and launching. Rather, it was as odd disc shaped affair that somehow used magnetism for propulsion and could travel faster than light!

As my interest in space exploration increased, I began, while still in junior high school, to become aware of the topic of "flying saucers" or, more scientifically speaking, "Unidentified Flying Objects" or "UFOs".

Our school library was fairly large and I discovered that it contained several books that dealt with the subject of UFOs. I wasn't long before, in addition to my regular school work, I had read the library's entire collection on the subject. From the dozens of cases I read about, two things eventually became quite clear to me. The first was that UFOs were as real as automobiles and, second, that their existence meant that there were already beings, usually "humanoid" in shape (meaning that they had a head, a torso, two arms, and two legs), but not exactly human, that were *already* engaging in interstellar and, perhaps, even intergalactic space exploration! Just as soon as I realized that, my interests in Earth's space programs dropped almost to zero. To me, it seemed like our scientists were only trying to play catch up with a far more advanced space propulsion technology and that, rather than continuing to build our clumsy chemical rockets like I was doing, we should be trying to reverse engineer the propulsion that was being demonstrated for us by our extraterrestrial visitors.

My interest in this subject did not begin and end in junior high school. Rather, that was just the beginning of it and, over the next several decades, I would spend thousands of hours researching cases and even, occasionally, investigating them. I won't devote any more space here to describing what my interest in UFOs led to except to say that it eventually resulted in my editing and publishing my own research journal devoted to the subjects of UFO propulsion and secondary effects and that was then the motivation for me to go on to author three volumes on the subject which,

I am confident, do resolve the mystery of how they operate and are able to travel far in excess of light velocity so as to make interstellar and even intergalactic travel practical.

In 1966 I graduated from the 9th grade at Grover Cleveland Junior High School and began the 10th grade at another even farther school from my home in Elizabeth, New Jersey named Thomas Jefferson High School. It was smaller in size and a bit more prestigious than my former junior high school and I immediately noticed several odd things upon beginning at the new school.

First, I noticed that most of my former male classmates were not at the high school. I learned that most of them hadn't done that well academically and, as a result, were sent to another school, the "Thomas A. Edison Vocational High School", on the other side of Elizabeth. Supposedly, they were going to learn less academic subjects there and focus on acquiring skills that involved working with their hands so they could eventually become carpenters, electricians, plumbers, etc. These were all things I was interested in, too, but, apparently, my high school was for those who were more academically inclined and would eventually go on to college and become a "professional" of some sort.

The second thing I noticed was that there were no girls at my new high school as there had been at my former junior high school. I was beginning to go through puberty and, just as my interest in the opposite gender was beginning to increase, they were all suddenly gone! I learned that they were all sent over to a girls only high school called "Battin High School" on the other side of town. When I inquired as to why there had been a separation of the genders someone told me that, in the past, girls had attended Thomas Jefferson High School, but there had been several pregnancies as a result and that the separation was necessary to prevent any more of them. It was no wonder then why so few of us were able to attend our high school's prom. Most of us didn't know any girls to ask to it!

The academic training at Thomas Jefferson High School was far more rigorous than it had been at my junior high school and I found the rigor quite to my liking. I was, literally, like a sponge that eagerly absorbed all of the subjects offered. I loved chemistry, math, and physics and even learned to read, write, and speak Spanish fluently. I also repeated my previous extracurricular reading by going through all of the available books on science and technology available at our school's library. They even had some books on UFOs! But, I learned that most of the better students were going over to Elizabeth's Main Public Library after we all got out at 3 pm in the afternoon and were then doing their homework over there. I'd never been inside of this library and decided to check it out. It was only about three blocks away from my high school.

On my first visit to the city's main library, I obtained a library card and began to explore their vast supply of books. They had the usual several sets of encyclopedias in a reference section, but it was in a back corner of the second floor that I discovered a rich cache of scientific and technical books. I was in heaven and began regularly checking out volumes on any subject I had even a passing interest in. I read books on electricity and electronics, clock and watch making, geology and mineralogy, and photography. Soon I was building simple AM radios, regularly repairing clocks and wristwatches for friends, collecting dozens of minerals and crystals, and even developing my own black and white film negatives and using a homemade enlarger I had constructed to print them in a darkroom that I had set up in our bathroom (my mother wasn't too happy about that!).

I also found the library's collection of UFO books. I'd read most of them already and only a few titles were new to me. These only further reinforced my belief that some of what people where reporting were, in fact, the "real thing"; that is, extraterrestrial air and spacecraft being operated in the skies of Earth by actual extraterrestrial beings which, when very rarely spotted in proximity to one of their craft, had a humanoid body shape. As I made my way through these books, I discovered another interesting topic that I was only slightly familiar with: the paranormal.

Like most of the kids of the 1950's and 1960's, I'd seen my share of horror movies. These were regular fare on Saturday night television and we'd watch the films with Lon Chaney Jr. portraying a werewolf, Boris Karloff as Frankenstein's monster, and Bela Lugosi as the bloodthirsty vampire Count Dracula. In the case of werewolves and vampires, the suggestion of these films was that strange things could happen to human beings as a result of mysterious forces influencing them. I even noticed that, despite the growing number of victims in their plots, I always felt a bit sympathetic toward the monsters. They were all originally just normal human beings who, through no fault of their own, were turned into hideous caricatures of what they had once been. These movies made me wonder if all of the "bad" humans throughout history might also have started out as normal people who, through very bad luck, had become evil due to forces acting on them over which they had no effective control.

But, the new books I found at the Elizabeth Main Public Library dealt with such things as ghosts, haunted houses, mysterious historical artifacts, and people who could demonstrate bizarre talents such as the ability to autolevitate, read the thoughts of others, know the future, and even "bilocate" or, literally, become two twins that would be spotted in widely separated locations at the same time. These effects were not being shown in the fictional horror films I was watching and I absorbed as much of the information available about them as possible. As in the case of my study of UFOs, I soon realized that, while the field of the paranormal was probably loaded with various hoaxers and the deluded, many of the events being described in these books had actually occurred. Again, as with UFOs, I believed that the effects being demonstrated in the cases were the result of new physics which was still not being discussed in the textbooks I was reading in school.

As I studied the various paranormal phenomena intensely, I noticed a rather interesting detail. Practically all of them were also seen in many of the UFO cases that had been recorded! I was not the only one to note this coincidence. There was an aerospace engineer and UFO researcher named Leonard G. Cramp who had written a book titled *UFOs and Anti-Gravity: Piece for a Jig-Saw* that was published in 1966 and a new copy of it showed up at the Elizabeth library just as I was being issued my own library card. In his classic treatment of the UFO topic, Cramp suggested that the ability of a UFO to hover without any visible means of propulsion might involve the *same* process that was taking place in the body of a human being who had the ability to autolevitate. I thought that, even without genuine extraterrestrial UFOs to study and reverse engineer, humanity might still be able to duplicate their propulsion if we could only find people today that were able to autolevitate and then get them into a physics laboratory where their bodies could be studied when they became airborne. Unfortunately, while this would probably provide a quick route to obtaining craft with the UFO-like ability to overcome gravity and inertia, no such individuals were being found. Skeptics simply said that meant that the effect was not real, but, to me, it seemed the actual reason was that it was a very rare talent and, perhaps, only one person in several centuries could demonstrate it at will.

I eventually absorbed so much information about paranormal phenomena that I was able to produce a book on the subject in the late 1970's which presented about three dozen of the effects I thought were most likely real along with a new theory to describe what was happening as the effects were manifesting. That book, unfortunately, spent over a decade in search of a publisher until, finally, quite by accident an upstate New York mail order bookseller found out about it and published it as a soft cover manual whose pages were cheaply photo offset printed and then bound together with two large staples in the spine. The bookseller had five hundred copies printed up and they all sold out in a matter of a few months. Then the title went out of print and remained so until, almost two decades later, when I rewrote and updated much of the book's original text and was able to have it published using the new "Print On Demand" digital printing technology that had been developed. Now that work, titled *The Physics of the Paranormal*, is available in soft and hard cover editions and also can be downloaded into various brands of ebook readers.

I found that I could not get enough of this new topic of the paranormal to study and then there came a day after school at the library when I came across a most unusual title. It was a thick volume by a British author named Rupert T. Gould titled *Oddities: A Book of Unexplained Facts* which had originally been published in 1928. The edition I found, however, was a later one published in 1944. Gould was a naval officer and, like myself, interested in clocks and watches. In 1920 he received permission to restore several of the original marine chronometers that had won John Harrison a huge prize back in the early 18th century. (The sum of the prize that Harrison had won, as mentioned in the previous chapter, most likely inspired Johann Bessler to set a price of 100,000 thalers for his secret imbalanced perpetual motion wheel mechanics because, at the time, 100,000 thalers was about equal in value to the British prize amount of 20,000 British pounds.) Also like me, Gould was deeply interested in various historical mysteries and rare exotic paranormal phenomena.

As soon as I held his book, I knew on the spot that it would be the next one I would read. The cover of the 1944 edition made available to me at the library had an image on it that also attracted me. On its right side there was a drawing of the left side of a strange looking wheel. An outer spherical metal weight attached to the end of a scissor-like set of arms was extended away from the center axis of the wheel's axle as another smaller inner weight sank toward the center axis of the axle. This action would, because of gravity, be repeated as each pair of mechanically connected weights swept along over the top of the counterclockwise rotating wheel.

From studying the book's cover illustration, it was obvious that, as the smaller inner weights passed under the wheel's 6:00 position, they would then fall away from the axle's center axis and, by so doing, cause the larger outer weights at the ends of their scissor mechanisms to be quickly pulled up closer to the axle's center axis. If this design worked, then it would keep the center of gravity of the wheel's 30 weights on the wheel's left or descending side at all times. As the wheel then rotated in a counterclockwise direction, this extension of the larger outer weights away from the axle at the 12:00 position of the wheel and their retraction at the 6:00 position would continue automatically and the maintained imbalance should then keep the wheel perpetually rotating in a counterclockwise direction.

I was intrigued and also noted on the cover that there was mention of "The Wheel of Orffyreus" which I assumed was what was depicted. I checked out the volume and had finished reading it before a week had passed and that was in addition to my required school work.

Even now, almost half a century later, I only vaguely remember the other chapters in Gould's book which dealt with various unexplained mysteries of a paranormal nature. What I do vividly still remember is chapter 5 which gave a brief and I later learned somewhat inaccurate account of Bessler and his marvelous self-moving wheels as when, for example, the *diameter* of the Draschwitz wheel was given as 5 feet and that of the Merseburg wheel as 6 feet! Obviously, he or the book's editor had mistakenly used the radii of those wheels for their diameters which then gave diameters which were only half of what they should have been (the correct radius of the Draschwitz wheel should actually have been 4.5 feet and its diameter 9 feet).

I learned from Gould's account that Bessler was mostly a self-taught craftsman who, in 1712, suddenly announced that he had invented a self-moving wheel or perpetual motion wheel. These wheels were then given various tests by the intelligentsia of the day and deemed to be genuine imbalanced perpetual motion wheels. Gould, however, could provide no explanation of how Bessler's wheels worked. I, of course, immediately began to wonder just how he had created such wheels.

My research into Bessler did not end with the material I found in the Gould book. I decided to look up the topic of perpetual motion in the library's various sets of encyclopedias. What I found in them was not very encouraging. They all suggested that there were laws of physics which implied that *any* sort of perpetual motion wheel was not physically possible. Since the validity of these laws had been firmly established by countless observations and experiments, that really meant that anyone claiming to have created a perpetual motion device, such as an imbalanced pm wheel, had to be either deluded or lying and perpetrating a hoax. The encyclopedia articles usually gave several examples of various types of perpetual motion devices that had been tried over the centuries with a quick explanation of why they were impossible.

Basically, all of these devices were impossible because, if they worked which they did not, they would have to be outputting energy that, literally, came from nowhere. In other words, in order to work, they would actually have to be *creating* energy out of nothing! That objection certainly made sense to me, but the fact remained that Bessler's wheels had been very carefully tested and, other than Bessler's brother, the only other man to ever view the internal mechanics of his wheels claimed that they were absolutely genuine. That man, the Count Karl described in the previous chapter, was a person of such highly respected integrity that the idea of him participating in a hoax with Bessler seemed to me to be an absolute impossibility. Bessler himself was a very pious and religious individual and, thus, it also seemed highly unlikely to me that he would lie about having found the long sought perpetual motion if he had not.

I remember one incident in my high school physics class when the subject of perpetual motion came up. Our teacher immediately dismissed it as total nonsense and said that, in modern physics, if one's theory predicted a result that indicated perpetual motion was taking place or that the 1st Law of Thermodynamics also known as the Law of Conservation of

Energy was being violated, then that theory automatically had to be false. He mentioned that, in formal logic, this was an example of *"reductio ad absurdum"* or, in English, "reduction to absurdity" in which a set of premises or a theory was automatically deemed to be false if the conclusions derived from it was known to be false. With respect to perpetual motion devices, this basically meant that any device that seemed to be producing energy in excess of what it required to overcome friction so that it could stay in continuous motion absolutely could not be genuine and must therefore be either fraudulent or not actually working the way some delusional inventor believed it was working. It was obvious to me that my teacher had never read about Johann Bessler and his wheels!

But yet, despite my teacher's dismissal of the subject, I could not get Bessler's invention out of my mind. The count who was his patron claimed that the wheels the inventor had made were so simple that a carpenter's apprentice could understand them and make one after having viewed a wheel's internal mechanics for a few minutes. In fact, the count even said he was surprised that, because the mechanics of the inventor's wheels was so simple, that no one else had ever thought of the design before Bessler. The inventor himself was also reported to have said that he feared that when one of his wheels finally found a buyer, the buyer might be disappointed that he had paid so much money for so simple a design!

Then I came across another book that contained a brief but tantalizing account of the Orffyrean wheel. I was a book that I encountered as I was researching the subjects of paranormal phenomena and UFOs.

That book was Frank Edwards' 1959 work, *Stranger Than Science*, which contained a chapter titled "Bessler's Wonder Wheel". Like Gould's treatment of the 18th century inventor, Edwards' book contained several errors in the descriptions he gave of the wheels and his account has been dismissed by many as somewhat fictional. For example, Edwards' account claims that the secret of Bessler's Kassel wheel was only revealed to the count *after* its testing was completed in early January of 1718. The reality, however, was that the secret of Bessler's wheels was revealed to Count Karl years earlier and was one of the conditions the inventor had to meet in order to obtain the count's funding for the construction of the Kassel wheel.

Edwards then went on to describe how, after finally having the secret mechanics Bessler used revealed to him, the count rushed back to his room in his castle and immediately wrote down what he had seen inside of the Kassel wheel. Supposedly, the wheel's imbalance was created because the weighted levers on the drum's ascending side eventually fell *inward* toward the axle and hit a pin or stop piece attached to the drum that then radially aligned them so their center lines pointed inward toward the axle's center axis. As the center axis of a weighted lever's pivot pin passed the 12:00 position of a clockwise rotating drum, the pin or stop piece that temporarily supported the lever would part company with it as the lever's end weight again began swinging outward toward the drum's peripheral wooden rim wall on its descending side and, eventually, made contact with another stop there or with the rim wall itself. In reading Edwards' account, one can easily get the impression that he had somehow gotten access to the count's written description of Bessler's secret imbalanced perpetual motion wheel mechanics and was accurately relating that description. Again, the reality is that such a written description by Count Karl does not exist as far as is known by anyone else. Apparently, the suggestion that it did was just fiction on the part of Edwards which allowed him to insert what he thought the secret mechanics might have been and to make his chapter on Bessler seem like it was solving the mystery of the inventor's wheels.

But, at the time all of this was enough to get me to thinking constantly about the subject and I even thought that Edwards had actually read a valid description of the Kassel wheel's mechanics and I used it as the basis for the first design I began working on which was a small wheel with weighted levers hanging from pivots attached to the wheel's outer rim. I had made sketches in a small notebook that showed that the center of gravity of the wheel's weights were, in the starting orientation of the wheel I drew, indeed located on the descending side of the wheel. On paper the design looked great and I was even a bit disappointed that solving the riddle of Bessler's wheels had been so easy for me! At the time I figured I'd have a running model wheel completed in, perhaps, two weeks time at the most.

That was about a half century ago!

What I did not realize in that pre-internet decade of the 1960's was that, since the time of Bessler, perhaps tens of thousands or even hundreds of thousands of young people just like me had also heard of the inventor and his wheels and had also thought that they would have the whole thing solved in a matter of weeks. I did not realize then that my starting design and practically every other design that I could imagine had probably already been tried *many* times by others without success. I did not realize that the quest to achieve a perpetual motion was really a sort of obsession that had made countless numbers devote large portions of their lives to the pursuit only to come up with nothing of value at the end of it all. I guess it was a good thing I was unaware of all of these things because, if I had been, then I might have found myself agreeing with the scientific orthodoxy that still, quite unfortunately, considers the entire pursuit nothing but a colossal waste of time for the deluded.

After my first few attempts to replicate one of Bessler's imbalanced perpetual motion wheels failed, I found the number of sketches I was making of potential designs for them steadily increasing. For every ten or so sketches I made that looked promising, I'd attempt one build of a physical model. These model wheels were made from things like tin cans, popsicle sticks, cut lengths of coat hangers, and steel washers glued in stacks onto wooden dowels to serve as metal weights. Only rarely did I ever visit a hardware store to purchase new parts.

After a year or so of these constructions, all I had to show for my work was a growing collection of notebooks filled with designs that did not work and a large collection of models that, once failing to work, had been cannibalized to provide parts for some future design that I was convinced would finally solve the problem. But, nothing worked despite the accuracy of my sketches or the precision with which I constructed my model wheels. Perhaps the scientists were right after all I would occasionally think, but, then, quickly deny that possibility because of the elaborate testing I knew Bessler's wheels had undergone and successfully passed. Constructing a working imbalanced perpetual motion wheel *had* to be physically possible. But how was it done?

To help me with my designs I found myself using more and more of the mechanics I was studying in my high school physics courses. Instead of just a quick sketch and determination of where the center of gravity of the wheel was located, I began to calculate the torque that would be produced and how it would vary as the wheel began to rotate. To my surprise, I found that no matter how clever my design, the center of gravity of its moving weights *always* dropped to the "*punctum quietus*" or equilibrium point located vertically below the center axis of a wheel's horizontal axle as the wheel rotated and, when that happened, the torque being produced immediately decreased to zero. I realized that, unless I could figure out how Bessler managed to prevent this from happening in his wheels, I would have no hope of ever duplicating his invention.

On one of my trips to the Elizabeth Main Public Library, I discovered that they had a copy of a somewhat rare tome on the history of perpetual motion devices. It was Henry Dircks' massive 1870 two volume book set titled *Perpetuum Mobile; or, A History of the Search for Self-Motive Power from the 13th to the 19th Century*. I found it to be an interesting compendium of the vain attempt to achieve a working perpetual motion device over the past centuries and it did briefly mention Bessler and his wheels, but pointed out his fired maid's accusation that the Kassel wheel had been fraudulent and quickly dismissed all of his wheels as such. Dircks also mentioned an English translation of Bessler's book, DT, which had appeared in England in 1770, but I was never able to locate that. Rather than being supportive of the subject, I was dismayed to realize that Dircks was actually a skeptic and his book appeared to me to really be an attempt to put the subject to rest in his own mind. Perhaps he had been frustrated by his own earlier efforts to achieve pm, concluded it was physically impossible, and felt it his moral duty to prevent others from wasting any of their lives pursuing it. I learned he had a prior history of debunking fraudulent spiritual mediums and his book was most likely just an extension of his debunking efforts into the realm of exotic mechanics. For example, one line from his book reads:

> "A more self-willed, self-satisfied, or self-deluded class of the community, making at the same time pretension to superior knowledge, it would be impossible to imagine. They hope against hope, scorning all opposition with ridiculous vehemence, although centuries have not advanced them one step in the way of progress."

Statements like that certainly don't portray pm seekers in a positive light!

During my years at Thomas Jefferson High School in Elizabeth, the "Vietnam Conflict" was slowly, but steadily escalating and in 1968 when

I was to turn 17 years old in May, I was required to report to our local draft board two weeks before my birthday so I could register for the draft that then existed and to which I would be subject to after I turned 18 years of age. If drafted and found medically fit, I would then be sent over to Vietnam to assist in the "police action" over there that was part of the US' cold war strategy to suppress the spread of Soviet style communism throughout the world. Not doing that was deemed undesirable because it would then result in the people of the capitalist countries losing their various freedoms and becoming slaves to an atheistic economic and political system.

On television the American people were able to follow the progress of the conflict on their nightly news shows. We saw regional maps of Southeast Asia that were referred to as "bleeding maps". First the map would only have North Vietnam colored red indicating communists were in control of the people of that country and then the color would slowly spread like a metastasizing cancer until all of South Vietnam was red. Then, one by one, the other countries of their peninsula turned red. But, the "cancer" did not just stop there. Soon it would spread to the Philippines, Indonesia, New Zealand, Australia, Japan and ever eastward to Hawaii and finally the west coast of the continental United States! From these maps I had visions of North Vietnamese landing craft coming aground on the beaches in California and discharging hordes of enemy soldiers that would be firing their machine guns in all directions as American beach goers ran for cover. Of course, this all sounds totally ridiculous nowadays, but these were the actual fears that were motivating most Americans to support the military effort in South Vietnam at the time.

In early January of 1968 I recall a school friend and myself standing in front of a dark wooden plaque attached to a wall in our high school that had several dozen little brass plates nailed to it. Each plate gave the name of a former student who, after graduating, had either enlisted or been drafted, went to Vietnam, and was then killed there. The plates gave the year the student graduated along with the year he died. This plaque sort of anticipated the much larger Vietnam Memorial Wall that was eventually erected in Washington, DC years after the conflict had ended.

As we gazed at this wall, I remember turning to my friend and asking him "Do think there will be enough room left for our names?" At that question his eyes just sort of bulged out of his face as a look of fear came over him. If one was drafted for this conflict there were only a few ways to avoid it. One either had to be found mentally or physically unfit for military service which had to be verified by a medical doctor in the form of a note to be presented during one's pre-induction physical, obtain a deferment from the draft because one was in some critically necessary occupation or was a student in a college or university, or he had to take flight to some country, like nearby Canada, which was harboring those who decided to "dodge" the draft. Without any of these, one either had to allow himself to be drafted or, if he refused, had to spend five years in a federal penitentiary as punishment for his crime of draft evasion.

Everyone in my family, however, was telling me not to worry about having to go to Vietnam because it was obvious from the television news broadcasts that the whole thing would probably be over before I went to register for the draft in May of that year. That calmed me down a bit.

Then on January 30th of 1968 the "Viet Cong", who were guerilla fighters inside of South Vietnam assisting the North Vietnamese military, launched hundreds of simultaneous attacks throughout the south and many people died. Some of these guerilla fighters managed to get inside the capital city of South Vietnam, Saigon, and even right inside our embassy there! When I saw this, I had no doubt that this "conflict" was not going to be over anytime soon. This unexpected and overwhelming attack was enough to make then President Lyndon B. Johnson decide not to run for re-election and was only the beginning of another seven years of fighting over there that would cost the lives of tens of thousands of additional Americans.

Like most of my fellow students, I was able to obtain a student deferment because in the fall of 1968 I began attending the Newark campus of Rutgers University. I filled out a single form at their administrative office and my deferment arrived in the mail a few weeks later. I was, however, only given a temporary "holding" status which could be removed at any time if the need was great enough. A few years later, they did away with these student deferments which were deemed unfair to those who were not going to college and did not have a medical exemption. The birthdays of those to be drafted were then randomly selected from a large drum and assigned a number from 1 to 365 and they would start by drafting those whose birthdays corresponded to number 1 and work their way down, if need be, to those whose birthdays corresponded to number 365. My birthday got number 147 so I figured I had about a 60% chance of being drafted. To my great relief on the *last* day that I was subject to this lottery selection method, the last number they reached was 145. That meant I did not have to worry about the matter again. If I'd not been so lucky, then I might not being writing his chapter now.

Unfortunately, the situation over there did not formally end until the Paris Peace Accords were signed in January of 1973. By then, about 58 thousand Americans would die from various causes during the conflict. The exact number of North and South Vietnamese that died is only an estimate, but is probably in the millions. In later years it was realized that the entire conflict had been a disastrous waste of human lives and money. It actually set in motion fiscal problems in the United States which laid the groundwork for much of the economic problems that continue to burden us even today.

I must admit that I found the academic work load at Rutgers rather easy compared to that of my high school in Elizabeth which is a nice testament to how rigorous the training and study habits encourage at that school were. I only had a few classes per day at Rutgers and my biggest problem was getting to a bus stop that was about a mile from my home. On many a morning I found myself speed walking / running that mile while carrying a heavy load of books in order to catch the bus to Newark so that I would get there in time for my first class. Sometimes I'd get to the bus stop just as the bus was pulling away and I'd have to run another full block to catch it at the next stop! This was a real challenge during the winter months or when it was raining. But, somehow I survived it all.

The major I pursued at college was chemistry and I'd eventually have enough credits for a minor in philosophy. Actually, I had so many interests when I arrived at Rutgers that I could have probably had any major and had success with it. During my "orientation day" that took place before my first classes began a month later, my advisor just happened to be a chemistry professor who was wearing a ring from the prestigious Max Planck Institute over in Germany. I told him I was interested in the subject of chemistry and so that is the major I pursued while there. If my advisor had been a professor of some other subject, then that would have been my major.

My interests in such things as paranormal phenomena, UFOs, and, of course, perpetual motion machines continued despite the academic work I did at the college. However, I noticed that the supply of books at their campus library that dealt with these subjects was quite limited. From informally chatting about such things with fellow students and an occasional professor, it became obvious to me that such topics were considered nothing more than irrational beliefs for which there was no "scientifically acceptable" evidence. I quickly got the message that they were not topics that "real" scientists were supposed to even think about! To do so might result in one's intellectual abilities and even mental health being called into question. However, I was so convinced in the reality of Bessler's wheels and the possibility of achieving a working imbalanced perpetual motion wheel that such negativity had little effect on me. I continued to work on my various imbalanced perpetual motion wheel designs, but, as usual I could not get their centers of gravity to persistently stay on their descending sides.

Sometime during my second year at Rutgers from the fall of 1970 to the spring of 1971 I purchased one of the first of the scientific calculators that became available, a "Bowmar Brain" handheld calculator. It was expensive, but I found it far easier to use and more accurate than the clumsy slide rule I was previously using. With this early electronic calculator I was able to perform the mathematical calculations of the torques of my imbalanced wheel models to an accuracy of 8 significant figures! Its tiny brightly glowing red LED digits were beautiful to behold, but the device drained its batteries so rapidly that I was obliged to use rechargeable nickel cadmium batteries with it and recharge them almost daily.

I actually used this device so much that its microchip failed in less than a year and I went on to purchase several more calculators by different manufacturers in the hope of finding one that would actually last. They, obviously, were not intended for the kind of "heavy duty" use they would receive in the hands of a "seeker after perpetual motion" which was how Leonardo da Vinci referred to my kind back in 1494.

As my handmade models continued to fail and my collection of filled up notebooks continued to increase in number, I wondered if Bessler somehow managed to incorporate magnetism into his wheels and that, perhaps, they were critical to their operation. I realized that the only magnets he would have had available would have been relatively weak ones called lodestones which are just the mineral magnetite that had become magnetized by the Earth's natural geomagnetic field. So, as a suitable substitute for this mineral, I used relatively inexpensive ferrite magnets that I obtained from a local electronics store.

I tried various designs in which the weighted levers approaching the 12:00 position of a wheel would use some magnetic repulsion to help them swing over toward the wheel's descending side. These designs, again, all looked great on paper, but never worked in practice. The problem was that whatever magnetic forces aided the shifting of a wheel's center of gravity onto its descending side as weighted levers approached the wheel's 12:00 position also tended to promote its shifting over to the ascending side as the weighted levers approached the wheel's 6:00 position so that there was no overall tendency of the design to keep its weighted levers' center of gravity on the descending side of the wheel at all times during wheel rotation. It was really a frustrating situation and I eventually concluded that this was not the method Bessler used. I realized that those investigating Bessler's wheels would probably suspect that he had hidden magnets in the floor below or the ceiling above their drums' outer rim walls and would have inspected those areas with a simple handheld compass to see if they could detect a strong external magnetic field there by the deflection in the compass needle that it would have caused.

The boost in my grades caused by my physics courses landed me on the prestigious "dean's list" for students that were doing exceptionally well. I also got involved in a tutoring program to help other students with the subject. Sometimes a small group of us would meet informally in an empty classroom after a physics class ended and I'd review that day's lesson with them. At other times I would have appointments set up with individual students in an effort to help them with their course work. Those I helped seemed to like my explanations and the approach to problem solving I used. One day, our physics professor called me aside after class and suggested that I should seriously consider changing my major from chemistry to physics because I seemed to have such a natural aptitude for the subject. I've often wondered how he would have reacted if he'd found out that I had developed my computational skills as the result of the many hours I'd spent trying to build a working perpetual motion machine. I suspect that if he had known that, then his recommendation would have been promptly withdrawn!

Finally, in June of 1973 I graduated from Rutgers with a Bachelor of Arts degree having majored in chemistry, but I had no working imbalanced perpetual motion wheel to show for the many hundreds of hours of "extra curricula" effort that I had invested in trying to construct one. It was time for me to begin seeking employment using my training so that I would have a more substantial income than had been provided by my summer jobs. That, however, proved more difficult than I thought it would be. The US was in a recession at the time due to the enormous cost of the Vietnam Conflict and a crippling oil embargo inflicted on the US and its allies by the Arab members of OPEC for our government's decision to re-supply the Israel military after the Arab / Israeli war ended with Israel victorious. That sent the cost of a barrel of oil sky high and reduced the supply in the US drastically.

With the number of jobs being offered in chemistry in the employment sections of our state's newspapers virtually disappearing, I was lucky to find a job as a shipping clerk in the basement of a nearby factory. So, there I was with my newly minted college diploma, an important part of the "American Dream", working for a low wage in a factory that was filthy, stank, and was contaminated with oil that actually dripped down on us from the constantly running machinery on the floor above mine. My job had absolutely nothing to do with chemistry and it was only then that I realized that, without an economy healthy enough to provide a college graduate with a suitable job, his diploma was virtually worthless as far as using his training was concerned. It was a very depressing thought, indeed. With required overtime, I was actually working about 60 or more hours per week and it was beginning to wear me out. After a few months of that misery, I managed to find a better paying and cleaner job at an electronics company not too far from where my college's Newark campus was located.

By then I had my first "previously owned" automobile which was a giant 3800 pound 1966 Chrysler "New Yorker" whose 8 cylinder, two barrel, 400 cubic inch engine, if kept well tuned up, managed to get about 8 miles to a gallon of gasoline. Shortly after I purchased it, our state instituted gas rationing due to the oil embargo and I was only allowed to visit a gas station every other day. When I got to the station there was usually a long line of cars ahead of me and I'd get in line and have to keep starting and stopping the engine every few minutes so as not to waste gasoline while idling. Then, when I finally got to the pump, I was limited to only a $3 purchase of gasoline. At that price at the time, I got about 3 gallons, but that did not fill up my car's 25 gallon gas tank. Some drivers, known as "tank toppers", would actually go from station to station spending hours in line until they had a full tank. I carried an emergency gallon of gasoline in my trunk just in case my tank ran dry when I was out on the road so that I could make it home with the car. It was a very stressful time and Americans, for the first time since WWII, had to start deciding if they really needed to make a trip and, if so, would they have enough fuel to complete it and get home again! This was devastating for the many businesses that relied upon drive in customers and many people found themselves unemployed as a result. This is what can happen when nations do not achieve energy independence and allow themselves to become dependent upon others for their energy needs.

Since this chapter is getting a bit longer than I had originally planned, I'll just compress my story up to the point where my Bessler research experienced a sort of renaissance due to the acquisition of information about him and his wheels which had not been previously available to me.

At the end of 1974 I finally found a job as a quality control chemist at a large pharmaceutical corporation in northern New Jersey. However, although the money was excellent, I found the work exceedingly boring. I did the same routine tests day after day and little creativity or problem solving was involved. I felt like little more than a glorified short order cook who just handled various chemicals instead of food. I finally left that position and began working for a smaller company in Newark that was involved in extracting certain metals from ores that were delivered by train to them. That was far more interesting to me and involved actual product research. As an assistant to the plant manager I had the run of the plant and was involved with the technical problems of its various departments. It was probably the most satisfying of my various careers.

Occasionally, I'd bring up the topic of perpetual motion devices and my co-workers, all with degrees in chemical, electrical, or mechanical engineering would just make fun of the subject. None of them had heard of Bessler and I thought it wise to keep quiet about my ongoing involvement in the subject or what was happening with my latest sure to succeed design. On several occasions I was tempted to see if I could get a machinist there who had become my friend to make up parts for me from stock rods, sheets, or tubes of aluminum or steel, but thought that would then lead to questions about what I planned to do with the items. Possibly, one or more of my co-workers was, like me, a "closet" perpetual motion seeker, but also decided to pretend they had no interest in devices that could achieve that type of motion.

One of the major problems I had noted with most of the exotic subjects that I had developed interests in was that an aura of ridicule had been established about them by the scientific community which then discouraged any serious scientific investigation of them. This, in my opinion, is not how real science is supposed to conduct itself. Researchers should always be free to investigate *any* phenomenon without fear of being socially or professionally stigmatized or financially punished by being denied employment or funding for research. Unfortunately, the situation has not significantly changed since the 1970's and persists to this day. It is for this reason that I hope this volume will, finally, lead to a successful duplication of Bessler's inventions and, with that, to a more open minded approach to the various "New Age" subjects of paranormal phenomena, sentient extraterrestrial life and UFOs, as well as self-motive, "free energy" devices.

I spent most of rest of the 1970's attending graduate school at the Newark campus of Rutgers at night in order to eventually earn a Master of Science degree in chemistry by 1980 with the hope that would lead to even more interesting employment and a higher salary. But, unfortunately, yet another recession had started by then and that was not possible for me. I eventually married and had two children and then was divorced. I attended several weekly support groups that were popular during the mid '80's and, eventually, found a new and enduring relationship at one of them that made me very happy and still does.

After several decades of frustrating attempts to achieve a working perpetual motion device, I finally had to put the matter to rest for a while and concentrated on the other topics I was interested in. I had been writing short technical articles on extraterrestrial UFOs for various civilian UFO magazines for several years and grew frustrated with the delays I experienced in seeing the material published. By 1985 I decided it was time for me to publish my own UFO research journal in which I could present as much of my research as I wanted to and to do so on my own time schedule. That publication, *Annals of UFO Research Advances* or *AURA* for short, was modestly successful and lasted for a decade during which time I explored almost every conceivable facet of the UFO phenomena and did so at a technical level that was unprecedented in the field. There is even some evidence that my one man effort stimulated one of the major civilian UFO research groups in the nation to follow suit and also try to duplicate my efforts. (A complete reprinting of all 24 issues of *AURA* that were published between 1985 and 1994 can be found in the 4th book I published, *The New Science of the UFO*, which came out in 2013. It is an interesting coincidence that the last two numerals in the years 1985 and 1994 each add up to a value of 13 and that the entire set of *AURA* was then republished in the 13th year of the 21st century or 2013.)

In 1987, as I previously mentioned, I made contact with that upstate New York mail order bookseller and he put out the first edition of my work titled *The Physics of the Paranormal*. It was later extensively edited, expanded, and updated and then republished in 2003 by another publisher.

Around the year 2000 I found my research in the subjects of the paranormal and UFOs beginning to come to an end. It was not because I had lost interest in them, but, rather, because I could not progress any further in them with the data I had available to me. I had, more or less, gone as far as I could with them and was searching for some other mystery to begin working on. That's when I remembered that I never really did find the secret of Bessler's wheels. But, then I recalled the many, many hours of work I had put into the subject with absolutely nothing real to show for it except that I still retained the belief that Bessler was not a hoaxer and had, somehow, managed to find a design that worked.

By 2002 I finally managed to join the growing numbers who were finding distraction on the newly revised and simpler to use internet. As I rummaged through a local church thrift shop in the early months of that year, by sheer chance I came across a box labeled "WebTV" and began reading the information on the box. Basically, it was a little electronic box that one could connect to his television set and then, after setting up an account, use an infrared keyboard or remote control to access the internet and even send and receive emails. At the time the only prior experience I had with computers was with an IBM 360 mainframe computer during my college years and a used Commodore 64 that I had picked up at a flea market in the early 1990's with which I had enjoyed writing little programs by using its simplified BASIC programming language. The Commodore, like the WebTV, had used a regular television screen as a monitor.

 I asked the lady in charge of the thrift shop how much she wanted for the WebTV unit and she said $5. I was leery about spending that much on something that was probably just a piece of discarded electronic junk that someone had donated to the church for free, but decided to take a chance on it. There was a toll free number on the side of the box and I used my girlfriend's cell phone to call the number. I did not expect anyone to answer, but to my great surprise the call was answered immediately and the representative said that the device I had was still being used and, if it worked, I could set up an account, which cost about $20 per month then, and immediately be surfing the web!

 Well, I got it home and found that one of its cables was missing so I took a trip to a nearby electronics store to get the right cable which only cost a few dollars. After that, however, it took only a half hour to attach it to our television set and use a credit card to set up the account. Once that was out of the way, I was suddenly on the internet for the first time in my life! I was 51 years young then and it was a bit exciting. All of our friends had been online for years and I was often embarrassed when someone asked me for my email address and I had to tell them I did not have one and did not own a personal computer.

 With that first WebTV unit I would eventually build a website and make contact with a publisher who would produce and distribute my first book on the paranormal as a professional looking title that was available on the internet as well as through about 25,000 bookstores worldwide! Needless to say, I soon became addicted to the new digital technology and began using it daily. Eventually, in order to go beyond the capabilities of an "internet appliance" like the WebTV unit, we purchased a much larger computer that cost a lot more, but had several drives in it that allowed it to play small floppy disks, and larger CDs, and even DVDs. Four of the six books I've authored were done on that first "tower" computer and, when finished, simply "burned" onto a blank CD and sent off to my publisher.

I was amazed that the entire process of writing a book could be done without using a single sheet of paper! The books I had published were done using a new technology called "Print On Demand". This meant that my book was first made into a digital format that was stored on a master disc. Then, when a customer wanted a copy, the disc would be inserted into one end of a locomotive sized machine made by IBM that cost about a million dollars. The operator of this machine could then press a few keys and one minute later a completely printed, bound, and packaged up for shipping book would pop out of the other end of the machine. A single volume could be produced this way or the machine could be allowed to run until it produced a thousand copies! The nice thing about this technology is that an author's work never goes out of print as almost all books produced by "mainstream" publishers eventually do and, even if a customer wants a copy a century from now, he will get a *newly* printed book and not one that has been sitting in a warehouse all of that time and which might be invested with mites.

Once online, I quickly discovered to my amazement that I was not the only person interested in perpetual motion devices and those of Bessler in particular. As I visited various "free energy" sites devoted to these topics, however, I found that most of those seeking to achieve mechanical perpetual motion weren't necessarily interested in Bessler's inventions although they did, generally, acknowledge that his machines were genuine. It seemed to me like those pm seekers had, basically, given up on trying to reverse engineer Bessler's particular design because the clues he had left behind in his writings were not sufficiently detailed enough for them to do so. They were, therefore, just "doing their own thing" when it came to the quest for pm. I realized that, within the pm seeking community, I was part of a purist subclass of individuals who were determined to resurrect the *exact* mechanics of the imbalanced pm wheels actually constructed by Johann Bessler. The successful physical duplication of his wheels that then soon followed would prove that his claims to having constructed such a machine were not lies and that his wheels' performances were not the result of some clever trickery.

It was after reviewing the available online information that Collins had obtained that I realized how filled with errors and serious omissions was the Bessler story I had gotten from authors like Dircks, Gould, and Edwards so many years ago. As one example there was the detail of Bessler's use of *springs* inside of his wheels that neither Gould nor Edwards mentioned because, like many others, they may have misinterpreted the meaning of the inventor's statements about springs not being used and, as a result, assumed no springs were actually used in any way. I eventually came to realize that this particular missing "little" detail in the Bessler story was the real reason why I had wasted about four decades chasing pm without any hope of finding success!

Indeed, I eventually realized there were really two reasons why Bessler only mentioned springs briefly in his writings about his wheels. The first and main reason was that he did not want his detractors to use that information to reinforce their claims that the inventor's wheels had to be cleverly powered by mainspring driven clockwork movements contained within their drums. The second reason was that, through very bitter experience, Bessler had learned that the achievement of a working imbalanced pm wheel was *not* possible *without* the use of springs! He even hints in his writings that a special material was needed to make his wheels a reality. I realized that this "special" material was the use of high quality steel springs for his wheels that were being developed for use in clock movements and other mechanical devices in his day. These springs were able to retain a constancy of their elasticity, when in use, far longer than was previously possible and that stability was critically necessary if one wanted to construct a continuously running imbalanced pm wheel whose ever moving internal weighted levers would alternately stretch and relax their attached extension springs with each wheel rotation.

As I continued to study the additional Bessler data that I encountered online, I came across many additional clues that showed me how I had been subtly mislead by the earlier works on the inventor's wheels, but this was not done by them on purpose. From those works I was convinced that Bessler used weights attached to the ends of levers, but I envisioned these levers as being simple wooden arms with a heavy metal weight at one end and a pivot attachment to a wheel's drum at the other end. These are the simple, hammer-like types of weighted levers usually shown for imbalanced pm wheel designs and I naturally assumed Bessler also used them.

But, it eventually became obvious to me that, aside from springs, Bessler's weights were attached to something more complicated in structure than a single straight lever arm. For example, in DT Bessler drops a hint about these levers when he writes "...the cause of it all being a humble tool which the famous scholars of the day have as yet seen in but an incomplete form." I realized that the "humble tool" mentioned was actually a lever, but its form was "incomplete" because those trying to use levers in their imbalanced pm wheel mechanics were only using levers that consisted of a single arm!

With this revelation, I decided to take another look at the Bessler wheel mystery and now, having settled my ongoing pursuit of solutions to the mysteries of the paranormal and extraterrestrial UFOs and fully published the results in book form, to redouble my efforts to see if I could, finally, after almost a half century of futile attempts, find a plausible solution to that three century old mystery. The solution would, of course, have to agree with the best translations of the various clues and hints Bessler left behind in his writings and would also, most importantly, have to lead to a *working* wheel. Any departure from those requirements, especially the second one, would not be acceptable to me. I was determined, before my generation as well as my own life ended, to be able to provide "the" solution that had been sought, fruitlessly, for the last three centuries.

But, I quickly ran into a major obstacle.

My enthusiasm for "hands on" construction of physical prototypes had long ago been killed off by dozens of useless wheel models over the decades and the idea of again trying to construct them no longer appealed to me. I needed a simpler and faster way to get the same results that would not require me to spend hours per day over weeks to complete and then test a single new wheel design or some modification of a previously failed one. Ideally, it had to be a method that was inexpensive and did not require special tools.

I found that method with the new computer modeling and simulation software that was becoming available at reasonable cost for the students of mechanical engineering. I had tried several free trials of this software available online from different publishers, but was not making any serious progress in using them. At one point, I almost gave up and was considering just forgetting about continuing the pursuit of a solution to the Bessler wheel mystery when, by chance, I came across a program that was truly as "user friendly" as possible and could serve as an introduction to more complicated computer simulation programs if I needed to use them.

I obtained a copy of this simulation program and began to practice with it. To my happy surprise, I found out that, after only a few hours of effort, I was able to "construct" simple virtual wheel models containing various numbers of spring tethered weighted levers of different masses which were interconnected by a network of cords of precise lengths which I could make have zero elasticity so that they would not stretch and change length when they were put under tension. I could also, in seconds, exactly adjust the elasticity and starting tensions of a model wheel's springs. It was, literally, a dream come true that would allow me to compress years and even decades of shop work into only a few months of work at a keyboard. The program only allowed one to construct two dimensional models, but that was fine with me because, basically, all of the mechanics in Bessler's wheels involved motions that only took place within the two dimensional planes contained within their drums. There was a three dimensional version of the simulation program available, but it would be more complicated to learn to use and I really did not need it for my research.

I immediately settled on making all of my wheels exactly 3 feet or 36 inches in diameter in an attempt to begin where Bessler had with his original prototype wheel in Gera, Saxony in late 1711. Once I had a model wheel completed after using various standard parts and connectors that were supplied from side panels near the program's workspace window, I would test it by simply left clicking on a single button above the workspace window. Depending upon the complexity of a design, the software's computation of the individual frames of its virtual reality simulation could take some minutes to complete, but, once done, the simulated test of the model wheel would then run in real time and I would very quickly know if I finally had "it" or if, seeing the model wheel fail to perform as it should, yet another modification was needed which was the usual case.

I had, like many, read all of the obvious clues that appeared in Bessler's writings including the ones that were noted in the various letters written by those who had witnessed or actually tested his wheels. These written clues, however, tended to be somewhat vague in nature and there always seemed to be some controversy as to whether or not their translations from 18th century German into contemporary English were accurate.

I had originally planned to cover all of the various written clues in Bessler's writings in this volume in detail, but eventually opted not to. This was because, as I pondered them, I realized that they really tell us little about what really matters, which is the mechanics that were contained *inside* of a wheel's drum.

However, I can briefly sum up where all of the obvious written clues take us.

Basically, we learn that Bessler's wheels consisted of relatively lightweight hollow wooden drums that were securely attached to a solid wooden axles whose inserted and securely fixed end steel pivot pins were free to rotate inside of semi-cylindrical depressions with "half moon" cross sections that had been cut out of brass bearing plates that were themselves embedded in solid wooden vertical support planks.

Each drum of a one-directional wheel also always contained eight weighted levers.

The above is, unfortunately, about all that one can determine from a careful reading and interpretation of the Bessler literature and, admittedly, it is not much to go on. This is all that Bessler was willing to divulge openly in his writings and I spent several years working with only this minimal amount of information and my simulation software in an effort to determine the exact design that Bessler had discovered and which allowed his 3 foot diameter Gera prototype wheel to finally display the pm effect in late 1711. During that time I probably produced around six hundred model wheels. Every single one of them "keeled" which is a term popular among pursuers of imbalanced perpetual motion wheels and refers to a failed design whose active weights' center of gravity, rather than remaining on a wheel axle's descending side during rotation, simply rotates along with the axle, passes from its descending to ascending side and through the *punctum quietus* equilibrium position, and then swings like a pendulum bob weight back and forth through that point several times until it finally comes to rest directly vertically under the center axis of the wheel's axle. Thus, the word "keel" describes the action of a failed imbalanced wheel's center of gravity which mimics the action of a weighted submerged keel attached to the bottom of a sail boat's hull that tends to drop to its lowest depth below the surface of the water and thereby automatically rotates the boat's hull so that its masts remain upright and vertically oriented.

It was sometime during the early months of 2011 that I made a discovery that, finally, put me on the still long road to successfully reverse engineering Bessler's wheels. This discovery occurred as I was convalescing from a very serious and nearly fatal illness!

For most of the year of 2010 I had been suffering from chronic bronchitis and, literally, became a walking phlegm factory. Antibiotics had virtually no effect on the problem and it was very annoying and would get worse after I ate a meal. The condition took a break during the summer months, but was then back again in the fall. While I was suffering from this, I was struggling to finish up what would be my second book on UFOs titled *The How and Why of UFOs* and get it off to my publisher. Then around the beginning of October I developed a low grade fever and flu-like symptoms. I made the mistake of ignoring these symptoms and continued to push on with my normal activities.

As the end of the month approached, I noticed that I was growing weaker and weaker and having trouble walking to the bathroom. I managed to visit a local walk-in clinic and got a prescription for what was described online as "the most powerful antibiotic in the world". The expensive supply I obtained contained only five tablets! I took them as directed over the next few days and…no effect on my symptoms whatsoever! Finally, I became completely bedridden, but thought that it had to be just another flu virus that would run its course and then I'd bounce back as I had done many times before during my life. Then as Halloween approached at the end of the month, I noticed one day that I was having a lot of trouble breathing. My partner decided that it was time for me to visit the hospital which I reluctantly agreed to do. I was so weak that two burly ambulance attendants had to pick me up and strap me to a stretcher and rush me to the hospital.

At the hospital, a chest x-ray revealed two huge shadows in the shape of lungs were occupying my chest cavity. I had severe double pneumonia and was put on more antibiotics and oxygen therapy. Over the next 48 hours my condition worsened and, at one point, I could hear some of my doctors quietly discussing how they did not think I was going survive! Needless to say, that was the last thing I needed to hear. I remember laying in the intensive care unit and being stunned to realize that people die every day and this was *my* day to do so even though I was only 59 years of age! I thought of all of the things I still wanted to do in life, of the people I'd leave behind who depended upon me, of my book that I'd worked on for years which was then at the publisher and which I'd never hold in my hands, and of the Sun rising the next day without me around to see it. At that point I blacked out.

In what seemed like only a few seconds, I awoke again feeling somewhat refreshed and I noticed my breathing was much easier. Then I noticed that my wrists were tied to the sides of my bed with Velcro straps and there were several plastic tubes coming out of my mouth. I was on a ventilator which assists the breathing of severely weak patients. After a few minutes a young Filipino nurse entered the room, hung her face over mine, and announced "You going to be okay" after which she promptly disappeared. I found out later that I had been in a drug induced coma for almost two days. To make a long story a bit shorter, I spent about 40 days in the hospital, half of them on a ventilator, lost about 40 pounds of body weight due to the chronic fever I had, and had, literally, quarts of antibiotic solutions of various types intravenously poured into my body. Even after I came home, it required another full six months of rehabilitation to get to the point where my atrophied leg muscles became strong enough to again support my upper body weight and I could walk normally again.

For the first two months of that time I just laid on a couch in our living room breathing 80% oxygen with the assistance of a cannula which is a strapped on plastic piece that fits into one's nostrils and is connected by a long plastic tube to an oxygen concentrating machine. Apparently, I did not have an infection from a single strain of the pneumonia bacteria, but from multiple strains and, based on the statistics I eventually researched, a person in my condition and on a ventilator has about a 50% chance of dying from such a severe infection in this country and far higher in Third World countries. So, I was still here and able to continue my Bessler research as the result of the metaphorical flip of a coin. Fortunately, heads came up! I had another bout of pneumonia a few years after that, but that time prompt treatment with a powerful antibiotic knocked it right out.

As I was recovering slowly over the course of the first half of the year 2011, there was little else to do but eat, sleep, watch television and DVDs, and think about all of my interests that were on hold. The author's copies of my second book on UFOs had arrived, but I had barely enough strength to open the package and flip through the pages of one of them. I was, however, able to review my research into Bessler's wheels and was dismayed at the large number of failed wheel models I had accumulated. Bessler, several times in his writings, hinted that, should his invention not sell, then he hoped others would find out how he did it and manage to duplicate them. He also suggested that he purposely left clues that would enable this to happen. But *where* were those clues, I wondered.

Then, for some strange reason, I could not get the two frontispiece portraits at the very beginning of Bessler's DT book out of my mind. I will deal with them in far greater detail later in this volume, but for the moment I will just state that there was something about the two portraits that seemed a bit odd to me.

Firstly, he placed them right at the beginning of his most professionally produced book on his wheels and he even cut a hole through the second portrait so that his face in the first portrait would show through the hole in the second portrait. I wondered why he would have done that. That's when it occurred to me that he wanted to make sure anybody reading DT knew who had authored the work and he also wanted the *same* face associated with the two portraits and, in particular, with the items shown in the *second* portrait which shows him standing behind a shop table covered with various, apparently randomly, arranged drafting tools, plumbing items, a telescope and microscope, and a globe with its South Pole facing the viewer. In the background of the second portrait, the reader of DT sees a collection of carpentry tools hanging on the wall. Why the need by the inventor to firmly establish that connection between himself and the second portrait's table items?

Then the answer became obvious to me. The secret of his wheels was actually contained *in* those two portraits! I, like many who had been mesmerized by the Bessler story Gould had presented in his 1928 book, *Oddities: A Book of Unexplained Facts,* had never seen these two engraved portraits before and when I finally did, my reaction was also to just give them a quick visual scan and then dismiss them as irrelevant decoration that Bessler put at the beginning of his book to decorate it, give readers a glimpse of the inventor's face, and, perhaps, amuse the reader with the gimmick of a fold over second portrait that used part of the first portrait's face in itself. I eventually realized how wrong that initial impression I had was. I discovered that the secret of Bessler's wheels was actually carefully hidden in the two portraits and was done so by using the truly universal language of mathematics. Bessler had published the text of DT in both German and Latin, but he also wanted to record the secret design of his wheels in it so that his discovery would not be completely lost in the event that he never found a buyer for it and then eventually died which, quite unfortunately, was exactly what happened. Only a truly universal language, mathematics, could make that possible.

The items shown in both portraits were also rather universal ones that would be recognizable in all cultures for many centuries to come. Today, three centuries after those portraits were made, we still have books, carpenter's hand tools, globes, drafting tools, rulers, plumb bobs (although, because of the toxicity of lead, they are now mostly made from brass), microscopes, telescopes, and thermometers. So, even today, it is still possible to decode the hidden wheel specifications in the two portraits and by so doing duplicate one of Bessler's wheels!

As I endured the long recovery from my near fatal illness I was able to spend a half hour or so per day perusing the two portraits and making notes on what I thought the various symbols in them meant and how they were related to the letters in the German words in the text under the portraits. Since I will be delving much deeper into the DT portrait clues in later chapters, I will not discuss them here.

By 2014 my number of computer models had soared to about 1,000 and still I could not find a design that would keep the center of gravity of its eight weighted levers on the axle's descending side throughout a 45º increment of drum rotation and which would have all of the starting orientations of the weighted levers repeated by the end of that short amount of rotation as the eight weighted levers' pivot pin axes again arrived at the drum's eight clock dial locations (in order to have a *perfect* match of the levers' orientation, however, one would have to allow the rotation to proceed at an infinitesimally slow rate in order to eliminate the effects of the centrifugal forces acting on the moving weighted levers). I continued to study the two DT portraits, but found that my several interpretations of a possible clue almost always proved to be erroneous and did not result in my computer model wheels performing better. However, perhaps one in a dozen possible interpretations would prove useful and I would then be overjoyed to see the performance improve even if only by a small amount.

I remember that as I approached model #1200 I was supremely confident that I was almost there and certainly would have the full details of the inventor's secret imbalanced pm wheel mechanics before I reached that model. I was so confident that I even proclaimed that, should I not have it solved by model #1200, I would then "retire" permanently from Bessler wheel research. Of course, I really believed that would never happen because I was so sure that I was finally within sight of victory!

Well, model #1200 came and went and still I found no success. It was a little embarrassing and taught me not to make optimistic projections or, as the farmers might say, not to count my chickens before they hatched!

In early 2015 I managed to publish my fifth book which, this time, did not deal with extraterrestrial UFOs in particular, but, rather, was an attempt to predict what established science might look like by the end of this our 21st century. In the introduction to that volume I lamented the fact that the book could not have been about Johann Bessler and his wheels because I knew that, for numerological reasons, the number 5 was one of his favorite numbers and I thought it would be a nice tribute to him to have my 5th book be the one in which his secret imbalanced pm wheel mechanics was finally revealed in enough detail to allow today's craftsmen to duplicate it. However, I had long ago decided that, unless I could actually provide my readers with a plausible and tested explanation for how Bessler's wheels worked, I would not just produce another book on the history of the inventor because that had already been done well enough by other authors.

Then, in early 2016 while studying the two DT portraits, several clues suddenly emerged from them to which I had previously been blind. It was amazing. In a matter of a few days I finally had all of the correct lever attachment points for the four types of coordinating cords that interconnected the weighted levers in the Gera prototype, Merseburg, and Kassel wheels!

I was rapidly approaching my 1500th computer model and immediately began using these recently interpreted clues. But, still, all I was getting for my efforts was one keel after another. This time, I vowed that if I did not find success by model #1500 that I would, indeed, finally quit the pursuit of the lost secret permanently. I had made nothing short of a heroic effort over the course of half of a century to solve the Bessler wheel mystery and the idea of continuing to endure the unrelenting frustration until I finally dropped dead without ever finding success was not too appealing to me. I had many other subjects I was interested in researching and, perhaps, authoring books on. My continuing fruitless search for Bessler's secret imbalanced pm wheel mechanics would only prevent me from moving on to those subjects with which, most likely, I would have a far better chance of finding success.

Around model #1480 I had a test that showed a model wheel undergoing unusually rapid initial acceleration. That was in the early spring of 2017. It was probably the most robust start that I had ever seen in one of my computer simulations. I was so impressed that I only allowed the model wheel to accelerate for about 10º or so of clockwise rotation. Surely, I finally had it and was beginning to announce to friends that I, indeed, had solved the Bessler wheel mystery! However, I did not have a complete 45º of clockwise drum rotation to fully confirm this. Despite that, I actually began writing this book based on that design and had it completed and illustrated in only about three months of intense effort. My intention was to have a book finished and published by the end of the summer of 2017 to commemorate the 300th anniversary of the completion of the construction of Bessler's 12 feet diameter, bidirectional Kassel wheel back in 1717.

By early summer of 2017 with the volume completed, I realized that it was time for me to stop procrastinating and finally complete the full 45º of clockwise drum rotation of my computer model wheel so I could send the volume in to my publisher. I still remember that night vividly.

As I ran the test, the simulation program slowly created each frame of the simulation and displayed it. The wheel had made it to 15º and was still accelerating smoothly and strongly. My heart began to pound with excitement. However, at about 20º something began to happen that made me wonder if I was seeing things. On the little added graph I used that displayed the rotational speed of the model wheel in rotations per minute versus time in seconds, the curve, rather than shooting almost vertically upward as it initially did, suddenly began to level off. At about 25º is was actually traveling horizontally which indicated that the torque on the model wheel's axle had dropped to zero and that the center of gravity of its eight little weighted levers was now located directly vertically below the center axis of the axle at the dreaded *punctum quietus* location. By the time the model wheel had rotated clockwise through 30º, the speed was actually decreasing which meant that the center of gravity of its eight weighted levers was now located on the *ascending* side and the resulting counterclockwise torque this created was actually braking the clockwise rotational momentum of the wheel. I let the simulation run long enough to see the typical rippling shape of the curve on the speed versus time graph that indicated I had yet another *failed* design that was just keeling!

So, there I was with a large heavily illustrated volume that had taken months to write and illustrate about a wheel design that was only yet another failure! I think I spent the next two weeks after that in a state of deep despair. I had promised many that I would deliver "the" secret wheel mechanics that Bessler had found and used and it was obvious that would not be happening despite my initial enthusiasm engendered by a partially tested computer model. If only I had taken the time of a few extra minutes to fully test the design, I could have spared myself months of unnecessary work, but when one is in a state of great excitement, his judgment can be impaired in such matters. I decided that the best thing to do would be to just archive a backup copy of the book and its illustrations on a CD and forget the matter. There would be no book and the secret of Bessler's wheels would remain lost, possibly forever.

About a month after that, however, I found myself growing more annoyed with the situation. I had probably spent thousands of hours during my lifetime researching Bessler and his wheels and had nothing of value to show for it. I felt like I had been personally cheated by Bessler! Yet, he had warned me and other hopefuls when, in AP, he wrote "…you'll soon find, you splendid mechanics, that this is a nut you can't crack!"

Well, not being a believer in uncrackable nuts, I decided that I would give it one final and supreme effort. Surely if I applied enough pressure to the Bessler wheel mystery, that "nut" would finally split open and reveal his wheel's internal mechanics.

Between the summer of 2017 and early April of 2018 I continued on with the most intense scrutiny of the two DT portrait clues that I had ever done. In the process I constructed an additional 500 computer model wheels using interpretations of the many new clues that were emerging from the portraits. I reached the point where my model wheels were no longer keeling and would accelerate throughout a full 45° of clockwise rotation, but the accelerations were not constant which would have been indicated by a speed versus time curve that was nearly a straight line because the center of gravity of the little 3 feet diameter computer model wheel's eight weighted levers was staying almost fixed at a particular location in space on the drum's descending side (*slight* movement of the center of gravity was, however, acceptable and entirely due to the increasing centrifugal forces acting on the weighted levers as their rotation rate around the axle's center axis increased). The curves I was obtaining were too curvy and that indicated centers of gravity that were excessively "wandering" about during the test increment of drum rotation. Often they would wander right over to the ascending side of the axle before whipping back again to the descending side. No, that would not do. I needed to find a design in which the center of gravity of the eight weighted levers remained almost fixed at a certain *horizontal* distance from the axle's center axis *throughout* the entire test increment of rotation and the larger that distance, the better since it would result in a greater start up torque for the wheel.

By the second week of April, 2018 all of my efforts had distilled down to two competing designs. Both had speed versus time graphs that were nearly perfectly straight lines and each had its eight weighted levers' center of gravity almost fixed at a certain horizontal distance from the center axis of the wheel's axle throughout the entire 45° test increment of rotation.

The first design which I had found, however, placed its eight weighted levers' center of gravity at a horizontal distance of about 0.035 inch from the axle's center axis, had a weighted lever's parallel pair of main arms in physical contact with its radial stop piece at the drum's 3:00 position, and had the angle of the parallel pair of the main arms of its weighted lever at the drum's 9:00 position at 53.5° away from an imaginary horizontal line passing through the lever's pivot pin's center axis. The second design, found a few days after the first, placed its eight weighted levers' center of gravity at a horizontal distance of about 0.0625 inch from the axle's center axis, had a weighted lever's parallel pair of main arms balanced a fraction of an inch above its radial stop piece at the drum's 3:00 position, and had the angle of the parallel pair of the main arms of a weighted lever at the drum's 9:00 position at exactly 62.5° away from an imaginary horizontal line passing through the lever's pivot pin's center axis.

Both designs worked and would display the pm effect if allowed to undergo a full 360° drum rotation and used angles of their main arms at the drum's 9:00 position whose digits summed to 13 which was a very important number for Bessler to incorporate into the design of his wheels (note that 5 + 3 + 5 = 13 and 6 + 2 + 5 = 13) for numerological and religious reasons of his own. I realized, however, that the *second* design was the one that Bessler would actually have used. It had double the start up torque and power of the first design and, for the 12 feet diameter Merseburg and Kassel wheels, would have put the center of gravity of their active internal one-directional wheel's eight weighted levers at a horizontal distance of about 0.25 inch from the center axis of their thick wooden axles upon start up. 0.25 inch is certainly a small distance, but we must remember that the Merseburg wheel placed a center of gravity with a total mass of 116 pounds at that distance and the Kassel wheel a total mass of 232 pounds at that distance! The startup torques created by this center of gravity horizontal displacement distance can then be used to rationalize the various test results gotten for these wheels.

Since I finally realized that it was the second design that was the one Bessler would have used on Friday, April 13th, 2018, I am going to take that as the official historical date of the rediscovery of the long lost secret of Bessler's self-moving wheel mechanics. All that this "nut" required for its cracking was about ten thousand or more hours of effort spread out over a half century and about 2,000 computer models and simulations of them running. Now that it is finally over with, I can move on to researching solutions to other mysterious subjects that, hopefully, will not present nearly as much of a challenge. From studying the life and works of Bessler, I think that the date of rediscovery of the secret of his wheel mechanics falling on a Friday the 13th would have pleased him greatly. Contrary to popular belief, the number 13 is actually a very lucky number in both numerology and theology. I consider its appearance here as a propitious omen of things to come!

It is my hope that the previously undiscovered information that will finally be revealed in this volume will eventually allow some skilled craftsman with much determination to one day duplicate either Bessler's bidirectional Merseburg or Kassel wheel. That will finally verify that my analysis of his design is correct and that Bessler was not a liar trying to defraud people with a fake wheel. Also, should someone actually manage to produce a working replica of one of Bessler's larger 12 feet diameter bidirectional wheels, it would be a very valuable collector's item and might, as a one of a kind item, be auctioned off for a considerable amount of money which would help repay the craftsman for the effort, money, and time he invested in the project. Indeed, he might even derive a tidy income by manufacturing these wheels for collectors around the world.

For the average builder, however, I have presented in this volume detailed plans for the duplication of Bessler's little 3 feet diameter, one-directional Gera prototype wheel with which he first found success in Gera, Saxony in late 1711. That wheel can probably be constructed within a few weeks of part time effort and at minimal expense. In fact, I don't recommend that anybody even attempt to construct a larger wheel until and unless he has first successfully constructed the 3 feet diameter Gera prototype wheel. That wheel should be considered as a basic training course for the future construction of a larger wheel. If one cannot get the little toy prototype wheel constructed and running reliably, then he will probably have considerable difficulty doing so with a larger and more expensive and difficult to construct version.

CHAPTER 3

Is Perpetual Motion Really Possible?

Before answering that question, I want to briefly review the history of the main component of an imbalanced perpetual motion wheel which is the wheel itself.

It was several tens of millennia ago that humans first discovered how to make fire and, in time, the many benefits that could derive from it such as warming caves, cooking meats to make them easier to consume, drying out materials like animal skins for clothing or mud to make pottery and bricks for small outdoor constructions, extracting metals from ores, and then softening them so they could be hammered into useful items such as tools and weapons. Unfortunately, the first tools made in this way used soft metals like copper and tin and would quickly dull in use.

Eventually, with the ability to generate higher temperatures and make stronger tools from hard metals like iron, it wasn't too long before humans were cutting large blocks of stone from natural geological deposits exposed in open, excavated pits known as quarries and using them to construct things like fortresses, pyramids, temples, and tombs. The major problem with these constructions was getting the heavy blocks of stone from their quarries to the construction sites.

At first, the stones were just dragged to the site along a track in the dirt. That required the expenditure of a lot of energy and if it rained and the dirt turned to mud, this method became useless.

Some time after methods for making fire were discovered, the wheel was also discovered. Primitive people noticed how things like boulders and tree trunks could easily roll downhill and, as these people attempted to construct large megalithic structures using the stone blocks that their stronger iron tools permitted them to cut from quarries, instead of just dragging them to construction sites, they started to put cut sections of tree trunks under the blocks of stone and used the resulting reduction in friction to, literally, push the stones along on a set of wooden rollers. However, this method required that the logs left behind the moving stone block needed to be physically picked up and then quickly carried around and placed on the ground again in front of the block in order for it to continue moving. While this approach made it easier to move the stones, it was still very energy and time consuming.

Then, someone had a bright idea. Why not construct a strong, but light wooden platform upon which the stones would be placed and then *permanently* attach several circular slices cut from a tree trunk to the platform to take the place of the logs that previously had to be continuously moved from the back to the front of a moving block? The slice of the tree trunks would, like the logs previously used, still roll along and carry the stone block, but it would no longer be necessary to constantly move them around the stone block because they would be permanently fixed to the platform and move along with it.

To do this, primitive people first just cut thick slices of wood from large diameter tree trunks, bored a hole in their centers, and then mounted these first manmade wheels on a wooden axle which was securely attached to the wooden platform. As long as the weight of the stone block being carried on top of the platform was not too great, this approached worked well and people found that they much preferred having the larger and stronger animals they herded for their meat and milk pull these stone carrying carts along rather than having it done by the humans themselves.

Purely wooden disc type wheels and axles have their problems, however.

As a load they carried increased in weight, such wheels could crack along their grain lines and it was necessary to lubricate with lard or vegetable oils the portion of the wheel's central hole which made contact with the end of the fixed wooden axle that carried a portion of the total weight of the platform (minus the weight of its other wheels) and its stone block load. Specialists arose whose job it was to construct more durable wheels and axles and these would eventually take different forms. The outer rims of solid wood wheels were reinforced with bands of iron and the wheels were made from separate sections of wood that were held together with metal plates and screws. The central hole was replaced with a metal "hub" that rotated around a fixed metal axle. Eventually, successful efforts were made to decrease the mass of a wheel by just having relatively narrow radial members or "spokes" support an outer iron clad wooden rim. The inward pointing ends of the spokes were then securely attached to an inner metal hub that rotated around the lubricated end of a fixed metal axle.

As wheels became lighter, stronger, and more reliable, people found them safe enough to use on small animal drawn carriages that they could use to make short trips between villages and towns. Eventually, in the 19th century, the animals previously needed to pull a human carrying carriage were replaced by an interesting device called the internal combustion engine. It used the force of the explosions produced by the sudden detonation of a mixture of fuel and air against movable pistons to turn a metal axle that was, through a mechanism known as a differential gear, connected to the hubs of a pair of wheels at the back of a carriage. The automobile was born and along with its larger cargo carrying version, the truck, is now a critically important part of modern human society.

Several millennia ago, however, as people began to construct wheels of various types they noticed that, in order to give a smooth ride, it was important that the wheel they used be as perfectly circular as possible, have its axle's center axis located as close to the exact center of the circular wheel as possible, and have its materials distributed around the axle as uniformly as possible. If not, the wheel would wobble in its plane when put into use and this undesirable motion would worsen as its speed of rotation increased. Many wheel builders began mounting each newly build single wheel on a well lubricated axle and checking to make sure that its outer rim remained at a constant distance from the center of its axle as the wheel was slowly rotated. Furthermore, they noticed that, if the materials of the wheel were not equally distributed around the axle, the entire wheel would begin to slowly rotate when released. The wheel builder would then have to compensate for this imbalance by securely attaching a small metal counter weight to another spot on the wheel until it balanced perfectly on the axle and would not begin rotating when released regardless of the particular orientation from which it started.

As ancient wheel builders studied their wheels, some of them wondered if, by cleverly attaching moving metal weights to their rims, it might be possible to create a situation in which the wheel would be imbalanced and would *remain* so as it rotated. It seemed intuitively obvious to them that, if this could be done, then the wheel should just continue to rotate forever if its hub's attachment to its axle was kept well lubricated in order to prevent excessive wear of the surfaces where the wheel touched the axle. Such wear would eventually cause the wheel and axle to separate and, obviously, the wheel would then fall to the floor or ground and stop turning.

Some with spare time on their hands would then go ahead and try to construct such a wheel. To their great surprise, however, they quickly discovered that, while achieving a chronically imbalanced wheel certainly sounded easy to do in principle, in practice it was proving impossible to achieve! No matter what design or variation of a design tried, nothing worked. They found that it really did not matter what they placed inside of their wheels to achieve the imbalance whether it was rolling metal balls inside of channels, liquid mercury flowing to the ends of glass vials, or metal weights fixed to the ends of lightweight wooden levers that would flop around as the wheel rotated. In every case the initial offset center of gravity of the wheel's various weights would just rotate down to a point below the center axis of the axle known as the "*punctum quietus*" (Latin for "point of rest") and, after a few oscillations through that point from one side of it to the other, the center of gravity of the wheel's weights would finally come to a stop there. It was all very disappointing and frustrating, especially for those who had spent large percentages of their lives trying to find a wheel design that would remain chronically imbalanced despite the rotation of the wheel and demonstrate to all that perpetual motion was possible.

By the 19th century, there had been so many failed attempts to produce an imbalanced perpetual motion wheel that serious scientists started to proclaim that such a wheel was a physical impossibility because it violated a fundamental law of nature. To operate, such a wheel would, they said, literally have to be *creating* energy out of nothing which no experiment ever performed up to then had shown could happen. Many of these scientists had heard of previous inventors who claimed to have invented such a wheel, one of them being a German / Polish carpenter and medicine man named

Johann Ernst Elias Bessler in the early 18th century, but they dismissed all of their claims as being either erroneous or fraudulent.

Surely, they reasoned, if such a device was possible, they would then have been using it instead of wondering if it such a wheel was possible!

This belief, quite unfortunately, still exists in the world of science to this very day. The vast majority of scientists and even average people have been so thoroughly "educated" with the belief that a working imbalance perpetual motion wheel is impossible that they almost immediately react to the story of Bessler and his inventions by just dismissing the whole affair as a fraud intended to fleece the gullible rich out of their money. Some believe that Bessler hoaxed his wheels' performances in some clever way so that he could obtain the funds he needed to continue his research with the hope of eventually replacing his fake pm wheels with genuine ones that actually worked.

I think, however, aside from not being more familiar with the details of the Bessler story as I described them in the first chapter of this book, many who have accepted that an imbalanced perpetual motion wheel is not possible are actually confused as to what is meant by the term "perpetual motion". Let me try to clarify this matter now.

Before proceeding to answer the question posed by this chapter's title, "Is Perpetual Motion Really Possible?", I must make the following somewhat surprising statement which is that the answer to this question can be *either* "yes" or "no" *depending* upon one's definition of "perpetual motion"!

For example, if one's definition of perpetual motion is simply the motion of an object which, while it moves along forever, performs *no* external work such as overcoming some form of mechanical drag along its path, then the answer is definitely "yes". In fact, this type of perpetual motion is actually *guaranteed* by the first of Newton's three famous laws of motion which says that the state of motion of an object, whether stationary or moving at a constant velocity (note that one can think of a stationary object as one moving with zero velocity), will persist *forever* unless the object is acted upon by an external force which then changes its state of motion by changing its velocity. Thus, if an object in outer space was moving along at a constant velocity and space was a perfect vacuum (which it is not), then the object would continue to do so for eternity unless it encountered some external force that changed its state of motion. It might, for example, slow to a near stop if it impacted against the surface of an asteroid. Or, if it encountered a larger body such as a moon toward which it directly moved, its velocity would be constantly increased by the moon's gravity field until it finally hit the moon's surface.

However, such perpetual motion is not limited to large objects moving through outer space. Indeed, every single atom within such an object contains a positively electrically charged nucleus around which many believe one or more negatively electrically charged electrons orbit at very high velocity (currently, atoms containing up to 118 electrons each have been discovered!). The electrons within such an atom of any element then tend to naturally "drop down" into "allowed" orbits that are as close their atom's nucleus as possible and which give them the lowest possible total energies that they can have inside of that atom (which orbits are "allowed" for the positively charged nuclei of all of the chemical elements has been determined by the science of quantum mechanics during the 20th century). Once all of the electrons inside of a particular atom are in their lowest possible energy "allowed" orbits about that atom's nucleus that they can occupy, the atom is said to be in its "ground state" and that tends to be a very stable state. The electrons in such ground state atoms will continue to orbit about their nuclei in their particular "allowed" orbits for all of eternity.

If fact, ground state atoms' electrons will continue to constantly orbit their nuclei forever even if the temperature of the object that contains them is lowered to absolute zero which is a temperature of −459.67° F. The only thing that can change this situation is if an object falls into a black hole in outer space and all of its atoms are then exposed to the black hole's extremely intense gravity field. As the object approaches the black hole's "event horizon" it would feel powerful gravitational forces acting on its atoms' nuclei and electrons that would tend to stretch the object out along its direction of motion while simultaneously compressing its cross sectional areas perpendicular to its direction of motion. The distorted object's various atoms' electrons would eventually actually be crushed down against and then finally fused with the protons contained in their respective nuclei to form neutrons and any electrostatic potential and kinetic energies those electrons previously had in their allowed ground state atomic orbits would be emitted as intense gamma radiation photons of various frequencies.

We see from all of this that perpetual motion of the first type in which neither an object nor its constituent atoms perform any external work in their environment as they move along, potentially forever, is something that is quite common in our cosmos and can exist from its megascopic and macroscopic levels right down to its submicroscopic levels.

This now brings us to the second type of perpetual motion being discussed in this chapter.

In this type of perpetual motion, an object continues to move along forever while *also* continuously performing work in its environment by overcoming any external drag forces that the environment provides. This situation can only exist if the object is able to continuously output energy *forever*. This kind of perpetual motion is most definitely *not* possible and the answer as to whether or not it can ever exist is "no".

The reason is simply that all objects of *finite* size and mass, regardless of their structural complexity, can only contain a finite amount of energy and, if the object continually loses portions of that energy to overcome external drag as it moves along, then eventually a point will be reached at which time the object will have no energy left to lose and will have to stop moving. In order to continue to move forever, the object would have to actually constantly create more energy out of nothing as it moved along in order to constantly replace the energy that was being depleted from it and that is absolutely impossible. I am also quite confident that no machine will ever be invented which will, somehow, be able to create energy out of nothing. However, while such a perpetual motion machine of the second type described here is not possible, it might be possible to construct a machine that would contain such a huge amount of energy compared to what it outputted per second that it would actually be able to run for years, decades, centuries, millennia, and even far, far longer. Such a machine would appear to exhibit the second type of perpetual motion described above, but that would only be an illusion which would not be dispelled unless the machine could be tested long enough to show that it eventually did stop running.

At this point the reader may be wondering exactly what type of perpetual motion wheels Bessler constructed.

Obviously, since Bessler's imbalanced perpetual motion wheels continuously rotated *and* were also able to continuously output mechanical energy, one is tempted to immediately assume that they were exhibiting the second type of perpetual motion described above which requires a wheel to be able to constantly create energy out of thin air in order to supply what it needed to overcome the various drags acting on its parts and also power any external machinery connected to its axle which represented yet another form of drag. However, that type of perpetual motion is absolutely impossible for any wheel to ever have and we are, therefore, forced to conclude that *all* of Bessler's wheels carried some unobvious and, most importantly, *finite* source of energy inside of their drums. Since that must have been the case, then they would only have been able to run for a *finite* amount of time and not for eternity. So, the inventor's wheels were not "perpetual" in the strictest sense of the word and, thus, would not have been able to run, nonstop for all of eternity if their various parts did not wear out.

But, wait a minute! Weren't Bessler's wheels tested at Weissenstein Castle in Kassel and shown to be able to run for almost two months? Didn't Bessler offer to let them run even longer to prove they were, indeed, truly perpetual in the absolute sense of the word? That certainly seems long enough to indicate they could run forever if their parts did not wear out or, if they did, were replaced as needed.

Yes, the testing of the 12 foot diameter, bidirectional Kassel wheel at Count Karl's castle that began on November 12th of 1717 and continued to January 4th of the next year was certainly impressive, but that test only lasted about 54 days and there is a huge difference between 54 days and eternity! However, although Bessler's wheels would, in reality, only have been able to rotate continuously for a finite amount of time because they had a finite amount of mass and therefore contained a finite amount of energy, that time interval could have been very, very long. Far, far longer than a single human's lifespan, in fact.

To begin to understand the physics of Bessler's wheels, one has to understand where they obtained the mechanical energy that they outputted.

When Leibniz first heard of the inventor's wheels, the former, being a scientist knowledgeable of the then most recent forms of storing energy, initially assumed that the wheels were powered by compressed air that would have been stored in tanks within a wheel's drum. Upon actually testing one of Bessler's wheels, however, he quickly realized that explanation was inadequate. He eventually just assumed that Bessler had discovered some previously unsuspected form of environmental energy and the wheels were able to harness that energy. To Leibniz, Bessler's wheels were like some new type of water or wind mill whose internal mechanics were able to interact with some other naturally occurring, momentum carrying substance which our normal human senses were incapable of detecting and which, most mysteriously, was still able to easily penetrate a drum's outer wooden rim wall or its wood slat or cloth clad sides.

To Christian Wagner, a mathematics professor and one of Bessler's most intellectual critics and detractors, the inventor's wheels were simply powered by mainspring driven clockwork mechanisms which had to be periodically wound up. As such, the total energy outputted by a wheel could be no more than was stored in its collection of mainsprings at each winding. In fact, Wagner was so obsessed with exposing Bessler's fakery that the former went to the trouble of constructing his own mainspring powered, bidirectional wheel and was able to use it to demonstrate weight lifting braking tests similar to those done with Bessler's wheels. He was quite confident that the inventor's 12 foot diameter wheels would not be able to run continuously for more than a few days even if their entire drums were filled with tightly wound mainsprings. This prediction was, of course, refuted after the Weissenstein Castle test.

To Johann Borlach, another of Bessler's more powerful enemies, the source of the energy Bessler's wheels outputted was simply provided by human muscles from an adjoining room and delivered to the wheels by means of a cleverly concealed crank and drive rod mechanism hidden inside of their wooden vertical support planks.

All of these hypotheses offered to explain away the performance capabilities of Bessler's wheels were inadequate and incorrect and a truly plausible one would not be available until the beginning of the 20th century even though it is still not even being considered by the scientific orthodoxy of our day, most of whom, quite sadly, are completely unfamiliar with the full details of the Bessler story.

At the beginning of the 20th century a remarkable discovery was made and promoted by scientists such as Albert Einstein. This discovery was that energy and mass, previously thought to be two different and separate entities, were actually the *same* thing!

This meant that wherever one had mass, he also had energy and wherever he had energy, he also had mass. It also meant that when one added energy to an object, he also increased its mass and that when he removed energy from an object, he also reduced its mass. Scientists even began to think of the mass possessing subatomic particles from which atoms were made as just "condensed" forms of energy. Previous to this important realization, mass was just pictured as a measure of the quantity of matter present in an object which determined things like its gravitational and inertial properties and energy was only considered as a measure of the amount of work that could be either absorbed by or released from matter as it changed from one energy state to another.

As simple examples of this new concept that is the most important one that helped classical physics evolve into modern physics, consider the following scenarios.

If one heats the water in a container, the temperature and the thermal energy content of the water will, of course, increase. What is not obvious to a casual observer is that the mass of all of the mass possessing subatomic particles in all of the water's atoms (that is the masses of all of the electrons, neutrons, and protons contained in the atoms of the water's molecules) also increases by a very tiny amount so that the water in the container weighs a slight amount more and also has slightly more inertia.

If one takes a steel helical coil spring, compresses it lengthwise by putting external energy into it, and can somehow lock it into its new shortened length, then the resulting increased proximity of the *same* polarity electrically charged subatomic particles in its atoms will raise their electrostatic potential energies with respect to each other as the repulsive forces acting between them also increase and try even harder to push them apart and return the spring to its original starting length. With that increase in electrostatic potential energies, there will also be an associated increase in the masses of all of the electrically charged subatomic particles involved and the spring will be a bit more massive and, as with the case of the water mentioned above, it will weigh slightly more and have slightly more inertia. (Note that in this example, I used a spring that had external energy added to it to compress it, but the exact same process happens if the added external energy is used to stretch the spring to a length greater than its starting length. This is because, regardless of whether a spring is compressed or stretched, there will always be subatomic particles of the same electrically charged polarity being forced to move closer to each other and their repulsions from each other will account for the restorative tension that increases in the spring.)

And, as a final example, if a weight is lowered at a constant velocity from one elevation above the ground of a planet's surface to another elevation that is lower in the planet's gravity field, then the weight will lose gravitational potential energy and there will be an associated loss of mass of all of the mass possessing subatomic particles in the atoms of the weight so that the weight will weigh slightly less and have slightly less inertia at the lower elevation than it would have at its original higher starting elevation.

One might wonder why, if these changes in the masses of objects with changes in their energy content were actually taking place, nobody ever noticed them before the beginning of the 20th century.

The reason is that the equivalence between energy and mass is a very lopsided one. The equation which describes this equivalence is $E = mc^2$ where, in the centimeter-gram-seconds system of units used in physics, the energy content of an object, E, in ergs, equals the mass of the object, m, in grams, times the velocity of light in a vacuum, c, in units of centimeters per second, which must then be squared.

The amount of energy typically applied to small objects to move them around on an everyday level of experience is in the range of a joule or so. One joule is the amount of energy that must be added to a stationary object with a mass of one kilogram to give it a final velocity of one meter per second. One joule is also equivalent to ten million ergs or, in exponential notation, 10^7 ergs.

Let us now determine the amount of mass associated with the amount of energy that would have to be added to a stationary object with a mass of one kilogram in order to accelerate it to a final velocity of exactly one meter per second. To do this we solve the energy mass equivalence equation above for m and get $m = E/c^2$. Inserting the values we have into this equation then gives $m = (10^7 \text{ ergs}) / (2.9979 \times 10^{10} \text{ centimeters per second})^2$. Solving this then gives $m = 1.1127 \times 10^{-14}$ grams which is equal to about 0.011127 *pico*grams or about a hundredth of a picogram!

This is, indeed, an extremely small increase in the mass of our one kilogram object as we supply it with the energy it needs to accelerate from a standstill to a velocity of one meter per second. It is so small a change in mass, in fact, that one would actually need special equipment to even detect and accurately measure it which was not available to scientists in the 19th century and is only recently beginning to become available in our 21st century. From simple calculations such as this, one quickly realizes that in order to be able to casually notice the changes in mass, weight, and inertia taking place as energy is either added to or subtracted from an object requires that the amounts of energy transferred be enormous and such enormous energy transfers simply do not take place at the usual level of human experience.

However, significant changes in mass can take place during nuclear reactions. In the case of the Hiroshima atomic bomb used in 1945, about 1 gram of mass was lost from the nuclei of the radioactive products created by the complete "chain reaction" fission of about 0.91 kilograms or 2 pounds of a rare isotope of uranium known as U^{235}. The energy released by that *nuclear* reaction was equivalent to the amount of energy that would be released by the explosive *chemical* reaction of 16 kilotons or 16,000 US *tons* of TNT! Obviously, it is not possible to confirm such large mass changes by simply collecting all of the radioactive fission products produced during a nuclear reaction detonation and then actually weighing them. Presently, this 0.11% loss of mass of such a weapon's original, prefission mass of U^{235} atoms' nuclei can only be determined by theoretical calculations.

In the case of our Sun, its luminosity requires that the nuclear reaction fusing of hydrogen and other nuclei in its core lose mass at the rate of about 2×10^{12} grams per second or about 2.2 million US tons per second! That is an enormous amount of mass eventually being constantly radiated away in the form of photons from its surface across all of the frequencies of the electromagnetic spectrum, but so small compared to the total mass of our Sun that the reduction in its total mass and the consequent reduction in the gravitation pull it exerts on our solar system's planets is virtually negligible over the course of billions years (the actual loss of mass per second is probably several times this when one includes the various subatomic particles constantly be ejected from the Sun's surface). Again, this loss of solar mass must be determined by theoretical calculations since we don't have a means of accurately measuring the reduction of the Sun's gravitational pull on the Earth over the course of only a few decades of time.

Now let us consider Bessler's wheels in light of these revelations about the equivalence of mass and energy.

In the case of the bidirectional Kassel wheel that Bessler constructed at Weissenstein Castle in 1717, my research indicates that its enlarged drum contained two internal, side by side, back to back, one-directional wheels each of which contained eight weighted levers with a mass of 29 pounds each. That means that *each* of the Kassel wheel's two internal one-directional wheels contained a total of 232 pounds of mass in its eight "perpetual motion structures" or weighted levers. If modern physics is correct about the equivalence of energy and mass which seems to be the case, then all of the mechanical energy Bessler's wheels outputted could only have come from one source. It came directly from the masses of their weighted levers themselves; that is, from the masses of the cylindrical lead end weights *and* the masses of the various components of the movable levers to which the end weights were attached. To state the matter more precisely, as his wheels turned and continued to output the mechanical energy and its associated mass that accelerated a wheel or operated some piece of "outside" machinery attached its axle, that expenditure had to be paid for by a continuing reduction in the energies and associated masses of the various mass possessing subatomic particles that composed the various atoms in the materials from which the active weighted levers of a wheel were made.

As noted previously, any change in the energy content of any object ultimately causes a change in the masses of its atoms' component subatomic particles which are just the electrons, neutrons, and protons of those atoms. This change of mass does not require that any sort of nuclear reactions take place in the atoms involved. Indeed, such mass changes are a natural and automatic process that the subatomic particles of all atoms are engaged in on a regular basis as their interactions with the environments outside of themselves cause them to either gain or loose small amounts of energy. An additional detail that should be mentioned here is that when energy is either added to or subtracted from all of the subatomic particles of an object's atoms at once, the associated mass that is also either added to or subtracted from the initial masses of all of those subatomic particles must be *equally* distributed among them so that each subatomic particle experiences the *same* percentage change in the initial mass it had before the energy transfer took place. It follows from this that the greater the initial mass of a subatomic particle in an atom of an object experiencing a change in the energy content of all of its atoms at once, the greater will be the change in the initial mass of the subatomic particle.

In the first chapter I stated that I had estimated that the Kassel wheel was expending energy at the rate of about 25.44 watts in order to run the Archimedean water screw attached to its axle when it was turning at 20 Rotations per minute and that this large 12 foot diameter, bidirectional wheel's maximum power output, at start up, would have been about 50 watts. Let us now see if we can compute how long the energy content associated with the masses of its weighted levers could power the wheel at that particular maximum rate of energy expenditure of 50 watts. Since this maximum rate was achieved immediately after the drum was given a push to start it turning, the rotation rate of the wheel *at that time* would be very low and far less than 20 revolutions per minute. In these calculations, it is assumed that the wheel is made from ideal materials which do not wear out from friction over time or in any other way deteriorate and lose their original structural strengths. This would then eliminate the need for any maintenance during a run time of any duration.

To begin, we first need to determine the energy content, in watt-seconds, of a mass of 232 pounds which was the total amount of mass of the eight 29 pound weighted levers contained in *one* of the bidirectional Kassel wheel's two internal one-directional wheels.

1 erg of energy is equal to 1×10^{-7} watt-seconds of energy and we calculated above that 1.1127×10^{-14} grams of mass was equivalent to 10^7 ergs of energy. After doing some additional simple mathematical calculations with the second equivalence, one can determine that one gram of mass is equivalent to 8.9871×10^{20} ergs of energy. Since 1 pound of mass is equal to 453.592 grams of mass, we can then calculate that 1 pound of mass is equivalent to 4.0765×10^{23} ergs of energy. We now multiply this by the number of watt-seconds of energy per erg of energy or 1×10^{-7} watt-seconds per erg and find that 1 pound of mass is equivalent to 4.0765×10^{16} watt-seconds of energy.

Since each of the two internal one-directional wheels inside of the 12 feet diameter, bidirectional Kassel wheel contained eight weighted levers with a total mass of 232 pounds, we can multiple this mass times the amount of watt-seconds of energy associated with a mass of 1 pound and determine that the eight weighted levers in that single internal one-directional wheel represented 9.4575×10^{18} watt-seconds of energy which, obviously, is an enormous amount of energy.

Now we can determine, *theoretically*, how long the Kassel wheel would rotate in one direction if it could continuously output energy at a *constant* rate of 50 watts.

We simply divide the total energy contained in one of its internal one-directional wheel's 8 weighted levers, 9.4575×10^{18} watt-seconds, by the constant rate of that one-directional wheel's power output, 50 watts, and we then get 1.8915×10^{17} seconds of time. Since there are 3.15576×10^7 seconds of time in one year of time, this means the Kassel wheel could turn in one direction and output energy at a constant rate of 50 watts for (1.8915×10^{17} seconds) / (3.15576×10^7 seconds per year) = 5.9938×10^9 years which is about 6 *billion* years!

The maximum time that the Kassel wheel could run, however, *in practice* would actually be longer than this. As the wheel continued to run over hundreds of millions of years, the total mass of its eight active weighted levers would slowly decrease from 232 pounds down to 0 pounds and, as this happened, its power output would slowly drop from 50 watts down to 0 watts. If the low start up rotation rate of the wheel remained constant, then, as the mass of its eight active weighted levers slowly decreased down to 0 pounds, so, too, would the torque of its axle and the power output of that axle. This means that the Kassel wheel's *average* power output during its run time would only be 25 watts. If, in practice, the wheel's average power output was only 25 watts or half of its start up power output of 50 watts, then that means that it would take *twice* as long for the wheel's imbalanced pm mechanics to output all of the energy content of its eight active weighted levers or about 12 billion years!

But, wait! That is the maximum run time, in practice, of the Kassel wheel in only *one* of its two possible directions of rotation that then outputs the total energy content of the mass of the eight active weighted levers of only *one* of its two internal one-directional wheels. If we could somehow use a time machine to travel ahead to the day, about 12 billion years hence, when the wheel finally stopped turning in the initial direction it was started in, we could exit our time machine and then give the wheel a gentle push in the opposite direction and it would then proceed to run for another almost 12 billion years in that direction. From this we see that, potentially, the Kassel wheel could, if its components were not subject to wear or deterioration, output mechanical energy at an *average* rate of 25 watts for almost 24 billion years!

24 billion years is a very long time period and, in fact, is greater than the presently theorized age of our visible universe which is about 13.7 billion years or the time that it has existed since the original Big Bang detonation created it. Most likely, 24 billion years from the present, our Sun will have already completed its catastrophic "red giant" phase almost 20 billions of years earlier and thereby destroyed the inner planets of our solar system including the Earth along with the Kassel wheel we were testing to determine its maximum run time starting with an initial energy output of 50 watts. But, even if this did not happen and the Earth and the wheel survived and the wheel continued to run, the fact is that the Kassel wheel would still have a *finite* run time and finally stop after about 24 billion years had passed. It would not run for eternity and, thus, would not be a perpetual motion machine in the strictest sense of the word. Despite this, however, I think that most people would not be too bothered by referring to Bessler's wheels as "imbalanced type perpetual motion wheels" even though they cannot actually meet the technical requirement for such a designation.

Just so there will be no confusion about what I have previously stated, let me emphasize that all of the atoms that composed the weights and levers inside of the Kassel wheel would still be present and have their usual chemical, electrical, magnetic, and optical properties after the wheel had been made to run in both of its directions for the 24 billion years. However, all of the formerly mass possessing subatomic particles such as the electrons, neutrons, and protons inside of those atoms would then be completely massless. Being now composed of massless atoms, the weights and levers would no longer have their normal gravitational and inertial properties and would also no longer have any energy that the wheel could output to its environment. If removed from the bidirectional wheel, the 48 cylindrical lead end weights and the 16 wooden levers and their various attached components to which the end weights were attached would actually float in the air and then slowly rise as though they were helium balloons! If one wanted, using his time machine, to continue to test the Kassel wheel for another 24 billion years, he would have to remove and replace all of its 16 massless weighted levers with "fresh" ones that had their normal masses and, most importantly, their normal energy contents.

In writing the above, I made a tacit assumption. I assumed that, as the Kassel wheel continued to run in each of its two possible directions of motion for 12 billion years, its weighted levers would just continue to lose mass as they outputted energy at an average rate of 25 watts until they all finally became totally massless. If that assumption is valid, then the scenario I presented would be valid. But, we must remember that when considering the operation of such a wheel, we are really into an entirely new realm of physics. It might be the case that, as wheel continually extracts and outputs the energy associated with the mass of its atoms' subatomic particles, the cosmos somehow metaphorically frowns upon this practice. There might be some lower limit below which the cosmos will not allow a subatomic particle's mass to *naturally* decrease.

Perhaps, after about 6 billion years had passed and the masses of the eight weighted levers inside one of the Kassel wheel's two internal one-direction wheels had dwindled to about 50% of their original values, some natural process would come into play that would prevent any further reduction in mass from taking place. At that point, the environment of the wheel might somehow actually start to replace, at the same rate, any future energy being outputted by the wheel. If this can be shown to be the case, then Bessler's wheels would, in fact, be displaying perpetual motion in the absolute and strictest sense of the word and one of them would be able to continue to run forever assuming that its parts were not subject to wear or deterioration and it had a stable planet available to provide the gravity field it required in order to operate. This is something for future physicists to ponder after we begin to successfully duplicate Bessler's wheels.

The final matter I want to treat in this chapter is that of exactly *how* Bessler's secret pm wheel mechanics managed to extract the energy associated with the masses of a wheel's eight active weighted levers that was needed to propel it whether the wheel was one of his one-directional, self starting types or was one of his bidirectional types requiring a push to start it turning so that only one of its two internal one-directional wheels would drive it at any time while the other opposed one-directional wheel inside of the enlarged drum was temporarily disabled. I will, however, here concentrate on the basic mechanics of the inventor's one-directional, self starting type wheels and leave the more complicated mechanics of his bidirectional type wheels to a later chapter where they will be discussed in considerable detail.

After working with what I have very good reason to believe is the actual imbalanced perpetual motion wheel mechanics that Bessler used, I found its method of operation to be fairly simple to understand and, like Count Karl, wondered why none prior to or since Bessler managed to find it despite many thousands of attempts over the centuries to do so. Again, I suspect that the major obstacle to rediscovery all along has been the failure by inventors and reverse engineers to achieve the precise balancing of a one-directional wheel's eight active weighted levers and the correct center axis to center axis cord lengths that coordinated the individual levers' motions which are critically necessary in order for their composite center of gravity to remain on the descending side of the axle's center axis as a drum rotation continues. Unless this precision, which involves many factors, is found, then it is just not possible for a Bessler type imbalanced pm wheel to operate.

To begin our analysis of how Bessler's imbalanced pm wheel mechanics worked, one can imagine that the center axes of the steel pivot pins of the eight weighted levers of a one-directional wheel are, at start up, arranged at locations inside of the drum which correspond to certain times on the face of a standard 12 hour round clock dial. Thus, a wheel's eight lever pivot pins' center axes can be imagined to be placed at regular angular intervals of 1.5 clock hours or 45° apart from each other all around this imaginary circular dial so that they are at the drum positions of 6:00, 7:30, 9:00, 10:30, 12:00, 1:30, 3:00, and 4:30. The eight pivot pins of the weighted levers were press fitted into holes drilled into the drum's radial frame pieces and, obviously, would rotate around the center axis of the wheel's axle as the drum rotated. (Interestingly, a lever's pivot pin's center axis was always located at a distance from the center axis of the wheel's axle that was exactly 0.77777 of the outermost radius of a drum *regardless* of the radius of the drum. Thus, for a 12 feet diameter wheel with a radius of 6 feet = 72 inches and having the center axis of a weighted lever's pivot pin at a distance of exactly 56 inches from the center axis of the wheel's axle, we see that 56 inches / 72 inches = 0.77777. For the small 3 feet = 36 inch diameter, one-directional Gera prototype wheel with a radius of 1.5 feet = 18 inches and having the center axis of a weighted lever's pivot pin at a distance of exactly 14 inches from the center axis of its small wooden axle, we see that 14 inches / 18 inches = 0.77777. It's highly likely that Bessler deliberately placed the center axes of each wheel's weighted levers' pivot pins at their particular locations on a drum's radial frame pieces so that this particular ratio value would exist with all of the lucky 7 numerals it contained. It would have been done in an effort to assure the successful outcome of his imbalanced pm wheel research and the eventual sale of the invention. It might also have been yet another tribute to God for having given Bessler the vivid dream which inspired him to make the modifications to his Gera prototype wheel that finally allowed it to work. The number 7 is associated with God since there is mention in the Bible of there being seven divisions of Heaven with the highest or 7th division being the abode of God and his angels.)

However, when referring to the eight important drum positions given above, we must imagine that the clock times are not drawn on the drum itself so that, when the drum rotates, the imaginary clock dial rotates with it. Rather, we have to imagine that the clock times have been drawn on a large circular sheet of transparent plastic that is placed near the circular face of the rotating drum, but is *not* attached to it and moving with it (this would, of course, require that a hole larger in diameter than the wheel's axle be cut out of the center of the circular sheet of plastic). This sheet of plastic is stationary with respect to the wheel's stationary wooden vertical support planks and the ceiling, floor, and walls of the room containing a wheel. In other words, when I refer to the drum position of a weighted lever's pivot pin's center axis and give a clock time for it, I will be referring to its position that is located on the 12 hour round clock dial drawn on this imaginary and, most importantly, *stationary* sheet of circular plastic. One can also just imagine these drum position clock times as floating fixed in space as the drum's eight weighted levers' pivot pins' center axes rotate past them on their trips around the center axis of the axle.

Each of the eight weighted levers inside of a one-directional wheel was attached by a single "main cord" to the weighted that led it by 45° so that all eight weighted levers were connected to each other. Note that by "led it" I mean the next lever moving clockwise 45° around the dial from a particular weighted lever. For example, the weighted lever whose pivot pin's center axis is located at a drum's 9:00 position is led by the weighted lever whose pivot pin's center axis is at the drum's 10:30 position. The weighted lever whose pivot pin's center axis is at a drum's 10:30 position is led by the lever whose pivot pin's center is at the drum's 12:00 position and so on around the drum for all of other weighted levers.

Next, each of the eight weighted levers was attached by a single "long lifter cord" to the weighted lever that led it in a clockwise direction by 90°. Thus, the weighted lever whose pivot pin's center axis was at the drum's 9:00 position was attached by a long lifter cord to the weighted lever whose pivot pin's center was at the drum's 12:00 position. The weighted lever whose pivot pin's center was at the drum's 10:30 position was attached by a long lifter cord to the weighted lever whose pivot pin's center was at the drum's 1:30 position and so around the drum for all of the other weighted levers.

Each weighted lever was attached by a single "stop cord" to a stationary steel pin, called a "stop cord hook attachment anchor pin", whose ends were held between a leading *inner* parallel pair of octagonal frame pieces that also served to separate and brace the two parallel pairs of radial frame pieces that held that weighted lever's pivot pin and the pivot pin of the weighted lever leading it by 45°. These eight stop cords were necessary to provide temporary support for each weighted lever as its pivot pin's center axis approached and then passed the drum's 9:00 position and to assure that the center lines of the lever's main arms did not form an angle with the center lines of the lever's radial frame pieces that exceeded a value of 62.5°.

Finally, and closest to the its pivot pin, each weighed lever had *two* parallel "spring cords" that attached the lever to the ends of a parallel pair of steel helical extension springs whose other ends were attached to a steel anchor pin, called a "spring hook attachment anchor pin". The ends of each spring hook attachment anchor pin were held between a leading *outer* parallel pair of octagonal frame pieces that served to separate and brace the two parallel pairs of radial frame pieces that held that weighted lever's pivot pin and the pivot pin of the weighted lever leading it by 45°.

Thus, each one-directional wheel that Bessler constructed, whether single and self-starting or *one* of an opposed pair within a bidirectional wheel whose enlarged drum required a push to start it turning, contained eight weighted levers each consisting of either a single cylindrical or ingot shaped lead end weight or a trio of cylindrical lead end weights held between the ends of a parallel pair of wooden main arms, eight steel pivot pins held by parallel pairs of radial frame pieces and around which the eight weighted levers were able to rotate through a maximum angle of only 62.5°, sixteen steel helical extension springs so that each of the eight weighted levers could have two springs attached to it, eight steel stop cord hook attachment anchor pins held by an inner parallel pair of octagonal frame pieces, eight steel spring hook attachment anchor pins held by an outer parallel pair of octagonal frame pieces, eight main cords, eight stop cords, eight long lifter cords, and sixteen spring cords. One can immediately see from this parts inventory that each of Bessler's one-directional wheels required a total of *forty* cords to precisely coordinate the motions of its eight individual weighted levers during drum rotation. In the case of a bidirectional wheel that contained *two* internal one-directional wheels, a total of *eighty* coordinating cords was required!

No doubt, this seems like a lot of components, but they were all necessary for Bessler to finally achieve an imbalanced pm wheel design that actually worked and he even found an ingenious way to keep all of the coordinating cords within his wheels separated from each other so that no two of them would rub together and fray as a wheel rotated and some of the cords periodically became taught while others became slack. Most likely, he had learned the method he used from his work as an organ maker. The precise arrangement of all of these various wheel components will become much clearer in the chapters to come where the reader will learn that, aside from a parallel pair of main arms, each weighted lever also had an additional *two* parallel pairs of wooden arms to which coordinating cords were attached. I have discussed these various wheel components here, somewhat prematurely, in order to give the reader a general idea of the mechanics Bessler employed and to make the following treatment easier to follow.

When a one-directional wheel was held in place so that it was stationary, six of its eight weighted levers were in a *very* delicate state of balance that was created by a combination of the contractive forces from the stretched extension springs attached to them and the torques applied to them by the other weighted levers in the drum to which they were connected by main coordinating cords. In this exquisitely balanced state the center of gravity of all of the eight weighted levers was located slightly onto what would become the axle's descending side when the drum was released and free to rotate.

If one was to reach into the opened side of the drum of one of the inventor's wheels that, if released, would begin to spontaneously rotate clockwise and placed a single finger under one of the main arms of its weighted lever whose pivot pin's center axis was located at the drum's 9:00 position, then he would find that he could cause it to begin to lift and rotate clockwise about its pivot pin with just the slightest amount of force from his finger. This would be because all of its weight, which would normally make the entire weighted lever rotate counterclockwise about its pivot pin, was completely counter balanced by the lifting force that the lever received from its own two stretched extension springs. However, one would not be able to push the weighted lever down so as to cause it to rotate counterclockwise about its pivot pin because this motion would be immediately prevented by the stop cord attached to that lever which was stretched taut.

As one carefully studied the weighted lever whose pivot pin's center axis was located at the 9:00 position of the drum, he would notice that the center lines of its parallel pair of main arms that held its single lead ingot end weight or trio of cylindrical lead weights hung below a horizontal line passing through the center axis of the wheel's axle and the center axis of the weighted lever's pivot pin so as to form an angle with that horizontal line which would have the largest such angle of all of the wheel's eight weighted levers. Careful measurement of this maximum main arm angle with a tool known as a "goniometer" would show that it was *exactly* 62.5°. This was an important number to Bessler for reasons that were discussed in Chapter 1. I'll just mention here again that this particular sequence of digits, 625, shows up repeatedly in many of the factors involved in the design of Bessler's wheels.

Now let's see what would happen when the drum of this one-directional wheel was released so that it could begin to rotate clockwise in response to the torque applied to it by its eight weighted levers' center of gravity which was *not* located below the center axis of the axle on a vertical line passing through that axis, but, rather, displaced horizontally away from that vertical line and projected slightly onto the descending side of the axle.

As soon as the wheel's drum was released and began to rotate clockwise, the weighted lever whose pivot pin's axis was leaving the drum's 6:00 position would begin, under the action of gravity, to swing its parallel pair of main arms counterclockwise about its pivot pin and away from its wooden radial stop piece that it was previously in contact with and, as this happened, that weighted lever's two extension springs would begin to stretch as its long lifter cord applied a small amount of clockwise lifting torque to the weighted lever whose pivot pin's center axis was beginning to leave the drum's 9:00 position. At the same time as that was happening, the weighted lever whose pivot pin's center axis was leaving the drum's 7:30 position would *continue* to swing counterclockwise about its pivot pin and further stretch its extension springs as its long lifter cord applied an additional small amount of clockwise lifting torque to the weighted lever whose pivot pin's center axis had just left the drum's 10:30 position.

Thus, every time a pivot pin's center axis left the 9:00 position of a clockwise rotating one-directional wheel's drum, its perfectly extension springs balanced weighted lever would immediately have a clockwise torque applied to it by the lagging weighted lever whose pivot pin's axis was just beginning to leave the drum's 6:00 position. (A "lagging" lever is one that is one or more clock time positions behind a particular lever in a clockwise rotating drum. Thus, the weighted lever which immediately lags the 9:00 weighted lever by one clock time position is the one whose pivot pin's center axis is located at the drum's 7:30 position.) That torque would then almost immediately begin to rotate the weighted lever whose pivot pin's axis was leaving the drum's 9:00 position *clockwise* about its pivot pin so as to bring center lines of its parallel pair of main arms that held the lead end weight(s) between its ends closer into alignment with the center lines of the parallel pair of radial frame pieces that held the lever's pivot pin. Thus, as a weighted lever's pivot pin's center axis traveled from the drum's 9:00 to 10:30 positions, its parallel pair of main arms would move closer to being in physical contact with their wooden radial stop piece that was attached to the weighted lever's parallel pair of radial frame pieces (final contact would not, however, occur until just after the lever's pivot pin's center axis passed the drum's 3:00 position on the descending side of the drum).

Even more interesting to observe would be that every time that the center axis of a pivot pin of a delicately counter balanced weighted lever passed the clockwise rotating drum's 9:00 position, that weighted lever, having completely stopped rotating *counterclockwise* around its pivot pin only about 6° of clockwise drum rotation earlier (because it had by then stretched its two extension springs as far as they were allowed to stretch by the lever's stop cord which was then taut), would *immediately* begin rotating *clockwise* about its pivot pin. This sudden reversal in swing direction was only possible because the weighted lever had dissipated all of its angular momentum relative to its pivot pin in the process of stretching the two extension springs attached to it to their maximum allowed lengths as the lever's pivot pin's center axis approached and reached the drum's 9:00 position.

During the 45° segment of clockwise drum rotation that carried this weighted lever's pivot pin's center axis to the drum's 10:30 position, its parallel pair of main arms would continue to rotate clockwise about their pivot pin through an angle of about 19.5° and the combination of this motion with the clockwise rotation of the drum would make their attached lead end weight(s) rise vertically at a rate about 43% faster than the drum could cause it to rise if there had been no additional clockwise rotation of the weighted lever's parallel pair of main arms about their pivot pin.

This rapid vertical rise of the weighted lever's end weight(s) was so noticeable, in fact, that Bessler even mentioned, possibly by accident in AP without realizing how much he was revealing, that in *his* perpetual motion wheels "the weights which hang below must rise in a flash". The implication of this, of course, is that the rapid clockwise rotation of a weighted lever's main arms about their pivot pin as that pin's center axis traveled between the drum's 9:00 and 10:30 positions was *critically* necessary to maintaining the center of gravity of all of the drum's eight weighted levers on the descending side of the axle so that the pm effect would manifest itself. After about 2,000 futile attempts that only finally ended successfully *after* I had also achieved this effect in my computer model wheels did I finally realized how very important this condition is.

Furthermore, as their individual pairs of extension springs contracted, all of the weighted levers whose pivot pin's center axes were moving between the drum's 9:00 and 3:00 positions would continually move their parallel pairs of main arms closer to their wooden radial stop pieces, but their approach speeds to those radial stop pieces would progressively diminish as a pair of main arms got closer to and finally made contact with its radial stop piece just after its pivot pin's center axis finally passed the 3:00 position of the clockwise rotating drum. The approach speed of a weighted lever whose pivot pin's center axis was traveling between the drum's 9:00 and 10:30 positions was, of course, the greatest.

As the one-directional wheel's drum continued to rotate clockwise through a 45° segment of rotation, it would, if its eight weighted levers were *not* free to rotate about their individual pivot pins in their various necessary directions, quickly cause the center of gravity of the levers to rotate clockwise around the center axis of the axle and, eventually, after oscillating back and forth for a while, come to rest directly below the center axis of the axle at its *punctum quietus* location at which time the axle's torque would drop to zero and the wheel would stop rotating. This, however, would not happen if the internal mechanics of the drum was able to rotate the center of gravity of all eight of the weighted levers around the center axis of the axle in the *opposite* counterclockwise direction and could do so at the same rate that the clockwise rotating drum was trying to rotate it down to the *punctum quietus* location below the axle's center. This is exactly what was happening inside of Bessler's successfully working imbalanced pm wheels.

Thus, the mechanics of his wheels had to produce two simultaneous motions in their eight weighted levers' center of gravity in order to keep it fixed at a certain location on the axle's descending side relative to imaginary horizontal and vertical lines passing through the center axis of the wheel's axle.

Aside from continually rising vertically toward the imaginary horizontal line passing through the axle's center axis, the center of gravity also had to constantly move horizontally away from the imaginary vertical line passing through the center axis of the axle and toward the axle's descending side in order for its location to remain stable during drum rotation.

The horizontal motion of the center of gravity away from the imaginary vertical line passing through the center axis of axle was not a problem because that motion did not require the expenditure of any significant amount of energy and mass from the weighed levers and was mainly the result of the counterclockwise rotation of the two weighted levers about their pivot pins as the center axes of those pivots left their starting locations at the 6:00 and 7:30 positions of the drum and stretched their pairs of extension springs. This horizontal motion of the eight weighted levers' center of gravity was also further assisted by the clockwise rotational motions of the weighted levers whose pivot pin's center axes were passing the drum's 10:30, 12:00, and 1:30 positions. In essence, the various rotational motions of the weighted levers mentioned about their pivot pins tended to continually "push" the center of gravity of the eight weighted levers horizontally away from the imaginary vertical line passing through the center axis of the axle and toward the axle's descending side.

However, the continuous *vertical* rising motion of the eight weighted levers' center of gravity toward the imaginary horizontal line passing through the center axis of a wheel's axle during any 45° segment of clockwise drum rotation *did* require the constant expenditure of energy by the wheel which, as previous mentioned, could only have been provided entirely by the mass of the weighted levers themselves (which included half of the masses of the seven coordinating cords and two unfixed extension springs attached to each lever). But, how was this energy extracted from the weighted levers as they each completed a trip around the wheel's axle from the drum's 6:00 position, up past its top 12:00 position, and then back down again to the 6:00 position only to repeat the trip over and over again with each drum rotation?

Answering this question is crucial to truly understanding how a Bessler wheel managed to produce the pm effect even though we know that effect would probably not be "perpetual" in the absolute sense of the word.

Most seeking to construct an imbalanced pm wheel simply assume that a wheel must run if they can just find a design that, during rotation, always keeps the center of gravity of its active weights on the descending side of the axle. This assumption, however, is one that is false because there have been many designs for imbalanced wheels that did keep their weights' center of gravity on their axle's or wheel's descending sides and yet did not run!

In such designs, despite the offset composite center of gravity of its various active weights, the wheel is actually in a state of balance regardless of what orientation it is rotated into by hand. However, the counter torque that prevents the imbalanced wheel's driving torque produced by its offset center of gravity from rotating the wheel is not obvious. This hidden counter torque results because, as the wheel begins to rotate, it must use some mechanical linkage between diametrically opposed weighted levers that uses the torque of a descending side weighted lever as it swings away from the axle's center axis and toward the wheel's outer rim to rotate the other linked ascending side weighted lever so that it's end weight moves closer to the outer rim of the wheel.

In practice, unfortunately, no shifting of the linked pair of weighted levers will take place because the torques of the two levers *always* have equal magnitudes, but are in opposite directions. Expecting them to shift each other every 180° of wheel rotation is somewhat like a person expecting to lift himself up off the ground and into the air by simply pulling up hard enough on his pants belt! Many a pm seeker has become excited over such a design and spent much effort, money, and time constructing a wheel using it only to be bitterly disappointed when the day came to give the device its first definitive test. For many, the resulting failure was sufficient to permanently end their continuing involvement in the construction of imbalanced pm wheels. From studying the pm wheels and devices illustrated in Bessler's MT, one finds that the majority of them are unworkable because of this subtle problem.

From studying the successful computer model for the 3 feet diameter Gera prototype wheel that I finally found, I realized that Bessler's design was doing something quite remarkable with the *average* vertical velocities of the centers of gravity of the four weighted levers on each side of its drum.

Imagine that we could take one of Bessler's one-directional wheels and, after removing its 40 coordinating cords and 16 extension springs, just securely tie down all of its eight weighted levers' parallel pairs of main arms so that they remained in physical contact with their radial stop pieces regardless of what orientation we manually rotated the drum into. In such a situation, all of the *individual* centers of gravity of its eight weighted levers' end weights would be as close to the drum's outer rim wall as possible, as far away from the axle's center axis as possible, and located at the vertices of a large imaginary octagon inside of the drum. As a result of this, the *composite* center of gravity of all eight of the weighted levers would be located exactly at the center axis of the axle (the reader will learn later in chapter 7 that this particular configuration of weighted levers was the one that existed inside of the *inactive* internal one-directional wheel of a bidirectional and was used to eliminate the torque of that inactive internal wheel and, thus, prevent it from cancelling any of the torque produced by the other and *active* internal one-directional wheel inside of the bidirectional wheel's enlarged drum).

We would then have four weighted levers on each side of the one-directional wheel's drum and, if we manually rotated the drum, the magnitude of the *average* vertical descent velocity of the four weighted levers' centers of gravity on the drum's descending side would exactly equal the magnitude of the average vertical ascent velocity of the four weighted levers' centers of gravity on the drum's ascending side (when using the "magnitude" of a number, we ignore its sign, whether positive or negative or, in the case of these vertical velocities, their up or down directions, and just concentrate on its numerical value).

As a consequence of this equivalence of vertical velocity magnitudes (but not directions), we would then find that, as the drum continued to rotate in either one of its two possible directions at a constant rate after we had given it an initial push, the magnitude of the average rate that the four descending side weighted levers *lost* gravitational potential energy would always exactly equal the magnitude of the average rate at which the four ascending side weighted levers *gained* gravitational potential energy. This would be because, as the four weighted levers each lost gravitational potential energy on the drum's descending side and converted it into kinetic energy, that kinetic energy would, metaphorically speaking, immediately "flow" out of those descending weighted levers and into the structures of the drum and its axle. However, just as quickly as that happened, the four weighted levers on the ascending side would immediately extract all of that kinetic energy from the structures of the drum and its axle and use it to increase their own gravitational potential energies. Because this process allows no kinetic energy to stay behind and accumulate in the drum and axle, they do not accelerate and their rate of rotation will remain constant (actually it will gradually slow over time due to aerodynamic and bearing drags that extract kinetic energy from all parts of the wheel).

With this particular symmetrical configuration of weighted levers, a wheel cannot accumulate any kinetic energy in the structures of its drum, its axle, or the weighted levers themselves. Under ideal conditions of zero aerodynamic and bearing drag, the drum of such a wheel, if manually brought up to a certain rotation rate and released, would continue to rotate for eternity at the same constant rate. However, the instant that the wheel was required to perform outside work, either by overcoming drag forces or operating some attached piece of machinery, the wheel would slow and eventually stop.

This situation, however, did not happen in Bessler's imbalanced pm wheels because, once allowed to start rotating, they could slowly accumulate kinetic energy in the structures of their drums, axles, and the weighted levers themselves. How could this be possible one might wonder?

It was only possible because the magnitude of the average rate that the four ascending side weighted levers' centers of gravity gained gravitational potential energy during drum rotation was always a little *less* than the magnitude of the average rate that the four descending side weighted levers' centers of gravity lost gravitational potential energy. This imbalance then led to a surplus of gravitational potential energy constantly being lost by the four descending side weighted levers which was not taken up by the ascending side weighted levers to increase their gravitational potential energies and, thus, this surplus remained behind as kinetic energy and was immediately distributed among all of the structures of the drum, axle, and even the eight weighted levers themselves which caused them to increase their rate of rotation around the axle's center axis. It was this constantly accumulating surplus kinetic energy during drum rotation which caused a wheel to accelerate if running freely or, while running at a constant rotation rate, to constantly overcome external drag or operate outside machinery attached to its axle by accelerating that machinery's moving parts and increasing their kinetic energies.

The reader has learned that the weighted lever whose pivot pin's center axis was traveling from a clockwise rotating drum's 6:00 position to its

9:00 position would swing counterclockwise about its pivot pin which caused the lever's center of gravity to move closer to the center axis of the drum. But the lever's rotation also resulted in a decrease in the vertical ascent velocity of the center of gravity of the weighted lever. On the other hand, once a weighted lever's pivot pin's center axis passed the drum's 9:00 position, the lever would begin rapidly rotating clockwise about its pivot pin and its center of gravity would begin increasing its separation distance from the center axis of the axle. This motion then caused the vertical ascent velocity of the lever's center of gravity to increase.

One might think that the decrease in the magnitude of the vertical ascent velocity of the center of gravity of a weighted lever whose pivot pin's center axis was moving between the drum's 6:00 and 9:00 positions would perfectly cancel out the increase in the magnitude of the vertical ascent velocity of the center of gravity of the a weighted lever whose pivot pin's center axis was moving between the drum's 9:00 and 12:00 positions so that the magnitude of the average vertical *ascent* velocity of a lever's center of gravity on the drum's ascending side would always exactly equal the magnitude of the average vertical *descent* velocity of the center of gravity of a weighted lever on the drum's descending side. That assumption, however, would be a *false* one to make and, if it was true, then Bessler would never have been able to construct a working imbalanced perpetual motion wheel!

The reality in Bessler's wheels and, indeed, in *any* successful, working imbalanced pm wheel is that the average vertical ascent velocity of the weights, whether they are attached to levers or not, on one side of the wheel *must* be *less* than the average vertical descent velocity of the weights on the other side of the wheel. And, this condition must exist despite there being an equal number of weights on both sides of the wheel.

When first presented with this critically necessary requirement, the reader may reject it and think it is an impossible condition to achieve. However, it must also be remembered that in Bessler's wheels the pivot pins were attached to and part of the drum and moved along with it while the weighted levers and their individual centers of gravity moved *independently* of their pivot pins *and* the drum everywhere around a clockwise rotating drum except *between* about 12° past its 3:00 position and its 6:00 position during which time a weighted lever's parallel pair of main arms were in physical contact with its lever's wooden radial stop piece and the weighted lever remained motionless with respect to the drum which carried it along around the center axis of the axle. It was these mostly independent motions of the drum's weighted levers that allowed for a difference in the magnitudes of the average vertical velocities of the centers of gravity of the four weighted levers on each side of the drum to exist. In order for this discrepancy in vertical velocities on each side of one of Bessler's wheels to be maintained during drum rotation, it was critically necessary that the composite center of gravity of *all* of the drum's eight weighted levers be kept on the axle's descending side during drum rotation.

Here's a simple analogy that may help make the physics of this situation clearer for the reader who has difficulty understanding why there should be any difference in the average magnitudes of the vertical velocities of a weighted lever's center of gravity on each side of the drum of one of Bessler's one-directional wheels which was rotating in a clockwise direction.

Imagine that a person lived on a river and had a friend who lived north of him on the same river, but who was 10 miles away from him. The friend decided one day to visit the person and had a boat with an outboard motor on it which was capable of a maximum speed in *still* water of 10 miles per hour. That friend then climbed into his boat, started the outboard motor on it, and began heading down the river. The water in the river was not moving at the time and it took the friend exactly one hour to reach the person so he could have a brief visit with him. On his way down the river to visit the person, the friend's boat had maintained a constant speed of 10 miles per hour relative to the banks of the river.

After the visit was over, it was time for the friend to return to his home which was 10 miles up the river in a northward direction. However, this particular river had a direction of flow that could be a bit unpredictable. When the friend first came down the river, the water was still. Then, however, as he began his return trip up the river, the water suddenly started flowing southward at 5 miles per hour and in the *opposite* direction that the friend wanted to travel which was northward. As a result, the friend's boat could only travel at a maximum speed of 5 miles per hour relative to the river's banks and it took him an entire hour to just travel a distance of 5 miles which was only half of the distance he needed to travel in order to get home again.

At the halfway point of his trip after he had traveled 5 miles up the river, the water suddenly changed its direction of flow. Now it was flowing in the *same* direction as the boat at a speed of 5 miles per hour relative to the banks of the river and the friend found that his boat's maximum speed of 10 miles per hour was added to that of the river water so that he was then traveling at 15 miles per hour relative to the river's banks. Because of his 50% increase in speed, he could then cover the remaining 5 miles of his return trip in only 20 minutes since 5 miles/15 miles per hour = 0.333 hours = 20 minutes.

Thus, the total time for the friend to return to this home, which was 10 miles away from the person he visited, was one hour and twenty minutes or 1.333 hours. If we divide the distance traveled, 10 miles, by his travel time, 1.333 hours, we can get the friend's average travel velocity during the return trip. Doing this, we get 10 miles/1.333 hours = 7.5 miles per hour which is 25% *less* than his average speed of 10 miles per hour when he began his trip down river to visit the person.

A somewhat similar situation existed inside of Bessler's one-directional wheels. A weighted lever on the descending side of the drum was similar to the boat traveling down the river when the friend decided to visit the person who lived 10 miles south of him on the river. However, a weighted lever on the ascending side of the drum that first swung its center of gravity toward the axle's center axis and opposite to the direction of drum rotation so that the magnitude of its vertical ascent velocity decreased and, after a brief pause of about 6° of clockwise drum rotation until its pivot pin's center axis arrived at the drum's 9:00 position, then began swinging its center of gravity away from the axle's center axis and in the same direction as drum rotation so that the magnitude of its vertical ascent velocity increased was similar to the friend's boat when it was traveling up the river to return him home and experienced a change in its speed relative to the river banks due to a change in the flow direction of the river at the halfway point in the return trip.

However, in the case of Bessler's wheels with their eight weighted levers, the situation was actually more analogous to having had the motor boats of four friends come down the same river with the still water for visits, one after another, while four other friends, again one after another, were each departing the person and traveling northward up four other *separate* parallel rivers each of which would change its flow direction when a single visitor's boat on it reached the halfway point in the return trip up the river. We see that all eight of the visitors' boats must travel the same distance of ten miles between their starting and final destinations, but the boats traveling south on the river always have a higher velocity, 10 miles per hour, compared to the average velocity of the boats traveling north on the river which was 7.5 miles per hour.

Additionally, if we imagine the boats arriving at the person's home, one after another, and each then immediately leaving for a return trip up the river on one of the four other separate parallel return rivers and those boats, upon reaching their starting location up the river, immediately making another trip down river to see the person using the single river whose water is always still, then we would, whenever we telescopically viewed the scene from, say, a helium filled observation balloon high in the sky, always count only eight moving boats on the five parallel rivers.

From the above we see that Bessler's one-directional imbalanced perpetual motion wheels could constantly accumulate a surplus of some of the lost gravitational potential energy of the centers of gravity of their descending side weighted levers as they continually traveled from a clockwise rotating drum's 12:00 position to its 6:00 position. Basically, this accumulation was solely due to the differences in the magnitudes of the average vertical velocities of the descending side and ascending side weighted levers. The descending side weighted levers were constantly losing more gravitational potential energy per second than could be taken up by the ascending side weighted levers per second. That resulted in lost gravitational potential energy that had to go somewhere. It could not be go into increasing the gravitational potential energy of the entire drum, its contained components, and the axle by making them rise to a higher elevation because the axle's center axis was restricted to a certain elevation above the floor of the wheel's room by the bearing plates that were firmly embedded in the wheel's two wooden vertical support planks. Once that method of storing the accumulating lost gravitational potential energy of the drum's descending side weighted levers was blocked, that energy had to be stored as an increase in the rotational kinetic energy of the drum, its contained components, and the axle which resulted in their rates of rotational around the axle's center axis increasing.

However, as the reader may recall from the beginning of this chapter, there is really no difference between energy and mass. I could have substituted the single word "mass" for "gravitational potential energy" and "rotational kinetic energy" in the above and it would still have been a valid description of the processes going on in Bessler's wheels when they were running. This means that, as long as one of his wheels was running, a weighted lever traveling from the drum's 12:00 to 6:00 positions would lose a tiny amount of its mass, only a *fraction* of a picogram (which is only one trillionth of a gram!), but it would then regain *less* than that amount of lost mass back on its trip from the drum's 6:00 to 12:00 positions. The tiny amount of mass lost by a descending side weighted lever that it did not gain back on the drum's ascending side would then be equally distributed among all of the wheel's moving parts such as its drum's various wooden structures, the materials covering its open sides, the eight weighted levers themselves, the 40 coordinating cords, the 16 helical extension springs, and the axle as they all rotated around the axle's center axis. If a piece of external machinery was attached to and moved by the wheel's axle, then some of that accumulating surplus mass being lost by the drum's descending side weighted levers would actually flow into the machinery to increase the masses of its moving parts.

How did Bessler conceive of such an approach to constructing a genuine imbalanced perpetual motion wheel? I've already mentioned the statements in his writings which involved his work in organ construction and a vivid dream he had one night. But, the real answer is that it just required a tremendous amount of effort on his part, an unwavering belief that it could be done, that he could do it, and, of course, a tremendous amount of luck.

At the time that Bessler found success in late 1711, there were probably hundreds or even thousands of rival perpetual motion seeking inventors throughout Europe who were just as skilled and intelligent as he. They, too, had studied the past failed designs offered in the then available books dedicated to mechanics and had spent years and, perhaps, decades in the pursuit of an imbalanced perpetual motion wheel. What they lacked was the phenomenal luck needed to finally find a design whose carefully balanced and coordinated weighted levers could, at any particular drum rotation rate up to its maximum terminal value, keep their composite center of gravity on the axle's descending side despite the rotation of the wheel. While the center of gravity of the active weighted levers of each of his rival inventors' wheels soon fell into its *punctum quietus* location below the center axis of the wheel's axle soon after rotation started, the center of gravity of one of Bessler's self-starting one-directional wheels remained away from that undesirable point at all times whether the wheel was stationary or allowed to rotate.

To return once again to the question posed by the title of this chapter, "Is Perpetual Motion Really Possible", the answer is "yes" if one has wheel mechanics that can slowly extract a small part of the energy and its associated mass from the wheel's descending side weighted levers during each of its drum rotations so that this energy and mass can then be used to accelerate all parts of the wheel about its axle's center axis or used continuously power at a constant rate any outside machinery attached to the wheel's axle. The answer is also "yes" if one is not bothered by having a wheel whose "perpetual" ability to constantly output energy and mass is limited to only a few tens of billions of years. Johann Bessler had such wheels and in the coming chapters I will show the reader how it may be possible for him to duplicate Bessler's first working 3 feet diameter Gera prototype wheel so that he can have one too!

As a postscript to this chapter, I want to briefly clear up a few additional misconceptions that some may have about Bessler's imbalanced pm wheels.

Aside from thinking that his wheels must be physically impossible because they would have to create energy out of thin air which would be an obvious violation of the universally accepted "Law of Energy and Mass Conservation", some may think that such a self-starting wheel is impossible because it would violate Newton's "Third Law of Motion" which states that the amount of momentum in a system must remain constant. That law specifically forbids an object from accelerating in one direction and, thus, constantly increasing its momentum, unless that increase is cancelled out by another object accelerating in the exact opposite direction. I believe that this chapter adequately explained why Bessler's wheels did *not* violate the energy-mass conservation law, but I did not adequately treat the other apparent violation of the momentum conservation law. Let me now address that matter.

This issue of violating the third law of motion was something that also bothered me for a long time during my research into Bessler's various wheels. It was finally satisfactorily resolved for me one day when I made a small computer model of a working imbalanced pm wheel (this was done years before I knew the exact internal mechanics of Bessler's wheels!) that was nothing more than a disc with a small amount of mass that could be kept in constant rotation by a small motor attached to it. I then made a second disc, much larger in radius and more massive, that represented the equatorial cross section of a small stationary planet on whose circumference I attached the small model pm wheel.

When the motor was turned off and the model pm wheel was stationary, there was no rotation in the larger planet disc. However, as soon as I started the model pm wheel spinning in *either* direction, the larger planet disc would immediately respond by rotating in the *opposite* direction, but with a far lower rate of rotation than that of the motorized model pm wheel. Angular momentum was being conserved and it also was conserved as Bessler's various imbalanced pm wheels were set into rotation in various towns in Saxony. In the case of his wheels, however, the opposite rotation of the Earth, although present, would have been imperceptible because of the huge mass of our planet compared to that of one of Bessler's wheels.

Also, some may believe that Bessler's wheels could only output energy if they were getting it from some source outside of themselves and that source must have been the Earth's gravity field.

This is impossible because a planet's *static* gravity field has no energy or mass which an imbalanced wheel can somehow extract and use. However, it is possible for a gravity field that is *varying* in intensity and radiating these variations outward from their source in the form of spherical waves to transmit energy and mass from one celestial body to another. This process, however, was not where Bessler's wheels' outputted energies came from. They came completely from the masses of their weighted levers and nowhere else.

To demonstrate that planetary gravity was not the source of the energies Bessler's wheels outputted, consider the following scenarios.

Imagine that we could construct a space station like those depicted in some of the 1950's and 1960's science fiction space movies. The station is, basically, a large diameter, wheel consisting of a hollow toroidal rim (with a core having a square or rectangular shaped cross section) attached by a collection of thinner spokes to a central axial body. Furthermore, the station is located far out in interstellar space so that there are no huge masses nearby it such as those of a star, planet, or moon. Using small rocket motors attached at different locations of the wheel's outermost perimeter, the entire station can be accelerated up to a rotation rate at which point the astronauts living inside of the hollow toroidal rim will feel centrifugal forces acting on their bodies that will hold them down against the interior walls of the outer curving surfaces of the rotating toroidal rim. If the rotation rate of this space station is properly adjusted, then every astronaut inside of the toroidal rim would have the same weight as he or she had if standing on the surface of the Earth.

If one was to then place a replica of one of Bessler's wheels into this rotating space station's toroidal rim and locate it on the same sections of its curving interior walls upon which the astronauts can stand and walk about, then the wheel would function exactly as it did on the Earth's surface. Most importantly, the wheel would be able to do this in the absence of a planet's natural gravity field.

Because of the huge difference in mass between the rotating imbalanced pm wheel and the much larger rotating space station that contained it, the amount of surplus energy and mass transferred to the station from the rotating imbalanced pm wheel would be small compared to those of the station and would have little effect on the rotational motion of the station after only a small amount of time had passed. However, over a longer period of time, the station's rotation rate would begin to be affected if the imbalanced pm wheel continued to run inside of its hollow toroidal rim.

Depending upon the direction of rotation of the imbalanced pm wheel's drum and the orientation of the center axis of the wheel's axle with respect to the center axis of the station's central axial body, the space station could either gradually slow to a stop or gradually speed up until it or the imbalanced pm wheel it carried tore themselves apart due to unsustainable centrifugal forces acting on their parts.

This experiment could even be performed without the need for the rotating space station!

One could simply take two identical wheels and attach them to the ends of two parallel extended support planks so that the parallel center axes of the wheels' two axles were rigidly held in space with respect to each other and 100 feet apart. Next, two small, identical rocket motors attached to opposite ends of this assembly would be briefly simultaneously fired in opposite directions in order to make the entire assembly continually flip end over end as the two wheels' drums rotated around the center of gravity of the assembly in the same direction, either both clockwise or both counterclockwise, as viewed from a distance along the assembly's center axis of rotation.

Each wheel's eight weighted levers would then feel centrifugal forces acting on them in a direction away from the center of gravity of the assembly which is located at the midpoint between the parallel center axes of the two widely separated imbalanced pm wheels' axles. The rotational rate of the assembly could then be carefully adjusted, using the rocket motors, until the average value of the centrifugal force acting on the weighted levels in each of the two wheels was identical to what the gravitational force acting on the weighted levers would be if they were all inside of the wheel when it was located on the Earth's surface.

When the brake bolts were finally released on both of the wheels (this could be done using a radio control activated electrical motor attached to each bolt to loosen its pressure on the steel pivot pin at the end of an axle), they would both begin rotating and accelerating up to their maximum terminal rotation rates, which, in the absence of air, would be solely due their various bearing drags. Again, this would all be happening without the need for the presence of a planetary gravity field.

Depending upon the directions of rotation of the two imbalanced pm wheels, the rotation rate of the entire assembly would either slow to a stop, reverse its direction of rotation, and then increase its rate of rotation in the new direction until the assembly was eventually torn apart by unsustainable centrifugal forces or the entire assembly would just continue to increase its rotation rate in the same direction until the assembly was again torn apart by unsustainable centrifugal forces.

CHAPTER 4
A Few Notes Before We Begin

Anyone seriously interested in reverse engineering Bessler's wheels and then replicating them must eventually decide what units of measurement for length and mass he was using and which were referred to in his writings and in the correspondence of others concerning his wheels. This is can be a daunting task because, prior to the introduction of the metric system of measurement in Germany on January 1st of 1872, practically every town in continental Europe used its own measurements. Traveling merchants arriving at a new town to sell their wares usually stopped at the town hall and found a chart on one of its walls that provided them with a drawing of what that town's citizens considered to be a certain standard length which, of course, was only "standard" in that particular town. The merchants would then sell things such as cloth and rope in multiples of those local standard lengths to the citizens and shopkeepers of that town and adjust their prices accordingly.

The standards for the measurement of weights were handled somewhat similarly only in their case different central governing bodies with authority over several towns in a region established the standards. Since these various authorities also each usually minted the coins used in a particular region whose towns they governed, they could make sure all of the coins were identical in weight and value in that region. They would also define other weights for use in commerce. Once established these units were also considered legally defined and any merchant misrepresenting the weights of the goods he sold could be fined or imprisoned for doing so which was considered a crime on a par with counterfeiting.

Some suppose that Bessler was using the Leipzig "ell" which was equal to about 22.3 English inches. (Note that the name "ell" derives from the same root as does the word "elbow" and was originally equal to a "cubit" or the distance from a person's elbow to the tip of his middle finger or about 18 inches. This distance was eventually lengthened for use in measuring the width of cloth coming off of rolls of cloth.) I am convinced, however, that Bessler actually used a *special* ell that was exactly equal to 24 inches in length with each of its inches also exactly equal to an English inch when he designed and constructed his wheels. Thus, a half of one of Bessler's special ells would have been 12 inches and corresponded to the English unit of distance of one foot. Why would he do this?

Bessler seems to have had an affinity for the British Isles. He traveled there as a young man and probably even spoke some English, later learned of a prize of £20,000 pounds offered by the British government in 1714 for a means to accurately measure the longitude of a ship at sea and then adjusted the amount he wanted in thalers for the secret of his imbalanced perpetual motion wheels to match the British prize amount, and eventually sent his brother over to London to try to find a buyer for his wheels. To interest English businessmen and the British Royal Society's members in his inventions, it is only logical to assume that he designed his wheels using the English units of measurement for length as well as mass since they would be more familiar to them. It is for these reasons that I have given the various dimensions of his wheels, the masses of the weights and levers he used inside of their drums, and other physical parameters in English units in this book and, indeed, used them exclusively during my research into the imbalanced pm mechanics he used. Those living outside of the United States and using the currently more popular metric system of units will have to make the conversions of the lengths, masses, and other physical parameters I provide in this book into that system's units themselves. It will also be noticed in the coming chapters that, whenever I give a distance measurement between two points or parallel planes in decimal form, I usually immediately follow it with the same measurement that provides the decimal portion of its value as sixteenths of an inch. I've done this on purpose in order to relieve craftsmen from the additional burden of having to make the conversions themselves as they attempt to replicate the various components of Bessler's wheels using rulers, yardsticks, and tape measures that have their inches subdivided into sixteenths of an inch.

There was an interesting clue that I came across that convinced me that my decision to exclusively use English units in my research was the correct one to make.

Although I will discuss it in far greater detail in the latter chapters of this volume, one of Bessler's most important books, DT, contains an engraved portrait at its very beginning that depicts the inventor in a workshop which was based, most likely, on a composite of several actual workshops in which he worked on his various perpetual motion wheel designs over the years (to see this portrait, the reader can look ahead to the end of Chapter 8 and find it in Figure 17). In that portrait he stands behind a shop table on which is located a long, thin measuring stick or ruler that is carefully balanced on the side of carpenter's plumb bob which rests of the table's top. Ordinarily, that is something one would expect to see in the workshop of a carpenter and inventor, although balancing a ruler like that is certainly not to be expected and, obviously, intended to attract a reader's attention. However, what really made this ruler seem unusual to me is that it shows 14 inches on it (even though part of the ruler is hidden behind a lamp on the table)!

I've used many kinds of rulers in my life, but every single one of them, even the larger yardsticks and both cloth and metal measuring tapes, assigned exactly 12 inches to a foot. So, why does he show us a ruler with 14 inches on it?

After doing some research, I found the answer and it actually further confirmed my belief that Bessler used the English units of length in the construction of this wheels.

Apparently, in certain parts of Germany in the early 18th century, they were using an ell that had 28 inches in it and those inches were identical to 28 English inches! Indeed, there is evidence that, after the period of the Roman occupation of Britain ended, the Anglo-Saxon period that followed resulted in many of the units of measurement being *exported* from northern Germany to the British Isles which is why that equivalence in the inches existed. The ruler that Bessler shows in his second DT portrait represents one *half* of one of those 28 inch ells and therefore only shows 14 inches and they were identical to 14 English inches at the time. But, didn't I just state that Bessler's "special" ell only contained 24 inches so that half of the length of that ell would equal 12 inches or one English foot?

Yes, I did, but the reader should take a closer look at that second DT portrait in Figure 17. The ruler has 14 inches on it, but *two* of them are obscured by the lamp so that the reader only sees 12 full inches on the ruler. That was Bessler's way of telling a future reverse engineer of his wheel's that he was, indeed, using a half ell of only 12 inches which corresponded to an English foot.

If one will grant that the linear measurements of single inches and feet containing 12 such inches that Bessler used for the designing of the parts in his wheels did correspond to the same English linear measurements, the next question one might ask is does this equivalence also apply to the masses associated with the various parts of his wheels? For example, did this equivalence apply to the single cylindrical lead end weight removed from the Merseburg wheel during its official test which was handled by witnesses and judged to have a mass of 4 pounds and also apply to the masses of the loads that his larger 12 foot diameter wheels could rapidly hoist which were given values of 60 or 70 pounds by other witnesses?

(Just so there will be no confusion as the reader proceeds through this chapter, I need to briefly digress and explain the difference between the mass of an object and its weight. Many people use these terms interchangeably and that is acceptable for an object located near the Earth's surface. However, there is a distinction which can become quite apparent in certain situations. Mass is a *property* of an object that depends upon the total number subatomic particles contained within its many atoms. The magnitude of this property then determines the gravitational and inertial properties of the object. Most importantly, the mass of an object does not vary with the strength of its local gravity field produced by some nearby massive planet or moon on which the object is located. The mass of an object on the surface of our Moon will be the same as it is when the object is on the surface of the Earth which has a much stronger gravity field. The weight of the object, however, *does* vary with the strength of the local gravity field affecting it and describes the amount of gravitation attractive force the object experiences. An object that is on the Moon will weigh significantly less there than it would when it is located on the surface of the Earth because the weaker gravity field of the Moon applies less attractive gravitational force to it than does the Earth's stronger gravity field. What blurs the distinction between these two terms is that both the mass and weight of an object are usually measured using the *same* units. So, when I use the word "mass" with respect to Bessler's wheels which were located on the Earth's surface, it means the same as "weight" and when I use the word "weight", it means the same as "mass". Thus, as far as Bessler's wheels are concerned, one pound of mass is equal to one pound of weight and one pound of weight is equal to one pound of mass.)

Again, I am quite convinced that the answer to these questions is "yes". Those pound units of mass mentioned in the Bessler writings and correspondence were identical to the ones being used over in the British Isles at the time and were English pounds and essentially the same as the pounds still used in the US today.

There is another important clue in that second DT portrait that establishes his use of English pounds for the masses of the parts in his wheels. It involves the carpenter's plumb bob weight upon which the above described 14 inch ruler sits perfectly balanced at its midpoint and is divided into two halves that are each 7 inches in length. Bessler, being a carpenter, would have used such a conical lead bob weight to make sure the floor to ceiling vertical wooden support planks for his more massive wheels were as close to being perfectly vertical as possible.

The left side of the carefully balanced ruler points to the center pivot of a large "duck's bill" goniometer which is a carpentry tool used to measure the angles between two adjacent flat sides of a piece of wood. In the second DT portrait the goniometer's two arms are opened to a precise inside angle of 45° and their pointed tips touch other items on the table so as to form a giant letter "M" which is rotated about 90° clockwise so that it appears to be lying on its right side. The lower arm of the goniometer is crossed by a drafting tool called an "L" square which forms the numeral "4" with the goniometer's lower arm.

The meaning of these various portrait symbols is fairly easy to interpret. The "M" indicates that Bessler is providing information about the Merseburg wheel whose first letter is "M" and the "4" tells us that information is that the wheel used 4 pound weights which, as was mentioned in Chapter 1, was what the examiners of the Merseburg wheel determined was the mass of the weights used in that 12 foot diameter, bidirectional wheel during its official test. (Indeed, if it was not for a single recovered letter, dated December 19th, 1715 that was sent by Christian Wolff to

Leibniz, we would not have any secondary verification of my interpretation of this important clue in the second DT portrait. Wolff was one of the people who conducted the testing of the Merseburg wheel on October 31st, 1715 and personally handled one of the weights Bessler extracted from its drum prior to its translocation to a secondary set of vertical support planks in the wheel's room. Wolff, aside from being Leibniz' good friend and probably the second most eminent philosopher in Europe at the time, was also a professor of mathematics and physics at the University of Halle.)

This letter from Wolff and the second DT portrait clue involving the duck's bill goniometer and the "L" square certainly seem to establish that Bessler's Merseburg wheel did, indeed, use 4 pound weights attached to its levers. But, how can we be sure that the 4 pounds that Bessler used had the same mass as a 4 pound weight used by the English then and in the US today? This, remarkably, is also assured by other symbols in the second DT portrait.

If one extends an imaginary horizontal line from the top edge of the ruler out to the left of it, that line will intersect the exact center of the pivot of the goniometer which is located very close to a semicircular protractor (which is a drafting tool used to measure angles from 0° to 180°) to its left that happens to look like a large letter "D" that is rotated a bit clockwise from its normal orientation. "D" is the fourth letter of both the English and German alphabets and, again, Bessler is telling us he will provide an important clue about the masses of the 4 pound weights used in the Merseburg wheel. If one extends another straight line from the 90° mark of the "D" shaped protractor up and through its origin point, then that line will intersect one of the Earth's poles that are indicated on a small globe on the leftmost end of the table visible in the second DT portrait (that pole is located at the center point of a collection of closely spaced concentric circles printed within the circular outer circumference of the globe).

What is Bessler trying to tell us with all of this?

From studying magnified images of that globe, it's evident to me that the pole being indicated is actually the Earth's *South* Pole and Bessler, in this second DT portrait, has actually provided us with what is known as a polar projection map of our planet's South Pole. I can make out most of the continent of South America on the globe on its upper left quadrant and part of the southern tip of Africa on its upper right quadrant. The continent of Antarctica is not shown because it was not officially discovered until about a half century later in the 1770's by an English captain named James Cook.

Obviously, this symbol of the globe is important, but to be properly interpreted, it must be combined with other symbols in the second DT portrait.

As previously mentioned the right side of the 14 inch ruler is partially hidden from view by an odd looking lamp that stands on the table on three feet, one of which is hidden from view behind the lamp. The lamp's cone shaped cover on its top points straight up. What is the symbolic significance of this lamp?

The lamp actually has multiple symbolic meanings.

The three feet at its base represent Bessler's first 3 feet diameter, prototype wheel that he constructed at Gera. The heavy conical cover at the top of the lamp points straight up like the *two* hands of a clock at 12 o'clock noon and that represents his *two* 12 foot diameter, bidirectional wheels which were the Merseburg and Kassel wheels. The vertical body of the lamp between its tripodal feet and its conical top then represents the other one-directional wheels he constructed that had drum diameters between 3 and 12 feet and were his 4.5 foot diameter second Gera wheel and 9 foot diameter Draschwitz wheel (and would also include the 6 foot diameter wheels he constructed years after the destruction of his 12 foot diameter, bidirectional Kassel wheel). Thus, the lamp symbol as used here tells us that he is going to provide us with information that pertains to *all* of his constructed wheels.

Again, recall that there is a 14 inch ruler perfectly balanced on the lead carpenter's plumb bob to the left side of the tripodal lamp. The ruler's *left* side points to the duck's bill goniometer pivot and that was near the protractor whose 90° radial line, when extended back through this drafting tool's origin point intersected the South Pole of the globe. That 14 inch ruler's *right* side also passes behind the lamp and that means there is some connection between the ruler, the carpenter's plumb bob weight, the tripodal lamp, and the globe that we must determine before we can obtain the information that Bessler hid in this portion of the second DT portrait. The right end of the ruler is also located near a group of four bars that look like metal ingots. Most likely, they were pieces of lead that, once melted, were used by plumbers to seal pipes or by gunsmiths to cast lead shot. They are also symbols for the various sized cylindrical lead end weights used in Bessler's wheels that had different masses.

The connection between these symbols was not an easy one to find and properly interpret, but I did eventually find it. One must imagine the lamp being miniaturized and then placed at the South Pole of the globe. When that is done, the lamp's conical cover will then, like a magnetic compass needle, be pointing due north (actually all directions from the South Pole point north!) and also, as a symbol for the hour and minute hands of a clock, be indicating 12:00 noon for some point on the globe. Although it's not obvious from just viewing the southern hemisphere of our planet on the globe in the second DT portrait, if one could somehow draw a line from the globe's South Pole upward, north, to the equator represented by the top portion of globe's circular circumference and then down and around to its North Pole which is hidden from view on the other side of the globe, he would discover that the line passed directly through *England*!

So, we see from the arrangement of these shop table items in the second DT portrait, that Bessler made a symbolic connection between the mass of the weights in his wheels and the country of England. That connection is made by the symbol of a ruler perfectly balanced on a carpenter's plumb bob weight made of solid lead. The ruler only remains balanced on the curving side of the conical bob because *both* of its ends are exactly 7 inches long and, thus, of *equal length* and because, being of uniform cross sectional area, *both* of these 7 inch lengths of the ruler are of *equal mass* (actually, both halves of the ruler are a fraction of an inch longer than 7 inches, but we may discount those fractional inches in considering this symbol and it does not affect the equality of their masses). We can further imagine that the circular cross section of the plumb bob weight that the ruler is precisely balanced upon represents a circular cross section of the Earth and that the balancing of the equal lengths and masses of the two 7 inch long sections of the ruler on the bob weight's two sides represents the use of the *same* measuring units for both length and mass in two separated regions of our planet. One of those uses was over in England and sanctioned by its government and the other use was in Saxony and employed by Bessler in the construction of his wheels.

Thus, the hidden information contained in the second DT portrait's related symbols of globe, protractor, duck's bill goniometer, "L" square, 14 inch ruler, carpenter's plumb bob weight, tripodal lamp, and lead ingots is now finally revealed. We see that the inventor is telling us that the units of length *and* mass he used in the construction of *all* of his wheels were *identical* to the same units as used in the England of his day. As previously stated, it is mainly because of their use by Bessler that I have adhered to these same units in my own research into the mechanics of his imbalanced perpetual motion wheels.

The use of the globe symbol, with the lamp superimposed on it that symbolizes the two hands of a clock pointing to 12:00 noon as well as a magnetic compass needle pointing north, to indicate the country of England also makes sense from the amount of money the inventor demanded for the secret mechanics he had found for achieving an imbalanced perpetual motion wheel. Bessler set the price at 100,000 thalers after learning about the prize of £20,000 being offered by the British Parliament under their "1714 Longitude Act" to anyone able to find a more accurate way to determine the longitude of a ship at sea. Bessler, being a clockmaker, might have considered trying to win that prize money himself because one method to determine the longitude of locations either east or west of London is to carry a highly accurate clock along on a sea voyage that always indicated the local time in London and that is how it was finally done.

A ship's navigator could then, when the time aboard his ship was exactly 12:00 noon (determined at the instant of the maximum elevation of the Sun above the horizon by using a precise optical instrument known as a "sextant"), note the difference between that time and what his reference clock aboard his ship said the time was in London, and then convert that time difference into degrees, minutes, and seconds of longitude on the Earth's surface. For every hour the ship's noontime was *ahead* of the time in London as shown on his reference clock, the ship would be 15° *east* of London and for every hour the ship's noontime was *behind* the time in London, the ship would be 15° *west* of London. After many decades of struggle, a self-educated English carpenter and clockmaker named John Harrison finally collected less than half of the promised prize amount in 1773 at the age of 80 with a highly accurate chronometer he designed and had constructed which was based on a design he had first developed in

1759 that used a mainspring powered oscillating balance wheel to precisely regulate its rate of timekeeping. Three years later he died.

In studying the second DT portrait, I also immediately noticed how many "V" shapes there are in it. There is the conical carpenter's plumb bob weight, the conical cover at the top of the lamp, the two visible feet of the lamp, the spread arms of the duck's bill goniometer, the drafting "L" square, the dividers or drafting calipers that rest on top of a metal file near the bottom of the portrait, and a larger version of this that Bessler holds against the upper right quadrant of the globe. What could all of these "V" shaped objects mean?

As I'm sure most readers know, the letter "V" is also used today as it was in Bessler's time for the number 5 because it happens to be the Roman numeral for the number 5. In fact, Roman numerals were almost exclusively used on the clock dials that Bessler, as a clockmaker, would have seen back then. Many have pointed out that this particular number was one of Bessler's "favorite" numbers and it crops up throughout his writings and illustrations and in the second DT portrait I'm referring to here there are even several pentagrams or five-pointed stars cleverly hidden within it which can be thought of as five "V's" attached to the vertices of a pentagon. The inventor's penchant for this number and the use of its Roman numeral was probably due to the various esoteric or numerological significances of this particular number to him.

Bessler had studied the Hebrew Cabbala which is a collection of various ancient Jewish mystical interpretations of the Old Testament that became of interest in Europe during the Middle Ages. From them one learns that the number 5 is associated with things like balance and motion as well as the fast moving planet Mercury (interestingly, this planet was associated with the Roman god of Mercury who, aside from being the swiftly moving messenger of the other gods, was considered the patron god of financial gain which is certainly something that Bessler was interested in!). 5 is also a number associated with both man and divine grace and suggests the state that humans can achieve after they overcome their lower animal selves. In the New Testament, this number is occasionally used and we read that Jesus had 5 wounds inflicted on this body during the Crucifixion which symbolize him being the divine in human form and, thus, subject to physical death, but also having the ability to eventually overcome it through bodily resurrection. So we see from all of this that 5 was a number that would have appealed to Bessler because of his desire to develop a wheel, which like the planet Mercury, would stay in rapid continuous motion and also because he may have done some immoral things in his youth due to bad associations he had made and wanted to atone for them with the help of God's divine grace and by trying to live a pious and religious adult life.

But, there is another important interpretation of this number that I feel is the real reason for its repeated use in the second DT portrait as its Roman numeral version.

The number 5 also represents the letter "E" in both the English and German alphabets because in both alphabets the *fifth* letter is "E". The word "England" also begins with the letter "E" and the German word for "England" is also "England"! Thus, this could be yet another way that Bessler was using symbols in the second DT portrait to tell his readers that the units of length and mass he was using in the construction of his wheels were identical to those being used in England at the time.

The unit of mass or weight known as the "pound" came into general use in Europe around the year 1300, but, like the ell, its value tended to vary from one region to the next although, generally, not by much. In England, during the reign of Queen Elizabeth (1558 to 1603), the pound was officially defined as being equal to the weight of exactly 7000 grains of wheat which, because a large sample of grains of very slightly varying mass were involved, tended to be quite uniform. This eventually became the "avoirdupois pound" which was divided into 16 ounces and, despite the revisions of many other units of mass over the centuries, it is still the same English avoirdupois pound weight that is commonly used today in the US.

After the introduction of the metric system of units, the pound was given a value of about 454 grams which is about 5/11ths of a kilogram. The word "avoirdupois" itself comes from the Anglo-Norman French words meaning, literally, "goods of weight" and refers to this unit of mass being used to measure bulk items that were placed on large balances or steelyards which were also known as Roman balances. Using the latter, the load to be weighed was placed in a suspended pan or was hung from a hook attached to the shorter end of a metal beam which had its pivot near the load. A metal counter weight was then slid along the top of the longer end of the beam which was located on the other side of the pivot and away from the load. The counter weight was slid away from the pivot until the beam was perfectly horizontal. At that time the position of the counter weight along the longer portion of the beam would indicate the weight of the load by markings on the beam itself.

So, because of Bessler used avoirdupois pounds and ounces that were virtually identical to the ones we use today, I am quite confident that any replicas of Bessler's wheels based on the designs I provide in this book will be, for all practical purposes, identical to the originals constructed by Bessler in terms of the physical dimensions of their parts as well as the masses of those parts.

I should point out, however, that, even if I am wrong about this matter and Bessler's inches and pounds were not identical to ours today, then it would still make no difference as far as the functionality of the designs for his wheels that will be found in the remainder of this volume. The successful operation of his wheels' imbalanced pm mechanics really depended upon the unique geometry of their parts and the various *ratios* of the lengths and masses of those parts with respect to each other and not upon the particular units of length and mass that he employed. This is because Bessler used the same Euclidean geometry in the design of his wheels' parts that we do today and the laws of mechanics and physics those parts obeyed as they worked inside of one of his wheels were identical to the same laws we use today and, most importantly, *independent* of the units used for lengths and masses.

In the unlikely event, however, that I am wrong and Bessler's inches, feet, and pounds were not the same as the English avoirdupois units being used over in England in his day and that we still use today, then the only result will be that the replica wheels made from the designs to be presented in the remainder of this book will vary a bit in terms of their sizes and masses from the wheels Bessler personally constructed, but these modern replicas should *still* work.

There are many more clues represented by the symbols of the globe, protractor, duck's bill goniometer, 14 inch ruler, carpenter's plumb bob weight, tripodal lamp, and metal ingots in the second DT portrait, but their analysis will have to wait until the last chapters of this book.

CHAPTER 5

DETAILS OF HIS ONE-DIRECTIONAL, 3 FEET DIAMETER, GERA PROTOTYPE WHEEL

In Chapter 3 I provided the reader with a qualitative overview of the principle that Bessler's wheels used in order to display the pm effect. In this, the 5th chapter (quite appropriate because, as was previously mentioned, the number 5 had special significance for the inventor), I want to give more details of the precise mechanics within his one-directional wheels which produced this effect. I also want to provide readers with enough details of his 3 feet diameter Gera prototype wheel, in particular, to allow for its duplication. That will involve an analysis of the weighted levers used inside of the drum of the prototype followed by an analysis of the arrangement of the various coordinating cords that controlled the motions of its eight weighted levers during drum rotation and which were essential to its operation.

Let us now begin with an analysis of the actual activity that took place within his "toy" Gera prototype wheel as its drum began its first 45° segment of clockwise rotation. For the following treatment, the reader should locate Figures 3(a) and 3(b) at the end of this chapter and refer to them.

As can be seen from these two figures, the 3 feet diameter, one-directional, Gera prototype wheel contained eight weighted levers made of wood that were free to rotate around stationary steel pivot pins that had outside diameters of 0.0625 inch = 1/16 inch. The ends of the eight pivot pins were pressed into and securely held by the drum's eight parallel pairs of wooden radial frame pieces all of which were securely attached to a central wooden axle (note that the *front* one of each parallel pair of radial frame pieces can be considered removed or present, but transparent in order to facilitate viewing the weighted levers). The pivot pins were all 4 inches in length and did not project beyond the outside facing surfaces of the radial frame pieces that held them. The two 3 arm side pieces of each parallel pair of wooden lever pieces (again, the *front* piece is not shown and can be considered removed or present, but transparent in order to facilitate viewing of the interior of the lever) also each had pivot holes drilled into them that then had thin metal sleeves pressed into them that helped protect the wood of the side piece from wear due to its cyclical and partial rotations around the lever's steel pivot pin which were just oscillations and not fully rotations about the pivot pin. The 16 tiny metal sleeve inserts used on all eight weighted levers would each have had an *inside* diameter 0.0625 inch = 1/16 inch and a length of 0.125 inch or 2/16 inch *after* being fastened in place. (One can use small "pop rivets", also known as "blind rivets" or "break stem rivets", for these little metal sleeves just as long as they all have the correct inside hole diameter. After insertion of a rivet into the hole drilled in the wood for it, a special hand tool is used to expand one the rivet's ends which will then tightly secure the rivet to the lever's 3 arm side piece. A well stocked craft supply store should have these rivets and the tool needed to fasten them.) The eight pivot pins' center axes were each located at an exact distance of 14 inches from the center axis of the axle and, geometrically, their ends formed the vertices of two imaginary octagons within the drum (if all of the points on all of the axes are considered, then they formed a single imaginary octagonal *prism* within the drum).

The two wooden parallel side pieces that gave form to each lever each had an odd 3 arm shape to it that made it look somewhat like the 25th letter of the English alphabet which is the letter "Y" and were only 0.125 inch = 2/16 inch thick. The flat *inside* facing surfaces of these 3 arm side pieces were spaced exactly 3 inches apart and that spacing was maintained by six 0.0625 inch = 1/16 inch diameter steel hook attachment pins that penetrated the wood of the arms of the side pieces and spanned the 3 inches distance between them. These hook attachment pins, however, unlike a weighted lever's single pivot pin which was 4 inches in length, were only

3.25 inches in length so that their ends did not project beyond the *outside* facing surfaces of a lever's parallel pair of 3 arm side pieces whose *combined* thickness was 0.25 inch or 4/16 inch. There were also six other 3 inch long thin wooden pieces called "hook keepers" glued in place between a parallel pair of 3 arm side pieces' inside facing surfaces that helped give a weighted lever rigidity and also kept its 3 arm side pieces at a fixed distance from each other (the wooden hook keepers are not shown in Figures 3(a) and 3(b) because of their small size relative to the 3 arm side pieces).

Located between the ends of the two *longest* pair of parallel arms or "main arms" of each weighted lever's parallel pair of 3 arm side pieces, there was a small end weight made from a cast ingot of lead that was 3 inches in length and had a trapezoidal cross section. It weighed 0.375 pound = 6 ounces and was attached by two small screws that penetrated the wooden ends of the lever's main arms and then penetrated about 0.5 inch = 8/16 inch into the trapezoidal ends of the ingot and along an axis that passed through the ingot's center of gravity which was in the middle of the ingot. This "center axis" for the ingot was located back 0.25 inch = 4/16 inch from the larger *outward* facing rectangular surface of the ingot so that the ingot's center of gravity, after the ingot was attached to the ends of the main arms, was located exactly 3.5 inches from the center axis of a weighted lever's steel pivot pin. (For an ingot with a *perfect* isosceles trapezoidal cross section, the center axis containing the ingot's center of gravity would have been located exactly 0.233 inches back from its larger, outward facing rectangular face. But, in order to a adjust the mass of the ingot and make it as close to 0.375 pound = 6 ounces as possible, Bessler would have rounded off the edges of an ingot's outward facing rectangular face and that would then have moved the ingot's center of gravity to almost exactly 0.25 inches back from its larger, outward facing rectangular face.)

Many craftsmen, however, may prefer not to have to melt, cast, and handle lead which is a toxic heavy metal. In that case, they may consider substituting a steel block of equal mass for each of the 3 feet diameter, Gera prototype wheel's eight lead ingot end weights. Obtaining eight steel blocks that are 0.5 inch = 8/16 inch thick x 0.88 inch = 14.1/16 inch wide x 3 inches long will do nicely. Each steel block will contain about 1.32 cubic inches of steel and have a mass of 0.375 pound = 6 ounces. The exact centers of the rectangular end faces of these steel blocks will need to have small pilot holes drilled into them so that they can be attached to the ends of each parallel pair of main arms of a lever using self-tapping machine screws. An additional, *notchless*, wooden hook keeper with a width of 0.5 inch = 8/16 inch can then be glued into place at an angle behind the trailing side of each steel block in order to function like an elongated roof that will keep the main cord attached to its lever's L1 hook attachment pin just behind the steel block from making contact with the block's sharp trailing edges when a weighted lever's pivot pin's center axis is approaching the drum's 9:00 position.

(Note: To prevent possible future confusion when I am providing values for the three physical dimensions of a three dimensional part, which is almost always a square or rectangular prism in shape, it should be remembered that I will always use the word "thickness" to identify the object's smallest physical dimension, the word "width" to identify the object's intermediate physical dimension, and "length" to mean the object's largest physical dimension. Often, I will only provide the values of two of a part's three physical dimensions, but the words attached to them will still give the reader an idea of what their relationship is to the missing dimensional value.)

Figures 3(a) and 3(b), when joined, show a cross sectional axial view of the weighted levers and the drum as viewed from the front side of Bessler's 3 feet diameter Gera prototype wheel when it was in clockwise rotation (I will also always refer to the circular side face of a drum undergoing clockwise rotation as being the drum's "front" side unless otherwise stated). We can see that there were actually six steel hook attachment pins that penetrated the two parallel 3 side arm pieces of each weighted lever (recall that in these figures the front 3 arm side piece of each weighted lever has been removed or made transparent in order to facilitate viewing the inside of the lever). It will also be noticed from Figures 3(a) and 3(b) that, in order to show as much internal detail as possible for the 3 feet diameter Gera prototype wheel's drum within the confines of the six inches by nine inches book pages of this volume, I was obliged to place the illustration of the drum's left *ascending* side on the left page of two adjacent pages that face each other and the corresponding illustration of the drum's right *descending* side on the right page of the two adjacent pages that face each other.

The exposed 3 inch lengths of the six 3.25 inch long hook attachment pins whose ends were embedded in the parallel pair of 3 arm side pieces that formed each weighted lever had coordinating cords (which were actually just strings or strong sewing thread for this first and smallest of Bessler's wheels) attached to them by little metal hooks. These hooks were confined to specific sections of the exposed length of a hook attachment pin by a cutout notch in the wooden hook keeper that was pressed up against that attachment pin (again, due to the small size of the metal hooks, they are not shown in Figures 3(a) and 3(b), but will soon be shown in Figure 4). The coordinating cords, as their name implies, coordinated or controlled the various partial rotational motions or oscillations of the wooden weighted levers about their steel pivot pins during drum rotation in the *exact* way necessary to keep the center of gravity of all of the eight weighted levers of the Gera prototype wheel on the descending side of the axle. This was because each cord and its attached hooks would only allow a maximum separation distance between the center axes of the two pins that they interconnected. These separation distances had to be as accurate as possible otherwise a wheel would not run as efficiently as possible or might not run at all!

The reader will notice that I have given names to each of a wooden 3 arm side piece's arms and can see the names for them indicated for the weighted lever shown at the 6:00 position of the drum in Figure 3(a). (As I previously mentioned, since the axial view of a drum's side face is circular and resembles a round 12 hour clock dial, I designate the different positions around its circumference by using the times, in hours and minutes, as found on a clock dial. These times should be imagined as remaining fixed in space as different parts of the drum rotate past them. Also, as is the case of the 3 feet diameter Gera prototype wheel shown in Figures 3(a) and 3(b), *all* of the wheel drums I illustrate should be considered to be in clockwise rotation unless otherwise stated.)

The longest arm of a 3 arm side piece is called the "main arm" or "L arm" (with the letter "L" standing for "Lever") and it has the end of only a single steel hook attachment pin penetrating it which is labeled the "L1" hook attachment pin. Located 45° clockwise around a lever's pivot pin's center axis from the main arm's center line (that's an imaginary line passing through the center axis of the pivot pin and the center of the straight cutoff end of one of a weighted lever's arms) is the center line of the "A arm" which is the shortest of the three arms. It has one of the ends of each of four hook attachment pins, labeled "A1, A2, A3, and A4" (with the letter "A" designating the A arm), penetrating it and the A1 hook attachment pin is closest to the weighted lever's pivot pin. Finally, there is the "B arm" which is longer than the A arm, but shorter than the main or L arm. The B arm is penetrated by the end of only a single hook attachment pin which is labeled the "B1" hook attachment pin (with, obviously, the letter "B" designating the B arm). Interestingly, the center line of the B arm is located exactly 157.5° degrees clockwise around the pivot pin's center axis from the A arm's center line and also located exactly 157.5° degrees *counter*clockwise around the pivot pin's center axis from the main arm's center line and this is what gives these 3 arm side pieces their resemblance to the letter "Y" with the main and A arms forming the top fork of the letter "Y" and the B arm forming the vertical lower section of the letter.

Bessler's one-directional, 3 feet diameter, Gera prototype wheel required a total of forty coordinating cords and sixteen little steel extension springs (two springs per weighted lever times eight levers = sixteen springs) in order to coordinate the various motions of its eight weighted levers so that their center of gravity would stay on the descending side of the axle during drum rotation. However, in order to prevent these cords from rubbing against each other during drum rotation as they alternately tightened when needed and slackened when not needed, then fraying, and eventually breaking, the inventor had to organize them into sets of cords of various lengths and functions and then confine these sets to five different "layers" inside of his prototype wheel's drum that spanned the distance of 3 inches between the inner facing surfaces of the parallel pairs of wooden 3 arm side pieces of the eight weighted levers.

We can call the front such layer "layer 1" (designated as "front" since it is *closest* to a person viewing the front circular side of the drum which is the side from which the drum would be seen to be rotating in a clockwise direction when released and allowed to spontaneously start rotating), the layer behind that "layer 2", then the next and middle of the five layers "layer 3", the next layer behind that "layer 4", and the final and backmost layer "layer 5". In the case of his Gera prototype wheel, each layer was only 0.6 inch thick! No doubt, Bessler's skills as a clockmaker aided him in compressing the five required coordination cord layers into a cylindrical volume of space only 3 inches thick. One should think of these five layers used inside of each of his one-directional wheels by their forty coordinating cords (which, as previously mentioned, were mere strings or threads in the case of the Gera prototype wheel) as occupying a *cylindrical* volume of space whose imaginary circular flat faces were in contact with the sixteen *inner* facing surfaces of the drum's eight weighted levers' parallel pairs of 3 arm side pieces and whose outermost cylindrical surface was in contact with the inner surface of the drum's outer rim wall.

For the purpose of analysis, in Figures 3(a) and 3(b) all five coordinating cord layers have been compressed into a single layer represented by the plane of the opened book's two facing pages and this gives the *false* impression that many of the cords are crossing each other and will come into rubbing contact as the weighted levers they are attached to swing or actually oscillate about their pivot pins during drum rotation. Furthermore, of the forty coordinating cords, only those fourteen which are visible and taut during any 45° segment of drum rotation are shown. This was done in order to make the explanation of how the individual weighted levers move during drum rotation easier to follow by removing the distraction of the cords which are not employed during any 45° segment of drum rotation. In the latter half of this chapter, all five of the coordinating cord layers will be *separately* illustrated and this information will be of great value to any craftsman intending to attempt duplicating Bessler's 3 feet diameter, Gera prototype wheel. Let us now, however, concentrate on briefly describing the five layers.

The width of each coordinating cord layer inside of the Gera prototype wheel assured that there would be an average separation distance of 0.6 inches between any two cords in adjacent layers or between a cord and a lever's 3 arm side piece. (I write "average separation" here because the notches in the wooden coordinating cord hook keepers did allow some axial movement of the hooks along their attachment pins which would then allow two adjacent hooks and their attached cords to come closer together than 0.6 inch. If the width of notch was 0.125 inch or 2/16 inch and the diameter of the wire used for the hooks was much smaller, then two adjacent hooks and their attached cords, assuming they had *theoretical* diameters of 0.000 inch, could only get as close as 0.6 inch − 0.125 inch = 0.475 inch to each other. A hook and its attached cord, however, could, again assuming they had theoretical diameters of 0.000 inch, get as close as 0.6 inch − 0.0625 inch = 0.5375 inch from the inside facing surface of a weighted lever's 3 arm side piece. These theoretical minimum separation distances would have been sufficient to prevent adjacent cords, whether both were bowed out by centrifugal forces or bowed in by gravitational forces, from touching each other. Of course, in the real world the hook wires and their attached cords would have had finite diameters greater than 0.000 inch and, the closer they got to a value of 0.125 inch without actually reaching it, which would have prevented the hooks from freely rotating around their attachment pins because they would be rubbing against the side walls of their hook keeper's notches, the closer the separation distances between adjacent hooks and between hooks and inner facing surfaces of 3 arm side pieces would be to a value of 0.6 inch.)

Layers 1 and 5 were identical and each of these layers contained eight small steel helical extension springs that each had a spring constant of k = 1.25 pounds per inch = 20 ounces per inch. Each spring had one of its ends attached to a little metal hook which was, in turn, attached to a 0.0625 inch = 1/16 inch diameter "spring hook attachment anchor pin" whose two ends were embedded at the midpoints of a pair of parallel *outer* octagonal frame pieces of the drum. The ends of these outer octagonal frame pieces, in turn, made contact with and braced two adjacent parallel pairs of radial frame pieces attached the prototype wheel's axle. In Figures 3(a) and 3(b), however, the reader will only count eight springs instead of sixteen. Where are the other eight springs? They are actually hidden behind the springs in layer 1 because layers 1 and 5 are identical and also superimposable.

The other lever pointing end of each extension spring had a small eyelet formed in it to which another little metal hook attached. That second metal hook then had a coordinating cord, the "spring cord", tied to it and the other end of that cord was then tied to a third little metal hook which was identical to the two attached to the extension spring (note that all of the forty-eight metal hooks inside of one of Bessler's one-directional wheels were identical in size, shape, and mass). That third spring cord hook was then, in turn, slipped onto the A1 hook attachment pin of a weighted lever's A arm. These little steel hooks were shaped so that they formed almost completely closed loops that were a bit springy so that they could easily be snapped into place when connected to a steel hook attachment pin inside of a weighted lever and would thereafter not come off of the pin unless manually removed which would only be necessary if, after prolonged wheel operation, one of the coordinating cords they were attached to or a spring itself broke. And, of course, occasionally one of the hooks itself would fail from metal fatigue and need to be replaced along with any coordinating cord that was attached to it.

The 16 steel helical coil extension springs located in layers 1 and 5 were unstretched for a weighted lever whose pivot pin's center axis was located at the drum's 4:30 and 6:00 positions (or traveling between those drum positions). At those positions, the weighted lever's parallel pair of main or L arms was in contact with their wooden "radial stop piece" that was attached to the lever's parallel pair of radial frame pieces which securely held the lever's pivot pin. The lead ingot end weight of each weighted lever, however, did *not* make actual physical contact with its lever's radial stop piece. Rather, only the parallel pair of main arms between whose ends each lead ingot end weight was located actually made physical contact with the lever's radial stop piece.

As a weighted lever's pivot pin's center axis orbited the center axis of the wheel's axle from the drum's 6:00 position, up to its 12:00 position, and then back down again to the 4:30 position, the *two* extension springs attached to the lever's A arms would, starting with unstretched lengths of at least 1 inch at the drum's 6:00 position, be stretched through a maximum distance of 0.5 inch as the weighted lever's pivot pin traveled clockwise to the drum's 9:00 position. However, from studying my successful computer models for the Gera prototype wheel carefully, I noticed that a weighted lever whose pivot pin's center axis was traveling toward the drum's 9:00 position actually had stretched its two extension springs to their maximum distance of 0.5 inch about 6° of drum rotation *before* reaching the drum's 9:00 position. Then as the pivot pin's center axis continued to travel clockwise from the drum's 9:00 position to about 12° past the drum's 3:00 position, the two extension springs would slowly contract back to their unstretched lengths. As they did this, the weighted lever would rotate clockwise about its pivot pin.

Layers 2 and 4 were also identical, but *not* superimposable in terms of the arrangement of the coordinating cords each contained. Each of these layers contained four "long lifter cords" each of whose two ends were tied to little metal hooks. One of the hooks on each long lifter cord was attached to a weighted lever's A3 hook attachment pin while the other was attached to the B1 hook attachment pin of a weighted lever whose pivot pin's center axis, going clockwise around the axle's center axis, was located 90° away from the first weighted lever's pivot pin's center axis. Bessler could only place four of the long lifter cords in each layer because, if he tried to put all eight into a single layer, they would have rubbed against each other, frayed, and quickly broken during wheel rotation.

Of the eight long lifter cords in coordinating cord layers 2 and 4, Figures 3(a) and 3(b) only show two of them. This is because, at the beginning of the 45° segment of clockwise drum rotation that started with the one depicted in the figures, only the two long lifter cords between the weighted levers whose pivot pins' center axes were located at the drum's 6:00 and 9:00 positions and between the weighted levers whose pivot pins' center axes were located at the drum's 7:30 and 10:30 positions were tight. As the drum then proceeded through 45° of clockwise rotation, the long lifter cord in layer 2 that started off between the weighted levers at the 6:00 and 9:00 positions of the drum would remain tight. However, the long lifter cord further back in layer 4 that started off between the weighted levers at the 7:30 and 10:30 positions of the drum would become slack after only 9° of clockwise drum rotation. Then, when the *next* 45° segment of clockwise drum rotation began (which is not depicted in the figures), it would be the long lifter cord in layer 4 that started off between the weighted levers at the drum's 6:00 and 9:00 positions that would remain tight throughout the entire next 45° segment of drum rotation while the long lifter cord in layer 2 that started off between the weighted levers at the drum's 7:30 and 10:30 positions that would become loose after the first 9° of clockwise rotation of that next segment of drum rotation. This clever *alternating* pattern of the long lifter cords tightenings and loosenings that Bessler conceived would be repeated four times during each complete drum rotation.

Finally, we come to coordinating cord layer 3 which was the middle layer of the 5 layers. It is unique because it actually contained *two* complete sets of eight coordinating cords and they were arranged so that, during drum rotation, none of the cords could make physical contact with each other. Of the eight main cords, only three are shown because they are tight at the beginning of the 45° segment of clockwise drum rotation depicted in the two figures. These three main cords interconnect the weighted levers whose pivot pin center axes were at the drum's 10:30, 12:00, 1:30, and 3:00 positions. Shortly after the 45° segment of clockwise drum rotation started, the main cord interconnecting the levers whose pivot pin center axes started were at the 1:30 and 3:00 positions of the drum would become slack after about 12° of clockwise drum rotation. The other complete set of coordinating cords in layer 3 consisted of eight "stop cords", but only one of them, the one that attached the weighted lever whose pivot pin's center axis was at the drum's 9:00 position, is shown because it is the only one of the eight cords that was tight. It actually got tight about 6° degrees of drum rotation *before* its weighted lever's pivot pin's center axis reached the drum's 9:00 position, but, as that pivot pin axis passed the drum's 9:00 position, the stop rope would almost immediately begin to slacken. It's primary and very important function was to assure that when a weighted lever's pivot pin axis reached the drum's 9:00 position, the center lines of the lever's main arms would form a precise angle of 62.5° with the center lines of that lever's radial frame pieces.

At this point the reader should have a fairly good understanding of the general arrangement of the 16 extension springs, forty coordinating cords (although only 14 of them that were tight at the beginning of the 45° segment of clockwise drum rotation shown in Figures 3(a) and 3(b)!), and some of the major structural details of the drum of Bessler's one directional, 3 feet diameter, Gera prototype wheel and its hidden set of eight weighted levers. Now it's time to put this information to use in providing a more detailed picture of how this wheel's mechanics functioned during its first 45° segment of drum rotation after its axle was released so that it could do the "impossible". To do this, the reader is again directed to Figures 3(a) and 3(b).

It will be noticed that in the two figures I have placed a small black arrowhead near the back main arms of each of the drum's eight weighted levers. These arrowheads indicate the directions, either clockwise or counterclockwise, that the three parallel pairs of arms, main or L, A, and B, of a weighted lever would rotate around their common pivot pin's center axis during a 45° segment of clockwise drum rotation began. As can be seen, the 9:00, 10:30, 12:00, 1:30, and 3:00 weighted levers all rotated clockwise around their respective pivot pins' center axes during the segment of drum rotation. Only the 6:00 and 7:30 weighted levers rotated counterclockwise.

The 4:30 weighted lever's back main arm has no black arrowhead near it because it underwent no rotational motion around its lever's pivot pin's center axis as that axis traveled clockwise around the center axis of the axle from the drum's 4:30 to 6:00 positions. During a 45° segment of clockwise drum rotation, this weighted lever's parallel pair of main arms remained in constant physical contact with its wooden radial stop piece which was attached to the parallel pair of radial frame pieces that held the steel pivot pin for the lever. I should again mention here that, in order to enhance the view of the Gera prototype wheel's mechanics, all of the *drum's* front frame pieces in Figures 3(a) and 3(b), such as the radial, inner octagonal, outer octagonal, and very short radial frame pieces have been removed or can be considered to be present, but transparent.

Let us now analyze these individual coordinated motions of the Gera prototype's eight weighted levers in more detail.

As the drum rotated through a 45° segment of clockwise rotation starting from a standstill after the axle was freed to rotate, the 6:00 and 7:30 weighted levers would immediately begin to rotate counterclockwise around their pivot pins' center axes. Thus, at the very beginning of the 45° segment of clockwise drum rotation, the 6:00 weighted lever's parallel pair of main arms separated from its wooden radial stop piece with which they were in physical contact. As the 6:00 weighted lever rotated counterclockwise about its pivot pin's center axis as that axis moved away from the drum's 6:00 position, the lever's center of gravity steadily swung closer to the center axis of the axle. This action then, through its long lifter cord, applied a torque to the B arm of the weighed lever whose pivot pin's center axis was leaving the drum's 9:00 position that would help that lever's three parallel pairs of arms rotate clockwise about their common pivot pin's center axis.

As the 6:00 weighted lever's parallel pair of main arms was rotating counterclockwise about their pivot pin's center axis, the 7:30 weighted lever was also rotating counterclockwise around its pivot pin's center axis as that axis approached the drum's 9:00 position and that weighted lever's center of gravity also steadily swung closer to the center axis of the axle.

Via its long lifter cord, that weighted lever would then apply a torque to the weighted lever whose pivot pin's center axis was leaving the drum's 10:30 position and help to rotate it clockwise about its pivot pin's center axis.

We see from this that, for the entirety of the 45° segment of clockwise drum rotation, the rapid clockwise rotations about their pivot pins' center axes of the two weighted levers whose pivot pins' center axes were leaving the drum's 9:00 and 10:30 positions was the result of the combined torques applied to them by the five weighted levers whose pivot pins' center axes were leaving the drum's 6:00, 7:30, 12:00, 1:30, and 3:00 positions. It was the resulting rapid vertical motion of these two weighted levers together with the horizontal motions of the five other swinging weighted levers which allowed the combined center gravity of all eight weighted levers to remain on the axle's descending side throughout the 45° segment of clockwise drum rotation and was an effect no other inventor aside from Johann Bessler had ever managed to achieve (that we know of, that is).

In an effort to truthfully describe the actions of the weighted levers inside of his wheels' drums during rotation without actually giving away enough details to aid rival inventors in duplicating them, Bessler simply states in AP:

"Anyone who wants to can go on about the wonderful doings of these weights, alternately gravitating to the center and then climbing back up again, for I can't put the matter more clearly."

Now that the reader is learning the details of the secret imbalanced pm wheel mechanics of Bessler's wheels, this vague description by the inventor can be more easily interpreted.

As the pivot pin's center axis of a weighted lever got closer to a clockwise rotating drum's 9:00 position, the center of gravity of that weighted lever was also getting closer to the center axis of the wheel's axle. By the time a weighted lever's pivot pin's center axis finally reached the drum's 9:00 position, the center of gravity of the weighted lever was as near to the center axis of the drum's axle as possible and this separation distance in the 3 feet diameter, Gera prototype wheel was about 15.6 inches. Since the maximum separation distance between a weighted lever's center of gravity and the axle's center axis occurred when the lever's main arms were resting on their wooden radial stop piece between the drum's 4:30 and 6:00 positions and was about 17.0 inches, we see that the center of gravity of a weighted lever got about 1.4 inches closer to the axle's center axis when the lever's pivot pin's center axis was at the drum's 9:00 position. Thus, the center of gravity of the weighted lever only came about 8.23% closer to the axle's center axis as the lever's pivot pin's center axis traveled from the drum's 4:30 to 9:00 positions. This same percentage would also be true for all of the imbalanced pm wheels Bessler constructed or planned to construct.

Now let us consider how the two extension springs that attached each weighted lever to the drum were stretched as the 45° segment of drum rotation took place.

As depicted in Figure 3(a), the two extension springs attached to each weighted lever would stretch a maximum distance of only 0.5 inch = 8/16 inch as the lever's pivot pin's center axis traveled from the drum's 6:00 to 9:00 positions. For the rest of the travel of the pivot pin's center axis from the drum's 9:00 to a location about 12° clockwise past the drum's 3:00 position, the two springs would be slowly contracting back toward their unstretched lengths and maintained those lengths as the lever's pivot pin's center axis traveled from that location down to the drum's 6:00 position to begin its next full clockwise rotation around the center axis of the drum. Since an extension spring used for this could not be allowed to stretch through a distance more than about 50% of its unstretched length, that meant that Bessler had to use springs that had a minimum unstretched length of no less than 1 inch. As indicated in Figure 3(b), *each* of the springs in the 3 feet diameter, Gera prototype wheel had a spring constant, k, equal to 1.25 pounds per inch = 20 ounces per inch. The reader should keep in mind that the two springs attached to each weighted lever's A1 hook attachment pin were not directly attached to that pin. Rather, each spring was attached by small metal hook to a short cord, a spring cord, which was then attached to the A1 hook attachment pin by the small metal hook at its other end.

If one could go back in time and open up the stationary drum of Bessler's 3 feet diameter Gera prototype wheel and insert a hand into it as Count Karl paid a sizeable amount of money to do, he would discover that the weight of the 9:00 weighted lever was so precisely counter balanced by the two small, but maximally stretched helical steel extension springs attached to it that he could cause it to immediately begin rotating clockwise about its pivot pin's center axis by just applying a clockwise torque to the end of one of its main arms with the lightest of touch. It was this near perfect state of balance of the 9:00 weighted lever by its two attached springs which allowed it to immediately be lifted by the long lifter cord from the 6:00 weighted lever as that lever's pivot pin's center axis began to pass the drum's 6:00 position and that lever began to swing counterclockwise about its pivot pin's center axis.

Bessler commented about this interesting effect in AP when, in a tirade against Christian Wagner, one of his most annoying and intelligent critics, Bessler described the motion of this particular weighted lever whose pivot pin's center axis was moving on its way from the drum's 9:00 position to its 10:30 position by writing:

"But the weights, which rest below must, in a flash, be raised upwards, and it is this, that Wagner cannot force himself to accept. But, crazy Wagner, just note that that is indeed the case with my device."

If one was to observe the internal mechanics of the inventor's 3 feet diameter, Gera prototype wheel as its drum reached a clockwise rotation rate in excess of 60 rotations per minute, he would see more than eight of its weighted levers' lead ingot end weights passing the drum's 9:00 position *every* second! However, the change of rotation from counterclockwise to clockwise of a weighted lever about its pivot pin's center axis would not be instantaneous. At the location where a weighted lever's approaching pivot pin's center axis still had to complete another 6° of clockwise drum rotation before reaching the drum's 9:00 position, the weighted lever would *already* have stretched its two attached extension springs to their maximal increase in length of 0.5 inches each, the stop cord attached to the lever would be stretched until it was tight, and, at this time, the weighted lever would be *motionless* with respect to its pivot pin. Only as the weighted lever's pivot pin's center axis passed the drum's 9:00 position would the lever begin rotating clockwise about its pivot pin's center axis. This brief 6° of drum rotation pause in the rotation of a weighed lever about its pivot pin as that pin's center axis approached the drum's 9:00 position was critically necessary in order to prevent the lever from overstretching and damaging its parallel pair of extension springs and possibly damaging the lever's parallel pair of B arms or the long lifter cord which connected its B arms to the main arms of the weighted lever whose pivot pin's center axis was approaching the drum's 6:00 position.

So, there we have it: the basic process by which Bessler's little 3 feet diameter Gera prototype wheel and, more importantly, *all* of his later and larger one-directional wheels managed to keep the centers of gravity of their eight weighted levers continuously displaced onto their axle's descending sides during clockwise drum rotation. The full explanation of exactly how his wheels managed to continuously output mechanical energy, however, had to await the discovery of new concepts in physics involving the equivalence of energy and mass at the beginning of the 20th century. Neither Bessler, Count Karl, Leibniz, nor Newton would have been able to understand exactly how Bessler's wheels worked. They simply did not have the physical concepts necessary to do that. All they would have instinctively known was that, if one could figure out a way to continuously keep the center of gravity of a wheel's weighted levers away from the center axis of the wheel's axle, then that wheel should continue to accelerate and, if it could supply enough torque, operate any outside machinery attached to it.

As Count Karl beheld the uncovered drum of Bessler's 3 feet diameter Gera prototype wheel when, in 1716, the inventor brought it to Weissenstein Castle in order to prove to the count that it was genuine, the former would, of course, have been unaware of all of the flows of energy and mass that were taking place among its diminutive wooden levers and their attached lead ingot end weights. The count would have just marveled at how intricately the wheel's eight weighted levers were interconnected by their main cords and long lifter cords. He would have realized that the design did, in fact, maintain the center of gravity of its axle's descending side because, every time he grasped and squeezed a section of its small wooden axle that was outside of the drum to stop the wheel and then released it, the little toy wheel would immediately have begun rotating again and slowly building up its rotation rate until it was in excess of 60 rotations per minute. He would have felt the continuous and cooling breeze coming from the drum at that speed and that would have been more than enough to convince him that Johann Bessler had finally done it. He had an actual, *working* imbalanced pm wheel design!

Next, I want to give further details of the little wooden levers used in Bessler's 3 feet diameter Gera prototype wheel followed by illustrations that provide the craftsman seeking to replicate this wheel all of the coordinating cords, both tight and loose in all five of the coordinating cord layers. Finally, the chapter will end with some general structural details of the drum and its support stand. My goal in this chapter is to provide the reader with sufficient detailed information so that he will not only be able to construct a small prototype replica that actually works, but one that is as close as possible to what Bessler created. Once a serious reverse engineer and craftsman manages such a successful duplication on the smallest scale that Bessler used, he should then be well prepared to go on to construct larger versions and, for the most skilled of craftsman, an exact replica of the mighty 12 feet diameter, bidirectional Kassel wheel. (While, for historical purposes, I give the details of the 3 feet diameter Gera prototype wheel in this chapter, it should be pointed out that this is not the minimum diameter that a one-directional Bessler type imbalanced pm wheel must have. Indeed, if one can achieve clockmaker precision in the fabrication of his replica's parts, there is no reason why a wheel only a single foot in diameter could not be produced. I suspect that, once the basic design in this book is fully verified as "the" one Bessler used, we may see inexpensive and small one to two feet diameter versions being produced for collectors and for use as decorative desk and table top conversation pieces.)

In order to begin the analysis of the structure of the 3 feet diameter Gera prototype wheel's diminutive weighted levers, the reader should now refer to Figure 4 at the end of this chapter which covers three pages.

The first page of Figure 4 shows a side view of one of the Gera prototype wheel's weighted levers on the top part of the page and a top view of the lever on the bottom part of the page. On the top part of the second page continuation of Figure 4, we see a collection of coordinating cord hook keepers. The top six of them were glued into place inside of the weighted levers between the inner facing surfaces of their parallel pairs of 3 arm side pieces. Below them are two longer hook keepers and these were glued into place inside of the drum between the inner facing surfaces of its parallel pairs of inner and outer octagonal frame pieces. The bottom half of the second page continuation of Figure 4 shows how each coordinating cord had its ends tied to two tiny metal hooks which could then be attached to either a weighted lever's or drum's hook attachment pin.

The purpose of three of the four different types of cords found in Bessler's wheels was simply to limit the maximum separation distance between the center axes of two coordinating cord hook attachment pins as the motions of the weighted levers caused these cords to tighten. The fourth type of cord, the spring cord, merely served to transfer spring tension to a weighted lever whenever the separation distance between the center axis of an A1 spring cord hook attachment pin inside of a weighted lever and the center axis of a spring hook attachment anchor pin inside of the drum exceeded a certain maximum distance which was determined for a spring cord and its attached extension spring by the separation distance that allowed neither slack in the spring cord nor stretching of the extension spring for a weighted lever that had its pivot pin's center axis at the drum's 6:00 position, had its parallel pair of main or L arm in contact with their radial stop piece, and was, thus, vertically oriented. At the bottom of that second Figure 4 continuation page there is a listing of the maximum separation distances between hook attachment pin center axes that a particular type of coordination cord and its two attached metal end hooks would permit. The reader should study these separation distance values carefully because, unless they are strictly adhered to, a replica of the 3 feet diameter, Gera prototype wheel may not run efficiently or even at all.

I recently added a second continuation page to Figure 4 that was prompted by a discovery of a clue in the DT portraits that provided information about the size and shape of the end weights that Bessler used in the weighted levers of the 3 feet diameter, Gera prototype wheel. As can be seen in the drawing on that page, they are not cylindrical as were the end weights in his much larger 12 feet diameter, bidirectional wheels, but, rather, were actually trapezoidal prisms which he would have cast by pouring molten lead into a ceramic mold. This then resulted in the production of identical lead ingots having fairly precise and uniform masses which, after having their ends filed down flat, were each mounted between the ends of a parallel pair of wooden 3 arm side pieces using two steel screws. Interestingly, it was the larger rectangular surface of each ingot that faced outward and away from the lever's pivot pin.

I should point out here that, while many of the construction details to be given can be found after a very careful analysis and interpretation of the various geometric and alphanumeric clues given in the two DT frontispiece portraits and which will be provided in the later chapters of this volume, many of them are solely the result of my work with about two thousand computer models of Bessler's Gera prototype wheel over the course of about a decade. My justification for the accuracy of any construction details *not* provided by the clues in the two DT portraits rests upon the fact that they did eventually lead to a computer model wheel the testing of which confirmed that it could maintain the center of gravity of its eight weighted levers on the axle's descending side *throughout* a complete 45° segment of drum rotation and, most importantly, finish the segment with the configuration of the eight weighted levers being identical to what it was at the beginning of the 45° segment of drum rotation. (Actually a drum rotation segment's ending weighted lever configuration will, because of the increasing effect of the centrifugal forces acting on the weighted levers during drum acceleration, only be *nearly* identical to what it was at the beginning of the segment and that difference will cause the center of gravity of all eight weighted levers to move slightly closer to its *punctum quietus* location below the center axis of the axle and thereby lower the torque that the drum and its axle receive. In order for the configuration of the eight weighted levers at the end of a 45° segment of drum rotation to be identical to what it was at the beginning of the segment, the drum must be constrained by an outside resistance so as to rotate at a constant rate.). With this qualification understood, let us now consider the various parts of Figure 4 in greater depth.

Upon inspecting the side view of the Gera prototype's weighted lever in the top half of the first page of Figure 4, one sees a black trapezoid at the end of its horizontal main arm (again, the lever's front 3 arm side piece has been removed or can be considered present, but transparent in order to facilitate viewing the interior of the lever). That trapezoid represents a cross section of the little lead ingot end weight attached between the ends of the weighted lever's parallel pair of "main arms" (note that I sometimes refer to these arms as the "L arms" which stands for "Lever arms"). That end weight was a cast lead ingot 6 inches in length that had an isosceles trapezoidal cross section with a height of 0.5 inch = 8/16 inch and two bases of different widths. The smaller base was 0.5 inch = 8/16 wide and the larger base was 0.75 inch = 12/16 inch wide. Each lead ingot's mass, after filing its ends down to turn it into a right isosceles trapezoidal prism was very close to 0.375 pound = 6 ounces in mass. One such ingot was held between the ends of each lever's parallel pair of wooden main arms by two small steel screws that penetrated the ends of the 0.125 inch = 2/16 inch thick wooden main arm pieces and then pierced the flat trapezoidal ends of the ingot to a depth of at least 0.5 inch = 8/16 inch. There would also have been two small pieces of wood glued to the inner facing surfaces of the weighted lever's parallel pair of main arms that were up against the back surface of the ingot. Their purpose was to prevent the ingot from rotating should its screws loosen during drum rotation.

The center axes of the two steel screws that held the ingot to the ends of the lever's main arms were both locate along the center axis of the ingot which was an imaginary line along its length on which its center of gravity was located. Thus, the center axes of the screws would have pieced each trapezoidal face of an ingot at a location about 0.25 inch = 4/16 inch away from each of its two different sized bases. This resulted in the larger outward facing base and rectangular side of the ingot being 3.75 inches = 3-12/16 inches from the center axis of the lever's two pivot pin holes and placed the center of gravity of the ingot at a distance of 3.5 inches = 3-8/16 inches from the center axis of the two pivot pin holes. The reason why Bessler had the larger parallel rectangular surface of the ingot face outward away from a lever's pivot pin rather than the smaller parallel rectangular surface will be explained later in the chapter.

The two steel attachment screws for each of the lead ingot end weights in the Gera prototype wheel's weighed levers as well as the longer bolts used to attach the end weights, both ingot and cylindrical, to the parallel pairs of main or L arms in Bessler's post Gera prototype wheels can, technically, be referred to as "L2 end weight attachment screws or bolts" since their center axes are farther from a lever's pivot pin's center axis than is the center axis of the L1 main cord hook attachment pin. Generally, however, in the schematics of the various cord layers for the Gera prototype wheel provided in this chapter I will not label them as L2 end weight attachment screws since, unlike true hook attachment pins, they did not have coordinating cord metal end hooks attached to them. However, whenever the reader sees an illustration that shows a lead end weight, whether ingot or cylindrical in shape, attached between the ends of a parallel pair of main or L arms of a weighted lever, he should remember that it is always held in that location by a pair of short screws or a single long bolt that passes through the entire length of the end weight so that the center axes of the weight and its attachment screw or bolt coincide.

Also located between the parallel pairs of main arms of the Gera prototype's weighted levers was the L1 main cord hook attachment pin and the distance of its center from the center of the pivot pin's hole was exactly 2.6 inches. Just above and to the left side of this cord hook attachment pin and in contact with it was its "hook keeper", a thin piece of wood with a single square notch cut into it that would keep the little metal hook attached to one end of a main cord restricted to the middle of the 3 inches of exposed length of the L1 hook attachment pin (its actual length was 3.25 inch = 3-8/16 inch and 0.125 inch = 2/16 inch of each of its two ends were embedded in the lever's parallel pair of 3 arm side pieces). Notice that the width of the wooden hook keeper for the L1 hook attachment pin was angled so that it formed an angle of 135° with the center line of the main or L arm. The different angles of the hook keepers of the various coordinating cord hook attachment pins were necessary in order to assure that, as the weighted levers rotated in various directions about their pivot pins during drum rotation, the hooks attached to their coordinating cord hook attachment pins would be able to freely rotate about them without hitting the wooden hook keepers and damaging them.

It should be noted here that all of the six coordinating cord hook attachment pins used in this lever as well as the two small steel screws used to secure the lead ingot end weight in place were 0.0625 inch = 1/16 inch in diameter. The pins would have been made from drawn steel wire and were in wide use by the upper classes of the early 18th century to secure garments, hats, wig pieces, etc. Bessler could have obtained them from any milliner's shop (basically, a hat shop) that catered to the wealthier citizens of Saxony and then, after carefully measuring the correct locations to place them, forced their sharpened points through slightly smaller diameter pilot holes previous drilled into the softer wood of the parallel pairs of 3 arm side pieces by gently tapping them into place with a small hammer. The wood he used for the weighted levers' 3 arm side pieces and their hook keepers, the drum frame pieces, axle, and support stand of the Gera prototype wheel would have been European white oak which was widely available and relatively inexpensive. But, if one is not too particular about making a duplicate of the prototype wheel with the exact same materials that Bessler used, then any wood with a density of about 0.025 pounds per cubic inch = 43.2 pounds per cubic foot will suffice. It should also be noted here that, because of their small size, Bessler probably cut the sixteen 3 arm side pieces he needed for the 3 feet diameter, Gera prototype wheel's eight weighted levers out of single pieces of wood that were 0.125 inch = 2/16 inch thick (and I must emphasize here that the *thickness* of the wood used for the 3 arm side pieces of the weighted levers was only *half* that of the wood used for all of the Gera prototype's drum frame pieces which was 0.25 inch = 4/16 inch thick with the exceptions of its wooden bridge pieces and outer rim wall). The 3 arm side pieces could also have been assembled by gluing three separate pieces of wood together, but that assemblage would not have been as strong as cutting the pieces from a single piece of wood.

In his writings, Bessler mentions that he purposely designed his wheels so that they would be as easy as possible for a craftsman to construct and using wood for their drum frame pieces which had the same cross sectional dimensions, but only varied by their cut lengths, was one way that he tried to achieve this. The other way to facilitate construction with his later 12 feet diameter, bidirectional wheels was to use cylindrical lead end weights with a standard diameter whose mass the craftsman could then vary by simply altering their cylindrical lengths.

Again with respect to the side view of the weighted lever of the Gera prototype wheel in Figure 4 whose pivot pin's center axis was located at the drum's 9:00 position, it will be noted that there was another parallel pair of arms, the A arms, which were rotated 45° clockwise about the weighted lever's pivot pin away from the main arms. As previously noted, these parallel arms held four coordinating cord hook attachment pins to which the coordinating cord hooks of the four different types of coordinating cords were attached along with their individual wooden hook keepers.

The hook attachment pin closest to the lever's pivot hole, the A1 spring cord hook attachment pin, had a center axis that was exactly 0.5 inch = 8/16 inch away from the center axis of the pivot pin's hole. It was to this hook attachment pin that the *two* little metal hooks with spring cords tied to them were attached for each of the eight weighted levers in the Gera prototype wheel's drum. Since the two spring cords of each weighted lever, as well as their attached extension springs, were located in coordinating cord layers 1 and 5, the wooden hook keepers for all eight of the A1 hook attachment pins of the prototype wheel's weighted levers also had two square notches cut into them that kept the two little metal hooks attached to each weighted lever confined to the sections of its A1 hook attachment pin that corresponded to coordinating cord layers 1 and 5. Note that the wooden hook keeper piece was in physical contact with the side of the A1 hook attachment pin such that the width of the hook keeper was an angle 90° to the center line of an A arm.

At a distance of 0.933 inch = 14.93/16 inch from the center axis of the weighted lever's pivot pin hole, one finds the center axis of the A2 stop cord hook attachment pin. For each of the eight weighted levers, this hook attachment pin only had a single hook attached to it to which one end of the stop cord was tied. Since the Gera prototype wheel's eight stop cords were all located in the middle layer of the five coordinating cord layers or coordinating cord layer 3, the wooden hook keeper piece for the A2 stop cord hook attachment pin also had its square notch located in the middle of its longest edge that was in contact with the A2 stop cord hook attachment pin. As was the case for the A1 spring cord hook attachment pin's hook keeper, the width of the hook keeper for the A2 stop cord hook attachment pin formed an angle of 90° with the center line of an A arm.

Next in the side view of the Gera prototype wheel's 9:00 weighted lever shown at the top in the first page of Figure 4, we come to the A3 long lifter cord hook attachment pin and its center axis was located between the parallel pair of A arms at a distance of 1.4 inches = 1-6.4/16 inches from the center axis of the weighted lever's pivot pin hole. The little metal hook at one end of a *single* long lifter cord was attached to this hook attachment pin. Since four of the drum's eight long lifter cords were located in coordinating cord layer 2 and the other four long lifter cords were in coordinating cord layer 4, the wooden hook keepers for the A3 long lifter hook attachment pins had *two* square notches cut into their longest edges that contacted the 3 inch exposed lengths of these cord hook attachment pins. However, one really only had to use a single notch of the two available for each of the eight A3 long lifter hook attachment pin hook keepers. But, providing two notches for all eight of these cord hook keepers eliminated the need during lever installation in the drum to keep track of which of the four weighted levers had the hook keepers with the single notches for the layer 2 long lifter cord hooks and which of the four weighted levers had the hook keepers with the single notches for the layer 4 long lifter cord hooks. This hook keeper design, obviously, made the final installation of the long lifter cords in Bessler's 3 feet diameter, Gera prototype wheel easier as it also did the longer and stronger coordinating cords of the inventor's later and larger wheels.

Finally, at the end of the weighted lever's parallel pair of A arms, we come to the A4 main cord hook attachment pin. The center axis of this cord hook attachment pin was located exactly 1.866 inches = 1-13.87/16 inches from the center axis of the lever's pivot pin hole. The little metal hook at the end of a main cord attached to this cord hook attachment pin and again, as for the other cord hook attachment pins held by a lever's parallel pair of A arms, its wooden hook keeper touched the side of this cord hook attachment pin so that the hook keeper's width would form an angle of 90° with the center line of an A arm. Since the other end of the main cord of a weighted lever whose pivot pin's center axis was at the drum's 9:00 position was attached by another little metal hook to the L1 main cord hook attachment pin of the weighted lever whose pivot pin's center axis was located at the drum's 10:30 position and all eight of the Gera prototype wheel's main cords were contained in coordinating cord layer 3, the wooden hook keeper for an A4 main cord hook attachment pin was identical to the one used for the L1 main cord hook attachment pin and both had a square notch in the middle of their longest edges that contacted their cord hook attachment pins. These central notches served to confine a main cord's two end hooks to the small exposed section of a cord hook attachment that was contained in coordinating cord layer 3.

The last of the 9:00 weighted lever's 3 parallel pairs of arms shown in the side view of the first page of Figure 4 that we must discuss is the pair of B arms. They had only a single cord hook attachment pin held between their ends which was the B1 long lifter cord hook attachment pin and its center axis was located exactly 2.1 inches = 2-1.6/16 inches from the center axis of the lever's pivot pin hole. Its wooden hook keeper, like all of those glued in between a weighted lever's parallel pair of 3 arm side pieces, had its 3 inch length pressed up against the exposed 3 inch length of the B1 long lifter cord hook attachment pin but with an angle that oriented its width 67.5° to the center line of a B arm. The B1 long lifter cord hook attachment pin's wooden hook keeper also had two square notches cut into its longest edge that contacted the cord hook attachment pin, but, again, only one of the notches would be used depending upon whether the long lifter coordinating cord hook it confined was located in coordinating cord layer 2 or 4 of the drum. And, of course, Bessler would have used the same hook keeper design in larger wheels that used stronger coordinating cords instead of the string-like or thread-like cords found in his Gera prototype wheel.

This now completes the treatment of the *side* view of the 9:00 weighted lever used in Bessler's 3 feet diameter prototype wheel with which he first found success in Gera, Saxony in late 1711, but there are some more details in Figure 4 that must be discussed.

Moving down to the bottom portion of Figure 4, the reader will find a *top* view of the prototype wheel's 9:00 weighted lever that shows how its six coordinating cord hook attachment pins would have appeared. Unfortunately, it would have too congested the drawing if I had tried to include the six wooden coordinating cord hook keepers which had different orientations relative to the center lines of the parallel pairs of arms they were glued between. However, despite this deficiency in the drawing, I have indicated which portions of the exposed 3 inch lengths of each steel hook attachment pin that its unseen wooden hook keeper's cutout notches would have confined a coordinating cord hook to and used small rectangles for this. I have also indicated the thicknesses of the five coordinating cord layers inside of the space between the inward facing surfaces of the lever's parallel pair of 3 arm side pieces. As can be seen, each of the layers was only 0.6 inch = 9.6/16 inch thick.

It will be noted that the mass of the lead ingot end weight held between the parallel pair of main arms of the weighted lever was as close to 0.375 pound or 6 ounces as possible. This mass *included* the mass of the two steel attachment screws that penetrated the ends of the parallel pair of main arms and then partially penetrated the two ends of the ingot in order to secure it in place. However, *before* the lead ingot and its attachment screws were added to a lever, we can refer to the lever as having been an "unweighted lever" and its mass was also very important. Actually, the larger mass of the lever's end weight(s) and the attachment screws or bolt(s) used to secure it (them) to the unweighted lever has to have a particular ratio with respect to the mass of the unweighted lever. That ratio is 4.8 meaning that the mass of the end weight(s) and their screws or bolt(s) must be 4.8 times the mass of the unweighted lever.

But, how is the mass of an unweighted lever to be determined?

Basically, for the 3 feet diameter, Gera prototype wheel's levers, it is the sum of the masses of the lever's two wooden 3 arm side pieces, its six wooden hook keepers, its six steel hook attachment pins, and the seven metal hooks attached to these hook attachment pins. The lever's pivot pin, being part of the drum and not the lever, does not have its mass included in that of the unweighted lever. (However, for larger wheels using heavier coordinating cords which are actually ropes or chains, *half* of the mass of each of the seven coordinating ropes or chains attached to a weighted lever's hook attachment pins must be included in the total mass of the *un*weighted lever. Also for larger wheels, one should include half the masses of the two extension springs which are attached, via spring ropes or chains, to the lever *if* these springs are only *loosely* attached to the drum's spring anchor pins and can rotate about them a bit. This need not be done if the ends of the springs are mechanically fixed to the drum so that they cannot rotate around whatever fastener is used to attach them to the drum. Finally, for each wooden lever of a larger wheel that is fitted with two metal sleeves through which the lever's pivot pin passes, the masses of these of these sleeves must be included in determining the mass of the unweighted lever.)

With the mass the end weight(s) and the screws or bolt(s) used to attach it(them) to an unweighted lever and the mass of the unweighted lever itself, we can now define the *total* mass of a weighted lever as simply being the sum of all of these masses.

If one can accurately duplicate the total masses of the weighted levers in any attempt to replicate the Gera prototype wheel, then the design should work well when all eight weighted levers, their 16 steel extensions springs, and the 40 coordinating cords are installed in the drum. These springs, although not depicted in Figure 4, should have a *minimum* length of 1 inch when *unstretched* (so as not to be damaged when achieving their maximum stretch of 0.5 inch = 8/16 inch when a weighted lever's pivot pin's center axis nears the drum's 9:00 position), weigh as little as possible, and have a spring constant value, k, of 1.25 pounds per inch = 20 ounces per inch of stretch.

The values for the various parameters that I give in this book are accurate to three decimal places and should be considered as ideal values that would assure that the finished replica wheel will work. However, in actual practice, one's handcrafted components will probably vary somewhat from these values. This is, unfortunately, unavoidable in any handcrafted item, but should not prevent the replica wheel from working unless the deviations are too great. Even Bessler mentioned that a *small* variation in mass of the weights was not "supercritical". Yet, I still think the craftsman should do as much as he can to minimize such variations. To do that, he might want to consider obtaining a sensitive digital balance upon which he can measure the mass of the weights that he will use on the ends of his levers as well as the masses of the unweighted levers themselves to a hundredth of an avoirdupois ounce.

As one can see from all of this, the construction of the eight weighted levers that were used in Bessler's 3 feet diameter Gera prototype wheel will be somewhat time consuming. The craftsman should not become discouraged if his first attempt is not successful. Indeed, he may have to make several attempts before he obtains eight weighted levers that have the correct mass. In fact, he may have to spend as much time working on these eight levers as he will on the wheel's drum and its support stand! But, such, unfortunately, is the price that must be paid when making a handcrafted imbalanced pm wheel for the first time.

In the *top* view of the 9:00 weighted lever shown at the bottom of the first page of Figure 4, the location of the center of gravity of the entire lever is also shown and it will be noted that it is actually *not* located inside of the lead ingot end weight. Rather, the center of gravity is located at a distance of about 3 inches from the center axis of the pivot pin hole which puts it about 0.5 inch = 8/16 inch back away from the center of gravity of the ingot (which is 3.5 inches away from the center axis of the pivot pin) and about 0.25 inch = 4/16 inch back away from the backward facing, smaller rectangular face of the ingot.

Let us now complete the discussion of Figure 4 by briefly commenting on the information contained in the two continuation pages of that figure.

On the top portion of the first continuation page of Figure 4 will be found all six of the wooden hook keepers used inside of *one* of the weighted levers of the prototype wheel (the eight weighted levers will, thus, require a total of 48 hook keeps be made for them!). These hook keepers were glued up against the sides of the coordinating cord hook attachment pins that spanned the gap between the parallel pairs of main, A, and B arms of a weighted lever. One of a coordinating cord's two end hooks could then only be attached to its designated hook attachment pin at the exposed sections of a cord hook attachment pin where the square notches in the sides of the hook keepers were located.

At the top of the collection is the L1 main cord hook attachment pin hook keeper with its single central square notch for a *lagging* weighted lever's main cord's end hook, then below that the A4 main cord hook attachment pin hook keeper which also had only a single central notch cut into it, but this time for a *leading* weighted lever's main cord's end hook. The central notch locations of both of these two wooden hook keepers confined a main cord and its two attached end hooks that interconnected two adjacent weighted levers to coordinating cord layer 3 inside the Gera prototype wheel's drum.

Next, moving down the collection of wooden hook keepers, there is the A3 long lifter cord hook attachment pin hook keeper which works with a 90° leading weighted lever's B1 long lifter cord hook attachment pin hook keeper to confine the two levers' interconnecting long lifter cord and its two end hooks to *either* coordinating cord layer 2 or 4 inside of the drum.

The A2 stop cord hook attachment pin hook keeper below that is for the end hook of a weighted lever's stop cord which, as explained earlier, assured that the center lines of the lever's main arms formed a precise 62.5° angle with the center lines of the parallel pair of radial frame pieces that held that lever's pivot pin as the lever's pivot pin's center axis approached the drum's 9:00 position. The A2 stop cord hook attachment pin hook keeper worked together with a special, slightly longer wooden hook keeper called a stop cord hook attachment anchor pin hook keeper which is shown below the collection of weighted lever coordinating cord hook keepers and was glued between a lever's nearby *leading* parallel pair of *inner* octagonal frame pieces of the drum. While the A2 stop cord hook attachment pin hook keeper was 3 inch wide, the stop cord hook attachment anchor pin hook keeper was 3.5 inches = 3-8/16 inches wide. Those attempting duplication of the 3 feet diameter, Gera prototype will have to make eight of these special stop cord hook attachment anchor pin hook keepers.

The A1 spring cord hook attachment pin hook keeper of a weighted lever worked with another special, slightly longer wooden hook keeper called a spring hook attachment anchor pin hook keeper. This special hook keeper, as shown at the bottom of the collection of wooden hook keepers in the top half of the first continuation page of Figure 4, also had a width of 3.5 inches = 3-8/16 inches and was glued between a weighted lever's nearby *leading* parallel pair of *outer* octagonal frame pieces of the drum. Each weighted lever used both of the two sections of its A1 spring cord hook attachment pin that were exposed by the two notches in that pin's wooden hook keeper. Each weighted lever also used both of the two sections of its spring hook attachment anchor pin that were exposed by the two notches of that drum anchor pin's wooden hook keeper.

Together, the four square notches in a weighted lever's A1 spring cord hook attachment pin hook keeper and its spring hook attachment anchor pin hook keeper located in the drum restricted a weighted lever's two extension springs and their attached spring cords and hooks to coordinating cord layers 1 and 5 inside of the drum. As in the case of the longer stop cord hook attachment anchor pin hook keepers, the craftsman attempting duplication of the Gera prototype wheel will have to make eight of these special, slightly longer anchor pin hook keepers.

I need to briefly mention a potential problem with the coordination cord hook attachment pin hook keepers at this point.

Aside from adding rigidity to the shape of a weighted lever, their main function is to confine the coordinating cords and their attached end hooks to certain of the coordinating cord layers within the drum. They do this by having a 3 inches long edge that is pressed tightly up against the 3 inches of length of a cord hook attachment pin and having small notches cut into their wood that will only allow the cord's end hook to be attached to a certain section of the cord hook attachment pin's length that is located in a particular coordinating cord layer within the drum.

In actual practice, however, the 0.0625 inch – 1/16 inch diameter steel attachment pin may begin to flex and pull away from its wooden hook keeper, especially when a weighted lever's pivot pin's center axis is located at the drum's 9:00 position and the entire weight of the lever is hanging on its stop cord and two spring cords. Such flexing could create a gap between the cord hook attachment pin and its wooden hook keeper that would then allow a lubricated metal hook to slide out of its assigned notch and wedge itself between the cord hook attachment pin and its wooden hook keeper so that the affected coordinating cord could be shifted over into *another* of the drum's coordinating cord layers! This could then produce two serious problems that would quickly lead to the cord breaking. One would have a metal cord hook that was not properly rotating around its attachment pin during the rotational motions of its weighted lever which would then puts stress on the cord where it was tied to the hook's end eyelet and, if the cord makes physical contact with another cord in the other cord layer, then the fraying caused by rubbing contact could make both of the cords break. This, obviously, is a highly undesirable situation which must be prevented from happening.

If this becomes a problem, it can be easily solved by wrapping some of the cord being used to make the coordinating cords around the hook keeper and its cord hook attachment pin after the hook keeper is glued into place and tightly pressed up against its cord hook attachment pin. The extra cord binding can then be tied in place and the knot coated with a drop of glue to keep it from coming undone. The best location for these extra cord bindings is near the middle of the 3 inches length of an attachment pin. If there is a notch in the middle of the hook keeper there, then the cord binding can be applied to either or both sides of notch.

The task for the serious craftsman of building a replica of Bessler's 3 feet diameter Gera prototype wheel will be a challenging one. He will, for his replica wheel, have to make eight cast lead ingots, sixteen wooden 3 arm side pieces, sixteen small wooden pieces to block ingot rotation, forty-eight steel hook attachment pins (each with a diameter of 0.0625 inch = 1/16 inch and a length of 3.25 inches = 3-4/16 inches and), forty-eight wooden coordinating cord hook keepers (each with a thickness of 0.0625 inch = 1/16 inch, a width of 0.21875 inch = 3.5/16 inch, a length of 3 inches, and one or more notches that are squares that are about 0.125 inch = 2/16 inch on a side), eight weighted lever pivot pins and sixteen drum anchor pins (each with a diameter of 0.0625 inch = 1/16 inch and a length of 4 inches), sixteen slightly longer wooden hook keepers for the drum's anchor pins (each with a thickness of 0.0625 inch = 1/16 inch, a width of 0.5 inch = 8/16 inch, and a length of 3.5 inches), forty coordinating cords, eighty little metal hooks to supply two for the ends of each of the forty coordinating cords plus another sixteen hooks for one end of each of the wheel's sixteen extension springs.

Once he has his eight weighted levers completed, he will need to complete the 3 feet diameter drum to hold them. That will require a wooden axle 1.5 inches = 1-8/16 inches in diameter x 1 foot in length that has steel pivot pins inserted into its ends to a depth of 1.5 inches = 1-8/16 inches which are each 0.1875 inch = 3/16 inch in diameter x 3 inches long. He will have to obtain enough long pieces of wood (0.25 inch = 4/16 inch thick x 0.5 inch = 8/16 inch wide x 17.125 inches = 17-2/16 inches long) to make 64 drum frame pieces of various lengths that he will have to carefully cut and glue together to produce each face of the drum. He will also need miscellaneous pieces of wood for the long lifter cord guide posts (made from 0.5 inch = 8/16 inch diameter x 3.5 inches = 3-8/16 inches long pieces of wooden dowels), the drum bridge pieces (0.0625 inch = 1/16 inch thick x 0.5 inch = 8/16 inch wide x 4 inches long), the outer rim wall which will be a thin sheet of wood that is 0.0625 inch = 1/16 inch thick x 4 inches wide x 112.9 inches = 112-14.4/16 inches long (the length can be the combined length of several smaller identical pieces), and the wedge shaped braces between the axle and the inside facing surfaces of the radial frame pieces.

The drum is not completed until it is attached to its axle and its eight weighted levers, their sixteen extension springs, all forty coordinating cords each with two little metal end hooks, the sixteen drum anchor pins and their sixteen slightly longer wooden hook keepers, and the sixteen long lifter cord guide posts made from cut sections of a wooden dowel have finally been installed in the drum.

However, prior to the final assembly of the drum and the installation of its various components, the craftsman should already have constructed a stable support stand for his replica Gera prototype wheel that will securely hold it when the support's two brake bolts are loosened and the axle and drum are allowed to reach a rotation rate well in excess of 60 rotations per minute. It's also a good idea to construct this wooden support stand, whose structural details will be described in greater detail later in this chapter, *before* the weighted levers and the drum are constructed. The stand is necessary for the initial balancing of the drum while it is still "empty" of its components and then again after the installation and alignment of the weighted levers to assure that the complete "full" drum is as perfectly balanced as possible which means that the center of gravity of the drum *and* its eight weighted levers (when each lever's parallel pair of main arms is radially aligned with the lever's parallel pair of radial frame pieces) is as close as possible to being located *exactly* on the center axis of the axle. Such delicate balance is necessary because, when the weighted levers are finally allowed to assume their starting configuration as shown in Figures 3(a) and 3(b) so that the drum can self-start because the center of gravity of its eight weighted levers is then located on the descending side of the axle, that center of gravity will only be projected horizontally about 0.0625 inch = 1/16 inch onto the axle's descending side!

While the design for the one-directional, 3 feet diameter Gera prototype wheel that I am providing and its operation is not really that difficult to understand (although it took me close to a half of a *century* to do so!), its construction will probably take the average craftsman one or more months to complete. It is far better to take one's time in making the parts for the replica prototype wheel as precisely as possible so as to greatly increase the probability that the final assembled model will work well on its first test, than to do a hasty job that produces parts with serious parameter errors that will doom the final model to running either inefficiently or not at all. I also recommend, as with any wood crafting project, that the craftsman first gather all of the materials needed for the replica Gera prototype wheel and have them all cut to their correct shapes and sizes prior to the beginning the construction phase.

Once the preshaped parts are ready, the best order for the assembly might be to make the support stand first, then made the drum with its attached axle which is carefully balanced while empty on the stand, then cast the eight lead ingot weights and carefully file their end surfaces and side edges until they have their correct masses, then make the eight weighted levers which are each carefully weighed to assure they have their correct masses, then make the forty coordinating cords each with its two attached little metal end hooks. With those parts made, the next step in the construction will be the attachment of the lead ingot weights to their levers with their attachment screws, installation of the weighted levers inside of the drum, the attachment of the 40 coordinating cords and 16 extension springs to the levers' coordinating cord hook attachment pins and the drum's hook attachment anchor pins, lubrication of all metal to metal contacting surfaces with a droplet of olive oil, and, finally, the placing of the full drum on the stand and, after locking the axle in place with the stand's two brake bolts, temporarily aligning each of the eight weighted lever's parallel pair of main arms with its lever's parallel pair of radial frame pieces, and again checking the balance of the full drum and correcting it if its center of gravity is not perfectly located at the center axis of the axle (which, if it was not located at the axle's center axis, would cause the drum to rotate into a "favored" orientation when released as its offset center of gravity moved down to the *punctum quietus* point below the center axis of the axle).

With all of these preliminaries completed, the weighted levers can finally be released and allowed to assume their starting orientations as shown in Figures 3(a) and 3(b) and the axle's two brake bolts can then be loosened. If one's parts have their correct physical parameters and have been precisely assembled, then the wheel should immediately self-start and accelerate to its rotation rate of over sixty rotations per minute! Only at that moment, will the diligent and persevering craftsman know exactly how Bessler felt that day in late 1711 when his Gera prototype wheel accelerated to such a high rotation rate that he could feel the breeze coming from its rapidly moving drum's radial frame pieces!

To complete the discussion of the first continuation page of Figure 4, the reader's attention is now directed to the bottom half of the page. It illustrates the shape of the little metal hooks to which the ends of the various coordinating cords were tied. One end of each hook had an eyelet or closed circle with a cord's end tied securely to it and the other end of the hook was like a squeezed closed fish hook with its end bent away from the point where it made contact with itself. This design allowed a little hook to be easily slipped over and secured to its assigned section of a weighted lever's coordinating cord hook attachment pin which was exposed by the square notch cut into the side edge of the wooden hook keeper that was pressed up against the attachment pin. This attachment method prevented a hook from coming off of its hook attachment pin unless it was purposely manually removed or had failed from metal fatigue while in use. It also assured that each coordinating cord and its end hooks would be restricted to one of the five coordinating cord layers inside of the Gera prototype wheel's drum.

In thinking about the simplest way to make these tiny metal hooks, it occurred to me that it might be possible to use metal staples for them. One would use regular pliers to straighten individual staples out and then a small pair of jeweler's needle nose pliers to form the eyelets and hooks in them. The craftsman will have to experiment with several different sizes of staples until he finds one that is strong enough so that its closed eyelet and will remain closed and its hook end will not significantly spread open despite the tension on the coordinating cord tied to the eyelet and yet is not wider than 0.125 inch = 2/16 inch so as to prevent it from *easily* rotating about the exposed section of the coordinating cord hook attachment pin to which it is attached (should this prove impossible to do, then it is permissible to *slightly* increase the width of a notch in the wooden hook keeper).

In the case of the 3 feet diameter, Gera prototype wheel's weighted levers, the coordinating cords and their attached hooks should be able to support a *minimum* weight equal to three *times* that of a weighted lever without the cord breaking or the metal hook significantly bending open at either its eyelet or hook end. This requirement also applies to the hooks and coordinating cords in all of Bessler's later wheels, whether they were one-directional or bidirectional, and is based on there being two spring cords and their attached metal hooks supporting the full mass of a weighted lever at the drum's 9:00 position and both being attached to a cord hook attachment pin whose center axis is at a distance from the lever's pivot pin's center axis that is 1/6 of the distance of the weighted lever's center of gravity from the pivot pin's center axis. Thus, in the case of the Gera prototype wheel whose weighted levers weighed 0.453125 pound = 7.25 ounces each, a coordinating cord and its attached metal hook should be able to easily support 3 x 0.453125 pounds = 1.359375 pounds = 21.75 ounces.

The bottommost portion of the first continuation page of Figure 4 is a very important one with two sketches there that the reader should study.

The one on the top shows how a coordination cord was attached to its two little metal end hooks. That was done by just tying the ends of the cord to the tiny eyelet of each of the end hooks. The lower sketch shows how the two ends of a spring cord were tied to two little metal end hooks and then one of those hooks was attached to the large end eyelet of an extension spring while the large eyelet at the other end of the spring was attached to another hook's tiny eyelet. This method allowed either a broken spring cord to be replaced without also replacing the spring or a broken spring to be replaced without also replacing the spring cord. And, of course, it also allowed both the spring cord and its attached spring to be replaced at the same time.

Below those two sketches is a heading that reads "Maximum allowed distances between attachment pin center axes". Below that heading is a list of the various types of coordinating cords found inside of Bessler's one-directional, 3 feet diameter, Gera prototype wheel. To the right of each different type of coordinating cord are the designations of the two cord hook attachment pins to which the two hooks at the two ends of a particular type of coordinating cord were attached (however, in the case of the spring cord, one of its end hooks was attached to the A1 spring cord hook attachment pin of a weighted lever while the other hook was attached to the eyelet at one end of an extension spring. The eyelet at the other end of that spring had a third hook's eyelet attached to it and the hook portion of that third hook was then attached the spring hook attachment anchor pin whose ends were embedded in a parallel pair of the drum's outer octagonal frame pieces). To the right of the list of pairs of cord connected cord hook attachment pins is a list of the separation distances between the center axes of the two cord hook attachment pins in each pair that ranges from 6 inches to 19.9 inches.

These distances, however, are *not* the lengths of the various coordinating cords themselves which were always a fraction of an inch shorter (and in the case of the spring cord, more than an inch or more shorter depending upon the length of its unstretched extension spring) than the actual separation distances given. Rather, each separation distance is the *maximum* distance that the *center axes* of a pair of cord connected cord hook attachment pins were allowed to separate during the Gera prototype wheel's rotation as the drum's eight weighted levers were undergoing their various rotational motions about their respective pivot pins and alternately tightening and loosening the various coordinating cords attached to each of the levers.

When and only when the center axes of two coordinating cord connected cord hook attachment pins reached the particular maximum separation distance permitted to them, would their connecting cord be tight. With any separation distance shorter than that, the coordinating cord was loose and, if the drum was in rapid rotation, that cord would just bow out away from the center axis of the axle due to the centrifugal forces acting on it. (The sole exception to this was for the spring cords. From the time that a weighted lever's pivot pin's center axis had reached a position 12° past the clockwise rotating drum's 3:00 position until it passed the drum's 6:00 position, the two extension springs attached to the weighted lever were unstretched and the lever's parallel pair of main arms was in physical contact with the lever's radial stop piece. During that portion of drum rotation, the weighted lever's two spring cords would not have been loose, but would have been slightly tight because they had the weight the two metal extension springs and the centrifugal forces acting on them applying tension to the two spring cords.)

The next question that may occur to the craftsman contemplating the duplication of Bessler's 3 feet diameter, Gera prototype wheel is how can he adjust the length of a coordinating cord tied between two of its little metal end hooks so that, when the two hooks are slipped into placed on the two hook keeper exposed sections of the cord hook attachment pins of the particular cord layer that they belong in, the result will be that the maximum separation distance they will allow between the center axes of the two cord hook attachment pins will be as close as possible to the values provided at the bottom of the first continuation page of Figure 4. This is an important issue because, if the builder does not achieve separations distances for the two center axes of the pair of cord hook attachment pins that are *very* close to the values I have listed, his wheel will either run inefficiently or not run at all.

There is, fortunately, a simple way to achieve this accuracy and it can be used even before the first weighted lever is completed for the replica Gera prototype wheel.

For this method, one will need a stiff wooden board that is about 24 inches in length, 4 inches wide, and at least 1 inch thick. At one end of the board, about 2 inches from the edge, one can then hammer in a nail to a depth of a half an inch whose diameter is 0.0625 inch = 1/16 inch. After that is done, the head of the nail is cut off with a pair of wire cutters and the sharp end rounded off with a file. This nail will serve as a reference cord hook attachment pin in the production of coordinating cord /metal hooks pieces of various types to use in the replica Gera prototype wheel.

Then, after making careful measurements *starting* from the center axis of this reference nail, one must hammer in three other identical nails along the length of the board so that the distances between their center axes and that of the reference nail near the end of the board will correspond as closely as possible to the maximum cord hook attachment pin center axes' separation distances allowed by the four different types of coordinating cords and their attached metal end hooks found in the Gera prototype wheel's drum as given at the bottom of the first continuation page of Figure 4 (only three additional nails are needed for the four cord/hooks pieces because the nearest nail to the reference nail will be used to make both the stop and spring cords). These three other nails should *not* be inserted into the board so that all of their center axes lie on a straight line, but, rather, should be spread out a bit from each other across the 4 inch width of the board. When all three of the these additional nails are securely embedded into the board to a depth of half of an inch, the heads of the nails must, as was done to the reference nail, be cut off and the sharp ends of the nails rounded off with a file.

This wooden board and its four nails can then be used as a template to accurately produce the four types of coordinating cord/metal end hooks pieces which, when actually attached to their various pairs of connected coordinating cord hook attachment pins inside of the replica Gera prototype wheel's drum, will then precisely restrict how far the center axes of those cord hook attachment pins will be allowed to separate from each other during drum rotation. It's very important that these maximum allowed center axes separation distances be as close to those shown in the list at the bottom of the first continuation page of Figure 4 as possible and that the cord and metal hooks, when under stress, not expand excessively so as to significantly increase the separation distances between the pin center axes that they restrict.

To use this method to quickly manufacture the main, long lifter, stop, and spring coordinating cords with their attached metal end hooks, one simple attaches one hook over the reference nail located at the end of the board and another hook to another nail along the length of the board whose center axis to center axis separation distance from the reference nail corresponds to the maximum separation distance between the center axes of two cord hook attachment pins allowed by a particular type of coordinating cord inside of the Gera prototype wheel's drum. The reference nail near the end of the board is used in making *all* of the four different types of coordinating cords with their attached end hooks; that is, the first hook of each cord/hooks combination is always placed on the reference nail near the end of the board and then the second hook of the combination is placed on the other nail on the board at the appropriate distance from the reference nail.

    A length of cord then has one of its ends first tied to the eyelet of the hook placed on the reference nail after which the other end of the cord is passed through the eyelet of the second hook placed on its nail. The cord is then pulled through the second hook's eyelet until it is tight at which point it must be securely tied to the eyelet. If, however, while tightening the cord, its tension begins to cause either of the hooks to bend open and risks pulling it off of the board's nail, then the cord is too tight and must be loosened a bit. If neither of the hooks can tolerate having the cord connecting them pulled taut, then it may be necessary to use a stronger piece of metal to make the little hooks and, for those using staples to make the hooks, a stronger staple. With the two ends of the coordinating cord finally securely tied to the eyelets of the two little metal hooks, the hooks can then be slid up and off of the two nails in the board.

    The only exception to this simple method is in the production of the replica Gera prototype wheel's spring cords. In this case, the hook end of a little metal hook is attached to the reference nail nearest the end of the board and the larger eyelet at the end of a small helical extension spring is then secured to that hook's eyelet. At that point the *hook* end of a *second* little metal hook is attached to the still unused larger eyelet at the *other* end of the extension spring. The hook end of a *third* little metal hook is then attached to the nearest of the three other nails in board (its center axis is only 6 inches away from the center axis of the reference nail) which is the nail that is also used to make the stop cord/hooks pieces. Finally, a piece of cord is tied to the eyelet of the second hook attached to the extension spring and the cord's other end is passed through the eyelet of the third hook which is attached to the first of the three nails which is nearest to the reference nail near the end of the board. Once that is done, this cord, now called a spring coordinating cord or just a spring cord, is pulled tight until it just *barely* begins to cause the little extension spring to which it is attached to stretch at which point the spring cord is securely tied to the eyelet of the third hook.

When one has completed this operation he should have an assemblage of parts that, starting from the reference nail near the end of the board will be: the first hook which will be attached to a spring hook attachment anchor pin inside of the drum, the extension spring, the second hook which is attached to the larger eyelet at the other end of the spring, a short length of spring cord, and, lastly, the third hook which will be attached to an A1 spring cord hook attachment pin inside of a weighted lever's parallel pair of A arms. Having two hooks attached to each of the Gera prototype wheel's 16 extension springs had the advantage of allowing only the spring and the hook attaching it to a drum's spring hook attachment anchor pin to be removed from the drum and replaced, should the spring break from metal fatigue after prolonged drum rotation, while the original spring cord that was attached to it could still be used with the new spring. The disadvantage of this way of making the spring cord and extension spring/hooks pieces was that it required an additional sixteen little metal hooks be used in a one-directional wheel's drum which could be eliminated if one end of each of the wheel's sixteen spring cords was tied directly to the larger eyelet at one end of its extension spring rather than being attached to it with a hook.

Since there were forty coordinating cords divided into four different types used inside of Bessler's one-directional, 3 feet diameter, Gera prototype wheel, using the above method to manufacture them should keep the amount of required work to a minimum. Because, inevitably, cords, hooks, and springs will fail when a replica wheel is allowed to run long enough, it's a good idea to manufacture additional cord/hooks pieces of the four different types to have on hand so that the broken pieces of a cord along with the hooks attached to them can be quickly removed from a replica wheel's drum and replaced with a fresh cord/hooks piece when needed. It might, therefore, be prudent to manufacture an entire extra set of all of the four different types of coordinating cord/hooks pieces so that they can be ready to use for quick repairs as needed. Bessler's basic imbalanced pm wheel mechanics always used 16 extension springs on its eight weighted levers in a one-directional wheel, so one might want to also keep at least an extra eight springs on hand to replace these when they occasionally succumb to metal fatigue and break.

Generally, however, one can expect the steel springs to far outlast the coordinating cords and their attached end hooks. To do this, however, the maximum *additional* distance through which a spring stretches, which occurs as its weighted lever's pivot pin's center axis approaches the drum's 9:00 position, should not exceed 50% of the unstretched length of the spring. For the Gera prototype wheel, since the maximum increase in the length of an unstretched spring was only 0.5 inch = 8/16 inch at the drum's 9:00 position, the extension springs used for a replica of this wheel must have an *unstretched* length that is at least one inch.

In the second continuation page of Figure 4, I have illustrated several views of one of the 0.375 pound = 6 ounce lead ingot weights that Bessler cast and indicated how its beveled ends were filed down flat so that it could be securely attached to the ends of a parallel pair of main arms in one of the Gera prototype wheel's unweighted levers which had a mass of 0.078125 pounds = 1.25 ounces *before* the lead ingot was attached to it. Note that the lower and smaller 0.5 inch = 8/16 inch wide rectangular surface of the ingot faced inward toward the lever's pivot pin while the upper and larger 0.75 inch = 12/16 inch rectangular surface of the ingot would have been very close to the drum's rim wall whenever a lever's parallel pair of main arms was in contact with the lever's radial stop piece. This orientation of the attached ingot is clearly indicated in a clue found in the second DT portrait as previously mentioned. Two small pieces of wood were also glued to the inside faces of the main arms and up against the inside facing rectangular surface of the ingot which assured that, should its small attachment screws loosen, the ingot would not begin rotating about the center axis of the two screws.

When constructing a replica of the Gera prototype wheel, the weighted levers must be precisely fitted between their supporting radial frame pieces or there is the possibility that the rounded ingot attachment screw heads projecting from the outside facing surfaces of a lever's parallel pair of main arms might rub against the lever's radial frame pieces near the drum's 3:00 and 6:00 positions during drum rotation and produce scratching sounds there. This apparently happened in the second Gera wheel and convinced many at public demonstrations of it that there was some sort of a trained animal inside of the wheel's drum whose claws were making the sounds as it ran along the bottom interior surface of the drum's thin outer rim wall in order to propel the wheel (imagine the difficult this would present to any small animal as he ran along a rapidly rotating drum's dark interior and had to constantly avoid tripping over the drum's wooden bridge pieces and colliding with its radial frame pieces!)

Bessler, in order to better conceal the internal mechanics of his larger bidirectional wheels, reduce drum mass, and make maintenance easier, had covered their sides with stretched sheets of cloth, most likely linen that is woven from thread made from the fibers of the flax plant, which had been dyed a dark color (possibly a mixture of blue and green dyes to produce a dark green color) and he probably began this practice with his little Gera prototype wheel so he could demonstrate it to his friends without having to reveal its internal mechanics to them. The craftsman replicating the 3 feet diameter, Gera prototype wheel might consider doing likewise in order to make it look as authentic as possible. The dyed linen sheets can be attached using small pressed in tacks to secure the stretched cloth to the drum's radial frame pieces and bridge pieces. Some builders, however, may choose to leave the drum uncovered or to cover its two round open side faces with a clear plastic material so that the mechanics can be viewed, but not touched by anyone.

Bessler probably also used stretch resistant linen strings or threads for the coordinating cords inside of his prototype wheel and, most likely, he would have used a different colored one for each of the four types of coordinating cords found in his prototype design. Aside from making things a bit more interesting to look at, different colored cords would have allowed him to be able to quickly visually identify the various types of weighted lever coordinating cords and make sure the two end hooks of each one was attached to its correct section of a cord hook attachment pin and, thus, within the notch of the attachment pin's wooden hook keeper that corresponded to that cord's correct coordinating cord layer within the prototype wheel's drum. Using colored coordinating cords would also make it easier to identify which type of cord to select from one's supply of replacement cords when it came time to replace a broken cord inside of the Gera prototype wheel's drum. The craftsman might consider using strings or threads dyed to have blue, green, red, and yellow colors. While colored thread is ready available, he may have to visit a local craft supply store for the colored string. An occasional builder might even consider obtaining white string or thread and dying it himself with colorfast dyes.

In an additional effort to enhance the understanding of the five coordinating cord layers and the various types of coordinating cords in each that were used in Bessler's successful one-directional, 3 feet diameter, Gera prototype wheel, the reader is at this point referred to Figures 5(a) through 9(b) at the end of this chapter which show, layer by layer, all five coordinating cord layers. As with Figures 3(a) and 3(b), the figures to be discussed are arranged so that each *pair* of them illustrating a *single* layer appears on two facing pages of this book. This, again, was done so that the figures would be as large as possible for the 6 inches by 9 inches page size of this volume. Unlike Figures 3(a) and 3(b), however, these illustrations show all 40 of the drum's coordinating cords, whether loose or tight, and will greatly aid the craftsman as he attaches these cords to the weighted levers and drum cord or spring hook attachment anchor pins inside of the drum of his replica prototype wheel.

When viewing Figures 5(a) through 9(b) the reader should imagine that he is located on the side of the 3 feet diameter, Gera prototype wheel's drum that, when released, will begin to spontaneously rotate in a clockwise direction. Consequently, using my convention, this side is designated as the "front" side of the drum. Next, he should imagine the cloth covering on this front side of the drum having been completely removed so that he can peer into the drum's interior and see its various coordinating cord layers. To further enhance his view of the coordinating cords, he should imagine that all of the eight weighted levers' front 3 arm side pieces and all of the drum's various front frame pieces have been removed or made transparent. The first layer he will see and the one nearest to him is layer 1, the one behind that is layer 2, the middle one of the five is layer 3, then layer 4, and finally near the cloth covering the back face of the drum (which would appear to be rotating counterclockwise when viewed from the *other* side of the drum) there is layer 5. Incredibly, all five of these coordinating cord layers were packaged into a drum that was only 4 inches thick!

Figures 5(a) and 5(b) show coordinating cord layer 1 of the Gera prototype wheel's drum. Figure 5(a) depicts the left ascending side of the drum when it and its attached 1.5 inches = 1-8/16 inches diameter axle were undergoing clockwise rotation around the axle's center axis.

Coordinating cord layer 1 contained eight of the Gera prototype wheel's 16 little steel helical extension springs. The springs Bessler used for his first working one-directional prototype wheel were all identical and each had a spring constant, k, of 1.25 pounds per inch or 20 ounces per inch. The eyelet at one end of each little extension spring had the smaller eyelet of a little metal hook *permanently* attached to it. That hook's hook end was then attached to the exposed layer 1 section of its weighted lever's spring hook attachment anchor pin that was held between the lever's leading parallel pair of *outer* octagonal frame pieces. The center axis of that spring hook attachment anchor pin was located exactly 16.5 inches = 16-8/16 inches from the center axis of the axle. The *other* end of the spring also had an eyelet formed in it to which the hook end of another metal hook was attached, but could easily be removed. The smaller eyelet of that second metal hook then had the end of a spring cord securely tied to it. The other end of that spring cord was then securely tied to the eyelet of a third hook. That third hook's hook end was then attached to a weighted lever's A1 spring cord hook attachment pin on the exposed section of the hook attachment pin that corresponded to coordinating cord layer 1.

I must mention something here that is important before proceeding further.

I have previously described the coordinating cord hook attachment pin hook keeper pieces or hook keepers, for short, that were glued into place between the parallel pairs of wooden 3 arm side pieces of the Gera prototype wheel's weighted levers. The six different hook keepers for each weighted lever are shown in the *side* view of the Gera prototype wheel's 9:00 weighted lever that was illustrated in the top half of Figure 4. I mentioned then that the wooden hook keepers' *widths* have different angular orientations relative to the center lines of the parallel pairs of arms containing them depending upon which particular pair of arms they are located between. I did not, however, previously mention the orientations for the longer coordinating cord hook *anchor* pin hook keepers which came in two types and were glued between either a parallel pair of outer octagonal frame pieces or a parallel pair of inner octagonal frame pieces. Now they must be mentioned.

For both the spring hook attachment anchor pins held between the parallel pairs of outer octagonal frame pieces of the drum and the stop cord hook attachment anchor pins held between the inner octagonal frame pieces of the drum, the larger 3.5 inches = 3-8/16 inches long wooden hook keepers were glued in place so that their wider 0.5 in. = 8/16 in. *widths* were in alignment with the center lines of the parallel pairs of octagonal frame pieces that contained them. Both types of hook keepers were also placed up against their hook attachment anchor pins so that they were located on the opposite side of the pin to which either a spring hook or stop cord hook was attached. Also, like the slightly shorter hook keepers used inside of the Gera prototype wheel's eight weighed levers, these longer drum hook keepers may have needed to be wrapped with cord near the middles of their lengths in order to keep the steel hook attachment anchor pins in contact with them from flexing away from them, especially when a weighted lever's pivot pin's center axis was nearing the drum's 9:00 position, and thereby opening up a gap between a hook attachment anchor pin its hook keeper that could allow the hook its notch confined to a particular coordinating cord layer of the drum to slide out of that notch and wedge itself between the hook attachment anchor pin and an unnotched section of the length of its wooden hook keeper. As previously mentioned, this is highly undesirable because it can lead to premature failure of the hook and its attached coordinating cord.

In Figure 5(a), the reader will note that the extension spring for the 6:00 weighted lever was unstretched and, consequently, applied no force to that lever's A1 spring cord hook attachment pin. (Since the spring was free to rotate about its anchor pin, any centrifugal force acting on the spring actually applied a very small amount of force to the A1 spring cord attachment pin, but it was negligible. However, in larger wheels with heavier springs, these forces can begin to become significant and the end of each spring attached to a spring hook attachment anchor pin may have to be prevented from freely rotating about them.) The distance between the center axes of the 6:00 weighted lever's leading spring hook attachment anchor pin and that lever's A1 spring cord hook attachment pin was exactly 6 inches. By the time a weighted lever's pivot pin's center axis reached the drum's 9:00 position, however, this distance increases to its *maximum* allowed value of 6.5 inches = 6-8/16 inches because of the counterclockwise rotation of the lever around its pivot pin due to the downward pull of gravity. Because the spring cords themselves did not significantly stretch, this means that, at the drum's 9:00 position, only each extension spring was stretched through a distance of 0.5 inch = 8/16 inch. Since a helical coil spring begins to experience damage to its coils on a microscopic level when it is stretched through a distance that is about 50% greater than its unstretched starting length, Bessler would have used small steel springs in his 3 feet diameter prototype wheel that were *at least* 1 inch in length when *unstretched*.

As the reader studies the eight extension springs of layer 1 shown in Figures 5(a) and 5(b) some mention needs to be made of the axle to which the prototype wheel's drum was attached. It was, like the drum frame pieces and the weighted levers' 3 arm side pieces, made of wood. To make the model wheel convenient to work on and move about on the top of a table, an axle length of 12 inches was ideal yet still allowed 4 inches of axle length to project out on each side of the drum so that a cord could, by using a small loop formed in its end, be quickly attached to a small screw projecting out from the surface of the rapidly rotating axle in order to lift a weight of a few pounds during a braking test of the prototype wheel. The axle was only 1.5 inches in diameter and had steel pivot pins probably projecting out 1.5 inches = 1-8/16 inches beyond both of its ends. These pivot pins, most likely, had a diameter of 0.1875 inch = 3/16 inch and were 3 inches in length. They will be illustrated later when we come to Figure 10.

Another important structural feature of the Gera prototype wheel's drum indicated in Figures 5(a) and 5(b) are the locations of the long lifter cord guide posts made from cut pieces of a wooden dowel which is just a cylindrical rod made of wood. There were two of these guide posts glued between the inward facing sides of each parallel pair of radial frame pieces and their lengths passed through all five coordinating cord layers. Consequently, they appear in all of the figures of the drum's coordinating cord layers provided in this chapter. Each cord guide post, like the width of a drum radial frame piece, was 0.5 inch = 8/16 inch in diameter and had a length less than the thickness of the Gera prototype's drum or 3.5 inches = 3-8/16 inches long. The guide post whose center axis was 9 inches from the axle's center axis was the "inner" guide post and the one whose center axis was 10 inches from the axle's center axis was the "outer" guide post.

The wooden dowel pieces used for the guide posts had their external curving surfaces sanded smooth and then varnished (the ends were sanded, but not varnished so that the glue would more firmly attach them to the inner facing wooden surfaces of the parallel pair of radial frame pieces that held them). This then gave them a hard surface finish that made them so slippery that any long lifter cord making contact with them did not bunch up on one side of a post due to friction between the cord and the post, but, rather, immediately slid over the post so as to equally distributed the cord's mass on both sides of the post. This slipping action was important during rapid drum rotation when a long lifter cord became slack and was bowed out by centrifugal forces so that it made contact with an outer cord guide post or when the drum was stationary and a slack long lifter cord near the top of the drum draped down over an inner cord guide post or a long lifter cord near the bottom of the drum draped down over an outer cord guide post there. In both cases, a loose long lifter cord was prevented from making contact with the any part of the weighted lever whose pivot pin was held by the same parallel pair of radial frame pieces as was the two cord guide posts and, possibly, becoming entangled with the lever's B arm. Although the guide posts passed through all five of the drum's coordinating cord layers, they were really only used by the long lifter cords located in layers 2 and 4.

This now adequately completes the discussion of coordinating cord layer 1. Let us quickly discuss the remaining four layers of Bessler's one-directional, 3 feet diameter Gera prototype wheel that were located behind layer 1 and slightly farther away from a person standing before of the drum's exposed front circular face.

In Figures 6(a) and 6(b) are shown the left ascending side and right descending side of coordinating cord layer 2 inside of the Gera prototype wheel's drum. This layer contains four of the drum's eight long lifter cords. They each connected the A3 and B1 long lifter cord hook attachment pins of two weighted levers whose pivot pins were separated by an angular interval of 90° from each other around the center axis of the wheel's axle. As was given in Figure 4, the maximum separation distance allowed between the center axes of the A3 and B1 long lifter cord hook attachment pins by a long lifter cord and its two attached little metal end hooks was 19.9 inches = 19-14.4/16. Of the four long lifter cords in layer 2, only the cord between the weighted levers whose pivot pins' center axes were at the 6:00 and 9:00 positions of the drum was very slightly loose at the beginning of each 45° segment of drum rotation. After only a fraction of a degree of clockwise drum rotation, however, as the 6:00 weighted lever's parallel pair of main arms swung clockwise away from their radial stop piece, this long lifter cord would become taut. The three other long lifter cords in layer 2 would remain loose during the entire remainder of the 45° segment of drum rotation.

In Figure 6(a) we see that, when the 6:00 to 9:00 weighted levers' long lifter cord was finally stretched tight, it did not touch the curving cylindrical surfaces between its own two guide posts or the outward facing surface of the 9:00 lever's outer guide post. And, when a long lifter cord is slack in a moving drum, centrifugal forces can cause the cord to "drape" around the curving inside facing surface of its own outer guide post. The situation is somewhat different for the two long lifter cords in layer 2 on the right descending side of the drum as illustrated in Figure 6(b).

Each of those two descending side long lifter cords, because they were even more slack than the long lifter cords on the drum's left ascending side and, thus, bowed outward farther from the axle by the centrifugal forces created by drum rotation, would not only be draped around the inside facing surfaces of their own outer cord guide posts, but the long lifter cord between the 3:00 and 6:00 weighted lever would also be able to make contact with the stop cord hook attachment anchor pin and its wooden hook keeper held between the 3:00 lever's leading parallel pair of *inner* octagonal frame pieces whose ends were attached to the drum's 3:00 and 4:30 parallel pairs of radial frame pieces. The repetitive contact of a long lifter cord with this obstruction during each complete 360° drum rotation was undesirable because it would, if the width of the stop cord hook attachment anchor pin and its hook keeper encountered was too small, tend to kink the cord at the contact area as the slack cord tried to drape itself around those parts and that would lead to the cord's premature failure. This is one of the reasons that the longer wooden stop cord hook attachment anchor pin hook keepers installed in the drum were twice as wide as the various coordinating cord hook attachment pin hook keepers used inside of the weighted levers. The drum's stop cord hook attachment anchor pin hook keepers were 0.5 inch = 8/16 inch wide and that more than doubled width compared to the lever cord hook keepers (which would, proportionately, be 2 inches inside the drum of one of Bessler's 12 feet diameter, bidirectional wheels) would act like a small shelf to provide a larger contact width for a long lifter cord that would help to minimize the kinking of the cord as it made contact with and draped itself about a drum's stop cord hook attachment anchor pin and its hook keeper.

While the draping of a long lifter cord about a stop cord hook attachment anchor pin's smooth steel surface allowed for enough slippage to minimize the rubbing wear of the cord, the rough, unfinished, axle facing surface of the stop cord hook attachment anchor pin's wooden hook keeper would not. To remedy this, the craftsman should lightly sand those rough surfaces to make them smooth and then varnish them so that they will allow the long lifter cords contacting them to slip easily over and around them. It's also a good idea to round off the side edges of the hook keepers opposite their notched edges that contact their stop cord hook attachment anchor pins. Ideally, if a coordinating cord makes any sort of contact with an obstructing part inside of the drum, the part should have a hard, smooth surface that will allow the cord to slide over and around it with as little friction as possible.

Now we come to the most complicated of the five coordinating cord layers found inside of the 3 feet diameter Gera prototype wheel's drum: layer 3 which is the middle layer of the five layers and contained two complete and *different* sets of eight coordinating cords or a total of sixteen cords. Both the left ascending and right descending sides of layer 3 are illustrated in Figures 7(a) and 7(b).

The eight main coordinating cords and their end hooks (the metal end hooks are not shown in any of these figures because of their small size relative to the cords tied to them) connected the A4 and L1 main cord hook attachment pins of adjacent weighted levers inside of the drum and restricted the maximum separation distance between their center axes to a distance of 11.3 inches = 11-4.8/16 inches which was previously indicated in the list at the bottom half of the first continuation page of Figure 4. Although not apparent in in Figures 7(a) and 7(b) (but quite apparent when viewing computer models of the Gera prototype wheel), the leading main cords of the weighted levers were only tight for weighted levers whose pivot pins' center axes were located between about 12° clockwise *past* the drum's 9:00 position and 12° clockwise *past* the drum's 3:00 position. For weighted levers whose pivot pin center axes were located outside of these six hours of drum face clock time during which time a pivot pin's center axis would travel 180° around the center axis of the axle from its descending to ascending sides, their leading main cords were loose.

Note that the leading main cords of the weighted levers whose pivot pin axes were at the drum's 3:00, 4:30, and 9:00 positions were slack enough so that the bowing of these cords due to the centrifugal forces acting on them during rapid drum rotation would cause them to *almost* touch those levers' spring hook attachment anchor pins and their hook keepers located between the levers' leading parallel pairs of outer octagonal frame pieces. For the weighted levers whose pivot pin center axes were located at the drum's 6:00 and 7:30 positions, however, their leading main cords were so slack that they actually would make physical contact with those levers' spring hook attachment anchor pins and their wooden hook keepers.

As previously mentioned when discussing the long lifter cords in layer 2, such repetitive contact and separation between a slack main cord and a drum's spring hook attachment anchor pin once during every 360° of drum rotation was undesirable because of the kinking of the cord caused by the centrifugal and gravitational forces acting on it that tried to drape it around the spring cord hook attachment anchor pin and its wooden hook keeper. And again for the main cords in layer 3, providing a wider spring hook attachment anchor pin hook keeper helped matters by serving as a shelf upon which a main cord could spread out a bit before centrifugal forces made it slip around the spring hook attachment pin and its hook keeper and thereby minimized the amount of kinking the slack main cord experienced. As with the drum's stop cord hook attachment anchor pin hook keepers held between the parallel pairs of a lever's leading *inner* octagonal frame pieces, the inward axle facing surfaces of the spring hook attachment anchor pins' hook keepers held between the levers' leading parallel pairs of *outer* octagonal frame pieces should have the edges of their lengths opposite their spring hook attachment anchor pins rounded off and then lightly sanded and varnished so that a main cord making contact with one of them can slide about it as easily as possible in order to better distribute the cord's mass around it while producing a minimum of friction in the process.

Of particular interest is what happened to the main cord of a weighted lever whose pivot pin's center axis was located at the drum's 7:30 position. This is shown in Figure 7(a) and one can see that the cord fully rests upon the lever's spring hook attachment anchor pin and its associated hook keeper. However, before the end of that main cord connected to the L1 main cord hook attachment pin of the weighted lever whose pivot's center axis was at the drum 9:00 position, the cord formed a small bowed out loop that is actually in contact with the lead ingot end weight of the 9:00 weighted lever! I believe it was this particular and undesirable contact that convinced Bessler to attach the ingot held between the ends of a weighted lever's parallel pair of main arms in the Gera prototype wheel's drum so that the *larger* rectangular side of the ingot (which, after the molten lead was poured into the mold to make it, became the top surface of the cooled and hardened ingot) faced *outward* and *away* from the lever's pivot pin. This then allowed the beveled trailing side of 9:00 weighted lever's lead ingot to serve as a shelf that minimized the kinking of a main cord from the 7:30 lever that was attached by a small metal hook to the 9:00 weighted lever's L1 main cord hook attachment pin.

The 3rd coordinating cord layer inside the Gera prototype wheel's drum also contained a complete set of eight stop cords each of which, along with its two little metal end hooks, restricted the maximum separation distance between the center axes of a weighted lever's A2 stop cord hook attachment pin and that lever's leading stop cord hook attachment anchor pin, as listed on the bottom first page continuation of Figure 4, to a distance of only 6 inches. Of the eight stop cords in layer 3, only the one whose weighted lever's pivot pin's axis was about 6° of clockwise rotation away from reaching the drum's 9:00 position would be taut. Then, as soon as that weighted lever's pivot pin's center axis passed the drum's 9:00 position, the lever's stop cord would become slack and remain so for about another 354° of clockwise drum rotation until the same lever's pivot pin's center axis was again approaching the drum's 9:00 position. As can be seen in Figures 7(a) and 7(b), none of the eight stop cords became loose enough to make contact with any of the main cords also occupying coordinating cord layer 3 even when the main cord was stretched tight in the top half of the drum.

Next, we must consider layer 4 whose left ascending and right descending sides are shown in Figures 8(a) and 8(b).

This layer contained the other four of the drum's eight long lifter cords and was identical to layer 2 with a few notable exceptions. First, this layer was *not* superimposable with layer 2 and was actually layer 2 rotated clockwise around the center axis of the drum's axle by 45°. The consequence of this was that, at the beginning of each 45° segment of clockwise drum rotation, the long lifter cord connecting the two weighted levers whose pivot pins' center axes were at the drum's 7:30 and 10:30 positions was initially tight, but it soon became loose after about 9° of clockwise drum rotation. The loosest of layer 4's long lifter cords, as seen in Figures 8(b), was located between the A3 long lifter cord hook attachment pin of the 1:30 weighted lever and the B1 long lifter cord hook attachment pin of the 4:30 weighted lever. This extreme looseness allowed this long lifter cord to be draped around its own outer guide post located between the drum's 3:00 pair of parallel radial frame pieces as well as the 1:30 weighted lever's spring cord hook anchor pin and its wooden hook keeper. Again, the use of the wider hook keeper piece there provided a shelf for the long lifter cord to slip around and then momentarily rest upon which helped minimized the kinking of that long lifter cord. Fortunately, there was just sufficient enough clearance to prevent that long lifter cord from getting entangled with the end of the parallel pair of B arms of the weighted lever whose pivot pin center axis was located at the drum's 3:00 position.

Because the long lifter coordinating cords in layers 2 and 4 were always 45° out of phase with each other, the long lifter cord from a weighted lever whose pivot pin's center axis was approaching the drum's 7:30 position that finally helped lift the main arms closer to their radial stop piece of a weighted lever whose pivot pin's center was approaching the drum's 10:30 position actually alternated back and forth between layers 2 and 4 as the drum moved from one 45° segment of clockwise drum rotation to the next!

Finally, we see layer 5 which is illustrated in Figures 9(a) and 9(b).

Coordinating cord layer 5 is identical to and superimposable with layer 1 and contains the *other* eight of the drum's 16 weighted lever extension springs as well as the eight spring cords attached to them by little metal hooks. The extensions springs in both layers 5 and 1 reached their maximum stretch distance of 0.500 inch = 8/16 inch when the pivot pin center axes of the weighted levers they were attached to each reached a position about 6° short of the clockwise rotating drum's 9:00 position. The extension springs in both layers 5 and 1 were unstretched when their weighed levers' pivot pins' center axes each traveled clockwise around the center axis of the axle from about 12° clockwise past the drum's 3:00 position to a fraction of a degree past the drum's 6:00 position during which time the parallel pairs of main arms of these weighted levers were in contact on their wooden radial stops.

This now concludes this chapter's treatment of the previously secret imbalanced pm mechanics of Bessler's 3 feet diameter Gera prototype wheel. The occasional reader seeking to duplicate this wheel should study with great care the five layers of coordinating cords that existed inside of its drum as depicted in Figures 5(a) through 9(b). Unless this cord configuration scheme is replicated *exactly* as illustrated using the correct maximum coordinating cord and attached end hooks allowed separation distances between the center axes of the various pairs of coordinating cord hook attachment pins, then, sadly, no success will come from one's efforts.

In the remainder of this chapter, I want to cover some additional important construction information for Bessler's 3 feet diameter Gera prototype wheel and the reader is now directed to Figure 10 which provides the various internal and external component dimensions for that wheel.

Figure 10 shows a profile view a weighted lever in the prototype wheel's drum whose pivot pin center is located at the drum's 12:00 position. The view is from the upper portion of the descending side of the drum, removes the intervening rim wall and 1:30 weighted lever, and shows what the lever would look like if the two main arms of the 12:00 weighted lever were in contact with the lever's wooden radial stop piece the leading surface of which is visible in the figure (note that the lever's 0.375 pound = 6 ounce lead ingot end weight does *not* make physical contact with the wooden radial stop piece).

Ordinarily, this orientation of a weighted lever can only be achieved when its pair of parallel main arms is perfectly aligned with and then clamped to that lever's pair of parallel radial frame pieces which hold the lever's steel pivot pin. Such an orientation then allows the radial distances of various parts of the lever from the drum's center axis to be accurately shown. In order to fit the top portion of the drum and lever into the illustration, however, it was necessary to remove a few inches from the lengths of the radial frame pieces which is indicated by the short diagonal lines crossing those frame pieces near where they are attached the wheel's axle.

As can be seen in Figure 10, the outside width of the 3 feet diameter prototype's wooden levers was exactly 3.25 inches = 3-4/16 inches and the inside width of the drum as measured between the inward facing surfaces of its pairs of parallel radial frame pieces was 3.5 inches = 3-8/16 inches. Thus, there was a *total* gap of 3.5 inches−3.25 inches = 0.25 inch = 4/16 inch between the weighted levers' parallel pairs of main arms and their parallel pairs of radial frame pieces. Dividing that total gap by 2 gives a gap on *each* side of a weighted lever of 0.25 inch / 2 = 0.125 inch = 2/16 inch. As indicated in the figure, I suggest that the craftsman trying to duplicate this prototype wheel insert *two* steel washers on *each* side of each weighted lever's pivot pin to fill in that gap. The washers should be 0.5 inch = 8/16 inch in diameter by 0.0625 inch = 1/16 inch thick with a center hole that is 0.0625 inch = 1/16 inch in diameter so that it will fit on a lever's 0.0625 inch = 1/16 inch diameter pivot pin. This fit should not be so tight as to prevent a washer from rotating about the pin, When each weighted lever has its four washers installed on its pivot pin then that should prevent any lateral swaying of the end of a weighted lever in either direction by more than 0.0625 inch = 1/16 inch. Obviously, this is a rather "tight" tolerance, but any excessive lateral swaying of a lever's lead ingot end weight might cause the heads of its attachment screws to begin to cut into the softer wood on the inward facing surfaces of the drum's radial frame pieces when the weighted lever's parallel pair of main arms again come into radial alignment with those frame pieces as the main arms come into contact with the lever's radial stop piece on the drum's descending side.

Thus, after the eight weighted levers are finally installed in the drum, each lever will have two of these washers on each side of it (that's 4 washers per weighted lever for a total of thirty-two washers installed on the drum's eight pivot pins). Before the initial test of the wheel, a small droplet of olive oil (or a modern silicone based lubricant if one it not overly concerned about using the same materials in his replica wheel as Bessler used in the original) should be placed on each pair of washers to minimize any torsional friction between them which would "steal" some of the kinetic energy accumulating in the accelerating drum and convert it into unwanted thermal energy in the surrounding air.

Another possible source of undesirable lateral swaying of the end of a weighted lever would be from the two thin metal sleeves inserted into the wooden lever's pivot holes should they be fitted too loosely on their pivot pin. This could then result in enough play existing near the end of a lever's main arms so that its lead ingot attachment screws heads would begin rubbing up against the portions of the wooden radial frame pieces near them. If this problem arises it would be necessary to remove the pivot pin from the axle and the lever and then replace one or both of the lever's metal sleeves. To avoid this, the craftsman should make sure that, as soon as he installs the sleeves in a weighted lever's pivot holes, he inserts a 0.0625 inch = 1/16 inch pivot pin into them to make sure the pin is as perfectly perpendicular to the planes defined by the various arms of the lever's 3 arm side pieces as possible and that there is no noticeable play between the pivot pin and the metal sleeves. Doing this precision check *before* final installation of the weighted levers into the replica Gera prototype wheel's drum can save the craftsman much extra later effort, time, and frustration.

On the bottom left side of Figure 10, the 12:00 weighted lever's pair of parallel radial frame pieces is shown attached to the middle of the length of the wheel's 1.5 inch = 1-8/16 inch diameter x 12 inch long wooden axle. While it is possible to attach the center of the drum more toward one of the ends of the axle than to the other, placing the drum's center plane in the middle of an axle assures that its weight will be evenly divided between its axle's two end pivots and that their supporting brass bearing plates held in a wheel's vertical support boards or planks will wear evenly over time and eventually both will need to be replaced together. However, a *possible* exception to this general rule *appears* to have been made in the case of the 12 feet diameter, bidirectional Kassel wheel as it was depicted in its DT illustration. Although this material more properly belongs in the next chapter, I shall include it here because of its relevance to the attachment of a wheel's drum upon its axle.

Careful analysis of the almost profile view of the Kassel wheel on the left side of Figure 2(b) at the end of Chapter 1 shows that the front projecting end of the bidirectional wheel's 6 feet = 72 inches long axle appears to be longer than the back projecting end and that this was achieved by moving the location of the *center* plane of the drum from the midpoint of the axle to a position on the axle that was about 3.6 feet = 43.2 inches from the front end of the axle and only 2.4 feet = 28.8 inches from the back end of the axle. Since the drum was 18 inches thick, that means that the front projecting section of the axle was 43.2 inches−9 inches = 34.2 inches long and the back projecting end of the axle was 28.8 inches−9 inches = 19.8 inches. This asymmetry would have put 0.6 times as much of the weight of the approximately 1100 pound Kassel wheel on the back vertical support plank and only 0.4 times as much on the front vertical support plank. Thus, the back vertical support plank would have had to have supported about 0.6 x 1100 pounds = 660 pounds and the front vertical support plank would have had to support about 0.4 x 1100 = 440 pounds.

If this axle placement was actually used for the Kassel wheel's drum, then the weight supported by the back vertical support plank's lower brass bearing plate would have been 1.5 times that supported by the front vertical support plank's lower bearing plate and one would expect the back lower bearing plate to wear out about 1.5 times faster than the front bearing plate. If, however, the center plane of the Kassel wheel's 18 inches thick drum had been located in the middle of its 6 feet = 72 inches long axle, then the front and back projecting sections of the axle would both have been (72 inches−18 inches) / 2 = 54 inches / 2 = 27 inches and each bearing plate would have had to have supported 0.5 times the weight of the drum and axle or about 0.5 x 1100 pounds = 550 pounds.

Using the asymmetrical drum placement suggested in the left side drawing of the Kassel wheel shown in Figure 2(b) would have made the front axle 34.2 inches long instead of the 27 inches that would have been provided if the drum's center plane was located at the axle's midpoint. That's only a gain of 7.2 inches. One might suppose the extra 7.2 inches was needed because, in addition to the four odd "Y"-shaped wooden forks that drove the rope that operated the Archimedean water screw, Bessler needed extra projecting axle length for the eight metal pieces near the drum that operated the two long levers that raised and dropped the two stamps of the stamping mill eight times during each axle rotation. However, I remain of the opinion that was not the case and that both projecting sections of axle from the drum were 27 inches in length which I believe would have been of sufficient length for the "Y"-shaped forks and the two groups of four metal pieces placed side by side on the axle nearer the drum's front side. The 27 inch length of projecting axle would have provided each of these three groups of four projecting pieces a full 9 inches of length of the axle and I don't believe Bessler would have wanted to have had an uneven loads placed upon the lower bearing plates in the Kassel wheel's two vertical support planks.

If that is the case, then why does the asymmetry in the axle location of the drum's center plane appear in Figure 2(b)? The simple reason is that the projecting length of the front end of axle needed to be exaggerated because Bessler decided to exaggerate the separation distance between the two lengths of rope connecting the square pulley on the water screw with the "Y"-shaped wooden forks on the front axle section in order to show that the two ropes were not rubbing against each other (which was assured by tilting the upper end of the water screw's axis of rotation away from the center axis of the Kassel wheel's axle while tilting the lower end of the water screw's axis of rotation toward the center axis of the Kassel wheel's axle) and, thus, subject to wear. In actual practice, however, the two lengths of rope only needed to be separated by a few inches from each other and not the foot or so suggested in Figure 2(b).

Now that this brief digression is completed, it's time to return to and complete the discussion of Bessler's 3 feet diameter, Gera prototype wheel.

For the 1.5 inch diameter wooden axle shown in Figure 10, one will have to cut 16 radial frame pieces from wood (Bessler used European white oak) that are each 0.25 inch = 4/16 inch thick x 0.5 inch = 8/16 inch wide x 17.125 inches = 17-2/16 inches long. The observant reader will note that a radial frame piece's width of 0.5 inch = 8/16 inch is equal to the width of the main arms of the 3 arm side pieces of the Gera prototype wheel's eight weighted levers. This equivalence then allows a miniature wood clamp to be used to lock each weighted lever's parallel pair of main arms into perfect radial alignment with the lever's parallel pair of radial frame pieces. Such alignment, temporarily maintained with eight small clamps, for all eight of a replica prototype wheel's weighted levers can make the installation of the wheel's forty coordinating cords easier during the finally assembly phase of the wheel.

Because of its small size and mass, it's possible to build a replica of Bessler's 3 feet diameter Gera prototype wheel without the use of such things as nails or screws and I believe that this is exactly how he did it. Thus, after having finished the construction of the replica wheel's wooden support stand, but before completing its eight weighted levers and the 40 coordinating cords with their attached metal hooks that would be attached to them, Bessler would have proceeded to assemble the drum's two side faces that consisted of a lattice work of their various radial and octagonal frame pieces by simply gluing them together.

The best way to assemble the two side faces of the drum for this prototype, based on my own past experiences with constructing hollow wooden drums from small pieces of wood, is to use a large table which, preferably, is circular and about a foot more than 3 feet in diameter. One must cover the surface of the table with sheets of taped down white paper to protect it. Next, find the exact center of the circular table top and place a dot on the paper there with a pencil. Use a small drafting compass to draw a circle around that dot with a radius of 0.75 inch = 12/16 inch. That circle will then be 1.5 inches = 1-8/16 inches in diameter which is the same diameter as the prototype wheel's axle.

Next use a protractor and *yardstick* to *very* carefully draw eight straight lines that will radiate out away from the center dot of the circle drawn on the paper in the middle of the circular table. Any two adjacent radial lines must form as close to an *exact* 45° angle with each other as possible. Extend each of the eight lines until it reaches the outer circular edge of the table. At a distance of exactly 14 inches from the pencil dot in the middle of the table, place a mark on each of the eight radial lines previously drawn. Using the yardstick again, measure the distances between these marks for all eight adjacent pairs of radial lines.

If the drawing of the eight radial lines was done accurately so that any two of the radial lines formed an exact 45° angle with each other, then the distance measured between the intersection points of the pencil marks and the radial lines on any two adjacent radial lines should be exactly 10.715 inches = 10-11.44/16 inches. If there are only *very* small variations in the eight distances measured, but they average out to this value, then this is sufficient accuracy and one may proceed to the next step in the drum's construction. If, however, there are big variations in the eight distances measured and they do not average out to this value, then these radial lines, which will serve as a guide to constructing an accurate and balanced drum, are unacceptable. The craftsman must then erase the radial lines and make another attempt at drawing them more accurately. Only when they are drawn accurately so that eight measured distances are nearly the same should he proceed to the next step in drum construction.

Once one has achieved an accurately drawn drum frame piece assembly guide on the paper covering his circular table's top, it's time to cut off a cylindrical disc piece that is 0.25 inch = 4/16 inch thick from the end of the wooden dowel piece left over after the 1.5 inches = 1-8/16 inches diameter x 12 inches length of the replica Gera prototype wheel's axle was previously cut off from a longer length of that dowel. That short cylindrical disc of wood is then glued on top of the 1.5 inches = 1-8/16 inches diameter circle drawn on the covering paper at the center of the table. It must be located so that the center of its circular side rests directly on the first center dot placed onto the paper drum frame piece assembly guide. When its glue is completely dry, the rest of the drum's construction will proceed quickly.

One then places the various previously measured and cut radial frame pieces of the prototype's drum onto the paper taped to the table. Their center lines, which can be lightly drawn on them using a pencil and yardstick, are then carefully aligned with the radial lines drawn on the paper and slid along those lines until one of their ends touches the outer surface of the cylindrical disc piece glued to the paper at the center of the table. Once the eight radial frame pieces are in place, they can be temporarily held in position with pieces of tape. At this time a point on each radial frame piece that is exactly 14 inches from the center axis of the cylindrical disc piece glued to the center of the paper on the table and also on the center line of the frame piece is marked off with a pencil. These points will be the locations at which the center axes of holes *slightly* less than 0.0625 inch = 1/16 inch in diameter will later be drilled and into which the drum's eight weighted levers' pivot pins will be inserted and pressed into place.

Once that is done, the remaining sixteen octagonal frame pieces, consisting of eight outer pieces of one length and eight inner pieces of somewhat shorter length are carefully positioned between the taped down radial frame pieces so that the ends of the octagonal frame pieces touch their adjacent radial frame pieces at the correct locations.

For the inner octagonal frame pieces, their locations and orientations will be determined by where the lightly drawn center line running along the length of an octagonal frame piece, if extended, would pass through the point marked on the adjacent radial frame piece it contacts that corresponds to where the center axis of that radial frame piece's weighted lever's pivot pin hole will be located and which was previously marked on the frame piece. For the longer outer octagonal frame pieces, the outermost corners of each one's beveled end should extend out 0.0625 inch = 1/16 inch beyond the corner of the straight cut off outermost end of the adjacent radial frame piece it contacts. The resulting indentation formed at the outermost end of each radial frame piece will later be filled in when a 0.0625 inch = 1/16 inch thick wooden bridge piece is glued to the two outermost straight cut off ends of each parallel pair of radial frame pieces during the final phase of drum construction.

I have not provided the exact lengths of these inner and outer octagonal frame pieces, but leave their determinations as an exercise for the craftsman to do and which he can accomplish by making careful measurements between the drum's adjacent radial frame pieces, as they lie on the layer of protective paper covering the assembly table, using the yardstick while also referring to Figures 3(a) and 3(b) for the positions of the drum's inner and outer octagonal frame pieces. For maximum structural strength when the drum is finished, the ends of the octagonal frame pieces should be cut at an angle so that the beveled ends formed will fit flat against the side edges of the radial frame pieces that they contact. That angle will be such that the beveled ends of all of the octagonal frame pieces, whether inner or outer, are inclined exactly 22.5° with respect to what would be, if it was so cut, an unbeveled, straight cutoff end that would be perpendicular to the center line of the octagonal frame piece.

With all sixteen of the correctly shaped octagonal frame pieces, eight inner and eight longer outer, almost in place and in contact with their radial frame pieces already lying on top of and taped to the paper on the table top, a drop of glue is applied to the lightly sanded wooden surfaces that will come into contact with each other and the octagonal frame pieces are then slid into their final locations that place them in contact with the eight radial frame pieces. When that glue has dried, the locations of the center axes of the holes (all of which will be *slightly* less than 0.0625 inch = 1/16 inch in diameter) that will be drilled out of the *midpoints* of the center lines of the eight outer octagonal frame pieces for the drum's eight 0.0625 inch diameter x 4 inch long spring hook attachment anchor pins and out of the midpoints of center lines of the eight inner octagonal frame pieces for the drum's eight 0.0625 inch diameter x 4 inch long stop cord hook attachment anchor pins should be located and marked with a pencil.

As was indicated in Figure 3(b), the center axis for a stop cord hook attachment anchor pin was located at the midpoint of an inner octagonal frame piece's center line and at distance of exactly 12.934 inches = 12-14.94/16 inches from the axle's center axis which, during the construction of the drum, will correspond to the center point of the 1.5 inch = 1-8/16 inch wooden disc glued to the paper at the center of the table. The center axis of an extension spring hook attachment anchor pin was located at the midpoint of an outer octagonal frame piece's center line and at a distance of exactly 16.5 inches = 16-8/16 inches from the axle's center axis which, again, also corresponds to the center of the wooden disc glued to the paper at the center of the table.

With the locations of the center axes of the eight weighted levers' pivot pins, the eight stop cord hook attachment anchor pins, and the eight extension spring hook attachment anchor pins clearly marked, it should then be possible to carefully lift the one complete side of the replica Gera prototype wheel's drum off of the paper if the glue holding its frame pieces together is thoroughly dried and all of the joints formed are strong (most likely, the paper will stick to some of the glued joints and possible tear as attempts are made to peel it away from those joints, but any resulting holes created in the paper template can be repaired by simply taping paper patches over them). This side of the drum will be fairly rigid, but it should be put aside by placing it down on a flat surface which could be either another table or the floor. Once it is out of the way, the above steps must be repeated in order to construct the *other* side of the drum which will be an exact copy of the first side previously constructed.

After *both* sides of the drum are finished, one must then very carefully drill out the sixteen lever pivot pin holes, sixteen stop cord hook attachment anchor pin holes, and sixteen extension spring hook attachment anchor pin holes at the various marked locations of their center axes. As previously mentioned, these holes should each be slightly smaller than 0.0625 inch = 1/16 inch in diameter to ensure that the pins to be pressed into them will, due to friction, remain fixed with respect to their frame pieces. One may refer to the glued together collection of frame pieces of *each* of the drum's two sides as a "latticework" or, simply, just a "lattice".

Each lattice, after having the central cylindrical disc around which it was constructed on top of the table removed, is then slipped onto the 1.5 inches = 1-8/16 inches diameter x 12 inches long wooden axle which has previously had its two 0.1875 inch = 3/16 inch diameter x 3 inch long steel end pivots installed in it by pressing them tightly into slightly smaller diameter holes that had been drilled out for them at the ends of the axle's center axis and to a depth of 1.5 inches = 1-8/16 inches each. The two *almost* 3 feet diameter lattices are then carefully slid along the length of the axle and rotated with respect to each other until one obtains eight parallel pairs of radial frame pieces separated from each other by exactly 3.5 inches = 3-8/16 inches. The outside facing surfaces of each parallel pair of radial frame pieces should also be exactly 4 inches away from the end of the axle nearest to it as illustrated in Figure 10.

After that three dimensional configuration of lattices is achieved, glue is applied to the inward facing end of each of the radial frame pieces to attach it to the axle. At this time one should also glue the small 0.5 inch = 8/16 inch thick wedge shaped wooden pieces shown in Figure 10 into place in the inside corners formed between the radial frame pieces and the axle's surface to enhance the strength of the attachment of the two parallel lattices to the axle (I have not specified the lengths of the perpendicular sides of these wedge shaped pieces, but, obviously, they must not exceed 1.75 inches = 1-12/16 inches or two of them will not fit between a parallel pair of radial frame pieces).

When the glue holding the drum's radial frame pieces to the axle is completely dry, the drum's sixteen long lifter cord guide posts, each of which is 0.5 inch = 8/16 inch in diameter x 3.5 inches = 3-8/16 inches in length, are glued between the eight parallel pairs of radial frame pieces so that the center axes of the eight inner posts are each located 9 inches from the center axis of the axle and the center axes of the eight outer posts are each located 10 inches from the center axis of the axle as indicated in Figure 10. When properly placed, the distance between the center axis of an inner guide post and the surfaces of the drum's axle nearest that post's parallel pair of radial frame pieces should be exactly 8.25 inches = 8-4/16 inches. The distance between the center axis of an outer guide post and the surfaces of the drum's axle nearest that post's parallel pair of radial frame pieces should be exactly 9.25 inches = 9-4/16 inches. The resulting gaps between the nearest inside curving surfaces of each pair of the long lifter cord guide posts should then be 0.5 inch = 8/16 inch (and if all of the drum's parts are properly sized and assembled together, then, when a long lifter cord is stretched tight between the pair of its guide posts, the cord will touch neither guide post). With the addition of these sixteen wooden dowel pieces used for long lifter cord guide posts, there will be a significant increase in the rigidity of the drum after the glue between the guide posts and the radial frame pieces has thoroughly dried.

The reader will notice in Figure 10 that the drum's 12:00 parallel pair of radial frame pieces are shown as each being 17.125 inches = 17-2/16 inches in length. The reason for this is that, after each radial frame piece is securely glued to the prototype wheel's axle which has a radius of 0.75 inch = 12/16 inch, the end of the radial frame piece will be located exactly 17.875 inches = 17-14/16 inches from the center axis of the axle. That is just 0.125 inch = 2/16 inch of an inch *shorter* than the 18 inches radius that the finished drum is supposed to have.

The replica prototype drum's radial frame pieces must have this length so that one can then attach, again via gluing, first a 0.0625 inch = 1/16 inch thick x 0.5 inch = 8/16 inch wide x 4 inches long wooden bridge piece to the cut ends of each parallel pair of radial frame pieces after which a circular band of wood that is the drum's outer rim wall and is also 0.0625 inch = 1/16 inch thick is attached by gluing it to the outer facing surfaces of the bridge pieces. When this is done, the drum of the Gera prototype wheel will have an outside radius of exactly 18 inches = 1.5 feet and an outside diameter of 36 inches = 3 feet. (Note that in Figures 3(a) and 3(b) there are also an *additional* eight identical wooden bridge pieces attached to the ends of eight parallel pairs of "very short" radial frame pieces whose other ends are attached to the midpoints of the drum's parallel pairs of outer octagonal frame pieces. The 16 very short radial frame pieces needed for the replica Gera prototype wheel's drum can be made from the same stock pieces of wood used for the drum's much longer radial and octagonal frame pieces. The very short radial frame pieces should be 0.25 inch = 4/16 inch thick x 0.5 inch = 8/16 inch wide x 1.125 inches = 1-2/16 inches long.)

The band of wood used for the outer rim wall of the replica Gera prototype wheel's drum can be made from a thin sheet of wood used for making furniture veneers and, if made as a single piece, would, as shown near the top of Figure 10, be 0.0625 inch = 1/16 inch thick x 4 inches wide x 112.9 inches = 112-14.4/16 inches long (the grain direction of the piece of wood used should be parallel to the center axis of the axle so that the piece will easily curl around the periphery of the drum from bridge piece to bridge piece without breaking). This length, which is about 9.4 feet, may be too long for a single piece of veneer wood so the builder can just use eight pieces that have the same thickness and width, but are each only 14.1125 inches = 14-1.8/16 inches long. If the craftsman opts to install this outer rim wall on his replica prototype Gera wheel, then its outermost curving surface will be exactly 18 inches from the center axis of the axle and the overall diameter of the drum will then be exactly 36 inches = 3 feet.

Even though it is not critically necessary for the operation of this little toy imbalanced pm wheel, the outer rim wall will help conceal the drum's internal mechanics just as it did on the original one Bessler constructed in late 1711 and, more importantly, it will reduce the aerodynamic drag that an open drum's various wooden pieces would experience during drum rotation which will then help the wheel accelerate as rapidly as possible up to its maximum terminal rotation rate after start up.

Before the eight weighted levers are finally installed in the replica Gera prototype's drum, however, its rigidity should be further enhanced by installing the eight 4 inches long stop cord hook attachment anchor pins and the eight 3.5 inches = 3-8/16 inches long wooden hook keepers in contact with these pins between the midpoints of the eight parallel pairs of *inner* octagonal frame pieces. After that, the eight 4 inches long spring hook attachment anchor pins and the eight 3.5 inches = 3-8/16 inches wooden hook keepers in contact with these pins are installed between the midpoints of the eight parallel pairs of *outer* octagonal frame pieces. When all sixteen of these hook attachment anchor pins and their associated hook keepers are in place and the glue used on the hook keepers has thoroughly dried, short lengths of cord can be tightly wrapped around the hook anchor pins and their contacting hook keepers near their middles, tied in place, and their knots secured with a drop of glue. As mentioned earlier, these extra cord wrappings will help prevent the hook attachment anchor pins from flexing away from the edges of their contacting hook keepers when a weighted lever's pivot pin's center axis approaches the drum's 9:00 position and the entire weight of the weighted lever is placed upon these hook attachment anchor pins by the stop cord and two stretched extension springs that connect the lever to the two anchor pins.

For this phase of the installation, the drum should be laid on a table so that a large portion of the drum *not* having the axle protruding from it lies on the table's surface. This then requires that the drum be laid on the table with a projecting section of the axle hanging out over the edge of the table. A weight should then be placed on several of the radial frame pieces touching the table so that the greater unsupported weight of the larger side of the drum overhanging the table does not cause the entire drum and its attached axle to fall off of the table and, possibly, get damaged.

Using a small hammer, the sixteen hook attachment anchor pins are *lightly* tapped into place in the slightly smaller holes drilled for them previously in the sixteen parallel pairs of octagonal frame pieces, but before doing that a piece of wood that is 3.5 inches = 3-8/16 inches long should be inserted between each parallel pair of octagonal frame pieces near where a hook attachment anchor pin is to be inserted. This wood piece will keep the hook attachment anchor pin's parallel pair of octagonal frame pieces separated from each other at their precise inside separation distance of 3.5 inches = 3-8/16 inches as the pin insertion takes place and also protect the various glued joints between the frame pieces in other locations of the drum from being damaged by any hammering done to insert the hook attachment anchor pins.

When a hook attachment anchor pin is fully inserted into the drum, its ends should not project out beyond the outside facing surfaces of the drum's various frame pieces. Also, as previously mentioned, these pins should fit snugly into their parallel pairs of octagonal frame pieces so that they do not rotate inside of the holes that were drilled for them. If they are loose, then a bit of glue on the ends of the hook attachment anchor pins, which have been lightly sanded to roughen their surfaces, should help immobilize them with respect to the parallel pair of octagonal frame pieces that hold them.

Once all of the eight hook attachment anchor pins are in place in the drum, their eight wooden hook keepers must be glued into place between their respective parallel pairs of inner and outer octagonal frame pieces with their correct orientations with respect to the center lines of the octagonal frame pieces to which they are attached. As previously mentioned, the notches in these hook keepers restrict the location of a cord or spring hook to a particular section of a hook attachment anchor pin that corresponds to a particular one of the drum's five coordinating cord layers. These hook attachment anchor pin hook keepers in the replica Gera prototype wheel are all 0.0625 inch = 1/16 inch thick x 0.5 inch = 8/16 inch wide x 3.5 inches = 3-8/16 inches long and have their square notches, each measuring 0.125 inch = 2/16 inch on a side, placed in them at the locations shown in the second continuation page of Figure 4.

The ends of a hook attachment anchor pin hook keeper need to be glued into place so that it has one of its 3.5 inches = 3-8/16 long edges, the edge with the notch(es) along it, in physical contact with its nearby hook attachment anchor pin and also, importantly, so that the width of the hook keeper is aligned with the center lines of the hook keeper's parallel pair of octagonal frame pieces. This alignment will then provide a flat, relatively wide surface for either a loose long lifter or main coordinating cord that contacts it to rest upon *after* the cord slides about a bit as the gravitational and/or inertial forces acting on it cause it to distribute its mass about the hook keeper and its hook attachment anchor pin. All sixteen of the wooden hook attachment anchor pin hook keepers should have their exposed axle facing surfaces lightly sanded and varnished in order to make them as slippery as possible which will then assure that the coordinating cords contacting them will be free to drape themselves as evenly as possible about them in response to the centrifugal forces acting on the cords during rapid drum rotation. The varnishing should be done after the cord wrappings are added to them that help keep their hook attachment anchor pins in contact with them.

With the drum now completed, it's time to install its eight weighted levers and their steel pivot pins. Obviously, this will only be possible if the craftsman has already completed the construction of the replica prototype wheel's eight miniature wooden levers as described earlier in this chapter.

To do this, the drum is again placed on the table so that one of its projecting axle ends hangs over the edge of the table. The piece of wood used to temporarily brace the pairs of parallel octagonal frame pieces while the sixteen hook attachment anchor pins were tapped into place is again used to insert the eight weighted lever pivot pins into the previously drilled out slightly smaller diameter holes for them in the drum's eight parallel pairs of radial frame pieces. These eight weighted lever pivot pins have the same dimensions as the thin metal rods used for the sixteen hook attachment anchor pins; that is, the weighted lever pivot pins are also 0.0625 inch = 1/16 inch in diameter and 4 inches in length.

The insertion of the eight weighted lever pivot pins into the drum is, however, somewhat more complicated than the previous insertion of the 16 hook attachment anchor pins. This is because each weighted lever pivot pin, aside from passing through the drilled holes in two parallel radial frame pieces, must also pass through the two thin metal sleeves that line the two pivot holes of each weighted lever's parallel pair of 3 arm side pieces. To make matters even more difficult, the craftsman must, as indicated in Figure 10, also be able to pass the pivot pin through the center holes of a set of two 0.0625 inch = 1/16 inch thick steel washers that act as spacers on *each* of the outward facing surfaces of a weighed lever's parallel pair of 3 arm side pieces and keep those surfaces (and especially the heads of the two attachment screws that secure the lead ingot end weight to the ends of the lever's parallel pair of main arms) from rubbing against the inward facing surfaces of the lever's radial frame pieces during drum rotation. I won't give specific instructions here on how to do this because it is a process that the craftsman will eventually figure out for himself after several initial failures. Obviously, he will have to work slowly and make sure all of the parts are in their properly aligned positions as a pivot pin is lightly tapped into place. When one has successfully inserted a weighted lever's pivot pin, then neither end of the pin should project out beyond the outward facing surface of either of the two parallel radial frame pieces which hold it. The weighted lever should, after a bit of lubrication is applied to its two thin metal sleeves that make contact with its pivot pin that is fixed to the drum and also to the two spacer washers on each side of the lever, be able to freely rotate about the pivot pin without any part of it contacting the pair of parallel radial frame pieces which hold the lever's pivot pin.

After all of the 3 feet diameter, Gera prototype drum's eight weighted levers have been installed, their eight parallel pairs of main arms should be clamped into place so that they align with their lever's radial frame pieces. If one cannot obtain sixteen *identical* miniature wood clamps that are small enough to use because the A arm of each weighted lever is in the way, he might try simply placing a 6 inch long cut piece of the same wood used for the drum frame pieces above and below the two parallel main arms and their nearby parallel pair of radial frame pieces and then using two rubber bands to pull the overhanging ends of these two 6 inch long pieces of wood together on both sides of the drum. This should then bring a weighted lever's parallel pairs of main arms and radial frame pieces into perfect alignment. One should make sure that all sixteen of these clamping pieces of wood have the same weight and are placed at the same distance from the surface of the wheel's axle which will be important later when one has to precisely balance the drum after it is mounted on its support stand.

When the main arms of all eight weighted levers are in perfect alignment with their radial frame pieces, their combined center of gravity should be located at a point on the center axis of the wheel's axle that is equidistant from the two parallel planes defined by the outward facing surfaces of the drum's two latticework faces which means in the exact geometric center of the 12 inch long axle. With all of the weighted levers clamped into position, one can then proceed to attach the eight wooden radial stop pieces required by a one-directional wheel.

I have not previously given the dimensions of these radial stop pieces and will do so now. They can be made by cutting off pieces of the same stock lengths of wood used to make the replica Gera prototype wheel's octagonal and radial frame pieces. One should, therefore, make each radial stop piece by cutting off a piece of wood that will be 0.25 inch = 4/16 inch thick x 0.5 inch = 8/16 inch wide x 4 inches long. After eight of these wooden radial stop pieces have been made, the lightly sanded ends of each must be glued to the leading surfaces (that is, the surfaces that face *toward* the clockwise direction in which the drum rotates) of each pair of a weighted lever's parallel pair of radial frame pieces so that the radial stop piece is in contact with the *leading* surfaces of a weighted lever's two main arms when they are aligned with the lever's parallel pair of radial frame pieces. It will be noted in Figures 3(a) and 3(b) as well as Figures 5(a) through 5(b) that I illustrated the cross sectional shape of each radial stop piece as being not that of a rectangle, but, rather, that of a right trapezoid with one side beveled so that the wood piece could fit snuggly into the 67.5° corner created where the end of an outer octagonal frame piece was attached to the end of a radial frame piece. The craftsman can, if he chooses, bevel one of the sides of the radial stop pieces described above so that they can also be so fitted. However, this is not critically necessary to the functioning of the replica Gera prototype wheel and the radial stop pieces I've described, which will have rectangle shaped cross sections will work just as well.

The reader should keep in mind that in none of Bessler's wheels, *starting* with his 3 feet diameter, Gera prototype wheel, did a weighted lever's end weight(s) actually make physical contact with the lever's wooden radial stop piece. Contact was made only between a weighted lever's two main arms and the radial stop piece. Consequently, while the replica Gera prototype's radial stop piece should be attached to its pair of parallel radial frame pieces so that it is as far from the center axis of the lever's pivot pin as possible in order to minimize the force applied to the radial stop piece as the weighted lever's pivot pin's center axis passes the drum's 3:00 position and the weighted lever's two main arms again make contact with their radial stop piece and rest their weight upon it, it is very important to make sure that the radial stop is not placed so close to the ends of the radial frame pieces so as to allow any physical contact to take place between the end weight(s) and the radial stop piece.

If such contact had been allowed, the metal weight to wood impacts taking place between the drum's 3:00 and 4:30 positions would have greatly increased the mechanics revealing sounds issuing from his wheels when they were running and that was something Bessler definitely did not want happening. In addition, metal to wood impacts, despite the softness of lead, tend to be more jarring than wood to wood impacts and might, as a result, have caused the wood of the radial stop piece to eventually shatter and thereby critically disable or damage a wheel and prevent it from running. Again, this would be an event Bessler would have wanted to avoid, especially during a private demonstration in which he was trying to convince a potential buyer how reliable his wheels were.

In the design I've given for Bessler's 3 feet diameter, Gera prototype wheel in this chapter, it should be possible to glue a beveled radial stop piece up against the parallel pair of outer octagonal frame pieces attached to the ends of the stop piece's parallel pair of radial frame pieces without the radial stop piece touching its weighted lever's lead ingot end weight when the lever's two main arms are in contact with the radial stop piece. If, however, this undesirable contact takes place, then one need only substitute radial stop pieces with rectangular cross sections which will not be able to fit snuggly into the 67.5° corners formed between radial and outer octagonal frame pieces.

As he proceeds to attach each radial stop piece to its two radial frame pieces' leading surfaces with glue and temporarily tapes them into place until the glue dries, the craftsman should measure the distances of all eight radial stops from the surface of the drum's axle to make sure that they are all equal. If not, then, before the glue is thoroughly dried, he should slide any misplaced wooden radial stop pieces about to make sure that their axle facing surfaces are all located at the same distance from the surface of their lever's pivot pin nearest. Once that is done, construction is paused until the glue holding the radial stop pieces to their radial frame pieces has completely dried.

At this point, the builder seeking to duplicate Bessler's Gera prototype wheel has a drum containing eight wooden weighted levers which are each clamped in place and in physical contact with their respective radial stop pieces. This assemblage of parts, however, despite how carefully it has been handcrafted, still needs to be precisely balanced *before* having its forty coordinating cords and sixteen extension springs installed. This is best done after the support stand for the wheel is constructed and that will now be briefly discussed before moving on to the final completion phase of a duplicated Gera prototype wheel.

In the continuation of Figure 10 which the reader should now find and study, I show four orthographic views of Bessler's 3 feet diameter prototype wheel mounted on its wooden stand. The upper left view is an axial view of what I, because of the drum's clockwise rotation, call the wheel's front side. The upper right view is a top view. The lower left view is a profile view as seen from the drum's descending side. And, finally, the lower right view is the bottom view of the wheel showing some additional details of the stand used to support it. (The two views which are missing are the axial view of the wheel's back side which would appear to a viewer there to be rotating counterclockwise and the profile view as seen from the drum's ascending side. These were not included because they are not really necessary for describing the overall structure of the drum, its axle, or the stand that supports the axle and its attached drum.)

Bessler mentioned in his writings that his Gera prototype wheel stood 3.5 feet = 42 inches high and some have assumed that he meant the wheel's drum was actually 3.5 feet = 42 inches in diameter. That was not the case. Rather, the drum was *exactly* 3 feet = 36 inches in diameter and the stand whose vertical supports held the drum's axle pivots kept the center axes of those two pivots and the axle at an elevation of 24 inches above the surface of a table's top or of a floor. As a result, there was a clearance of 6 inches between the bottom of the prototype's drum and the surface of the table top or floor and that resulted in the top of the 36 inch diameter drum being located 6 inches + 36 inches = 42 inches = 3.5 feet above the tabletop or floor.

As can be seen in the continuation of Figure 10, the Gera prototype wheel's stand can be quickly fabricated using only fourteen pieces of wood. There are two pieces that are 1 inch thick x 2 inches wide x 22.75 inches = 22-12/16 inches long which are used for the two vertical supports that will contain the bearings that will hold the pivots at the ends of the axle. There are two pieces of wood that are each 1 inch thick x 2 inches wide x 20 inches long and are used on the two sides of the drum to prevent it from falling over onto one of its axle's ends. There is a single piece that is 1 inch thick x 2 inches wide x 12.375 inches long, oriented parallel to the axle, and located below the drum. This wood piece connects the two horizontal pieces on the sides of the drum together to form an "H" shape as seen in the bottom right illustration of the continuation of Figure 10.

There are also an additional ten wedges used in the construction of the base that are isosceles triangle shaped pieces of wood that help increase the rigidity and stability of the stand. Six of them are 1 inch thick with two equal sides that are each 4 inches long and brace the vertical supports that hold the axle's end pivots and four of them are 2 inches thick with two equal sides that are each 4 inches long and which brace the horizontal pieces on the sides of the drum with the shorter piece that interconnects them. The second group of four wedge shaped pieces also adds mass to the bottom of the stand which can help prevent it from falling over on its side should there be some resonant oscillation that builds up in the drum when it is running at its maximum terminal rotation rate. Because of the light weight of the Gera prototype wheel of only a little more than 6 pounds, the various pieces of its wooden stand can just be glued together after their contacting surfaces have been lightly sanded.

On the bottom right side of Figure 10, the reader will see some of the details of the bearing that supported the Gera prototype wheel's *back* axle pivot pin. The bearing on the wheel's front axle pivot pin is not shown, but was identical.

To place the axle's two projecting pivot pins onto their "half moon" brass bearing plates (note that I will continue to refer to these as "plates" even though they were not flat on both sides, but, rather, had one flat side and one side that had a semi-cylindrical shape to it which together suggested the appearance of the Moon in its first or third phase in which it is referred to as being a "half Moon") requires one to first place *two* 0.0625 inch = 1/16 inch thick x 1 inch diameter steel washers, each having a 0.1875 inch = 3/16 inch diameter center hole, onto *each* of the axle's 0.1875 inch = 3/16 inch diameter end pivot pins (thus, four washers must be used).

Next, one must place one of the axle's steel pivot pins into the open semi-cylindrical space of one of the stand's vertical supports. The other vertical wooden support piece is then *gently* pulled away from the other end of the axle until it is possible to slip the end of the pivot pin on it into the semi-cylindrical space in the pulled away vertical support piece. Once that second axle pivot pin is inserted, the second vertical support piece is released and one must then make sure both of the axle's pivot pins are resting in the hollows that have been drilled out for them in the half moon brass plates.

(Note that in Bessler's later and larger wheels that had far more massive weighted levers than found in the Gera prototype wheel's drum, he would have switched over to using brass bearing plates that were square blocks with semi-cylindrical hollows drilled into them. These have the advantage over the semi-cylindrical "half moon" bearing "plates" used in the Gera prototype wheel of eliminating the possibility of the bearing plate itself being rotated around in a semi-cylindrical nest cut into the wood of a vertical support plank for it by the rotating axle's end pivot pin it supported should there be a lubrication failure which caused the rotating pivot pin to bind up inside of its semi-cylindrical hollow. If that was to happen because a larger semi-cylindrical bearing plate had been used, then the axle's pivot pin could roll out of its bearing plate and then out and off of the support plank! Such an accidental dislocation of a wheel's axle would then allow its rapidly rotating drum to crash to the floor of the wheel's room and result in the destruction of months of work and possible serious injury to anyone near the wheel at the time.)

The half moon bearing plates used in the Gera prototype wheel may be easily fabricated by simply using a hacksaw to cut a cylindrical brass bushing clamped in a shop table vice in half and then gluing one of the half pieces into the circular hole bored into the top end of each of the stand's two vertical support pieces. It is acceptable to use a bushing whose center hole is slightly larger than 0.1875 inch = 3/16 inch in diameter. This will then allow the lubricant placed on the bearings to more easily coat the rotating surfaces of the two steel axle pivot pins. As shown in the bottom right axial view of the top part of the prototype's vertical support piece in Figure 10, the center axis of the half moon bearing plate and the air gap above it should be located 0.75 inch = 12/16 inch below the top of the wooden vertical support piece.

The tops of both of the stand's vertical support pieces also have wing head bolts installed in them. Each bolt can be hand tightened and, as this happens, its end will descend down through the upper half of the cylindrical hole at the top of the vertical support piece which also contains its half moon bearing plate in its lower half. Eventually, the end of the bolt will contact the axle's pivot pin and begin pressing it against the half moon bearing plate that supports it. When both bolts have been tightened, the friction on the two axle pivot pins will be sufficient to prevent the wheel from rotating even though its imbalanced pm wheel mechanics are providing the axle with maximum torque.

With the axle and drum being supported by the wooden stand, it's time to balance the drum. I cannot overemphasize here how important is this phase of the replica Gera prototype wheel's construction. The center of gravity of the eight weighted levers in this wheel, *after* they are placed in their starting configuration, is only displaced about 0.0625 inch = 1/16 inch horizontally onto the descending side of the axle and away from an imaginary vertical line passing through the center axis of the axle (in his 12 feet diameter, bidirectional wheels, the horizontal displacement distance of the active internal one-directional wheel's weighted levers' center of gravity was only 0.25 inch = 4/16 inch!). The resulting start up torque of the Gera prototype wheel was, consequently, very low and if the replica drum's center of gravity with its eight weighted levers installed and locked into contact with their radial stop pieces is not as close as possible to being located *exactly* on the axle's center axis, then the wheel may not run efficiently or even at all when its forty coordinating cords and sixteen extension spring are installed, its weighted levers are finally released, and start up is attempted. Thus, the craftsman must *very* carefully balance the drum and its eight locked down levers using the instructions below. After the initial construction of all of his wheels, Bessler would spend several days meticulously balancing them so they could deliver maximum torque at start up.

After construction of the replica Gera prototype wheel's drum and its attachment to its axle is completed, the eight weighted levers installed, and their parallel pairs of main arms have been carefully clamped against their levers' parallel pairs of radial frame pieces, one will notice that, when the vertical support piece brake bolts are loosened, the drum will have a tendency to rotate into a particular favored orientation whenever one tries to move it to a different orientation. This indicates that there is a small amount of imbalance in the distribution of mass around the center axis of the axle. Most likely, it will be due to very small misalignments of the radial or octagonal frame pieces around the axle which do not cancel each other out so as to put the wheel's center of gravity exactly on the axle's center line.

To compensate for this, the craftsman must add a little extra mass to the drum *directly* above the axle of the wheel as its axle sits on the two vertical supports of the stand *after* the drum has come to a complete stop in its favored orientation (this extra mass will compensate for the drum and levers' center of gravity being slightly *below* the center axis of the axle).

Using tape, he can try temporarily taping one or more 0.0625 inch = 1/16 inch diameter x 0.5 inch = 8/16 inch thick steel washers to various places on the inward facing surfaces of the outer octagonal frame pieces near the drum's 12:00 position until he finally achieves a state of perfect balance for the drum. "Perfect balance" means that the drum, while mounted on its two well lubricated half moon brass bearing plates with the two brake bolts not in contact with the axle's end pivots, will remain stationary *regardless* of whatever orientation it is rotated into by hand. When one finds the locations for the extra washers that balance the drum, the locations can then be marked with a pencil, the tape holding them removed, and the washers finally glued permanently to those exact locations.

With the 3 feet diameter prototype wheel's axle supported on its wooden stand and the drum's eight weighted levers installed and clamped against their radial stops, it's time to begin installing the forty coordinating cords into their proper layers.

Place the wheel and its stand on a table and then unfasten its two brake bolts so that the drum can be easily rotated to any orientation. Next, on another portion of the table arrange the four different sets of coordinating cords.

As an example, if one has opted to use color coded coordinating cords as I suggested earlier, he may have eight main cords that are dyed green, four long lifter cords dyed orange, four long lifter cords dyed yellow, eight stop cords dyed red, and sixteen spring cords dyed blue. There should be forty cords in all. Next to the cords, which each have little metal hooks attached to their two ends, one should place the sixteen little steel extension springs to be used, each of which has an *un*stretched length of a least 1 inch and has a little metal hook attached to *one* of its ends. At this time, the builder can proceed to attach one blue colored spring cord, via *either* of its unused end hooks, to the eyelet at the end of a spring that does not already have a hook attached to it. Do this until all sixteen of the extension springs have spring cords attached to them and put them near the other sets of coordinating cords.

At this time, it is also a good idea for the craftsman to keep the five coordinating cord layers of the drum as depicted in Figures 5(a) through 9(b) in this volume at hand and to make reference to the various layers mentioned as he reads the following instructions for installing the coordinating cords and extension spring's in the replica Gera prototype drum's five cord layers.

As one faces the front side of the drum (which is the side from which, when released, it will be observed to begin to spontaneously rotate in a clockwise direction), the coordinating cord layer between the inward facing surfaces of the parallel pairs of 3 arm side pieces of its eight weighted levers that is *farthest* from him will be layer 5 and he should install the correct coordinating cords into that layer first. As shown in Figures 9(a) and 9(b), that layer requires eight of the spring cords and their attached springs be installed. The craftsman should, one weighted lever at a time, attach a blue colored spring cord's unused end hook to the lever's A1 spring cord hook attachment pin that is exposed by the square notch in its hook keeper for layer 5. As soon as that is done, the extension spring attached to that spring cord should have its unused hook attached to that lever's spring hook attachment anchor pin at the section exposed by the square notch of its drum's hook keeper for layer 5. One can then slowly rotate the drum and repeat this action until all eight of layer 5's spring cords and their attached extension springs are installed. Since all eight of the weighed levers' parallel pairs of main arms are clamped up against their wooden radial stops, none of the layer 5 extension springs attached to the levers' A arms will be stretched significantly (I wrote "significantly" because the weight of a spring that has one end which is free to rotate around an spring hook anchor pin may cause some small and unavoidable amount of spring stretching).

After layer 5 has all of its eight blue colored spring cords and their attached extension springs installed, it's time to work on coordinating cord layer 4 which is the next closest layer to the front of the side of the drum that faces the craftsman. As shown in Figures 8(a) and 8(b), layer 4 receives four of the orange colored long lifter cords.

One of each orange long lifter cord's end hooks will be attached to the hook keeper exposed layer 4 section of the A3 long lifter cord hook attachment pin held between the parallel pair of A arms of the first weighted lever in each pair of connected levers while the other hook of the long lifter cord will be attached to the hook keeper exposed layer 4 section of the B1 long lifter cord hook attachment pin held between the parallel pair of B arms of the second weighted lever in each pair. The four orange long lifter cords of layer 4 will, therefore be installed between the pairs of weighted levers whose pivot pin center axes are located at the drum's 7:30 and 10:30 positions, the drum's 10:30 and 1:30 positions, the drum's 1:30 and 4:30 positions, and, finally, the drum's 4:30 and 7:30 positions.

It is very important that each installed long lifter cord pass through the 0.5 inch = 8/16 inch gap between the inner and outer long lifter cord guide posts located between the parallel pair of radial frame pieces for the weighted lever that is located *between* the two weighted levers of a pair connected by a long lifter cord. Thus, the long lifter cords in layer 4 will pass between the inner and outer guide posts located between the parallel pairs of radial frame pieces for the weighted levers whose pivot pins' center axes are located at the drum's 6:00, 9:00, 12:00, and 3:00 positions.

Next, coordinating cord layer 3, the middle one of the five cord layers, must have its coordinating cords installed. There will be sixteen cords to install, eight green colored main cords and eight red colored stop cords as shown in Figures 7(a) and 7(b).

Firstly, install all eight of the green dyed main cords. These are, as the name implies, the primary coordinating cords in the drum. Each main cord must be installed between two adjacent weighted levers so that it will have one of its end hooks attached to the hook keeper exposed layer 3 section of the first lever's A4 main cord hook attachment pin held between its parallel pair of A arms and its other end hook attached to the hook keeper exposed layer 3 section of the second lever's L1 main cord hook attachment pin held between the parallel pair of main arms of the second weighted lever.

Secondly, install all eight of the red colored stop cords. Each is placed so that it will have one of its end hooks attached to the hook keeper exposed layer 3 section of a weighted lever's A2 stop cord hook attachment pin held between the lever's parallel pair of A arms and its other end hook attached to the hook keeper exposed layer 3 section of the drum's stop cord hook attachment anchor pin that is held between the weighted lever's *leading* parallel pair of *inner* octagonal frame pieces.

Next, the craftsman must move on to installing the yellowed dyed long lifer cords in coordinating cord layer 2 of the drum. As shown in Figures 6(a) and 6(b), there are, as was the case with layer 4, only four long lifter cords in layer 2.

One of each yellow long lifter cord's end hooks will be attached to the hook keeper exposed layer 2 section of the A3 long lifter cord hook attachment pin held between the parallel pair of A arms of the first weighted lever in each pair of connected levers while the other hook of the long lifter cord will be attached to the hook keeper exposed layer 2 section of the B1 long lifter cord hook attachment pin held between the parallel pair of B arms of the second lever in each pair. The four yellow colored long lifter cords of layer 2 will, therefore be installed between the pairs of weighted levers whose pivot pin center axes are located at the drum's 6:00 and 9:00 positions, the drum's 9:00 and 12:00 positions, the drum's 12:00 and 3:00 positions, and, finally, the drum's 3:00 and 6:00 positions. Also, as was the case in layer 4, each of the layer 2 long lifter cords must pass between the two long lifter cord guide posts located between the parallel pair of radial frame pieces that hold the pivot pin of the weighted lever located between the two levers that a long lifter cord connects. In the case of layer 2, the long lifter cords should pass between the gaps of the long lifter cord guide posts located between the drum's 7:30, 10:30, 1:30, and 4:30 parallel pairs of radial frame pieces.

When layer 2's four yellow long lifter cords have been installed, the builder should observe that the eight pairs of long lifter cord guide posts around the drum will each have one long lifter cord located between its inner and outer posts that is either yellow and in layer 2 or orange and in layer 4. It's important here to again emphasize that, while both layers 2 and 4 each contain four long lifter cords, these two layers are *not* superimposable. If a builder accidentally makes the arrangement of long lifter cords in layers 2 and 4 identical, then it will be impossible for the 3 feet diameter, Gera prototype replica wheel to run.

Finally, this brings us to coordinating cord layer 1 which is shown in Figures 5(a) and 5(b) and is the layer closest the viewer of the open front side of the drum of the replica Gera prototype wheel. It contains eight blue colored spring cords with their attached extensions springs and is identical to and superimposable with layer 5 except, of course, that the spring cords and springs are all contained in layer 1 instead of layer 5. To install the extension springs and their attached spring cords into layer 1, just attach a spring cord's unused end hook to the hook keeper exposed layer 1 section of a weighted lever's A1 spring cord hook attachment pin held between the lever's parallel pair of A arms and then attach, by its unused end hook, the extension spring of that cord to the drum hook keeper exposed layer 1 section of that lever's spring hook attachment anchor pin which is held between the lever's leading parallel pair of outer octagonal frame pieces. Do this for all eight of the weighted levers.

At this point, the replica of Bessler's one-directional, 3 feet diameter, Gera prototype wheel has all eight of its weighted levers installed, each of which has a wooden radial stop piece glued to its parallel pair of radial frame pieces, and also has all forty of its coordinating cords and its sixteen extension springs installed. It is now structurally finished and *almost* ready for final testing!

Prior to that, however, one should tighten both of the axle brake bolts to immobilize the axle and its attached drum and then place a *small* droplet of oil, either olive oil or a modern silicone based oil, on the axle's two end pivot pins that rest on their brass half moon bearing plates including their spacing washers, the sixteen spots within the drum where the thin metal sleeves that line the insides of the weighted levers' pivot holes make contact with their pivot pins whose ends are embedded in the drum's radial frame pieces, the spacing washers on the weighted lever pivot pins which are located on each side of a lever, and all of the places where a little metal hook is attached to a cord hook attachment pin within a weighted lever or to a stop cord or spring hook attachment anchor pin whose ends are embedded in the drum's octagonal frame pieces, and, finally, to the larger eyelets at the ends of each extension spring. That works out to well over one hundred spots that will require lubrication!

In order to make sure that all places where there is metal to metal contact receive lubrication, one should start with the axle pivots and their axle end spacing washers and then proceed to the drum and begin by lubricating all of the spots associated with the 6:00 weighted lever. Without rotating the drum, since it is now locked in place by its axle pivots' two brake bolts, work clockwise around the drum until all of the metal to metal contacts associated with each of the eight weighted levers have received their small droplet of oil that will help to reduce frictional wear there to the minimum possible and thereby allow the prototype wheel to run continuously for a long time before its metal components will have worn enough to need replacement.

But, even though the wheel is now fully constructed and well lubricated, the craftsman is not yet ready for the final test of his craftsmanship. This is because the drum's eight weighted levers must still be placed into their correct starting configuration before sustained drum rotation can occur.

Without freeing the axle by loosening its two brake bolts, one must remove all of the clamps that were used to lock the parallel pairs of the main arms of the eight weighted levers into alignment with their nearby parallel pairs of radial frame pieces. Again, start by removing the clamps from the 6:00 weighted lever. With its clamps removed, this weighted lever's main arms will remain vertically oriented and in light contact with their wooden radial stop piece that is attached to the lever's pair of parallel radial frame pieces.

Next, remove the clamps from the 7:30 weighted lever and carefully hold and lower its parallel pair of main arms as they swing away from their wooden radial stop piece and counterclockwise about their pivot pin until the tension supplied by their stretching extension springs prevents anymore motion of the lever. Then, do the same with the 9:00 weighted lever and so on around the drum until all eight weighted levers have been completed unclamped and have assumed the start up configuration that places their combined center of gravity onto the descending side of the axle.

When final equilibrium is achieved, one should notice that only the main arms of the 4:30 and 6:00 weighted levers are in contact with their wooden radial stop pieces and their extension springs unstretched while the other six weighted levers' pairs of parallel main arms are rotated through various angles counterclockwise around their pivot pins and away from their radial stop pieces and their extension springs are stretched by various amounts. The weighted lever whose pivot pin's center axis is at the drum's 3:00 position should *not* have its parallel pair of main arms in contact with that lever's radial stop piece. Rather, the center lines of its main arms should form an angle of less than only 2° with the center lines of the lever's parallel pair of radial frame pieces. Final contact between the leading surfaces of this weighted lever's main arms and their radial stop piece will not occur until the lever's pivot pin's center axis has rotated clockwise about 12° past the drum's 3:00 position.

The two most important angles that the craftsman should measure in his replica wheel will be those of the center lines of the main arms of the weighted levers whose pivot pin center axes are located at the drum's 9:00 and 3:00 positions relative to the center lines of these levers' parallel pairs of radial frame pieces. Ideally, the angle for the 9:00 lever's main arms should be 62.5° and for the 3:00 lever's main arms should be less than 2°.

While it should be possible for the wheel to run with some *small* variation in these angles, the closer that the angles in one's replica prototype wheel match them, the more efficiently the wheel will run. If large discrepancies are noted, however, one will need to carefully check the maximum hook attachment pin center axes separation distances allowed by the main coordinating cord/end hook combinations in layer 3 which were listed at the bottom of the first continuation page of Figure 4.

Most likely, the problem will be with the little metal hooks tied to the ends the coordinating cords or attached to one end of each of the extension springs. Perhaps, the metal staples used to make the metal hooks are too thin and, consequently, have hook or eyelet ends that are flexing open excessively from the tension applied to them by the weighted levers in the replica wheel's drum. This problem can only be remedied by using staples for the hooks that are thicker or wider. The width of any substitute hook must not exceed 0.125 inch = 2/16 inch, however, or the new hook will not fit into the notches in the prototype wheel's wooden hook keepers (these, however, may be enlarged if necessary). Another potential problem might be the excessive stretching, under tension, of the coordinating cords themselves. One must make sure he uses cords which are as inelastic as possible.

When the craftsman reaches this point, then he will have a fairly accurate duplicate of Bessler's one-directional, 3 feet diameter Gera prototype wheel standing before him on a table's top. Weeks, perhaps months, of work now come down to a single final test. With one last visual check to make sure that all of its forty coordinating cords are in their proper layers and none are crossing or even touching another, it's time to loosen the two wing head brake bolts at the top of the wheel stand's two vertical wooden support pieces and see what happens.

If my computer simulations of Bessler's imbalanced pm wheel mechanics are free of glitches which I, of course, am quite convinced they are, and my interpretations of the many clues Bessler left concerning the parameters and geometry of parts used in those mechanisms are valid which I am also quite convinced is the case, and the craftsman's skills are sufficient to follow the construction details described above so as to produce an accurate model, then, upon unloosening both of the stand's brake bolts, the wheel produced should very slowly begin to rotate clockwise as viewed from its front side. As it does so, the successful craftsman will be able to see the exact same sight that amazed Count Karl almost three centuries ago: a genuine and undeniably working chronically imbalanced type perpetual motion wheel!

As the drum continues to rotate and pick up speed, one will become aware of the obvious rapid clockwise rotation of its 9:00 weighted levers' three parallel pairs of arms (its main, A, and B arms) as the levers' pivot pins' center axes travel clockwise about the drum axle's center axis and rapidly move from the drum's 9:00 to 10:30 positions.

The center lines of a lever's parallel pair of main arms, at the 9:00 drum position, hang down counterclockwise away from the center lines of the lever's supporting pair of parallel radial frame pieces with an angle of 62.5°. By the time that lever's pivot pin's center axis reaches the drum's 10:30 position, however, the center lines of that weighted lever's main arms will be located about 43° clockwise away from the center lines of that lever's pair of parallel radial frame pieces. This means that, during the 45° of clockwise drum rotation that took place, this weighted lever's three parallel pairs of arms *all* rotate clockwise through an angle of about 62.5°–43° = 19.5°. This represents the largest clockwise rotation among the drum's five clockwise rotating weighted levers as their main arms move closer to their radial stops during a 45° segment of drum rotation.

To an outside observer of the open and rapidly clockwise rotating drum of the 3 feet diameter, Gera prototype wheel replica, the little 6 ounce lead ingot end weight held between the ends of the 9:00 weighted lever's main arms will appear to make a very rapid vertical ascent. This motion is due to a combination of the drum's clockwise rotation about its axle's center axis and the rapid clockwise rotation of the weighted lever's lead ingot weight about the lever's pivot pin's center axis as that axis travels 45° clockwise around the axle's center axis from the drum's 9:00 to 10:30 positions.

After a minute or so, the replica Gera prototype wheel will have reached its maximum rotation rate which should be in excess of 60 rotations per minute. Even if the craftsman has equipped his replica with the thin wooden outer rim necessary to give the side lattice pieces of the drum a circular shape in an axial view, there will still be enough aerodynamic drag produced by the drum's radial and octagonal frame pieces to produce a noticeable draft. This draft indicates that energy and mass are constantly being extracted from the model wheel's lead weights and their levers and is flowing into the air molecules surrounding the wheel's drum.

Since this book openly reveals the secret mechanics of Bessler's imbalanced pm wheels, there is really no need for one who has constructed a duplicate of his self-starting, one-directional, 3 feet diameter Gera prototype wheel to conceal its internal mechanics with coverings on its drum's sides. However, some craftsmen may wish to do so just to make their reproduction look as close as possible to what Bessler originally built. To do so, one might consider covering both sides of the drum with single sheets of 36 inch wide x 36 inch long linen that have been dyed a dark bluish-green color and then stained with olive oil.

The purpose of the oiled cloth coverings the inventor would have used on his original little toy Gera prototype wheel was to hide any oil stains that might appear on dry cloth from oil oozing out of the various weighted lever pivot pins and drum hook attachment anchor pins and making its way around the radial and octagonal frame pieces to the cloth. Aside from revealing the locations of the drum's eight weighted levers' pivot pins and its radial, inner, and outer octagonal frame pieces and thereby providing clues about the arrangement of the parts inside of the drum, if the cloth used to cover the sides of the drum was dry, then, via capillary action or the so-called "wick effect", enough oil could have been removed from the drum's internal mechanics so that the metal to metal contacts taking place there would become virtually free of lubrication over time and then subject to maximum wear during prolonged rotation of the prototype wheel's drum. If enough of the oil was finally extracted, then the thin metal sleeves secured inside of the pivot holes of the weighted levers' 3 arm side pieces might begin to excessive heat up, expand, and then seize up so they could no longer easily rotate around their snuggly fitted pivot pins. If that did not happen, then so much metal might wear away until the metal sleeves *and* the sections of a lever pivot pins they surrounded that microscopic gaps between their contacting surfaces would increase in size and thereby begin to loosen up or have "excessive play" in them. Such a problem would eventually require new sleeves to be installed in the 3 arm side pieces along with installing new pivot pins in the drum's parallel pairs of radial frame pieces. In a worst case scenario, it might actually be necessary to manufacture new weighted levers!

The square sheet of linen used for each side of the replica Gera prototype wheel's 36 inch diameter drum will need to have a circular hole 1.5 inches = 1-8/16 inches in diameter cut out of its geometric center so that the entire sheet can be placed over a section of the drum's axle that projects out 4 inches from the center of the circular side of the drum. The sheet can then be pulled tight at the 12:00 position of the drum and a small tack pressed into the wood at the end of the radial frame piece there (it can be inserted into the seam between the end of the radial frame piece and the bridge piece glued onto that end). Next, the same is done at the drum's 1:30, 3:00, 4:30, 6:00, 7:30, 9:00, and 10:30 positions. Finally, another eight tacks are inserted at the ends of the very short radial frame pieces located between the above eight drum positions. After this is done, only sixteen tacks will have been used on one side of the drum to hold its cloth covering in place. Once completed, the drum must be turned over and the same done to its other side.

Since the drum's faces are circular and the sheets of linen are square, one will have to use scissors and cut around the two circular edges on both sides of the outer rim wall of the drum to remove the excess cloth. Even with the excess cloth removed and both sheets now circular, there will still be a problem. Between the tacks at the ends of the radial frame pieces and the very short radial pieces, there will be 22.5° arcs of cloth which remain unattached to the drum because there are no additional frame pieces there to receive tacks and the rim wall itself is only 0.0625 inch = 1/16 inch thick and, thus, probably too thin to hold any tacks. These excess flaps of cloth can just be tucked inside of the rim wall and that should be sufficient to prevent outside observers from seeing the prototype wheel's internal mechanics.

In the event that one of the coordinating cords or a extension spring should fail inside of the replica Gera prototype wheel's drum, which will inevitably happen after so many hours of continuous operation at its maximum free running rotation rate, one can then pull the tucked in flaps of cloth out of the drum and peer in through the resulting opening to inspect the drum's interior. Once the broken cord or spring is located, the tacks securing the linen covering near it can be pried out to expose enough of the drum's interior for the cord or spring to be manually replaced. This is made easier when one uses the minimum number of tacks to secure the cloth coverings to the sides of the prototype wheel's drum. For these times, having a supply of spare coordinating cords with their attached end hooks and springs each with their single attached end hook will be very convenient and allow the repair to take place as quickly as possible.

Aside from just letting the duplicate Gera prototype wheel run, feeling the breeze it puts out if its sides are uncovered, and hearing the rapid tapping sounds as the leading surfaces of its descending side's weighted levers' parallel pairs of main arms gently impact their wooden radial stop pieces as their lever's pivot pin's axis reaches a drum position that is about 12° clockwise past the drum's 3:00 position, one can also use the replica wheel to perform various demonstrations of its miniscule 0.2 watt power output.

For example, one can place a small, lightweight screw into the middle of the wheel's front 4 inch projecting axle section and then use that to hoist weights of several pounds when the wheel is turning at its maximum speed. To do this, one needs only to attach a cord to a weight of several pounds and position the weight directly under the part of the axle's projecting section that contains the screw (because some of the wheel stand's triangular shaped wooden reinforcing wedges are located under each of the prototype wheel's projecting axle sections as can be seen in the lower right drawing of the continuation page of Figure 10, one will need to construct a small wooden platform that can be placed over these wedge pieces and upon which the test weight can rest prior to being lifted).

Next, the cord from the weight is drawn out vertically until it passes the elevation of the axle by about a half foot. Then a small loop is formed at the end of the cord. This loop is then quickly slipped onto the rapidly rotating screw head so that it can begin wrapping the cord clockwise around the wheel's rapidly turning axle. When the slack in the cord is taken up by it being wrapped about the rotating axle, the weight will rise and, as it does so, the drum will rapidly slow down. By trial and error, starting with a weight too massive to be lifted all the way to the axle, one can, by gradually lowering the mass of the weight, find a mass that allows the weight to be raised until it *almost* touches the bottom curving surface of the axle as the drum comes to a complete standstill. At that point, one can grab and hold the weight and then detach the cord from it and manually unwind it from the axle and, finally, remove its loop from the screw head.

Another possible test is to see what happens if one does *not* grab the weight as it reaches its maximum elevation and is almost touching the bottom of the axle as the axle and drum come to a standstill. In this case, the drum and axle will then begin rotating counterclockwise in the direction opposite to that which they normally turn as the attached weight unwinds its cord from the axle and begins to descend back to its starting elevation below the axle and finally lands on the top of the small platform provided for it. During the weight's descent, the drum's former descending side will become its new ascending side and its former ascending side will become its new descending side.

However, this change in direction of drum and axle rotation does not affect on what side of the axle the center of gravity of the drum's eight weighted levers is located. Their offset center of gravity now remains located on the axle's new ascending side (its former descending side) and *opposes* its forced counterclockwise direction of rotation as the descent of the weight takes place. In fact, one may notice that the weight does not accelerate as it descends toward its starting elevation, but, rather, does so at a fairly steady rate. This is because the counterclockwise rotation of the drum causes the center of gravity of the eight weighted levers to also rotate a small amount counterclockwise around the center axis of the axle and thereby places that center of gravity horizontally farther away from an imaginary vertical line passing through the center axis of the axle. That repositioning of the eight weighted levers' center of gravity then actually makes the torque *opposing* the counterclockwise rotation of the drum and axle *greater* in magnitude than is the clockwise torque acting on them when they are freely rotating in the wheel's preferred clockwise direction. Once the weight reaches its starting elevation on its little elevated platform below the projecting section of the axle and comes to a rest, its cord will momentarily become slack and then immediately tighten again as the drum and axle continue to rotate a short amount in their unpreferred counterclockwise direction which will then raise the weight a short distance. But, the drum and axle will soon come to a stop again and, as the weight descends back toward its platform, they will again be rotating in their preferred clockwise direction of rotation and accelerating.

However, when it comes time to again begin lifting the weight off of its platform by the drum and axle which are again turning in their preferred direction of clockwise rotation, the prototype wheel will be trying to lift the weight with only a fraction of the rotational kinetic energy and its associated mass that wheel had when it was running at its maximum terminal rotation rate of 60 or more rotations per minute *before* the *first* lift of the weight took place. As a result of this lesser rotational kinetic energy, the weight will only rise a fraction of the distance it did the first time before the drum again stops and begins rotating counterclockwise as the weight sinks back at a steady rate toward its little platform again. As the above scenario is repeated, the weight will rise to a lower and lower elevation each time that the drum turns in either direction. Eventually, the weight will just remain stationary on its little platform and its cord attachment to the screw on the axle will be tight and serve as a tether that prevents the low torque acting on the replica Gera prototype wheel's drum from self-starting and accelerating the drum in its preferred direction of rotation.

What happened to the accumulated rotational kinetic energy and mass that had originally been in the wheel when it was running at its maximum terminal free running rate before it lifted the weight the first time? It was eventually completely dissipated by the aerodynamic drag acting on the rising and falling weight and which was experienced by the drum's eight weighted levers as they were suddenly forced to change direction and begin stirring the air inside of the drum in the opposite direction that its original preferred motion had caused it to flow along in inside of the drum. That aerodynamic drag actually heated the air molecules surrounding the rising and descending external test weight and the weighted levers inside of the drum and resulted in their thermal energies (which are just the sum of the kinetic energies of the vibrating atoms inside of the weights) and masses increasing very slightly.

There are also some other simple demonstrations that can be done with this replica Gera prototype wheel despite its low power output.

Using small screws, one could attach thin sheet aluminum metal fan blades to both of the wheel's projecting axle sections. One set would push air toward the drum and the other set on the other side of the drum would pull air away from the drum. This will result in a flow of air toward and around and then away from the drum and the latter flow of air allows the wheel to function as a continuously running fan that can help circulate the air in a room. One might even leave the linen cloth coverings off of the faces of the drum and then allow the air flow to pass directly through the *interior* of the drum so it does not have to flow around the drum's outer rim wall! The additional aerodynamic drag on the axle caused by the presence of the fan blades will also serve to greatly reduce the maximum rotation rate of the wheel. The benefit of that is that the slower the replica Gera prototype wheel rotates, the longer it can operate before it experiences an internal part failure that disables its mechanics enough to make it stop running.

If one uses a single axle pivot several inches longer than the 3 inch long one I suggest in Figure 10, then he can attach a small grinding wheel to its end. When the wheel is turning at its maximum rate, small metal tools in need of sharpening can then be pressed against the topmost part of the grinding wheel and will be sharpened. This will cause the wheel to decelerate and when its rate becomes too slow, one should remove the drag of the tool being sharpened and only continue once the wheel has reached its maximum rotation rate again.

If a craftsman manages to successfully construct a working replica of Bessler's one-directional, 3 feet diameter Gera prototype wheel, then he should be justly proud of his accomplishment. With the help of this volume, he will have achieved what it took Johann Bessler about a decade to achieve. For most craftsmen, this achievement will probably suffice and they will enjoy showing off their handiwork to their family and friends. A small percentage of craftsmen, however, will not be content to stop with a successful duplication of the little toy prototype wheel.

They will want to go on to construct larger and more powerful editions of the prototype wheel and, in order to meet their needs, I have provided the next two chapters. In them I discuss the inventor's 12 feet diameter Merseburg and Kassel wheels in sufficient detail for a one-directional only version of them to be constructed by a diligent craftsman which only uses one of their two internal one-directional wheels. For the most skilled of craftsman, even such a wheel will still not suffice. He will want to go on to construct a fully operational copy of the *bidirectional* Merseburg or Kassel wheel and be able to duplicate all of demonstrations attributed to it. That requires a detailed analysis of the novel additional mechanisms that Bessler used to provide his two 12 feet diameter wheels with their remarkable property of bidirectionality. The details of these necessary additional mechanisms will also be treated in order to allow duplication of these wheels.

Eventually, the reader will reach the latter part of this volume where I will devote some space to explaining the many geometric and alphanumeric clues in the two DT portraits that allowed me to find what I am *very* well convinced is *the* secret of Bessler's wheels and to also be able to write this long needed book about it.

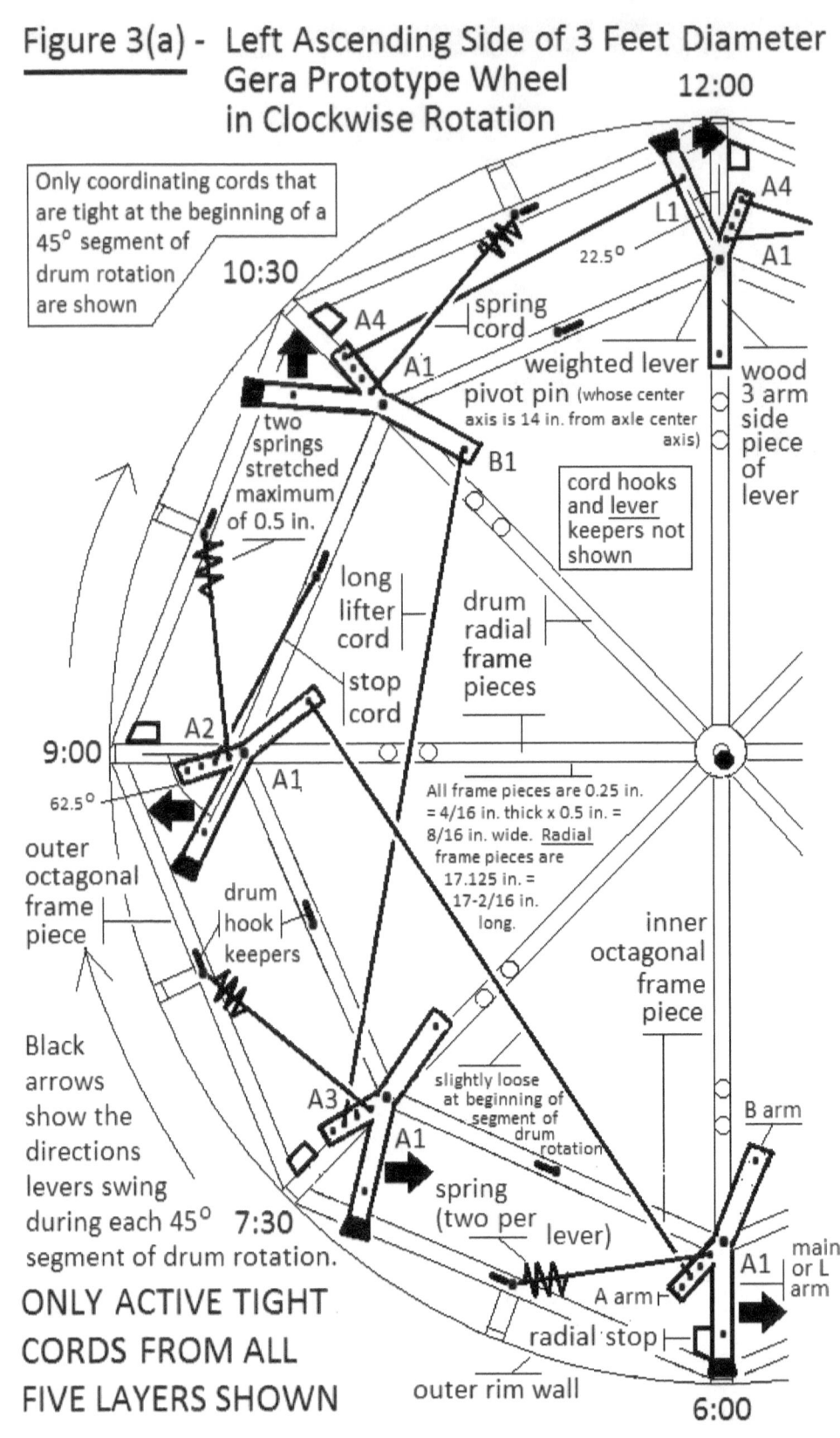

Figure 3(a) - Left Ascending Side of 3 Feet Diameter Gera Prototype Wheel in Clockwise Rotation

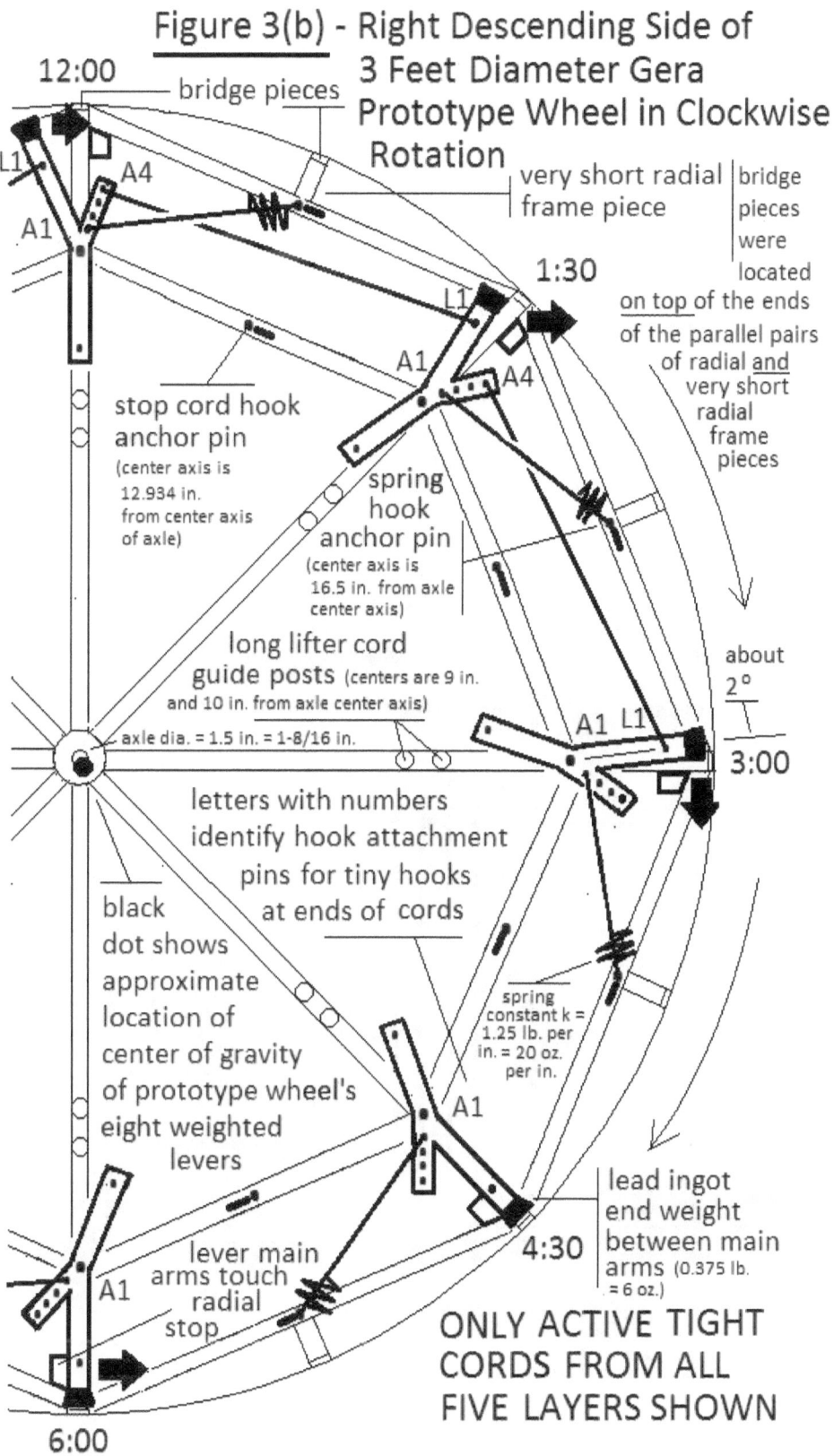

Figure 4 - Details of the 3 Feet Diameter Gera Prototype Wheel's Weighted Lever

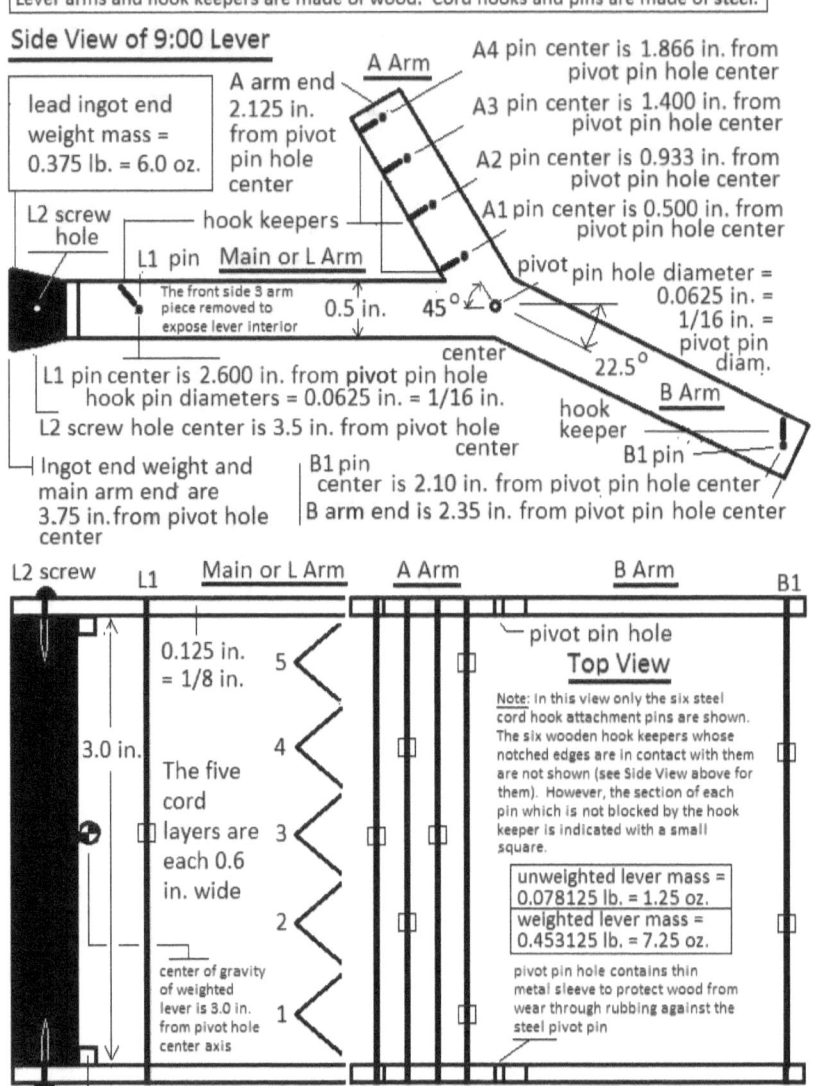

Figure 4 - continued

Details of the Coordinating Cord Hook Attachment Pin Hook Keepers

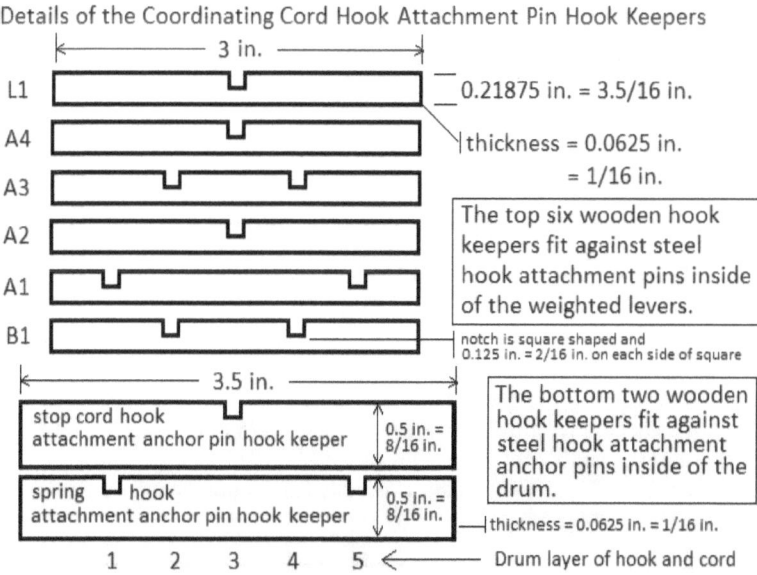

Details of the Coordinating Cords and the Metal Hooks Attached to Them

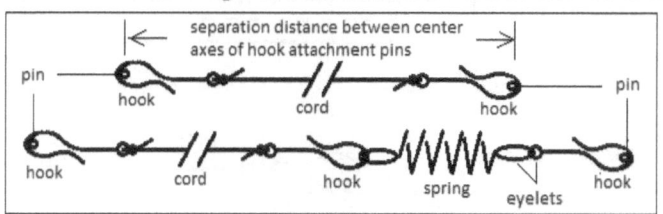

Maximum allowed distances between attachment pin center axes

Main Cord	= A4 to L1 = 11.3 in. = 11-4.8/16 in.
Long Lifter Cord	= A3 to B1 = 19.9 in. = 19-14.4/16 in.
Stop Cord	= A2 to stop cord hook attachment anchor pin between drum's <u>inner</u> octagonal frame pieces = 6.0 in.
Spring Cord*	= A1 to spring hook attachment anchor pin between drum's <u>outer</u> octagonal frame pieces = 6.0 in.

* <u>includes</u> the length of the <u>unstretched</u> extension spring

Figure 4 - continued

Details of the lead ingots Bessler cast to use for the end weights in his one-directional, 3 feet diameter Gera prototype wheel's weighted levers

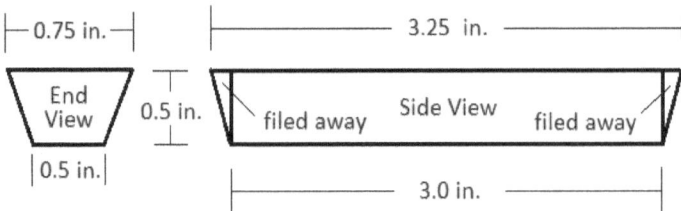

After casting, the ends of an ingot were filed flat. This produced a ingot that weighted slightly more that 0.375 pounds or 6 ounces. Additional filing away of any sharp edges would have made the ingot weight almost exactly 0.375 pounds or 6 ounces.

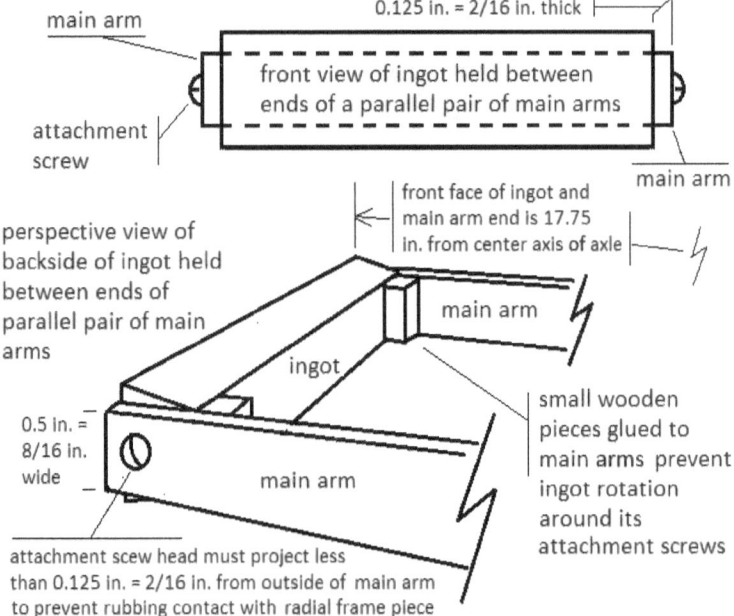

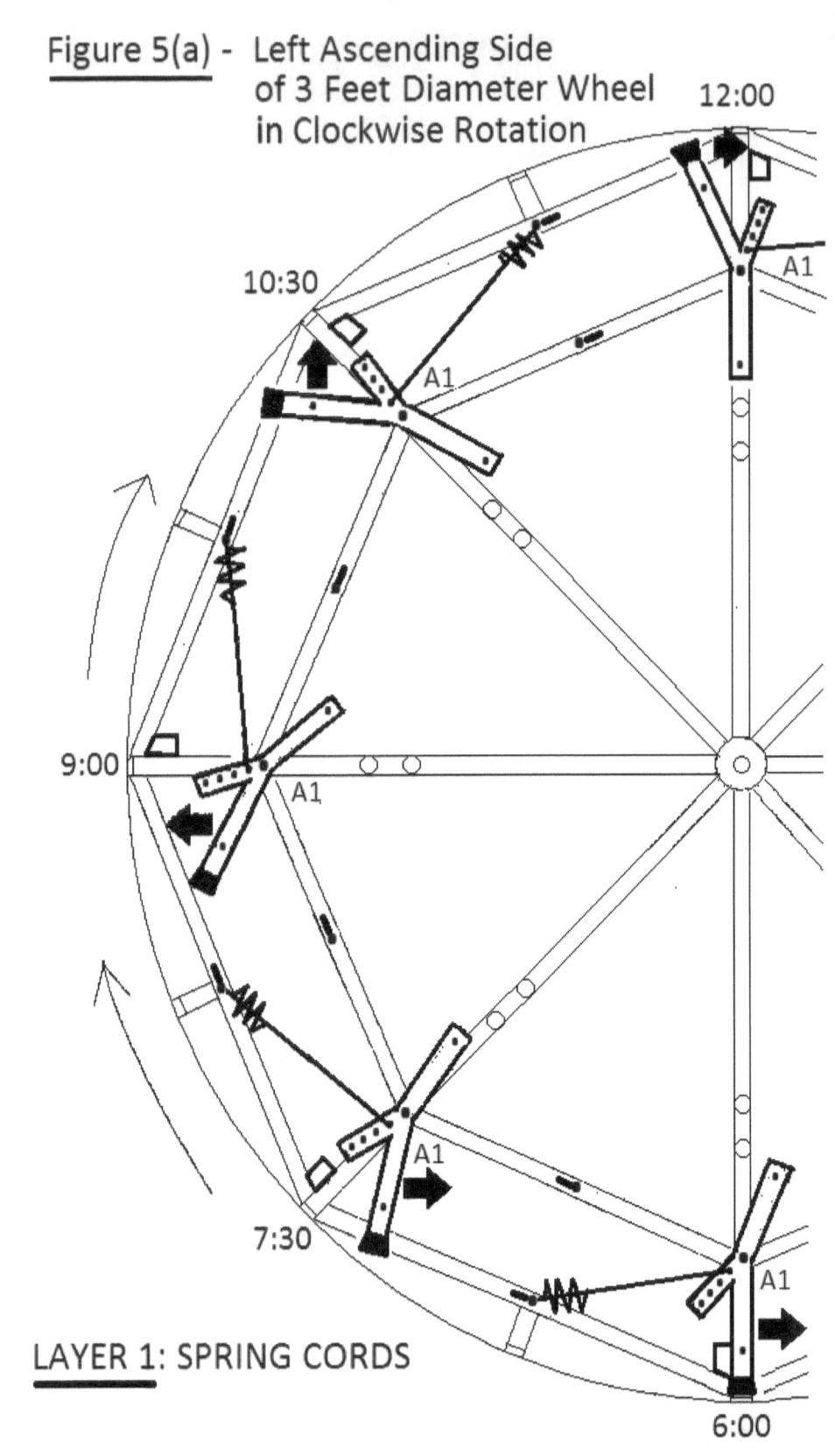

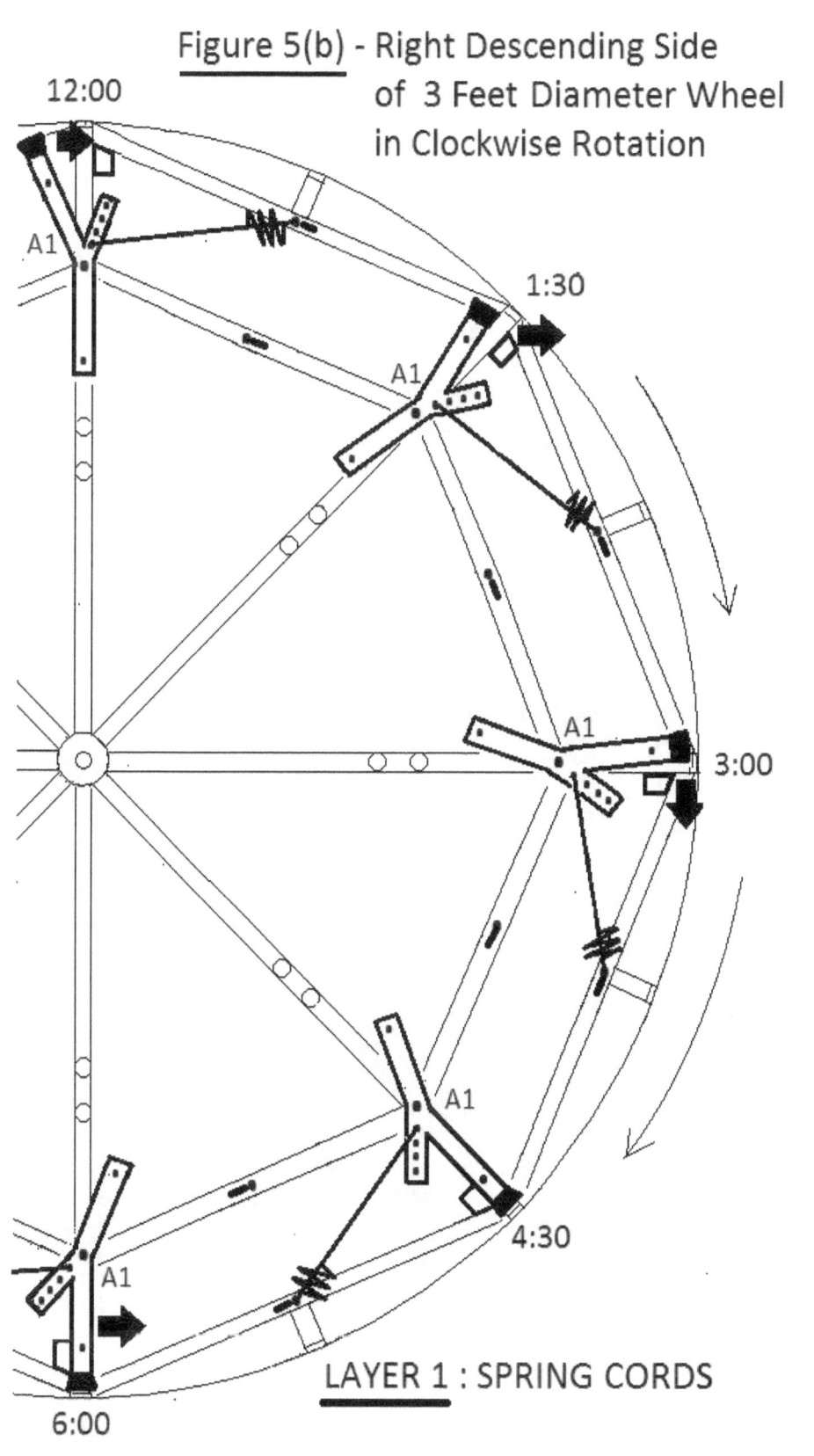

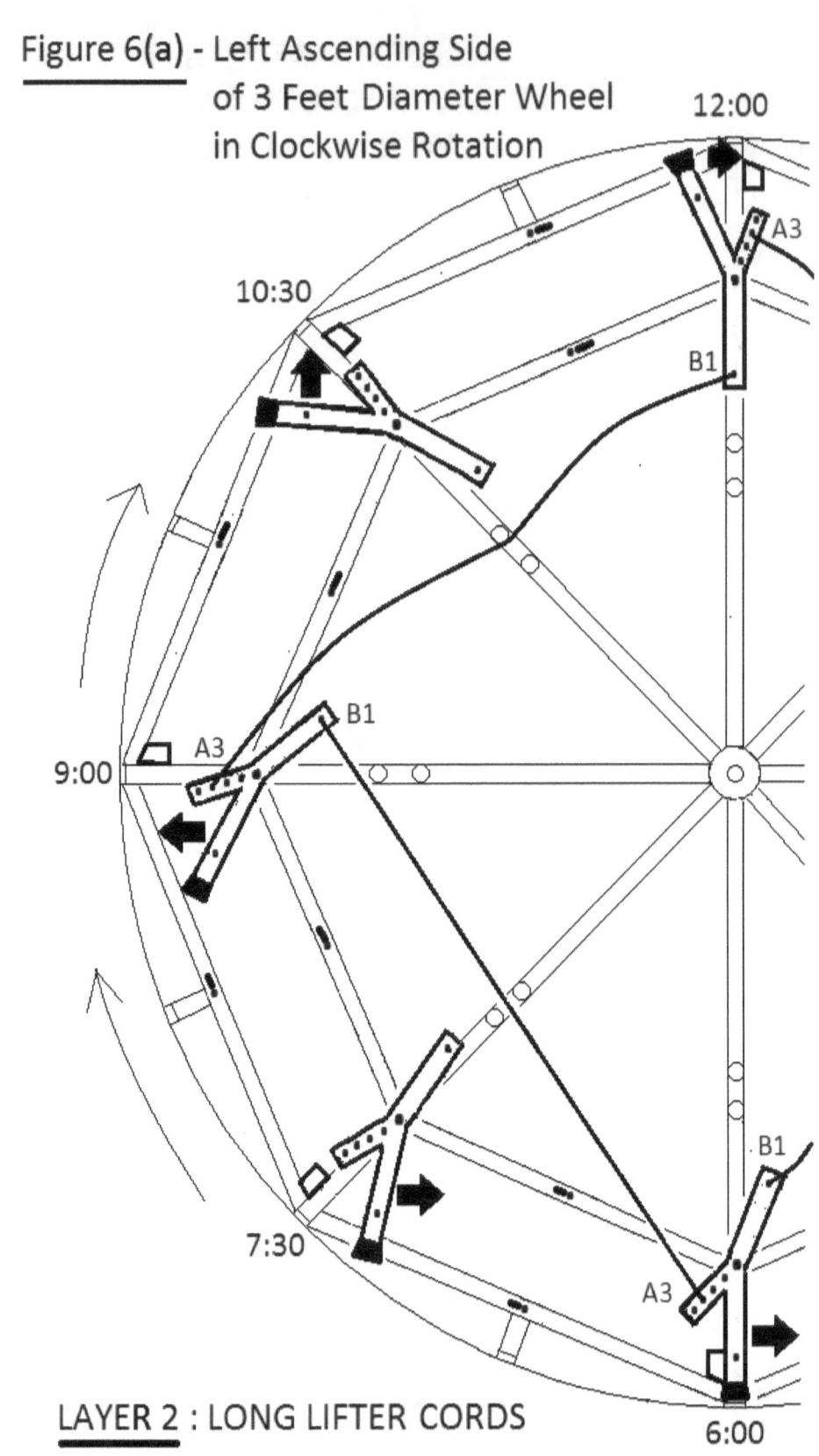

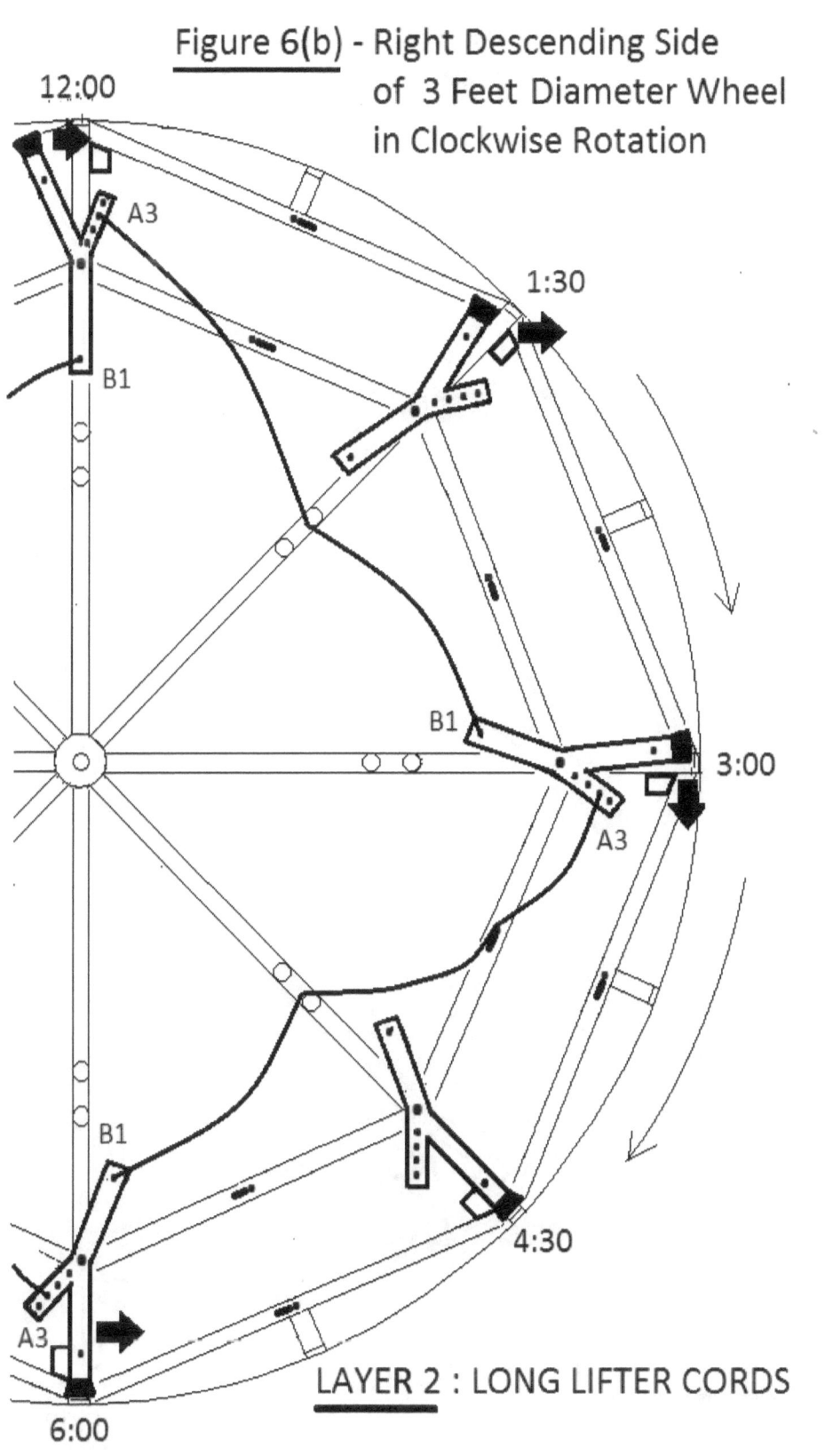

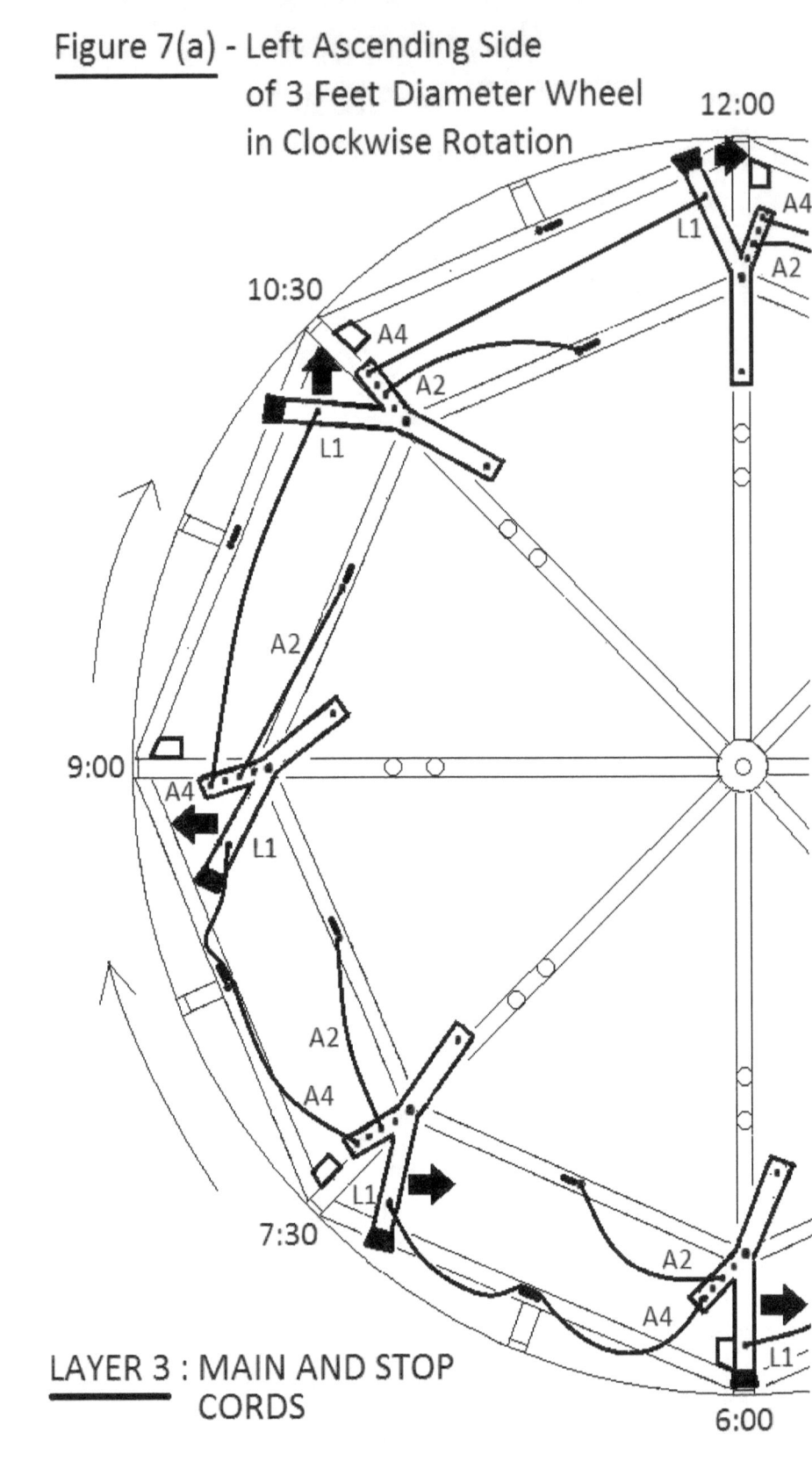

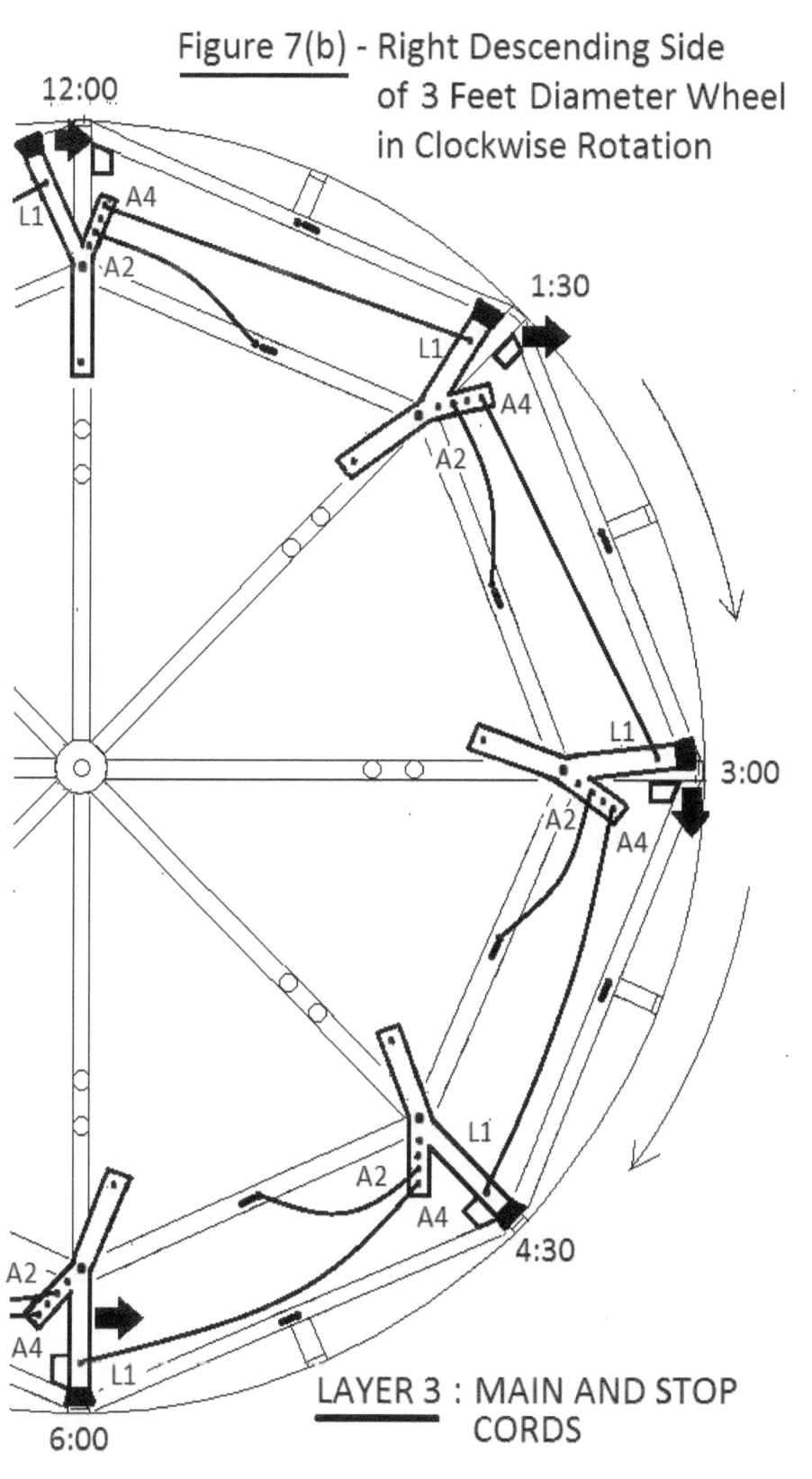

Figure 7(b) - Right Descending Side of 3 Feet Diameter Wheel in Clockwise Rotation

LAYER 3 : MAIN AND STOP CORDS

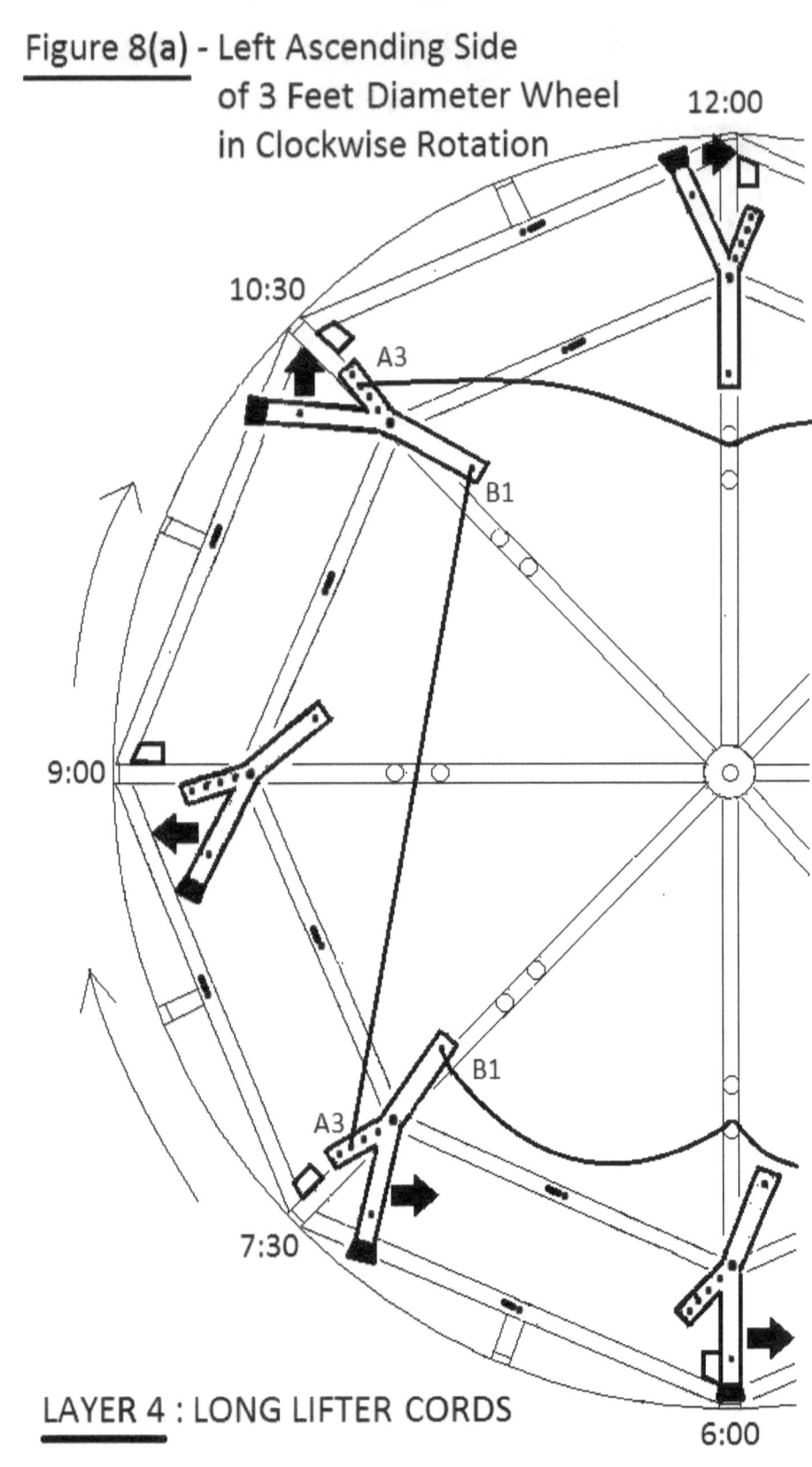

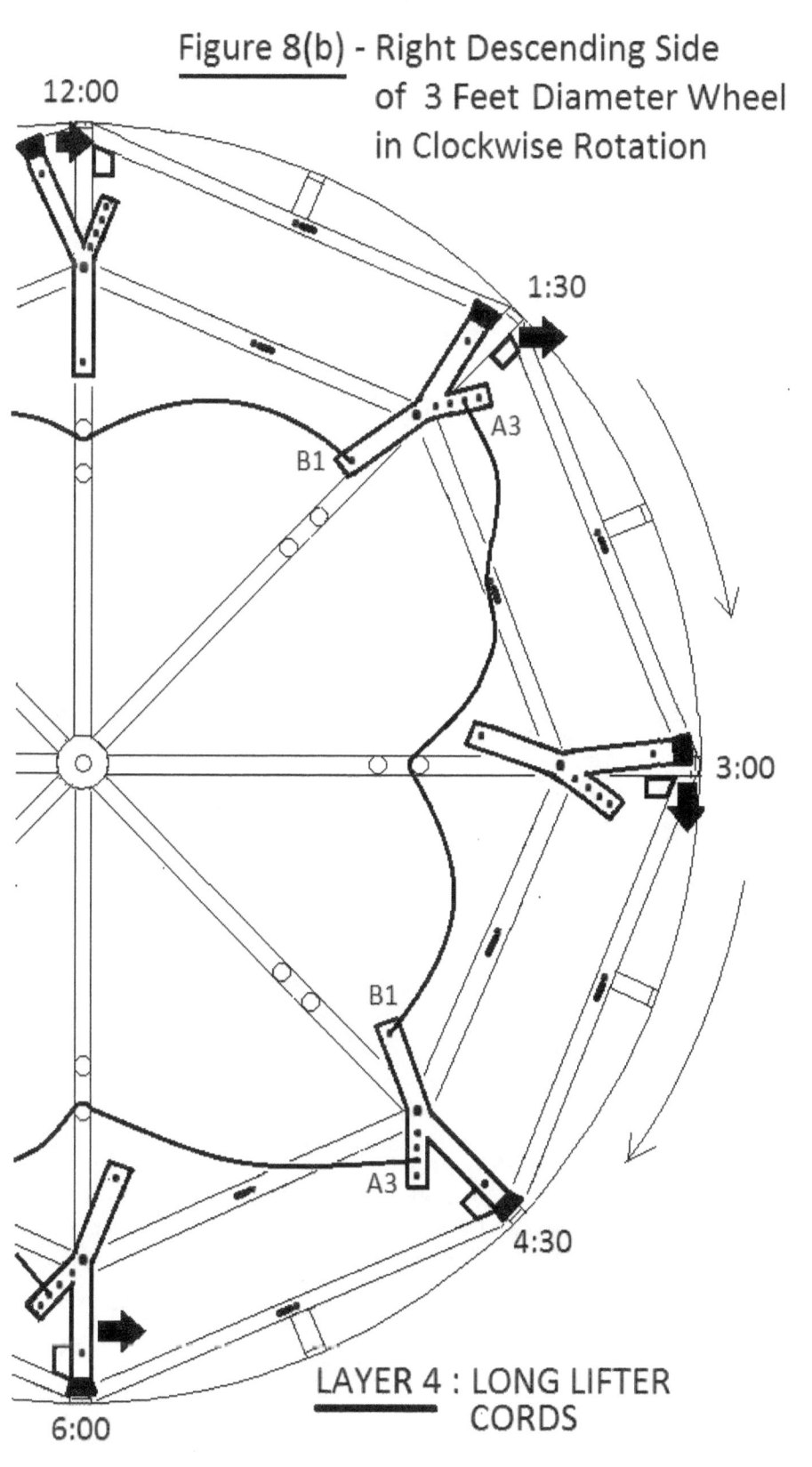

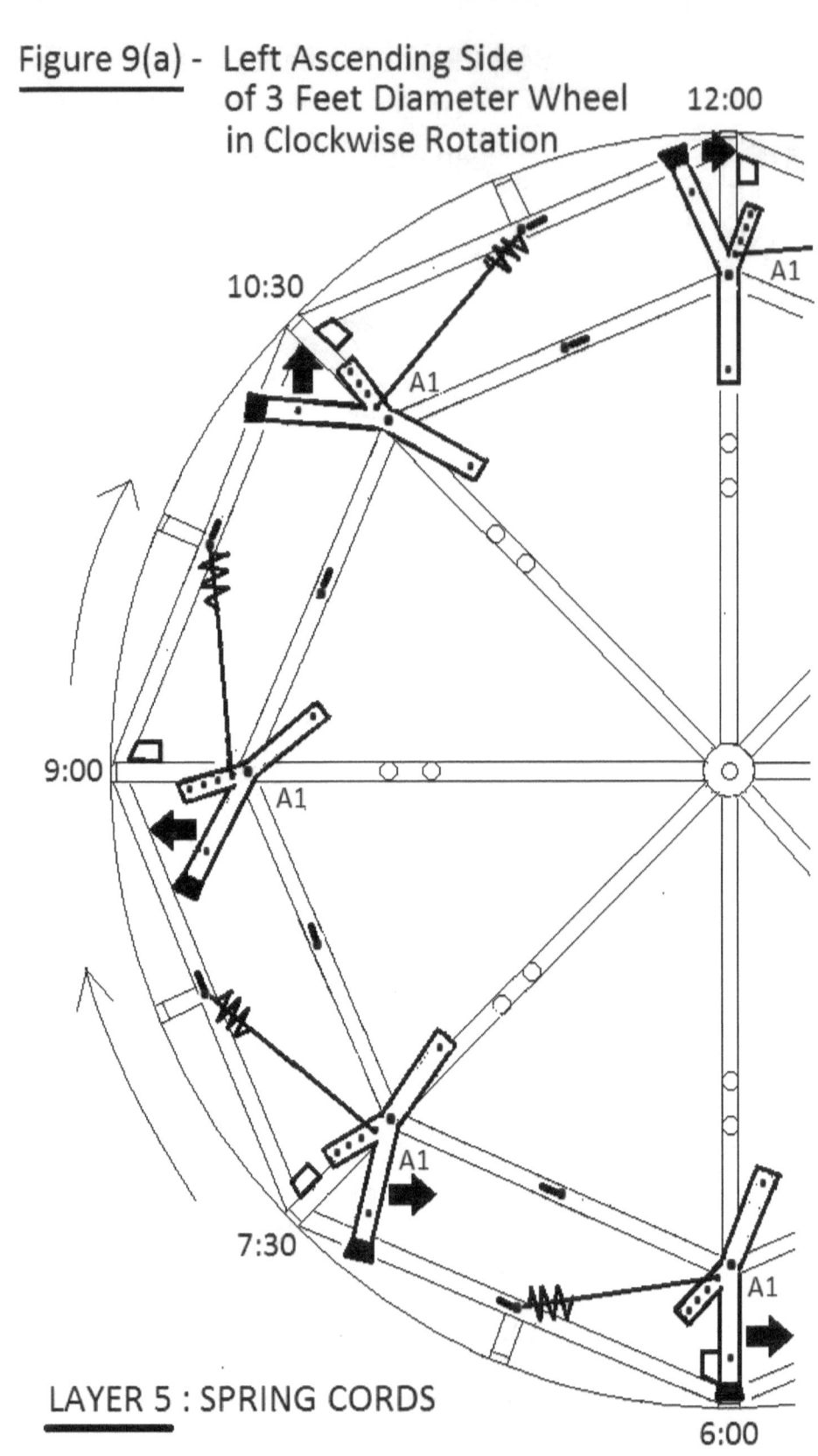

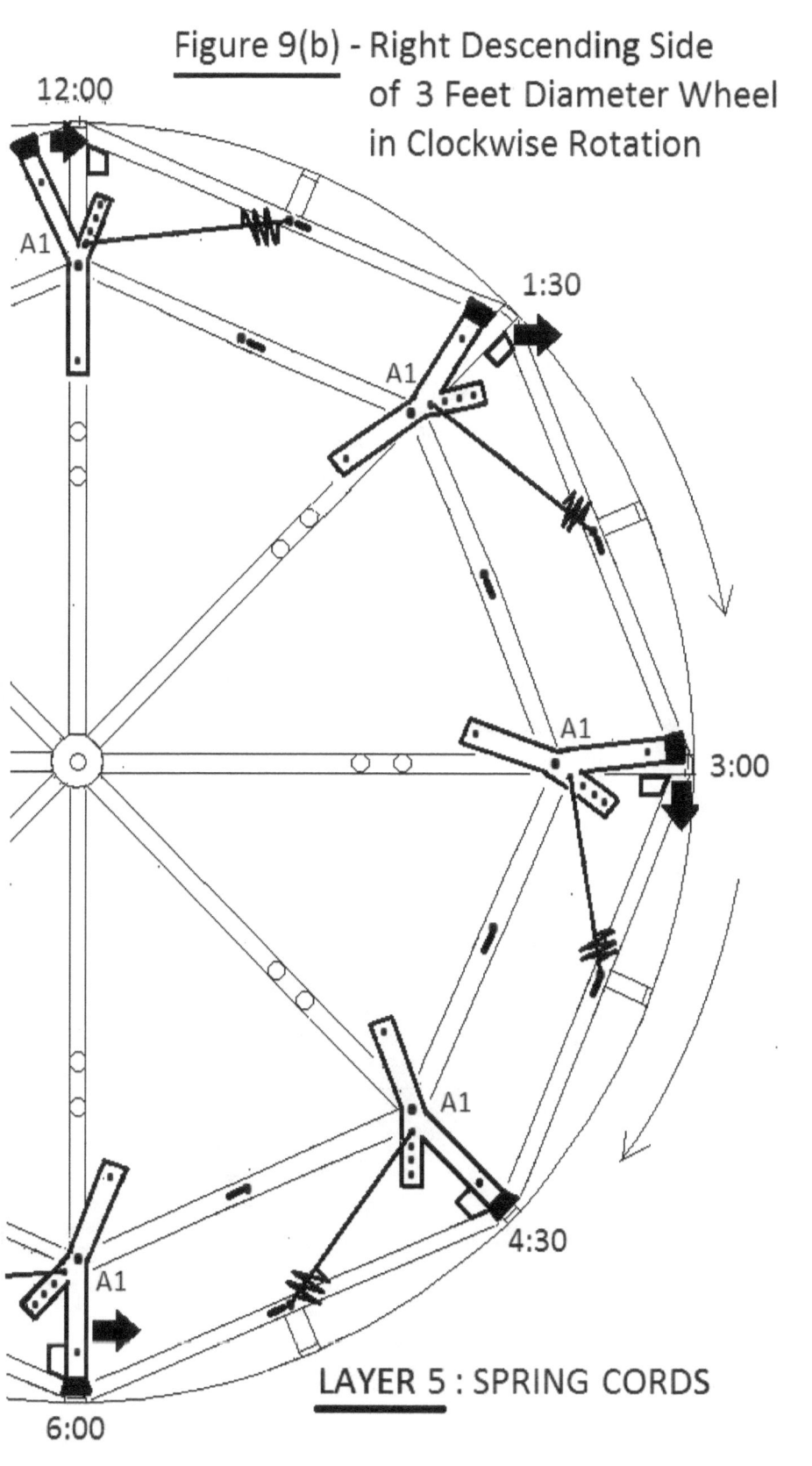

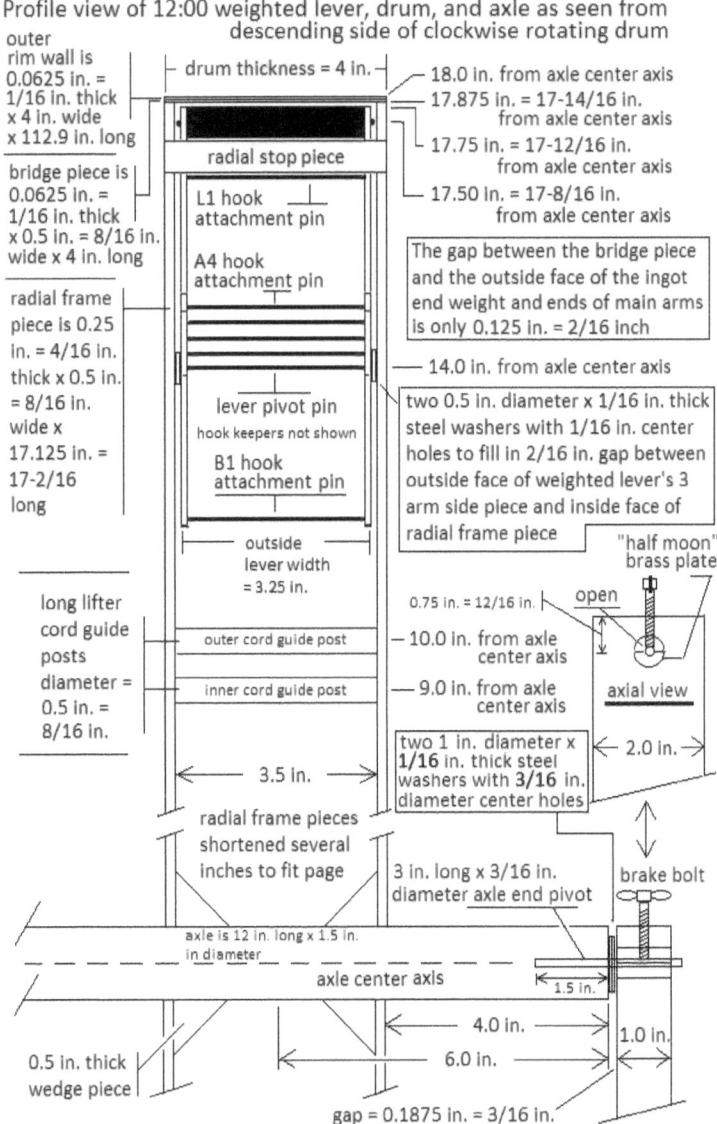

Figure 10 - More Details of the 3 Feet Diameter Gera Prototype Wheel's Construction

Figure 10 - continued

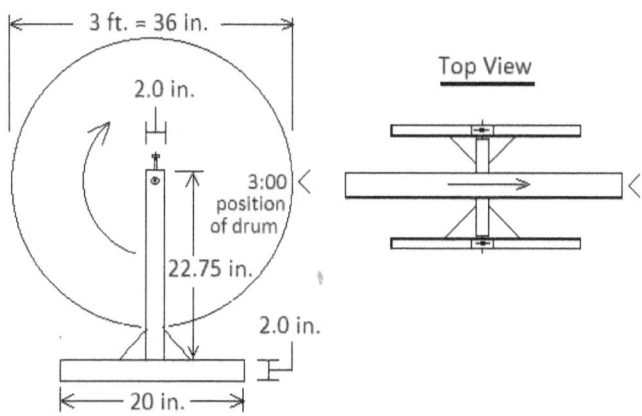

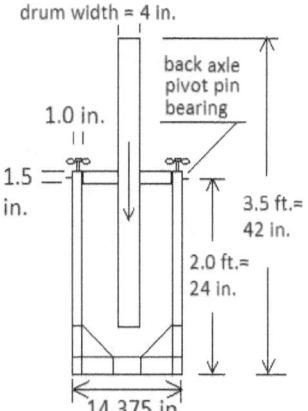

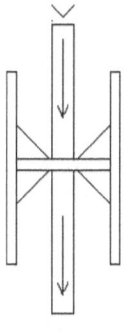

CHAPTER 6
THE BIDIRECTIONAL 12 FEET DIAMETER MERSEBURG AND KASSEL WHEELS

As noted in Chapter 1, Bessler's efforts at constructing working imbalanced perpetual motion wheels did not end with his first successful 3 feet diameter, one-directional, prototype wheel constructed in Gera, Saxony near the end of the year 1711. Realizing that he had finally found a design that worked and which no one else had previously stumbled upon, his next task became trying to sell his invention and to do so for an enormous amount of money with which he hoped to found an engineering school, his "Fortress of Wisdom", that, aside from teaching various forms of crafting, science, and technology, would actually revolutionize life on Earth by instilling what he considered to be the correct Christian values in its students. His was a very ambitious plan, but, without the sale of his invention, it would remain an unrealized dream that, at best, might only result in a few published books on his religious philosophy which would not enjoy wide circulation much less actual practice by anyone.

To Bessler, the quickest route to the fortune he needed seemed obvious. He had to convince some rich buyer, such as a member of the aristocracy or a consortium of businessmen, that his wheels could, if made large enough, be a more powerful and versatile power source than one being provided by an animal, water wheel, or windmill. He was, of course, aware of the developments being made in harnessing the power of expanding steam to drive machinery, but they all required the constant burning of fuel that cost money because it had to be manually collected either from a forest as wood or as fossilized prehistoric wood, also known as "coal", from deep within the bowels of the Earth. To Bessler it seemed that his wheels had so many clear advantages that it would only be a matter, at most, of a few years before he sold their secret and could move on to his ultimate goal: improving human life and its religious values worldwide. Indeed, he may have envisioned the ultimate creation of a single planetary religion which would, finally, result in the moral perfection of the human race. While his wheels were capable of perpetual motion on Earth, his new worldwide religious order would assure everyone living on Earth of an eternal spiritual existence in Heaven after the death of their physical bodies because they would, finally, be living exactly the way God wanted them to, at least according to Bessler's analysis of the Scriptures.

So, to interest the business community of his day, he needed wheels far more powerful than the table top toy he had constructed at Gera. Yes, it worked, but it was feeble and businessmen wanted a form of rotary power that could run stamping machines or turn the massive stone wheels that ground grain or squeezed the oils from various nuts and seeds. Those that owned mines were always bothered by flooding from ground water and needed pumps powerful enough to quickly drain a flooded mine so that the extraction of its stony treasures could continue. To gain their attention for his wheels, Bessler began to construct a series of wheels of ever increasing size and power.

Increasing his wheels' power outputs was actually very simple. He needed only to increase the masses of their weighted levers and then move the centers of gravity of those individual levers farther from the center axis of the wheel's axle. This, unfortunately, also required housing the wheel's imbalanced pm mechanics in larger and heavier drums attached to larger and heavier axles. But, by using all of the carpentry skills he had acquired, the inventor could keep the mass of a wheel's drum and axle to a minimum. They could be made just strong enough to safely contain the weighted levers as they sped around their axle and slowly gave up and equally distributed their energy and mass content to the wheel's drum, its contents, and the axle to which they were attached to accelerate their motions. If, however, those components already moved with constant rotation rates, then to any external operating machinery attached to a wheel's axle or to just the air outside of the wheel if it was running at its maximum terminal rotation rate.

By early 1714, over two years after he first found success with his Gera prototype wheel, Bessler and his family were living in the nearby town of Draschwitz where he had completed the construction of a one-directional wheel whose drum was 6 inches thick and 9 feet in diameter. Because of its size and mass, he had to have it permanently installed in his rented home there and then allowed the public to enter the first floor room where the wheel was kept in order to view its capabilities. What probably amazed people the most was the dramatic moment when he first released the short piece of rope which kept the rim of its drum tied down to the floor beneath the drum. As soon as he undid the lock from the drum and pulled the rope away, the drum would slowly begin to rotate. Over the course of several minutes it would build up to its maximum free running speed of about 50 rotations per minute.

At that point Bessler had a variety of amazing demonstrations he would perform with his wheel for the public. It would make stamping machines run and pendulums attached to its axle's end cranks swing to and fro. Heavy loads of bricks would be hoisted up from the floor until they nearly touched the ceiling by a rope running through a ceiling mounted pulley whose other end wound rapidly around the wheel's axle. As all of this was going on, Bessler would have been enthusiastically singing the praises of his invention and telling everyone how it could revolutionize their lives and its use in commerce would help lower their costs for manufactured items while increasing profits for factory owners.

Everyone was fascinated by what they were hearing and seeing, but, oddly, although Bessler was starting to interact with more and more businessmen, noblemen, intellectuals, and craftsmen, no one was interested in purchasing his wheel for the enormous price he demanded. Indeed, many began to say that it was just too much money to risk on something that might be a hoax, especially with a stranger who had just arrived in town and had a history of moving out of small villages in a hurry. Some began to suggest that his wheels were nothing more than cleverly constructed mainspring driven machines and the fact that they could, like mainspring powered clock movements, only rotate in one direction proved that in their minds. Some doubted that the wheels would ever be powerful enough to meet the ongoing and escalating needs of the Industrial Revolution that was sweeping across Europe at the time.

In September of 1714 Bessler had a visit from Gottfried Wilhelm Leibniz who had heard of the inventor's wheels and wanted to personally see one. Leibniz was as famous for being a scientist in continental Europe as Newton was in England and Bessler was only too glad to give such a notable personage a private demonstration of his Draschwitz wheel. The older scientist spent several hours having the inventor and his assistant, probably his brother, conduct various tests on this large one-directional wheel.

At the conclusion of the tests, Leibniz was quite convinced that Bessler did, in fact, have something very unusual although he thought it must somehow be tapping a previously unsuspected environmental source of power. Perhaps, he thought, it was being propelled by the same gravitational and inertial forces that made the ocean tides possible or that it was coupled to some moving source of magnetism within the bowels of the Earth or that the inventor had found some wondrous substance for the wheel's weights that, after falling a certain distance and performing work, could be raised back to its starting point by requiring less work to do so. That latter mechanism was, however, in light of the developing science of mechanics of that time, considered a physical impossibility. But, Leibniz was aware that many things previously thought to be impossible were being achieved on an almost yearly basis. In any event, Leibniz was convinced that the Draschwitz wheel was real and not being propelled by any known source of either external or internal power including metal tanks filled with compressed air that might have been used to shift metal pistons about inside of metal cylinders so as to keep their collective center of gravity on the wheel's descending side during drum rotation. The amount of work the wheel could do continuously, Leibniz noted, was still far less than could be delivered by even a small water or windmill, but, he reasoned that, with further development, it might be able to compete successfully with those.

During his visit, Leibniz made several suggestions to Bessler that influenced his future plans for wheels and their exhibitions. The older scientist impressed upon the younger inventor that he needed to have a wheel that was sturdy enough to run continuously for at least several weeks while also continuously performing work while it did so. Next, he suggested that it would be nice if Bessler could somehow modify his one-directional wheels so that they could be started up rotating in either of two directions. These two changes, Leibniz assured Bessler, should finally remove any doubts about the inventor's wheels being genuine. And, finally, Leibniz suggested that any future wheels Bessler constructed be made much more powerful. Since steam engines were being developed to operate pumps for draining flooded mines, he would have told Bessler that it would be very advantageous for the sale of his invention if he could come up with a wheel that could raise water to a height of several hundred feet and do so at a rate that would make it suitable for draining a flooded mine.

Before he parted company with Bessler, Leibniz promised that he would alert the many scientists he was in contact with throughout Europe about his impressions of the wheel he had just tested and, hopefully, they would then pass the word on to their associates and that might help Bessler finally sell his marvelous invention.

Bessler was so impressed by Leibniz' advice that, by the time the former and his family moved to another neighboring town of Merseburg in 1715, the inventor already had complete plans for a wheel that would meet most of Leibniz' requirements. It would be the largest and most powerful he had ever constructed and its stronger parts and coordinating cords would allow it to run for weeks, perhaps months, without having to be repaired and could do so while continuously operating machinery attached to its axle. Most importantly, it would incorporate an ingenious system of gravity activated latches that the inventor had conceived that would allow it to be started up, from a standstill, in either of two directions after being given a gentle push in that direction.

Bessler was quite convinced that his Merseburg wheel would, indeed, be *the* one which he would finally sell. Any complaints that it was still not powerful enough could be answered by just telling a prospective buyer that any level of power could be obtained by simple mounting several of the wheels, up to four, on a single axle and, if necessary, using several of these ganged quartets of wheels together.

Finally recovering from a severe bout of depression, Bessler had his Merseburg wheel completed in time for the Whitsun Holiday which is the Christian festival of Pentecost that takes place on the seventh Sunday after Easter to commemorate the descent of the Holy Spirit upon Christ's disciples as described in the New Testament's Acts of the Apostles, chapter 2. This religious holiday was celebrated sometime in late May of 1715 and the wheel, originally planned for exhibition at an earlier Easter fair in the town of Leipzig, was, however, like the Draschwitz wheel, publicly exhibited in one of the ground floor rooms of the inventor's apartment in the town of Merseburg. While Bessler knew he could draw larger crowds to his public exhibitions by having them conducted out of doors, he also distrusted such an approach to getting publicity because of the security risk it posed. He would not be able to personally guard the wheel at all times and that meant he would have to hire strangers to do so for him. Any such stranger could, as far as he was concerned, be someone out to steal the secret of his wheel's mechanics and sell it to a rival inventor or, possibly, steal the entire wheel when the inventor was not around!

In this chapter, however, I will not again recount all of the details of the construction and exhibition of the Merseburg wheel and its even more powerful version constructed at Count Karl's castle in Kassel because this was done adequately in Chapter 1. This chapter's goal will be to provide that rare craftsman who might want to attempt to duplicate one of these last two exhibited wheels of Bessler with some additional details of their internal construction and mechanics that might make that possible, but most of the treatment will deal with the Merseburg wheel because, while not his first wheel, it was his first bidirectional wheel and, therefore, a sort of prototype for wheels with that capability.

From my study of Bessler's various wheels, it eventually became obvious to me that, basically, he only had a single design that worked and that design was the one used in his original one-directional, 3 feet diameter Gera prototype wheel which was extensively treated in the last chapter. If one understands the mechanics of that toy prototype wheel and can successfully duplicate them, then he can, as Bessler did, just scale up the design to create wheels of greater size and power. In fact, that is exactly what Bessler did and all of his wheels constructed after the Gera prototype were simply enlarged versions of that wheel with a few extra mechanical enhancements to give them added capabilities.

In one of the last chapters of this volume, I will provide the serious student of the inventor's wheels with two convenient tables of data that provide the various constructions parameters for all of the imbalanced pm wheels Bessler built or planned to build. With these tables is should be possible for a craftsman to duplicate all of Bessler's wheels. And, again, I would strongly encourage anyone considering attempting to duplicate any of Bessler's later wheels to only begin on such a project *after* he has successfully constructed a duplicate of the inventor's one-directional, 3 feet diameter Gera prototype wheel. That toy wheel's successful construction will serve as "basic training" for what will need to be done as one pursues a larger edition. But, in the present chapter, let us begin by considering some of the general details of the wooden drum used in the Merseburg wheel.

We can begin by discussing Figure 11(a) whose center part shows a *front* axial view of the 12 feet diameter bidirectional Merseburg wheel's drum after its front and backside dyed and oiled linen cloth covers have been removed and then placed to the left and right of the 12 inch thick drum.

As the reader will notice, the front lattice of the drum is somewhat familiar and contains all of the radial, inner and outer octagonal, and rim wall support bridge frame pieces of the much smaller one-directional, 3 feet diameter Gera prototype wheel's drum. For the four times larger diameter Merseburg wheel's drum, however, many extra wooden frame pieces were added for extra strength and rigidity.

There was an additional set of eight short "innermost octagonal frame pieces" in *each* lattice (one of which is labeled "f" in Figure 11(a)) whose beveled ends' centers contacted the drum's eight main radial frame pieces (one of which is labeled "a") at a distance of about 28 inches from the center axis of the wheel's 6 inch diameter axle or 25 inches from the surface of the axle to which the radial frame piece they contacted was attached. From the ends of each of these innermost octagonal frame pieces, two "diagonal frame pieces" (one of which is labeled "g") extended outward away from the axle, but came together and then contacted an "inner octagonal frame piece" (one of which is labeled "e") at its midpoint near where the stop cord hook attachment anchor pin's end was located. Then, near each stop cord hook attachment anchor pin end embedded in an inner octagonal frame piece there was an additional "short radial frame piece" (one of which is labeled "b") that extended radially outward away from the center axis of the axle until it contacted the midpoint of an "outer octagonal frame piece" (one of which is labeled "d") near where the spring hook attachment anchor pin end was located. From the midpoint of that outer octagonal frame piece, an additional "very short radial frame piece" (one of which is labeled "c") finally extended outward until it contacted a wooden bridge piece (two of which are labeled "h" in Figure 11(a)).

The drum's outer rim wall (labeled "i"), most likely in sections with lengths only $1/8^{th}$ or $1/16^{th}$ of that needed to cover the drum's full circumference with a single long piece, was then nailed or screwed down to the outside facing surfaces of the drum's sixteen wooden bridge pieces whose *lengths*, like the Merseburg wheel's drum's thickness, were exactly 12 inches each. The various types of wooden lattice frame pieces are labeled in the lower left quadrant of the drum's front lattice shown in the middle of the Figure 11(a) with the letters "a" through "h" and the types of frame pieces they correspond to are horizontally listed below the floor of the wheel's room. The letter "i" which stands for the drum's outer rim wall is not included because I do not consider it to be part of a lattice, but, rather something added to a drum's lattice after its completion.

Thus, *each* of the *three* lattices of the 12 feet diameter, bidirectional Merseburg wheel's drum contained eight radial frame pieces (a), eight outer octagonal frame pieces (d), eight inner octagonal frame pieces (e), eight innermost octagonal frame pieces nearest the axle (f), sixteen diagonal frame pieces (g), eight short radial frame pieces (b) located between the middle and outer octagonal frame pieces (e and d), and eight very short radial frame pieces (c) located between the midpoints of the lengths of the outer octagonal frame pieces (d) and the bridge pieces (h) to which the drum's outer rim wall was attached. So, a total of sixty-four wooden frame pieces of various lengths and angular end cuts were required for *each* of the Merseburg drum's three lattices to make a grand total of one hundred and ninety-two frame pieces needed for *all* three lattices in this wheel's bidirectional drum!

From this, we see that this total number of sixty-four frame pieces per lattice for the Merseburg wheel was double the thirty-two frame pieces per lattice found in the design for Bessler's 3 feet diameter Gera prototype wheel given in the previous chapter. These counts of thirty-two and sixty-four frame pieces *per* lattice, however, do *not* include the extra sixteen 0.0625 inch = 1/16 inch thick x 4 inch long wooden bridge pieces that were attached to the ends of the sixteen radial frame pieces and sixteen very short radial frame pieces of the one-directional Gera prototype wheel's *two* parallel lattices that were mounted on the same 1.5 inches diameter axle. Nor do they include the extra sixteen 0.5 inch = 8/16 inch thick x 12 inch long bridge pieces that were attached to the ends of the twenty-four radial frame pieces and the twenty-four very short radial frame pieces of the bidirectional Merseburg wheel's *three* parallel lattices that were mounted on the same 6 inch diameter axle. (Note that the heavier and wider drum of the later constructed Kassel wheel also used sixty-four frame pieces per lattice and three lattices mounted on the same 8 inch diameter axle whose ends were also held together by sixteen wooden bridge pieces, each 0.5 inch = 8/16 inch thick x 18 inches long, that were attached to the ends of its three lattices' twenty-four radial frame pieces and twenty-four very short radial frame pieces.)

Most of Figure 11(a) shows the front axial view of the Merseburg wheel's uncovered front lattice, but the middle and rear lattices would also be present, but hidden from view behind the front lattice because all three lattices were parallel. Since this might confuse some readers as to how the 12 feet diameter bidirectional wheel's other two hidden parallel lattices were arranged on its axle, I added a small side view of the top portion of the drum in the upper left hand side of the figure. This is what one would see if the outer rim wall was removed from the drum, one then ascended a ladder placed near the periphery of the drum until he was level with the top end of the front lattice's 12:00 radial frame piece, and then finally looked toward that frame piece. In that inserted side view of the drum one can see the ends of the 12:00 radial frame pieces of all three of the parallel lattices that were attached to the wheel's axle (which is not shown) and the *single* wooden bridge piece that is attached across the ends of *all* three of the 12:00 radial frame pieces.

As I did for the Gera prototype wheel design, I will only tell the craftsman here that the twenty-four radial frame pieces of the three lattices that were attached to the Merseburg wheel's 6 inch diameter oak axle were 0.5 inch = 8/16 inch thick x 2 inches wide x 68.25 = 68-4/16 inches in length and he will have to determine the correct lengths of the other frame pieces for lattices (their thicknesses and widths will be identical to those of the radial frame pieces) using the skills he has already acquired in constructing the parallel pair of lattices that formed the two sides of the replica 3 feet diameter, Gera prototype wheel's drum if he followed my suggestion to construct that replica before attempting a replica of one of Bessler's larger self-moving wheels.

Also, as mentioned above, the same basic lattice design used in the Merseburg wheel's drum as depicted in Figure 11(a) was also used for the Kassel wheel's drum, but, due to the doubling of the mass of the latter's weighed levers, the twenty-four radial frame pieces of the three lattices that were attached to the Kassel wheel's 8 inch diameter oak axle were 1 inch thick x 2 inches wide x 67.25 = 67-4/16 inches in length (and, again, the thicknesses and widths of the other frame pieces within a Kassel wheel's lattice were the same as those of its radial frame pieces).

In AP Bessler mentions how the axial view of his uncovered Merseburg wheel's drum was as glorious as the spread tail feathers of a peacock and when one views the exposed front lattice of frame pieces shown in the center portion of Figure 11(a), they do have a bit of a resemblance to that particular bird's tail plumage. Bessler must have been filled with a tremendous sense of awe and pride as he completed each of the three lattices of the Merseburg wheel and finally got their planes into a vertical orientation and mounted onto that wheel's 6 inch diameter wooden axle into whose ends he had previously installed two 0.75 inch = 12/16 inch diameter x 12 long steel pivot pins. Once in place and carefully aligned so that their corresponding frame pieces were perfectly parallel to each other, he would then have securely affixed the ends of all twenty-four radial frame pieces to their 6 inch diameter axle. One almost immediately wonders how that was done.

Before the three lattices were attached to their axle, however, he had to carefully assemble each lattice's sixty-four separate wooden frame pieces of various lengths and end cut angles into a single rigid piece that could then be placed onto the wheel's axle.

The simplest way to do this, and the one most likely employed by Bessler, was to use the exact same method I suggested for the construction of the replica 3 feet diameter Gera prototype wheel's two parallel lattices in the previous chapter, but to use it with a far larger flat surface than a table top. The sixty-four pieces of each of the Merseburg wheel's lattices were simply laid out on an area of the floor of the room in his rented home in which the 12 feet diameter bidirectional wheel would be displayed after its axle with its inserted steel end pivot pins were mounted onto the lower brass bearing plates of the two vertical support planks that spanned the 14 feet distance from the same room's floor to its ceiling. This floor area, obviously, had to be larger than 12 feet x 12 feet or larger than 144 square feet in area. Bessler, in order to keep his landlord happy by not marking up the floor boards of his rented home's room, would have covered a 12 feet x 12 feet section of the floor with 144 precisely cut and fitted 1 foot x 1 foot square sheets of paper that were temporarily pasted to the floor and then have drawn a small spot in the exact center of the resulting larger square assembly of individual sheets of paper. This center spot would have been located exactly where the corners of four sheets met at a location six sheets in from *each* side of the large square assembly of paper sheets.

Around this center point he would have used a drafting compass to draw a circle exactly 6 inches in diameter which represented a cross section of the axle of the Merseburg wheel with the center point representing the center axis of the axle. Then using a large protractor to carefully measure angles, he would have attached and then stretched dark threads between pins that had been pressed into the wood of the floor boards to hold them in place. The dark threads would then have defined the locations of the straight center lines of the various wooden frame pieces of a lattice. Bessler would have used the resulting configuration of dark threads as a template to determine the exact lengths, widths, and arrangements of the sixty-four diagonal, octagonal, and radial frame pieces that were required for each of the large 12 feet diameter Merseburg wheel's three drum lattices. Once all of the threads were in place and in contact with the paper, he would have placed small drops of glue along their lengths to hold them to the paper and then removed all of the pins. In the exact center of this template he would have glued down a 0.5 inch = 8/16 inch thick circular slice cut from the end of the stock cylindrical piece of wood he used for the axle of the Merseburg wheel so that its center was directly above the dark spot placed at the center of the 12 feet x 12 feet array of square paper sheets.

Once all of the sixty-four wooden frame pieces were cut for each lattice, they would have been carefully laid down onto the dark threads glued to the paper covering the floor that surrounded the glued down section of axle and had glue applied to their previously lightly sanded contacting surfaces and then been carefully pushed together. To further enhance the rigidity of the complete lattice, the inventor may have also used small metal plates and screws on its various joints. None of the eight radial frame pieces of the lattice, however, would have been glued to the circular wooden disc in the center of the template, but were merely in physical contact with its curving surface.

After the glue had dried, the complete 12 feet diameter lattice piece could then be lifted up off of the floor into a vertical orientation by Bessler and, due to its weight of only about 65 pounds, carried over and placed leaning up against a nearby wall by him. But, his work was not done yet. In order to construct the bidirectional Merseburg wheel, he had to construct a total of three of these 12 feet diameter drum lattices!

All three of the completed lattices then had to be precisely mounted on the wheel's 6 inch diameter x 6 feet long wooden axle with two of the lattices forming the circular outside surfaces of the drum and the third lattice placed between these two and located precisely at the midpoint of the drum. Once properly located in their positions of the axle, the lattice pieces were finally glued to the axle and isosceles triangle shaped wood pieces were glued into the corners between the bottom surfaces of the radial frame pieces and the outer surfaces of the axle. Possibly, he also used small wood screws at the axle contacting ends of the radial frame pieces to attach them to the wedge shaped wooden pieces and then used extra screws to attach the wedge shaped pieces to the axle's surface for greater joint strength. He would have previously drilled small pilot holes into the wood pieces in order to minimize the chance of them spitting when the screws were fully tightened.

How these lattices could then have been attached to their common axle by one man working alone is a bit of a mystery, but the method that seems easiest to me would be for him to have used a system of three strong ropes to temporarily suspend the wheel's 6 feet long axle from the ceiling of the wheel room as the three lattices were attached.

While the first two ropes held the center of the axle at an elevation of greater than 7 feet off of the floor, he would have slipped the opening formed by the converging radial frame pieces at the center of a lattice onto one end of the axle and slid it close to the nearest of the two extension ropes which we can call the "second rope" while the other rope was the "first rope". Once that was done, a "third rope" was added to support that end of the axle to which the lattice was added and the second rope nearest the lattice was removed. The lattice was then slid a little farther toward the first rope at the other end of the axle until the lattice was at its correct location on the axle. The second rope was then reattached between that lattice and the third rope so that the third rope could then be removed. The next and second lattice was then placed onto the same end of the axle as the first had been and this procedure was repeated until it too was in its correct position on the axle next to the first lattice that had been previously mounted on the axle. Finally, one more repetition of this procedure added the last and third of the three lattices to the Merseburg wheel's 6 inch diameter x 6 feet long axle. This method only slides lattices to the center of the axle from *one* of its ends and this attachment method would absolutely have been necessary if, prior to mounting the lattices on the axle, Bessler had already installed the eight steel pins on one end of the axle that were used to power the portable stamping mill he used during demonstrations of the wheel.

Once the three lattices were properly positioned and aligned so that all of their various frame pieces were perfectly parallel to each other on their common axle, they would have been securely attached to the axle. At this point, the axle, its two steel end pivot pins, and the three lattices would have weighted about 250 pounds!

Next, a special transport cart Bessler had constructed for the wheel would have been carefully placed under the projecting ends of the axle and the ropes released so as to lower the axle's bottom surfaces onto the cradles on the top ends of the cart's vertical support planks. With the use of this cart, the inventor and his assistant could then have easily moved the axle and its three attached lattices to the wheel's permanent set of vertical wooden support planks in another part of the wheel's dedicated room for public viewing and, using two overhead block and tackle hoists, lifted and pushed the axle about until its end pivot pins finally rested in the hollows of their brass half moon bearing plates installed in the wheel's two vertical wooden support planks. These hollows would have been previously well lubricated with either tallow or lard.

I shall discuss the mechanics of Bessler's two 12 feet diameter, bidirectional wheels in greater detail in the next chapter, but for the moment let it just suffice for me to mention again that both his Merseburg and Kassel bidirectional wheels each contained *two* independent, one-directional wheels, each containing eight weighted levers, that were next to each other, side by side and back to back, and which actually *shared* the same middle lattice of the three mounted on their common axle. These two "internal" one-directional wheels were arranged so that the directions they each would spontaneously begin rotating in, if released, were opposite to each other. Obviously, their opposed torques cancelled each other out so that a bidirectional wheel, when released, would remain stationary.

After the Merseburg wheel's drum's three parallel lattices were firmly secured to their common axle, the sixteen bridge pieces which were *not* previously part of the completed lattices were installed. These sixteen bridge pieces, to which the sections of the rim wall would later be glued and nailed, were attached to the ends of the eight sets of three parallel radial frame pieces whose other ends were attached directly to the axle and to the ends of the eight sets of three parallel *short* radial frame pieces whose other ends were attached to the midpoints of the eight sets of three parallel outer octagonal frame pieces that were located in the drum's eight 45° octants.

The drum at this time was still completely "open" and had to have the sections of its outer rim wall attached. The sixteen curved wooden sections of the outer rim wall to be attached each measured 0.25 inch = 4/16 inch thick x 12 inches wide x 28.18 inches = 28-11.25/16 inches long and were, in the case of Merseburg wheel, probably glued and also nailed to the drum's sixteen bridge pieces one section at a time until the entire perimeter of the drum was covered. For the Kassel wheel which needed to later be disassembled and moved to another location in Weissenstein Castle, the sixteen outer rim wall sections were the same thickness and length as those used for the Merseburg wheel's drum, but were 18 inches wide instead of 12 inches wide and would have been screwed down to the drum's sixteen bridge pieces to make their later removal easier.

In order to make their attachment to the sixteen wooden bridge pieces previously glued and nailed to the ends of the two types of radial frame pieces of the drum's three lattices as easy as possible, each of the outer rim wall sections would have been previously soaked in water and then, once softened, forcefully warped and tied in that shape with rope so as to form a piece with a curvature suitable to precisely fit 1/16th of the outer circular circumference of either the Merseburg or Kassel wheel's 12 feet diameter drum. When the wood pieces completely dried by being left out and exposed to the warmth of the Sun, the ropes would been removed and the pieces would then retain much of their new curvature so that they could more easily be attached in order to form the complete outer rim wall of the drum.

Next, a bidirectional wheel's eight shared stop cord hook attachment anchor pins, its eight shared spring hook attachment anchor pins, and their thirty-two wooden hook keepers (these were not shared and sixteen were located in each of the drum's two internal one-directional wheels) were installed in the drum.

For the Merseburg wheel's drum each of its sixteen hook attachment anchor pins was 0.25 inch = 4/16 inch in diameter x 12 inches long. In the case of the Kassel wheel, these hook attachment anchor pins were all 0.375 inch = 6/16 inch in diameter x 18 inches in length. Note that the different lengths of these hook attachment anchor pins were the same as the thicknesses of their particular bidirectional wheel's drum because each type of anchor pin had to pass through *three* parallel octagonal frame pieces from the outward facing surface of the drum's front lattice to the outward facing surface of its rear lattice. Thus, we see that each hook attachment anchor pin was actually shared by a front and back weighted lever each of which was located in one of their enlarged drum's the two internal one-directional wheels. As a consequence of this, each spring hook attachment anchor pin had four spring hooks attached to it, but each stop cord hook attachment anchor pin only had two stop cord hooks attached to it.

With the Merseburg wheel drum's two types of hook attachment anchor pins in place and the long notched sides of their wooden hook keepers glued up against them, it was time to install the 12 feet diameter wheel's sixteen weighted levers and their pivot pins. The two ends and middle section of each pair of side by side weighted levers' shared pivot pin was press fitted into three parallel radial frame pieces so that its center axis was at a distance of exactly 56 inches from the center axis of the axle and this same distance was also used in the Kassel wheel. For the Merseburg wheel, the eight shared weighted lever pivot pins were, like the drum's two types of hook attachment anchor pins, 0.25 inch = 4/16 inch in diameter x 12 inches long. For the Kassel wheel, its eight weighted levers' pivot pins were, like its drum's two types of hook attachment anchor pins, 0.375 inch = 6/16 inch in diameter x 18 inches in length.

As the reader will notice, for both of these 12 feet diameter, bidirectional wheels' drums, the weighted lever pivot pins used were identical in their physical dimensions to those used for the drum's two types of hook attachment anchor pins. As Bessler intended, this would make drum constructions as easy as possible so that, as he wrote in AP, "…even a poor workman could put the thing together without a lot of head-scratching and get it completed almost before you could notice."

I used the term "weighted lever" above, but, most likely, at this stage of the drum's assembly, the Merseburg wheel's forty-eight cylindrical lead end weights and their steel attachment bolts with each weight and bolt combination having a mass of four pounds, had not yet been attached in trios to the ends of the drum's sixteen levers when the latter had been installed. So, from this point on in the assembly description, I will refer to the levers being installed in the drum as "unweighted levers" until the point in the drum's assembly is reached where their end weights were finally attached.

Installing the sixteen unweighted levers into a bidirectional wheel's drum was, however, a bit of a challenge despite them being less massive due to the absence of their attached cylindrical lead end weights. Not only did each of the unweighted levers' steel pivot pins have to pass through the *three* snug holes previously drilled into the wood of each set of *three* parallel radial frame pieces and the *four* aligned thin metal sleeves inserted into the wood of the four 3 arm side pieces of *two* separate unweighted levers, but Bessler had to also make sure that the correct number of washers were located on the single 18 inches long pivot pin so that the *same* number of washers filled the *four* gaps between the two unweighted levers and the three radial frame pieces nearest them. This task had to then be repeated for all *eight* pairs of side by side unweighted levers around the drum!

As each side by side pair of unweighted levers was installed, their two wooden radial stop pieces would also each be attached via gluing and perhaps small nails across its particular section of each unweighted lever's pair of parallel radial frame pieces. When the glue was completely dry, each unweighted lever's two main arms would be clamped up against their respective radial stop piece using a small carpenter's wood clamp. When this was done, the center lines of the four parallel main arms of each pair of side by side unweighted levers would be perfect alignment with the two levers' three parallel radial frame pieces. This would have to be done for all eight of the side by side pairs of unweighted levers found within the bidirectional wheel's drum. The little wood clamps used would have all been identical in mass and shape so that, when all sixteen were in place holding their respective unweighted lever's main arms in contact with their lever's radial stop piece and also in perfect alignment with their nearest parallel pair of radial frame pieces, each of the sixteen clamps' individual centers of gravity would have been the exact same distance from the center axis of the wheel's axle.

At this point it was time to check the balance of the drum.

The still incomplete drum, however, no matter how carefully Bessler had tried to construct it, install its sixteen unweighted levers, their sixteen radial stop pieces, and the sixteen identical wood clamps, when released would not remain stationary in *all* orientations because its center of gravity was not located exactly at the center axis of the axle. It would have slowly rotated until its initial slightly horizontally offset center of gravity was finally located directly under the center axis of the axle at its *punctum quietus* location. This inherent imbalance then made it necessary to add one or more steel (and perhaps even lead in case of severe imbalance) washers to the outer octagonal frame pieces to move the center of gravity upward until it was located exactly at the center of the axle. When this balance was achieved, the drum could then be manually rotated into any orientation and it would stay there when released.

But, this was only an *initial* balancing. There would have to be two additional balance checks made after the extra mechanisms were added in order to give the wheel its bidirectionality. I will cover these extra mechanisms in detail in the next chapter.

With the initial balancing completed, it was time to install the bidirectional wheel's eighty coordinating cords. As with his 3 feet diameter, one-directional Gera prototype wheel, Bessler probably would have used dyed cords so as to keep better track, visually, of which of the four different types of coordinating cords he was installing and to make sure they were placed into their correct layers in *each* of the drum's two side by side, internal one-directional wheels.

If we call one of the Merseburg wheel's internal one-directional wheels the "front wheel" and the one behind it the "back wheel" then, most likely, the following procedure would have been used to install their eighty coordinating cords.

Because the drum was still uncovered, he would have installed forty coordinating cords in each of the drum's two one-directional wheels separately.

He probably first worked on the back internal one-directional wheel (the one on the side of the drum facing away from the members of the public who came to view the Merseburg wheel at his home) and did so from the *back* side of the drum. As was done with the one-directional Gera prototype wheel, he first installed the farthest of its five layers from his location or layer 5 which contained its eight blue colored spring cords and their attached extension springs. Then he moved closer to himself to layer 4 and installed its four orange colored long lifter cords. After that he moved even closer to himself to what was layer 3, the middle layer of the five layers, and installed the eight green colored main cords and the eight red colored short lifter cords. Next, he moved forward another layer to layer 2 and installed four yellow dyed long lifter cords paying special attention to make sure their arrangement did *not* superimpose with the orange long lifter cords in layer 2. Finally, he would have come forward to layer 1 which was the closest to him and installed another eight blue spring cords with their attached springs.

With all forty of the coordinating cords installed in the bidirectional drum's back internal one-directional wheel, Bessler then went around to the front side of the 12 feet diameter drum (the one on the side of the drum facing toward the members of the public who came to view the Merseburg wheel) and, treating its internal one-directional wheel like the back internal one-directional wheel he had just completed, proceeded to install, layer by layer, the same four sets of colored coordinating cords and springs into it. However, with the front internal one-directional wheel he would have *reversed* the order of the layers. That is, the *farthest* of the 5 layers from his location would be treated like layer 1 and the closest to him would be treated like layer 5.

Why would he do this? The answer is that, after the drum's dyed linen cloth side coverings were attached, he would have been inspecting and replacing broken coordinating cords and extension springs from inspection holes cut into the cloth covering of the drum's *backside* which meant that, in order to reach the broken cords and springs of the drum's front internal one-directional wheel, he would have to reach through the entire thickness of the back internal one-directional wheel. By using the arrangement of colored coordinating cords I suggest above, his hands would, as they moved through the inspection holes cut into the drum's back cloth covering to the intact cloth covering on the front side of the drum, have passed through the *ten* coordinating cord layers of two separate internal one-directional wheels in the order of layers 1, 2, 3, 4, and 5 and then 1, 2, 3, 4, and 5 again. This ordering of the coordinating layers makes the arrangement of the two sets of five layers in the bidirectional drum's two side by side internal one-directional wheels identical which aided Bessler, in addition to the use of colored cords, in identifying the particular layer that contained a broken cord or spring.

While the sixteen unweighted levers were still clamped to their wooden radial stop pieces, an additional sixteen structures I refer to as "cat's claw gravity latches" were installed (these will be discussed in far greater detail in the next chapter). After their installation and the temporary immobilization of their moving parts in identical orientations relative to the unweighted levers whose rotations about their pivot pins they would prevent, the drum was again checked for any imbalance introduced into it by the additional structures. If there was any, it was again compensated for by attaching one or more steel (or lead) washers to various places near the drum's 12:00 position after the imbalanced drum came to rest in its equilibrium orientation which would place its newly introduced slightly offset center of gravity under the axle's center axis at its *punctum quietus* point. After balancing with the added washers, the center of gravity of the drum and its contents was again returned to the exact center axis of the axle and the drum would remain stationary regardless of what orientation it was placed into.

At this point, it was time to check the wheel to see if it was displaying the pm effect in *both* of the drum's two possible directions of rotation. But first, the drum's forty-eight identical 4 pound cylindrical lead end weights and their forty-eight steel attachment bolts had to be attached to the ends of the parallel pairs of main arms of their wooden unweighted levers inside of the drum (each unweighted lever had a mass of 2.5 pounds in the Merseberg wheel and 5 pounds in the Kassel wheel). This required undoing the clamp that held each unweighted lever's pair of main arms in contact with its wooden radial stop piece, pulling the main arms down and out of alignment with their lever's parallel pair of wooden radial frame pieces, inserting the three cylindrical lead end weights, one at a time, between the ends of the main arms so that each weight's center end holes aligned with the holes at the ends of the main arms for it, inserting the weight's steel attachment bolt first through the hole in the *back* main arm, then through the snuggly fitting shaft for the bolt that had been drilled through the center axis of the cylindrical lead end weight, and then finally screwing the threaded end of the attachment bolt into the threaded hole of a nut for it that was embedded in outward facing surface of the *front* main arm. Once an end weight and its attachment bolt was in place, Bessler then used a small screwdriver to tighten the bolt's slotted head so that the end weight was securely held between the ends of its parallel pair of main arms and would not rotate relative to them. He would have repeated this procedure until all three of the 4 pound cylindrical lead end weights were attached to each lever.

Because all servicing of the 12 feet diameter, bidirectional Merseburg wheel (and also the Kassel wheel) had to be done only from the backside of the drum through the inspection and servicing holes cut into the sheets of stretched, dyed linen cloth that covered the back circular face of the drum in order to minimize the chance of prying eyes seeing the shapes of the levers the inventor used in his wheels and how the coordinating cords were attached to them, Bessler had to attach each trio of cylindrical lead end weights to a lever's parallel pair of main arms from the backside of the drum. This is why the steel attachment bolts, as mentioned above, had to be inserted *first* into holes drilled into the outward facing surfaces of the *back* main arms of each lever. Later, when it became necessary to remove the forty-eight end weights in order to lighten the drum during a test required translocation of the wheel, he would then have been able to reach through the backside of the drum and unscrew and remove all forty-eight end weight attachment bolts from the backside of the drum. This required that the *all* forty-eight of the attachment bolts' slotted heads face toward the backside of the drum. The craftsman attempting to duplicate one of

Bessler's 12 feet diameter, bidirectional wheels should keep this in mind and make sure that all of the slotted heads of its forty-eight cylindrical lead end weight attachment bolts, when inserted to secure their weight to the ends of a lever's parallel pair of main arms, face whatever side of the drum that will be its backside and will later be covered with the sheets of linen that will have the eight inspection and servicing holes cut into them.

Starting with an unweighted lever of the drum's *front* internal one-directional wheel whose pivot pin's center axis was located at the drum's 6:00 position and using a ladder, Bessler slowly worked his way clockwise around the drum (which was prevented from rotating during this procedure by attaching a tether rope to it from the floor beneath the drum) until he had unclamped all eight of its unweighted levers, attached their three cylindrical lead end weights with their steel attachment bolts between the ends of their parallel pairs of main arms, and then manually supported each weighted lever's parallel pair of main arms as it *slowly* settled into the particular orientation relative to the lever's parallel pair of radial frame pieces as allowed by the various coordinating cords and stretched springs attached to the lever's A arms. As that was done, the moving parts of the cat's claw gravity latches for all eight weighted levers in the drum's front internal one-directional wheel would also be free to engage in their normal motions during drum rotation. The eight unweighted levers without their end weights attached to them and the eight cat's claw gravity latches of the *back* internal one-directional wheel, however, were all still locked into their various orientations that kept their center of gravity exactly at the center axis of the drum of the 6 inch diameter axle of the Merseburg wheel.

At this point, assuming that the front internal one-directional wheel's spontaneous direction of rotation was clockwise as viewed from the front side of the drum (which is the "standard" spontaneous start up direction I use for one-directional wheels in this volume), the drum, when released, would immediately start rotating in that direction and accelerating slowly up to its maximum free running rotation rate which for the Merseburg wheel was about 40 rotations per minute, but this would only happen if all eight weighted lever's of the drum's front internal one-directional wheel had the correct masses for their cylindrical lead end weights and wooden unweighted levers, their extension springs had their correct spring constant values, and the weighted levers were connected to each other by coordinating cords that allowed only the correct separation distances between the center axes of the hook attachment pins to which they were attached. If there were small deviations from the ideal values of the parameters these parts were supposed to have by only some of the parts in the front internal one-directional wheel, then the acceleration rate of the drum and its maximum rotation rate would be somewhat reduced. If, however, there was too much deviation by too many parts or a broken cord or extension spring in the front internal one-directional wheel, then the drum might have little or no acceleration after being released. Because of the precision of Bessler's craftsmanship, this probably never happened.

The above describes what needed to be done to check only one of the two internal one-directional wheels inside of the enlarged drum of the bidirectional Merseburg wheel. After determining that its front internal one-directional wheel was fully functional, Bessler had to again immobilize all eight of the then *weighted* levers of the front wheel using eight small clamps, immobilize the motions of their eight pairs of cat's claw gravity latches (in the next chapter the reader will learn that each weighted lever in a bidirectional wheels requires two small gravity latches be assigned to it), and finally repeat the same steps described above for the drum's *back* internal one-directional wheel. Then, if that back internal one-directional wheel proved to be running correctly, he released all of the moving parts of the front internal one-directional wheel's mechanisms again and, finally, had a *fully* functional bidirectional wheel. Obviously, all of these balancing and performance tests would have taken Bessler and his brother Gottfried, working together, several days to a week or more to complete depending upon how much time per day they could both devote to the assembly of the Merseburg wheel. The balancing and testing procedures for the more massive Kassel wheel the inventor later constructed would have been identical, but more arduous due to the doubling of the masses of the weighted levers that 12 feet diameter bidirectional wheel used.

Now it's time to finish the discussion of Figure 11(a) by mentioning the two cloth coverings shown on each side of the drum's front lattice of frame pieces which is located in the center of the illustration. In order to reveal the intricate workmanship used to construct the front lattice of the drum's three identical lattices in the figure, the front and back coverings were removed from the drum and placed next to it. The covering on the left side appears as it would from the front side of the drum and the covering on the right side appears as it would from behind the drum. These large pieces of cloth were attached to the drum's two circular sides as a last step after Bessler had completely finished the balancing, adjusting, and testing the 12 feet diameter, bidirectional Merseburg wheel. The linen cloth used for these coverings was dyed a dark green color by mixing blue and green dyes together and Bessler probably just purchased the cloth already dyed so that he did not have to perform the messy task of dying it himself.

On the left side of the figure is shown the two 6 feet wide sheets of colored linen that covered the front side of the Merseburg wheel's drum and hid direct viewing of the mechanics of its front internal one-directional wheel by members of the general public who were kept several feet away from the wheel's front vertical support beam by a guard rail that surrounded three sides of the wheel. On the right side of the figure we see the two 6 feet wide sheets of linen that covered the backside of the wheel's drum and hid direct viewing of the mechanics of its back internal one-directional wheel by anyone Bessler allowed inside of the guard rail through its locked front gate. That privilege was reserved only for people both known to Bessler and trusted by him or for those who could prove that they represented serious potential buyers of his invention. Being able to view the wheel's axle from behind the plane of its drum allowed them to better observe the way the eight steel pins projecting out from the rear section of the axle there could operate the stamping mill's heavy wooden stamps when it was slid up against the side of the axle. However, during such examinations, Bessler would have always been present to make sure that they did not start inspecting any of the cloth patches covering the inspection holes cut into the drum's back cloth covering lest they see too much!

To begin attaching these four large sheets of linen to the drum's two circular sides, Bessler probably started by attaching one 3 feet wide x 12 feet long sheet to the front side of the drum after it had been immobilized. He might have tacked one corner of the sheet of cloth to the lower half of the end of the 2 inch wide 9:00 radial frame piece and then, after stretching the cloth tightly, tacked the other corner of it, which was 12 feet away from the first corner he tacked down, to the lower half of the end of the 2 inch wide radial frame piece at the drum's 3:00 position.

After making a small semi-circular cut in the edge of the cloth near the axle, he hammered two tacks into the cloth on each side of the axle and then added a few more tacks along the top edge of the cloth to secure it to the lower halves of the drum's 2 inch wide 9:00 and 3:00 radial frame pieces. He would have added all of tacks symmetrically with respect to an imaginary vertical line passing through the axle's center axis so as not to introduce any imbalance into the drum. With the top edge of the 12 feet wide sheet of linen securely fastened to the drum, he would have then pulled the sheet down at its bottom midpoint edge and tacked this edge to the center line of the drum's 2 inch wide 6:00 radial frame piece. Next, the same was done for the sheet at the ends of the drum's 7:30 and 4:30 radial frame pieces. He would then have used a sharp knife to trim off most of the excess cloth around the lower circumference of the drum, but left a few inches of excess material hanging past the outer circular edge of the wooden outer rim wall.

With the lower half of the front side lattice of the drum thus covered, the inventor freed the wheel, rotated its drum 180°, again immobilized it, and then proceeded to attach a second 6 feet wide x 12 feet long sheet of dyed linen to what had previously been the upper half of the drum's front lattice in a manner identical to what he had done for the first sheet.

(I should again mention here that the various clock time drum positions I constantly refer to must be imagined as *fixed* in space regardless of whichever side of the drum one is viewing. They do not change their locations as a drum rotates about its axle's center axis or when the direction of rotation of a drum is reversed. In fact, they do not change even when one walks around from the drum's front to back side. When viewing the back side of a drum, the clock time locations will be same as though one was viewing the front side of the drum and no horizontal mirror image reversal of the times takes place because one is viewing the back side of the drum. In the case of a one-directional wheel, its spontaneous start up direction of rotation will make it appear to be rotating clockwise when viewed from one of its drum's two circular sides, but counterclockwise when viewed from the drum's other side.)

With the front side of the drum of the Merseburg wheel fully covered by two long sheets of darkly dyed linen, Bessler then moved on to the backside of the drum and also covered it with two identical sheets of linen using the method previous described.

At this point, however, he had something that looked a bit sloppy. Both of the circular edges of the drum's rim wall had a few inches of fabric hanging out all the way around them except for two places which, when the drum was immobilized, were near the drum's 6:00 and 12:00 positions. To tidy this up, he would, as suggested in Figure 1(a) at the end of Chapter 1, have pulled the few inches of excess linen tightly over the circular edge of the rim wall all the way around the outer circumference of the drum on each of its front and backsides and tacked this material down to the *outer* curving surface of the rim wall and close to its curving edge.

It is difficult to determine from Figure 1(a) what Bessler used to dress up the outermost circumference of each circular cloth side of the drum near where the cloth was folded over and tacked to the rim wall's *outside* curving surface. Possibly, he might have just taken strips of thick leather a few inches wide by several feet long and then, after folding each strip lengthwise over the edge of the drum's outer circumference and onto the rim wall's outside curving surface, attached the leather strips with small tacks.

The side of the leather strip that was folded over onto the circular face of the drum then sandwiched the layer of linen cloth between itself and the 0.25 inch = 4/16 inch thick layer of wood used for the outer rim wall. As he did this, he proceeded to hammer very small tacks with large heads along the length of the leather strip so as to attach it to the 0.25 inch = 4/16 inch thick curving *side* of the thin outer rim wall. When he finished tacking down that strip of leather, he then added another strip by placing one of its ends in contact with the end of the first strip and then continued to tack down the length of that second strip of leather to the curving side of the outer rim wall. Bessler then continued this process until he had gone all the way around the outer circumference of one circular face of the drum and, by carefully studying Figure 1(a), it looks like he used a total of about one hundred and twenty tacks to do this! These tacks were not temporary, but meant to be left in place permanently.

Once the strips of leather had been tacked on so as to cover the entire outer rim wall's rim circumference of the front side of the Merseburg wheel's 12 feet diameter drum, Bessler would have then cut any excess linen off beneath the flaps of leather that were folded over onto the outer curving surface of the drum's rim wall and removed the tacks holding them down. Then, using far less than 120 tacks, he would have added the minimum number of permanent tacks necessary to hold those flaps of leather down to the outside curving surface of the rim wall. With the complete 360° circular rim edge of the front side of the drum completed, he would have gone through the same procedure on the backside of the drum.

The finished product would have been a drum that if one, such as a prospective buyer's agent, grasped its rim wall's opposite circular edges in an effort to start or stop the motion of the 12 feet diameter bidirectional wheel's drum, there was the least chance of him cutting his hands on the edges of the outer rim wall or picking up a painful splinter from one of them. Bessler wanted anyone representing a potential buyer to be able to go back him and not have to report that his hand had been severely slashed by the sharp edges of the rapidly rotating drum. If that had happened, then the buyer might decide that the invention, although interesting, would be too dangerous to have around his workmen.

However, in Figure 2(a) Bessler seems to have used a different approach to dressing up the circular outer rim wall edges of the Kassel wheel's drum. That wheel still had the sheets of dyed linen stretched across the front and back circular faces of its drum, but now Bessler actually nailed curved sections of wood molding along the curving rim wall edges of the drum. Underneath those attached and closely fitted sections of wood he would still have used many small tacks to permanently attach the stretched sheets of dyed linen to the 0.25 inch = 4/16 inch thick curving edges of the drum's attached wooden rim wall sections, but all of those tacks would have been placed so that they were carefully concealed from view by the molding pieces. To impress the count with the Kassel wheel, Bessler wanted it to look as much like a work of fine furniture making as possible and he would have sanded, stained, and varnished the outside curving surface of the rim wall as well as its two circular rims of attached molding pieces. Because of the preference for darkly stained wood in Europe at the time, one can safely assume the visible woodwork of the Kassel wheel was stained a dark brown color.

A drum with completely cloth covered sides posed several problems for Bessler, however.

In order to frustrate the reverse engineers among the crowds of admiring visitors who came to see Bessler's demonstrations of the Merseburg wheel, he had decided, as previously mentioned, to use coverings of stretched linen cloth that were dyed a dark greenish color to conceal the drum's open lattice sides. This was because, if he had used the uncolored natural linen which is almost pure white, then, from handling over time, it would have began to accumulate unsightly dirty spots and stains on it and, worst of all, anyone viewing one side of the cloth covered drum while the other side was illuminated by bright sunlight, might actually see the shadows of its three lattices' frame pieces as well as the locations, shapes, and orientations of its sixteen weighted levers, their forty coordinating cords, and their thirty-two extension springs. Using white linen would, therefore, under certain lighting conditions, actually be like publishing a schematic of his wheel's secret internal mechanics for everyone to see! That, obviously, was unacceptable and so very dark colored linen had to be used.

But, aside from keeping the prying eyes of those that wanted to steal the secret of his genuine working imbalanced pm wheels without paying for it, the cloth coverings also prevented Bessler from seeing what was going on inside of the drum and being able to reach and repair any parts that had failed during an extended period of operation. The simple solution to that problem is illustrated by the removed backside cloth covering shown on the right side of Figure 11(a). (Note that Figure 11(a) shows the drum *before* its weighted levers, coordinating cords, steel extensions springs, and other mechanisms had been installed. One may refer to this as an "empty drum".)

Bessler most likely cut eight isosceles trapezoidal inspection holes into the linen cloth covering the backside of the drum. Each hole was made with *three* straight cuts that were along the center lines of the hole's two adjacent radial frame pieces and the inner octagonal frame piece that joined and helped brace them so as to form a flap of material that could be opened up by folding it back away from the axle in order to expose the most important parts of the wheel's internal mechanics which were the three parallel pairs of arms of two adjacent weighted levers and the coordinating cords attached to them.

The three edges of each flap that was formed could be temporarily pinned down with tacks that penetrated the wood of the frame pieces yet had heads large enough to keep the linen from slipping off of them.

Perhaps Bessler found and purchased tacks made specifically for attaching cloth to wood or he actually made them by soldering small washers' center holes to the shafts of nails whose tips he had filed to a sharp point so that they acted like a larger version of the modern push pins used to affix pieces of paper to the surfaces of bulletin boards. Once he pinned a flap into place to cover an inspection hole, he would always have put the tacks back into the same holes again. This would have made it easier to reinsert the tacks and prevented the wooden frame pieces from having multiple holes formed into them which might, conceivably, eventually have weakened the wood and caused it to split from the varying internal stresses created in it during drum rotation.

It wasn't long before Bessler constructed the Merseburg wheel and began running it with its newly attached dark linen cloth coverings that he noticed another disturbing effect.

As the wheel ran during its many demonstrations, Bessler, as part of routine maintenance so as to minimize the wear of its metal to metal contacting surfaces and thereby allow the wheel to run longer before a part such as a cord or spring broke, would periodically open all of its drum's backside cloth flaps and then apply a drop of a vegetable oil like olive oil to all of the places were such wear could occur. After a while, however, he noticed that some of the excess oil, due to centrifugal force and capillary action, was beginning to show the outlines of the radial and inner octagonal frame pieces near the pivot pins of its various weighted levers.

Bessler quickly realized that if nothing was done about this problem, then an outside viewer of the 12 feet diameter, bidirectional Merseburg wheel restricted to viewing its drum's front side would be able to plainly see and estimate that its levers' pivot pins' center axes were located about 56 inches from the center axis of the wheel's axle and that the distance from a weighted lever's pivot pin's center axis to the outer rim of the drum was 16 inches. If one then assumed that there was a separation of about 2 inches between the outer surface of the drum's rim wall and the center of gravity of a weight attached to the end of each of the drum's levers, he could quickly calculate that the levers that the Merseburg wheel used were about 14 inches long between a lever's pivot pin's center axis and the center of gravity of its attached end weight.

Counting the number of oil spots on the front cloth covering would also tell an outside observer of the bidirectional wheel that there were weighted levers in the front half of the drum whose pivot pins' center axes were arranged to form the vertices of large octagon and, if the viewer managed to see the identical pattern of oil stains on the colored linen covering on the backside of the drum, he would also have counted another eight weighted levers there whose pivots pins' center axes also formed the vertices of a large octagon. It would not have taken too much thinking on the part of an observant reverse engineer in a crowd viewing the bidirectional Merseburg wheel to realize that its drum actually contained a total of sixteen weighted levers that were divided into two groups of eight weighted levers with each group of eight located near one face of the drum and responsible for driving the wheel in one of its two possible directions of rotation.

Having this minimal amount of information about the wheel's internal mechanics would then have motivated a reverse engineer viewing the Merseburg wheel to begin wondering about exactly how Bessler's design managed to coordinate the motions of the weighted levers in each of its internal one-directional wheels so as to keep their centers of gravity on the descending side of the wheel during drum rotation in either direction.

If the viewer looked carefully at the drum's front side before the inventor gave the 12 feet diameter drum a gentle push to start it rotating, he might have noticed that there were other fainter oil stains located around the drum and positioned so as to also form octagonal patterns whose vertices were all located *midway* between the stains resulting from the lubrication of adjacent weighted levers' pivot pins. He might have realized that these secondary stains were due to the excess lubrication of hooks attached to additional pins whose ends were then press fitted into parallel pairs of octagonal frame pieces used for the bracing of the drum's radial frame pieces and that the lengths of those extra pins passed through the entire 12 inch thickness of the drum. Next, he would have concluded that some of those additional pins, like the ones farthest from the center axis of the axle, served to anchor one end of a spring whose other end, via a short piece of cord, was attached directly to each of the drum's weighted levers.

As the reader can well surmise, this was information that Bessler, obsessed with secrecy and security, did not want being made available to any of the visitors to his home who came to see demonstrations of his first bidirectional wheel. To make sure they did not have these clues, Bessler did the only thing he could. He hid the oil stained spots in the dark colored linen coverings on the drum caused by excess oil seeping onto them at those places by simply applying a quart or more of oil to the entire cloth covering on each side of the drum. That perfectly obscured the oil spots he did not want the public to see and the only negative feature was that it made the cloth a bit oily to the touch. But, aside from Bessler when he was opening a flap to expose a servicing hole, no one else would be touching the drum's cloth coverings. Agents of prospective buyer's that personally handled the wheel's drum in the inventor's presence would push start it by pushing on the outer surface of its finished wooden rim wall or by grasping the layer of leather that was folded over the outer rim of its drum and would not be touching the cloth covering material itself. And, even if the oil got onto anyone's hands and caused concern, Bessler would simply tell them that is was just olive oil and completely harmless. This method of lightly oiling the linen cloth coverings on a 12 feet diameter, bidirectional wheel's two drum faces worked so well that Bessler also used it on the Kassel wheel. This method of saturating the linen cloth covering also had the advantage of greatly reducing the "wick effect" of dry cloth extracting oil from the drum's various internal bearing surfaces via capillary action. That meant that the metal to metal contact between the cord and spring end hooks and the various lever and drum anchor pins would retain their lubricant coatings longer and help reduce wear to those contacting metal surfaces.

Using oil soaked, darkly dyed cloth side covers on the Merseburg wheel's drum finally put Bessler's mind at rest about the reverse engineers visiting his home and getting too much information about the Merseburg wheel's internal mechanics, yet he still had to contend with the occasional skeptic who claimed that, despite the bidirectional wheel's ability to turn in both directions, it must still have some sort of weights hanging from ropes wrapped around the section of axle that was hidden from view inside of the drum. To silence these skeptics, Bessler cut a small slit into the front cloth covering of the drum as depicted on the left side of Figure 11(a) next to the black silhouette of a 6 feet tall man.

The skeptic, with a small group of other members of the public looking on, was then invited up to the front side of the bidirectional wheel's drum through a gate in the guard rail that kept the visitors several feet away from the axle's front end vertical support plank. At that point, the skeptic could insert his raised hand into the slit in the cloth covering and begin groping around near its axle.

No doubt, Bessler, seeking some extra money for himself and the various church poor boxes to which he contributed, would have made a bet with the doubter. The skeptic would have to wager a thaler (which was about a day's pay for the average person back then) that, when he reached through the slit in the linen to touch the hidden section of the axle inside of the drum, he would feel ropes or chains wrapped around the axle. If he did, then Bessler would pay him ten thalers and admit to the group that he, Bessler, was a sinful hoaxer and also refund, perhaps, double the price of their admission to the demonstration. If, however, the skeptic felt no ropes or chains, then he would lose his thaler and have to turn to the group and publicly confess to them that he felt nothing on the wheel's hidden axle section and that Bessler was an honest man.

Upon reaching over his head and inserting his outstretched arm into the drum, the doubter would have only felt the various radial frame pieces of the three lattices attached to the axle and the small wedges that helped secure those pieces to the axle. There were no chains or ropes to be felt either wrapped around the hidden axle section or hanging anywhere near them. How embarrassing this must have been for those jeering skeptics who were absolutely positive that the 12 feet diameter, bidirectional wheel had to be a fake powered by the same kind of falling weights that similarly powered the pendulum clocks of that time. Bessler must have looked forward to these opportunities to teach people the risk and expense of jumping to conclusions without having all of the facts first which, of course, the inventor would not be divulging until far more thalers were paid to him.

While there is some evidence that Bessler may have allowed visitors to peer into the drum of the one-directional Draschwitz wheel through a palm sized gap in its drum's wood slat cladding in order to verify that there were no trained animals running about inside of it or weights suspended by ropes wrapped around its hidden section of the axle, with the Merseburg wheel this practice ended.

Direct visual inspection of the 12 feet diameter, bidirectional wheel's internal mechanics could only be made through inspection holes in the linen covering on the backside of the drum to which the general public did not have access. Only special visitors invited by Bessler to inspect the wheel were allowed there and only if the inventor was present and they gave their sacred words honor as gentlemen not to try to open a flap over one of the holes to see what the internal mechanics looked like. That was a privilege reserved only for the person or group that finally paid Bessler the huge sum he demanded and did so in a single, upfront payment. The only known exception to this policy was with Count Karl who paid Bessler 4,000 thalers for the privilege of examining the internal mechanics of the 3 feet diameter Gera prototype wheel prior to becoming Bessler's patron and funding the construction of his most powerful Kassel wheel in 1717.

For numerological and theological purposes, Bessler wanted the Merseburg wheel's drum to be exactly 12 inches in thickness x 12 feet in diameter. The number 12 has special significance in theology and appears in the Judeo Christian Bible at several places in reference to a prophesied coming of the Kingdom of God to Earth. The inventor firmly believed that the sale of the Merseburg wheel would provide all of the money he needed to found a new global religion that would revolutionize human morality on Earth which would then actually help prepare the way for the Second Coming of Jesus and the eventual establishment of God's Kingdom on Earth! Bessler, therefore, had to make sure that this wheel had that particular number of 12 incorporated into the dimensions of its drum's thickness and diameter and so he made the thickness 12 inches and the diameter 12 feet in this, his first bidirectional wheel.

Since each foot contains 12 inches, this means that the diameter of the Merseburg wheel's drum was 144 inches. This is yet another important number that appears in the Bible.

In Revelation 21, verses 10 to 27 it states that there was a holy city being prepared for occupation by those faithful to God which would physically descend from Heaven when God finally took up permanent residence on planet Earth after Jesus' Second Coming and temporary reign as king for a period of one thousand years. That city is depicted as a huge cube whose internal structures were made of pure gold and which would be12,000 furlongs or 1,500 miles long on each of its 3 edges. With that size, it would cover most of the Indian subcontinent and extend out beyond our atmosphere! Obviously, there was an error made in this figure when the Book of Revelation was originally being written in Hebrew or when it was later translated into Latin and then into English. The correct figure probably should have been only 120 furlongs or 15 miles on an edge for this odd cube-shaped city. The city walls of this giant cube were to be made out of the mineral known as jasper and exactly 144 cubits or 216 feet thick assuming a cubit with a length equal to exactly 18 inches.

How many would be occupying that holy city with God? They were referred to as "The Elect" and, out of Earth's entire population, only 144,000 people would be deemed by God and Jesus to be worthy to be part of this select group as is mentioned earlier in Revelation 14, verses 1 to 5. This white robed group would be composed of 12,000 very righteous people or "saints" from each of the twelve tribes of Israel.

Being Bessler's largest wheel after the Draschwitz wheel with weighted levers having a total mass of 14.5 pounds each, the inventor had to become more concerned about the centrifugal forces that these weighted levers and, more importantly, their pivot pins and the radial frame pieces they were embedded in would have to withstand.

I won't burden the reader with the not too difficult math here, but for a 14.5 pound weighted lever whose center of gravity (which located 2 inches *closer* to the center axis of the lever's pivot pin than was the center of gravity of its end weights which then puts the center of gravity of the weighted lever only 12 inches from the center axis of its pivot pin) was revolving at a speed of 40 rotations per minute (which was the Merseburg wheel's maximum terminal rotation rate when running freely) and at a distance of 56 inches plus 12 inches or 68 inches from the center axis of the drum's axle, the centrifugal force was about 44.8 pounds acting on the center of gravity of the weighted lever when the lever's parallel pair of main arms was in contact with its radial stop piece. This force acted constantly *outward* away from the center axis of the wheel's axle on each weighted lever as it traveled around the drum at 40 rotations per minute.

At the drum's 6:00 position, the downward acting force applied to a weighted lever's pivot pin when the drum was *stationary* was just the lever's normal downward acting weight due to gravity or 14.5 pounds. However, at the drum's maximum terminal rotation rate of 40 rotations per minute, the downward acting force applied to a weighted lever's pivot pin and the radial frame pieces holding it at the drum's 6:00 position was equal to the lever's normal downward acting weight of 14.5 pounds *plus* the downward acting centrifugal force of about 44.8 pounds that the lever experienced because of its motion around the axle's center axis which was then equaled to total of about 59.3 pounds. At the drum's 12:00 position, the downward acting force that would be applied to a weighted lever's pivot pin and its radial frame pieces when the drum was *stationary* was again just the lever's normal downward acting weight or 14.5 pounds. At the drum's maximum terminal rotation rate of 40 rotations per minute, however, the *upward* acting force a weighted lever's pivot pin and its radial frame pieces experienced at the drum's 12:00 position was equal to the *upward* acting centrifugal force of about 44.8 pounds on the lever *minus* the lever's normal downward acting weight of 14.5 pounds which was then equal to about 30.3 pounds.

So, when the Merseburg wheel's drum was turning at its maximum terminal, free running rotation rate of 40 rotations per minute, the steel pivot pins embedded in the drum's wooden radial frame pieces would experience *outward* directed radial forces acting on them from a single weighted lever that varied between a minimum of about 30.3 pounds at the drum's 12:00 position and a maximum of about 59.3 pounds at the drum's 6:00 position which is almost a doubling of the outward applied force between its minimum and maximum values.

But, we must not forget that the Merseburg wheel was a bidirectional wheel containing *two* side by side, internal one-directional wheels which means that when center axis of one of its 0.25 inch = 4/16 inch diameter x 12 inches long steel pivot pins passed through the drum's 6:00 position as the drum was rotating at its maximum terminal rate of 40 rotations per minute, that single pivot pin actually experienced a downward force applied to it by *two* weighted levers or a force of about 2 x 59.3 pounds = 118.2 pounds. This outward force, however, was also simultaneously applied to the wood of *three* radial frame pieces with a total cross sectional area of 3 x 0.5 inch = 8/16 inch x 2 inches = 3 square inches from which we must subtract the amount of cross sectional area lost from the wood of the frame pieces due to the presence of the pivot pin. Since each pivot pin was 0.25 inch = 4/16 inch in diameter and completely passed through each radial frame piece from one of its sides to the other which was through a distance of 0.5 inch = 8/16 inch, we must, therefore, *subtract* an area of 0.25 inch = 4/16 inch x 0.5 inch = 8/16 inch or 0.125 square inch from the cross sectional area of *each* radial frame piece which is then 3 x 0.125 square inch or 0.375 square inch for *all* three of the lattices' radial frame pieces that a pivot pin's two ends and middle section were embedded in.

Thus, at the drum's 6:00 position during the maximum terminal drum rotation rate of 40 rotations per minute and because of the weight and centrifugal forces applied by the *two* 14.5 pound weighted levers attached to their single shared pivot pin, each pivot pin would experience a total force of 118.2 pounds acting on it which would then be applied to the drum's three oak radial frame pieces that held the pivot pin. These three radial frame pieces had an *effective* cross sectional area of 3 square inches *minus* 0.375 square inch which is equal to an effective cross sectional area of 2.625 square inches. This force would always try to tear the 3 radial frame pieces apart at the locations where the single 12 inch long pivot pin penetrated them which was at a distance of 56 inches from the center axis of the axle for each of the radial frame pieces.

A total effective cross sectional area of wood of only 2.625 square inches may not seem to have sufficient tensile strength to withstand such a force, but if one obtains a short length of oak with a cross sectional area of only 1 square inch and then uses a piece of rope to attach a weight with a mass of, for example, 75 pounds to the end of the piece of wood, he will find that he can lift the weight by holding onto the piece of wood without the wood breaking. In fact, he could probably increase the mass of the weight lifted up to 100 pounds or more and still be able to lift it with the piece of wood without the wood breaking. So, we see from this that the stock wood which had a rectangular cross sectional area of only 0.5 inch = 8/16 inch x 2 inch = 1 square inch that Bessler used for each of the three parallel radial frame pieces of the 3 lattices of the Merseburg wheel's drum that contained each steel pivot pin was, even after reducing that area by 0.375 square inch for the cross sectional area that was removed from *each* radial piece by the steel pivot pin that penetrated it, *more* than sufficient to withstand the maximum force of about 118.2 pounds applied to the 3 radial frame pieces at the drum's 6:00 position when the drum was turning at its maximum possible rate.

Of course, it was also very important that all of the radial frame pieces of any the drum's three lattices be very securely attached to the surface of the wheel's wooden axle *and* to each other by the other octagonal and diagonal frame pieces of that lattice which they were.

Each front and back lattice radial frame piece inside the drum of the Merseburg wheel only had *half* of the maximum downward vertical force of a weighted lever or about 59.3 pounds / 2 = 29.65 pounds applied to it by the *end* of a steel weighted lever pivot pin embedded in it at the drum's 6:00 position when the drum was rotating at its maximum terminal rotation rate. However, the middle lattice's radial frame piece was shared by *two* side by side weighted levers and, thus, would have had the full maximum downward vertical force of about 59.3 pounds applied to it. That force was applied to a cross sectional area of wood that was 0.5 inch = 8/16 inch thick x 2 inches long = 1 square inch. Subtracting the area removed for the pivot pin that penetrated it requires us to reduce its area by 0.125 square inch. Thus, the middle lattice radial frame pieces of the Merseburg wheel's 12 feet diameter drum had to support a maximum downward vertical force of about 59.3 pounds with only 0.875 square inch of wood. Since 1 square inch of wood can lift up to 100 pounds or more without the wood breaking, the middle lattice radial frame pieces of the Merseburg wheel, made from the same stock wood as its front and back lattices' radial frame pieces, could easily withstand this force and there was no need to double the thickness of this middle lattice radial frame piece to 1 inch. However, in the case of the Kassel wheel with its weighted levers having double the mass of those in the Merseburg wheel or 29 pounds each compared to the 14.5 pounds for the Merseburg wheel, it was necessary to double the thickness of all of the drum lattices' radial frame pieces to 1 inch each although the *width* of the frame pieces in the Kassel wheel was kept the same as that of the frame pieces in the Merseburg wheel or 2 inches.

Now let us move on to treating the details of the Merseburg wheel's sixteen weighted levers, eight of which were located in each of the two internal one-directional wheels contained within that wheel's enlarged 12 inch thick drum. One of these weighted levers is shown in Figure 11(b) and has the orientation it would have if its parallel pair of main or L arms was clamped into contact with its wooden radial stop piece (which is not shown in the figure) at the drum's front internal one-directional wheel's 9:00 position.

As can be seen in Figure 11(b), each of the two wooden side pieces of the weighted lever had three parallel pairs of arms (the *front* 3 arm side piece of the weighted Merseburg wheel lever in the top view of Figure 11(b) can be considered to have been removed or made transparent in order to better view the lever's interior) that were, as far as geometrical orientation proportionality was concerned, identical to the ones used in the small wooden levers of the 3 feet diameter Gera prototype wheel. And, as one might expect, since the drum of the Merseburg wheel was four times the diameter of the Gera prototype wheel's drum, the lengths of the Merseburg wheel's weighted levers' three parallel pairs of arms were also four times as long as those of the prototype because the distances between the center axis of a Merseburg wheel's weighted lever's pivot pin hole and the center axes of the lever's various cord hook attachment pins and cylindrical lead end weight attachment bolts were quadrupled. As would also further be expected, the center of gravity of this larger weighted lever was four times farther from the center axis of its pivot pin's hole as was the center of gravity of the Gera prototype wheel's weighted lever from the center axis of its pivot pin's hole or 12 inches (again indicated by a small circle divided into alternating black and white quadrants) compared to 3 inches for the Gera prototype wheel. But, aside from these details, there were also some other major differences in the construction of the Merseburg wheel's weighted levers.

The biggest difference occurred in the *total* mass of a Merseburg wheel's weighted lever. This total mass was the sum of the mass of its three cylindrical lead end weights including their three steel attachment bolts and the mass of an unweighted lever. The mass of an unweighted lever (this is the mass of lever *before* its three end weights and their three attachment bolts were added to it) was simply the sum of the masses of its two parallel wooden 3 arm side pieces, its six press fitted steel hook attachment pins, its six wooden hook keepers, the seven metal coordinating cord hooks attached to its six hook attachment pins, and *half* of the masses of the seven coordinating cords and two extension springs attached to the lever *if* the two springs were free to rotate about their spring hook anchor pins in the drum's axle (if they were fixed to the drum and could not rotate independently of it, then their masses would not be included).

The mass of *each* of the three cylindrical lead end weights held between the ends of a weighted lever's pair of parallel main arms was *about* 4 pounds as was determined by those weights that the inventor allowed some to handle during the translocation of the Merseburg wheel when it was being officially tested at his home. However, it was only with the addition of a weight's steel attachment bolt that the combination had a mass of almost exactly 4 pounds.

Each cylindrical lead end weight was 1.78 inches = 1-12.5/16 inches in diameter and 4 inches in length or about the size as a small can of juice as indicated in Figure 11(b). A weight was held to the ends of a lever's parallel pair of main arms by a slotted round head bolt that passed through a snuggly fitting shaft drilled out along the center axis of the cylindrical lead weight. These attachment bolts had a diameter of 0.1875 inch = 3/16 inch and a smooth shank length of 4 inches. At the end of the shank length, the treaded section of a bolt would have extended an additional 0.5 inch = 8/16 inch. The treaded end of the bolt would have completely filled the center hole of a 0.1875 inch = 3/16 inch thick *square* nut that was sunk 0.0625 inch = 1/16 inch into the wood of the outward facing surface of its 3 arm side piece so that the nut only projected out 0.125 inch = 2/16 inch above the surface of the wood. These nuts were held in place by friction and some glue and remained in place on a lever's 3 arm side piece whenever their bolt and the cylindrical lead end weight it penetrated were removed from the lever.

At the top of the continuation page for Figure 11(b), one will find the various wooden hook keepers used inside of the weighted levers of the 12 feet diameter, bidirectional Merseburg wheel and, as one might expect, they are larger than those found in the 3 feet diameter, one-directional Gera prototype wheel because the width of an arm of a 3 arm side piece in a weighted lever in the Merseburg wheel, which was 2 inches was four times larger than the analogous width of an arm of a 3 arm side piece in a weighted lever in the Gera prototype wheel which was 0.5 inch = 8/16 inch.

The wooden coordinating cord hook keepers that were in physical contact with the steel hook attachment pins that joined the parallel pairs of wooden 3 arm side pieces of the Merseburg wheel's sixteen weighted levers were each 0.1875 inch = 3/16 inch thick x 0.90625 inch = 14.5/16 wide x 4 inches long (thus making each of the five coordinating cord layers inside of the Merseburg wheel's weighed levers 4 inches / 5 = 0.8 inch = 12.8/16 inch thick). As in the case of the Gera prototype wheel, the edges of the hook keepers in the Merseburg wheel's weighted levers had to have thicknesses that matched the diameters of the metal hook attachment pins that they were in contact with inside of each lever or 0.1875 inch = 3/16 inch and the unusual width of the these lever hook keepers was necessary so that, when their notched edges were placed up against the steel hook attachment pins on the lever's A arm in the necessary orientations, they would not project out beyond the lower trailing edge of the arm. Also, as it the case of the wooden hook keepers inside of the Gera prototype wheel, cord bindings may have been wrapped around the central regions of the Merseburg wheel's wooden hook keepers and the steel hook attachment pins they contacted in order to prevent any flexing of the pins from taking place which could then allow a metal hook to slip along the length of the pin and away from its notch restricted section of the pin and thereby become wedged between the steel pin and the wooden hook keeper.

Both the stop cord and spring hook attachment anchor pin hook keepers at the bottom of the vertical collection of wooden hook keepers used in the Merseburg wheel as shown in the continuation of Figure 11(b) are 0.25 inch = 4/16 inch thick x 0.875 inch = 14/16 inch wide x 5.25 inch = 5-4/16 inch long. Their thicknesses and widths are different from the hook keepers found inside of the weighted levers of the Merseburg wheel because the steel hook attachment anchor pins of the drum that they contact have diameters that are 0.25 inch = 4/16 inch and are larger than the diameters of the hook attachment pins found inside of the Merseburg wheel's weighted levers which were 0.1875 inch = 3/16 inch.

Other than dimensional size differences, by comparing the collection of Merseburg coordinating cord hook attachment pin hook keepers in the continuation of Figure 11(b) with those shown in the first continuation of Figure 4 in the previous chapter for the Gera prototype wheel, the reader will notice an interesting difference. It is found in the L1 main cord hook keeper in the continuation of Figure 11(b).

This hook keeper has the square center notch that receives the hook at one end of a main cord, but there are now two additional notches, longer and wider rectangular ones, on each side of the square center notch. These extra rectangular notches are each 0.375 inch = 6/16 inch wide x 1 inch long. The purpose of these additional notches will be explained in the next chapter. For the moment, however, let it just suffice to say that they are there to allow a special gravity activated latching mechanism Bessler invented to grab onto an L1 main cord hook attachment pin and thereby hold a weighted lever's parallel pair of main arms against their wooden radial stop piece so that they could not swing inward toward the center axis of the axle when their one-directional wheel was undergoing retrograde motion or motion opposite to its normal direction of spontaneous start up motion around the center axis of the axle. Basically, the latches kept a weighted lever's main arms aligned with their parallel pair of radial frame pieces as the weighted lever's pivot pin went up and over the drum's 12:00 position. When this alignment was achieved for all of the eight weighted levers in the internal one-directional wheel of a bidirectional wheel undergoing retrograde motion, the center of gravity of those eight weighted levers would then be moved to and, more importantly, be kept at the location of the center axis of the axle until that internal one-directional wheel was again turning in its normal spontaneous start up direction of motion.

Finally, at the bottom of the continuation of Figure 11(b) we see the larger and stronger metal hooks to which the ends of the various coordinating cords were tied. As one might reasonably suppose, they can no longer be made from something as delicate as a steel staple that would be suitable for a replica of Bessler's 3 feet diameter Gera prototype. For the Merseburg wheel they were most likely made from a short length of 0.1875 inch = 3/16th inch diameter steel rod that had a closed eyelet formed at one of its ends and an almost closed loop hook at the other end. These coordinating cord and spring hooks had to be springy enough so that the loop hook end could spread far enough to allow it to be easily attached onto a notch exposed section of a 0.1876 inch = 3/16th inch diameter weighted lever hook attachment pin or one of the drum's 0.25 inch = 4/16th inch diameter stop cord or spring hook attachment anchor pins.

The cords found in the Merseburg wheel were very strong and probably also color coded for ease of identification during installation into the drum's two side by side, back to back, internal one-directional wheels found inside of that 12 feet diameter bidirectional wheel. Obviously, the more massive weighted levers in the Merseburg wheel would have necessitated Bessler using cords and attachment hooks that were stronger than used in his previous wheels with their less massive weighted levers. By apply the same formula used to determine the strength of the cords and hooks used in the

3 feet diameter Gera prototype wheel, the craftsman can get an idea of how strong the Merseburg wheel's cords and hooks needed to be.

From the Gera prototype wheel we learned that each of its cords and one of its attached hooks were strong enough if they could be used to vertically lift and support a mass equal to three times the mass of a single weighted lever or 3 x 0.453125 pounds = 1.359375 pounds = 21.75 ounces without the cord breaking or the hook significantly bending open. In the case of a replica of the Merseburg wheel, its cords and hooks will have sufficient strength if a single cord and its attached hook can be used to vertically lift and support a mass that is three times that of one of that wheel's weighted levers which had a total mass of 14.5 pounds without the cord breaking or the hook significantly bending open. Thus, a cord and hook combination used in any replica of the Merseburg wheel should be strong enough to lift and support a mass of 3 x 14.5 pounds = 43.5 pounds.

Since all of the arm lengths of the Merseburg wheel's weighted levers were four times longer than those of the levers found in the Gera prototype wheel, this means that all of the distances between the center axes of the coordinating cord hook attachment pins and the center axis of a Merseburg wheel's weighted lever's pivot pin's center axis (or the pivot pin hole's center axis if the lever was not yet installed in the drum) were also quadrupled and this can be seen by noting the various distances labeled in the weighted lever's side view in the top portion of Figure 11(b) and comparing them to the corresponding distances given for the Gera prototype wheel's weighted lever shown in Figure 4.

Because the distance between the center axis of a weighted lever's pivot pin and the center axis of the axle in the Merseburg wheel was exactly four times that of the corresponding distance found in the Gera prototype wheel *and* the distances between the center axes of the coordinating cord attachment pins and the center axis of their weighted lever's pivot pin in the Merseburg wheel were also four times those of the corresponding distances found in the Gera prototype wheel, this results in *all* of the allowed maximum separation distances for the various coordinating cords in the Merseburg wheel being four times those of the corresponding maximum allowed distances for the various coordinating cords found in the Gera prototype wheel. These quadrupled maximum allowed separation distances for the Merseburg wheel's four different types of coordinating cords are listed at the bottom of the continuation of Figure 11(b).

In manufacturing the sets of coordinating cords for wheels larger than the Gera prototype each of which has a metal hook attached to its two ends, the craftsman can still use the method I described in the last chapter for the Gera prototype wheel.

For the 12 feet diameter Merseburg wheel, however, one will not be able to use a small length of wood with nails inserted into it at various distances from a common reference nail near one of its ends. He will have to use large 0.25 inch = 4/16 inch diameter nails or, preferably, wood screws (this diameter is identical to that of the coordinating cord hook attachment pins of the Merseburg wheel's levers and their stop cord hook and extension spring hook attachment anchor pins in the drum) that are inserted into a board that is at about 7.25 feet = 7 feet and 3 inches in length. One of the nails or screws will serve as the common reference one at the end of the board while the others are placed along the length of the board and fanned out a bit so that the distances to their center axes from the center axis of the shaft of the reference nail or screw correspond to the maximum allowed separation distances shown at the bottom of the continuation of Figure 11(b) for the four different types of coordinating cords used in the Merseburg wheel.

The heads of the nails or screws are then cut off and the projecting shafts filed down a bit to round off any sharp edges on their ends. Next, one attaches one previously made metal hook to the shaft of the reference nail or screw at one end of the plank (the craftsman need not actually manufacture these hooks himself if he can find a source that sells what he needs), attaches another hook to a second shaft whose center axis is located at a specific maximum allowed separation distance from the reference shaft's center axis so as to correspond to the maximum distance permitted between the center axes of a pair of coordinating cord hook attachment pins of the Merseburg wheel replica for one of its four types of coordinating cords, and then, finally, ties the two ends of the correctly colored coordinating cord to the eyelets of both hooks so that the cord is *tightly* stretched between them. Once finished, the colored coordinating cord and its two attached metal hooks are then easily slid off of the two nails or screw shafts which they were stretched between and any excess cord dangling from the cord's two attached metal hooks' eyelets is cut off.

Up to this point I have not mentioned the parallel pairs of steel extension springs that were attached to each of the Merseburg wheel's sixteen weighted levers. One might suppose that, since the masses of the weighted levers in the Merseburg wheel were exactly 32 times those of the weighted levers used in the Gera prototype (because 14.5 pounds / 0.453125 pound = 32), he need only multiply the k value of each of the Gera prototype's springs by a factor of 32 to derive the spring constant, k, of each of the springs used in the Merseburg wheel. Thus, since *each* of the two extension springs attached to the A1 spring cord hook attachment pin of one of the Gera prototype's weighted levers had a spring constant, k, of 20 ounces per inch = 1.25 pounds per inch, then one might conclude that the k value for each Merseburg wheel's extension springs must have simply been 32 x 1.25 pounds per inch = 40 pounds per inch and that would then mean that the *combined* k value for the *two* parallel springs in coordinating cord layers 1 and 5 attached to each of the 12 diameter bidirectional Merseburg wheel's sixteen weighted levers would have been 80 pounds per inch.

This conclusion, however, would be false.

The reason is that, although the mass of the Merseburg wheel's weighed levers was 32 times that of the little wooden levers used in the Gera prototype wheel, the *distance* that a Merseburg wheel's extension spring was stretched by its weighted lever at *any* position of the Merseburg wheel's drum was always four times greater than the corresponding stretch distance for a smaller spring at the same drum position in the Gera prototype wheel. Unless Bessler had compensated for this quadrupling of stretch distances in his Merseburg wheel's extension springs at any of its drum positions compared to a spring's stretch distance at the same drum position in the Gera prototype wheel, he would have had the Merseburg wheel's springs applying four times too much force to the weighted levers to which they were attached and the center of gravity of drum's eight weighted levers in the active internal one-directional wheel that drove the 12 diameter, bidirectional drum would not have remained on the descending side of the axle. To prevent this problem, Bessler had to compensate for the quadrupled stretch distances when selecting the k value of the springs he would use in the Merseburg wheel.

Fortunately, the compensation was simple and merely consisted of dividing the expected increased k value needed for the Merseburg wheel's 32 times more massive weighted levers of 40 pounds per inch by a value of 4. Thus, *each* of the two extension springs attached to a weighted lever inside of the drum of the Merseburg wheel had to have a k value equal to that of a Gera prototype wheel's extension spring, 1.25 pounds per inch, that had been multiplied by a factor of 32 as one would expect, but then, in order to compensate for the quadrupling of stretch distances in the larger bidirectional Merseburg wheel, this product was divided by 4. We can write this as (32 x 1.25 pounds per inch) /4 = (40 pounds per inch) /4 = 10 pounds per inch for *each* of the Merseburg wheel's 32 steel extension springs. For each parallel *pair* of extension springs attached to each weighted lever inside of the Merseburg wheel's drum, the *total* k value was, therefore, 20 pounds per inch. Since the Kassel wheel's weighted levers had twice the mass of those in the Merseburg wheel, but its drum had the same 12 feet diameter, we can quickly obtain the k value for each of the Kassel wheel's sixteen extension springs by simply doubling the k value of each of the springs used in the Merseburg wheel. Thus, the k value of a single extension spring in the Kassel wheel was 20 pounds per inch and the total k value for a parallel *pair* of such extension springs attached to a single lever's A1 spring cord hook attachment pin was 40 pounds per inch.

As a final note on the Merseburg wheel's extension springs, it should be mentioned that when the pivot pin center axes of the weighted levers they were attached to reached the drum's 9:00 position, each parallel pair of springs had its maximum stretch from the unstretched length it had as a weighted lever's pivot pin's center axis traveled from the drum's 4:30 to 6:00 positions in a clockwise rotating drum. For the Gera prototype wheel's springs, we saw in the last chapter that this maximum stretch distance was only 0.5 inch = 8/16 inch at its drum's 9:00 position and that the *unstretched* length of that wheel's diminutive extension springs needed to be a *minimum* of 1 inch in order to prevent the additional maximum 0.5 inch = 8/16 inch stretch distance they experienced from exceeding about 50% of a spring's unstretched length and thereby permanently damaging the spring by stretching it beyond its elastic limit. Due to the quadrupling of the stretch distances of all of the stretched extension springs in the Merseburg wheel, a parallel pair of extension springs whose weighted lever's pivot pin's center axis reached the drum's 9:00 position would have been stretched a distance 4 times as great as the corresponding distance for a parallel pair of springs in the Gera prototype wheel or 4 x 0.5 inch = 2 inches and this then required that Bessler use extension springs for the Merseburg wheel's weighted levers whose *unstretched* lengths were *at least* 4 inches long so as to prevent the permanent damage to their coils that could be caused whenever a steel spring is stretched through a distance that exceeds about 50% of its unstretched length.

I can also mention here that the descriptions given above for the basic geometry of the Merseburg wheel's drum and weighted levers can also be applied to the Kassel wheel that Bessler and his brother Gottfried constructed at Weissenstein Castle under the patronage of Count Karl.

While the three lattices of the drum of the also 12 feet diameter, bidirectional Kassel wheel used the same lattice design found in the Merseburg wheel, in order to compensate for the doubling of its sixteen weighted levers' masses, Bessler had to double the thickness of the Kassel wheel's drum's radial and octagonal frame pieces so that they had a *square* cross section that was then 1 inch thick x 2 inches wide = 3 square inches (it's possible that he may have used the same 0.5 inch = 8/16 inch thick x 2 inches wide stock wood that he used for the frame pieces of the Merseburg wheel's three lattices and just glued and clamped them together to double their thickness).

I might mention here that, while the various frame pieces of the three drum lattices inside of the Merseburg wheel were attached to each other with a combination of glue and metal plates, I am of the opinion that, in the construction of the Kassel wheel, Bessler stopped using glue and held the frame pieces of the drum together exclusively using small metal plates of various shapes that were attached to two or more touching frame pieces using small screws. I say this because he would have known in advance that he would be required to disassemble the wheel from where it was being constructed in the Weissenstein castle's garden shed and then transport it up several flights of stairs and through somewhat narrow doorways to a room on the castle's second story above its ground story or three stories above the level of the shed. He would have needed to be able to completely disassemble the drum without the risk of damaging its various frame pieces which might have happened if they had been glued and nailed together as was his practice with his earlier wheels and he had to tear their joints apart.

The *total* mass of a Kassel wheel's weighted lever was 29 pounds which was the sum of the masses of its three cylindrical lead end weights and their three attachment bolts (each weight and bolt having a combined mass of 8 pounds to make a total of 24 pounds for the three weights and their attachment bolts) and an unweighted wooden lever having a mass of 5 pounds (this mass included the masses of the seven coordinating cord hooks attached to a lever, half the masses of those hooks' seven coordinating cords, and half of the masses of its two extension springs *if* they were not fixed by one of their ends to the drum). This is in contrast to the total mass of a Merseburg wheel's weighted lever which was 14.5 pounds and was the sum of the masses of its three cylindrical lead end weights and their three attachment bolts (each weight and bolt having a combined mass of 4 pounds to make a total of 12 pounds for the three weights and their attachment bolts) and an unweighted wooden lever having a mass of 2.5 pounds (and, again, this mass included the masses of the seven coordinating cord hooks attached to a lever, half the masses of those hooks' seven coordinating cords, and half of the masses of its two extension springs *if* they were not fixed by one of their ends to the drum).

Although the side view of a Kassel wheel's unweighted lever (which I have not illustrated, but which a skilled craftsman should be able to deduce) was identical to that of the Merseburg lever with a parallel pair of wooden 3 arm side pieces whose arms were 2 inches wide, its top view would reveal a lever that measured 7 inches between the *outside* facing surfaces of its two parallel 3 arm side pieces which were each 0.5 inch = 8/16 inch thick. Because of the doubling in mass of the Kassel wheel's weighted levers, its cylindrical lead end weights' attachment bolts would have each been 0.25 inch = 4/16 inch in diameter x 7.125 inches = 7-2/16 inches in length with a slotted round head adding another 0.125 inch = 1/8 inch to the bolt's overall length and its six coordinating cord hook attachment pins would have been a larger 0.375 inch = 6/16 inch in diameter and exactly 7 inches in length so that their ends would not project out beyond the outside facing surfaces of a lever's parallel pair of 3 arm side pieces. However, the eight shared weighted lever pivot pins and sixteen drum hook anchor pins used in the Kassel wheel's drum would also have been 0.375 inch = 6/16 inch in diameter, but 18 inches in length.

At this point in the chapter the reader should have a fairly detailed understanding of how Bessler's 12 feet diameter, bidirectional Merseburg wheel's drum was constructed and of the parallel pairs of 3 arm side pieces from which its sixteen weighted levers were constructed. Now it's time to move on to some of the external features of this wheel that Bessler hoped would so impress a rich businessman that the inventor then could quickly sell it and begin working on his far grander plan to unify and pacify humanity globally by instilling in them what he considered to be proper Christian values.

The reader can begin by locating and studying Figure 12(a) at the end of this chapter. In that figure, two views of one end of the Merseburg wheel's 6 inch diameter, oak axle are shown and the construction shown would be identical for the axle's front and back ends.

At the top of the figure we see a cross sectional view of the axle, the embedded steel pivot pin that extends out of its end, and the two brass "half moon" bearing plates embedded in one of the 12 diameter, bidirectional wheel's 14 feet tall, vertical wooden support planks. The only "half moon" aspect of these brass bearing pieces was the cross sectional appearance of the semicylindrical depressions drilled out of them into which the axle's end pivot pin was inserted and turned. To form the two required half moon bearing plates for his wheels, Bessler simply took a cube of solid brass, drilled a hole through one of its three center axes having a diameter *slightly* larger than a wheel's axle pivot pin, then cut the cube into two identical halves, and, finally, drilled smaller holes into the midpoint of the length of each of the semicylindrical depressions formed in each of the two plates of brass that had been made from the original whole cube.

These smaller holes allowed him, by using a very short shaft screwdriver, to use two wood screws to secure these brass bearing plates into the nests he had carved out of the portions of the vertical support planks for them. The *slight* gap between the outer surface of an axle's pivot pin and the inner surface of a brass bearing plate's semicylindrical depression served as a reservoir for the grease that lubricated the axle pivot pins. Because of the weight of the Merseburg wheel pressing its two axle pivot pins down into their lower bearing plates, these bearing plates would not have been lubricated with a runny liquid vegetable oil such as olive oil, but, rather, with a thicker grease made from tallow which is rendered animal fat that is composed mostly of triglycerides and is a hard, waxy solid at room temperature. It would be applied to the axle pivot pins before they were place upon the semicylindrical recesses of their lower brass bearing plates.

It was *very* important that the Merseburg wheel's four half moon brass bearing plates did not move about as the axle pivot pins they embraced continuously rotated. If, for example, one of the lower plates did detach and somehow came out of its nest in its wooden vertical support plank when that bearing was "opened" for inspection, then the end of the axle near that dislocated bearing could have rolled out of the plate and, along with its axle and its attached drum, have dropped a distance of about a foot toward the floor. If the drum and axle of the rapidly spinning Merseburg wheel weighing 550 pounds or those of the Kassel wheel weighing 1,100 pounds was to suddenly hit a stationary floor while the other end of its axle was still being loosely held in its vertical support plank, then the results would have been catastrophic and resulted in the drum and the other vertical support plank being torn apart. This was definitely something Bessler did not want to risk happening as he was demonstrating a wheel to a potential buyer or his agent. He always tried to portray his inventions as close to being problem free as possible which was exactly what he knew a buyer, bothered by the high price the inventor was asking, wanted to hear.

In the side view of the axle pivot pin encased by its two half moon bearing plates nested in the wheel's vertical support plank shown in the top part of Figure 12(a), I show that a 6 inch diameter oak axle and a 0.75 inch = 12/16 inch diameter steel pivot pin were used in the Merseburg wheel and this pivot pin would have been 12 inches in length with half that length or 6 inches hammered into a tight fitting hole bored into the end of the axle to receive it. However, for the Kassel wheel an 8 inch diameter oak axle and two 1.125 inches = 1-2/16 inches diameter steel pivot pins had to be used because the mass of that wheel was about double that of the Merseburg wheel. The steel pivot pin used on the Kassel wheel's axle would have been 18 inches in length with half of that length or 9 inches hammered into a tight fitting bored hole in the end of that wheel's 8 inch diameter axle. There is some suggestion in left side profile illustration of the Kassel wheel shown in Figure 2(b) at the end of chapter 1 that Bessler may have attached metal collars to the very ends of that wheel's axle. They would have been intended to reinforce the ends of the axle where their pivot pins were inserted so that the ends would not split apart when the weight of the drum was rested upon the two axle end pivots as they came to rest on their lower half moon brass plates.

As a steel pivot pin projected from the end of the Merseburg wheel's axle, it immediately entered the inward facing side of a vertical oak support plank that was *slightly* less than 14 feet long x 12 inches wide x 3 inches thick. This single vertically oriented plank spanned the distance from the floor of the room containing the Merseburg wheel to its 14 feet high ceiling. Making each of the axle's vertical support planks slightly less than 14 feet in length was necessitated by the fact that when one tries to vertically fit planks between a floor and ceiling that are exactly 14 feet apart, the length of the plank must be a little shorter than 14 feet because the *diagonal* distance measured between *opposite* thickness or width edges of a plank that is exactly 14 feet in length will always be a little *more* than 14 feet which will then prevent the plank from being fitted into position vertically. Thus, making a plank that is a little less than 14 feet in length is the only way that it can be fitted between a floor and ceiling that are exactly 14 feet apart.

Each plank, once carefully positioned so that the 6 feet long axle would fit between two of them without the ends of the axle rubbing against the inside facing surfaces of the planks, was then secured in place to the floor and ceiling with a sort of box-like frame of molding pieces that tightly embraced the ends of the plank and were themselves nailed to each other and to the wood of the floor boards or ceiling beams. Once in place, the molding pieces finally would have been nailed to the end widths of the vertical support planks to prevent any wobbling motion of the rapidly rotating axle caused by a slightly imbalanced drum from lifting, through an axle pivot pin, the bottom end of a plank off of the floor until its top end traveled a fraction of an inch and hit a ceiling beam. Such an undesirable "hopping" motion of a plank might eventually cause one of the vertical support plank's molding frame boxes to come apart and thereby allow a plank to slip out of its vertical orientation resulting in the axle pivot pin it supported falling to the floor and also destroying the drum and the other vertical support plank in the process.

Bessler probably chose this approach to securing the wheel's vertical support planks in place because it would allow him to more easily remove and relocate the planks if needed.

As the Merseburg wheel's two axle end pivot pins were placed upon the 3 inch thicknesses of the wheel's two vertical support planks, they would only momentarily touch the wood of the planks before they were pushed farther in and finally fell into place in the semicylindrical recesses of the two planks' embedded lower half moon brass bearing plates. Since the two bottom half moon bearing plates in the wheel's two vertical support planks were at least 2 inches thick each, they could easily withstand the load placed upon them by the two axle pivot pins attached to the axle and drum which had a combined weight of about 550 pounds. Thus, each bottom half moon bearing plate had to withstand a downward force on it of about 275 pounds. Both bottom bearing plates would have been firmly fixed into their nests by long wood screws to assure that they could not be worked out of their nests by any vibrations transmitted to them from the axle through its pivot pins.

Each upper half moon bearing plate's primary purpose was to keep a rapidly rotating axle pivot pin from riding up and out of the lower bearing plate's semicylindrical depression should there be a sudden increase in pivot pin to lower bearing plate friction due to a lubrication failure. That, as previously noted, would have been catastrophic and the resulting damage would destroy weeks or even months of work. However, an upper half moon bearing plate could only do its job if the bottom bearing plate and the pivot pin resting or turning in its semicylindrical depression were "closed" to visual inspection because the upper half moon bearing plate was in place with the upper semicylindrical portion of the pivot pin securely contained within the upper bearing plate's semicylindrical depression.

As each axle pivot pin of the Merseburg wheel projected beyond the outside facing surface of its vertical support plank, it passed through the center hole of a large washer which was held near the outside surfaces of that axle pin's two touching half moon bearing plates by a split pin that was inserted through a small hole drilled through the axle pivot pin and which then had one of its emerging ends bent in order to keep the split pin locked into the small hole and the washer from moving away from the bearing plates along the axle's end pivot pin. The split pin and the axle pivot pin washer whose outward axial motion it blocked also served the purpose of preventing the axle pivot pin from withdrawing into the plank as the *other* end of the axle then moved closer to the *inside* surface of its vertical support plank and began rubbing against it. If such rubbing was allowed to occur, it would create energy wasting drag that would prevent the wheel from rotating at its maximum rate and, if severe enough, might even slow the wheel to a stop.

The exterior end of each axle pivot pin was "squared off" by using a file to give it a square cross section and then had a threaded hole tapped into the center of its outward facing square face that would receive the threaded section of a small bolt. Bessler, having been an apprentice clockmaker, would have been familiar with this technique because it was a method commonly used to securely attach the hour and minute hands to the posts projecting out of the center of a clock's dial.

Once the exterior ends of the axle pivot pins were squared off, it was then possible to attach two somewhat unusual cranks to them. One of these cranks is shown at the bottom of Figure 12(a). It had a square hole in its central cylindrical piece that fitted onto the square cross section at the end of an axle pivot pin after which it was locked into place by a tightened bolt with a large slotted head on it. Then, from the cylindrical piece attached to the end of an axle pivot pin, a handle forming a semicircular arc extended away and its end was attached to a smaller diameter cylindrical piece with a hole in its center. The resulting radial distance between the center axis of the axle's pivot pin and the center axis of the hole in the smaller cylindrical piece at the end of the crank handle was about 3 inches.

Why this unusual shape instead of just using a crank with a straight handle? It's hard to tell, but I suspect that Bessler found them to be decorative and, perhaps, liked them because he thought their curves would distract the reverse engineers viewing his wheels from noticing the exact way that the two pendulums were attached to the ends of the Merseburg wheel's axle and for reasons that will become obvious before the end of this chapter.

As previously mentioned, at the end of each crank's curving handle, there was a small cylindrical piece with a hole through which a bolt could be inserted. In the side view shown at the top portion of Figure 12(a), we see how the inventor would have attached one of the Merseburg wheel's pendulums to the crank. The attachment was made by simply passing a bolt through a hole at the bottom end of a pendulum's drive rod, whose circular cross section of 0.75 inch = 12/16 inch diameter had been heated and flattened until it was about 0.35 inch = 5.6/16 inch thick x 1.25 inches = 1-4/16 inches wide, and then through the hole in the small cylindrical piece at the end of the crank's curving arm, and, finally, through a short length of a metal spacing sleeve and a pair of square nuts that were then tightened to lock against each other.

This would assure that the two nuts would stay securely attached to the bolt, yet there would be some looseness so that the rotating semicircular crank handle could easily move the pendulum drive rod up and down in order to move the pendulum it was attached to at its other end. The important thing about this design is that it allowed Bessler to quickly detach a pendulum from the end of an axle pivot pin. To do this, the inventor simply removed the single bolt that attached the curving crank to the squared off end of the axle pivot pin, pulled the crank's central cylindrical piece off of the pivot pin, and then just let the cylindrical piece hang down away from the end of the pivot pin while the smaller end of its curving handle remained loosely attached to its pendulum's drive rod by the bolt and two square nuts. Once this was done, the offset downward pull caused by the weight of its attached drive rod and crank on one side of the pendulum's pivot would cause its plumb bob weight to swing a few degrees to the other side of the axle's pivot pin and stay there.

Bessler could then quickly reattach the little curved crank to the squared off end of the axle pivot pin whenever he wanted to by just slowly rotating the drum until the orientation of the squared off end of the axle pivot pin properly aligned with the square hole in the central cylindrical piece of the crank, pushing that piece back onto the end of the axle pivot, and, finally, inserting and tightening the bolt to lock the central cylindrical piece of the crank in place again. Thus, detaching each pendulum from the end of an axle pivot pin might have taken about a half minute and reattaching it again, after the Merseburg wheel's axle pivot pin's orientation had been properly adjusted, might have taken a full minute at most. (Shortly in this chapter, where I discuss what the real purpose of the two pendulums attached to the Merseburg wheel's axle pivot pins were, the reader will realize what I mean by "properly adjusting" an axle pivot pin's orientation prior to attaching its curving end crank.)

Lastly for Figure 12(a), the reader's attention is directed to the perspective axial drawing of the section of one of the Merseburg wheel's vertical support planks which occupies most of the lower half of the figure.

In order to allay suspicions about hidden drive belts attached to pulleys or drive rods attached to crankpins located on the sections of the pivot pins inside of the wooden vertical support planks of the Merseburg wheel and normally out of sight, Bessler had a means of exposing the entirety of bottom brass half moon bearing plates upon which the axle pivot pins rested. In this portion of the figure I show what I believe is the most probable way he did this. Basically, just above each pivot pin and the bottom bearing plate there were two tightly fitted sections of the plank that could be easily removed. Firstly, the upper wedge shaped wood piece was removed by unscrewing a long wood screw that attached it to the main plank and sliding it out of its section of the plank. Next, the lower wood block, which contained the upper brass half moon bearing plate, was lifted up vertically a short distance to clear its two pins from their holes in the plank and then removed by sliding it out of the cutout section of the plank.

This exposure of the Merseburg wheel's axle pivot pins was done by Bessler for anyone representing a buyer who might actually be in a financial position to purchase the wheel and its secret mechanics and needed to be convinced that there was absolutely nothing attached to the hidden sections of the pivot pins inside of the vertical support planks that was driving the axle. Since the bearing plates and pivot pins were at an elevation of 7 feet above the surface of the wheel room's floor, Bessler and another person viewing the exposed bearings would have reached them by using a portable wooden platform with attached steps and, most likely, a railing around three sides of it whose top level was about 2.5 feet off of the floor.

Once the visual inspection was completed, the lower block containing the top half moon bearing plate would be lowered back into position on top of the axle pivot pin so that the four steel pins in the lower block went into their matching holes and slots in the main plank. Next, the upper wedge block was slid into place above the lower block until the two steel pins at its smaller end engaged their matching holes in the main plank. Finally, the long wood screw was inserted into its hole in the upper wedge block and tightened. When this was done, both blocks would be securely locked into their places and the axle pivot pin located within the semicylindrical depressions of both brass half moon bearing plates would be securely held by the lower and upper half moon bearing plates.

No doubt, Bessler was often asked to let his Merseburg wheel run at full speed while the axle bearings in both of its vertical support planks were exposed. While he probably obliged the most serious of those examining his wheel, this was not something he would have routinely done. The reason why is that when a lower half moon bearing plate and its supported pivot pin were completely exposed because the upper wedge and lower block pieces of a main plank had been removed, there was a cross sectional area of wood of only about 3 inches x 3 inches connecting the 7 feet long length of the main plank below the center axis of the axle's pivot pin with the slightly less than 7 feet long length of plank above that center axis. Normally, if the lower block and upper wedge block were in place and secured, then the cross sectional area of wood connecting the lower length of the main plank with the upper length would be 3 inches x 12 inches. From this we can surmise that the tensile strength of the main plank with regard to *lateral* forces applied to it, which was proportional to the cross sectional area of the plank, was reduced by a factor of 4 whenever the lower and upper wedge blocks were not in place. That is a 75% reduction in strength and considerable despite the thickness and width of the planks Bessler used for the 12 feet diameter, bidirectional Merseburg wheel.

This, however, would have no effect on a vertical support planks' ability to *vertically* support the weight of the wheel's mass. But, it posed a risk in the event that any unanticipated *horizontal* forces were suddenly applied to the wheel's axle. For example, if one of the coordinating cords somehow got wrapped around the end of a weighted lever's arm so that the eight weighted levers of the Merseburg wheel's active internal one-directional wheel that was driving the axle could not smoothly shift during drum rotation, then this might cause the displaced center of gravity of the disabled one-directional wheel drop below its *punctum quietus* and then begin rapidly rotating about the center axis of the axle as the drum continued to rotate while gradually losing speed.

The imbalanced centrifugal force from this would, during each complete drum rotation, then have caused the two axle pivot pins to each first press against one side of the semicylindrical hollow of the lower half moon bearing plate it rested in and then against the other side of the hollow. This might, if the pivot pins were not adequately greased, then cause them to ride up and roll out of the two lower half moon bearing plates. Depending upon the direction in which the Merseburg wheel's drum was rotating, the two axle pivot pins might then roll right out of the vertical support planks and allow the drum to fall to the floor and be destroyed or they might both roll in the *opposite* direction and slam into the reduced cross sectional area connecting the upper and lower halves of the two vertical support planks. In the latter scenario, the impact might, for the 550 pound bidirectional Merseburg wheel when rotating at its maximum free running rate of about 40 rotations per minute, actually apply enough horizontal force to the narrowed connection between the upper and lower portions of the two support planks to cause them to fracture and then tear loose from the wooden frame pieces holding their ends in place near the floor and ceiling.

Even if this did not happen and the axle pivot pins both stayed seated in their lower half moon bearing plates' semicylindrical depressions, then any suddenly induced violent horizontal oscillation of the axle might, via resonance, have caused the entire slightly less than 14 feet long lengths of the vertical support planks to begin flexing from one side to the other and to increase the amplitude of their oscillations with each drum rotation until a point was reached where the two planks split into two pieces and the wheel's drum was then allowed to crash to the floor and destroy itself. The risk of plank fracture due to unexpected horizontal forces being applied to them would also be made greater if the top ends of the planks had not been nailed to the molding pieces that jacketed them and secured them to the ceiling beams. This is because the reduced area of the section of wood connecting the upper and lower portions of a plank would also have to bear all of the weight of the upper portion of a plank. For the Merseburg wheel, that upper portion of each plank had a weight of about 75 pounds!

Obviously, with all of these potential risks to an invention that had taken him months to construct, Bessler would only have let the wheel run with open bearings for the minimum amount of time necessary to satisfy an examiner or potential buyer's agent that there was nothing inside of the two large oak planks that was propelling the wheel through its axle's two steel end pivot pins.

Now it's time to move on and discuss another important feature of both Bessler's Merseburg and Kassel wheels: the pendulums attached to the ends of their axles' pivot pins through the curved handle cranks previously described. The serious student of both of these wheels needs to understand what their purpose was and how they achieved this. For this treatment, the reader is directed to Figure 12(b) at the end of this chapter.

In the upper left hand corner of the figure is an inserted image of the drawing of the *front* pendulum of the Merseburg wheel found in DT (See the left hand side of Figure 1(a) at the end of Chapter 1 for the complete drawing). Of all of the perspective errors I've found in Bessler's various book illustrations, this is probably the worst. From studying this DT image one gets the impression that all of the pieces of the pendulum, that is, its vertical bob weight support, its horizontal cross beam, its two diagonal braces, and its oddly placed "U" piece that attaches the end of the vertical bob support to the crossbeam, are all in the same plane. Then, one notices that the fulcrum of the pendulum, a supposedly fixed rod that extends from near the top of the axle's front vertical support plank to a wooden cross piece held between two smaller wooden vertical support boards, does not pass through the inside of the "U" shaped piece that attaches the pendulum's vertical bob weight support to the pendulum's cross beam as one might expect, but, rather to the right side of it! This is also done with the back fulcrum rod which passes to the left side of its "U" shaped piece. The actual use of the two illustrated pendulum fulcrums shown in DT would only result in pendulums that were sloppily slipping and sliding about as they were driven by the drive rods connecting them to the ends of the curving axle pivot pin cranks of the Merseburg wheel. How can we rationalize Bessler using these bizarre attachments for the pendulums in his DT illustration?

One answer is that it was a simple mistake he made while engraving the copper printing plates for the illustrations in his book. We have to remember that to make such a plate, he had to take a sketch he'd previously made on paper and then engrave the horizontally *reversed* mirror image of it onto the plate so that, after printing, the printed image would appear as it did in his original sketch. Although he was certainly a gifted craftsman, it's possible that he got confused while making the mirror image engraving and only later, after a test printing was made, did he realize how erroneous the final image was and by then, perhaps because he was under time pressure to get what would be his largest and most professionally done work on his wheels completed, decided to use the erroneous illustration details anyway.

Yes, that is one possible explanation for this obvious mistake, but Bessler, being somewhat of a perfectionist, was not the kind of man who would just let something like that go without correcting it. If it was not corrected, then there must have been another very good reason for that. I think that other reason that he *purposely* left the errors in was because he hoped that any of the more knowledgeable reverse engineers studying DT would see them and conclude that such a mistake proved that his wheel had to be a hoax and therefore not worthy of their time to further study and try to duplicate or, at a minimum, would dissuade them from trying to determine exactly how the pendulums were suspended and then, through their drive rods, attached to the axle's curving end cranks. Surprisingly, the details of how the bottom ends of the drive rods were attached to the axle's end pivot pins was *very* important and necessary because of the relatively low power output of the wheel even at start up. This was some additional information about his wheels that Bessler did not want anyone obtaining. (Interestingly, a more finely engraved illustration of the Merseburg wheel was actually first published in GB in December of 1715 that seems to show the correct attachment of the "U" shaped piece to the front pendulum's movable pivot rod. This changed, however, when the second, less finely engraved image of the 12 feet diameter, bidirectional wheel was used in his DT book published in October of 1719. Again this suggests a conscious decision by Bessler to try to confuse the reverse engineers studying the printed images of the Merseburg wheel in DT.)

However, before I give the details of how the compound pendulums were attached to the semicircular axle cranks and what that suggests about the wheel's performance, let me first provide the reader with the actual details of how the pendulums were suspended for use with the Merseburg wheel. This is shown in the upper right hand corner of Figure 12(b).

There we see that the supposedly "fixed" fulcrum rod depicted in the left side insert of the Merseburg wheel's pendulum attachment from DT is actually a *movable* iron pivot rod that had been forge welded at a right angle to and in the middle of another iron rod which served as the cross beam of each of the compound pendulums that Bessler could attach to the Merseburg wheel. This horizontal movable iron pivot rod, which had the vertical bob weight support rod attached to it by the "U" shaped piece of iron, partially rotated clockwise, then counterclockwise, then clockwise, and then counterclockwise again and again repeatedly about its lengthwise center axis as the bob weight at the bottom end of the vertical bob weight support rod swung from one side of the 12 feet diameter, bidirectional wheel's axle center axis to the other side.

As shown on the left side of Figure 1(a) at the end of Chapter 1, the two tapered ends of the *front* pendulum's movable pivot rod of the Merseburg wheel were supported by two bearing plates one of which was embedded in the wheel axle's front wooden vertical support plank above the half moon bearing plates that contained the axle's front end pivot pin and the other one which was located in the middle of a horizontal wooden cross piece that connected two additional wooden vertical support boards which all served to support the front end of the front pendulum's movable pivot rod.

Note, however, that this arrangement was not done with the *back* end of the movable pivot rod of the wheel's *back* pendulum. If the reader takes a careful look at the left edge of Figure 1(a), he will notice that there is a iron bracket bolted to the wall behind the Merseburg wheel. Apparently, the back of the wheel was so close to the rear wall of the room Bessler kept it in for public exhibitions in his home, that he was able to do away with using two extra wooden vertical boards and a wooden cross piece between them to support the back end of that rear pendulum's movable pivot rod. He just slipped its tapered back end into a slot in the wall mounted bracket and lubricated the iron to iron contact there. This same simpler approach to mounting axle driven pendulums was used in four places for the Kassel wheel which can be seen in Figure 2(b) at the end of Chapter 1 (note that the pendulums in that figure of the Kassel wheel are not the same type of compound pendulums used on the Merseburg wheel, but are really just fans whose side to side sweeping motion was meant to apply aerodynamic drag to the 12 feet diameter bidirectional wheel's axle in an attempt to reduce its maximum free running rotation rate of 26 rotations per minute to a substantially lower value).

Again in Figure 12(b), we note that a pendulum's vertical bob weight support rod was not attached directly to the cross beam rod or the movable pivot rod, but, rather, was attached to the bottom of a "U" shaped iron piece whose two ends were then attached, again via forge welding, to the movable pivot rod. Why this odd way of attaching the vertical bob weight support rod to the movable pivot rod?

Possibly, it was done this way because it was too difficult to forge weld three iron rods together in one place and have all three of them oriented at right angles to each other. The solution was to just weld two together at right angles to each other, the movable pivot rod and the cross beam rod in this case, and then to weld the end of the vertical bob weight support rod at right angles to the middle of a short piece of iron rod after its ends had been heated and bent up to form the "U" shape. Finally, the two ends of the "U" shaped iron piece were welded to different spots on the movable pivot rod.

In order to prevent metal fatigue from finally breaking the two arms of the "U" shaped piece that were attached to a pendulum's movable pivot rod after the pendulum had been in use for a long time and its vertical bob weight support rod had repeatedly slightly flexed and then relaxed the "U" shaped piece's attachment arms thousands of times, the inventor further equipped each pendulum's vertical bob weight support rod with two diagonal brace rods to prevent this flexing action. They each had one end forge welded to the same location on the vertical bob weight support rod and their other ends welded to opposite sides of the cross beam rod as shown in the upper right hand corner of Figure 12(b).

One of the two diagonal brace rods on each pendulum, however, was longer than the other and its excess length extended above the point at which it was welded to the cross beam rod. This extended section was bent back a little *toward* the nearby vertical support plank and had a small hole drilled in its end. Then, again via a small bolt and two locked together square nuts as was used to connect the end of the curved crank handle to the bottom end of a pendulum drive rod, the top end of this extended section of a diagonal brace rod was attached to the top end of the pendulum's drive rod. In order to make attachment with a straight bolt as easy as possible, the end of the extended section of a diagonal brace rod, which was through most of its length 0.75 inch = 12/16 inch in diameter, was flattened so that it was about 0.35 inch = 5.6/16 inch thick x 1.25 inches = 1-4/16 inches wide and, thus, was the same width as the flattened ends of the drive rod. Also, the flattened end of the extended section of a diagonal brace rod was bent *away* from the drive rod so that the flattened ends of both rods would be coplanar and, thus, fit flat against each other when the bolt and its two square nuts were installed to hold the ends of these two rods together.

I might add here that, based on my measurements made using the DT illustrations, all of the iron rods used in the Merseburg wheel pendulums had a circular cross sectional area, unless flattened for drilling holes for connecting bolts, and a diameter of 0.75 inch = 12/16 inch. The total length of the stock iron rod used to make *each* pendulum including its movable pivot rod was about 26.5 feet or 318 inches. Using the density of iron of about 0.284 pounds per cubic inch, one can determine that the total mass of just the rods alone (that is, with no extra weights attached to them) for each pendulum, including its drive rod, was about 40 pounds.

Now it's time to discuss the various types of weights attached to the Merseburg wheel's two compound pendulums, their masses, and their locations. Some very interesting information about how they worked will emerge from this.

I've concluded, again based on measurements made from the DT Merseburg wheel illustrations, that the distance between the individual centers of gravity of the two iron block weights at the ends of a pendulum's cross beam was about 8 feet and this is shown by the distance labeled "A" in the left of center inset in Figure 12(b) which was taken from DT and shows a section of the axial view of the 12 feet diameter, bidirectional wheel's drum and the front pendulum. Each iron block weight had a mass of about 15 pounds and the two weights were either directly welded to the ends of the cross beam rod or there was a thinner plate welded to the ends of the cross beam which had holes drilled into it and to which the inventor could bolt on block weights of different masses. The cross beam rod itself, without the two iron blocks attached to its ends, the two upper ends of the diagonal rods attached to it, or the movable pivot rod welded to its middle, weighted about 13 pounds.

From the middle of bottom curve of the "U" shaped piece of iron attached to each movable pivot rod another rod, the vertical bob weight support rod, extended downward and was attached to a spherical brass pendulum bob weight with a diameter of about 3 inches and a mass of about 5 pounds. The main purpose of these shiny brass pendulum bob weights was to dress up the end of its vertical support rod and also mesmerize the viewers of the Merseburg wheel with its swinging motion. Most likely, each bob weight had a threaded hole cut into it and it was then just screwed on to the end of the vertical support rod which had also been threaded to receive it. The distance between the center axis of the movable pivot rod and the center of gravity of the spherical brass bob weight, labeled "B" in the figure, was about 9.7 feet.

Adding up all of the masses of its weights and rod pieces, one finds that the total mass of each of the Merseburg wheel's compound pendulums, *without* their drive rods attached, was about 75 pounds. The drive rods were about 6 feet or 72 inches in length and weighed about 9 pounds each. I do not count a drive rod's mass as part of a pendulum's mass because a drive rod's mass was mostly supported by the wheel's axle pivot pin and was not, therefore, actually part of the pendulum.

As soon as I had what I considered to be fairly accurate estimates for the masses of the Merseburg wheel's front and back pendulums, I decided to make a computer model of one of them and see what its natural frequency of oscillation was.

I was not too surprised to discover that the pendulum completed almost exactly 16 oscillations in one minute where each complete "oscillation" consisted of the bob weight swinging from one side of a vertical line passing down through the center axis of the pendulum's attached movable pivot rod to the other side and then back again to the side from which it had started. I expected that the oscillation frequency of these compound pendulums would be much less than the axle's 40 rotations per minute rate since Bessler had connected them to the end cranks of the Merseburg wheel's axle in order to reduce its free running rotation rate to a lesser value in order to extend the running time of the wheel before an internal mechanical failure occurred that necessitated a repair. Perhaps with the pendulums attached, the 12 feet diameter bidirectional wheel would only attain a simple average of the drum's maximum free running rotation rate of 40 rotations per minute and the two pendulums' oscillation rate of 16 oscillations per minute or a final drum rotation rate of only 28 rotations per minute.

What I found next was a bit of a surprise, however. Using my computer model pendulum, I decided to see where its center of gravity was located and then try to locate that point on the section of the front side of the Merseburg wheel's drum shown in the left center inset image of Figure 12(b).

According to my model pendulum, the center of gravity of a pendulum used in the Merseburg wheel was located on the vertical bob weight support rod and 2 feet below the center axis of its movable pivot pin. When I then located this point on Bessler's DT illustration showing the front side of the drum which is what the public usually was only allowed to view, I noticed that the inventor had already marked the spot! In Figure 12(b) it is the point labeled "C" where the vertical bob weight support rod is crossing in front of the compound pendulum's drive rod. Notice that this location is also exactly located on the left edge of the front vertical support plank for the wheel's axle.

Bessler put some interesting shading on the left half of that front vertical support plank. He actually used vertical lines that run the entire length of the plank. There are six strips formed by these lines and, since the plank was 12 inches wide and there are six strips that cover half of the width of the plank, making each strip exactly one inch wide. By placing the overlap of the two rods on the left edge of the vertical support plank, Bessler was telling the future reverse engineer with this clue that the center of gravity of one of the Merseburg wheel's compound pendulums was displaced only 6 inches horizontally from the center line of its nearest vertical support plank when the pendulum's bob weight had swung to its maximum distance on either side of the center line of that vertical support plank.

Thus, the inventor literally marked the spot where the center of gravity of a compound pendulum was located when at its maximum horizontal displacement and that mark is actually a large letter "X" formed by the overlapping of the drive and vertical bob weight support rods! Bessler, in his youth, tried unsuccessfully to be a treasure hunter and, interestingly, in most of the treasure maps I've ever read about, the location of the treasure was always marked on the map with a large letter "X". I consider this clue as being Bessler's way of telling the astute reader of DT that he will learn something valuable about the Merseburg wheel by studying the arrangement of the two overlapping rods in the illustration he provided that shows the front axial view of his 12 feet diameter bidirectional wheel.

If one thinks of the axial view of the Merseburg wheel's drum as a treasure map of sorts, then the "X" formed by the two rods means that one will find treasure or something of valuable in the illustration. That something of value is really the secret information its many clues contain, but, of course, can also refer to the money that a future re-discoverer of Bessler's imbalanced pm wheel mechanics could make by selling replicas of the inventor's wheels.

Next, notice the two small black arrowheads I've inserted onto the section of the front of the Merseburg wheel's drum shown in Figure 12(b). Each arrowhead points to the number 8 which Bessler put on the axial view of the Merseburg wheel's drum with one of the numbers on the left side of the vertical bob weight support rod and the other one on the rod's right side. What could the meaning of this be? The locations of the two number 8's indicates that the clue tells us something about the left to right swinging motion of the vertical bob weight support rod. If we add the two numbers together we get the number 16 which represents the oscillation frequency for each of the Merseburg wheel's two pendulums or 16 oscillations per minute for each pendulum which was confirmed by my computer model using one of the compound pendulums. In other words, during a time period of one minute, each pendulum swings sixteen times to the right of the vertical support plank's center line and sixteen times to the left of that center line. But, how do we know that we have 16 oscillations in a time period of one minute?

The letter "X" formed by the crossing pendulum drive and vertical bob weight support rods as shown in Figure 12(b) can also represent an hourglass which is an ancient device for measuring time by seeing how long it takes for a given quantity of fine grain sand to pour through a small orifice connecting an upper glass bulb containing the sand with a lower bulb where it would collect. When all of the sand had poured from the top to bottom bulbs, a particular unit of time had elapsed which was usually an hour as the device's name suggests.

If one carefully measures the two angles formed above and below the location where the pendulum's drive rod and vertical bob weight support rod cross or the center of the "X" in Figure 12(b), he will see that each angle is exactly 30° as I've indicated in that figure. If we then add the two angles together we get 60° which then represents a time interval of 60 seconds or one minute. So, this clue, combined with the two number 8's, very cleverly tells the reverse engineer that the Merseburg wheel's pendulums had a natural oscillation frequency of 16 oscillations per *minute*. Once again, however, this was not necessarily the rotation rate that the Merseburg wheel, freely running, would have been constrained to by its pendulums. Its rotation rate with the two pendulums attached was probably somewhere around the average value of the pendulums' natural oscillation rate and the wheel's maximum terminal rotation rate when running freely *without* the pendulums attached to its axle pivot pins' oddly curved semicircular cranks.

From making precise measurements off of the two DT illustrations of the Merseburg and Kassel wheels, it appears that each of the oscillations of the Merseburg wheel's spherical brass bob weight carried it twice through an arc of 25° around the center axis of its movable pivot rod and each of the oscillations of the Kassel wheel carried its football shaped lead bob weight twice through an arc of 52° around the center axis of its movable pivot rod. (The arcs measured relative to the center axes of these wheels' *axles*, however, would have appeared much larger and this was another illusion Bessler used to make his wheels seem more powerful than they actually were.) This difference in bob weight arc swings between the two 12 feet diameter, bidirectional wheels would have been due to the larger semicircular curved cranks attached to the ends of the pivot pins of the Kassel wheel's 8 inch diameter axle that increased the distance between the center axes of the square holes in their central cylindrical portions and the little circular holes at the ends of their curving arms.

To complete the treatment of the Merseburg wheel's pendulums (and this will also apply to the Kassel wheel's pendulums), we must now discuss the diagram at the bottom of Figure 12(b). This is actually the most important part of the figure.

In that diagram, I illustrate how the axle's *two* semicircular cranks that drove the two pendulum's two drive rods were attached to the ends of the two steel pivot pins that projected from the opposite ends of the wheel's axle. We saw previously in Figure 12(a) that the end of each axle pivot pin was filed down on four sides so as to give it a square cross sectional area onto which one could fit a curving crank which had a square hole located at the center of its central cylindrical portion. The central cylindrical portion of the crank was then secured to the end of the axle pivot pin with a bolt having a large round head on it.

Bessler, mostly likely, would have squared off the ends of his axle pivot pins and then drilled out and tapped the threads into the resulting holes, before hammering the pivot pins into the holes he had bored into the ends of the wheel's 6 foot long wooden axle for them. But, when he did finally insert the axle pivot pins, he would have done so very carefully so that the two square cross sectional shapes of the pivot pin ends were perfectly matched up with each other. In other words, if one had imaginary x-ray vision and could look through the entire length of the axle from the squared off end of the front axle's pivot pin all the way to the squared off end of the back axle's pivot pin, then he would notice that the two square shaped cross sectional areas of the ends of the two pivot pins perfectly overlapped each other and their sides and corners aligned so that they appeared to be a single square. Let's further imagine that each of the squared off axle pivot pins was inserted into its end of the axle so that the four corners of its square shaped cross section would point to the clock dial times of 3:00, 6:00, 9:00, and 12:00 if one was to also imagine that the cutoff circular end of the axle containing that pivot pin was a round clock dial.

Next, because of the square holes in the central cylindrical portions of the semicircular curved cranks, it was possible to attach a crank to the squared off end of an axle pivot pin with one of four possible orientations with respect to the pivot pin and to the cutoff circular end of the wooden axle in which half of the length of that pivot pin was embedded and could not, because of friction, rotate around in. Again, if the hours of a clock dial are used for position references, then one could attach the central cylindrical portion of a crank to the squared off end of the *front* axle pivot pin so that the little circular hole at the end of the crank's extended curving arm was located at either the 3:00, 6:00, 9:00 or 12:00 position of the cutoff circular end of the axle. One could also do the same with the semicircular curved crank attached to the squared off end of the axle pivot pin embedded in the *back* end of the axle and, again using imaginary x-ray vision to look through the entire length of the axle and then see the back end of the axle and its embedded pivot pin and attached crank from a viewpoint located *inside* of the axle, one could attach the central cylindrical portion of that back crank to the back axle pivot pin so that the little circular hole at the end of the crank's curving arm was located at either the 3:00, 6:00, 9:00 or 12:00 position of the cutoff circular face of the back end of the axle.

In the drawing at the bottom of Figure 12(b) I have labeled, with very small hour numbers just outside of the front and back flat circular ends of the Merseburg wheel's axle, four clock time locations for the center axis of the little circular hole at the end of a crank's semicircular curving arm. (I did not so label the front and back cutoff circular axle ends of the axle in the rest of the diagram because it would have made the diagram too crowded. The reader will have to imagine the hour numbers surrounding those cutoff circular axle ends as being identical to the two ends that are labeled.) It should be noted that the assignment of the clock time locations for the center axes of these little circular holes, whether discussing a one-directional or bidirectional wheel's axle as in the cases of the Merseburg and Kassel wheels, is always done using only a *front* axial view of the wheel's axle.

With four possible ways to attach *each* of the two cranks to the squared off ends of the two axle pivot pins at the ends of a wheel's axle, it then becomes possible to attach the *two* cranks so that imaginary radial lines extending from the center axes of their central cylindrical portions to the center axes of the little circular holes at the ends of their curving arms can *appear* to form only three possible angles with respect to each other as viewed, again using imaginary x-ray vision, along the length of the axle from the front of the axle. (The two imaginary radial lines that form these angles would actually be located in two different planes and separated by a distance greater than 6 feet which is the length of the Merseburg and Kassel wheel's 6 feet long axles plus the additional distances from the cutoff circular ends of their axles to the planes containing the semicircular curved arms of their two axle pivot pin cranks.)

Firstly, the two cranks could be attached to their respective pivot pins so that the center axes of the two little circular holes at the ends of their curving arms appeared to be at the *same* clock time locations on their respective cutoff circular axle ends as viewed through the length of the axle from its front to back ends and, in that case, the imaginary radial lines connecting the center axes of the two cranks' central circular portions with the center axes of their little circular holes at the ends of their curving arms would appear to be superimposed and, thus, 0° apart. This is what would happen if the center axes of the two little circular holes of the two cranks were *both* located so as to appear to be located at either the 3:00, 6:00, 9:00, or 12:00 locations of their respective cutoff circular end of the axle.

Or, secondly, the two cranks could be attached to their respective pivot pins so that the center axes of the two little circular holes at the ends of their curving arms appeared to be located at diametrically opposite clock time locations in the superimposed view of their two cutoff circular axle ends of the axle as seen through the length of the axle from its front to back ends and, in this case, the imaginary radial lines connecting the center axes of the two cranks' central circular portions with the center axes of their little circular holes at the ends of their curving arms would appear to form a straight line and, thus, be 180° apart (but, again, remember that this is only an optical illusion because both imaginary radial lines would actually be over 6 feet apart). This situation would exist if the center axes of the two little circular holes at the ends of their curving arms appeared to be located *either* at the 3:00 and 9:00 locations or at the 6:00 and 12:00 locations of the superimposed view of the two cutoff circular ends of the axle.

Since we are dealing with a superimposed view of the front and back cutoff circular ends of the axle, it does not matter if the center axis of the front crank's little circular hole appears at the 6:00 location of the cutoff circular end of the front end of the axle and the center axis of the back crank's little circular hole appears at the 12:00 location of the cutoff circular end of the back end of the axle or if the center axis of the front crank's little hole appears to be at the 12:00 location of the cutoff circular end of the front end of the axle while the center axis of the back crank's little hole appears to be at the 6:00 location of the cutoff circular end of the back end of the axle. Similarly, it does not matter if the center axis of the front crank's little circular hole appears to be at the 3:00 location of the cutoff circular end of the front end of the axle while the center axis of the back crank's little circular hole appears to be at the 9:00 location of the cutoff circular end of the back end of the axle or if the center axis of the front crank's little hole appears to be at the 9:00 location of the cutoff circular end of the front end of the axle while the center axis of the back crank's little hole appears to be at the 3:00 location of the cutoff circular end of the back end of the axle.

Or, thirdly and finally, the two cranks attached to the squared off ends of their two pivot pins at the front and back cutoff ends of the axle could be attached so that the center axes of the two little circular holes at the ends of their curving arms appeared to be *adjacent* to each other and, thus, the two imaginary radial lines from the center axes of their central cylindrical portions to the center axes of the little circular holes at the ends of their curving arms would appear to form an angle of 90° with respect to each other. In this situation the center axes of the two little circular holes would appear to be located at either the 3:00 and 6:00 locations, 6:00 and 9:00 locations, 9:00 and 12:00 locations, or 12:00 and 3:00 locations in a superimposed view of the cutoff circular ends of the front and back ends of the axle as viewed along the length of the axle using imaginary x-ray vision.

Again, in this case, it does not matter if, for example, the center axis of the front crank's little hole appears to be at the 3:00 location of the cutoff circular end of the front end of the axle while the center axis of the back crank's little hole appears to be at the 12:00 location of the cutoff circular end of the back end of the axle or if the center axis of the front crank's little circular hole appears to be at the 12:00 location of the cutoff circular end of the front end of the axle while the center axis of the back crank's little circular hole appears to be at the 3:00 location of the cutoff circular end of the back end of the axle. This allowed exchange of locations also applies to the other three superimposed views of the pairs of clock time locations for the two center axes of the two little holes of the curving cranks which would make them seem to form a 90° angle using imaginary radial lines extending to them from the center axes of their central circular portions.

Now that the reader understands the three options Bessler had for attaching the two semicircular curved cranks to the squared off ends of the two axle pivot pins of his 12 feet diameter, bidirectional Merseburg wheel, it's time to analyze what the mechanical consequences of each type of attachment option would be. As the reader will soon learn, Bessler only used *one* of them because of its unique benefit for the performance of his wheels.

Let's start by analyzing what would happen if the *first* optional method of axle pivot pin crank attachment described above was used by Bessler.

In this first method the lower ends of the drive rods of the two pendulums of the Merseburg wheel (whose *upper* ends, when viewed using imaginary x-ray vision from the front side of the wheel in order to see through the drum to the back pendulum, always appeared attached to their superimposed cross beams on opposite sides of their movable pivot rods) were both attached to the little circular holes at the ends of the extended curving arms of the cranks that appeared to have the *same* clock dial locations on the superimposed front and back cutoff circular ends of the axle. As a result, when the axle rotated in *either* of its two possible directions, the forces applied to the cross beam rods of both pendulums by their attached drive rods would cause the two pendulums' spherical brass bob weights to always swing in *opposite* directions at the same time. This would then cause the superimposed individual centers of gravity of the two compounds pendulums, which were each actually swinging from one side of a vertical line passing through the center axis of its own movable pivot pin, to *appear* to continuously move horizontally away from and then back toward each other, meeting at the superimposed center lines of the wheel's two vertical support planks, over and over again while those centers of gravity continuously both moved vertically upward as they appeared to separate and then downward again as they appeared draw closer together.

The *combined* center of gravity of the two compound pendulums (actually located at an imaginary point inside of the middle lattice of the bidirectional wheel's enlarged drum and, when the pendulum bob weights had both swung as far away horizontally from the superimposed center lines of the vertical support planks as they could, at an elevation above the center axis of the axle indicated by the point labeled "C" in the middle inset image of the front of the Merseburg wheel's drum in Figure 12(b) where the front drive rod and front vertical bob weight support rod appear to cross each other to form the letter "X") would then undergo no *net* horizontal motion, but would simply rise and fall through a short vertical distance directly above the center axis of the wheel's axle.

During *alternating* half rotations of the axle, the wheel would have to expend energy to lift the combined center of gravity of the two compound pendulums while during the remaining alternating half rotations of the axle, the pendulums would return their increased gravitational potential energy to the wheel. This means that, when starting up from a standstill with the two pendulums attached by their curving cranks to the squared off ends of the axle pivot pins using this first method, the bidirectional wheel would, within a single axle rotation before it had accumulated significant rotational kinetic energy, be required to lift the pendulums' combined center of gravity which had a mass of about 150 pounds! And, surprisingly, this lift would have been required *regardless* of the direction in which the Merseburg wheel's drum was started rotating. Thus, we see that this first method of attaching the pendulums to the Merseburg wheel's axle would have almost immediately caused the 12 feet diameter, bidirectional wheel to stall and stop moving and, possibly, even sheared off the squared off ends of its pivot pins. Such an event was certainly not something Bessler wanted a potential buyer's agent or a witness during an official test to see happen!

Let us now move on to analyzing what would happen if the *second* optional method of axle pivot pin crank attachment described above was used by Bessler.

In the second attachment method, the bottom ends of the drive rods of the two compound pendulums are attached to the little holes at the ends of the curving arms of their respective cranks which appear to be located at opposite clock times on the superimposed view of the two cutoff circular ends of the axle when, by again using imaginary x-ray vision, they are viewed from the front of the axle through its length to its back. As a result of this arrangement of the cranks, as the 12 feet diameter, bidirectional wheel's axle rotated in *either* of its two possible directions, the forces applied to the cross beam rods of both compound pendulums by their two drive rods would cause their spherical brass bob weights to always swing in the *same* directions at the same time. Thus, the motions of the two pendulums become synchronized as do their centers of gravity.

In this scenario, the *combined* center of gravity of the two pendulums simply swings repeatedly to the right and then back to the left side of its equilibrium location or *punctum quietus* which is again located inside of the drum's middle lattice and above the hidden section of the axle inside of the enlarged bidirectional drum. With each swing toward one side of the axle, the combined center of gravity of the two pendulums moves horizontally 6 inches away from its equilibrium location nearest the axle's center axis and, as it does so, it must rise through a certain vertical distance. This second method of attaching the two axle cranks to their respective squared off pivot pins' ends suffers from the same problem that the previously discussed first method had. It requires the wheel, upon start up and having little accumulated rotational kinetic energy available, to lift a combined center of gravity with a mass of about 150 pounds. As with the previous scenario, this arrangement would quickly stall the Merseburg wheel and could, possibly, shear off the squared off ends of the two axle pivot pins.

With all of these troublesome problems, one might conclude that attaching the two compound pendulums to the Merseburg wheel's axle would have been most undesirable and that means that their depiction as shown in Chapter 1's Figure 1(a) must have been some fiction Bessler included in the DT illustration to confuse and frustrate the reverse engineers trying to steal the secrets of this wheel based on hints in that drawing. This, on the surface, seems to make some sense because, other than the two wheel speed "modifying" pendulums illustrated in his drawings of the Merseburg and Kassel wheels and mentioned by Bessler in his writings, there is no mention of their use being reported by anyone else. Some may even assume that the pendulums were included in the figures and mentioned by Bessler because he only *thought* they might work and intended to employ them some time after he published his books about his wheels. In other words, he did not realize as he introduced these fictional pendulums that they would make his wheels unworkable. This, however, would be a false assumption to make.

Fortunately, the *third* axle crank configuration described earlier above *does* allow two compound pendulums to be attached to a bidirectional wheel's axle without impairing the wheel's start up in either of its two possible directions of rotation and so we can have faith in the validity of the presence of the pendulums shown in both the GB and DT illustrations of the Merseburg wheel and of their use being mentioning by Bessler in his writings.

In this remaining third attachment method, which my research indicates was actually used by Bessler on his Merseburg and Kassel wheels, the bottom ends of the drive rods of the two compound pendulums were attached to the little holes at the ends of the curving arms of their respective cranks after the cranks' central cylindrical portions were attached to the squared off ends of their respective axle end pivot pin as shown in the diagram at the bottom of Figure 12(b). If one was to use imaginary x-ray vision to look through the entire length of the 6 feet long axle of either wheel so as to have a view of the two superimposed cutoff circular ends of the axle, then the small holes at the ends of the two cranks' curving arms would appear to be at clock time locations that were always three hours apart with the clock time location of the front crank's small hole always being three hours *behind* that of the back crank's small hole. If we imagine a radial line drawn from the center axis of each crank's central cylindrical portion to the center axis of the little hole at the end of its curving crank arm, then the superimposed view of the two radial lines would form an angle of 90°.

This particular method of attaching the two cranks to the squared off ends of the axle's two projecting pivot pins allowed the two drive rods of a *clockwise* rotating axle, as viewed from the front end of the axle, to apply forces to the two compound pendulums' cross beam rods so that the *individual* centers of gravity of the two pendulums moved horizontally in *opposite* directions for a *quarter* of an axle rotation, then moved in the *same* direction for the next quarter of an axle rotation, then moved in *opposite* directions again for the next quarter of an axle rotation, and then moved in the *same* directions again for the last quarter of an axle rotation as the axle completed a single full rotation (for clockwise axle rotation, each 360° full axle rotation can be considered to finish and then start again when the little hole at the end of the curving arm of the front crank appears to be at the 12:00 location of the cutoff circular end of the front end of the axle). This alternating pattern of horizontal motions of the two individual centers of gravity of the two compound pendulums was then repeated during each complete axle rotation over and over again. As this happened, the center of gravity of the back compound pendulum *always* had a location, as seen in an imaginary x-ray vision axial view from the drum's front side, that was a quarter of a complete pendulum oscillation *behind* the location of the center of gravity of the front compound pendulum.

In the bottom portion of Figure 12(b) I have illustrated this third method of attaching the two lower ends of pendulums' drive rods to the curving crank arms attached to the squared off ends of the axle pivot pins of the Merseburg wheel and it was also used with the Kassel wheel. The bottom row shows a complete rotation of the *front* cutoff circular end of the axle as a series of 90° segments of clockwise rotation of the little circular hole at the end of the curving arm of the crank attached to the squared off end of the axle's *front* pivot pin. Above each of these is a small representation of the wheel's *front* compound pendulum which shows the orientation it had for the particular position of the front crank to which its drive rod was connected. The top row shows the corresponding complete rotation of the *back* cutoff circular end of the axle as a series of 90° segments of clockwise rotation of the little circular hole at the end of the extended curving arm of the crank attached to the squared off end of the axle's *back* pivot pin as viewed through the length of the axle from its front end using our imaginary x-ray vision (the pendulums in the figure were grossly reduced in size and also moved above their actual locations relative the cutoff circular ends of the axle in order to more clearly show how their orientations related to those of the little holes at the ends of the curving arms of the front and back axle pivot pin cranks).

In this bottom portion of Figure 12(b), the "follow the leader" motion of a bidirectional wheel's two pendulums (with the front pendulum being the "leader") becomes more obvious. What is less obvious is that the *vertical* elevation of the *composite* center of gravity of the front and back pendulums above the floor of the wheel's room does not significantly change throughout a complete axle rotation. (Note that in this bottom portion of Figure 12(b) I have not tried to indicate the location of the center of gravity of *each* compound pendulum in order to keep the drawing from becoming overly cluttered, but it would be roughly located on a pendulum's bob weight support rod just below the point where the two diagonal brace rods are attached to that rod.)

To give the reader a better understanding of how this third method of attaching the bidirectional Merseburg wheel's two cranks to the squared of ends of their respective axle pivot pins worked in practice, let me now describe the resulting motions of the individual centers of gravity of the wheel's two compound pendulums in detail for a complete axle rotation as viewed from the front end of the axle. What I'm about to describe is illustrated by the five front and back vertically arranged *pairs* of pendulums and cranks shown at the bottom of Figure 12(b) and which are labeled "D", "E", "F", "G", and "H". I will be starting with the leftmost pair which is labeled "D".

At "D", when the center of gravity of the front compound pendulum had reached its maximum displacement of 6 inches to the left of its *punctum quietus* or equilibrium point located directly under the center axis of its *pendulum's* movable pivot rod and stopped, the back pendulum's center of gravity was moving to the left and at its maximum horizontal velocity, but was only located at its equilibrium point directly under the center axis of its pendulum's movable pivot rod at that time. Then, at "E" as the front pendulum's center of gravity reversed direction, moved to the right, reached its equilibrium point directly under the center axis of its pendulum's movable pivot rod, and was moving with its maximum horizontal velocity to the right, the back pendulum's center of gravity was then located at its maximum displacement of 6 inches to the left of its equilibrium point located under the center axis of its movable pivot rod, had stopped moving, and was about to change its direction and move to the right.

At "F" as the front pendulum's center of gravity continued to move to the right, it eventually reached its maximum displacement of 6 inches to the right of its equilibrium point located directly under the center axis of its pendulum's movable pivot rod and stopped at which time the center of gravity of the back pendulum was moving to the right at its maximum horizontal velocity and was then located at its equilibrium point directly under the center axis of its pendulum's movable pivot rod.

Then at "G", when the front pendulum's center of gravity had again reversed direction and moved left until it was again located at the equilibrium point directly under the center axis of its pendulum's movable pivot rod and was moving with its maximum horizontal velocity to the left, the back pendulum's center of gravity was located at its maximum horizontal displacement of 6 inches to the right of its equilibrium point directly under the center axis of its pendulum's movable pivot rod, had stopped, and was about to begin moving to the left again.

Finally, at "H" as the fourth and final quarter or 90° of a single 360° rotation of the axle took place, the front pendulum's center of gravity would continue to move to the left until it again reached its maximum horizontal displacement of 6 inches to the left of its equilibrium point directly under the center axis of its pendulum's movable pivot pin and stopped. At that instant, the center of gravity of the back pendulum would be located directly under the center axis of its pendulum's movable pivot rod and would be moving at its maximum horizontal velocity to the left.

At this time, the axle's complete rotation was finished and the four quarters of unique horizontal motions of the individual centers of gravity of the bidirectional Merseburg wheel's two compound pendulums would then be repeated during the next full rotation of the wheel's axle and every rotation thereafter as long as the two pendulums both remained attached to the wheel's drum by their respective semicircular curved axle end cranks.

Somewhat amazingly, this pattern of pendulum motions was maintained regardless of which direction a bidirectional wheel's drum was given a push to start it rotating. For the clockwise axle rotation I show in Figure 12(b), the front pendulum is the leader which is followed by the back pendulum. However, if the bidirectional drum was given a push to start it rotating counterclockwise as viewed from the front side of the drum, then the back pendulum would be the leader that was followed by the front pendulum (for counterclockwise axle rotation, each 360° full axle rotation, consisting of four 90° quarters of rotation, can be considered to finish and then start again when the center axis of the little hole at the end of the curving arm of the *back* crank appears to be located at the 12:00 location of the cutoff circular end of the *back* portion of the axle).

The interesting thing about this "follow the leader" motion of the two compound pendulums' individual centers of gravity was that their *composite* center of gravity, located inside of the inner middle lattice of the Merseburg wheel's drum, only moved about 3 inches to the left and then 3 inches to the right of that composite center of gravity's equilibrium point which was located on a vertical line passing through the center axes of the two pendulum's movable pivot rods and the wheel axle's center axis (for the Kassel wheel, however, the horizontal displacement would have been larger due to the larger arcs that its two compound pendulums swung through). But, because one pendulum's center of gravity was always falling vertically as the other pendulum's was rising vertically, there was virtually no change in the elevation of the *composite* center of gravity of the two pendulums relative to the Earth's surface or the floor of the wheel's room. As indicated in the center portion of Figure 12(b), that composite center of gravity would have remained in the plane of the drum's middle lattice and have been located slightly below the elevation indicated by the letter "C" in the inserted illustration from DT showing the front side of the Merseburg wheel's drum which would have placed it at a location slightly more than about 2 feet below the center axes of the two pendulums' movable pivot rods.

Thus, since the 12 feet diameter, bidirectional Merseburg wheel used this particular crank to axle end pivot pin configuration, it could easily start up with the two compound pendulums attached to its axle's end cranks because the wheel did not have to vertically lift the two pendulums' 150 pound composite center of gravity during its first quarter of complete axle rotation and before the wheel had accumulated sufficient enough rotational kinetic energy to do so.

However, both the wheel's rate of rotational acceleration and its maximum terminal rotation rate would definitely be significantly reduced as its axle, during the first 90° quarter of *every* 360° of full rotation, made the two pendulums swing in *opposite* directions from each other as their composite center of gravity remained stationary on one side of an imaginary vertical line passing through the center axis of the axle. That was then followed by the next and second 90° quarter of axle rotation during which time the two pendulums were made to swing in the *same* direction by the axle as it *horizontally* accelerated the 150 pound composite center of gravity of the two pendulums through a distance of 6 inches so that it passed through its equilibrium point located inside the plane of the drum's hidden middle lattice to the other side of the imaginary vertical line through the axle's center axis and then just as rapidly was made by the axle to decelerate to a standstill again as the end of the second quarter of axle rotation drew near. Then, during the next and third 90° quarter of axle rotation, the two pendulums would again be made by the axle to swing in *opposite* directions as their composite center of gravity remained stationary on the opposite side of the imaginary line passing through the center axis of the rotating axle. Finally, during the next and last fourth 90° quarter of the complete 360° rotation of the axle, the axle made the two pendulums swing in the *same* direction again as the axle horizontally accelerated their composite center of gravity *back* through the same 6 inches to the original side of the imaginary vertical line passing through the center axis of the axle that the composite center of gravity was momentarily stationary on during the *first* 90° quarter of each full 360° of axle rotation and, once there, again rapidly decelerated the composite center of gravity of the two pendulums to a standstill as the last and fourth 90° quarter of axle rotation was completed. This same cycle of motions of the two pendulums' composite center of gravity would then be repeated over and over again during every successive full 360° of axle rotation.

It was this periodic, every other 90° quarter of axle rotation *inertial* drag acting on a wheel's axle due to its effort to accelerate and then decelerate the composite center of gravity of the two pendulums attached to the axle's end cranks in either a one-directional or bidirectional wheel that Bessler relied upon to reduce the wheel's free running maximum terminal rotation rate and thereby reduce the total amount of wear to its drum's various internal components that would have occurred during any prolonged period of wheel rotation as compared to the amount of wear that would take place during the same period of wheel rotation at a higher rotation rate because the two pendulums had not been attached to the wheel's axle. As with all of Bessler's inventions, this method of inertial braking was simple, efficient, yet amazingly novel.

Most importantly, it completely eliminated the need to have any type of mechanical brake shoes being pressed against a metal collar wrapped around and attached to a wheel's rotating axle which, eventually, would have failed as their constant rubbing contact with the metal collar wore away the contacting surface material of the brake shoes. The inventor's inertial braking system using dual pendulums also eliminated the need for any sort of steel bolts located in the axle's two wooden vertical support planks that would have had their ends tightened against the axle's turning steel end pivot pins and would have, eventually, cut grooves into those axle pivot pins and thereby weakened them. Also, unlike mechanical brakes that would apply their maximum braking drag to the axle at start up and would prevent such a start up, the inertial braking system's drag would increase slowly as the axle rotation rate increased and would, thus, not interfere with start up when the wheel's rotational kinetic energy was at a minimum.

Since this chapter also deals with the Kassel wheel, I need to give some details of the pendulums it used, an illustration of which the reader can find in the right hand side of Chapter 1's Figure 2(a).

Although this DT illustration of the 12 feet diameter, bidirectional Kassel wheel only shows us a view of the wheel's front compound pendulum, we can safely assume that there was an identical one attached to its back axle end's pivot pin crank because the left hand side of Chapter 1's Figure 2(b) shows that there was a back axle end's pivot pin crank and movable pivot rod support brackets attached to the wheel's back vertical support plank and the wall behind the wheel (note again that, as in the case of the Merseburg wheel, Bessler placed the back of the Kassel wheel near a wall which was intended to discourage the average visitor to Weissenstein Castle from having a good view of the backside of the drum where the eight pinned down flaps of cloth were located that covered the inspection and servicing holes that had been cut into the sheets of dyed linen covering that side of the drum). So, why can we not see the back compound pendulum of the Kassel wheel in the front axial view of its drum shown on the right side of Figure 2(a)?

One reason would be because Bessler used the *second* method described above to attach the crank to the projecting squared off end of the pivot pin at the *back* cutoff end of the axle so that the center axis of the little hole at the end of its curved arm, as seen in a superimposed x-ray view of the two cutoff circular ends of the axle, was located *opposite* to the clock time position of the center axis of the little hole at the end of the *front* crank's curved arm which then resulted in the back pendulum's foot ball shaped lead bob weight and its vertical support piece, its cross beam piece including its two diagonal support rods, and its two cross beam end spherical weights being in alignment with the corresponding pieces of the front pendulum as the back pendulum's foot ball shaped lead bob weight swung from one side of the wheel's 8 inch diameter axle to the other. Thus, the back pendulum's foot ball shaped bob weight and its cross beam's two spherical weights were always hidden behind those of the front pendulum and would not be seen in a front axial view of the Kassel wheel's drum. But, as we saw earlier, this second method of attaching the two pendulums to a wheel's axle would have made it impossible for the wheel to self start or, if bidirectional as the Kassel wheel was, be started with a gently push in either of its two possible directions. Therefore, this was not why we only see one of the compound pendulums in the right hand side of Figure 2(a).

Another reason and the actual one why we do not see any part of the back compound pendulum in the right side of Figure 2(a) is because, as we saw above, Bessler had to attach the two pendulums to a wheel's axle so that they would operate in the "follow the leader" fashion as the axle turned and its attached pendulums swung in various directions relative to each other during each of a full 360° axle rotation's four quarters of rotation. By studying the diagram at the bottom of Figure 12(b) which depicts the motion of the two pendulums during a complete clockwise rotation of the Merseburg wheel's axle, it will be noted that whenever the bob weight of the front pendulum is at the extreme of its swing to either the right of left side of the axle, the back pendulum's bob weight support rod is perfectly vertically oriented. Such would also be the case for the Kassel wheel's back pendulum's wooden bob weight support piece and, in that orientation, the back pendulum's foot ball shaped bob weight and the two spherical weights at the end of its iron cross beam would all be hidden behind the drum and out of sight. Since it was the third pendulum to axle pivot pin attachment method that Bessler had to use to allow the Kassel wheel to start up with the two pendulums attached to its axle, this is exactly what we are seeing depicted in the right hand side of Figure 2(a). Bessler then leaves it to the reader of DT to mistakenly think that the back compound pendulum is either missing or in alignment with the front pendulum using the second of the pendulum to axle end pivot pin attachment methods.

In the right hand side drawing of the front side of the Kassel wheel's drum in Figure 2(a), the reader will notice that Bessler again used several number 8's near the compound pendulum, but this time he only placed a single number 8 *on* the wooden vertical bob weight support piece and tilts it so that it almost looks like the heads of two closely spaced bolts that might be attaching the top end of the vertical bob weight support piece to the cross beam piece (I write "piece" here rather than "rod" because the structures on the Kassel wheel's pendulums, other than the iron diagonal rods, do not have circular cross sections). There are also three other 8's in the illustration near each of the three weights of the Kassel wheel's compound pendulum. I think Bessler may have used these four 8's in an attempt to obscure their use in the earlier Merseburg wheel illustration he included in DT. In any event, there were four number 8's used in the axial view of the front side of the Kassel wheel's drum on the right side of Figure 2(a) and this does have some significance in terms of the natural oscillation frequency of each of its more massive compound pendulums.

As one might suspect, being twice as powerful as the 12 feet diameter

Merseburg wheel due to its weighted levers having twice the mass of those used in that bidirectional wheel, the mass of each of the Kassel wheel's compound pendulums would have been at least double that of a compound pendulum used in the Merseburg wheel. However, after making many attempts to accurately determine the total mass of all of the component parts of a Kassel wheel's pendulum (minus the mass of its drive rod), my best estimate is that its two iron cannon balls at the ends of the cross beam piece weighed about 30 pounds each, the iron cross beam piece itself was about 60 pounds, the iron movable pivot rod was about 70 pounds, the wooden vertical bob weight support piece was about 30 pounds, the two iron diagonal brace rods were about 60 pounds each, and, finally, the lead foot ball shaped bob weight was about 60 pounds. Thus, the masses of all of these parts actually totaled 400 pounds which is about 5.333 times as massive as one of the Merseburg wheel pendulums!

Surprisingly, the center of gravity of this massive pendulum was also located about 2 feet under the center axis of its movable pivot piece (which had a square shaped cross section to minimize the flexing it did when the weight of the rest of the pendulum was placed upon it) and its natural oscillation frequency was also about 16 oscillations per minute just as was the case for the Merseburg wheel. Since Bessler placed four number 8's on the front side of the bidirectional Kassel wheel's 12 feet diameter drum shown in the right side of Figure 2(a), their total of 32 can be used to obtain the natural oscillation frequency of *each* of the two compound pendulums by simply dividing 32 by 2 to obtain 16 oscillations per minute. Perhaps we are supposed to perform the division by 2 because there are two pendulums attached to the wheel's axle even though one only sees a single pendulum in the DT illustration. Also, in the case of the right side view of the front side of the Kassel wheel's drum provided in Figure 2(b), the location of the front compound pendulum's center of gravity was not located at the elevation of point where the pendulum's diagonal brace piece appears to cross in front of the drive rod and form the letter "X" as was noted for the Merseburg wheel in the middle inset illustration of Figure 12(b), but, rather, about a half foot above that point.

The Merseburg wheel's axle, when turning with its end pendulums attached, was forced to horizontally accelerate and decelerate a *composite* pendulum center of gravity for *two* compound pendulums with a total mass of 150 pounds from one side of an imaginary vertical line passed through the center axis of its 6 inch diameter wooden axle twice per complete 360° axle rotation while the axle of the Kassel wheel, when turning with its axle's end pendulums attached, was forced to horizontally accelerate and decelerate a composite pendulum center of gravity for two compound pendulums with a total mass of 800 pounds from one side of an imaginary vertical line passed through the center axis of its 8 inch diameter wooden axle twice per complete 360° axle rotation! That was a big increase in the amount of mass used by the inventor's compound pendulum inertial braking system in the Kassel wheel.

If we denote the power output at any drum rotation rate of the Merseburg wheel by 1 and that of the Kassel wheel at the same rotation rate by 2, then we can determine the ratio of the composite pendulum center of gravity mass to the power output for both wheels and compare them. We then write, for the Merseburg wheel, 150 pounds / 1 = 150 pounds. For the Kassel wheel we write 800 pounds / 2 = 400 pounds. This shows us that the Kassel wheel had to horizontally accelerate 400 pounds / 150 pounds = 2.666 times as much mass for its power level than did the Merseburg wheel. As a result of this we can probably safely assume that the reduction that the two compound pendulums made in the Kassel wheel's maximum free running terminal rotation rate of 26 rotations per minute was not just the average of that bidirectional wheel's pendulum oscillation rate of 16 oscillations per minute and its free running terminal rotation rate of 26 rotations per minute, which would be 21 rotations per minute, but was probably much closer to a value of 16 rotations per minute. Perhaps it was as low as 18 rotations per minute.

While the cross sectional area of the Merseburg wheel's vertical support planks was 3 inches thick x 12 inches wide = 36 square inches, the cross sectional area of each of the Kassel wheel's vertical support planks had to be doubled to support that wheel's doubled mass of about 1100 pounds and was 6 inches thick x 12 inches wide or 72 square inches and, therefore, might be more properly referred to as a vertical support *beam*. The axle of the Kassel wheel was 8 inches in diameter compared to the 6 inches diameter of the Merseburg wheel's axle. As was previously mentioned, the cutoff circular ends of the Merseburg wheel's axle were each fitted with a steel pivot pin 0.75 inch in diameter x 12 inches long that weighed 1.5

pounds and was inserted to a depth of 6 inches into the end of a 6 inch diameter axle, the cutoff circular ends of the Kassel wheel's axle were each fitted with a pivot pin 1.125 inches in diameter x 18 inches long that weighed 4 pounds and was inserted to a depth of 9 inches into the end of an 8 inch diameter axle.

For the Merseburg wheel, each of its vertical support planks was slightly less than 168 inches in length = slightly less than 14 feet in length (this was so it could be raised into a vertical orientation between a floor to ceiling beam distance that was exactly 168 inches = 14 feet). It is fairly easy to estimate the length of the beams used to support the Kassel wheel and its attached compound pendulums because the ends of each beam were secured to a specially shaped rectangular metal plate, which I referred to as a "bridge piece" in Chapter 1, that had its four corners bent down to give it the appearance of a small table. Because of these bridge pieces, the two ends of each of the Kassel wheel's wooden vertical support beams were probably separated from the surface of the wheel room's floor or an overhead beam by a distance of at least 4 inches. That would then mean that length of each beam was 160 inches = 13.333 feet.

To install each vertical support beam, a solid iron bridge piece was first securely attached to the lower end of the beam and it was then raised into a vertical orientation on the correct spot of the floor for it so that its bottom end was elevated above the surface of the floor by 4 inches. The second bridge piece was then slipped horizontally between the top cutoff rectangular end of the vertical support beam and the bottom surface of a ceiling beam. The final step would have been to secure the second bridge piece to the ceiling beam and then to the wheel's vertical support beam. It's even possible that there was a 6 inch wide x 12 inch long rectangular hole formed in the center of the beam's top bridge piece that was used to slip the latter over the top rectangular shaped end of the wheel's vertical support beam and then, after it was raised into a vertical orientation, the bridge piece was simply raised up to the end of the wheel's vertical support beam and the corners of the top bridge piece were then attached to the ceiling beam with large wood screws. There would have been no need for fasteners to be used between the bridge piece and the top end of the wheel's vertical support beam whose remaining two inches of length would have been prevented from horizontal movement by the inner edges of the rectangular hole in the 2 inch thick bridge piece. If this was the method for installing the Kassel wheel's two vertical support beams, then each beam would have been 162 inches = 13.5 feet in length.

One of the problems with the more massive Kassel wheel's two pendulums was that, weighing 400 pounds each, they would each have added another 200 pounds to the weight that *each* of this wheel's two oak vertical support beams had to support (400 pounds is not added to the load on each beam because half of the weight of each pendulum was supported by structures *other* than the vertical support beam, so that only 200 pounds of extra pendulum weight was applied to each of the axle's vertical support beams). Thus, *each* of the Kassel wheel's two vertical support beams, when the two compound pendulums were attached to the wheel's axle, had to be able to support half of the weight of the wheel and its axle, about 550 pounds, plus half of the weight of a pendulum, about 200 pounds, plus its *own* full weight which would have been about 288 pounds. Thus, each of the Kassel wheel's vertical support beams needed to be able to support a mass of 1038 pounds which is certainly easy for a beam with a cross sectional area of 72 square inches to do.

And, finally, we can end our discussion of the dual pendulums used in the Merseburg and Kassel wheels and Figure 12(b) by noting that, when attached via drive rods to the ends of a wheel axle's curved arm end cranks that were properly oriented with respect to each other, these compound pendulums did exactly what Bessler said they did. They "modified" or reduced the free running, maximum terminal rotational rate of his bidirectional wheels, regardless of which direction their enlarged drums were made to turn via an initial push start. They did this by applying intermittent inertial drag to the axles which exerted a braking torque and caused the wheels to slow down, but, unlike a brake that uses friction and ultimately converts rotational kinetic energy into an increase in the thermal energy of the air surrounding the brake, the swinging compound pendulums attached to Bessler's wheels actually dissipated little of the rotational energy that the bidirectional wheels were accumulating from the loss of the mass of their active internal one-directional wheel's weighted levers by using it to overcome the aerodynamic drag of the air their parts moved through. They mainly just put an upper limit upon how much rotational kinetic energy a wheel could accumulate.

However, the semicircular flanged planks, one of which is shown near the Kassel wheel's back axle on the left side of Figure 2(b), would have worked contrary to the manner in which the compound pendulums did. These two special devices, which were not true pendulums, were intended to siphon off as much of the Kassel wheel's rotational kinetic energy as possible just as soon as it was generated and immediately use it to increase the thermal energy and mass of the air near the ends of the axle. Bessler may even have preferred using these flanged planks instead of the 400 pounds compound pendulums for the obvious reason that they could be far easier to handle during demonstrations due to their low weight. I can even imagine that, during the hot summer months when Bessler was demonstrating the Kassel wheel to the curious visitors to Weissenstein Castle, the groups entering the wheel's room would have found the breezes generated by the oscillatory sweeping motions of the flanged planks to have been quite refreshing!

Now we move on to Figure 12(c) which shows some of the finer details of how the two internal one-directional wheels, both front and back, were housed inside of the Merseburg wheel's enlarged 12 inch thick drum. As can be seen from the figure, Bessler used all of his carpentry skills to place the two separate one-directional wheels into a drum that was as thin as possible. Again, the reason was to allay the suspicions of those who might think that his bidirectional wheel was being propelled by a small trained animal located inside of the drum at its 6:00 location.

At first glance, Figure 12(c) may seem a bit confusing, but with a brief explanation, it should become easy to understand. First, the reader must realize exactly what he is seeing in the figure.

To begin, the reader should imagine that he has just entered the room containing the Merseburg wheel and is then looking directly at its front vertical support plank. Perhaps Bessler had removed the front and back pendulums including their two curved arm axle cranks that were attached to the squared off ends of the axle's two pivot pins. The reader must now imaging that he walks around to the right side of this 12 feet diameter, bidirectional wheel's drum and then, after turning to his left side, stands and faces the outermost portion of the drum's rim wall so that he then sees a profile or side view of the 12 inch thick drum. If the reader quickly reviews the derogatory illustration circulated by Borlach that I provided in

Figure 1(b) at the back of Chapter 1, he will realize that he is now standing near that large middle window of Borlach's drawing that is partially hidden by the Merseburg wheel's drum and that he would be behind the railing.

From this vantage point, the reader must, once again, imagine that he has x-ray vision and can view an interior *profile* cross section of the enlarged drum from its bottommost 6:00 to topmost 12:00 positions. In Figure 12(c), however, I only show this interior cross section view from the *top* surface 12:00 position of the bidirectional wheel's *axle* up to the 12:00 position of its attached *drum*. Unfortunately, because the radii of Bessler's drums were a lot larger than their thicknesses, I was not able to show all of the length of the drum's radius in this figure if I also wanted to make the drum's thickness large enough to show its interior structural details with clarity. Thus, as will be noted near the bottom of Figure 12(c), at a distance of only a few inches above the upper surface of the wheel's 6 inch diameter wooden axle, I was forced to omit nearly all three of the lengths of its 12:00 radial frame pieces. However, despite this, I was still able to include most of the important construction details of this bidirectional wheel's drum near its outer rim wall and that information will be of great benefit to anyone attempting to duplicate this wheel since these structural details, basically, will be replicated for the front and back internal one-directional wheels' side by side, back to back pairs of weighted levers at the drum's seven other clock time locations.

At the top of Figure 12(c) we see the outer ends of the two parallel pairs of main arms of the front and back internal one-directional wheels' weighted levers whose shared pivot pin is located at the drum's 12:00 position. The two black rectangles each represent the *middle* one of the three 4 pound cylindrical lead end weights (which includes the mass of each weight's steel attachment bolt) attached between the ends of the parallel pair of main arms the front and back internal one-directional wheel's weighted levers at the 12:00 position of the Merseburg wheel's drum. As was shown in Figure 11(b), each of a weighted lever's three cylindrical lead end weights was 1.78 inches = 1-12.5/16 inches in diameter x 4 inches long. Each of these three cylindrical lead end weights was attached to the ends a weighted lever's parallel pair of wooden main arms using a steel bolt that was 0.1875 inch = 3/16 inch in diameter and had a total slotted head, smooth shank, and treaded end section length of 5 inches.

The middle weight on the left side of Figure 12(c) is held between the ends of the main arms of the bidirectional wheel's front internal one-directional wheel and the cylindrical lead end weight on the right side is held between the ends of the main arms of the wheel's back internal one-directional wheel. Since the wooden 3 arm side piece arms of this wheel's weighted levers were each 0.375 inch = 6/16 inch thick, this means that the two parallel main arms of each weighted lever and the 4 inch long cylindrical lead end weight that they held had a total thickness of 4.75 inches = 4-12/16 inches which the reader can verify by adding up the thicknesses of the two main arms and the length of the cylindrical lead end weight they held for each of the two weighted levers shown in Figure 12(c).

Using steel attachment bolts that had an overall length of 5 inches on each of the cylindrical lead end weights attached to the ends of these lever's main arms then resulted in the bolt's slotted head and the opposite threaded end of its length *each* projecting out 0.125 inch = 2/16 inch beyond the *outside* facing surfaces of the main arms of each weighted lever. That then resulted in there being only a 0.125 inch = 2/16 inch clearance between the surfaces at the ends of an attachment bolt and the *inside* facing surfaces of a weighted lever's parallel pair of radial frame pieces.

If the weighted end of a lever's parallel pair of main arms was to axially move only that short distance so that it's cylindrical lead end weights' attachment bolts made physical contact with the inside facing surface of the nearest radial frame piece, then the bolt heads or threaded ends could have began cutting into the wood of the radial frame piece. This rubbing contact would have made a loud scrapping sound and, over time, might even progress to the point where one of the wooden ends of a main arm also began rubbing against the radial frame piece or even hitting it and coming to a sudden stop which could actually disable or damage the drum's internal mechanics and prevent the wheel from running until the problem was corrected.

To minimize the chance of this happening, it was critically important that the metal sleeve lined pivot pin holes of a weighted lever's two parallel 3 arm side pieces not be able to slide back and forth along the length of their allotted section of the 12 inch long, shared pivot pin whose two ends and middle had been press fitted into three of the drum's parallel radial frame pieces each of which was in one of the drum's three lattices. Since the width of the space between an adjacent parallel pair of radial frame pieces that was available to a weighted lever was 5.25 inches = 5-4/16 inches and the width of a weighted lever, *not* including the protruding ends of its cylindrical end weights' attachment bolts, was 4.75 inches = 4-12/16 inches, that means that the two gaps between the outside facing surfaces of the lever's main arms and the inward facing surfaces of the lever's parallel pair of radial frame pieces was 0.5 inches = 8/16 inch which only allowed a distance of 0.25 inch = 4/16 inch *per gap*. This is only double the width of the gap between the end of the weighted lever's cylindrical end weight attachment bolt's ends and their nearest radial frame piece. For each of the bidirectional drum's eight pairs of front and back, side by side weighted levers, all four of these gaps would have to be filled in to prevent the unwanted sliding of the two levers along their shared pivot pin during drum rotation.

These four gaps would have been filled in by placing *two* steel washers, each of which was 0.125 inch = 2/16 inch thick with a center hole that was 0.25 inch = 4/16 inch in diameter and an outside diameter that was 1 inch in diameter into *each* of the four gaps. Each pair of washers would have been positioned on their correct section of 0.25 inch = 4/16 inch diameter x 12 inch long common pivot pin (which, for reasons stated above, I could not show in Figure 12(c)) that was shared by each pair of front and back, side by side weighted levers in the bidirectional wheel's 12 inches wide drum. When this was done during the initial installation of the drum's sixteen unweighted levers, the weighted levers of each active internal one-directional wheel would then later be free to rotate, within certain limits, about their respective half of their shared pivot pin during drum rotation (actually through an arc of no more than 62.5°) and no part of a weighted lever, including the protruding ends of the bolts used to attach its three cylindrical end weights, would come into rubbing contact with its nearest parallel pair of radial frame pieces. A total of 64 washers were needed for this and, of course, their contacting surfaces would have been routinely lightly lubricated with olive oil.

That Bessler could achieve such a high degree of precision in the construction of his sixteen weighed lever containing, bidirectional Merseburg wheel is a true tribute to his skills as both a carpenter and a mechanic. It is even more remarkable that it was all achieved with only the use of simple hand tools!

The 16 *unweighted* levers used in the Merseburg wheel's drum each had a mass of 2.5 pounds as indicated in the lower right corner of Figure 12(c). By "unweighted lever", in the case of this 12 feet diameter, bidirectional wheel, I mean a lever *without* its three cylindrical lead end weights and their three attachment bolts attached to the end of its parallel pair of main arms. (Since the square nut that each end weight's attachment bolt screwed into was partially embedded in and glued to the end of one of a lever's wooden main arms and, thus, could not be removed along with the end weight and its steel attachment bolt, we can consider the mass of the three square nuts to be part of the mass of an unweighted lever.) Thus, the 2.5 pound mass of an unweighted lever used in the Merseburg wheel was the remaining *sum* of the masses of the lever's parallel pair of wooden 3 arm side pieces, its six wooden coordinating cord hook keepers, its six steel coordinating cord hook attachment pins, the seven steel coordinating cord attachment hooks, half of the lengths of their seven coordinating cords whose ends were tied to the steel hooks attached to the unweighted lever's various coordinating cord attachment pins, half of the mass of the two extension springs if their attachments to their drum's spring hook attachment anchor pin were free to rotate around the pin and this would be the case if the springs are attached via steel hooks to the anchor pins, and the three steel cylindrical end weight square nuts into which the threaded ends of the three end weights' three steel attachment bolts were tightened when the three cylindrical end weights were attached to the ends of the weighted lever's parallel pair of main arms.

The masses of the steel coordinating cord attachment hooks and half lengths of the cords only needed to be included in the 2.5 pound mass of an unweighted Merseburg wheel lever *if* they were significantly massive and exceeded more than, perhaps, a quarter of a pound (that is why their masses can be safely ignored in the construction of a replica of the 3 feet diameter, one-directional Gera prototype wheel which used little steel staples for hooks and strong thread or string for coordinating cords). I suspect that Bessler *did* include the masses of these components in the construction of the Merseburg wheel and this is why anyone attempting to duplicate this wheel should also take their masses into account during its construction.

If a craftsman's Merseburg wheel duplicate's unweighted lever mass begins to exceed 2.5 pounds by more than a few ounces despite using the lightest cord attachment hooks and coordinating cords that it is possible to use safely, then he will have to compensate for the excessive mass by reducing the mass of the unweighted levers themselves. One way to do that would be to drill out holes along the center lines of each of the bidirectional wheel's 16 weighted levers' parallel pairs of arms, that is, their main, A, and B arms. The removal of the wood (a process referred to as "skeletonization" in clock and watchmaking when it is used to remove metal from movement plates so that their enclosed gear train can be better viewed) must be carefully done so that its absence does not seriously compromise the structural strength of the arms when the drum is in rotation and various torques are being applied to these arms by the tight coordinating cords attached to them.

In Figure 12(c), I've also given the distances, in inches, measured from the center axis of the Merseburg wheel's 6 inch diameter oak axle to various points near the outer rim wall of the drum.

We saw, previously, that the 0.25 inch = 4/16 inch diameter x 12 inch long common pivot pin used by a pair of front and back, side by side weighted levers in the Merseburg wheel's enlarged drum had its center axis located exactly 56 inches from the center axis of the axle which, as might be expected for a drum that is four times the radius of the 3 feet diameter Gera prototype wheel's drum, is four times the distance of the center axis of a Gera prototype's weighted lever's pivot pin from the center axis of that relatively diminutive wheel's axle which was 14 inches. Since the distance from the center axis of a shared pivot pin's center axis to the center of gravity of the three cylindrical lead end weights in a Merseburg wheel's weighted lever (which was located on the center axis of the *middle* of the three weights) was 14 inches, this means that, when one of the Merseburg wheel's weighted lever's parallel pair of main arms was in contact with its wooden radial stop piece, the center of gravity of the lever's three cylindrical lead end weights was exactly 70 inches from the center axis of the Merseburg wheel's 6 inch diameter axle. This 70 inch distance is indicated on the right side of Figure 12(c). The actual center of gravity of the *entire* weighted lever, however, was only located about 68 inches from the center axis of the axle.

Since the radial frame pieces that were attached to the Merseburg wheel's axle were each 68.25 inches = 68-4/16 inches in length (this distance is *not* shown on the right side of Figure 12(c) since only distances relative to the axle's center axis are indicated there, but it is mentioned in the lower left side of the figure), that means that, after being securely attached to the surface of the bidirectional wheel's 3 inch *radius* wooden axle, the outer *ends* of the radial frame pieces would have been located at a distance of 71.25 inches = 71-4/16 inches from the center of the axle (this distance is shown on the right side in Figure 12(c)). Then, across all three outermost ends of *each* set of three parallel radial frame pieces of the drum's three lattices (there were eight of these sets of three radial frame pieces attached to the axle), a single 0.5 inch = 8/16 inch thick x 1 inch wide x 12 inch long piece of oak wood that I call a "bridge piece" was attached with glue and nails to help increase the drum's rigidity and Bessler used the same stock wood for these bridge pieces as he did for all of the other frame pieces of the drum's lattices. That added another 0.5 inch = 8/16 inch to the end of the radial frame pieces so that the outside facing flat surfaces of these bridge pieces were then 71.75 inches = 71-12/16 inches from the center axis of the axle while their *inside* facing flat surfaces only cleared the outer curving surfaces of the 0.89 inch radius, cylindrical lead end weights by 0.36 inch = 5.75/16 inch! All of these dimensions are shown toward the right side of Figure 12(c). This clearance might seem too small, but in every computer model wheel that I have made, that gap is more than sufficient to allow the Merseburg wheel's weighted levers to swing away from the bridge pieces without touching them.

Finally, thin previously warped sections of wood were attached to the sixteen bridge pieces of the drum's three lattices to change their outer shape into that of a circular drum shaped one (note that eight of the bridge pieces were attached to the ends of the radial frame pieces and the other eight bridge pieces were attached to the ends of the very short radial pieces as indicated in Figure 11(a)). Only one of these bridge pieces is shown in Figure 12(c) attached to three parallel radial frame pieces that are vertically oriented. That thin outer layer of wood was only 0.25 inch = 4/16 inch thick, but with its attachment the drum's *outer* rim wall surface was formed and was then exactly 72 inches from the center axis of the axle or 6 feet. And, of course, this meant that the diameter of the bidirectional Merseburg wheel's enlarged drum was then exactly 12 feet.

The reader will notice in Figure 12(c) that the two wooden radial stop pieces for the bidirectional wheel's two 12:00 weighted levers shown appear different from each other. The reason for this is that with the radial stop piece for the front one-directional wheel's 12:00 weighted lever (on the left side of the drum's middle radial frame piece), we see the *back* surface of the radial stop piece as the lever's two parallel main arms make contact with its hidden front surface. For the back internal one-directional wheel's 12:00 weighted lever (on the right side of the drum's middle radial frame piece), however, we see the *front* surface of the radial stop piece against which that lever's two parallel main arms make contact while that radial stop piece's unused back surface is hidden from our view (the back surfaces of the radial stop pieces never had a weighted lever's parallel pair of main arms in contact them and, thus, were unused). From this is becomes obvious that the adjacent front and back, side by side pairs of weighted levers in a bidirectional wheel's drum were designed to swing away from their respective radial stop pieces in opposite directions which is to be expected since the "preferred" directions of rotation for the front and back internal one-directional wheels were always opposed to each other.

There is also a prominent rectangular opening in the wooden radial stop pieces of the 12 feet diameter Merseburg wheel that did not appear in the far smaller radial stop pieces of the 3 feet diameter Gera prototype shown in Figure 10 of the last chapter. These cutouts were required for use by the mechanisms (which are not shown in Figure 12(c)) that allowed the Merseburg wheel to display the property of bidirectionality and which will be discussed in greater detail in the next chapter. As the reader looks through the rectangular opening in the radial stop piece for the 12:00 weighted lever of the *front* internal one-directional wheel shown in Figure 12(c), he will notice the L1 main cord hook attachment pin for that lever's parallel pair of main arms and the three sections of the notched edge of the wooden hook keeper that is in contact with the L1 main cord hook attachment pin.

The rectangular opening in the radial stop piece for the 12:00 weighted lever of the *back* internal one-directional wheel shown in Figure 12(c) is a bit more difficult to make out. While the lower *beveled*, steel sheet clad inside edge of that opening is visible, the upper flat steel sheet clad inside edge is obscured by the wooden hook keeper whose three notches expose particular sections of the back weighted lever's L1 main cord hook attachment pin. The middle notch exposes the section of the pin to which one of the end hooks of the main cord is attached and the two larger notches on each side of the pin expose sections that are used by the mechanisms that are responsible for the Merseburg wheel's bidirectionality and which, as previously mentioned, will be discussed in the next chapter in detail.

As a final comment on Figure 12(c), I can briefly mention the method used to attach the three parallel 12:00 radial frame pieces of the drum's three lattices to the surface of the axle as shown at the bottom of the figure. After their final assembly into a completed lattice, Bessler would then have suspended the axle on ropes using a method described in a previous chapter, slid each lattice onto the axle, maneuvered it into its proper position and orientation with respect to a previously attached lattice, and then glued the lightly sanded ends of the added lattice's eight radial frame pieces to the lightly sanded surface of the axle. He would then have glued in addition wedge shaped wooden pieces (most likely shaped like isosceles right triangles) between the inside corner that each end of a radial frame piece made with the axle in order to reinforce the initial glue bond that had been made. He may also have used small nails for further reinforcement of the glue bond between the wedge shaped pieces and the axle. For the Merseburg wheel a total of thirty-two wedge shaped pieces of wood would have been required and they each would have been about 1 inch thick.

Before leaving this chapter's discussion of the Merseburg wheel there is one additional matter concerning its drum which needs to be mentioned.

During the many years that I was an active builder of physical model wheels, I made various attempts to construct hollow wooden drums with *ratios* of dimensions similar to those found in Bessler's wheels even though their diameters were never larger than about 4 feet. Thus, like Bessler, I tried to construct drums with diameters that were much larger than their thicknesses. These drums were constructed of thin pieces of wood that had been glued together to form two identical polygonal lattices (usually having only eight sides) that were then each attached separately by their eight converging radial frame pieces to a wooden axle and, in order to make them more rigid, had additional pieces of wood glued between them at various positions around the drum with the most being located at the drum's outer polygonal vertices.

Considering how thin their individual wooden frame pieces were, these drums had exceptional rigidity and I remember, in particular, one stress test I did in which I laid a portion of a axle mounted drum flat on a table top and then proceeded to pile weights on it to see if it would collapse under the load. To my amazement, it easily held a load of 30 pounds wherever I placed it without the drum being damaged in any way. Also, after I installed variously shaped weighted levers into these drums which had a mass of less than a pound each, I found that the total mass of all of the weights did not seem to damage the drum or cause the radial frame pieces to tear away from the places on the axle to which they had been glued even when the drums were given a spin to start them rotating. Sadly, without the same design that Bessler used, none of my wheels ever displayed the pm effect. What a different matter that would have been if I'd then had the same design presented in this volume!

However, after constructing several drums that were about 4 feet in diameter and 4 to 6 inches in thickness, I noticed a disturbing tendency in them after they were mounted on their axles.

If, with both hands, I grasped *one* end of the axle to which an empty drum had been attached (that is, a drum without its weighted levers having yet been installed) and then rapidly moved the axle toward and then away from myself repeatedly with a motion one can refer to as "axial oscillation", I could get the entire drum to sort of wag, but the wagging action was not just confined to one of its parallel pairs of radial frame pieces. Rather, the wagging involved the *entire* circumference of the drum. In essence, it would behave under these conditions like a giant cymbal and, if I suddenly stopped rapidly moving the axle back and forth along its center axis, the entire outer circular rim of my drum would, for a second or so, continue to whip back and forth with respect to the then stationary axle that I was holding onto. I refer to this secondary effect as "tympanic oscillation". In fact, every hollow drum, regardless of the materials from which it is constructed, will have a natural tympanic oscillation frequency that will be determined by its particular structural parameters.

One might then wonder how, if the axle of one of Bessler's wheels was prevented from undergoing axial oscillation because its pivot pins were securely seated in their top and bottom brass half moon bearing plates in each of the vertical support planks, it would be possible for the outer rim wall of a drum to begin undergoing any sort of tympanic oscillation. There is, however, a way that this can happen.

For one of Bessler's one-directional wheels it could happen if the center axis of one of the weighted lever pivot pins embedded in its parallel pair of radial frame pieces was not sufficiently parallel with respect to the center axis of the wheel's axle. Then, as that weighted lever's cylindrical lead end weight(s) and main arms experienced a rapid ascent as they rotated rapidly clockwise around their misaligned pivot pin as that pin's center axis, in turn, traveled from the drum's 9:00 to 10:30 positions, the outward acting centrifugal force produced on the clockwise swinging weighted lever would, indeed, have a small axial *component* that would be *momentarily* applied to the drum, through the lever's pivot pin, every time that weighted lever's pivot pin's center axis moved between those two drum clock time positions. When the wheel's accelerating drum finally reached a rotation rate that matched its natural tympanic oscillation frequency, each time the weighted lever's misaligned pivot pin passed between the drum's 9:00 and 10:30 positions, the axial component of its cylindrical lead end weight(s) and main arms' centrifugal force applied to the pin would beginning "pumping" that part of the drum. These repeated pulses of axial force applied to the lever's pivot pin as it traveled between those two drum locations would keep the tympanic oscillations going and, in fact, adding together so that there would be a steady increase in the amplitude of the drum's tympanic oscillations with every additional complete drum rotation.

This situation would not be that serious if the drum's natural tympanic oscillation frequency was greater than the wheel's maximum terminal rotation rate because, even if allowed to run freely, the wheel would never reach the particular rotation rate that would allow the drum's tympanic oscillations to become additive. Also, this situation would not be that serious if the drum's natural tympanic oscillation frequency was below the minimum rotation rate of the drum when its axle was performing some sort of external work via a machine attached to its axle. This is because the natural tympanic oscillation frequency of the drum would be one that was only momentarily experienced as the drum accelerated past it to some higher operational rotation rate at which the tympanic oscillations did not become additive. The most dangerous situation, however, would occur if the drum's natural oscillation frequency just happened to be equal to the maximum terminal rotation rate of the drum. In this case, when the drum was spinning at its highest possible rate, each axial pulse delivered to the drum through the misaligned pivot pin whose ends were embedded in a parallel pair of its radial frame pieces would be in phase with the drum's natural tympanic oscillations and these oscillations would be additive and increase in amplitude with each drum rotation.

As the drum's natural tympanic oscillations continued to increase in amplitude, a point might be reached at which the constant axial flexing of the drum's lattice pieces could actually become severe enough to cause the glue bonding their separate frame pieces together to fail. At that point the drum's structural integrity would be compromised and it could then be torn apart by the centrifugal forces acting on its other weighted levers that were all attached to the drum by their steel pivot pins. Obviously, this was a potentially dangerous effect and Bessler needed to come up with a way of preventing if from appearing in any of the hollow wooden drums he was building.

My solution to this problem for the drums I was constructing was to just glue in additional thin wooden frame pieces using "X" patterns across the middle sections of the parallel pairs of radial frame pieces inside of a drum. This approach worked well and I was able to construct light wooden drums that, when still empty of their weighted levers and attached to their axles, I could even hold by one end of their axles with a drum's two side lattice planes horizontal and the other end of the axle pointing vertically downward without having the two octagonal outer perimeters of the lattices sag noticeably downward and out of the planes of the central portions of two lattices that were securely attached to the axle (I did *not*, however, try this test with a drum that had all of its weighted levers installed and would not risk doing so!).

However, I do not think that my method is the one that Bessler would have employed to suppress these potentially catastrophic oscillations in the drums of his wheels. Using large wooden pieces to suppress tympanic oscillation adds internal structures into a drum that, by their stirring of the air inside of the drum and the rising of its temperature, tends to waste the energy and mass that a wheel extracts from its weighted levers and makes available to perform outside work. The best solution, then, would be to use thin metal pieces, but to use the minimum necessary because they only serve as dead weight inside of a wheel that adds to its gross weight and places additional stress on its axle's end pivot pins and their supporting bearing plates.

The method I believe Bessler used is illustrated in Figure 12(d) at the end of this chapter. Unfortunately, I can not prove that he used this method from the clues he left us in his writings and illustrations, but it is a solution that I think a clockmaker like he would have tried because the materials for it would have been readily available to him.

In Figure 12(d) I show another cross sectional profile or side view of the Merseburg wheel's drum except that this time the view is from the *left* side of the wheel which is the side opposite to which it was viewed in Figure 12(c) and shows both the 12:00 and 6:00 weighted levers of its front internal one-directional wheel (on the right half of the cross section of the drum) and its back internal one-directional wheel (on the left half of the cross section of the drum). All structures in the figure are in the correct proportions with respect to each other as the corresponding structures found in the actual Merseburg wheel that was illustrated on the left side of Bessler's illustration in DT (the complete version of which is shown in Figure 1(a) at the end of Chapter 1).

As can be seen in Figure 12(d), the 6:00 and 12:00 side by side, back to back pairs of weighted levers of the Merseburg wheel's two internal one-directional wheels share common 12 inch long steel pivot pins that, along with their four long lifter cord guide posts, help to reinforce the outer portions of the drum and dampen any tympanic oscillations that might begin to occur in them. However, other than the four small wedge shaped blocks of wood that help to secure each triplet of parallel radial frame pieces to the surface of the axle, there is no bracing for the inner portion of the drum near the axle where runaway tympanic oscillations could tear the radial frame pieces away from the axle and, perhaps, break the glue bonds between the radial frame pieces and the innermost octagonal frame pieces between adjacent radial frame pieces near the axle's surface.

To remedy this problem, Bessler could have used a total of sixteen thin strips of steel he obtained by simply unwinding and then flattening out clock mainsprings (which had begun to appear in European clocks during the 15th century). He would have used strips made from 0.5 inch = 8/16 inch wide mainsprings that were each then cut to a length of 20 inches. As shown in Figure 12(d), each of the eight triplets of parallel radial frame pieces within the Merseburg wheel's drum would have had two of these strips of steel attached to it so that they formed an "X" shape when viewed at an angle perpendicular to the lengths of the radial frame pieces. But, the two pieces of steel mainspring were not in the same plane. Rather, they were attached so that one 20 inches long strip connected the thicknesses of the three radial frame pieces on one side of the pieces (which would be a leading or trailing side depending upon which direction the 12 feet diameter bidirectional wheel would be rotating) while the other 20 inches long strip connected the thicknesses of the three radial frame pieces on the other side of the radial frame pieces.

After drilling holes close to the two ends and in the middle of a 20 inches long strips made from a straightened out clock mainspring, the inventor would have attached one end of a strip to an outside lattice's radial frame piece at a distance of 24 inches from the center axis of the axle by attaching it to the frame piece with a small wood screw. The middle of the strip would then, by the small hole drilled in it, be similarly attached to the middle lattice's radial frame piece at a distance of 16 inches from the center axis of the axle. Finally, the remaining end of the strip of clock mainspring steel would be attached with a screw to the other outside lattice's radial frame piece at a distance of 8 inches from the center axis of the axle which then put that attachment point at a distance of only 5 inches from the surface of the axle to which that radial frame piece was attached. The same attachment of a 20 inches long steel mainspring strip would then be made on the three opposite sides of the three parallel radial frame pieces, but, as shown in the figure, this would be done with a reversed order that resulted in the two strips appearing to form a letter "X".

The sixteen strips of steel mainspring would have been installed in the Merseburg wheel's drum shortly after its three lattice pieces were attached to their common oak axle. Their attachments to the radial frame pieces were far enough away from the surface of the axle so as to not prevent a skeptic's probing hand, after being slipped through the slit in the wheel's front linen clothe covering, from groping the section of axle that supported the drum's front internal one-directional wheel or from reaching over to feel the section of axle that supported the back internal one-directional wheel in order to verify that there were no ropes or chains wrapped around either of those hidden axle sections. In the event that a skeptic's hand actually came into contact with the end of a mainspring strip attached to a radial frame piece at only a distance of 5 inches from the axle's surface, there would be little danger of him cutting his hand because, although thin, the edges of mainsprings are generally not sharp enough to cut skin.

This method of suppressing tympanic oscillations in the Merseburg wheel could also be used in the more massive Kassel wheel. However, in that wheel the width and thickness of the sixteen required mainspring strips used would have been slightly larger although the lengths of the strips, 20 inches, would remain the same. In reconsidering this method of suppressing potentially catastrophic oscillations in his rapidly rotating post Gera prototype wheels' drums, I suppose, in order to minimize aerodynamic drag inside of a drum's interior volume, Bessler could also have used lengths of thin stretched steel wire instead of straightened out steel clock mainsprings. Those seeking to duplicate such a post Gera prototype wheel should also consider this alternative option.

Finally, before ending this chapter, I want to briefly disclose some of the important structural features that a craftsman attempting to duplicate the Kassel wheel will need to know. For this purpose, the reader is now directed to Figure 12(e) at the end of this chapter.

The first thing to know about Figure 12(e) is that it shows the view of the Kassel wheel's drum from the same viewer location as was used in Figure 12(c) for the Merseburg wheel's drum; that is, the reader must imagine that he has entered the room into which the Kassel wheel was moved inside of Weissenstein Castle and sees the front axial view of the wheel as shown in Figure 2(a) in Chapter 1. He then walks to the right side of either of the two drums shown in that figure and, as he reaches the rim wall of the drum, turns to his left so that he can see a profile view of this 12 feet diameter, bidirectional wheel's drum which was 18 inches thick.

Using imaginary x-ray vision, the reader can then view a cross sectional profile view of the 12:00 position of the drum. Because of the extra thickness of the Kassel wheel's 18 inch thick drum, I was not able to include the surface of the axle to which its three vertical parallel radial frame pieces were attached in Figure 12(e), but it would look much the same as the attachments made for the Merseburg wheel's three parallel radial frame pieces shown in Figure 12(c). Unlike Figure 12(c), however, I had to omit practically all of the lengths of the Kassel wheel's radial frame pieces and the lengths of the parallel pairs of main arms of its side by side, back to back pair of weighted levers at the drum's 12:00 position and just concentrate on the outer portions of those two levers' parallel pairs of main arms that each held three cylindrical lead end weights (we only see one of the three cylindrical lead end weights between the main arms of each weighted lever, but that is because the two other identical cylindrical weights are hidden behind the one that is visible). Again, as was the case with the Merseburg wheel's drum illustrated in Figure 12(c), the Kassel wheel drum's front internal one-directional wheel's 12:00 weighted lever is on the left side of Figure 12(e) and the back internal one-directional wheel's 12:00 weighted lever is on the right side of the figure.

As one might expect, in order to double the start up torque and power output of his Kassel wheel, Bessler equipped it with cylindrical lead end weights that had double the mass of the ones used in the Merseburg wheel and with their steel attachment bolts weighed 8 pounds apiece. Each of these lead weights was 2.04 inches = 2-0.64/16 inches in diameter x 6 inches in length and was attached to the ends of a weighted lever's parallel pair of main arms with a steel bolt that was 0.25 inch = 4/16 inch in diameter and whose combined rounded slot head thickness, shank length, and threaded end section length was 7.25 inches = 7-4/16 inches in length. The unweighted levers were also made to have double the mass of those used in the Merseburg wheel by increasing the thickness of each unweighted lever's parallel pair of wooden 3 arm side pieces so that they were 0.5 inch = 8/16 inch thick, increasing the thickness of the wooden hook keepers to 0.375 inch = 6/16 inch, and using coordinating cord hook attachment pins that were 0.375 inch = 6/16 inch in diameter x 7 inches long. As a result, the unweighted levers in the Kassel wheel would have weighed 5 pounds each. Doubling both the masses of the three cylindrical lead end weights with their attachment bolts and the unweighted lever between whose parallel pair of main arms they were bolted resulted in a Kassel wheel's weighted lever with a mass of 29 pounds as compare to a weighted lever in the Merseburg wheel with a mass of 14.5 pounds.

But, again, I must remind the craftsman that the mass of an unweighted lever inside of a replica of the Kassel wheel will, unless one is using extremely light and strong materials for the coordinating cords, their metal end hooks, and also using extension springs that are fixed to the drum so that they do not rotate freely, via hooks, around the drum's spring hook attachment anchor pins, have to include half of the masses of the cords, hooks, and springs to the determine the actual final mass of an unweighted lever which must be as close to 5 pounds as possible. If one's replica Kassel wheel's unweighted levers deviate too much from this mass, then the resulting wheel will not run as efficiently as possible and may not even run at all! For this reason, the craftsman should carefully and accurately determine the masses of the lengths of the four different types of coordinating cords he is using, the hooks that the cords will be tied to, and the extension springs being used. It's also a good idea to check the masses of the cylindrical lead end weights being used along with their steel attachment bolts to assure that their combination is as close to 8 pounds as possible. A variation of more than a single ounce is undesirable and should be compensated for by adding or removing some lead from the cast cylindrical lead weight.

The most important aspect of Figure 12(e), however, is what it reveals about the way that Bessler packaged the bidirectional Kassel wheel's two internal one-directional wheels inside of its 18 inch thick drum.

We see from the figure that, in order to withstand the doubling of centrifugal forces applied to each of the drum's eight parallel triplets of radial frame pieces by the eight pivot pins that carried more massive weighted levers, the thickness of the individual radial frame pieces had to be doubled from the 0.5 inch = 8/16 inch used in the Merseburg wheel to a value of 1 inch for the Kassel wheel. Since it is possible that Bessler continued to use the same stock wood pieces when making the Kassel wheel as he did in constructing the Merseburg wheel, he may have made the radial frame pieces for the Kassel wheel's three lattices by simply gluing together, lengthwise, two of the 0.5 inch = 8/16 inch thick lengths of the wood used for the Merseburg wheel's three lattices to form lengths of wood that were 1 inch thick. Or, he may have just switched over to using stock wood that already had a 1 inch thick x 2 inch wide = 2 square inch rectangular cross sectional area for the various frame pieces of the Kassel wheel's drum lattices.

Since I cannot determine exactly what he did use based on the clues he left in this drawings, I decided to show the radial frame pieces in Figure 12(e) for the Kassel wheel's drum as being formed from two frame pieces of the smaller 0.5 inch = 8/16 inch thick frame pieces which were used in the Merseburg wheel. In any event the dimensions of each radial frame piece in the Kassel wheel were 1 inch thick x 2 inches wide x 67.25 inches = 67-4/16 inches long. The lengths of the radial frame pieces in the Kassel wheel's drum are an inch shorter than those in the Merseburg wheel's drum because the radius of the Kassel wheel's axle was one inch larger than the radius of the Merseburg wheel's axle.

We can now quickly calculate how much gap space would have been left between *each* side of a Kassel wheel's weighed lever's main arm's outside surface and the nearest inside surface of a radial frame piece. We know that the weighted levers themselves were 7 inches wide as measured between the *outside* surfaces of their two wooden parallel 3 arm side pieces. Each parallel triplet of double strength radial frame pieces was made from three frame pieces that were 1 inch thick (or six that were 0.5 inch = 8/16 inch thick). Each parallel triplet of radial frame pieces, therefore, took up 3 inches of the drum's 18 available inches of thickness which would leave 15 inches for an side by side, back to back pair of front and back internal one-directional wheels' two weighted levers. Since there were two weighted levers that were each 7 inches wide (this does not include the projecting ends of the cylindrical end weights' attachment bolts), they account for another 2 x 7 inches = 14 inches of drum thickness. Subtracting 14 inches from the previously available 15 inches leaves only 1 inch for the four gaps inside of the drum between the outside facing surfaces of the weighted levers' main arms and the nearest inside facing surface of a radial frame piece. Thus, there was only a gap or clearance of 1 inch / 4 gaps = 0.25 inch per gap = 4/16 inch per gap. If, as was done with the Merseburg wheel, steel washers were used to fill in these gaps, then each gap in the Kassel wheel would again have been filled in by placing two of the 0.125 inch = 2/16 inch thick steel washers on each of the exposed ends of a weighted lever's steel pivot pin during the installation of the unweighted levers in the drum. Since the eight shared weighted lever pivot pins were 0.375 inch

= 6/16 inches in diameter x 18 inches long, these washers would also have center holes with that diameter and an outside diameter of 1 inch as was used in the Merseburg wheel. Because of a gap width that was identical to that of the Merseburg wheel, a total of sixty-four washers would be needed for the sixteen weighted levers contained within the bidirectional Kassel wheel's 12 feet diameter drum.

As in the case of the Merseburg wheel, the center of gravity of a weighted lever's three attached cylindrical lead end weights inside of the Kassel wheel's drum was exactly 70 inches from the center axis of its axle when the leading thickness of a parallel pair of main arms whose ends held the three weights was in contact with its wooden radial stop piece. But, since the axle was increased to a diameter of 8 inches and, thus, had a radius of 4 inches, this location of the three cylindrical lead end weights' center of gravity required a radial frame piece that was exactly 67.25 inches in length so that when an 0.5 inch = 8/16 inch thick x 2 inches wide x 18 inches long bridge piece was added to the ends of a triplet of parallel radial frame pieces to hold them in place relative to each other and an additional wooden outer rim wall section 0.25 inch = 4/16 inch thick was then put on top of the bridge pieces, the outer surface of the rim wall would have been exactly 72 inches from the center axis of the axle and the drum's overall diameter would then be exactly 144 inches = 12 feet. With the configuration I show in Figure 12(e), the reader will notice that there was only a 0.23 inch = 3.7/16 inch gap between the outer curving surface of one of the Kassel wheel's cylindrical lead end weights and the inner facing surface of a bridge piece, but it was still sufficient to prevent direct end weight to drum contact as a weighed lever's parallel pair of main arms swung out of alignment with their nearest parallel pair of radial frame pieces and then, later, swung back into alignment with them again.

When looking at the portion of the Kassel wheel's *front* internal one-directional wheel's 12:00 weighted lever shown in Figure 12(e) on its left side, one notices that, again, only the backside of the weighted lever's wooden radial stop piece is visible and that lever's two parallel main arms are actually in contact with this radial stop piece's *unseen* front side. For the *back* internal one-directional wheel's 12:00 weighted lever on the right side of Figure 12(e), we see that weighted lever's parallel pair of main arms are again in contact with the front side of the wooden radial stop piece and that we cannot see the radial stop piece's back side. We can now clearly see this back internal one-directional wheel's L1 main cord hook attachment pin and the wooden hook keeper near it whose three notches expose certain sections of the pin. The center notch exposes the section of the L1 main cord hook attachment pin that receives the hook at the end of a main cord while the two larger side notches expose sections of the hook attachment pin that will receive a special sort of hook that will be the subject of the next chapter.

With the information provided in this chapter, a skilled craftsman is now in a position to construct a 12 feet diameter, one-directional wheel that uses the same imbalanced pm wheel mechanics as was employed in Bessler's wheels. However, he still does not have the technical information necessary to make such a wheel bidirectional. The mechanics that Bessler needed to bestow that property to his largest constructed and publicly demonstrated wheels will be the subject of the next chapter.

Figure 11(a) - 12 Feet Diameter Wheel with Its Cloth Covers Removed

Each bidirectional wheel's drum contained three lattices attached to its axle

The drum has none of its internal mechanics installed at this time and can be referred to as "empty"

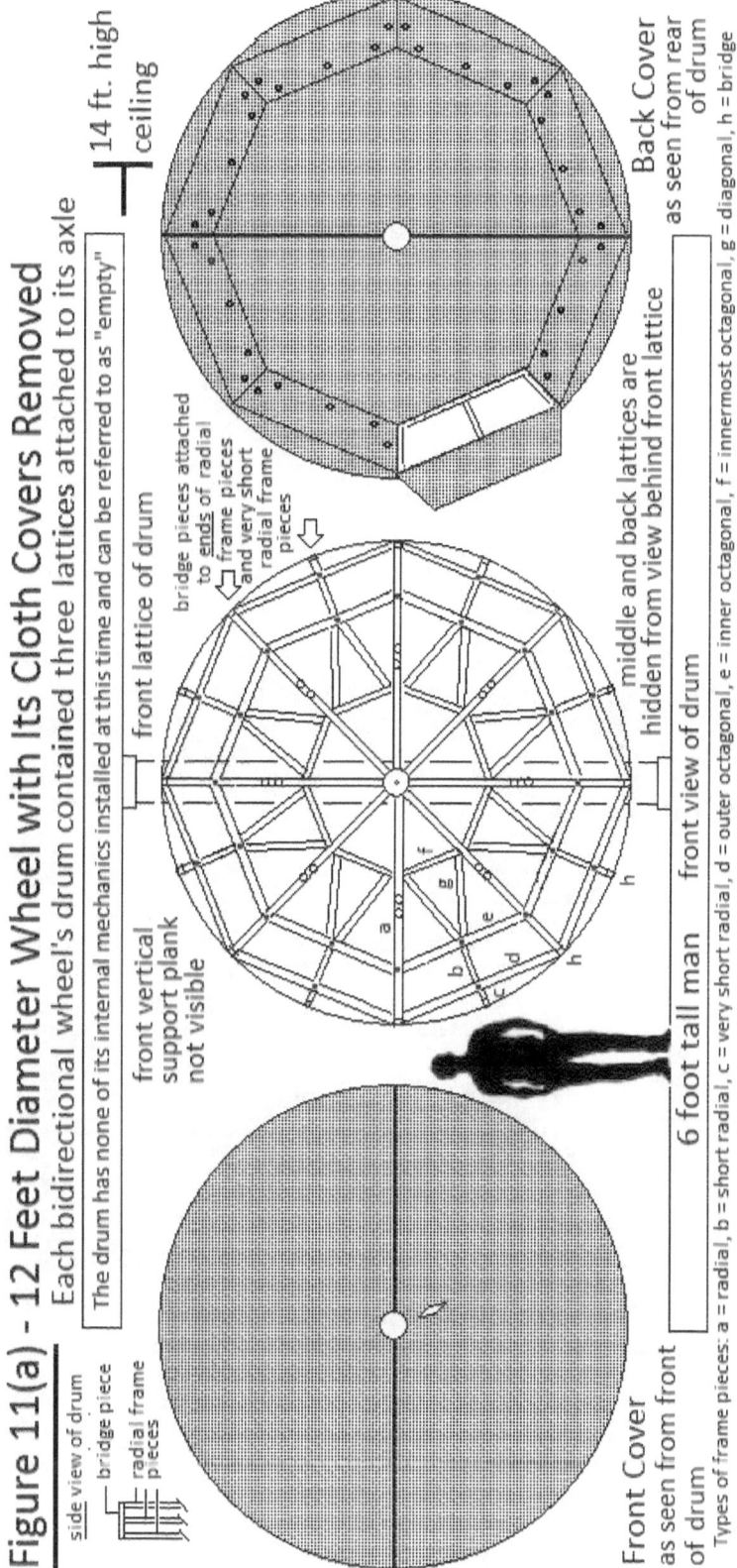

Types of frame pieces: a = radial, b = short radial, c = very short radial, d = outer octagonal, e = inner octagonal, f = innermost octagonal, g = diagonal, h = bridge

Figure 11(b) - Details of the 12 Feet Diameter Merseburg Wheel's Lever

Side View of 9:00 Lever

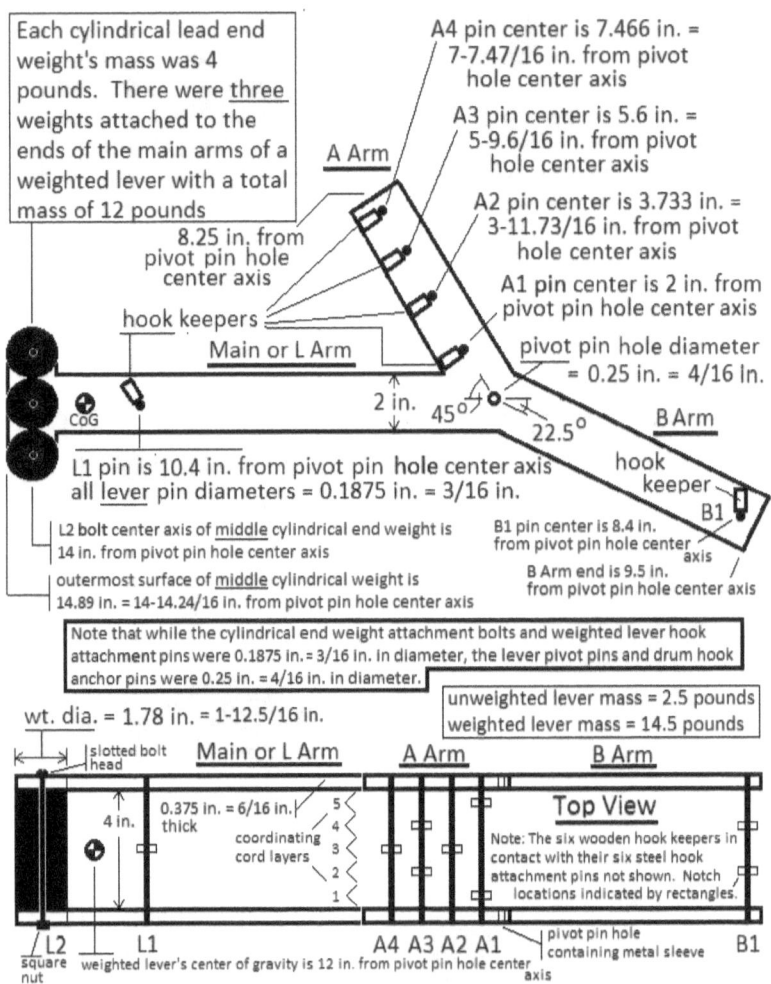

Figure 11(b) - continued

Merseburg wheel weighted lever cord hook attachment pin wooden hook keepers

Drum stop cord and spring hook attachment anchor pin wooden hook keepers

coordinating cord layers

All weighted lever hook keepers are 0.1875 in. = 3/16 in. thick x 0.90625 in. = 14.5/16 in. wide x 4 in. long.

All square notches on hook keeper edges are 0.25 in. x 0.25 in. = 4/16 in. x 4/16 in.

The two rectangular notches of the L1 hook keeper are each 0.375 in. = 6/16 in. wide x 1 in. long.

All drum anchor pin hook keepers are 0.25 in. = 4/16 in. thick x 0.875 in. = 14/16 in. wide x 5.25 in. = 5-4/16 in. long.

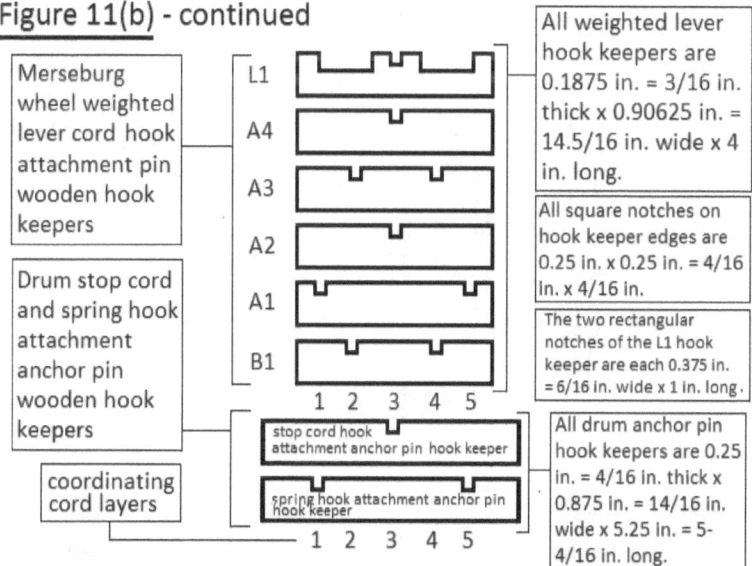

Values for the various Merseburg wheel coordinating cords

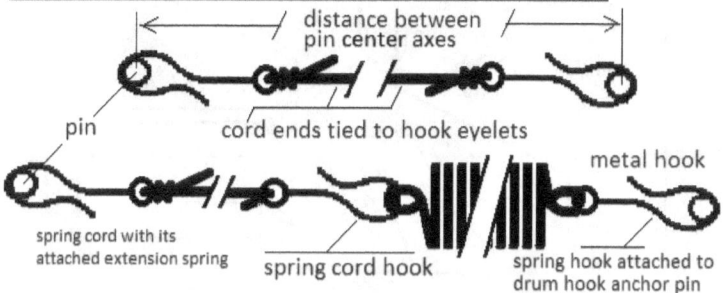

Maximum allowed distances between pin center axes

```
Main Cord       = A4 to L1 = 45.2 in. = 45-3.2/16 in.
Long Lifter Cord = A3 to B1 = 79.6 in. = 79-9.6/16 in.
Stop Cord       = A2 to stop cord hook anchor pin between drum's
                  leading parallel pair of inner octagonal frame pieces
                  = 24 in.
Spring Cord*    = A1 to spring hook anchor pin between drum's
                  leading outer octagonal frame pieces = 24 in.
```
* includes the length of the unstretched extension spring

Figure 12(a) - Merseburg Wheel Axle Pivot Pin and Bearing

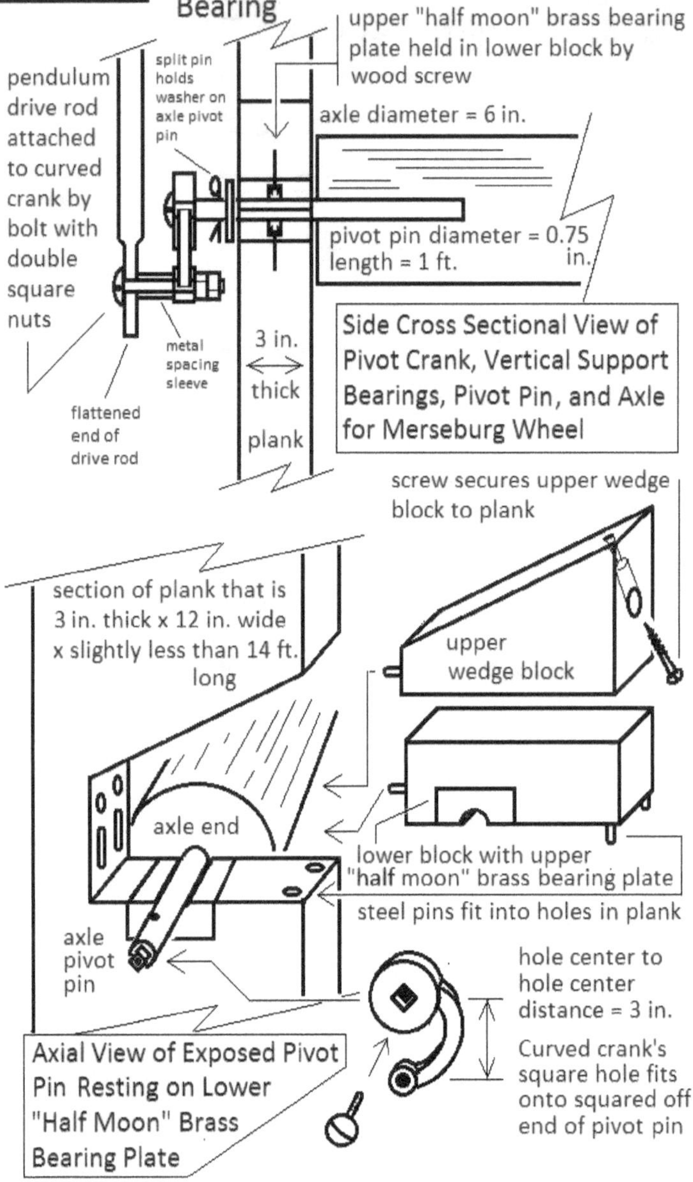

pendulum drive rod attached to curved crank by bolt with double square nuts

split pin holds washer on axle pivot pin

upper "half moon" brass bearing plate held in lower block by wood screw

axle diameter = 6 in.

pivot pin diameter = 0.75 in.
length = 1 ft.

metal spacing sleeve

3 in. thick plank

flattened end of drive rod

Side Cross Sectional View of Pivot Crank, Vertical Support Bearings, Pivot Pin, and Axle for Merseburg Wheel

screw secures upper wedge block to plank

section of plank that is 3 in. thick x 12 in. wide x slightly less than 14 ft. long

upper wedge block

axle end

lower block with upper "half moon" brass bearing plate

steel pins fit into holes in plank

axle pivot pin

hole center to hole center distance = 3 in.

Curved crank's square hole fits onto squared off end of pivot pin

Axial View of Exposed Pivot Pin Resting on Lower "Half Moon" Brass Bearing Plate

Figure 12(b) - Pendulum Details

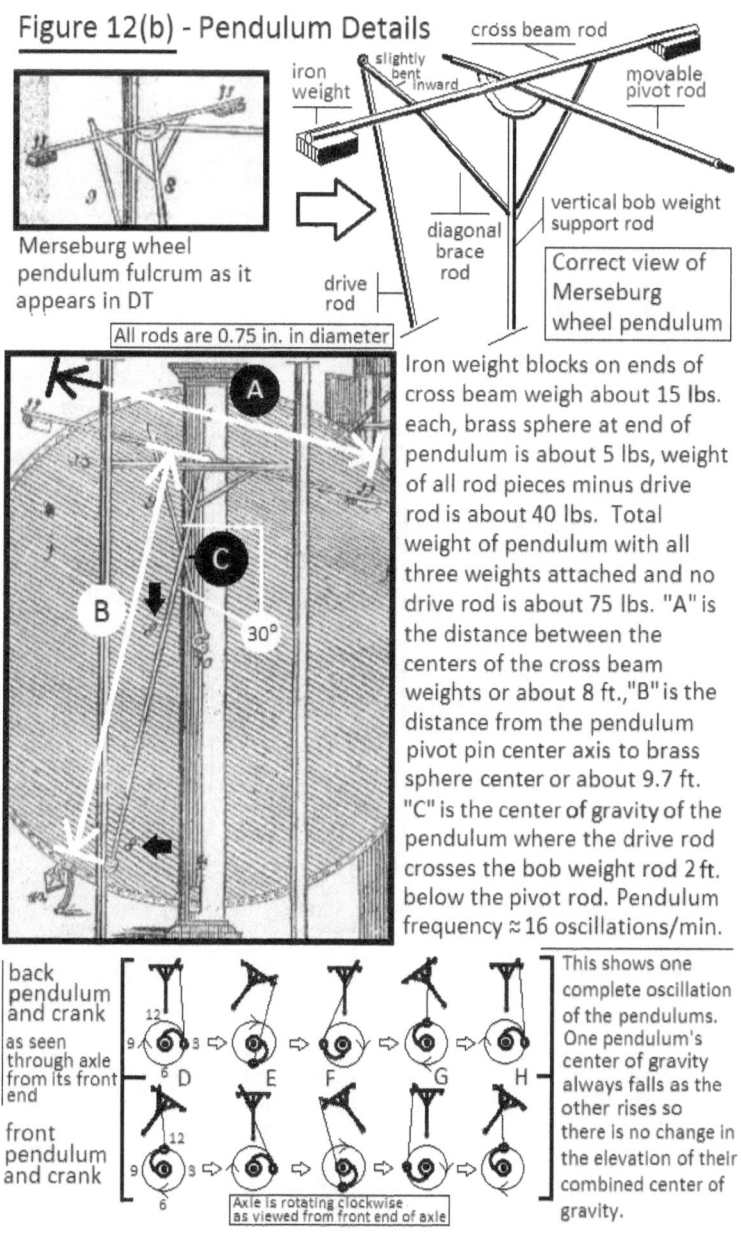

Merseburg wheel pendulum fulcrum as it appears in DT

All rods are 0.75 in. in diameter

cross beam rod
iron weight
slightly bent inward
movable pivot rod
diagonal brace rod
drive rod
vertical bob weight support rod

Correct view of Merseburg wheel pendulum

Iron weight blocks on ends of cross beam weigh about 15 lbs. each, brass sphere at end of pendulum is about 5 lbs, weight of all rod pieces minus drive rod is about 40 lbs. Total weight of pendulum with all three weights attached and no drive rod is about 75 lbs. "A" is the distance between the centers of the cross beam weights or about 8 ft.,"B" is the distance from the pendulum pivot pin center axis to brass sphere center or about 9.7 ft. "C" is the center of gravity of the pendulum where the drive rod crosses the bob weight rod 2 ft. below the pivot rod. Pendulum frequency ≈ 16 oscillations/min.

back pendulum and crank as seen through axle from its front end

front pendulum and crank

D E F G H

Axle is rotating clockwise as viewed from front end of axle

This shows one complete oscillation of the pendulums. One pendulum's center of gravity always falls as the other rises so there is no change in the elevation of their combined center of gravity.

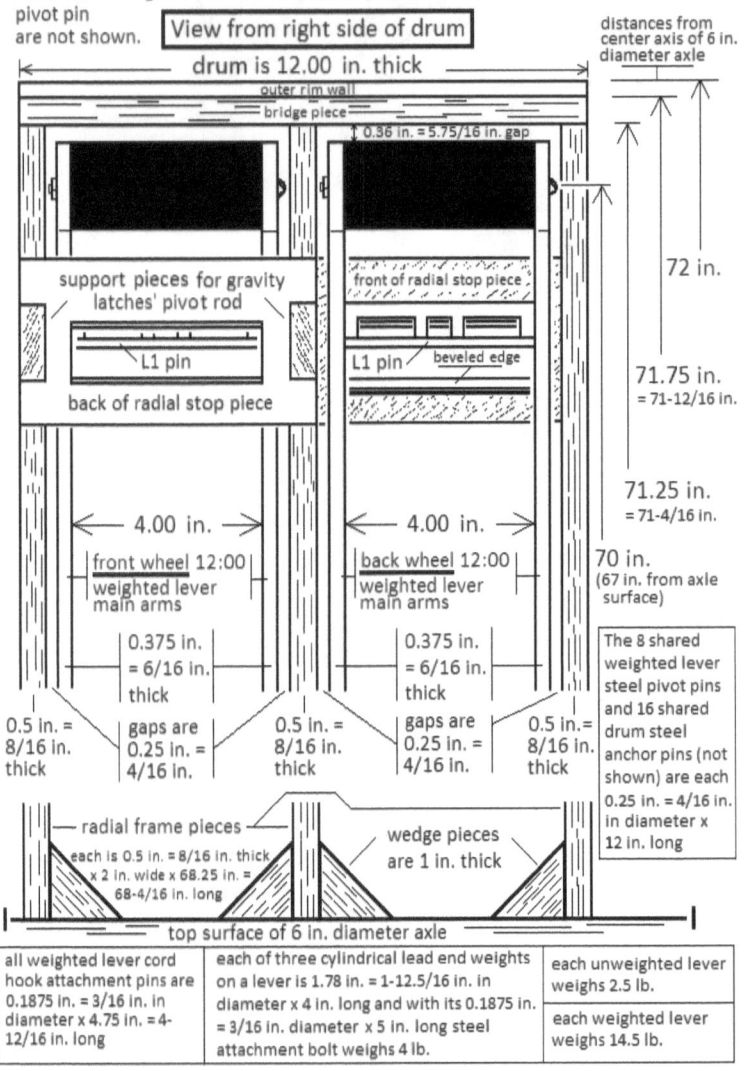

Figure 12(c) - More Details of the Merseburg Wheel's Construction

Profile view of front internal one-directional wheel's 12:00 weighted lever (left side) and back internal one-directional wheel's 12:00 weighted lever (right side). Most of the lengths of the 3 radial frame pieces that hold the two levers' common pivot pin are not shown.

Figure 12(d) - Method Used to Prevent Drum Tympanic Oscillation

0.5 in. = 8/16 in. wide x 20 in. long lengths of clock mainsprings are used as bracing to prevent tympanic oscillation of drum. There is one length of mainspring on each side of three radial frame pieces for the adjacent pair of weighted levers of a bidirectional wheel's internal front and back one-directional weighted levers

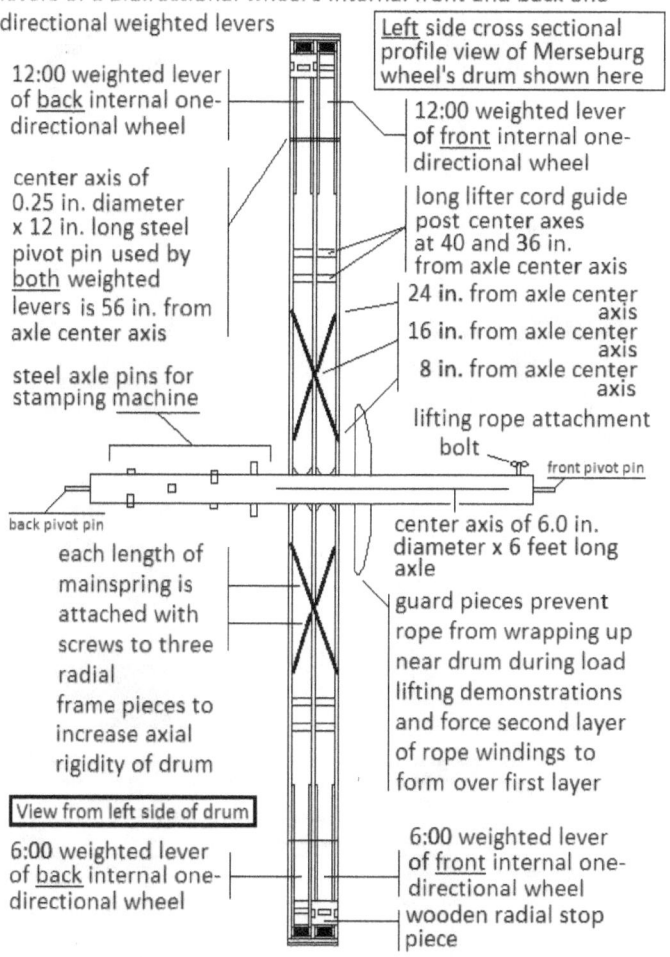

Left side cross sectional profile view of Merseburg wheel's drum shown here

12:00 weighted lever of <u>back</u> internal one-directional wheel

center axis of 0.25 in. diameter x 12 in. long steel pivot pin used by <u>both</u> weighted levers is 56 in. from axle center axis

steel axle pins for stamping machine

back pivot pin

each length of mainspring is attached with screws to three radial frame pieces to increase axial rigidity of drum

View from left side of drum

6:00 weighted lever of <u>back</u> internal one-directional wheel

12:00 weighted lever of <u>front</u> internal one-directional wheel

long lifter cord guide post center axes at 40 and 36 in. from axle center axis

24 in. from axle center axis
16 in. from axle center axis
8 in. from axle center axis

lifting rope attachment bolt

front pivot pin

center axis of 6.0 in. diameter x 6 feet long axle

guard pieces prevent rope from wrapping up near drum during load lifting demonstrations and force second layer of rope windings to form over first layer

6:00 weighted lever of <u>front</u> internal one-directional wheel

wooden radial stop piece

This method also used in the Kassel wheel

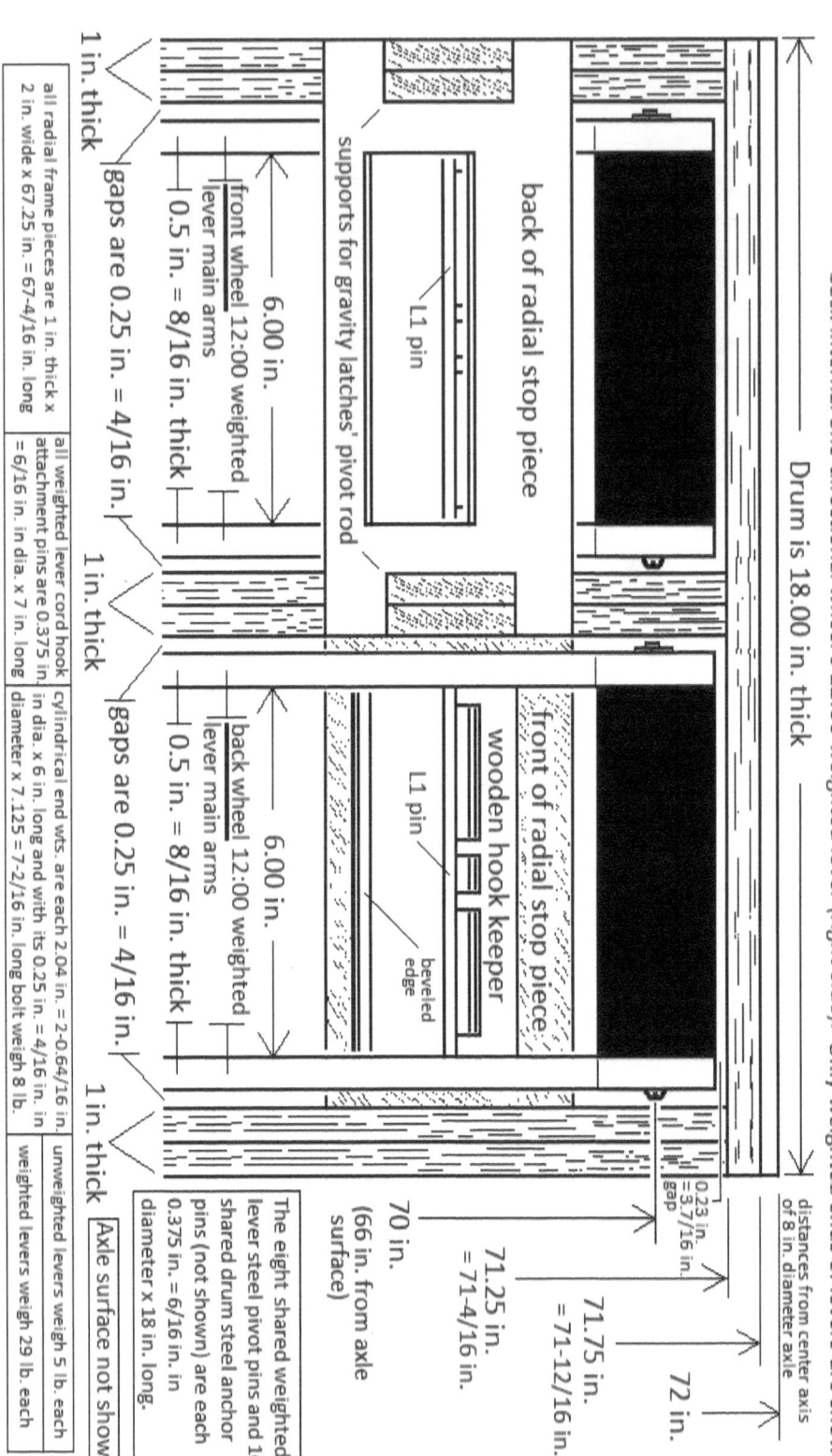

Figure 12(e) - More Details of the Kassel Wheel's Construction. View from right side of drum. Interior profile view of drum's front internal one-directional wheel's 12:00 weighted lever (left side) and back internal one-directional wheel's 12:00 weighted lever (right side). Only weighted ends of levers are shown.

CHAPTER 7
HOW BESSLER MADE THE MERSEBURG AND KASSEL WHEELS BIDIRECTIONAL

As was mentioned in Chapter 1, by the time Bessler was publicly exhibiting his one-directional Draschwitz wheel in 1714, he was beginning to hear skeptics claiming that his wheels' motions were merely the result of cleverly concealed mainspring powered clockwork mechanisms and, to reinforce their beliefs, they pointed to their relatively weak power outputs, that they could only turn in one direction upon having their drums released, and that they only ran for relatively short time periods during his demonstrations of their capabilities.

Then, in September of that year, the inventor received a visit from one of the most famous scientists in Europe, Gottfried Wilhelm Leibniz himself. Leibniz directed the testing of the Draschwitz wheel while carefully observing it and making various measurements of its ability to accelerate and lift loads of various masses. He also listened to Bessler's laments about his inability to convince many of the average people at his public demonstrations that his wheels were, in fact, genuine imbalanced perpetual motion wheels without having to resort to the unacceptable measure of actually opening up their drums so that everyone could see that the drums contained no gears, mainsprings, or anything else related to clock movements.

The solution was simple Leibniz said. All the inventor had to do was make a larger and more powerful wheel and let it run constantly for a period of a few weeks at a minimum. That, the scientist believed, should be sufficient to convince even the most skeptical of the men of science that Bessler's wheels were not frauds. But, it would also be nice if the Bessler could figure out a way to make his wheels bidirectional; that is, somehow modify their internal mechanics so that they could be started up and would run in *either* direction of a wheel's two possible directions of rotation although the older man could offer no advice on how to achieve this last enhancement.

We must keep in mind that Leibniz had no idea of how Bessler's wheels actually worked and soon realized that it was not a subject that the latter wanted to discuss in any great detail. At most, the young inventor probably mentioned to Leibniz that his wheels were simple imbalanced wheels whose centers of gravity always remained on their descending sides because of the unique mechanics they employed. Leibniz would have accepted that description as valid, but then assumed that, somehow, Bessler's secret mechanics were able to use some previously unsuspected force of nature to maintain the imbalance during their rotation of which, perhaps, even Bessler was unaware. Leibniz, like most of the scientists of his day, did not believe that an isolated piece of machinery could just suddenly start operating by itself unless it obtained power from outside of itself in some way for immediate use or, like a clock that is wound up, could obtain the external power at one time, store it internally in some way, and then use it later for its operation. In reality, Bessler wheels were actually doing just that, but it would take another two centuries before the very unobvious process involved would be fully understood.

Bessler did not take Leibniz' advice lightly and almost immediately began to ponder making the suggested changes in his next wheel which, as we saw, would be constructed in the town of Merseburg, Saxony. It would be larger in diameter than the Draschwitz wheel with heavier weighted levers and sturdy enough to hold up during weeks of continuous operation. But, how could he make it bidirectional so as to finally remove that lack of capability from the arsenal of his detractors' objections?

The first approach that would have occurred to Bessler, in order to keep the Merseburg wheel's drum as thin as possible, was to make a bidirectional wheel using only a *single* one-directional wheel. How could this be done?

The only practical method would be to have a wheel within a wheel type of design in which one mounts a single one-directional wheel inside of the bidirectional wheel's larger external drum so that the internal one-directional wheel can rotate independently of the outer drum *and* its attached axle. To do this requires that the radial frame pieces of the internal one-directional wheel be attached to a metal collar that fits around the hidden section of the bidirectional wheel's axle that is inside of its drum and can rotate around it. Ideally, the collar should snuggly fit around a metal sleeve that is attached to a portion of the hidden section of the axle inside of the larger external drum. The contact area between the metal collar and axle's attached metal sleeve would be well lubricated.

Then, when Bessler wanted this type of bidirectional wheel's external drum and its attached axle to turn in the same spontaneous start up direction of its internal one-directional wheel, he would just need to make sure that the collar of the internal one-directional wheel was securely attached to the drum's axle and then release the external drum so that the internal one-directional wheel's eight weighed levers could start rotating the axle and its attached drum in the same direction that the internal one-directional wheel was designed to start up.

Before releasing the external drum, however, the inventor might reach through a small slit cut into its back side layer of concealing linen near the axle and make sure that the pointed end of a hand tightened bolt on the single internal one-directional wheel's metal collar was pressing tightly against an exposed wooden surface of the hidden section of the axle inside the larger external drum that extended beyond the end of the metal sleeve attached to the axle. This bolt would securely attach the internal one-directional wheel to the drum's axle and they could then begin spontaneously rotating together just as soon as the external drum was released.

However, making the axle and its attached external drum turn in the *opposite* direction to the spontaneous direction of rotation of this type of bidirectional wheel's single internal one-directional wheel would be more complicated.

In order to do that, Bessler would have to stop the drum's rotation and then again reach into the drum through the slit in its back linen cloth covering near the axle, but this time he would activate an additional mechanism built into the drum that, like the internal one-directional wheel already there, also had a metal collar that snuggly embraced the metal sleeve that surrounded and was fixed to *most* of the length of the hidden section of wooden axle inside of the external drum and could rotate around it, but was also able to slide a little along the length of that metal sleeve. This additional mechanism would then be slid toward the metal collar of the external drum's internal one-directional wheel and, once reaching it, would then mesh the teeth of an idler gear it carried with the teeth of a metal ring gear attached to and surrounding the end of the internal one-directional wheel's metal collar.

Once that was done, the idler gear would then be able to transfer any torque produced by the internal one-directional wheel to an identical metal ring gear securely attached to an exposed portion of the external drum's *wooden* axle. When the teeth of these various gears were firmly enmeshed with each other, Bessler would then completely *loosen* the bolt he had first tightened in order to directly affix the internal one-directional wheel's metal collar to the hidden wooden section of axle inside of the larger external drum and then release that drum. The imbalanced internal one-directional wheel, without changing its preferred start up direction of rotation, would then drive the surrounding external drum and its attached axle in the *opposite* direction of the rotating single internal one-directional wheel and the metal collar to which it was attached.

The obvious problem with this approach is that, in order for it to work and allow a wheel to display bidirectionality, it would require that the additional mechanism carrying the idler gear be one that was mechanically independent of *both* the internal one-directional wheel and the external drum and the axle to which it was attached and would also be able to serve as a "fixed" point within the external drum. The additional idler gear carrying mechanism, which must be able to slide along the axle's metal sleeve for a short distance, could only be somewhat fixed in space by attaching a heavy ballast weight to a vertical metal rod that extended downward from its bottom so that the ballast weight was located at the 6:00 position of the external drum.

Stopping such a bidirectional wheel when its external drum and axle were rotating in the opposite direction of the internal one-directional wheel, would result in the internal one-directional wheel causing the ballast weight on the additional mechanism to then swing away from the external drum's 6:00 position until its gravitational retarding torque equaled, in magnitude, the driving torque produced by the internal one-directional wheel's eight weighted levers' offset center of gravity.

At that point the external drum would be tethered to prevent any rotation and, if Bessler wanted to then make the external drum rotate in the opposite direction, he would have to reach into the external drum and manually disengage the additional mechanism's idler gear from the ring gear attached to the metal collar of the internal one-directional wheel which would immediately cause the ballast weight to swing back toward the external drum's 6:00 position and oscillate about its equilibrium point directly under the axle's center axis until it came to rest, manually prevent the internal wheel from self-starting, and then *tighten* the bolt that secured that internal wheel to the wooden section of the bidirectional wheel's axle hidden inside of the external drum. Then, the external drum could finally be released from its tether and it, its axle, and the internal one-directional wheel would all start rotating again in the latter's spontaneous start up direction.

To make this system work, Bessler would have had to invent a novel and rather thin manual mechanical transmission with a 1:1 reverse gear ratio that could efficiently transfer and reverse the torque provided by the internal one-directional wheel to an external drum and its axle which would then turn in the opposite direction. And, this mechanism would have to be one that could be engaged and disengaged in a matter of seconds preferably using only one hand! Not really an impossible task for a skilled clockmaker like Bessler, but certainly a task that would make the duplication of a bidirectional wheel something the average craftsman would not want to undertake and which would have increased the construction cost of such a wheel due to need to use precisely cut idler and ring gears.

Another problem of a manual mechanical transmission being used to change the direction of a bidirectional wheel's rotation is that it would require Bessler to insert his hands into the interior of the wheel's drum from the back side of the wheel. Those members of the public then viewing a demonstration of the wheel could then claim that while he was adjusting the gearing inside of the drum to reverse it rotation direction, he might also be winding up the mainspring(s) of a clockwork type moment that would, as it slowly unwound, propel the drum during the next phase of the demonstration and up to the point at which the inventor decided another direction change was necessary or he decided the demonstration should end because it was time for his small audience to leave the wheel's room so that the next group could come in to view the wheel. And, of course, he would then be suspected of winding the drum's mainspring(s) up before the next group of viewers was allowed into the room.

An obvious solution to this problem would have been for Bessler to equip the bidirectional wheel's drum with an *automatic* mechanical transmission which would be able to reverse the external drum's direction of rotation without Bessler having to reach into the drum and touch any of the mechanics there. The automatic mechanism would be activated whenever the external drum and its axle were stopped so that the next time it was released it would start rotating in the opposite direction. While such an automatic mechanical transmission is physically possible, it would have greatly increased the complexity of the mechanics inside of the external drum and required a great deal of time and money to invent, test, and install in his bidirectional wheels. Most likely, Bessler produced many sketches of variations of such a mechanism before he realized its negative features.

The next possible approach, which also uses a single internal one-directional wheel with eight weighted levers, requires that one find a way to make the cylindrical lead end weight carrying mains arms of a weighted lever capable of swinging from one side of its nearest pair of parallel radial frame pieces to the other side. That means that he would not be able to just use a simple flat piece of wood, a radial stop piece, to span the distance from one radial frame piece to the other to provide a resting place for a weighted lever's main arms which, when moving clockwise about the axle, make contact with this wooden stop piece by the time that their weighted lever's pivot pin's center axis reaches a point about 12° past the 3:00 position of the drum.

This could be achieved by using a wooden support piece that would be attached between the parallel pair of main arms of a weighted lever using an extra pivot pin so that the support piece could swivel between the main arms and then have its free end rest on carefully located spots inside the drum depending upon which direction the drum was made to rotate. As far as the extension springs on each weighted lever were concerned, it would also be possible to have two pairs of springs attached to opposite sides of a lever so that when the lever was stretching one pair of springs, the cords connecting the other pair of springs to the other side of the lever would be slackened.

However, while these modifications might be possible, there would still be the problem of the various coordinating cords between the levers. Basically, the number of long lifter and stop cords would need to be doubled and each weighted lever would then have to have two parallel pairs of A arms and two parallel pairs of B arms. To keep the four types of coordinating cords from contacting and rubbing against each other, the parallel pairs of A and B arms on each side of a weighted lever's single parallel pair of main arms would have to be put into different planes which would then result in a significant increase in the width of the single bidirectional weighted lever.

After considering all of these problems and realizing how complicated they would make the construction of a bidirectional wheel, an alternative solution finally occurred to Bessler. He would simply place two complete and thin as possible one-directional wheels, each containing eight weighted levers, side by side and back to back, inside of a thicker drum. They would be placed as close to each other as possible and, in fact, would even share the sixty-four wooden frame pieces of a middle drum lattice between them as well as eight weighted lever pivot pins, eight extension spring hook attachment anchor pins, and eight stop cord hook attachment anchor pins embedded in the three lattices of an enlarged drum. However, each of the two internal one-directional wheels would be arranged so that its spontaneous start up direction of rotation was *opposite* to that of the other wheel. Then, in order for the bidirectional wheel to rotate in either of its two possible directions, all the inventor would have to do was figure out some way of temporarily locking down *all* of the main arms of the eight weighed levers of *one* of the internal one-directional wheels against their radial stop pieces and thereby inactivating that wheel whenever it was the one turning *opposite* to its normal spontaneous start up direction of rotation and, therefore, undergoing what is referred to as "retrograde rotation" in mechanics.

Once that was done, that inactive internal one-directional wheel's eight weighted levers would all have the center lines of their parallel pairs of main arms in alignment with the center lines of their particular pair of nearby parallel radial frame pieces. With this symmetry achieved, the composite center of gravity of the inactive internal one-directional wheel's eight weighted levers would then be located at the exact center axis of the hidden section of the wheel's axle inside of the drum to which that inactive internal one-directional wheel's two lattices were attached. At this point, the entire enlarged drum, the wooden axle to which it was attached, *and* the inactive internal one-directional wheel would be completely driven by the other *active* internal one-directional wheel whose weighted levers were still free to rotate about their respective pivot pins. This active internal one-directional wheel would be rotating in the spontaneous start up direction it was intended to rotate in and its weighted levers would be maintaining their composite center of gravity on the descending side of the portion of the drum's interior axle to which the two lattices of the active internal one-directional wheel were attached.

But, how could the inventor come up with a simple mechanism that would be able to align all of the inactive internal one-directional wheel's weighted levers' main arms with their respective pairs of parallel radial frame pieces quickly and reliably and could do so without adding too much extra weight to the wheel and further thickening its drum?

Bessler was the type of person who, once presented with a mechanical problem to solve, would obsess about it until he finally found a desirable solution as, obviously, was well demonstrated by his decade long search for an imbalanced pm wheel design that actually would remain imbalanced as it rotated.

As a result of this work ethic, by the beginning of June, 1715, after less than only two months of effort, he had, with the assistance of his brother, finished the 12 feet diameter, bidirectional Merseburg wheel and equipped it with a remarkably simple and reliable mechanical system that allowed it to run in either direction and did so without him having to insert his hands into its drum's interior using the inspection holes that had been cut into the linen covering the back side of the drum's circular face. The details of that mechanism are the subject of this chapter and its operation will be of great importance to any craftsman who plans to duplicate one of Bessler's largest bidirectional wheels.

We can begin by referring to Figure 13(a) which shows a front perspective view of a bidirectional wheel. The wheel's enlarged drum contains two internal one-directional wheels each of which contains eight weighted levers. However, in order to make it easier to see how the individual centers of gravity of each of the two internal one-directional wheels behaved when the single drum containing both wheels was turned in either of its two possible directions of rotation, I have separated the two internal one-directional wheels from each other by splitting the middle lattice they share through its plane into two pieces and then moving both wheels to the ends of the wheel's axle and fixing their radial frame pieces to those locations. (In fact, this could actually have been done with either the Merseburg or Kassel wheel without affecting its operation if each of the internal one-directional wheel's half slice of the middle drum lattice was replaced with a whole lattice, but then it might have been apparent to viewers at public demonstrations that a wheel's bidirectionality depended upon the use of two internal one-directional wheels that took turns rotating their shared axle in opposite directions. That was information that Bessler did not want any reverse engineer having.)

Let us start with a bidirectional wheel in the orientation shown in the *center* of Figure 13(a). As can be seen, the center of gravity of the front internal one-directional wheel is on the right side of the axle's center axis and that wheel, therefore, is designed to normally rotate in a clockwise direction and produce torque in that direction as indicated by the portion of a circle with an arrowhead on it. The center of gravity of the back internal one-directional wheel, however, is on the left side of the axle's center axis and that wheel is, therefore, designed to normally rotate in a counterclockwise direction and produce torque in that direction as indicated by the portion of a circle with an arrowhead drawn on it.

Because the individual centers of gravity of the two internal one-directional wheels are located on opposite sides of the axle's center axis, the two torques they produce are in opposition to each other and the axle, its attached drum, and the two internal one-directional wheels themselves will remain motionless as a consequence of this. If I had not split the bidirectional wheel's middle lattice into two pieces as shown in Figure 13(a), but had allowed the wheel's two internal one-directional wheels to remain together near the middle of the axle and share the middle lattice there, then their *composite* center of gravity would be located directly under the axle's center axis and in the plane of the middle lattice. That location would then be the *punctum quietus* for the composite center of gravity of the two internal one-directional wheels inside of the bidirectional wheel's enlarged drum.

Now we must turn our attention to the right side of Figure 13(a) which shows what happens to the centers of gravity of the drum's two contained internal one-directional wheels when the drum is given a push that makes it rotate clockwise as viewed from the front end of the axle.

As can be seen, the front internal one-directional wheel is rotating in the direction, clockwise, that it was designed to rotate in and its center of gravity remains on the right side of the axle's center axis to produce a clockwise torque on the axle. The back internal one-directional wheel also turns clockwise, however, it is forced to rotate *opposite* to the direction it was designed to rotate in which is counterclockwise. The consequence of this forced retrograde rotation of the back internal one-directional wheel is that, after a certain amount of rotation, all of its eight weighted levers will have their main arms securely locked down against their wooden radial stop pieces and the composite center of gravity of all of *that* back wheel's weighted levers will be located at the exact center axis of the axle where it can contribute no torque to the axle. As a result of this, the bidirectional wheel and its axle, including the inactive back internal one-directional wheel contained inside of its enlarged drum, are propelled in a clockwise direction solely by the front active internal one-directional wheel.

Although the back internal one-directional wheel's center of gravity is located at the center axis of the axle, the composite center of gravity of *both* internal one-directional wheels' sixteen weighted levers is still located to the right side of the axle's center axis and, if I had not split the middle lattice in two and moved the two internal one-directional wheels to opposite ends of the axle, would be located in the plane of the middle lattice and to the right side of the axle's center axis.

Finally, we must consider the left side portion of Figure 13(a) which shows what happens when the bidirectional wheel in the center of the figure is given a push that makes its drum rotate in a counterclockwise direction.

Now we have a situation in which it is the back internal one-directional wheel that is rotating in the counterclockwise direction that is was designed to rotate and, as a result, the center of gravity of its eight weighted levers remains on the left side of the axle's center axis.

The front internal one-directional wheel, however, is now forced to rotate in the direction opposite to the clockwise direction that it was designed to rotate and this, after a certain amount of drum rotation, results in all of the main arms of its eight weighted levers being securely locked down against their wooden radial stop pieces. Because of this, the center of gravity of the front internal one-directional wheel's eight weighted levers is moved up to the center axis of the axle where it can provide no torque to the axle. When this happens, the entire bidirectional wheel and its axle, including the front, now inactive, internal one-directional wheel, will be propelled solely by the still off set center of gravity of the active back internal one-directional wheel.

Although the center of gravity of the front internal one-directional wheel is located at the center axis of the axle, the composite center of gravity of *both* internal one-directional wheels' sixteen weighted levers is still located to the left side of the axle's center axis and, again, if I had not split the middle lattice in two and moved the two internal one-directional wheels to opposite ends of the axle, the composite center of gravity of both internal one-directional wheels would be located in the plane of the middle lattice and to the left side of the axle's center axis.

After studying Figure 13(a), one might wonder what would happen if one of Bessler's bidirectional wheels, either the Merseburg or Kassel wheel, was allowed to run for a while in one of its two possible directions of rotation and was then forced to stop by applying drag to its enlarged drum's outer rim wall or its axle in order to quickly dissipate all of the wheel's rotational kinetic energy by converting it to thermal energy and the drum or axle was then released?

The answer is that the axle and its attached drum would, immediately upon being released, start turning and accelerating in the *same* direction that they had been rotating before they were stopped. The reason for this is that all eight of the parallel pairs of the main arms of its *inactive* internal one-directional wheel's eight weighted levers would still be in contact with and locked down against their particular wooden radial stop pieces. With this configuration the inactive internal one-directional wheel can still contribute no torque to the axle because the center of gravity of its eight weighted levers would still be located at the center axis of the axle. However, all eight of the weighted levers of the *active* internal one-directional wheel contained within the bidirectional wheel's enlarged drum would still be free to rotate about their pivot pins and the composite center of gravity of all sixteen of the bidirectional wheel's weighted levers would still be located on whichever side of the axle's center axis that had been the descending side prior to the wheel's rotation being stopped.

The above situation would persist every time that the rotation of the drum and axle of the bidirectional wheel was stopped and then released. Next, one might wonder how, once started in either direction, a bidirectional wheel could be brought to a stop so that it would remain motionless and not spontaneously begin rotating again once released. The method for achieving this is actually simple.

In order to completely stop one of Bessler's bidirectional wheels after it had been running for a while in one of its two possible rotational directions or rotation, one would have to stop it completely and, once that was done, immediately *counter* rotate the drum through an angle of between a quarter and a half of a complete drum rotation which is about 135°. That action would then cause some of the weighted levers of the *active* internal one-directional wheel inside of the drum to have their main arms *locked* into contact with their wooden radial stop pieces while it would simultaneously cause some of the locked up weighted levers of the *inactive* internal one-directional wheel to become *unlocked* from contact with their radial stop pieces and, thus, free to again rotate about their pivot pins.

As a result, the center of gravity of the eight weighted levers of the active one-directional wheel would still be displaced onto the drum's *previously* descending side while the center of gravity of the eight weighted levers of the inactive one-directional wheel would be moved away from the center axis of the axle and now displaced onto the drum's *previously* ascending side. Since the displacements of the individual centers of gravity of both of the drum's internal one-directional wheels are again on opposite sides of the axle and at the same horizontal distances from the center axis of the axle, there will be no net torque acting on the axle and the entire bidirectional wheel will remain motionless when its drum is released. The *composite* center of gravity of these two internal one-directional wheels' centers of gravity is then located in the plane of the drum's middle lattice and at a point, the *punctum quietus*, located vertically below the center axis of the axle.

At this time, the bidirectional wheel's two internal one-directional wheels can *both* be considered to be 50% active and also 50% inactive with either one having the potential to become fully active if the drum is given a push to make that one-directional wheel rotate in the direction it was designed to spontaneous start up in while the other wheel then becomes fully inactive as it is forced to move in the direction *opposite* to which is was designed to spontaneously start up in. Thus, once so stopped, one of Bessler's bidirectional wheels would remain completely stationary until it was again given a push in either of its two possible directions of rotation.

No doubt, carefully rebalancing the opposing torques of the two internal one-directional wheels of one of his 12 feet diameter, bidirectional wheels was a skill that had to be learned from much handling of the wheel's drum. After one of these wheels' huge drums had been stopped, Bessler would have quickly back rotated it through 90° and then slowed down as the drum approached having been back rotated by about 135°. He would then release the drum and see if it remained stationary. If not, he would have continued to back rotate it slowly and then released the drum again to see if the counter torque equilibrium between the bidirectional wheel's front and back internal one-directional wheels had been achieved. If not, more rotation followed until, finally, he could release the drum and it would remain stationary. But, this equilibrium was a delicate one and, as noted in a certificate given to Bessler which was signed on October 31st, 1715 after the official testing of the Merseburg wheel had been completed, the inventor could make the drum of the 550 pound wheel begin rotating in either of its two possible directions by merely pushing on its outer rim wall with only two fingers!

Achieving the capability of bidirectionality required, as was mentioned above, that special mechanisms be added to Bessler's Merseburg wheel and, later, his Kassel wheel and it is now time to begin discussing these special mechanisms. For that purpose, the reader is now directed to Figure 13(b) at the end of this chapter.

In Figure 13(b) we see the two separate types of latches that had to be used with *each* of the bidirectional Merseburg wheel's sixteen weighted levers and each latch is actually a "gravity latch" because it depends upon its orientation in the Earth's gravity field in order to function. There is the "outer latch" shown at the top of the figure and the "inner latch" below that one (the reason for these names will become apparent when Figure 13(c) is discussed next). Also, for each of the two types of latches shown in Figure 13(b), there is an axial view on the left side and a profile view on the right side. In order for the Merseburg wheel to have been bidirectional, each of its sixteen weighted levers needed to have one inner latch and one outer latch located near the outer end of its parallel pair of main arms.

Each of these two types of latches may be referred to as a "cat's claw latch" because of the unusual shapes at the ends of their longer arms which somewhat resemble a cat's forepaw claws. Each latch, for the Merseburg wheel, basically consisted of two thin pieces of steel that were shaped somewhat like an "L" square, but, instead of an interior angle of 90°, had an angle of 95°. The two "L" shaped pieces of steel were held together by a single bolt located at their shorter ends which passed through a cylindrical lead counter weight and was then secured into place with a square nut. It will be noted that, for the outer latch, the shorter arms and their connecting lead counter weight pointed down relative to the longer arms, while for the inner latch, the shorter arms with their connecting lead counter weight pointed up relative to the longer arms. Thus, although both types of cat's claw latches shown are the same size and mass, they are neither superimposable nor mirror images of each other. As will soon be seen, this asymmetry was critical to their functioning inside of Bessler's bidirectional wheels.

The parameters that I've given for these two types of latches used in the Merseburg wheel are based upon many hours of careful computer modeling that used the same materials that would have been readily available to Bessler. My intention was to keep the design as simple, reliable, and as close to his as possible. Thus, what is provided in Figure 13(b) is the particular design that I believe Bessler would have eventually used. There is an obvious clue in the Bessler literature that points to these types of latches being used, but its description will have to await a later chapter.

The two parallel steel "L" arms used in either of the two types of latches were each only 0.0625 inch = 1/16 inch thick x 0.5 inch = 8/16 inch wide with longer arms whose beveled tips were exactly 5 inches from an imaginary vertical line passed through the center axis of a 0.25 inch =

4/16 inch diameter pivot rod hole located at the corner in the metal piece (note that I'm referring to this as a "pivot rod hole" instead of a "pivot pin hole" in order to distinguish it from the pivot pin hole of a 3 arm side piece used to make a wheel's weighted levers and thereby avoid confusion in the following descriptions). The two parallel shorter arms of an inner latch were extended a bit below the ends of its two parallel longer arms. This was done in order to provide some extra metal about this type of latch's pivot rod holes and, thus, more securely attach the latch to its pivot rod which, in turn, was fixed to the drum via several wooden pieces.

The straight ends of the two shorter arm of either type of latch were exactly 4 inches from an imaginary horizontal line passed through the center axis of the latch's pivot rod hole. At the end of each longer arm of either type of latch, there was a square shaped notch that was 0.25 inch x 0.25 inch = 4/16 inch x 4/16 inch whose center was located about 4 inches horizontally away from the center axis of the latch's pivot rod hole. These notches on *both* the inner and outer cat's claw latches were designed to fit over and enclose the 0.1875 inch = 3/16 inch diameter steel L1 main cord hook attachment pins that spanned the distance between the parallel pairs of main arms of the Merseburg wheel's weighted levers. As either an inner or outer latch's *two* parallel longer arms engaged a single L1 main cord hook attachment pin at the same time, *both* of their square notches would grab onto the wooden hook keeper exposed section of the L1 main cord hook attachment pin which was located on *one* side of the L1 pin's central exposed section (that central section was reserved for the metal hook attached to the end of a main cord).

As shown on the right side profile views in Figure 13(b), each cat's claw latch consisted of two parallel steel "L" arm pieces that were bolted together with a disc-like counter weight of lead held between the ends of their shorter arms. These cylindrical lead counter weights were only 0.5 inch thick by 1.78 inches in diameter and weighed 0.5 pounds = 8 ounces each. Bessler would have made them by simply sawing off slices from the 4 pound cylindrical lead end weights he used at the ends of the parallel pairs of main arms of the Merseburg wheel's weighted levers. Each cat's claw latch's counter weight had a 0.1875 inch = 3/16 inch diameter x 0.5 inch = 8/16 inch long hole drilled through its center axis and was held between either type of latch's parallel pair of shorter arms using a small steel bolt that had a slotted screw head and whose threaded section was 0.1875 inch = 3/16 inch in diameter x 0.75 inch = 12/16 inch long. This bolt was then secured with a small square nut.

Both types of cat's claw latch, inner and outer, had two "L" square shaped steel pieces that *each* weighted 0.078125 pounds or 1.25 ounces. The cylindrical lead counter weight held between the parallel pair of shorter arms in a latch weighed 0.5 pounds or 8 ounces *including* its small attachment bolt. Thus, the total weight of a *single* cat's claw latch of either type used in the Merseburg wheel was (2 x 0.078125 pounds) + 0.5 pounds = 0.65625 pounds = 10.5 ounces. Since each of this bidirectional wheel's sixteen weighted levers required two latches, an inner and outer one, a total of thirty-two latches had to be installed in the wheel's enlarged 12 inches thick drum and they added an extra 32 x 0.65625 pounds = 21 pounds to the total weight of the wheel.

This value, however, does not include the additional weight of the latches' extra sixteen fixed steel pivot rods, each one 0.25 inch = 4/16 inch in diameter x 6.25 inches = 6-4/16 inches long around which the cat's claw latches independently rotated, the sixty-four steel washers soldered to the pivot rods, or the extra wooden support pieces needed to hold the center axes of these sixteen pivot rods at their proper locations inside of the drum. These wooden cat's claw latch pivot rod support pieces would have been made from the same stock pieces of wood used for the other frame pieces inside of a one of the drum's lattices and were attached in parallel pairs between the parallel pairs of a weighted lever's radial and outer octagonal frame pieces. A total of sixty-four of these extra wooden pieces were needed for a bidirectional wheel's sixteen weighted levers.

The sixteen steel pivot rods only added 1.39 pounds, their sixty-four soldered on washers added about 0.45 pound, and their extra sixty-four wooden support pieces added about 5 pounds of weight to the 12 feet diameter bidirectional wheel's total weight or about 5.45 pounds of weight was added. Thus, the *total* extra weight due to the presence of the thirty-two cat's claw latches inside of the bidirectional Merseburg wheel was the sum of the weight of the thirty-two cat's claw latches, 21 pounds, plus the weight of their supporting structures which was about 5.45 pounds and that makes a total weight of about 26.45 pounds. This added weight was a rather small price to pay for achieving the bidirectionality that Bessler was convinced would help him finally silence his skeptics whose doubts were interfering with him finding a buyer for his invention.

As Bessler was constructing the three lattices that would form the width and thickness of the 12 feet diameter drum of the Merseburg wheel, he included the extra wooden support pieces needed to hold the pivot rods for each side by side pair of the two types of cat's claw gravity latches used by each of the wheel's sixteen weighted levers. Only after the three lattice pieces were mounted on the wheel's 6 inches diameter oak axle did he then install their sixteen steel pivot rods and, as he did that, the two types of latches, one inner and one outer, on each of the pivot rods. Unlike the weighted lever pivot pins and hook attachment anchor pins that were embedded in the radial and octagonal frame pieces of a bidirectional wheel's three drum lattices and were shared by two weighted levers from the front and back internal one-directional wheels inside of the enlarged drum, the cat's claw gravity latch pivot rods were only shared by the two types of latches used by *each* weighted lever. Thus, each weighted lever had its own pivot rod and did not share it with another weighted lever in the drum's other internal one-directional wheel and this is why sixteen rather than eight gravity latch pivot rods were required.

While the two ends of each pivot rod were firmly embedded in their two wooden support pieces for each weighted lever and did not rotate, the two cat's claw latches, an inner and outer one, could freely rotate around their common fixed pivot rod. However, there was still the problem of restricting each of the two side by side gravity latches to certain sections of their shared pivot rod.

I'm not convinced that Bessler used the same type of wooden hook keepers that he used to restrict the hooks at the ends of the various types of coordinating cords to certain sections of a coordinating cord hook attachment pin inside of a weighted lever or the extension spring hooks or stop cord hooks to certain sections of their hook attachment anchor pins that were fixed to the drum. Rather, I think he simply added four small steel washers onto each pivot rod as he was adding the inner and outer latches to it and then soldered the washers to the pivot rod so as to restrict each latch to a certain section of the pivot rod which would then align each latch's pair of parallel longer arms and their square end notches with one of the sections of the L1 hook attachment pin exposed by one of the *larger* rectangular side notches in that pin's wooden hook keeper. These washers were probably 0.0625 inch = 1/16 inch thick with a center hole that was 0.25 inch = 4/16 inch in diameter and an outside diameter of 0.75 inch = 12/16 inch in diameter. How they were soldered to the pivot rods is shown in the right side profile views of the two types of latches in Figure 13(b). Since this design uses steel "L" square pieces rotating around a stationary steel pivot rod, the inventor would have kept their contacting surfaces well lubricated with a drop of a vegetable oil like olive oil.

As one might expect, the cat's claw gravity latches required to make the Kassel wheel bidirectional would have been larger and stronger because that wheel's weighted levers had twice the mass of those used in his Merseburg wheel. Figure 13(b) only illustrates the two types of cat's claw gravity latches, inner and outer, used in the Merseburg wheel, but those used in the Kassel wheel would have been geometrically similar with only a few changes made to them.

For the Kassel wheel, Bessler probably only needed to double the widths of the gravity latches that were used in the Merseburg wheel and also the masses of their counter weights.

Thus, a Kassel wheel's steel "L" square piece would have still had shorter and longer arms that were the same lengths as measured from their pivot rod hole's center axis as those in the Merseburg wheel, 4 and 5 inches respectively, and the piece was also still 0.0625 inch = 1/16 inch thick, but it would have been 1 inch wide and weighed 0.15625 pounds or 2.5 ounces each. The cylindrical lead counter weights held between the ends of their shorter arms would still have had a diameter of 1.78 inches, but the thickness of these weights would have been increased to 1 inch which gave them a weight of 1 pound = 16 ounces each and that was double what was used in the Merseburg wheel's latches. These changes then gave each of the Kassel wheel's latches a total weight of 1.3125 pounds or 21 ounces.

The square notches at the ends of the long arms of the Kassel wheel's gravity latches would have been 0.5 inch x 0.5 inch = 8/16 inch x 8/16 inch square so that they could securely enclose an L1 pin that was 0.375 inch = 6/16 inch in diameter. Each of the sixteen steel pivot rods needed for the thirty-two cat's claw gravity latches of the bidirectional Kassel wheel would have been 9.5 inches long by 0.375 inch = 6/16 inch in diameter. The thirty-two cat's claw latches including their counter weights and their attachment bolts, their sixteen pivot rods and their soldered on sixty-four steel washers, and the sixty-four wooden support pieces needed to make the Kassel wheel bidirectional added about 57 pounds to the total weight of that wheel. This value is a little more than double that of the added weight of the Merseburg wheel's thirty-two cat's claw latches and their support structures which was 26.45 pounds mainly because of the larger 0.375 inch = 6/16 inch diameter steel pivot rods used for the Kassel wheel's thirty-two cat's claw latches.

At this point, the reader may be wondering why the cat's claw gravity latches needed to have the precise parameter values that I've described above and illustrated in Figure 13(b). The reason is simple. These values agree with the few clues Bessler provided that describe these latches and also allow each latch's two parallel *longer* arms to remain perfectly horizontal when the latch is free to rotate about its pivot rod. This precise balance of a latch was critical to its proper functioning which is to be shortly described.

Now that I have covered the basic structural features of the cat's claw gravity latches that Bessler needed to use in order to make his largest 12 feet diameter wheels bidirectional, it's time to move on to seeing exactly how these odd shaped latches did this.

But, firstly, I need to provide some extra details about how the two types of latches, both inner and outer, were installed near the rim wall end of each weighted lever's parallel pair of wooden radial frame pieces so that the square holes near the ends of the two longer arms of each latch could engage their particular exposed section of the L1 main cord hook attachment pin that was held between that weighted lever's parallel pair of main arms.

As the reader will recall, the exposed center portion of the length of the L1 main cord hook attachment pin provided an attachment location for a metal hook tied to one of the ends of a one-directional wheel's main cord (the hook tied to the other end was then attached to the B1 main cord hook attachment pin held between the parallel pair of B arms of the leading weighted lever). The main cords were the most important coordinating cords found in all of Bessler's wheels and, literally, functioned like a backbone to connect all eight of a drum's weighted levers together. Three of a one-directional wheel's eight main cords were always stretched tight between the four weighted levers whose pivot pin center axes were located at a clockwise rotating drum's 10:30, 12:00, 1:30, and 3:00 positions. But, with the introduction of bidirectionality to Bessler's wheels, the L1 main cord hook attachment pins served another critically important purpose. They each provided *two* additional sections of the L1 main cord hook attachment pin for the two square notches at the parallel longer ends of *each* weighted lever's two types of gravity latches to grab onto so that, under certain specific conditions, that weighted lever's parallel pair of main arms would be held in contact with its wooden radial stop piece. This will become clearer as the discussion proceeds, but, for the moment, the reader should find and study Figure 13(c).

In this figure I have provided three separate interior views of a weighted lever's parallel pair of main arms and their three attached cylindrical lead end weights which are located at the 6:00 position of a bidirectional wheel's drum. When I presented the reader with the internal structures of the 3 feet diameter, one-directional Gera prototype wheel, I provided diagrams showing all eight of its weighted levers for each of its five layers so that the arrangement of its four types of coordinating cords could be better studied. In the cases of the Merseburg and Kassel wheels descriptions to follow, however, I will only illustrate the most important aspects of their construction which should be sufficient for a skilled and well equipped craftsman to use to construct an entire 12 feet diameter, bidirectional wheel.

We begin with the view shown on the left side of Figure 13(c) which is an interior axial view showing only the end of the parallel pair of main arms of the weighted lever of the *front* internal one-directional wheel of the 12 feet diameter, bidirectional Merseburg wheel whose pivot pin's center axis is located at the 6:00 position of the drum. Note that the front main arm of the parallel pair of main arms of the weighted lever can be considered to be removed or made invisible so as not to obstruct the view of the two types of cat's claw gravity latches' parallel pairs of longer arms that could engage the weighted lever's L1 main cord hook attachment pin. Furthermore, only the front arm of each type of latch's parallel pair of longer or shorter arms can be seen because it hides the back longer or shorter arm from view.

The first thing that is noticeable is that there are two of the cat's claw gravity latches, an inner and outer latch, which are both mounted on the same pivot rod near the front internal one-directional wheel's 6:00 weighted lever and free to rotate about it within limits. Their common pivot rod's ends are embedded in a parallel pair of short wooden support pieces (but, we only see the *back* support piece in Figure 13(c) because the front piece has been removed so as not to hide the cat's claw levers) made from the same wood stock as the drum's radial frame pieces that each have a rectangular cross section that is 0.5 inch = 8/16 inch thick x 2 inches wide and a length of about 3.5 inch = 3-8/16 inches and which is attached to the *back* side of the 6:00 weighted lever's wooden radial stop piece which is 0.5 inch = 8/16 inch thick. (I previously stated that the back surface of a radial stop piece was unused. This is only true for one-directional wheels, but for bidirectional wheels that surface can be used with the support pieces for a weighted lever's gravity activated cat's claw latches). That parallel pair of pivot rod support pieces is then attached to another even shorter parallel pair of wooden brace pieces with the same rectangular cross section whose beveled ends are, in turn, attached to the weighted lever's leading parallel pair of outer octagonal frame pieces (but, again, in Figure 13(c), we only see the back brace piece and the back middle lattice's outer octagonal frame piece to which the brace piece is attached so as not to hide the two latches).

It should now be apparent why I have labeled one of the gravity latches as the "inner latch" and the other as the "outer latch". The inner latch is the one whose two parallel shorter counter weight carrying steel arms point toward the interior of the drum (but not directly toward the center of the axle), while the outer latch's two parallel shorter counter weight carrying steel arms point away from the drum's interior. Another way to think of the difference between the two types of cat's claw gravity latches is that an inner latch keeps its counter weight's center of gravity closer to the center axis of the drum's axle than does the outer latch.

While the shorter counter weight holding arms of both types of cat's claw gravity latches point in nearly opposite directions, the longer arms of both types of latches have square notches at their ends that open on the bottom edges of the longer arms. This is so that the two parallel notches of both types of latches can engage the L1 main cord hook attachment pin of a weighted lever from the same curving side which is its side that faces the weighted lever's pivot pin.

In the left side drawing of Figure 13(c) it will be noted that the outer cat's claw gravity latch is positioned in front of the inner latch and obscures the portion of that latch's two pivot rod holes that ride on the pivot rod. It will also be noted that the two longer arms of the outer latch (again recall that we only see the front longer arm in the figure which blocks our view of the parallel back longer arm) are horizontally oriented and that their two square end notches have engaged the L1 main cord hook attachment pin held between the main arms of the 6:00 weighted lever whose left and leading surfaces are in contact with that lever's wooden radial stop piece (also again recall that we only see the back main arm because the front one can be considered removed or made invisible).

In this situation, the 6:00 weighted lever's two parallel wooden main arms can be said to be "locked down" against their wooden radial stop piece which, of course, then means that the center lines of the main arms are both parallel to the center lines of their lever's nearest radial frame pieces. When this happens, the bottom edges of the outer latch's two parallel longer arms do not touch any portion of the lever's L1 main cord hook attachment pin and it is only the deepest edge within the square notch that makes any contact with the L1 main cord hook attachment pin. At this time, these longer arms may or may not be in physical contact with the upper surface of the lower side of the rectangular hole cut into the wooden radial stop piece. Since this contact can take place, this surface as well as the beveled surface of the upper side of the rectangular hole in the radial stop piece, as indicated in Figure 13(c), should have a thin layer or lining of steel attached to them to prevent wear whenever contact is made.

For the inner cat's claw gravity latch, however, the situation is different. The square notches at the ends of its two parallel longer arms (only the front one of which is visible in the Figure 13(c)) do not engage the L1 main cord hook attachment pin of the front internal one-directional wheel's 6:00 weighted lever and the latch's two parallel longer arms have rotated a little counterclockwise and away from the L1 main cord hook attachment pin. Their motion, however, stopped just as soon as they made contact with the inner surface of the upper edge of the rectangular hole cut into the 6:00 weighted lever's wooden radial stop piece. That surface they contact is beveled at the angle needed (about 14° away from an imaginary perpendicular line passing through the thickness of a 0.5 inch = 8/16 thick radial stop piece) so that the upper edges of the two parallel longer arms of the inner gravity latch will distribute the force they apply to the beveled surface across the largest area possible. As mentioned above, the upper beveled inside surfaces of the rectangular holes cut into the radial stop pieces should be lined with thin layers of steel to prevent wear to them by the constant impacts from both types of cat's claw gravity latches during prolonged drum rotation.

The wooden radial stop pieces for the Merseburg wheel or a replica of it should measure 0.5 inch = 8/16 inch thick x 4 inches wide x 6.25 inches = 6-4/16 inches long. For the Kassel wheel they should measure 1 inch thick x 4 inches wide x 9.5 inches = 9-8/16 inches long.

As far as the rectangular *holes* cut into the radial stop pieces are concerned, for the Merseburg wheel or a replica of it, they should be 0.5 inch = 8/16 inch thick (or deep) x 1.15625 = 1-2.5/16 inches minimum width on the radial stop piece's back side surface beveled out to 1.28125 inches = 1-4.5/16 inches maximum width on the radial stop piece's front side surface (these widths include the two thicknesses of the thin metal lining) x 4 inches long. For the rectangular holes cut into the radial stop pieces of the Kassel wheel or a replica of it, they should be 1 inch thick (or deep) x 2.3125 inches = 2-5/16 inches minimum width on the radial stop piece's back side beveled out to 2.5625 inches = 2-9/16 inches maximum width on the radial stop piece's front side surface (these widths include the two thicknesses of the thin metal linings) x 6 inches long.

For the Merseburg wheel replica, the rectangular holes in their radial stop pieces should be cut so that their lower flat inside surfaces, *including* the thicknesses of the metal linings on these surfaces, are exactly 0.15626 inch = 2.5/16 inch offset from the exact center of the 5 inch wide radial stop piece and lie along its width. For the Kassel wheel replica, the rectangular holes in their radial stop pieces should be cut so that their lower flat inside surfaces, *including* the thicknesses of the metal linings on these surfaces, are exactly 0.3125 inch = 5/16 inch offset from the exact center of the 4.5 inch wide radial stop piece and lie along its width. The reader may notice that the exact center of each wooden radial stop piece also has another interesting geometric property. If one was to extend an imaginary line through it that was perpendicular to the parallel front and back surfaces of the radial stop piece, that line would pass directly through the center axes of both the shared pivot rod of the two types of gravity latches and their weighted lever's L1 main cord hook attachment pin.

When the upper beveled inside surfaces of the rectangular holes in the radial stop pieces have been formed and *before* their metal lining pieces have been attached in *either* a Merseburg or Kassel wheel replica, they should be checked with a goniometer to verify that they slope away from an imaginary perpendicular line drawn from the front to back surfaces of a radial stop piece by an angle of exactly 14°. If not, then the surfaces must be carefully sanded away until this angle is achieved and then the metal lining can be attached. This particular angle is critical to the upper edges of the parallel longer arms of both types of gravity latches making contact along their full contacting areas with the metal lined inside surface of a bidirectional wheel's wooden radial stop pieces and thereby reducing damage to both the longer arms and the metal lining. Unfortunately, there is not much that can be done to reduce the tapping sounds associated with these metal to metal contacts constantly taking place during drum rotation, but Bessler probably welcomed them because he knew they were mainly issuing from near the enlarged drum's 6:00 and 12:00 positions which helped mask the thumping sounds of weighted levers landing their parallel pairs of main arms on their radial stop pieces as the center axis of a weighted lever's pivot pin reached about 12° past the 3:00 position of a clockwise rotating drum as viewed from the front of the axle.

The craftsman seeking to replicate either a bidirectional Merseburg or Kassel wheel should first attach each pair of cat's claw gravity latches, minus their counter weights, to its weighted lever's wooden radial stop piece before finally attaching the radial stop piece to the lever's two radial frame pieces. That will allow him to slide the radial stop piece up and down a fraction of an inch along the length of the radial frame pieces to make sure that the notches in both types of gravity latches will securely enclose the lever's L1 main cord hook attachment pin while the bottom edges of the two parallel pairs of longer arms of the two latches just barely touch the surface of the metal lining covering the flat bottom inside surface of the rectangular hole cut into the radial stop piece. At that point, the two wooden brace pieces can be inserted and glued into place between the two gravity latches' pivot rod's wooden support pieces and the weighted lever's leading parallel pair of wooden outer octagonal frame pieces. Because of slight misalignments in positioning from one radial stop piece to another, it's a good idea to check the overall balance of the drum after all sixteen of its radial stop pieces and their thirty-two gravity latches, including their thirty-two attached counter weights have been installed in the drum. If any imbalance is found, it should be compensated for by the addition of small lead weights to the appropriate locations on the drum's outer octagonal frame pieces until perfect balance is achieved.

Due to space limitation in the left side drawing of Figure 13(c), I did not illustrate the two types of gravity latches that were attached to the wooden radial stop piece of the *back* internal one-directional wheel's 6:00 weighed lever. However, those latches' wooden support pieces would be attached to the radial stop piece of that back internal wheel's 6:00 weighted lever which was, in turn, attached to the *right* sides (as viewed from the front side of the drum) of the parallel pair of 6:00 radial frame pieces of the drum's middle and back lattices. Those unseen back gravity latches' brace pieces would then be attached between the latches' wooden support pieces and the nearest parallel pair of outer octagonal frame pieces of the drum's middle and back lattices. As a result of this, the two types of gravity latches *and* their wooden support structures, including their radial stop piece, for the back one-directional wheel's 6:00 weighted lever would appear to be the mirror image of the two latches for the one front one-directional wheel's 6:00 weighted lever which is illustrated in the figure.

(Although one cannot see the back internal one-directional wheel's 6:00 weighted lever in the left side drawing of Figure 13(c) because the lower ends of its main arms and their attached cylindrical lead end weights are hidden behind the drum's middle lattice 6:00 radial frame piece and two of the cylindrical end weights attached to the front internal one-directional wheel's 6:00 weighted lever and also its parallel pairs of A and B arms are not shown, that back internal one-directional wheel's 6:00 weighted lever would also be mounted so that, when viewed through the front lattice from the front side of the drum, it would appear to be a mirror image of the front internal one-directional wheel's 6:00 weighted lever. We see from this that Bessler used the same design for both of a bidirectional wheel's two internal one-directional wheels, however, they were mounted on their common axle so that they were, essentially, in a "back to back" relationship to each other. The reader may be better able to visualize this arrangement by taking two identical coins and then pressing them together so that their two "tail" sides are in contact. Then, by using imaginary x-ray vision so that he can look through one of the "head" sides of the two joined coins to see the head side of the other coin, he would notice that, if properly aligned, the two head sides would appear to be mirror images of each other.)

Now it's time to move on to a discussion of the center drawing in Figure 13(c).

There we again see a cross sectional view of the Merseburg wheel's front one-directional wheel's 6:00 weighted lever, however, this time the view is a profile view from the right side of the drum. That is, this is the view one would see if he walked into the room containing this 12 feet diameter, bidirectional wheel and, facing the front side of the drum which is the first side he would see, slowly walked around to the right side of the drum and then stopped when in the plane of the drum. He would then turn to his left until he faced the outer rim wall of the drum so as to have a profile view of it. Next, after dropping down onto his knees and using his imaginary x-ray vision, he would look into the bottom of the drum so that he could see both the front internal one-directional wheel's 6:00 weighted lever on the left side of the 6:00 radial frame piece of the drum's middle lattice and the 6:00 weighted lever of the back internal one-directional wheel on the right side of that middle lattice's 6:00 radial frame piece. Again, because of space limitations, there was only enough room in the center drawing of Figure 13(c) to include a small portion of the bidirectional Merseburg wheel's back internal one-directional wheel's 6:00 weighted lever and its radial stop piece.

As can be seen, the 6:00 weighted lever of the front internal one-directional wheel on the left side of the drum's middle lattice has its two parallel main arms in contact with the front side of their wooden radial stop piece. In this view, we can see the full 4 inch length of one of the weighted lever's three attached 4 pound cylindrical lead end weights (which includes the mass of its attachment bolt) that is held between the ends of the lever's two parallel main arms as well as those arms' L1 main cord hook attachment pin. Also visible is the angled wooden hook keeper just below the L1 main cord hook attachment pin whose upper notched edge is in contact with the pin and exposes three sections of its 4 inch length.

There is a square notch in the center of the wooden hook keeper and it is flanked by two larger rectangular notches. The center square notch is for the end hook of a main coordinating cord that would be attached to the L1 main cord hook attachment pin at the center section of the pin exposed by the notch (note that the metal hook as well as the main cord attached to it are not shown in order to prevent the drawing from becoming cluttered). The larger rectangular notches on each side of the center square notch of the wooden hook keeper expose sections of the L1 main cord hook attachment pin that are used solely by the two types of cat's claw gravity latches assigned to that front internal one-directional wheel's 6:00 weighted lever.

The reader will notice in the central drawing of Figure 13(c) that the square notches at the ends of the two parallel longer arms of the *outer* cat's claw gravity latch are engaged with the L1 main cord hook attachment pin and that the bottom edges of these two longer arms almost or actually do touch the lower, thinly metal clad inner surface of the bottom edge of the rectangular hole cut into the weighted lever's wooden radial stop piece. The notches of the two parallel longer arms of the *inner* cat's claw gravity latch, however, are not engaged with the L1 main cord hook attachment pin but the *upper* edges of those two arms are in contact with the rectangular hole's thinly metal clad, upper *beveled* surface. Since the inner surfaces of the two parallel longer arms of each type of gravity latch are 0.5 inch = 8/16 inch apart and the metal arms are each 0.0625 inch = 1/16 inch thick, this means that the rectangular notches in the L1 main cord hook attachment pin's wooden hook keeper for them must each be at least 0.625 inch = 10/16 inch long. In practice, however, the notches would have been longer than this and, most likely, 1 inch long in order to compensate for any unwanted axial motions of a weighted lever's parallel pair of main arms that could cause the metal at the ends of the longer arms of the two types of gravity latches to make contact with the wooden hook keeper and wear it away over time from repeated contacts during prolonged drum rotation.

Finally, we come to the right side drawing in Figure 13(c) which shows a *top* view of what was shown in the center drawing of the figure. Here we see the two side by side, outer and inner cat's claw gravity latches mounted on their common pivot rod which is fixed to the drum via their parallel pairs of wooden support and brace pieces (the brace pieces cannot be seen because they are hidden by the support pieces). The two latches are free to rotate around the pivot rod within an arc of only about 14° that is imposed on them by the widths (which appear as heights in the figure) of the rectangular holes cut into the radial stop piece of the front one-directional wheel's 6:00 weighted lever. The four steel washers soldered onto the pivot rod on the sides of the two types of latches help restrict their longer arms to their allocated sections of the pivot rod's length.

The parallel pairs of longer arms of each type of cat's claw gravity latch both pass through the rectangular hole cut into the wooden radial stop piece that is attached to the Merseburg wheel's front internal one-directional wheel's 6:00 front and middle lattice radial frame pieces. The square notches at the ends of the parallel pairs of longer arms of the two types of latches can then swing down into the plane of the page and engage the L1 main cord hook attachment pin of the front 6:00 weighted lever.

However, as shown in the right side drawing of Figure 13(c), only the two parallel longer arms of the *outer* gravity latch (the right one of the side by side pair of latches) have come down so that their square notches are engaging the L1 main cord hook attachment pin. The inner gravity latch's parallel pair of longer arms is raised up above the L1 hook attachment pin and its square holes do not engage the pin. In this drawing we can see the wooden hook keeper located just below, yet in contact with sections of the length of the L1 main cord hook attachment pin. The black rectangles on it represent the notches in this hook keeper. There is one center notch that exposes the section of the L1 main cord hook attachment pin to which the metal hook of a main coordinating cord is attached (again, neither the metal hook nor the main cord are shown in this figure). The larger left side rectangular notch of the hook keeper exposes the section of the L1 main cord hook attachment pin which is engaged by the square notches in the parallel pair of longer arms of the *outer* cat's claw gravity latch *when* it is in use. And, the larger right side rectangular notch of the hook keeper exposes the section of the L1 main cord hook attachment pin which is engaged by the square notches of the parallel pair of longer arms of the *inner* cat's claw gravity latch *when* it is in use.

Again, the reader should keep in mind that Figure 13(c) only shows one pair of inner and outer cat's claw gravity latches that are assigned to a single one of the sixteen weighted levers inside of Bessler's bidirectional

Merseburg wheel which happens to be the weighted lever of the front internal one-directional wheel whose pivot pin's center axis is located at the drum's 6:00 position.

In order to make that wheel bidirectional, however, the inventor had to have such a *pair* of inner and outer gravity latches on *each* of its 16 weighted levers! That's a total of thirty-two latches contained within a bidirectional wheel's drum. The effort needed to manufacture all of these latches, install them, and then individually adjust them to function reliably must have been enormous. These adjustments would have involved carefully filing away some of the metal of the three inner edges of the square notches at the ends of the longer arms of the latches in order to enlarge them a little as well as removing some of the metal at their beveled ends to make sure a latch's two square notches could, when in use, completely and securely engage and then quickly disengage from an L1 main cord hook attachment pin at the right times during drum rotation. Most likely, he would have had to carefully bend the longer arms of a latch to bring them into parallel alignment with each other and assure that they were not making contact with the sides of the rectangular notch of their L1 main cord hook attachment pin's wooden hook keeper. He may also have had to have added or subtracted small amounts of lead from the cylindrical counter weights of the latches in order to assure that, if free to rotate completely around their pivot rods (which the radial stop pieces would never actually permit), the parallel pairs of longer arms of both types of latches would remain perfectly horizontally oriented when released.

But, Bessler was motivated by a burning desire to prove to people witnessing his wheels that they were not just fakes driven by clockwork movements because, unlike the individual gears found in the gear trains of such movements, his wheels could turn in either of two possible directions. A mechanical clock's hands can only move clockwise as a result of the one-directional rotations of the individual gears in the gear train to whose extended pivots the hands are attached, but the drum's of a Bessler's bidirectional wheel could, depending upon the choice of its operator, turn either clockwise or counterclockwise.

At this point the reader should have a good idea of how the individual cat's claw gravity latches were shaped and what their various physical parameters were. However, I have still not covered in detail exactly *how* they functioned in order to allow one of Bessler's 12 feet diameter wheels to achieve bidirectionality. To understand that, the reader is now directed to Figure 14 which shows the two latches of a single weighted lever in operation as the lever's main arms and their trio of cylindrical lead end weights traveled around a wheel's axle in either of its two possible directions (unfortunately, due to space constraints, only the end of the lever's main arms is shown and we do not see the rest of the main arms, the A and B arms, or the lever's pivot pin). Once again, the figure only shows the back main arm of a weighted lever and the three end weights attached to it. The front main arm has been removed or made invisible so as not to hide the positions of the longer arms of the inner and outer gravity latches as their square notches engaged and disengaged from the L1 main cord hook attachment pin of the weighted lever during drum rotation.

The important thing to remember about a pair of gravity latches, one inner and one outer, assigned to any of a bidirectional wheel's sixteen weighted levers is that they were carefully designed to only release a weighted lever's L1 main cord hook attachment pin so that the lever could then freely rotate around its pivot pin when that lever's internal one-directional wheel was rotating around its axle's center axis in the direction it was designed to rotate in which means that the internal wheel is rotating in its spontaneous start up direction. If, however, the internal one-directional wheel was forced to rotate counter to the direction it was designed to rotate in and, thus, undergoing retrograde rotation, then the weighted lever's pair of gravity latches would, through the use of the square notches at the ends of their parallel pairs of longer arms, actually take turns engaging the lever's main arms' L1 main cord hook attachment pin and thereby keeping that lever's parallel pair of main arms in *constant* contact with their wooden radial stop piece. When this "locked down" condition affected all eight of the weighted levers inside of an internal one-directional wheel undergoing retrograde rotation inside the enlarged drum of a bidirectional wheel, their collective or composite center of gravity would then be located at the exact center axis of the wheel's axle as was illustrated in Figure 13(a) where it could provide no torque to the axle. This should become clearer as we continue the discussion of Figure 14.

In order to concentrate solely on the interactions of the two types of cat's claw gravity latches with a single weighed lever's L1 main cord hook attachment pin, Figure 14 does not show an entire internal one-directional wheel with all of its eight weighted levers. Also, because of the difficulty of drawing small versions of the parallel pairs of "L" square shaped side pieces of the gravity latch levers, I have simply depicted them with stick-like figures with circles at the ends of their shorter arms to represent their attached cylindrical lead counter weights and hooks at the ends of their longer arms to represent the two square notches of an actual latch's two parallel longer arms.

For the inner latches, the circles representing their shorter arms' lead counter weights are "unfilled" and, thus, appear as black circles on the white page with their small unfilled black center circles representing their small attachment bolts. For the outer latches, the circles representing their lead counter weights are "filled" and, thus, appear solid black with the exception of their small white center dots that represent their small attachment bolts. It will also be noticed that the lines representing the longer and shorter arms of the *outer* cat's claw gravity latches are thicker and thus bolder than those representing the arms of the inner gravity latches which was done to make it easier for the reader to visually distinguish between the two types of gravity latches' parallel pairs of longer arms only the front arms of which are visible in the figure. The end portion of a weighted lever's back main arm is represented by a short segment of parallel lines. Between these lines there is a small unfilled black circle that represents the L1 main cord hook attachment pin and three solid black circles with white center dots that represent the circular cross sections of the weighted lever's three attached cylindrical lead end weights. The weighted lever's radial stop piece is represented in the figure by a long, thin rectangle in contact with the side of the lever's back main arm. The longer arms of the two types of latches pass through a rectangular cutout in the radial stop piece which limits their swinging motions to an arc of only 14°.

In the top two thirds of Figure 14 the reader can see how the two types of cat's claw gravity latches, inner and outer, behaved when the weighted lever to which they belonged was forced to rotate around the wheel's axle in the *opposite* direction, in this case counterclockwise, to which its internal one-directional wheel was intended to spontaneous start.

Just before a weighted lever crossed the 6:00 position of the drum, its main arms were held against the lever's wooden radial stop piece by the *inner* cat's claw gravity latch whose square notches at the ends of its longer arms were engaging the L1 main cord hook attachment pin located between the weighted lever's main arms. At this time, the square notches of the longer arms of the outer cat's claw gravity latch were raised up off of the L1 main cord hook attachment pin and the upper edges of that pair of parallel longer arms were resting against the interior beveled surface of the rectangular hole cut into the lever's wooden radial stop piece. Thus, we see that the weighted lever's parallel pair of main arms was held against its wooden radial stop piece and in alignment with the weighted lever's parallel pair of radial frame pieces (which are not shown) by only *one* of the lever's two cat's claw gravity latches which was the *inner* latch.

As the pivot pin center axis of the weighted lever left the drum's descending side on the left side of an imaginary vertical line passing through the center axis of the axle and crossed the drum's 6:00 position on its way to the ascending side of the drum which was located on the right side of that imaginary vertical line through the axle's center axis, the situation changed. The counter weight of the outer gravity latch continued to rotate clockwise about that latch's pivot rod and that caused the outer latch's longer arms to come down on the L1 main cord hook attachment pin from above so that, eventually, the square notches at the ends of the longer arms engaged the L1 main cord hook attachment pin. Shortly *after* that happened and as the weighted lever's pivot pin continued along on the ascending side of the drum, the counter weight of the inner gravity latch, because it was leaning so much to the left of its latch's pivot rod, finally fell over to the left and caused the square notches of that latch's two parallel longer arms to suddenly disengage from the L1 main cord hook attachment pin. Those two longer arms then rose away from the L1 main cord hook attachment pin through an arc of 14° until their upper edges made contact with the upper beveled inner surface of the rectangular hole cut into the weighed lever's wooden radial stop piece. As the weighted lever's pivot pin's center axis continued to rotate around the axle's center axis on the ascending side of the drum, the lever's parallel pair of main arms was then only held against its wooden radial stop piece by one of the two cat's claw gravity latches which was then the *outer* gravity latch.

This situation prevailed until the center axis of the pivot pin of the retrograde rotating internal one-directional wheel's weighted lever began to approach the drum's 12:00 position. Then, as the pivot pin center axis of the weighted lever passed over from the drum's ascending side and back onto its descending side again, the *opposite* sequence of events to those described above took place!

As this happened, the inner gravity latch's counter weight continued to rotate clockwise about its shared pivot rod and, eventually, the two notches at the ends of this latch's pair of parallel longer arms engaged the weighted lever's L1 main cord hook attachment pin. Shortly *after* that happened and as the weighted lever's pivot pin continued on its way along the drum's descending side, the counter weight of the *outer* cat's claw gravity latch was leaning so much to the left of its latch's pivot rod that it fell suddenly to the left and the two square notches at the ends of that latch's parallel pair of longer arms then quickly disengaged from the L1 main cord hook attachment pin as their longer arms *dropped* away from it and, after moving counterclockwise through an arc of 14°, came to rest on the beveled, thin metal clad inside surface of the rectangular hole cut into the lever's radial stop piece.

As the weighted lever continued to rotate about the bidirectional wheel's axle and its elevation on the descending side of the drum decreased, the weighted lever's parallel pair of main arms was only held against the lever's wooden radial stop piece by *one* of its two cat's claw gravity latches which was then the *inner* latch. This situation persisted until the center axis of the weighted lever's pivot pin had rotated enough counterclockwise about the center axis of the wheel's axle to again carry it past the drum's 6:00 position at which time the notches at the ends of the parallel pair of longer arms of the inner latch would lift up and disengage their two square notches from the lever's L1 main cord hook attachment pint. By the time that happened, however, the parallel pair of longer arms of the *outer* latch would *already* have engaged the weighted lever's L1 main cord hook attachment pin and would be securely holding the lever's parallel pair of main arms in contact with the lever's wooden radial stop piece.

The cycle described above then began again and continuously repeated itself as long as the internal one-directional wheel containing the particular weighted lever discussed was undergoing *retrograde* rotational motion or rotation opposite to the direction that the wheel was designed to spontaneously start. As a weighted lever rose on the drum's ascending side, its parallel pair of main arms was held against its radial stop piece by the two square notches of the *outer* cat's claw gravity latch's parallel pair of longer arms. However, as a weighted lever lost elevation on the drum's descending side, its parallel pair of main arms was held against its radial stop piece by the two square notches of the *inner* cat's claw gravity latch's parallel pair of longer arms. During the time that all eight weighted levers of an internal one-directional wheel had their parallel pairs of main arms locked down against their respective wooden radial stop pieces, the composite center of gravity of the eight levers was located on the center axis of the section of the wheel's axle to which those eight weighted levers' internal one-directional wheel was attached and would be in the plane that contained the individual centers of gravity of the eight levers' twenty-four cylindrical lead end weights.

We see from this that, in practice, any internal one-directional wheel inside the drum of a bidirectional wheel, whose eight weighted levers were each equipped with one inner and one outer cat's claw gravity latch, had all of those weighted levers' main arms securely locked into contact with their wooden radial stop pieces whenever the wheel was forced to counter rotate through a complete 360° of rotation. If, however, half of an internal one-directional wheel's eight weighted levers were *already* locked down against their wooden radial stop pieces, then the wheel only needed to be counter rotated through an additional 180° in order to lock down its four remaining unsecured weighted levers.

But, what good would a bidirectional wheel's internal one-directional wheel be if all of its eight weighted levers' parallel pairs of main arms had been locked into contact with their wooden radial stop pieces and then *stayed* that way with their collective center of gravity was *always* located at the center axis of the wheel's axle where it could produce no torque to accelerate the wheel? The obvious answer is that such a wheel would be useless for the purpose of constructing a bidirectional wheel. That is why the bidirectionality mechanics Bessler employed had to provide some means by which the internal one-directional wheel's locked down eight weighted levers would again be released for rotations about their respective pivot pins *after* the wheel's drum had been stopped and then given a push to start it turning in the *same* direction in which that internal one-directional wheel was designed to spontaneously start rotating. Once that happened, the imbalanced pm mechanics of that internal one-directional wheel would, once again, be able to maintain the location of the center of gravity of its eight weighted levers on the descending side of the center axis of that internal one-directional wheel's section of axle inside of the bidirectional wheel's enlarged drum.

Amazingly, the system of automatically activated cat's claw gravity latches that Bessler used in his 12 feet diameter bidirectional wheels met this requirement because it only worked in one direction of rotation which was the direction that made a particular internal one-directional wheel undergo retrograde rotation. Should the retrograde rotation of the internal one-directional wheel being discussed above be stopped and the wheel subsequently given a push in the direction in which it was intended to spontaneously turn, then, *after* a full 360° of rotation was completed, all of that internal one-directional wheel's eight weighted levers were unlocked and their parallel pairs of main arms were then again free to swing away from their respective wooden radial stop pieces and they would retain this status as long as the internal wheel continued to rotate in its normal spontaneous start up direction. As that happened, the center of gravity of the eight weighted levers returned to its normal horizontally (and vertically) offset location on the descending side of the internal one-directional wheel's section of axle located within the bidirectional wheel's drum.

To understand how the weighted levers were automatically unlocked when an internal one-directional wheel equipped with these two types of cat's claw gravity latches again began rotating in its designed spontaneous start up direction when its bidirectional drum wheel's drum was stopped and then given a push start in that direction, the reader must now focus his attention on the lower third of Figure 14 which shows a *single* locked down weighted lever as it began to move back across the drum's 6:00 position from the its *previous* ascending side (which was normally its internal wheel's descending side) to its *previous* descending side (which was normally its wheel's ascending side).

As the center axis of the weighted lever's pivot pin moved clockwise around the wheel's axle center axis and approached the drum's 6:00 position, the square notches of the parallel pair of longer arms of its outer cat's claw gravity latch were engaged with the weighted lever's L1 main cord hook attachment pin and securing the lever's main arms against their wooden radial stop piece which was attached to the weighed lever's pair of parallel radial frame pieces. The parallel pair of longer arms of the weighted lever's other or inner cat's claw gravity latch was not engaging the L1 main cord hook attachment pin, but, rather, was still in contact with the upper inner beveled and thin metal clad surface of the wooden radial stop piece's rectangular cutout hole.

As soon as the weighted lever's pivot pin began to pass the drum's 6:00 position, however, the counter weight of the outer cat's claw gravity latch swung counterclockwise about its pivot rod as that latch's parallel pair of longer arms swung up and their two square end notches disengaged from the weighted lever's L1 main cord hook attachment pin. When this happened, the weighted lever was then immediately free to begin swinging counterclockwise about its pivot pin and stretching its two extension springs in the process as would normally happen when the weighted lever's pivot pin's center axis reached and then passed the drum's 6:00 position and continued its *clockwise* rotation around the center axis of the clockwise rotating drum's axle. Note, however, that the parallel pair of longer arms of the lever's *inner* cat's claw gravity latch was still in contact with the beveled and thinly metal clad upper and inner surface of the rectangular cutout hole in the lever's wooden radial stop piece.

The center axis of the weighted lever's pivot pin actually had to be approaching the drum's 7:30 position before the center axis of the inner gravity latch's cylindrical lead counter weight was far enough to the right of the center axis of the latch's pivot rod to make the counter weight suddenly fall over to the right and rotate the entire latch clockwise so that its parallel pair of longer arms quickly dropped down toward the flat and thinly metal clad inside surface of the bottom of the rectangular hole cut into the weighted lever's wooden radial stop piece. By this time, however, the weighted lever's L1 main cord hook attachment pin was not in position so that the two square notches at the ends of the inner latch's longer arms could engage it. The L1 main cord hook attachment pin was to the right of these square notches of the inner gravity latch and the latch's longer arms just continued to rotate clockwise about their pivot rod until their lower edges finally made contact with the flat and thinly metal clad inside surface of the rectangular hole cut into lever's radial stop piece. As previously mentioned, this surface had been covered with a thin lining of steel in order to protect the wood of the radial stop piece from being worn away by the repeated impacts from the longer arms of the inner cat's claw gravity latches during prolonged drum rotation.

By the time the center axis of the weighted lever's pivot pin reached the drum's 7:30 position, the pairs of longer arms of its inner and outer gravity latches were both in contact with the thinly metal clad inside surfaces of the rectangular hole cut in the lever's wooden radial stop piece and the weighted lever's L1 main cord hook attachment pin that they could engage was far to the right of them. The bottom edges of the inner cat's claw gravity latch's two parallel longer arms were in contact with the thinly metal clad, flat lower inside surface of the radial stop piece's rectangular cutout hole and the upper edges of the outer cat's claw gravity latch's two parallel longer arms were in contact with the also thinly metal clad, beveled upper inside surface of the radial stop piece's rectangular cutout hole. Needless to say, during rapid drum rotation these metal to metal impacts taking place between the latches' longer arms and the thin linings of steel that protected the inside upper and lower surfaces of the rectangular cutout holes in the wooden radial stop pieces contributed much to the sounds emanating from the linen cloth covered sides of a bidirectional wheel's drum.

From all of this we can see that, when *either* internal one-directional wheel of a bidirectional wheel was undergoing *retrograde* rotation, its weighted levers had their parallel pairs of main arms kept locked into contact with their respective wooden radial stop pieces by their *outer* cat's claw gravity latches whenever the center axes of the levers' pivot pins were on the *ascending* side of the center axis of the drum's axle. But, whenever their pivot pins' center axes were located on the *descending* side of the center axis of the drum, the parallel pairs of main arms of the weighted levers were kept locked in contact with their radial stop pieces by their *inner* gravity latches. The transition or what I refer to as "the handover" (perhaps it should be referred to as the "pawover" since *cat's* claw latches are involved!) took place as a retrograde rotating internal one-directional wheel's weighted lever approached and then passed through either the drum's 6:00 or 12:00 positions.

It's important to realize that, while this transition was taking place, there was enough of an overlap in the engagements of the two types of gravity latches with their L1 main cord hook attachment pin so that the pin was at no time left simultaneously disengaged from the square notches at the ends of the longer arms of *both* of the weighted lever's two types of latches. There was always one cat's claw gravity latch securing a weighted lever so that its parallel pair of main arms stayed in physical contact with its wooden radial stop piece and, thus, the center lines of these main arms always remained perfectly aligned with the center lines of the weighted lever's parallel pair of radial frame pieces.

This alignment of the *inactive* internal one-directional wheel's levers was critically necessary in order to keep the center of gravity of the eight weighted levers located *exactly* on the center axis of the bidirectional wheel's axle. And, at that location, their center of gravity could produce no torque which opposed that being produced by the offset center of gravity of the eight weighted levers that were inside of the other *active* internal one-directional wheel that was not undergoing retrograde rotation and, therefore, was rotating in the direction that it had been designed to spontaneously rotate if free to do so.

When, however, a one-directional wheel was undergoing rotation in its designed spontaneous start up direction, assumed to be clockwise in the above treatment, each of its eight weighted levers' parallel pairs of main arms normally came into contact with its wooden radial stop piece as a lever's pivot pin's center axis reached a location that was about 12° past the drum's 3:00 position. Just as contact with the radial stop piece was about to be made, however, the weighted lever's L1 main cord hook attachment pin would hit the ends of the parallel pair of longer arms of that lever's outer cat's claw gravity latch which had their bottom edges with the square notches in them pressed against the lower inside *flat* metal clad surface of the rectangular cutout hole in the weighted lever's wooden radial stop piece. The L1 main cord hook attachment pin, however, was *not* blocked by the ends of the outer latch's two longer arms so that the weighted lever's pair of parallel main arms that held the L1 main cord hook attachment pin could not make contact with their wooden radial stop piece as they also aligned (temporarily until the lever's pivot pin's center axis passed the drum's 6:00 position) with the lever's parallel pair of radial frame pieces. How was that possible one might wonder?

If the reader looks back at Figure 13(b), he will notice that the ends of the longer arms of both types of cat's claw gravity latches, inner and outer, were not flat (actually very slightly beveled) like the ends of their shorter arms. Rather, the ends of the longer arms were cut at a slant or beveled with an angle of about 30° to the upper edge of an arm. The reason for that was so that when the longer arms of either type of gravity latch might block the L1 main cord hook attachment pin of a weighted lever whose main arms were trying to come into contact with their wooden radial stop piece so that the main arms and the lever's parallel pair of radial frame pieces would be perfectly aligned, the L1 main cord hook attachment pin would be able to easily lift the end of the latch's parallel pair of longer arms out of the way and continue moving so that the alignment could finally take place.

Thus, as the pivot pin center axis of a normally turning internal one-directional wheel's weighted lever reached a location about 12° past the 3:00 position of a clockwise rotating drum, the lever's L1 main cord hook attachment pin would contact the beveled ends of the longer arms of the lever's *outer* gravity latch and cause those arms to move far enough toward the axle of the drum so that the L1 main cord hook attachment pin could continue on to reach the position which allowed its two parallel main arms to make contact with the surface of the lever's wooden radial stop piece. After that happened, the square notches at the ends of the outer latch's longer arms would drop down around and enclose the L1 main cord hook attachment pin. This condition would then persist until the weighted lever's pivot pin center axis reached the drum's 6:00 position at which time the outer latch's parallel pair of longer arms would immediately begin rotating counterclockwise about their pivot rod and the square notches again begin disengaging from the weighted lever's L1 main cord hook attachment pin so that the lever could rotate counterclockwise about its own pivot pin and its parallel pair of main arms could part company with its radial stop piece as the weighted lever's pivot pin's center axis continued to rotate clockwise about the wheel's axle and rise on the drum's ascending side.

So, we see from this that, for a bidirectional wheel's active internal one-directional wheel rotating in its normal spontaneous start up direction, its weighted levers would only be locked down against their wooden radial stop pieces by their outer cat's claw gravity latches as the levers' pivot pins' center axes traveled from just past the drum's 3:00 position to its 6:00 position. This was, however, irrelevant because these levers always normally had their main arms in contact with their radial stop pieces between these locations of the drum whether or not each lever was equipped with the two types of gravity latches. Despite this, one can still refer to the weighted levers of an active internal one-directional wheel as being "locked down" between these two drum positions because, if one was to stop the bidirectional wheel's drum and then reach into it and try to pull such a locked down weighted lever's parallel pair of main arms out of alignment with its parallel pair of radial pieces, he would not be able to do so unless he applied enough force to actually tear the ends off of the lever's outer latch's parallel pair of longer arms.

To test the strength of such a latching system, I conducted several tests using a piece of steel that was 0.0625 inch = 1/16 inch thick x 0.25 inch = 4/16 inch wide x several inches long. I placed this metal bar under the hand grip portion of a 30 pound exercise dumbbell in order to see if it would break as I tried to lift the dumbbell with the bar's 0.0625 inch = 1/16 inch thick edge. The piece of metal was able to lift the iron weight off of the floor and experienced virtually no flexing along its length with the load applied to its middle. Considering that a weighted lever used in the Merseburg wheel only had a total mass of 14.5 pounds and an actual *single* cat's claw gravity latch of either type would have held it in place with *twice* as much steel as I used, I think that the two square notches in an *inner* latch's two longer arms would have been more than adequate to support the *entire* mass of a weighted lever at the 9:00 position of an internal one-directional wheel undergoing retrograde rotation. However, since the weighted lever at that drum position had the other ends of its parallel pair of main arms supported by a pivot pin fixed to the drum, the inner cat's claw latch holding it in contact with its radial stop piece actually was about four times as strong as it needed to be to lock down the weighted lever at that drum position!

This method of using inner and outer cat's claw gravity latches on each of the sixteen weighted levers of a bidirectional wheel to achieve its bidirectionality is certain ingenious. It is fairly simple to construct and install, does not add too much mass to the wheel, and, as Bessler's

Merseburg and Kassel wheels amply demonstrated can, with careful adjustment, be highly reliable.

Achieving that reliability, however, was no easy feat. Bessler complained in his writings that getting these gravity activated latches to function smoothly and reliably in the Merseburg wheel gave him *many* a headache. I can certainly appreciate that statement. When I finally decided to reverse engineer the method he used to achieve bidirectionality (a task actually completed several years *before* I had finally found the correct weighted lever shape and the many other parameters that made a one-directional imbalanced pm wheel work!), it took me dozens of computer models and weeks of effort to, finally, find a design that worked flawlessly. Any reader deciding to computer simulate this latching system or actually make a physical version of it had better be prepared to expend a *lot* of effort and endure much frustration before he has it functioning flawless as it did in Bessler's two 12 feet diameter bidirectional wheels.

Let us now very briefly see how the above analysis of the functioning of the cat's claw gravity latches can be applied to the Merseburg wheel. For this, the reader must imagine that he can go back in time and has joined a small crowd of the curious who have lined up at the door of Bessler's rented home in order to view his first bidirectional wheel in the town of Merseburg.

Upon entering the room and standing behind the guard rail so as to have an axial view of the huge 12 feet diameter drum's circular front side, imagine that the designed spontaneous start up direction of its unseen internal *front* one-directional wheel is clockwise and that the spontaneous start up direction of its also unseen back one-directional wheel located immediately behind the front internal wheel is counterclockwise as viewed *from your location* in front of the bidirectional wheel's drum. Now imagine that Bessler and his assistant have entered the room and, after greeting your group and the inventor telling you some of the capabilities of his new wheel, begin to demonstrate its bidirectionality. Assume further, that for this demonstration, the two axle end curved crank driven pendulums had been previously removed.

Before his demonstration, however, Bessler orders his assistant to remove a lock on the rim of the drum which attaches a short length of rope to it whose other end is tied to the head of an eyelet that is screwed into a wooden floor plank directly beneath the drum. As the lock and rope are removed from the drum, you notice that the drum remains stationary. This is a bit surprising to you because you had heard that his previous somewhat smaller 9 feet diameter wheel constructed in the town of Draschwitz started up spontaneously as soon as the rope tether to its drum was removed.

Bessler then explains that because his new wheel is bidirectional that proves that it is not driven by any kind of concealed mainspring powered clock type movement. The inventor then asks someone in your group to call out a direction in which he or she would like to see the wheel start turning. Someone calls out "Clockwise!" and the inventor then gives the rim of the drum a gentle push using only two fingers of his hand so that its front circular side begins to rotate in a clockwise direction. The 12 feet diameter drum's speed of rotation is very slow at first and there is an audible clattering and thumping sound coming from the interior of the drum as it turns. Over the course of several minutes, the drum finally completes a few full rotations and is moving at its maximum rate. A wealthy person with a pocket watch in your group says he was carefully timing the drum's rotation rate by noticing how many times a slit in the front cloth covering on the drum traveled around the axle in a single minute and concludes the rate is about 40 rotations per minute. You notice that, as the wheel's rotation rate slowly increased, so too did the rate of the various sounds coming from its drum.

Bessler then orders his assistant to stop the wheel and he obediently does so by reaching over his head and pressing his hands against the speeding outer rim wall on the drum's left ascending side while he wears thick leather gloves to protect their skin from any injury. Using this method, he manages to slow the wheel to a stop in less than 30 seconds. You notice that once the wheel's drum is stopped, Bessler takes the place of this assistant and then grasps the curving edges of the drum's outer rim wall on the left side of the drum and, by slowly pulling down on them, rotates it counterclockwise by about 135° after which he releases the drum and it again remains stationary.

He then says that he will now show the wheel's bidirectional capability and gives the huge drum a nudge that causes its front circular face to begin rotating about the center axis of its axle in a counterclockwise direction as viewed from your location in front of the drum. Again, you notice its minutes long acceleration up to a terminal speed of about 40 rotations per minute and the increasing frequency of the clattering and thumping sounds emanating from the drum as this acceleration takes place. Finally, Bessler orders his assistant to stop the huge drum and he does so by again pressing the leather gloves he wears against the outer rim wall on the drum's right ascending side. This time, however, something unexpected happens.

After the assistant stops the drum's counterclockwise rotation, he accidentally releases it before Bessler can reach and take hold of it. You notice that, as soon as the assistant releases the drum, it immediately starts rotating in the *same* counterclockwise direction as it previously did and does so *without* having to be given a push! The inventor, also noticing this, immediately rushes over to the drum to stop it and then back rotates it in a clockwise direction as viewed from the front of the drum by about 135°. Upon releasing the drum this time, it again remains stationary.

Bessler appears a bit embarrassed by this incident perhaps because he did not want anyone in your small group seeing this happen. You wonder why as he then proceeds to demonstrate another of the bidirectional Merseburg wheel's abilities, once it had again reached its full terminal rotation speed of 40 rotations per minute again in a clockwise direction, which is to quickly lift a 60 pound load of bricks located outside of the wheel's room by using a rope suddenly attached to a wing shaped head of a projecting bolt near the front end of the wheel's rapidly rotating axle. Once attached, that rope runs through a pulley bolted to the floor under the front end of the wheel's axle and then through an open window pane of the wheel's room to another pulley mounted on the wall outside of the house and finally down to the load of bricks itself.

This entire imaginary demonstration lasts about 45 minutes during which time Bessler tells your group that his new invention, because it requires no expensive fuel like the steam engines being developed and is far more reliable than water or wind power, will help to greatly reduce the cost of the manufactured products you buy. He even hints that he has plans for carts and ships that will be powered by his wheels and which will help make the delivery of products faster and less expensive. Everyone in your group is amazed by what they have seen and been told and you leave his home eager to tell others of what you have seen there.

With what the reader has learned so far about the construction of Bessler's bidirectional wheels, it is now an easy matter to rationalize everything that was observed during the above hypothetical visit to the inventor's home to see a demonstration of the Merseburg wheel.

When the bidirectional Merseburg wheel's drum was first detached from the floor rope to which it was secured by a lock, it remained motionless because the *composite* center of gravity of its *two* individual and separate internal one-directional wheels, the front and back ones, was located immediately below the center axis of the axle section inside of the drum and in the plane of the drum's middle lattice of frame pieces. In that *punctum quietus* location, the two internal one-directional wheel's composite center of gravity could provide no net torque to accelerate the 12 feet diameter drum in either of its two possible directions of rotation, either clockwise or counter clockwise, as viewed from the location of the small group of visiting onlookers which was facing the front circular side of the drum.

If Bessler had cut away the sheets of darkly dyed and oiled linen covering the front circular side of the drum which had been rendered stationary after running by being stopped and then carefully back rotated, then the existence of these two separate, side by side and back to back internal one-directional wheels and the symmetrically reversed configurations of the eight weighed levers in each of them would have become quite apparent.

The group would have noticed that in the *front* one-directional wheel, which was closest to them, most of the four weighted levers on the drum's right side had the longest of their three parallel pairs of arms, their main arms, in alignment with the drum's radial frame pieces that held those levers' pivot pins and that was because these main arms were all in physical contact with the wooden radial stop pieces that were attached to the radial frame pieces of each weighted lever.

However, on the left side of the drum they would have noticed that some of the front internal one-directional wheel's four weighted levers were separated from their radial stop pieces and rotated a little counterclockwise about their respective pivot pins so as the bring their main arms' trios of cylindrical lead end weights a little closer to the center axis of the wheel's axle. This difference between the weighted levers on both sides of the drum then resulted in the center of gravity of the eight weighted levers of the Merseburg wheel's front internal one-directional wheel being displaced, horizontally, a fraction of an inch onto the *right* side of the center axis of the drum's interior 6 inches diameter section of the axle to which the sixteen radial frame pieces of the front internal one-directional wheel were attached. These sixteen radial frame pieces were provided by the front and middle lattices of the drum.

If the group could see deeper into the 12 inch thick drum of the bidirectional Merseburg wheel, they would have noticed that the configuration of the eight weighted levers of the drum's front internal one-directional wheel was exactly reversed by the configuration of the eight weighted levers in the drum's back internal one-directional wheel.

Thus, in the *back* internal one-direction wheel, some of the parallel pairs of main arms of the four weighted levers on the drum's right side were rotated clockwise around their respective pivot pins by varying amounts and separated by varying distances from their wooden radial stop pieces so as to bring their trios of cylindrical lead end weights closer to the center axis of the axle and, consequently, did not have their main arms in alignment with their levers' parallel pairs of radial frame pieces. Most of the four weighted levers on the left side of the drum, however, had their parallel pairs of main arms in physical contact their wooden radial stop pieces and were aligned with their respective radial frame pieces. As a result of this difference between the weighted levers on both sides of the drum, the center of gravity of the eight weighted levers of the Merseburg wheel's back internal one-directional wheel would have been offset, horizontally, a fraction of an inch onto the *left* side of the center axis of the section of the drum's interior axle to which the back one-directional wheel's sixteen radial frame pieces were attached. These sixteen radial frame pieces were provided by the middle and back lattices of the drum.

The net effect of the bidirectional Merseburg wheel's two internal one-directional wheel's having their weighted levers' centers of gravity on opposite sides of the axle's center axis was a *composite* center of gravity for the two internal one-directional wheels sixteen weighted levers that was located in the plane of the drum's middle circular lattice and less than an inch below the axle's center axis at its *punctum quietus* location.

It was because of this location of the bidirectional wheel's weighted levers' composite center of gravity that the 12 feet diameter drum did not spontaneously start rotating as soon as Bessler's assistant unfastened its rim from the floor's tether rope.

When Bessler then gave the Merseburg wheel's enlarged drum a push so as to cause it to begin rotating in a clockwise direction as viewed from the location of the spectators in front of the wheel, something most remarkable happened to the eight weighted levers in *each* of the bidirectional wheel's two internal one-directional wheels as the drum rotated through its first 180° of such clockwise rotation.

During that first half turn of clockwise drum rotation, all of the previously locked down main arms of the weighted levers on the right side of its front internal one-directional wheel were released as their pivot pins' center axes sequentially passed the drum's 6:00 position so they could then swing first counterclockwise and then clockwise about their pivot pins as they ascended the left side of the drum. Simultaneously, all of the unlatched and free weighted levers on the right side of the back internal one-directional wheel (as viewed by the spectators located in front of the drum), that was now forced to undergo retrograde rotation, had their parallel pairs of main arms securely locked down against their respective wooden radial stop pieces as their pivot pins' center axes sequentially passed the drum's 6:00 position and began ascending the left side of the drum and could, therefore, not swing freely about their pivot pins any longer.

After the 180° of clockwise drum rotation was completed, all eight weighted levers of the Merseburg wheel's front one-directional wheel would be free to rotate about their pivot pins so as to keep their center of gravity fixed on the right descending side of the axle's center axis. But, after the completion of the same 180° of clockwise drum rotation was completed, all eight weighted levers of the bidirectional wheel's back one-directional wheel would have their parallel pairs of main arms locked down against their radial stop pieces and in alignment with their parallel pairs of radial frame pieces and their center of gravity, if it could be seen, would be located on the center axis of the 12 feet diameter wheel's 6 inch diameter axle in the section of the drum's interior axle to which the back internal one-directional wheel's sixteen radial frame pieces were attached.

At this point, the drum would be driven completely by the offset center of gravity of its front internal one-directional wheel's eight weighted levers while the back and now disabled or inactive internal one-directional wheel's locked down eight weighted levers acted only as flywheel mass that could store some of the rotational kinetic energy and its associated mass that was extracted from the front internal one-directional wheel's eight weighted levers during drum rotation.

After the bidirectional Merseburg wheel had finally achieved its terminal free running rotation rate of about 40 rotations per minute, Bessler then ordered his assistant to bring it to a stop as previously mentioned. The assistant did this by applying the drag of his two leather gloves to the outer rim wall of the left ascending side of the 12 inches thick drum. What the group did not notice was that, as he did that, the leather gloves he wore heated up considerably. They did this because most of the rotational kinetic energy and its associated mass extracted from the front internal one-directional wheel's eight weighted levers had then been transferred to the gloves. However, a certain percentage of this energy and mass was also transferred to the air surrounding the wheel's drum, its axle, and the axle's end pivot bearings as well as the air contained within the drum. What mass was gained by the assistant's gloves was, therefore, only *almost* equal to the amount lost by the front internal one-directional wheel's eight weighted levers.

Just as soon as the great bidirectional wheel's drum was finally brought to a stop by Bessler's assistant, the inventor quickly grasped its outer rim and then slowly back rotated it counterclockwise by about 135°. Why did he do that? The answer is that, if he had not done that and then released the drum, it would have immediately and spontaneously begun turning in a clockwise direction again and would eventually accelerate up to its terminal free running rotation speed of about 40 rotations per minute. This would happen because the drum's front internal one-directional wheel's center of gravity was still located on the previous descending side of the axle's center axis and the its back one-directional wheel's center of gravity was still located at the center axis of the axle. However, the about 135° counterclockwise back rotation that Bessler gave the drum prevented that from happening.

As that almost 135° counterclockwise back rotation of the drum took place as viewed by the small group of spectators located on the front side of the drum, most of the previously free weighted levers of the front internal one-directional wheel located on the left side of the drum were locked down by the outer cat's claw gravity latches assigned to them as their pivot pin center axes passed the drum's 6:00 position while some of the previously locked down weighted levers of the back internal one-directional wheel located on the drum's left side were, as their respective pivot pins' center axes sequentially passed the drum's 6:00 position, *unlocked* and again free to rotate about their pivot pins because the outer latches that engaged their L1 main cord hook attachment pins lifted up their longer arms and disengaged their square notches from the L1 main cord hook attachment pins.

After the almost 135° counterclockwise back rotation of the drum had been completed, Bessler could again release the rim of the Merseburg wheel's drum and it would remain stationary because some of its front internal one-directional wheel's left side weighted levers were still unlocked, most of the back internal one-directional wheel's left side weighed levers were still locked down, most of the front internal one-directional wheel's right side weighted levers were locked down, and, finally, some of the back internal one-directional wheel's right side weighted levers were unlocked.

In essence, the 135° counterclockwise back rotation that the inventor gave the bidirectional Merseburg wheel's drum restored the configuration of its sixteen weighted levers to what it had been when he first had to give the drum a push to start it rotating in a clockwise direction. In this starting configuration, the front internal one-directional wheel's center of gravity was located a fraction of an inch horizontally onto the right side of the center axis of its section of axle and the back internal one-directional wheel's center of gravity was located a fraction of an inch horizontally onto the left side of the center axis of its section of axle. The two centers of gravity were, thus, located on opposite sides of the axle's center axis and at the same horizontal distances from an imaginary vertical plane passing through the axle's center axis and so could produce no net torque on the axle which then allowed it to remain stationary when it was finally released.

At this point in his demonstration, Bessler decided to amaze his audience by demonstrating the Merseburg wheel's bidirectional property and then gave the 12 feet diameter drum a gentle push to get it rotating, as seen from the viewpoint of the spectators in front of its drum, in a counterclockwise direction.

During the first 180° of that counterclockwise rotation, the front internal wheel's remaining left side unlocked weighted levers became locked up as their pivot pins passed the drum's 6:00 position and the back internal one-directional wheel's remaining left side locked down weighted levers became unlocked and freed as their pivot pins passed the drum's 6:00 position. After its first half of a complete counterclockwise rotation, the drum was then solely propelled by the offset center of gravity of the eight unlocked weighted levers of its *back* internal one-directional wheel which was located on the left side of the center axis of that wheel's axle section while the front internal one-directional wheel, with its eight locked down weighted levers' center of gravity drawn up less than an inch vertically into the center axis of that wheel's axle section, served only as flywheel mass to accumulate some of the rotational kinetic energy and its associated mass that was being extracted from the back internal one-directional wheel's eight weighted levers with each drum rotation.

As the bidirectional 12 feet diameter Merseburg wheel's drum continued to rotate in a counterclockwise direction and accelerate, energy and mass continually flowed from the back one-directional wheel's eight unlocked weighted levers to the eight locked up weighted levers of the front internal one-directional wheel as well as all of the other rotating structures of the drum and axle including even some small percentage that flowed back into the back internal one-directional wheel's eight unlocked weighted levers themselves. But, the transfer of this energy and its associated mass was not 100% efficient. Some of it was always transferred and, thus, "lost" to the air surrounding the drum, axle, its end pivot pins, and even some to the air inside of the drum.

After Bessler was convinced that the quick change in rotational direction of the Merseburg wheel's drum convincingly proved to his small group of observers that his wheels could not possibly be powered by a simple one-directional, spring wound clock type movement, he again directed his assistant to slow the 12 feet diameter bidirectional Merseburg wheel to a standstill and this the latter quickly did by, again, using his leather gloved hands to apply continuous drag to the outer rim wall on the drum's *right* ascending side. It was then that something occurred that Bessler did not want the group to see. Because the inventor had not reached the drum's rim in time to grasp it when his assistant released it, the drum immediately began to spontaneously turn again in the same counterclockwise direction it previously had as soon as the assistant released his grip on its drum's outer rim wall.

This occurred because, although brought to a standstill, all of the front internal one-directional wheel's eight weighted levers were still locked down with their center of gravity located at the center axis of that wheel's axle section and all of the back internal one-directional wheel's eight weighted levers were still unlocked and free to rotate about their pivot pins and, thus, keeping their center of gravity on the left descending side of the center axis of their axle section. Because of this situation, after the assistant had stopped the drum's counterclockwise rotation and it was released, it had to immediately and spontaneously begin rotating and accelerating in a counterclockwise direction again. In fact, no matter how many times Bessler's assistant might have stopped the counterclockwise rotating drum (as viewed from the front side of the drum), it would have immediately begin rotating in a counterclockwise direction again as soon as he released it.

Knowing this could happen, Bessler had to quickly reach the drum's rim, stop any further drum rotation, and then slowly back rotate the drum about 135° *clockwise* in order to again restore the configuration of its two internal one-directional wheels' sixteen weighted levers back to the one they initially had when the group of spectators first arrived to witness the demonstration of the bidirectional Merseburg wheel in Bessler's home. That initial configuration perfectly counter balanced the drum's two internal one-directional wheels' horizontally offset centers of gravity against each other so that no net torque in either direction would be applied to the bidirectional wheel's axle.

Well, I'm sure by now that the reader realizes that, although understanding the construction and basic operation of the two types of cat's claw gravity latches assigned to an individual weighted lever inside of one of Bessler's bidirectional wheels is fairly easy, understanding *how* all thirty-two of the cat's claw gravity latches, sixteen inner and sixteen outer latches, worked together in order to enable the drum to be able to start up, *after* a gentle initial, two finger push in *either* direction, and keep accelerating thereafter until the wheel's terminal rotation rate was reached is not quite as easy to understand.

This is because, to achieve bidirectionality, required the use of two side by side and *back to back*, identical one-directional wheels whose spontaneous start up directions of rotation were always in opposition to each other. Consequently, whenever one of the two internal one-directional wheel's weighted levers was passing the drum's 6:00 position and its parallel pair of main arms was being unlocked from contact with their radial stop piece by its outer cat's claw gravity latch lifting its two square notches off of the lever's L1 main cord hook attachment pin, the other internal one-directional wheel's weighted lever was also passing the drum's 6:00 position and its parallel pair of main arms was being locked down into contact with their radial stop piece by its outer gravity latch's two square notches coming down on and engaging its L1 main cord hook attachment pin. Basically, it was the use of a side by side pair of inner and outer cat's claw gravity latches, working together on each weighted lever near the top and bottom positions of a rotating bidirectional wheel's drum, was critically necessary to achieving bidirectionality.

It is a real tribute to the ingenuity and craftsmanship of Johann Bessler that he could have conceived of and installed such a unique gravity powered latching system into his 12 feet diameter, bidirectional Merseburg wheel during the few months needed to complete its construction. Now, hopefully, with the information provided in this chapter, Bessler's method for achieving imbalanced pm wheel bidirectionality will not be lost and, perhaps, some day soon will be fully replicated by the most skillful of craftsmen.

The craftsman seeking to duplicate the same automatic gravity latching system employed by Bessler in his bidirectional wheels should reread the material in this chapter concerning the operation of the latches *several* times until he thoroughly understands their function and can, in his mind's eye, actually "see" how the two types of gravity latches interact with a weighted lever's L1 main cord hook attachment pin as the lever completes full rotations about the center axis of the drum's axle in both clockwise and counterclockwise directions. Only a craftsman who thoroughly understands the functioning of these gravity activated latches can ever hope to successfully duplicate them.

One should keep in mind that I have arbitrarily taken the spontaneous start up direction of a bidirectional drum's front internal one-direction wheel as *always* being clockwise as viewed from the front side of the drum which is the side that average visitors to Bessler's public demonstrations ordinarily would have been restricted to viewing a bidirectional wheel from upon entrance to the room containing it. I think, once having been a clockmaker's apprentice, he started using this as a "standard" spontaneous start up direction, beginning with the construction of his 3 feet diameter, one-directional Gera prototype wheel, because it reminded him of the direction that the hands of a clock turn when viewed from the front of the dial (unfortunately, this similarity only later served to encourage some of his detractors to start suggesting his wheels were frauds containing windup mainspring powered clock movements!). Unfortunately, confusion can often arise when discussing the spontaneous start up direction of rotation of a bidirectional drum's *back* internal one-directional wheel and how it relates to the spontaneous start up direction of the front internal one-directional wheel.

The spontaneous start up direction of the back internal one-directional wheel would, when viewed *through* the front internal one-directional wheel from the *front* side of the drum have been counterclockwise, but it would have been clockwise when viewed from the *back* side of the drum. This was because the two internal one-directional wheels were actually *identical*, but mounted so that their spontaneous start up directions were always opposite to each other when *both* were viewed from *either* the front or back side of the drum.

In the design Bessler used, the inventor simple took the arrangement of weighted levers, radial stop pieces, and side by side pairs of cat's claw gravity latches used in the front internal one-directional wheel and rotated them 180° around an imaginary vertical axis between the drum's 12:00 to 6:00 locations and used the resulting arrangement to determine the placement of those components in the back internal one-directional wheel as viewed from the *back* side of the drum. The only exception to this rule was that he reversed the locations of all of the slotted head bolts and their square retaining nuts on the back internal one-directional wheel's eight weighted levers and sixteen gravity latches so that their end weights and counterweights could all be loosened and tightened through the inspection holes cut into the linen sheets on the *back* side of a bidirectional wheel's drum. This then allowed all servicing of a bidirectional drum's hidden internal mechanics to be done solely from the back side of the drum though a collection of eight trapezoidal holes cut into the linen sheets covering the back circular side of the drum.

Finally, before ending this chapter, we must briefly discuss an unusual problem in the operation of the bidirectional Merseburg wheel which sometimes occurred.

On occasion, when someone claiming to be the representative of some rich person interested in purchasing the wheel made an appointment with Bessler to come and personally test it, the visitor would do something the inventor preferred he not do. While Bessler's back was turned, the representative would give the wheel's drum a sudden downward yank by reaching up and grasping two of its leather covered outer rim wall perimeters and quickly pulling down on them to more quickly start the wheel's rotation instead of just providing a gentle upward nudge to the rim wall's outer curving surface. Most likely, he hoped that this would shorten the time needed for the drum to reach its maximum terminal rotation rate of about 40 rotations per minute which some of my computer models indicated could take several minutes.

Most of the time this trick, only applied upon the start up of a stationary bidirectional wheel, would help to shorten the acceleration time somewhat, but, every once in a while, the yank, if excessive, would actually completely disable the wheel and prevent it from running in one of its directions.

At these times Bessler became quite annoyed that some stranger he had allowed access to the wheel had done something he was not supposed to do with it, but, out of fear of having the representative return to a potential buyer with a negative report on the testing of the Merseburg wheel, the inventor had to contain his anger and simply tell him that there was a problem with the bidirectional wheel that only very occasionally occurred and which the inventor was working to permanently solve. The representative was then allowed to continue his testing once the problem was fixed and he had been admonished to only give the wheel a gentle push to start it in either direction.

What was that "problem" and how did Bessler quickly fix it while a representative impatiently waited to continue his testing of the wheel?

Imagine that the representative is standing near the *left* side (as viewed from the front of the drum) of the Merseburg wheel's 12 inch thick bidirectional drum and, while Bessler is preoccupied with some other matter, the former decides to reach over his head and yank down hard on the rim wall edges of the initially stationary and untethered drum. Since the drum is stationary, that means the *front* internal, normally clockwise starting, one-directional wheel inside the drum has most of the parallel pairs of main arms of its *right* side of the axle weighted levers locked down against their wooden radial stop pieces and only some of the levers locked down on the left side of the axle.

The main arms of the front internal one-directional wheel's weighted lever whose pivot pin's center axis is located at the drum's 9:00 are still locked down and in contact with their radial stop piece and they and their three attached cylindrical lead end weights and their attachment bolts with a mass of 12 pounds are, literally, holding a horizontal orientation there because they are being completely supported by the two square notches at the ends of the longer arms of the lever's *inner* cat's claw gravity latch that hold onto the lever's L1 main cord hook attachment pin. The two longer arms of the lever's outer gravity latch, however, do not have the two square notches at their ends holding on to the L1 main cord hook attachment pin. The upper edges of the longer arms of the outer gravity latch are actually in contact with the thin metal lining on the beveled inside surface of the rectangular hole cut into the front 9:00 weighted lever's radial stop piece.

Ordinarily, the inner gravity latch would have no problem supporting the 9:00 weighted lever's main arms and their attached trio of cylindrical lead end weights. However, this weighted lever's inner cat's claw gravity latch is subject to a vulnerability that only becomes apparent when a sudden counterclockwise accelerating torque is applied near the drum's 9:00 position. That torque will cause the drum, as viewed from the front side, to suddenly rotate in a counterclockwise direction which is opposite to the normal direction that the front internal one-directional wheel was supposed to start. However, if the torque was excessive, then, due to inertia, the 0.5 pound = 8 ounce disc-like lead counterweight attached between the ends of the shorter arms of the inner cat's claw gravity latch in the Merseburg wheel would tend to stay in place as the surrounding drum dropped vertically relative to them. Because of this, the shorter arms of this type of latch would then actually rotate counterclockwise around their shared pivot rod and the two notches at the ends of the arms would suddenly disengage from and release the 9:00 weighted lever's L1 main cord hook attachment pin allowing the lever's main arms and their three attached cylindrical lead end weights to momentarily be in a state of near free fall!

The resulting sudden dropping of the 9:00 weighted lever's main arms would then also cause the lever's parallel pairs of A arms and B arms to rapidly rotate counterclockwise around the lever's pivot pin. Since the main arms of the 10:30 weighted lever are still locked down against that lever's radial stop piece (which is located above the main arms), the sudden counterclockwise rotation of the 9:00 weighted lever would be abruptly halted as the main cord connecting its A arms' A4 main cord hook attachment pin to the L1 main cord hook attachment pin of the 10:30 weighted lever suddenly became taut and this would happen well before the center lines of the main arms of the 9:00 weighted lever's main arms reached an angle of 62.5° relative to the center lines of that lever's parallel radial frame pieces. In fact, it would happen after they only reached an angle of about 8.9°.

If the briefly applied torque of the representative's two handed yank on the drum's rims was not too great, then the taut main cord would be able to stop the counterclockwise rotation of the 9:00 weighted lever around its pivot pin and the wheel would continue to run normally in the direction that the representative had tried to give its drum some assistance. As the weighted 9:00 weighted lever's pivot pin's center axis then reached and passed the 6:00 position of the drum, the lever's L1 main cord hook attachment pin would be locked down against the lever's wooden radial stop piece by the two square notches at the ends of the longer arms of the lever's outer cat's claw gravity latch and the problem would be resolved as the drum continued to accelerate in a counterclockwise direction up to its maximum terminal rotation rate. The drum could then be stopped and, after being back rotated in a clockwise direction, as viewed from the front of the wheel, by about 135° in order to again unlock some of the front internal one-directional wheel's left of axle weighted levers while simultaneously locking down most of the back internal one-directional wheel's left of axle weighted levers, would remain motionless when released.

If, however, the briefly applied torque was excessive, then the sudden tension applied to the main cord interconnecting the front internal one-directional wheel's 9:00 and 10:30 weighted levers might be sufficient to break one of the main cord's metal attachment hooks or even break the cord itself. If this scenario occurred, the drum would still run normally in the counterclockwise direction that the representative was trying to speed up its acceleration. However, the problem would become obvious when the drum was finally stopped and back rotated clockwise in an attempt to balance out its two internal one-directional wheel's opposing torques. If the broken main cord was high up on the right side of the drum, then back rotation by 135° would be possible and the drum would remain stationary when released. But, when an attempt was then made to push start the drum so that it began rotating in a *clockwise* direction, as viewed from the front of the wheel, the drum would probably only turn through about 180° before it stalled and came to a stop. The wheel would only run normally as long as its drum, after balancing, was push started in the same counterclockwise direction it had before being stopped and balanced so that it remained stationary when released. Such damage, therefore, immediately removed a drum's bidirectional capability.

This would be due to the broken main cord finally reaching a location near the upper left side of drum where an intact main cord was critically needed to rotate the 9:00 weighted lever clockwise about its pivot pin as its center axis traveled to the 10:30 position of the drum. That failure to rapidly rotate the 9:00 weighted lever and its main arms' attached trio of cylindrical lead end weights would then result in the center of gravity of the front internal one-directional wheel's eight weighted levers dropping to a *punctum quietus* location directly below the center axis of the section of the axle to which the front internal one-directional wheel was attached. Since the center of gravity of the back internal one-directional wheel's eight locked down weighted levers was then located at the center axis of its internal wheel's section of the axle, that center of gravity could provide no torque to drive the axle. The drum would probably only oscillate from side to side or "keel" briefly before coming to rest and any attempt to then gently push start it in *either* direction would be futile.

To quickly repair such a catastrophic failure, Bessler would have to open some of the cloth flaps on the inspection and serving holes cut into the linen on the back side of the drum, locate the damaged main cord, replace it with a spare he had ready, and then, finally, close up the cloth flaps that were pinned in place to cover the holes. Needless to say, the representative would be instructed to keep well away from the back linen covered circular side of the drum as Bessler performed this repair.

This exact same problem could also happen to the bidirectional Merseburg wheel's *back* internal one-directional wheel's 3:00 (as viewed from the front side of the drum) weighted lever if the representative were to grab the outer leather covered rims of the rim wall on the right side of the drum and yank down excessively hard on them. I suspect, however, that this unusual problem did not manifest itself with the Kassel wheel. That wheel had double the mass of the Merseburg wheel and that would have required the application of a brief torque of double the magnitude applied to the Merseburg wheel's drum to produce the effect. If the same representative who caused the problem to manifest in the Merseburg wheel tried to give the Kassel wheel an acceleration boost at start up, he would probably find his yanking on the drum caused his entire body to quickly lift up off of the floor of the wheel's room in Weissenstein Castle while the massive drum barely moved!

With the conclusion of this chapter, the reader finally has all of the insights I've gained into the internal imbalanced pm mechanics of Bessler's wheels after many years of trying to reverse engineer them both unsuccessfully with physical models and then eventually and successfully with computer models. I believe that the information presented so far should be adequate to allow the serious Bessler researcher with average crafting skills to duplicate the inventor's 3 feet diameter, one-directional Gera prototype wheel without too much difficulty. For the craftsman of greater skill, the construction of self-starting one-directional wheels with drum diameters up to 12 feet should be possible and for the most skilled and diligent of craftsmen, even the construction of a fully functional replica of the bidirectional Merseburg or Kassel wheel should be possible.

To assist him, I will provide the various critical part parameters necessary to do this in the form of several easy to read tables at the end of Chapter 11 in this volume.

In the next several chapters, however, I want to first provide the reader with a detailed analysis of the many clues that Bessler carefully hid in his writings and which, upon being *properly* interpreted, finally enabled me to find the details of the mechanics he used in his imbalanced pm wheels. The finding of these previously unsuspected clues was the major turning point in my own research into the mystery of the inventor's "lost" secret and without their discovery this book could not have been written.

Figure 13(a) - Effect of Latching Mechanisms on Each Internal Wheel's CoG

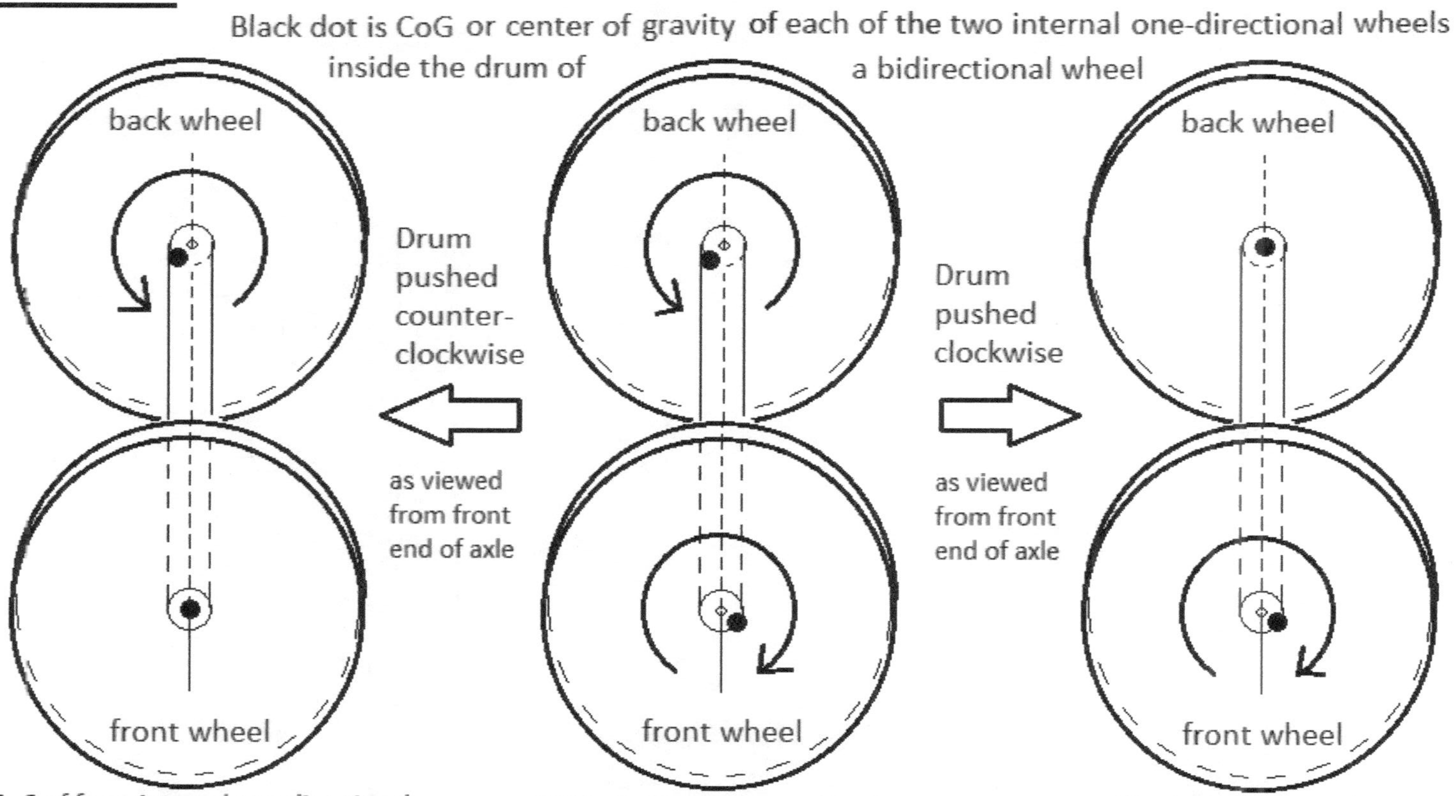

Black dot is CoG or center of gravity of each of the two internal one-directional wheels
inside the drum of a bidirectional wheel
back wheel
front wheel
Drum pushed counter-clockwise
as viewed from front end of axle
back wheel
front wheel
Drum pushed clockwise
as viewed from front end of axle
back wheel
front wheel

CoG of front internal one-directional wheel at center axis of axle. Only back internal one-directional wheel provides axle torque.

START - CoG's of front and back internal one-directional wheels on opposite sides of axle center axis. No net axle torque.

CoG of back internal one-directional wheel at center axis of axle. Only front internal one-directional wheel provides axle torque.

Figure 13(b) - Details of the "Cat's Claw" Gravity Latches
(used in the Merseburg wheel)

After the two steel "L" square pieces are attached to their counter weight, their 5 in. longer arms should remain perfectly horizontal when the complete latch is mounted on a pivot rod and is free to rotate about it.

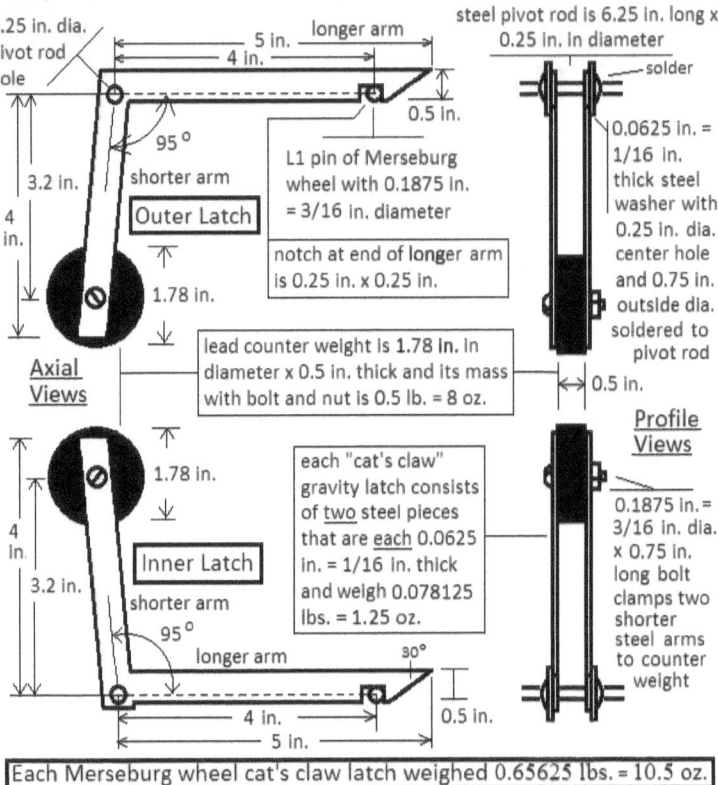

Each Merseburg wheel cat's claw latch weighed 0.65625 lbs. = 10.5 oz.

These types of gravity latches were also used in the Kassel wheel to make it bidirectional, but were heavier and stronger. They would have had arms that were 1 in. wide and weighed 2.5 oz. each. The lead counter weights would have been 1.78 in. in diameter and 1 in. thick, with a mass of 1 lb. = 16 oz. The pivot rod would have been 0.375 in. in diameter x 9.5 in. long. The notch at the end of the longer arm would have been 0.5 in. x 0.5 in. to accomodate an L1 pin that was 0.375 in. in diameter.

Figure 13(c) - The Gravity Latches Installed on Merseburg Wheel Weighted Lever
(L1 to A4 main cord and hook not shown)

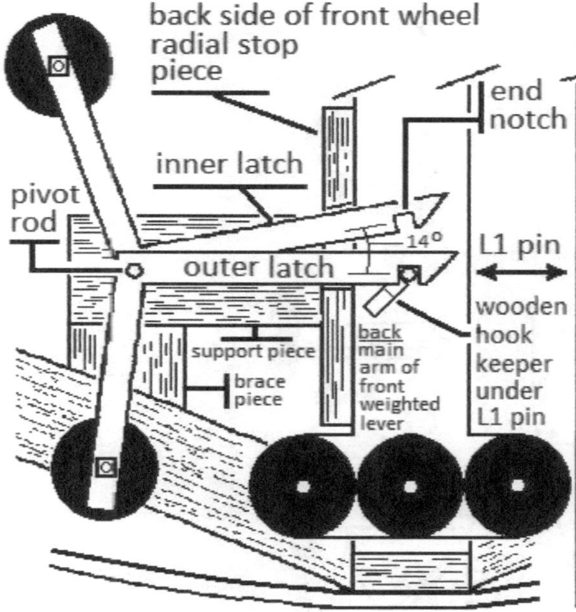

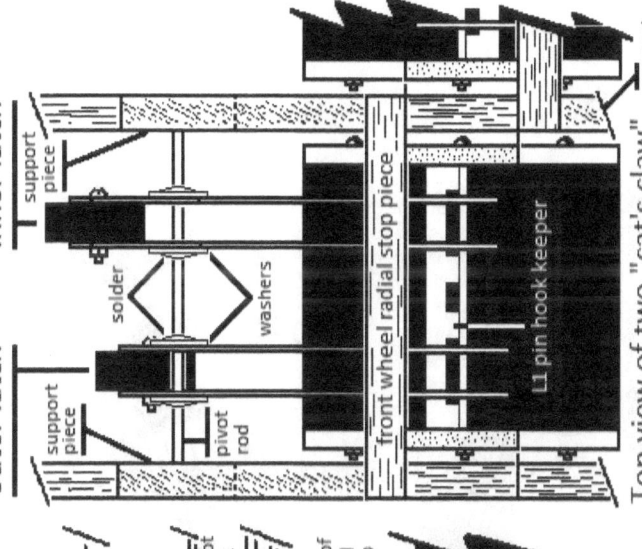

Axial view of two "cat's claw" gravity latches for 6:00 lever of front internal one-directional wheel of a 12 foot diameter bidirectional wheel.

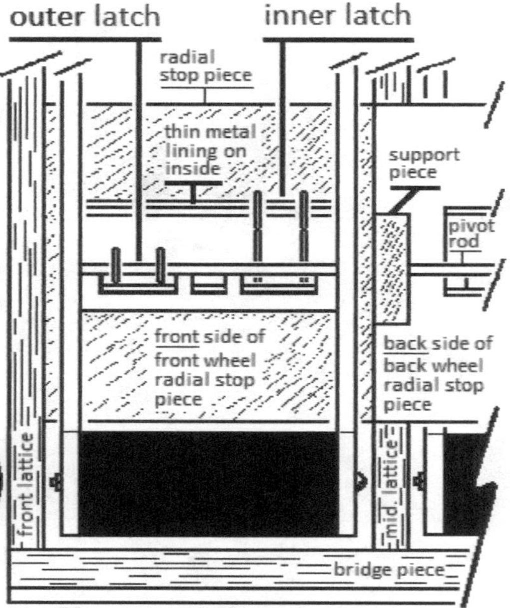

Profile view of front internal one-directional wheel's 6:00 lever from right side of drum showing its outer "cat's claw" gravity latch engaging the lever's L1 pin. Inner "cat's claw" gravity latch is disengaged.

Top view of two "cat's claw" gravity latches with their end notches over cutouts in the hook keeper and L1 pin of the front internal one-directional wheel's 6:00 lever.

Figure 14 - The Gravity Latches in Operation

Interaction of two "cat's claw" gravity latches with weighted lever of an internal one-directional wheel rotating counter (i.e., counterclockwise) to the direction it normally turns in (i.e., clockwise).

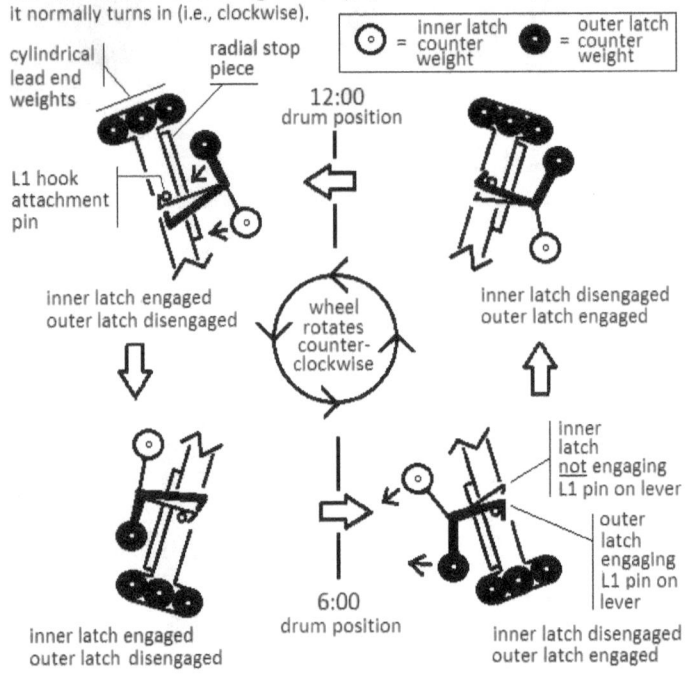

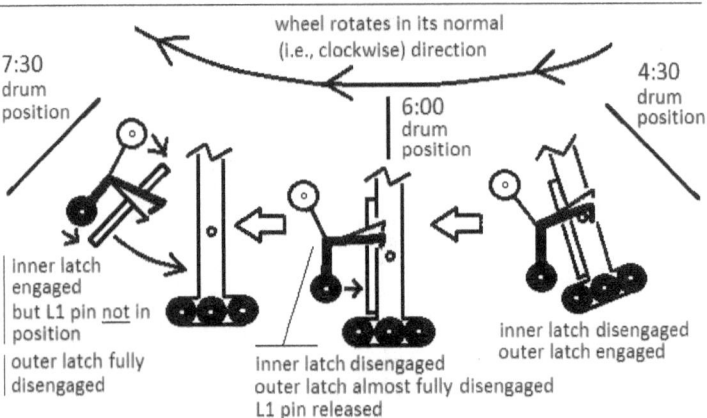

CHAPTER 8

Introduction to the DT Portraits

It was mentioned in Chapter 1 that Bessler was subject to occasional bouts of severe depression which seemed to coincide with the onset of the dark and dismal winter months of northern Europe. When one of these episodes began in late 1717 and extended into 1718 during his stay at Count Karl's luxurious Weissenstein Castle and after the completion of his most powerful 12 foot diameter bidirectional wheel there, the count, in an effort to help the inventor cope with his melancholia, inspired him to author his most professional effort ever to describe his invention to potential buyers. As with his earlier literary efforts, the goal of the work was to convince those who had heard about his wheels, but not personally witnessed them, that they were genuine perpetual motion machines and could be very profitable for anyone who acquired and used them.

That work was titled *Das Triumphirende Perpetuum Mobile Orffyreanum an Alle Potentaten, Hohe Häupter, Regenten und Stände der Welt* or, in English, "The Triumphant Perpetual Motion Machine of Orffyreus, to All the Potentates, High Leaders, Regents and Ranks of the World" and Bessler completed and published it in 1719. This book, unlike his earlier ones, was not just aimed at the masses, but also at the more elite members of the then growing scientific community. To achieve that goal, it had the same text written in German on the left column of each page in Gothic type for the average reader and also in Latin on the right column of each page in plainer type so as to make it appeal to scientists who were then using Latin as a sort of international language.

One of the suggestions that the count made to Bessler was that he provide enough details of how his secret mechanics worked so that, should something happen to him such as an accidental death or murder (and the count may also have feared the possibility of suicide considering the severe nature of Bessler's depressive episodes!) and the Kassel wheel also somehow destroyed, the secret of its perpetual motion mechanics would not be lost to humanity.

While the inventor would have done nearly anything that the count requested in most matters, this request was very near the limit of what he would try to fulfill. Yes, he owed the ample funding of his research to the count and also his employment and residence in a castle for himself and his growing family to this generous man, but Bessler also did not want to risk revealing too much because that might then allow someone to duplicate his invention and to do so without paying him a single thaler for it. Reconciling these two diametrically opposed desires would have given the inventor many a sleepless night, no doubt, as he planned the contents of the book in his mind.

But, he did manage to do it and using what I believe is one of the cleverest techniques ever devised.

As one reads through an English translation of DT, he finds that it presents three basic themes. The first is overwhelming praise to God for giving the inventor the secret of perpetual motion and to Count Karl for funding the construction of his latest and most powerful wheel. The second theme is a scathing condemnation of all of the people that Bessler considered his enemies and that included nearly all of humanity. And, finally, his third theme deals with the matter of his wheels. There are illustrations of the exteriors of the Merseburg and Kassel wheels and descriptions of the various official tests they were given which, he believed, provided ample proof that he had, in fact, managed to find a perpetual motion wheel design that actually worked. However, he never quite gets into the details of what was going on *inside* of the drums of his wheels. The most the reader learns is that, somehow, the weights on the ascending side of a one of his rotating pm wheels would, momentarily, move closer to the axle and this constantly repeated action as the drum rotated then allowed a wheel to maintain a chronic state of imbalance.

There are no drawings of the internal mechanics to be found anywhere in DT and when it came time in the text to describe what was happening to the weights on the descending side of a wheel's rotating drum, Bessler merely states:

> "...these weights themselves provide the perpetual motion of the device and are the essential constituent parts which must of necessity continue to exercise their motive force indefinitely as long as they keep away from the center of gravity of the wheel...they are enclosed in a structure or framework and coordinated in such a way that not only are they prevented from attaining their desired equilibrium or 'point of rest', but they must forever seek it, thereby developing an impressive velocity which is proportional to their mass and to the dimensions of their housing."

All that one learns from this is that Bessler had a working imbalanced wheel design and there are no details about *how* that imbalance was achieved and, most importantly, maintained during drum rotation.

But, as I thought over this and other relevant quotes from DT, I realized that there had to be some very specific details about the actual mechanics he used somewhere in the text. Without such details, it seemed to me that he would not be complying with the count's request and, indeed, it might be impossible for any reverse engineer in the future to find the same design Bessler found so that working replicas of his wheels could be constructed. Somewhere in DT, I was convinced, those details had to exist. But, where could they be?

After much searching, I finally found those specific details and, incredibly, they have been out in plain sight for nearly three centuries! They were very carefully hidden in the two frontispiece portraits and their underlying text that appear at the very beginning of DT.

One of the main reasons that the resurgence of interest in Bessler's wheels that took place after the publication of Rupert T. Gould's *Oddities: A Book of Unexplained Facts* never led to a rediscovery of these portrait clues is because, while Gould refers to DT in his book's chapter on Bessler, he never mentions the portraits or provides illustrations of them. If he had done that, then I believe we would not have had to wait an extra almost century for a solution to finally be found. As far as my own research was concerned, I only learned of the existence of these two portraits about a decade ago and it was only from that time onward that my own research began to make serious progress.

It's possible, however, that the copy of DT Gould used for his book may have been missing both portraits. Currently, there are probably less than a dozen copies of DT in private libraries and, perhaps, as many as another two dozen copies scattered about in the rare book sections of university libraries throughout Europe. Of those in existence, it seems some are missing the second portrait.

Figure 15 shows how Bessler arranged the two frontispieces in DT. One the left side of the figure we see the first portrait which shows him wearing an elaborate wig (the fashion in those days) and standing near a table on which several items are carefully arranged. Attached to the right edge of the first portrait is a folding flap with an oval hole cut out of it whose center is located about a quarter of the way down from the top of the flap. The flap is blank on the side facing the reader, but on its other side one finds the second portrait of the inventor in which he appears to be standing behind a shop table covered with various drafting tools and other small items. On the right side of the figure we see what happens when the flap containing the second portrait is folded over onto the first portrait. Bessler's face from the first portrait appears through the hole in the second portrait.

As one views the two printed portraits in DT, the first question that comes to mind is why did Bessler decide to include the portraits in this manner? Why not simply have printed them on separate facing pages? That would have seemed a far simpler thing to do then to go to the extra trouble of cutting a hole in the second portrait and then having, book by book, to carefully align the flap of the second portrait so that the inventor's face in the first portrait would show through the hole when its right edge was finally pasted down against the page on which the first portrait was printed.

When it came time to include his own portraits in DT, Bessler, based on the recommendation of an acquaintance, decided to hire a young artist named Christian Fritzsch to do the engravings of the images on the copper plates used in the printing process which had to be made as accurate reversed mirror images from an original sketch. Fritzsch was born in Saxony in 1695 (he died in Hamburg in 1769) and was twenty-four years old when Bessler hired him. He was based in Dresden and found steady work producing illustrations for book publishers in the area.

From the woodcut printing blocks used for the test images in his incomplete and unpublished MT, it's obvious that Bessler was capable of producing simple reversed relief images of various perpetual motion machines on wood blocks that could then be used in a printed book. But, when it came time for the far finer detail involved in engraved copper printing plates, although he had some experience with this, he decided to turn to a professional artist whose refined skill could give him the accuracy that he required in the two portrait illustrations especially since they would have to contain some very precise angles, ratios, and alphanumeric clues.

In the lower left corners of the two frontispiece portraits used in DT, Fritzsch signed his name as "*C. Fritzsch Sc.*" The abbreviation "*Sc.*" stood for the Latin word "*Sulpsit*" which means, in English, "Sculpted by". This was an indication that the engraving work had been performed by an artist who had mastered special techniques that allowed the engraved images of certain objects to actually appear similar to ancient sculptures or masonry. This might, as an example, be done by using pumice stone or some other abrasive substance to roughen up the surface of the copper printing plate which would then result in the printed image having a grainy or pitted appearance like actual stone material. Needless to say, engravers who could legitimately add this title after their names on their artwork earned higher fees than those who could not.

Bessler contacted Fritzsch and supplied him with enlarged sketches of what the inventor wanted his two DT portraits to show and, of course, the younger artist was not aware of the true significance of the precise engravings he was required to produce. I believe that, initially, the two portraits including the text under them, which measured only about 6 inches wide x 9 inches high, were to be placed on separate facing pages; that is, they were not originally intended to be attached to each other by a pasted hinge.

However, when Fritzsch finished the work on the two portraits and then supplied Bessler with some sample prints of what they would look like in his *magnum opus* about his inventions, there was a problem with one of them that was unacceptable to Bessler.

One must realize that placing the specific clues about his imbalanced pm wheel mechanics into the portraits and then having them in the same image with his face was the inventor's way of symbolically establishing his claim to have been the first person to actually achieve a working imbalanced pm wheel. When Bessler saw the first portrait, he must have been satisfied with it because it shows him as a cultured man relaxing in his study and surrounded by the symbols of intellectualism. With the second portrait, however, this was not the case. The more Bessler stared at the second portrait's face, the more it must have seemed to him that it did not look enough like the depiction of his face in the first portrait.

Things like that can happen when one attempts to draw the same person's face and head from two different angles or from the same viewpoint after the person has turned his face toward a different direction. Bessler, realizing that the most important clues about his wheels' secret mechanics were contained in the second portrait might have thought that, because the second portrait's face did not closely enough resemble the first portrait's face, some unscrupulous person might emerge later, after the wheel was sold and its secret finally publicly revealed, and claim that the person in the second portrait was not actually Bessler and that, in fact, someone other than Bessler had really been the wheel's inventor. Indeed, that person might then even claim that *he* and not Bessler was the inventor and that, due to his closer resemblance to it, the face in the second portrait actually belonged to him! He could then claim that the one hundred thousand thalers that Bessler received should really have been paid to him rather than Bessler! With enough luck, he might even get a court to agree with him and be able to legally take the money from Bessler.

Bessler, obviously, after all of the sacrifices he had made to finally find a working imbalanced pm wheel design had no intention of letting that happen. He would have contacted Fritzsch and said that he wanted the second portrait redone with a face that more closely resembled the one in the first portrait. At that time Fritzsch probably explained to him that he would have to charge extra to completely redo the second portrait and, even so, he could not guarantee that the second portrait's new face would look enough like the one in the first portrait to satisfy Bessler. Fritzsch then most likely suggested that the least expensive solution would be to just make a few minor alterations to the engraved copper plate for the second portrait which would obliterate the face in it and then, after an image of the revised version of the second portrait was printed, simply cut out a hole in it so that the face from the first portrait would show through when the two portraits were overlapped. This solution, however, required printing the second portrait on a second smaller than page size sheet of paper, manually cutting or punching out the hole, and then carefully gluing down the right edge of the smaller sheet of paper to the page containing the first portrait so that, when the second portrait's sheet was folded over the first portrait, Bessler's face from the first portrait would show through and appear as part of the second portrait.

This approach would guarantee that Bessler's face from the first portrait was also associated with the second portrait and, I suspect, that the novelty of this feature in his book also intrigued the inventor. It would make the book stand out a bit from other works and that might result in the extra attention needed to finally sell his invention.

From studying the second portrait, I believe that the problem with it was that it originally showed an almost profile view of Bessler's head with him looking to *his* right side and toward the globe on the left side of his shop table. To Bessler, that face just did not look enough like the frontal view shown in the first portrait and had to be eliminated. He wanted it blotted out completely and then replaced with something into which an oval hole could be cut through which the inventor's face from the first portrait could show.

To hide the original face in the second portrait, Fritzsch decided to just add a thick, almost solid black, furry head covering which was tied in a knot under the inventor's chin and had ends that then hung down over the front of the bulky coat he is shown wearing. The viewer would then, because of the coat, automatically assume that the head covering was a natural addition to the coat and helped keep Bessler's head warm when he was working away in his shop during a long frigid winter month when, despite a small stove for heat, the temperature was still uncomfortably cold. On the upper left hand side of the second portrait a thermometer is attached to the wall and this further creates the impression that his workshop temperatures might get so cold that Bessler would have wanted to be able to measure them in order to know how much wood to put in the stove or how much clothing to wear.

Now that the reader understands some of the circumstances that probably led Bessler to include the two portraits in DT so that a single face was used for both, it's time to look a little closer at each of the two portraits. In the two chapters following this one, we will scrutinize each of the portraits in far greater detail in order to extract the embedded clues from them. But, for the remainder of this chapter, an analysis of some of the more obvious symbolisms in these two crucially important sources of clues will suffice.

To begin this preliminary analysis, the reader should now refer to Figure 16 which shows the first portrait.

One immediately notices that there are German words, often shortened, written under the portrait and below them I give the English translation for those words as:

> "J(ohann). E. E. B. or ORFFYREUS; His Highness's Counselor of Commerce, Medical Practitioner, Mathematician, and Inventor of Perpetual Motion."

With this, the reader learns that Bessler was appointed to an important government position and, thus, had the approval of Count Karl. The former was also a medical doctor, mathematician, and, most importantly, had found the secret of what he believed was true perpetual motion!

Looking at the portrait itself, we see Bessler apparently standing behind a table. He wears an expensive handmade wig, is well dressed for the period, and has either a blanket or cloak covering his left side from his shoulder down to his wrist. About his neck is a cravat and attached to its lower end is what was probably a silver badge of office that he was required to wear while performing his duties as the count's commerce counselor. Both of Bessler's hands are visible.

The thumb and index finger of his left hand (which appears on the right side of the portrait) are bracing the bottom end of a large book that rests upon the table and thereby keeps the book's upper raised end from sliding off of the face of a jawless skull that it rests upon which would then let the skull, precariously balanced on its rounded back portion, fall over to the right so its upper teeth would hit the table. The crown of the skull further has a tilted vase leaning against it. Bessler's right hand (which appears on the left side of the portrait) touches nothing, but its outstretched index finger points to the book.

Bessler's right forearm appears to be resting upon a set of three books, but there is a dark space to the right of the set of books and the implication is that the book on the table was removed from the set which actually consisted of four books.

Behind the inventor's right arm we can see four shelves that each hold four jars although the right sides of the three top shelves are partially obscured by a heavy drape. The jars have diagonal labels on them and are of the type used to store various herbs and dry powdered medications. There is also a large tassel on the end of a cord and the implication is that, by reaching up with his right hand, Bessler could easily pull down on the cord which would then result in the drape being drawn toward the right side of the portrait so that what is hidden behind it would be revealed.

Analyzing the obvious symbolism of the first portrait is fairly easy.

Bessler wanted the reader to appreciate that the inventor was an intelligent man who was comfortable in the world of the aristocrat and intellectual. Books are symbolic of acquired knowledge and his right arm resting on the three books indicates that he had spent about three decades gaining knowledge of the world from his infancy to his present age in the portrait. His right arm is also a symbol of manual work and, because it is resting on the books, this tells us that much of his accumulated knowledge was the result of hard physical work. But, one of the books, the fourth one, was removed and then placed on the table in a rather precarious state of balance. This indicates that it was in his fourth decade of life, when he was almost thirty-two years of age in the year 1711, that he found the unique mechanics that allowed a wheel to remain chronically out of balance as it rotated and accelerated.

His wheels, which could remain permanently out of balance, are represented by the nearly circular shaped skull on the table which is supported by the fourth book so that it cannot lose its imbalanced orientation and fall over onto its upper teeth. The use of the forth book as a support tells us that it was, in particular, the knowledge Bessler gained from a combination of intellectual study and physical work during the fourth decade of his life that finally allowed him to make a wheel that would remain in a permanent state of imbalance whether it was stationary or rotating. The four visible upper teeth in the skull also suggest the fourth decade as the one in which the inventor found success. But, what about the symbolism of the missing lower jaw of the skull?

The lower jaw of a living person is used for talking and communication and its absence in the first portrait indicates that Bessler intended to keep the secret of his wheels' internal mechanics confidential by not talking too much about their details. It also indicates that, unless he sold his invention, he was prepared to keep the secret right up to the point of his own death!

There are thirteen jars visible on the four shelves behind Bessler (as determined by the number of diagonal labels on them that can be counted) and four books in the portrait (three under his right elbow and one on the table in front of him). Adding the number of jars and books gives us a total of seventeen items. The index finger of his right hand points to the book on the table which is the seventeenth of the seventeen items if one starts counting from the leftmost jar on the top shelf.

Then, we notice that his right hand's middle finger crosses behind his index finger and this represents the "+" sign in arithmetic. Thus, this portrait clue is telling us that we must perform some sort of an addition or joining together that involves the number 17 which is represented by the book on the table and some other number. To find that other number, we notice that the thumb of his right hand points directly to the lowest button on his *shirt* which is a bit difficult to see because half of it is hidden by his jacket. If one counts the buttons, starting from the leftmost one on his jacket's right sleeve, then up along the right side of his jacket, and, finally, down the length of his shirt starting from its collar, he will discover that the lowermost button on his shirt is the twelfth button in the portrait. However, that twelfth button is only partially visible indicating that it represents a number *less* than 12. This particular combination of clues in this first DT portrait therefore tells us that we must add or join together the number 17 and another number that is *less* than 12. If we place the number 17 and a number that is less than 12 such as 11 next to each other in decreasing order, then we get the string of digits 1711 which just happens to be the year that Bessler first found success with his one-directional, 3 foot diameter Gera prototype wheel!

Although it is very difficult to see in Figure 16, the twelfth button on his shirt only appears as a thin crescent that peeks out from behind the edge of his jacket. This suggests that the working design he found in 1711 was on a small wheel and that original Gera prototype was, indeed, the smallest one-directional wheel he ever constructed. (There is also another important interpretation of the number of shelf jars and books on the table which I will go into in detail in the upcoming chapter that treats the first DT portrait items in depth.)

The closed drape behind Bessler was his way of symbolically telling the reader that there was something important being hidden from his view and that information was contained in the first DT portrait itself. However, the drape is partially pulled open near the inventor's head. The message from this is that, in order to discover what is hidden behind the drape *and* in the portrait, will require much thought on the part of the reader of DT. The drape appears to be opening from the left to the right of the first portrait and there is a major fold in the drape, the corner of which is near the right side of the second shelf from the top on the left side of the portrait.

If one extends an imaginary vertical line downward from that fold's corner, it will connect up with the place of contact between the two diagonal jar labels on the right side of the fourth shelf from the top. These two labels are oriented with respect to each other to form a downward pointing arrow head whose tip appears in the portrait to be touching Bessler's right arm. Thus, we have a sort of downward pointing arrow that extends from the corner of the drape's fold and points directly toward Bessler's right arm.

Since, as previously mentioned, the right arm is a symbol of physical work, the inventor is telling the reader with this arrow clue that to finally open the curtain or obtain all of the hidden information in the first and also, by extension, the second portrait, will require the reverse engineer of his imbalanced pm wheels to perform *much* physical as well as mental work. In other words, it will be a very difficult task to extract all of the information about his wheels' secret mechanics from the portraits alone, but it can be done if someone is willing to do the necessary mental and physical work. The mental work will involve using initial interpretations of the clues in the two DT portraits to guide the planning of a possible replica wheel and then the physical work will involve the actual construction and testing of that wheel and, when it fails, again going back to the DT portrait clues to see if an alternative interpretation of them can suggest modifications of the possible replica wheel which will improve its performance. The two diagonally oriented labels on the jars forming the arrowhead can also symbolize the side by side, back to back arrangement of the two internal one-directional wheels found inside the enlarged drums of the inventor's bidirectional Merseburg and Kassel wheels.

So, in essence, Bessler was symbolically telling the reader of DT and future reverse engineers of this wheels that the clues contained in the two portraits would only be of use to him if he is also continuously building and testing possible physical replica wheels as he was simultaneously trying to correctly interpret the clue symbols. Thus, the existence of these clues shows us that Bessler was willing to give the secret of his imbalanced pm wheels away, but only to the person who was patient and devoted enough to receive the information in the exact way that the inventor intended to deliver it to him.

That method was a slow process that would involve continuous feedback between the possible correct interpretations of the DT portrait clues and the result of their incorporation into a physical wheel by a craftsman that could, in practice, actually take him decades or even an entire lifetime to complete! Indeed, in my own research into the inventor's wheels, I doubt if I would have made any sort of significant progress, even after a lifetime of effort, if I had been limited to just constructing and testing potential physical replica wheels by hand. What allowed me to finally succeed where many, many others had failed was having access to modern computer simulation software and, even then, it took almost a decade of using it as well as a tremendous amount of sheer luck!

But what about that vase that is leaning against the top of the skull both of which are propped up by the book which, in turn, is braced by Bessler's left hand?

The vase is a symbol for the womb and birth and the skull, aside from being a symbol for one of the inventor's wheels as previously noted, is also a symbol for death, dying, and extreme old age. Here the book is again a symbol for knowledge and medical knowledge in particular. The book is preventing both the vase and the skull from falling over to the right side of the portrait. Falling over is something that animals with feet do when they are dying and are too weak to stand upright against the pull of gravity. Thus, the items on the table in front of Bessler are symbolically telling the viewer of the portrait that, through the application of his medical knowledge, Bessler had prevented the deaths of people of all ages from their birth to their old age.

Most importantly, however, we see here that Bessler could use *multiple* meanings for the *same* symbol in the first DT portrait and, indeed, we will see in the next chapter that this also applies to the symbols in the second DT portrait as well. The analyst of the DT portrait symbols must always keep the potential *multiple* interpretations of a symbol in mind when trying to extract the mechanical clues they contain. It certainly makes interpreting them more difficult, but it was not Bessler's intention to make the job of the reverse engineer an easy on. Bessler wanted to make sure that whoever finally obtained the secrets of his imbalanced pm wheel mechanics would, literally, have to work as hard for them as had the inventor!

This now concludes the analysis of some of the more obvious clues in the first DT portrait, but it is only the barest tip of the iceberg when it comes to the entire symbolic content of this portrait. I will have much more to say about the many other symbols and clues that are also present, but that will have to wait for another two chapters after this one.

Now it's time to move on to the second DT portrait and for this the reader's attention is directed to Figure 17 at the end of this chapter.

In Figure 17 we see a scene with Bessler that is located in a workshop environment and we can begin with a translation of the German text under the portrait. That text is not an easy one to translate, but my best effort, in English, is:

> "Note well that a useless invention can be said to be perfected when something (i.e., a modification) is found that causes it to be praised."

Here he is expressing exactly what happened to him as he pursued a working imbalanced perpetual motion wheel design. Every design he had previously tried proved totally useless and then, one day, after several weeks of working with a modification suggested to him in a particularly vivid dream he had one night, he found success that, eventually, led to larger and more powerful wheels which were praised by many for their amazing self-motive abilities.

In the second portrait, Bessler is again positioned behind a table, but this time it is one found in a workshop that holds a diverse collection of objects.

There is a globe that would be of interest to a person studying geography, a telescope for someone studying astronomy, various drafting tools of use to a craftsman, and even a small microscope which would be of interest to the student of biology. In the lower right hand corner of the portrait we see an upright metal lamp that stands on tripod feet and, to its right, a group of four small metal ingots which are probably of lead. There is also a curious little piece of paper with curved and straight lines drawn on it to the left of the lamp as well as a carpenter's plumb bob with a 14 inch ruler carefully balanced upon it.

Unlike the first portrait, Bessler appears more active in the second portrait and this was intended to depict him as a man who was physically active and, thus, did not have a lot of time available to be sitting around and relaxing. The lamp on the table suggests that he would work long after sundown in his pursuit of a working perpetual motion wheel. In the inventor's left hand is a divider whose arms' pointed ends make contact with two points on the surface of the globe. In his right hand he holds a pair of armless eyeglasses. Bessler himself is dressed in a heavy coat and has a furry covering tied around his head. As previously mentioned, the face we see in the second portrait is really the one from the first portrait underneath it that is showing through an oval hole cut out of the paper that the second portrait was printed on.

The items that appear in the background of the second DT portrait are just as important in terms of symbolism as are the various foreground tabletop objects as shown in Figure 17.

Behind Bessler's left shoulder we see an organ with a clock attached to its side and hung on a hook or nail. On the front top edge and just below the front upper right corner of the organ there is some decorative woodwork which shows the heads of two cherubs or angels complete with their wings. These beings are described in Christian angelology as members of the second highest order of angels who attend God and their distinguishing gift was *knowledge*. Their presence, of course, is entirely appropriate on an organ which, most likely, would have been used to play music during a church ceremony. Bessler includes the organ in the second portrait to show the reader that he was also a skilled organ maker.

On the other side of Bessler's head and attached to the wall in the background one finds a device known as an armillary sphere or spherical astrolabe. It is a collection of metal, usually brass, rings that form a sphere. The Earth is often represented in the center of such a sphere by a small hollow metal ball. The various rings around the outside of an armillary sphere represent certain main latitudinal and longitudinal lines that can be imagined to exist on the surface of the apparent sphere of the sky, both daytime and nighttime, that surrounds and is visible from the Earth's various locations.

Other rings attached to the outside of the spherical form of the armillary depict things like the apparent yearly path of the Sun through the various constellations of the Zodiacal "belt" which was sometimes shown as a band with the pictures of the twelve constellations painted onto it. (This apparent yearly path of the Sun through the sky surrounding Earth is known as the "ecliptic" because lunar and solar eclipses can occur when the path of the Moon crosses it.) An armillary allowed one to determine with some degree of accuracy which of the brightest stars and constellations would be visible in the night sky for any particular time of the night and month of the year and what their positions would be relative to the horizon. Because it also allowed one to know what constellation the Sun was in on any given day, it was an important tool for use by astrologers in producing horoscopes and Bessler may have so used it.

Placing the organ cherubs and armillary sphere in proximity with his head in the second portrait was Bessler's way of symbolically telling the reader that his invention of a working perpetual motion wheel was the result of divine inspiration because both the cherubs and sphere are associated with God and Heaven. In other words, these items tell the reader of DT that Bessler was actually *given* the secret of achieving mechanical perpetual motion by God Himself!

In Chapter 1 it was mentioned that Bessler only found success after he had a particularly vivid dream that stimulated him to make a small change in a useless design that he was working on and which then resulted in it becoming operational. Angels, such as cherubs, were considered to be the messengers for God. Since cherubs were also considered to be responsible for imparting knowledge to the devout, the inventor uses them in the second portrait to suggest that they were responsible for the dream he had. Perhaps he thought that, as he slept, two little cherubs appeared in his bedroom and whispered into his ears what changes he had to make to his nonfunctional Gera prototype wheel in order to make it finally work and that information took the form of a vivid dream. These little angels, however, would only have done that if directed to do so by God.

The symbol of the armillary sphere represents the heavens above and, in particular, the seventh heaven which, according to religious texts like the Talmud, was the highest and farthest out heaven and the place where God dwelled. This also fits in with the number of vertical pipes visible on the organ behind the inventor's left side. We see that there are seven of them that the lower cherub hovers over the smallest pipe and the barely visible and largest seventh pipe is nearest to Bessler's covered left ear. This suggests sound or a voice that was traveling from the smallest pipe over to the left and largest pipe visible. Thus, Bessler is telling the observant reader of DT with these clues that the secret of achieving perpetual motion was actually delivered into his ear by cherubs dispatched from the seventh heaven which was the dwelling place of God himself!

Moving over to the left side of the portrait, one finds a variety of woodworking tools such as a saw, two carpenter's planes, and a collection of chisels as well as a large thermometer and even a shotgun. Because the reader already possesses an understanding of the basic design Bessler used in his wheels, interpreting the meaning of these items in the second portrait is now actually fairly easy.

We start with the saw and notice that it is shaped and oriented somewhat like the number "9". It represents a weighted lever whose pivot pin's center axis has reached the 9:00 position of a clockwise rotating drum and that is indicated by the proximity of the top part of the saw to the 9:00 position of one of the circular rings of the armillary sphere hanging on the back wall to the right of the saw. The wooden handle of the saw just touches the back of the wooden stock of the shotgun below the saw whose barrel points downward toward the globe. A shotgun is a device which, because of the triggered action of its firing mechanism, can suddenly and forcefully expel the lead shot ammunition it carries from the end of the gun's barrel.

With this symbol Bessler is telling the reverse engineer of one of the inventor's clockwise rotating one-directional, imbalanced pm wheels that there was rapid motion of the cylindrical lead end weights at the end of a weighted lever *after* its pivot pin's center axis passed the drum's 9:00 position. We know that the drum's motion he refers to is clockwise because the shotgun's barrel points toward the globe, a symbol for one of his one-directional wheels (whether it was a single self-starting wheel or one member of an opposed pair of internal one-directional wheels within the drum of a bidirectional wheel that had to be push started) and we see the South Pole of the Earth on the globe. When viewing the Earth from a position out in space above its South Pole, our planet will appear to be rotating clockwise.

The connection between the shotgun and globe symbols is further reinforced when it is noticed that an imaginary line drawn from the *top* of the thumb grip on the shot gun's hammer through the point where the *border* between the portion of the wooden stock under the barrel and the barrel intersects the upper edge of his coat's right sleeve will, if extended farther downward, pass directly through the South Pole of the globe! On the upper half of the globe, we can see the southern tips of South America and Africa. Antarctica is not shown because it was not discovered until later in the 18th century.

Very prominent in the background are the two carpenter planes. These tools are used to shave off a thin layer of wood from the side of a board and must be used with and not against the grain of the wood. They both point downward toward the *top* of the globe or its equator and are arranged one on top of the other. This tells us that, if we were very tall and were to look down into top of a bidirectional wheel's enlarged drum at its 12:00 position while standing on its axle's front end, then we would see its two internal one-directional wheels located next to each other and the back internal wheel, represented by the upper carpenter plane, would appear to be above and behind the front internal wheel which is represented by the lower carpenter plane. It will be noted that a vertical line extended downward along the side lengthwise edges of the two hung carpenter planes passes directly through the center axis of the upper lens of the pair of armless eyeglasses that Bessler holds in his right hand. Eyeglasses are worn by people with vision problems that do a lot of reading and studying. The message in this symbolism is that the reader should study these symbols carefully. As seen above, that is because the inventor is here revealing how the two internal one-directional wheels were arranged inside of the larger drum of a bidirectional wheel.

Another important feature of the second DT portrait is the presence of a large number of optical systems which do not appear in the first DT portrait. I've already mentioned the two lenses in Bessler's armless eyeglasses, but there are also lenses, although unseen, in the telescope near the left side of the bottom border of the portrait, the screw barrel microscope to the right of the conical top of the lamp and near the right border of the portrait, and even Bessler's two eyes. The important feature of these various optical systems is not the lenses, but, rather, their optical *axes* which can be thought of as imaginary straight lines that project out of the optical system and then point to some other item in the second DT portrait. When such an optical axis is found, it is a signal from Bessler that one must intensely study what the optical axis is pointing toward as would happen when, for example, one points the optical axis of a telescope toward some distant object to more closely view it.

For example, consider Bessler's right eye (which is the left one in the portrait). If one drops a vertical line downward from the center of the pupil of that eye, it will pass directly through the center axis of the pivot pin that holds the two arms of the large duckbill goniometer together. This indicates that one must study this goniometer symbol and try to extract the information it contains. While I will do this in far greater detail in the next chapter, for the moment let it just be mentioned that the two arms of the goniometer have an interior angle of exactly 45°. This tells us that we are going to find some information in this portion of the portrait that relates to what happens during each 45° segment of a wheel drum's rotation. We are also told that this information will involve other items near the arms such as the "L" square drafting tool that the goniometer's lower arm rests on top of, the strange looking piece of paper with the concentric arcs and straight lines on it, the conical lead carpenter's plumb bob weight with the 14 inch ruler precisely balanced on it, and even the tripod metal lamp with the end of a steel file resting on one of its feet.

What are the precise meanings of all of these symbols? Again, a more detailed treatment must await the next chapter.

Finally, I want to discuss the large thermometer that is attached to the back wall in the upper left hand corner of the second DT portrait. While it is possible that it was the type that used mercury in its glass bulb and capillary tube and would have been one of the original ones manufactured by Daniel Fahrenheit that was calibrated using the scale he developed, I'm convinced that it is just the less accurate type that used dyed alcohol and was used for decades before Fahrenheit became a glass blower in 1717 and began making scientific measuring instruments such as altimeters, barometers, and thermometers.

The possibility that it could have been an original Fahrenheit mercury thermometer is, however, strengthened by the fact that Fahrenheit maintained correspondence with other notables in the Bessler story such as Leibniz and s'Gravesande. No doubt, Count Karl would also have had several of his handmade and calibrated scientific instruments in his large collection of these. However, the second portrait was intended to show the reader the type of devices that Bessler handled during his quest to achieve mechanical perpetual motion and, during that decade of research, he would most likely only have used thermometers containing colored alcohol instead of the more toxic liquid mercury.

I'm sure that many have looked at this thermometer and automatically assumed that, since the thermometers Fahrenheit manufactured used liquid mercury, this was a clue that Bessler's wheels used flowing mercury contained in vials attached to their inner rim walls or in hollow spokes to achieve the imbalanced center of gravity that drove them. However, the thermometer in the second DT portrait was most likely one of the pre-Fahrenheit types and its presence, again like the fur head piece Bessler wears, was intended to tell the reader of DT that Bessler continued to work on his imbalanced perpetual motion wheels despite the cold temperatures in his workshops during the frigid winter months in Saxony.

But, the thermometer has far more significance than that. Indeed, the clue it contains is actually the place where the successful reverse engineer must *begin* in his quest to unravel the complex web of clues contained in the two DT portraits! I will delve far more deeply into this in the next chapter, but, for the moment, let me just point out a few things that focused my attention on the thermometer.

I mentioned above that there were two cherubs carved into the decorative woodwork of the background organ on the right side of the second portrait. Each of those cherubs has somewhat distinctive facial features, especially their pointed little noses. In thinking about those pointed noses, I decided, using a graphics program on my computer, to draw a straight line from the upper cherub's nose to the lower cherub's nose and then keep extending it to the lower right side of the portrait. To my great surprise, I discovered that the line passed directly through the center of the dial of the clock that is hung on the right side of the organ.

In a sense, the center pivots that carry the hands of a clock's dial are analogous to the nose on a face and, even today, it is not unusual to refer to a clock's dial as its "face". So, upon inspecting the alignment I had found, it became obvious to me that Bessler had carefully positioned the three noses, two belonging to the cherubs and one to the clock's face, so that they would be collinear or all lying on the same straight line. Were there any other "noses" in the second DT portrait? I could find another two. One, obviously, belonged to Bessler whose printed face from the first DT portrait was showing through the oval hole cut into the folded over piece of paper upon which the second portrait was printed and the second belonged to the pair of armless eyeglasses Bessler holds in his right hand. The "nose" of the eyeglasses, however, would correspond to the empty space in the center of the frame which would be occupied by the inventor's actual nose when he wore the glasses.

I then drew a straight line from the center of the clock's dial to the tip of Bessler's nose and decided to measure the angle between the two lines I had drawn. Careful measurement indicated it was exactly 62.5°. Seeing how precisely this angle was formed in the second portrait seemed to be no accident to me. Bessler purposely put these alignments of nose points into the portrait to provide the observant reader with a very important starting clue as he attempted to solve the mystery of how the inventor had achieved a working imbalanced perpetual motion wheel.

At this point I decided to draw a horizontal line through the center of the clock dial and extend it to the left of the clock. Since that horizontal line immediately passed through the 9:00 position of the clock's dial, I realized that Bessler was saying with this that the angle of 62.5° was an angle what would be found for all of the weighted levers as their pivot pins' center axes arrived at a clockwise rotating drum's 9:00 position. After having studied the earlier chapters of this book, the reader will realize that this clue was intended to show the orientation of a one-directional wheel's 9:00 weighted lever's parallel pair of main arms when the center axis of their lever's pivot pin was located at the 9:00 position of the drum. The center lines of the main arms there are rotated exactly 62.5° counterclockwise about their pivot pin away from the center lines of the lever's supporting parallel pair of radial frame pieces which are then perfectly horizontally oriented. That orientation then brings the center of gravity of the weighted lever as close to the center axis of the wheel's axle as possible.

This particular angle is critically important in the construction of one of Bessler's one-directional wheels because it presents the angle that a weighted lever's main arms' center lines will make with respect to the lever's radial frame pieces' center lines at the drum's 9:00 position when the lever is being *entirely* suspended by the tension in its two attached and maximally stretched extension springs. This angle is only achieved when one has weighted levers the masses of whose end weight(s) and wooden levers are in a precise ratio with respect to each other and are, when attached to each other, suspended by two extension springs that have the correct spring constant value for the total mass of the weighed lever used.

To state the matter as bluntly as possible, if the center lines of the main arms of one's extension springs suspended weighted levers whose pivot pin's center axes are located at the drum 9:00 position do *not* make this particular angle with the center lines of the lever's horizontal parallel pair of radial frame pieces, then he has failed to exactly duplicate the mechanics that Bessler used in his wheels and whatever final wheel he manages to construct will not produce its maximum possible magnitude of torque at any given axle rotation rate. Indeed, if the angle he achieves with his drum's suspended 9:00 weighted levers' main arms' center lines departs too much from this value, then his wheel will not work at all! It is because of the importance of this angle that it is the first clue that the Bessler wanted a future reverse engineer of his wheels to find.

The angle of 62.5° also would have fascinated Bessler because its digits just happen to add up to 13 and the Hebrew name for this number is "adonai" which is also their name for God. This surely would have convinced the inventor that this very important value was given to him by God Himself.

It is now time to end this brief introductory chapter on the two DT portraits. Its purpose was to make the reader aware of the great importance these two printed copper plate engravings have in terms of solving the Bessler wheel mystery and to give a very general overview of the symbols used in the portraits.

I have focused primarily on the portraits themselves and have not really delved much into the text below each portrait other than providing English translations for them. The reality, however, is that there is an abundance of precise values for the physical properties of the various components used in Bessler's imbalanced perpetual motion wheels that were carefully hidden in the text below the portraits. That information is encoded using the simple alphanumeric system that Bessler preferred and much of it will be decoded in the following two chapters. And, I feel obliged to point out now that, while I made enormous progress in decoding the many clues Bessler left in the two DT portraits for future reverse engineers to use in duplicating his inventions, most likely, there are *still* many clues that I have not found despite years of effort! No doubt, as interest in Bessler's wheels escalates following their successful replication, future researchers will find these additional clues and properly interpret them.

When I became aware of these previously unknown clues in the two DT portraits, I also realized that they are additional and very strong evidence that Bessler's wheels could not have been the product of trickery. What swindler would go to the lengths of so carefully hiding the design parameters of fake imbalanced pm wheels into two portraits that would only finally be successfully analyzed three centuries later and long after his death? That possibility seemed vanishingly small to me. I can only conclude, like many have over the centuries, that Bessler's wheels were 100% genuine and, if one is not too rigorous with his definition of "perpetual", were real perpetual motion machines. Their existence shows that such devices are possible and that Bessler found one way of building such a device although it is obvious from his writings that he thought he had found the only way possible.

Someday, hopefully soon, when self-motive machinery continuously outputting the energy associated with the masses of their active components are in widespread use on our planet, Johann Bessler will be remembered more kindly as the "father" of this subject and finally achieve his rightful place in the history of science.

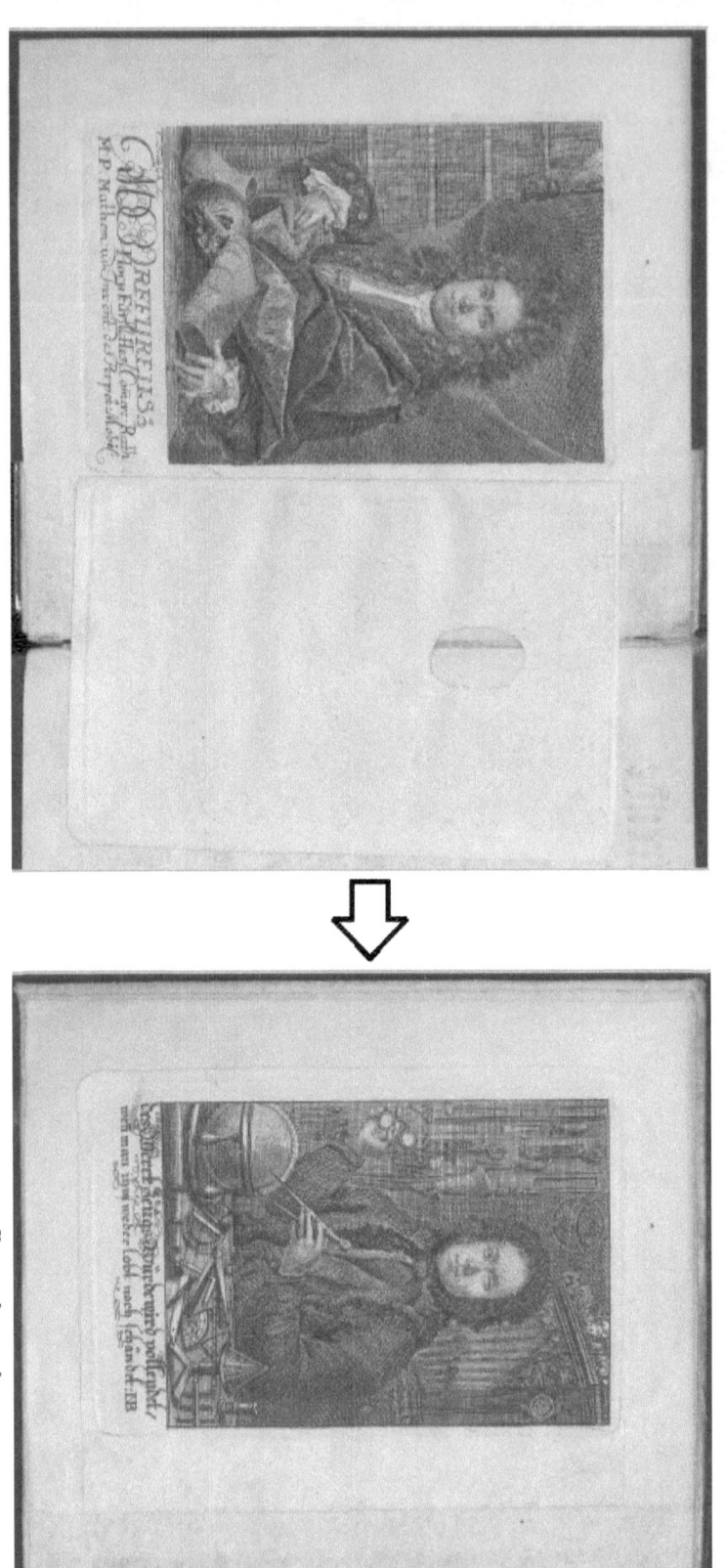

Figure 15 - The Two DT Portrait Frontispieces

The second portrait is printed on a flap which, when folded over to the left, allows Bessler's face from the first portrait to show through an oval hole in the second portrait.

Figure 16 - The First DT Portrait

"J(ohann). E. E. B. or ORFFYREUS: His Highness's
Counselor of Commerce, Medical Practioner,
Mathematician, and Inventor of Perpetual Motion"

Figure 17 - The Second DT Portrait

"Note well that a useless invention can be said to be perfected when something (i.e., a modification) is found that causes it to be praised."

CHAPTER 9

CLUES FOUND IN THE SECOND DT PORTRAIT

The reader may be a bit surprised that this chapter, the first of the two that deal with a more in depth treatment of the clues in the two DT portraits, is starting with what, in the previous chapter, was referred to as the "second DT portrait" and which shows Bessler standing behind the table in his workshop. Why, one might wonder, am I not starting with what was previously referred to as the "first DT portrait" that shows the inventor standing near a table, fashionably dressed, and adorned with an expensive wig?

There are several reasons for this.

Firstly, the number of clues about the values of the various construction parameters involved in his secret imbalanced pm wheel mechanics is actually much greater in the second portrait than it is in the first one. Once a reader has been made aware of those clues and their locations in the second portrait, identifying the same clues in the first portrait becomes much easier.

Secondly, I realized that it was actually Bessler's intention that a reader of DT, upon opening the book, would first see what we have been calling the "second portrait". That is, the reader would first see the printed engraving showing Bessler in his workshop and then, only after flipping the pasted on paper flap containing that image to the right, would he finally and *secondarily* see the "first portrait" with the inventor standing near the corner of a table and in proximity to the symbols of a learned man.

Why did Bessler arrange the portraits in this manner? As was pointed out in the last chapter, one of the motivations was that Bessler was not happy with the appearance of the differently oriented face and head that was originally done for the second portrait showing him standing behind his workshop table. It just did not look enough like the version of his face in the first portrait with him located in his study and wearing the wig and he feared some envious rival inventor, whose face just happened to somewhat resemble the face in the second portrait, might then be able to come forth later and claim that the invention was not really made by Bessler, but, rather, by that rival inventor. He might then be in a legal position to demand that he and not Bessler should be paid the money for the invention after it had been sold for such an enormous sum.

Bessler could not even allow the possibility of that happening, so he then decided to have oval cutouts made in all of the second portraits in the workshop and have them attached to the first portraits in the study so that the face of the first portrait could appear on the second portrait. This arrangement then guaranteed that the reader of DT would *first* see Bessler in his workshop where, obvious, his 3 feet diameter, one-directional Gera prototype wheel was conceived and constructed and there would then be absolutely no doubt that it was the exact same man who appeared in the first portrait and had been hired by Count Karl as the commerce counselor for the principality of Hesse-Kassel in Saxony.

But, there is also some symbolic value in this arrangement of the portraits.

Bessler had a certain amount of disdain for people who thought they could merely intellectualize their way to finding a working design for an imbalanced perpetual motion wheel. He knew from his years of arduous construction and sacrifice that the only way to achieve success was by physically working hard for it. This work ethic is symbolized by the arrangement of the two portraits in DT. The second portrait of him in his workshop is placed on top of the first portrait in his study and that was Bessler's way of telling the reader that the secret of success in finding a perpetual motion lies in hard physical work and not mere armchair philosophizing. It is a message that particularly resonated with me after it had taken me around two *thousand* computer models to finally find the details of the mechanics used in his wheels!

So, in keeping with Bessler's wishes, the subject of this chapter will be the clues concerning the parameter values of the various parts used in his imbalanced wheel mechanics that he carefully hid in what is usually referred to as the "second portrait" even though it was really the first portrait that a reader of DT would see.

Before we begin, however, I need to mention a few of the methods I will be using as I reveal these previously unknown clues.

As I've done in earlier chapters, all of the illustrations showing the locations of the second DT portrait clues have been placed at the end of this chapter. In many cases, I needed to draw a line from a dark portion of the portrait to either a label or piece of text that was off of the portrait and located on the white page background outside of the portrait itself. In these cases, since this book does not use color printing, I was forced to start with a white line on the dark portion of the portrait that then turns into a black line as it extends off of the portrait. This may seem a bit odd at first, but the reader should get used to it as he continues to view the various portrait clues.

With these details in mind, let us begin with Figure 18 which shows an enlarged section of the tabletop in the second portrait.

One of the first things I noticed about the second portrait was that if I drew a straight line vertically downward from the pupil of Bessler's right eye (which is the left eye in the portrait), it passed directly through the center of the pivot of the duckbill goniometer. Because that pivot is on an optical axis, that of Bessler's eye, it indicates that the inventor wants the reverse engineer to carefully study the goniometer. This instrument is used by carpenters to measure the angles between two flat adjacent surfaces and careful measurement of the angle between the *inner* edges of the two arms of the device shown on the tabletop shows its angle is exactly 45°. The extended sweep of the two arms also includes several of the many items on the tabletop and Bessler wanted us to study all of them carefully because each can provide valuable clues to help someone replicate one of his imbalanced pm wheels.

The particular items we are interested in are: the carpenter's plumb bob upon which the 14 inch ruler is carefully balanced, the "L" square upon whose longer lower arm the lower arm of the duckbill goniometer rests, the small divider whose upper arm rests on top of the tip of the goniometer's lower arm, the steel file whose end rests on top of one of the lamp's tripod feet, the three vertically oriented metal ingots that support a fourth larger horizontally oriented ingot, and, finally, the upright lamp with the conical cover. The angular spread of the goniometer's two arms tells us that there is an angle of 45° associated with these items.

The analysis of this duckbill goniometer clue is actually rather simple.

The eight items represent the eight weighted levers found in one of Bessler's one-directional wheels. The items themselves each represent one of the eight weighted levers whose pivot pin's center axis was located at one of the wheel drum's eight clock time positions *after* its pivot pin's center axis had traveled *clockwise* through an arc of 45° degrees around the center axis of the wheel's axle from its *previous* clock time position.

Thus, starting from the upper left hand corner of Figure 18 and moving counterclockwise about the figure, the carpenter's plumb bob represents the 10:30 weighted lever after its pivot pin's center axis had traveled 45° clockwise around the center axis of the wheel's axle from the drum's 9:00 position to its 10:30 position, the "L" square represents the 9:00 weighted lever after its pivot pin's center axis had traveled 45° clockwise around the center axis of the axle from the drum's 7:30 position to its 9:00 position, the divider whose upper arm rests on the tip of the lower arm of the duckbill goniometer represents the 7:30 weighted lever after its pivot pin's center axis had traveled 45° clockwise around the center axis of the axle from the drum's 6:00 position to its 7:30 position, the metal file whose pointed end rests on the lamp's visible left tripod foot represents the 6:00 weighted lever after its pivot pin's center axis had traveled 45° clockwise around the center axis of the axle from the drum's 4:30 position to its 6:00 position, the right vertically oriented lead ingot represents the 4:30 weighted lever after its pivot pin's center axis had traveled 45° clockwise around the center axis of the axle from the drum's 3:00 position to its 4:30 position, the middle vertically oriented lead ingot represents the 3:00 weighted lever after its pivot pin's center axis had traveled 45° clockwise around the center axis of the axle from the drum's 1:30 position to its 3:00 position, the smallest left vertically oriented lead ingot represents the 1:30 weighted lever after its pivot pin's center axis had traveled 45° around the center axis of the axle from the drum's 12:00 position to its 1:30 position, and, finally, the metal lamp with the cone shaped cover represents the 12:00 weighted lever after its pivot pin's center axis had traveled 45° around the center axis of the axle from the drum's 10:30 position to its 12:00 position.

From the above, we see that Bessler had, with the items shown on the right side of the shop table in the second portrait, provided the reverse engineer with symbols for all of the eight weighted levers found in one of this one-directional wheels. Their presence also emphasizes that his one-directional wheels only used eight weighted levers. As our analysis of the clues continues, it will be seen that it becomes possible to use these weighted lever symbols to obtain precise values for the parameters used in their construction.

Before leaving Figure 18, I must mention a few more details of the eight weighted lever symbols which help to reinforce my interpretations of them. Again, I will begin with the symbol for the 10:30 weighted lever which is represented by the carpenter's plumb bob.

That the plumb bob represents the 10:30 weighted lever is obvious from the orientation of this cone shaped lead weight. It sits on the surface of the shop table on its large circular base and the small steel eyelet screwed into the center of the circular base to which a tangle of suspension string is attached. The apex or pointed tip of the plumb bob is raised up off of the table's top and that inclines the center axis of the cone at a 45° angle to the table's surface. The angle that the cone makes with the table's top is the same angle that the hour hand of a clock makes with a horizontal line passed through the center of the dial when the time shown is 10:30! Thus, the carpenter's plumb bob represents a weighted lever whose pivot pin's center axis is located at the 10:30 drum position of a clockwise rotating one-directional wheel.

If one draws a horizontal line from the center of the ball of string near the eyelet at the base of the conical bob weight (in the highest resolution images of the second portrait, that center of the ball of string is located at the end point of a dark and slightly curved curl that ends *inside* of the dark outer perimeter of the ball of string's shape) leftward until it reaches the curving edge of the protractor immediately to the left of the duckbill goniometer's pivot pin, that line will pass through the 9:00 position of an imaginary circle of which the protractor forms a visible half. This indicates that the carpenter's plumb bob is a symbol for the weighted lever whose pivot pin's center axis is at the 10:30 position of a one-directional wheel's drum *after* its pivot pin's center axis had moved from the drum's 9:00 position to its 10:30 position.

The "L" square represents the weighted lever whose pivot pin's center axis is located at the drum's 9:00 position and there are several clues that verify this.

One may immediately notice that the "L" square forms the number "4" with the lower arm of the large duckbill goniometer and the number appears to have been rotated over 90° clockwise for it normal orientation. The goniometer also looks like the letter "V" which has also been rotated about 90° clockwise from its normal orientation. The letter "V" is also the Roman numeral for the number 5. Note that the place where the goniometer's lower arm passes over the longer arm of the "L" square looks somewhat like a slightly clockwise rotated "+" sign which is the arithmetic symbol for addition. If we perform the indicated addition, we get 4 + 5 = 9 which indicates that the "L" square represents a weighted lever whose pivot pin's center axis is located at the 9:00 position of a clockwise rotating one-directional wheel's drum. It should also be pointed out that the "V" shaped goniometer can, due to the presence of its pivot which looks like a decimal point, also be a symbol for 0.5.

More evidence that the "L" square represents the 9:00 weighted lever is gotten by drawing a horizontal line from the tip of the drafting tool's upper *shorter* arm leftward until it reaches the curved edge of the protractor that is located near the pivot pin of the goniometer. As was done in the case of the carpenter's plumb bob, we imagine the curved edge of the protractor as being 180° of arc of a full circle, then, if the horizontal line is extended through the protractor's curved edge to the other *missing* part of the circle, the horizontal line will intersect that circle at its 9:00 position.

The place where the line intersects the missing part of the circle suggested by the protractor is not visible in Figure 18 and the reader is directed back to Figure 17 of the previous chapter in order to "see" it. If the reader obtains a high resolution image of the second portrait from an internet source, then he can either use a graphics program to draw in the horizontal line and the missing part of the circle that the protractor is part of or he can just directly use a ruler on his computer's monitor screen to see where a horizontal line from the tip of the small arm of the "L" square reaches when it is extended to the left and passes through both the curved edge of the protractor and the origin of the protractor.

But, the symbol of the "L" square also represents a weighted lever *after* its pivot pin's center axis had traveled 45° from the 7:30 to 9:00 drum positions of a clockwise rotating one-directional wheel. We can also interpret the "L" square as being a symbol for the number 7 and, since its larger arm crosses under the lower arm of the duckbill goniometer and the goniometer can stand for the number 5 or 0.5 we can, after performing an addition, use the combination of the "L" square and the goniometer to represent the number 7.5 which, in clock time, becomes 7:30 since 0.5 of an hour is 30 minutes. So, we see from this that we can also interpret the "L" square to represent a weighted lever whose pivot pin's center axis is at the 9:00 position of a clockwise rotating one-directional wheel's drum after it had traveled from the drum's 7:30 to 9:00 positions.

The divider whose upper arm covers the lower arm of the goniometer represents the 7:30 weighted lever. The divider looks like a mirror image of the number 7 which we can take it to represent and, since one of its arms touches the "V" shaped goniometer's lower pointed arm and the goniometer, as mentioned above, can represent the number 0.5, the suggestion is that we simply add the 7 and the 0.5 together. This gives us 7.5 which can also be written as the clock time 7:30 and verifies that the divider is a symbol for a one-directional wheel's 7:30 weighted lever.

There is a dark wooden knob at the top of the divider by which it is held and can be employed to make circles of various radii. If one extends a horizontal line from the center of that knob leftward and then past the curved edge of the protractor, he will discover that it will intersect the lower corner of the protractor that corresponds to the 7:30 clock time position of a wheel's drum when the protractor is considered to be half of a full circle.

That the divider also symbolizes a weighted lever after its pivot pin's center axis had traveled through an arc of 45° around the center axis of the axle of a clockwise rotating one-directional wheel's drum from its 6:00 to 7:30 positions is indicated by what the upper arm of the divider points to. It points to the same visible left foot of the lamp that the end of the steel file rests upon. As will be shown next, that foot of the lamp and another item it touches (which is not the steel file) represent the number 6.

Moving on we reach the steel file which represents a weighted lever whose pivot pin's center axis is located at the 6:00 position of a clockwise rotating one-directional wheel's drum. This symbol is easily interpreted.

Note that the lower end of the file with the hole in it (intended to allow it to be hung on a nail in a wall or to be secured by a bolt to a wooden handle) rests on one of the two visible tripod feet of the lamp. The two visible feet of the lamp form an inverted "V" and the right side foot touches the smallest of the three vertically oriented lead ingots. If we consider the smallest metal bar to represent the letter "I", then, ignoring the inversion of the letter "V" formed by the two visible lamp feet, we have the letter group "VI" which the reader will recognize as the Roman numeral that was used for the number 6. From this we can conclude that the steel file represents a weighted lever whose pivot pin's axis is located at the 6:00 drum position of a clockwise turning one-directional wheel.

The upper end of the steel file has the longer arm of the "L" square resting on it and, above, I mentioned that the "L" square and the lower arm of the duckbill goniometer formed the number 4. If we combine the number 4 with a value of 0.5 derived from the Roman numeral "V" represented, after clockwise rotation by more than 90° from its usual orientation, by the shape of the divider that sits astride the file, we can get 4.5 and use it as 4.5 hours which, in clock time, can be written as 4:30. Thus, we also have the two clock times associated with the steel file of 4:30 and 6:00 and this tells us that the symbol of the steel file represents a weighted lever whose pivot pin's center axis is located at the 6:00 position of a clockwise rotating drum after that axis had traveled from the drum's 4:30 position to its 6:00 position.

The next three weighted levers with pivot pin center axes located at the 4:30, 3:00, and 1:30 positions of a clockwise rotating one-directional wheel's drum, are represented by the three vertically oriented lead ingots on the lower right side of the second portrait. These symbols for the weighted levers are a little more difficult to interpret.

The largest, rightmost vertical lead ingot on the table represents the 4:30 weighted lever after its pivot pin's center axis has traveled from the drum's 3:00 to 4:30 positions.

The drum location of 3:00 is symbolized by the ingot being the third largest vertical one. The 4:30 position interpretation is found by extending a vertical line segment downward and out of the bottom border of the portrait. Although not shown in Figure 18, it will touch the right leaning slash mark or "virgule" at the end of the first line of German text under the portrait. The lower end of that slash mark is shaped somewhat like an arrowhead and points downward to the Latin abbreviation of "*NB*" on the second line of text which means, in English, "Note Well". Bessler fused together the separate letters in the abbreviation so that the letters "N" and the "B" both share a common vertical line. One can then imagine this single fused abbreviation to be pulled apart into three separate numerals, two Roman and one Arabic: I, V, and 3. Recombining the I and the V gives IV which is the Roman numeral equivalent of the Arabic numeral 4 which is then taken as representing four hours and the Arabic numeral 3 is then taken as representing 30 minutes. By adding the symbolic time values of these Roman and Arabic numerals, we obtain the clock time of 4:30 which then tells us that the rightmost vertical lead ingot represents a weighted lever whose pivot pin's center axis was located at the 4:30 position of the clockwise rotating drum of a one-directional wheel after that axis had rotated through an arc of 45° around the center axis of the drum's axle.

The middle vertically oriented lead ingot represents the 3:00 weighed lever after its pivot pin's center axis has traveled from the drum's 1:30 to 3:00 positions. Correctly interpreting this symbol requires one to begin to delve more deeply into the alphanumeric coding system that Bessler used to hide the various physical parameters of his secret imbalanced wheel mechanics into the two DT portraits.

While the coding system is rather simplistic because it merely assigns a number to each letter of the 26 letters of the German alphabet which are the same as found in the English alphabet, in order to confuse and frustrate the reverse engineer looking for quick answers, Bessler regularly varied how he used this coding system to represent various parameter values. This will become clearer as the reader continues to study the many DT portrait clues revealed in this and the next chapter. For now, however, showing how he used it to give the drum positions for the weighted lever represented by the middle vertical lead ingot is a good time to start learning some of the different ways he used this simple alphanumeric coding system.

If one draws a vertical line downward from the *right* side of the middle vertically oriented lead ingot and out of the bottom border of the portrait, that line will run along the *right* side of the vertical portion of the letter "t" at the end of the word "*vollendet*" at the end of the first line of German text under the portrait. This alignment is not visible in Figure 18 and the reader will need to refer to Figure 17 to see it.

Of importance here are the two last letters of the word "*vollendet*" which are "et". "e" is the 5th letter of the alphabet and "t" is the 20th letter. Why these two particular letters are important for determining which weighted lever is represented by the middle ingot in the second portrait has to do with their order in the word. Bessler, as we saw, already used the "N" in the "*NB*" at the end of the second line of text to stand for the Roman numeral IV which is gotten by subtracting the value of "I" which is 1 from the value of "V" which is 5 to obtain a value of 4. This is a convention used in determining the overall value of a Roman numeral made up from two or more individual Roman numerals. If there are one or more Roman numerals with a smaller summed value to the *left* of a larger valued Roman numeral, then the summed value of the numeral(s) on the left must be *subtracted* from the larger valued numeral to obtain the overall value of the *group* of individual Roman numerals. On the other hand, if there are one or more Roman numerals with a smaller summed value to the *right* of a larger valued Roman numeral, then the summed value of the numeral(s) on the right must be *added* to the larger valued numeral to obtain the overall value of the group of individual Roman numerals.

We can now apply this method of Roman numeral value evaluation to the alphanumeric values of the two letters at the end of the word "*vollendet*". Doing this requires us to subtract the value of the letter "e" or 5 from the value of the letter "t" or 20 and obtain the number 15. What does the number 15 have to do with a weighted lever whose pivot pin's axis is located at the 3:00 position of a clockwise turning one-directional wheel's drum? Well, if one starts counting the hours around a clock's dial from midnight, he will find out that the 15th hour is 3:00! But, how do we get the starting location of 1:30 for the pivot pin axis of the 3:00 weighted lever *before* that axis completed its 45° of clockwise rotation about the center axis of the wheel's axle?

This value can easily be gotten by simple taking the 15 and moving its unwritten decimal point one place to the left to obtain 1.5. In clock time this corresponds to one hour plus one half of an hour which equals one hour plus 30 minutes or 1:30 in this case. How can we justify adding the decimal point? Again, one must look to see if there is a symbol, a dot or point, nearby that tells us to add the decimal point. As we look at the top of the "t" at the end of *"vollendet"*, we notice that, indeed, it resembles a dot!

I'm sure that the reader may be a bit surprised by this rather bizarre and confusing method that Bessler used in order to encode his wheel's component parameters into the two DT portraits. But, one must remember that, while he wanted to preserve the secret for eventual rediscovery in the future, he did not want to make finding the details of his imbalanced pm wheel mechanics too easy. Those details had to be difficult enough to find so that only the most persistent of reverse engineers would ever solve the mystery and be in a position to duplicate his wheels. As I continue to provide correct interpretations for the clues given in the two portraits, the reader will begin to see the logic in his approach and, quite possibly, will even be able to identify clues and parameter values that I've missed despite my years of studying the portraits. Let us now continue with Figure 18's full description.

The remaining vertically oriented lead ingot is the smallest and leftmost of the set of three. It is symbolic of the weighted lever whose pivot pin's center axis was located at the drum's 1:30 position after it had traveled 45° clockwise around the center axis of the axle from the drum's 12:00 to 1:30 positions. Its interpretation is little more difficult to do in comparison to the interpretation of the middle ingot symbol.

A vertical line drawn *upward* from the *left* side of the smallest lead ingot will intersect one of the major inch marks on the 14 inch ruler above the ingot (which is visible in both Figures 17 and 18) and this is the division mark that ends the 13th inch starting from the left side of the ruler. Starting from midnight, the 13th hour of a day is 1:00 pm. Next, notice what the bottom left side of the small ingot is touching. It's the right foot of the two visible feet of the tripodal lamp's base. Those two visible feet, when inverted, become the Roman numeral "V" which is equal in value to 5. If we could use the 5 as 0.5, then we could add it to the 13th hour to get 13.5 hours or 1:30 in clock dial time. The justification for this change of the 5's value is that the small ingot only touches the lamp's right foot at one *point* on the ingot's *left* side which then directs us to move the decimal point one place to the left of 5 to turn it into 0.5.

As far as the starting location for the weighted lever whose pivot pin's axis is at the drum's 1:30 position being the drum's 12:00 position before the 45° of clockwise drum rotation carried that axis around the center axis of the axle until its location was at the drum 1:30 position is concerned, it can be justified by the small ingot being the only one of the three that actually physically touches the upright metal lamp which is a symbol for both a weighted lever whose pivot pin's axis is at the drum's 12:00 position and for that drum position itself. Thus, we see from this that the smallest vertically oriented lead ingot of the set of three represents a weighed lever whose pivot pin's axis is at the drum's 1:30 position *after* it has traveled there from the drum's 12:00 position.

Finally, in discussing Figure 18, we come to the metal lamp with the conical cover that stands upon three feet only two of which are visible in the second DT portrait. I've already mentioned that this lamp represents a weighted lever whose pivot pin's axis is located at the 12:00 position of a drum and that interpretation is obvious from the vertical orientation of the lamp and its pointed conical cover which points vertically upward as do *both* the hour and minute hands on a clock's dial at either 12:00 noon or 12:00 midnight. But how do we associate the starting drum position of 10:30 with the lamp symbol for a weighted lever whose pivot pin's axis is located at the drum's 12:00 position?

One could just say that, since the ruler balanced on the carpenter's plumb bob passes only behind the lamp's vertical support base and the bob, as previously shown, is a symbol for a weighted lever whose pivot pin's axis was at the drum's 10:30 position, this indicates that the lamp symbol represents the 12:00 weighted lever *after* its pivot pin's axis had traveled from the drum's 10:30 to 12:00 positions. But, there is another way to get the drum position of 10:30 associated with the lamp symbol.

Note that if one counts the inch marks, starting from the left end of the ruler, then the 10th inch mark is the only one hidden from view behind the lamp's vertical support post. This then gives us the number 10. Since the ruler is balanced on the carpenter's plumb bob and that weight is "V" shaped, we can take the bob's shape as representing the Roman numeral for the number 5 and then use it as 0.5 with the left pointing tip of the bob weight justifying the movement of the unwritten decimal point one place to the left of the number 5. Adding the number 10 gotten from the ruler with the 0.5 gotten from the carpenter's plumb bob then gives us 10.5 which, in clock time, is 10:30. Thus, again, we see that the drum position of 10:30 is associated with the lamp symbol for a weighted lever whose pivot pin's axis is located at the drum's 12:00 position.

At this point, the observant reader may be wondering why I have not yet mentioned the large *horizontally* oriented lead ingot that rests on top of the three smaller vertically oriented lead ingots. Let me now give its correct interpretation.

Since the largest horizontal ingot is lifted off of the surface of the table by the three vertical ingots that are in contact with the table, this indicates that the three weighted levers, the ones whose pivot pins' center axes are, during each 45° segment of drum rotation, traveling from the drum's 12:00 to 1:30, 1:30 to 3:00, and 3:00 to 4:30 positions are playing an important role in lifting up some other lever. But, in this case that "other lever" symbolized by the horizontal ingot is actually the *two* levers whose pivot pins' center axes are traveling from the drum's 9:00 to 10:30 and 10:30 to 12:00 positions. This is confirmed by noting that the center axes of both the carpenter's plumb bob and the lamp converge on a point inside of the lamp's vertical support base and that the center axis of the largest horizontally oriented lead ingot, if extended to the left of the ingot with an imaginary horizontal line, will pass exactly through that convergence point of the bob weight and lamp's center axes.

With this clue, Bessler is telling the reverse engineer of his wheels that the five weighted levers whose pivot pins' center axes are located from the drum's 9:00 to 3:00 positions are interconnected by *only* their main cords and very precisely counterbalanced against each other during each 45° segment of clockwise drum rotation.

From working with precise computer models of Bessler's one-directional wheels, however, I found that this interconnection only actually begins simultaneously moving the five weighted levers when the pivot pin axis of the 3:00 weighted lever reaches a drum location about 11° clockwise past the drum's 3:00 position and ends when that weighted lever's pivot pin's axis reaches a location about 33° clockwise past the drum's 3:00 position during which time the weighted lever whose pivot pin's center axis is moving from the drum's 1:30 to 3:00 positions begins to rotate clockwise about that pivot pin and the *three* main cords between that weighted lever and the weighted lever whose pivot pin's axis is moving from the drum's 9:00 position to 10:30 position also begins to rotate clockwise about its pivot pin so that it moves its parallel pair of main arms closer to their wooden radial stop piece. Thus, the interconnection implied by the intersection of the center axes of the carpenter's plumb bob, the lamp's vertical support, and the large horizontal lead ingot is one that only lasts during about 22° out of each 45° segment of clockwise drum rotation.

I have spent considerable time on interpreting the symbols for the eight weighted levers of a one-directional wheel that appear in the second DT portrait because, aside from convincing the reader of their presence, they will become of additional importance as the analysis of other symbols in the portrait proceeds.

Now it's time to move on to pointing out and further evaluating some of the second portrait clue symbols that were briefly discussed in the last chapter, but were not illustrated there. For these, the reader is directed to Figure 19 at the end of this chapter.

One of the most important breakthroughs that started me down the path to determining the type of mechanics Bessler used in his wheels, involved an unusual alignment I noticed that occurs in the second DT portrait. That alignment is now presented in Figure 19 where the reader will find an image of the second portrait with the various alignments indicated by white lines and labeled with letters.

"A" is the cherub located in the middle of the front top edge of the background organ. Close examination shows him to be looking to *his* left side and it even looks like his left wing is also being used to point to his left! This, of course, immediately caught my attention. Then I noticed the second cherub, labeled "B" in the figure, in the upper right corner of the organ's front side and he, too, seemed to be looking toward the right side of the second portrait. This second cherub was less distinctly engraved than the middle one at "A", but I noticed he seemed to have his little mouth opened in an expression of surprise. Surprise is an emotion associated with making a discovery. I wondered if this symbol suggested that a closer examination might provide such a surprise. But, where should I look for it?

I then noticed how pointy the nose of the first cherub at the middle of the top front edge of the organ looked. The nose of the second cherub, however, was difficult to see even under extreme magnification.

Remembering one of the basic postulates of Euclidean geometry that states that through any two points only a single straight line can be drawn, I decided to draw a line through the tips of the noses of the two cherubs and see where it went. To my surprise, when extended downward and to the right, it passed *exactly* through the center of the clock's circular dial, "C", which is another point that is somewhat analogous to the nose of the clock's "face"! I then had three points that all lain on the same line segment. That seemed a bit improbable to me and I looked for yet another nearby nose point through which to extend another line from the clock face's "nose". The only other nearby nose in the portrait was that on Bessler's face, so I drew a straight line segment from the center of the clock's dial to his nose which is labeled "E".

After this lengthening line segment reached the tip of the inventor's nose, I decided to continue extending it farther and it eventually reached the nose space under the bridge of the armless eyeglasses, "F", that Bessler holds in his right hand. So, once again, I had three nose points which lain on a single line segment that belonged to the clock's dial, Bessler's face, and the imaginary one that would fit into the nose space of the armless eyeglasses when the inventor wore them.

These two straight line segments, the one from "A" to "C" and the other from "C" to "F", both had something in common: they each connected three noses or nose symbols and were, obviously, intentionally arranged in that way. As I viewed them, I kept thinking about an old saying about something being "as plain as the nose on one's face" which meant it was very obvious and difficult to overlook and, most likely, there was a similar expression in the German language that Bessler was thinking of when he used these symbols in the second DT portrait. Since the two line segments formed an angle about the center of the clock's face or its "nose", I decided to measure this angle and it turned out to be exactly 62.5° and this angle is indicated above Bessler's head in Figure 19.

I then decided to start a horizontal line segment at the center of the clock's dial and extend it toward the left side of the portrait. This horizontal line segment did not precisely bisect the 62.5° angle, but, when extended, I noticed that the it immediately passed through the 9:00 position of the clock and then, on the left side of the second portrait, the line segment passed through the saw hanging on the background wall.

That saw, because of its shape and its orientation, represents the Arabic numeral 9 and, along with the 9:00 position on the clock dial, indicates that we are receiving information that concerns the weighted lever whose pivot pin's axis was located at the 9:00 position of a clockwise rotating one-directional wheel.

When the horizontal line segment finally reached the markings near the capillary tube of the thermometer at "D", I noticed that the *temperature* indicated there corresponded to *exactly* 62.5°!

To get that temperature, one must start with the first major temperature mark just above the colored alcohol filled bulb at the bottom of the thermometer's capillary tube and assign to it, as the manufacturers of these early devices did, a value of 0° and then label the *major* temperature marks upward *after* that starting mark of 0° as 10°, 20°, 30°, 40°, 50°, and then, finally, 60° just below the point where the horizontal line segment extended from the center of the clock's dial intersected the thermometer's glass capillary tube.

Careful inspection of the engraving work of the second DT portrait under magnification shows that the 10° temperature change between any two consecutive major 10° temperature marks on the thermometer is further subdivided into *eight* less prominent smaller temperature changes that are each equal to 1.25°. Noting this, one then discovers that the point of intersection of the horizontal line segment with the thermometer's glass capillary tube is two of these smaller temperature changes or 2 x 1.25° = 2.5° *greater* than 60° which means that the horizontal line segment actually indicates a point on the capillary tube for the height of its column of dyed alcohol that would be achieve when the air temperature in the workshop was 62.5°. In his thermometer symbol, Bessler obviously used degrees of temperature to represent degrees of an angle. (Unfortunately, the little 1.25° temperature changes on Bessler's second DT portrait thermometer that I'm discussing here are just barely possible to discern from studying Figure 19. Because of this, I would urge the serious student of Bessler's wheels to obtain and study enlargements of the highest resolution images of Bessler's two DT portraits that he can find so that he can see these minor temperatures changes for himself. Such higher resolution images may be obtained from various sources on the internet.)

This was just too much of a coincidence to actually be just a coincidence as far as I was concerned. Here were two very carefully inserted clues in the second DT portrait and they both indicated an angle of 62.5°. In fact, they both associated that angle with a one-directional wheel's whose weighted lever's pivot pin's center axis was located at the 9:00 position of a clockwise rotating drum. We saw above that the horizontal line segment extended to the left of the portrait from the center of the clock's dial immediately passed through the dial's 9:00 position, but, then, noted that as that line segment was farther extended over to the left side of the second portrait, it passed through the saw that hangs on the same back wall to which the thermometer is also attached. That saw, as I mentioned previously, is, because of its shape and orientation on the wall, also a symbol for the 9:00 position of the drum of a clockwise rotating one-directional wheel.

Can we draw any other conclusions from these symbols of cherubs, noses, clock, and thermometer that appear in the second DT portrait? Remarkably, we can.

The horizontal line segment from the center of the clock's dial to the thermometer passes through the fur headpiece Bessler wears and just above his head. The line segment from the center of the clock's dial to the inventor's nose, however, passes though the portion of the headpiece that covers his left ear. The symbolism of this is quite obvious.

Here Bessler is telling us that, in order to begin to duplicate his imbalanced one-directional pm wheels, one must have the correct angular relationship between the center lines of the main arms of a weighted lever whose pivot pin's center axis is located at the 9:00 position of a clockwise rotating drum and the center lines of that lever's parallel pair of radial frame pieces. He also tells us that this information was actually given to him by God, symbolized by the armillary sphere attached to the back wall to the left of his head, and conveyed to him by angels, symbolized by the two cherubs on the organ. A second line segment is then started at the center of the clock dial extended from the clock to Bessler's nose and, as that happens, the line segment passes directly through the place where the inventor's left ear would be under the fur headpiece he wears. (There is also another interesting symbol present which indicates this important information came to him from the seventh and highest level of heaven where God resides. That symbol is provided by the seven visible flue pipes of the organ in the background on the right side of the portrait, the *upper* edges of whose "mouth" openings form a straight line pointing directly toward Bessler's left ear.)

Thus, we are told that the two cherubs whispered this secret into Bessler's ear and this fits in with the belief in the early 18th century that certain vivid dreams that provide important information could be due to spirit beings, acting as messengers for God, whispering it into the ears of a sleeping person that God intended to receive that knowledge.

We will, unfortunately, probably never know the exact details of the vivid dream that inspired Bessler to make the modifications to his inoperative 3 feet diameter Gera prototype wheel that, finally, allowed it to display perpetual motion, but I believe that in his dream, the inventor actually saw an axial view of the prototype's drum with the center lines of its 9:00 weighted lever's parallel pair of wooden main arms rotated through an angle of exactly 62.5° clockwise from a horizontal line passing through the center axis of the wheel's axle. He then saw the five weighted levers whose pivot pins' center axes were located at angular intervals of 45° going clockwise from the 9:00 position of the drum over to its 3:00 position interconnected by the main coordinating cords with each lever also having the force of two stretched helical extension springs applied to it. Perhaps a cherub appeared in the dream and fluttered close to the 9:00 weighted lever and then gently touched its little lead ingot end weight at which point the weighted lever instantly partially rotated clockwise about its pivot pin because it was so perfectly counter balanced by the tension of the two extension springs attached to it.

So, to end this digression and finally to return to the analysis of the lines in Figure 19, I can simply state that I had two equal numerical values, both in degrees, although one was in angular degrees and the other in temperature degrees, and two symbols for a weighted lever whose pivot pin's center axis was located at the drum's 9:00 position. This was just too emphatic to be dismissed as a coincidence by me. It is a powerful clue as to an angle associated with the main arms of a weighted lever whose pivot pin's center axis is located at the 9:00 position of the drum. But, are there any additional clues in the second portrait to support this? The answer is a definite "yes"!

If one extends a vertical line segment downward from the tip of Bessler's nose in Figure 19 (which in the figure appears to slant to the right a little because the portrait image I used was similarly slanted), it will arrive at "G" which is the place where the longer arm of the "L" square passes under the lower arm of the duckbill goniometer. As we saw in the discussion of Figure 18 above, the "L" square is a symbol for the 9:00 weighted lever of a clockwise rotating one-directional wheel. Here, the longer arm of the "L" square represents the main arms of this weighted lever. If one then draws a horizontal line segment whose right end touches the *outside* corner formed by the "L" square's long and short arms, that line segment will then represent the center lines of the horizontal parallel pair of radial frame pieces which contain the embedded ends of the pivot pin of the 9:00 weighted lever. If one then measures the angle between that horizontal line segment and the outer edge of the longer arm of the "L" square which represents the center lines of the parallel pair of main arms of the 9:00 weighted lever, he will discover that the angle is *exactly* 62.5° as is indicated in the figure.

I even recently found another line segment to add to Figure 19 which I've drawn in as a thinner white line labeled "H". In studying the fingers on Bessler's left hand that holds the divider, I noticed that they could be compared to the features on a human face. The thumb and pinky would be the ears, the index and ring fingers would be the eyes, and the middle finger would correspond to the nose of a face. I then wondered what would happen if I extended a line segment from the bridge space of his armless eyeglasses through the tip of his left hand's middle finger which represented a nose. To my great surprise, when extended diagonally downward and to the right of the second portrait, that line segment eventually and exactly intersected the outer corner of the "L" square which represents the weighted lever whose pivot pin's center axis is located at the 9:00 position of a clockwise rotating drum.

So, we see from Figure 19 that Bessler provides multiple clues in the second portrait as to the exact angle that the center lines of the main arms of the 9:00 weighted lever make with the center lines of that lever's parallel pair of radial frame pieces.

Just as soon as I found Bessler's clues revealing the correct angle for the center lines of the main arms of a 9:00 weighted lever, I immediately wondered if he also provided clues for the correct angles of the other seven weighted levers in one of his one-directional wheels. Only years later, after

I had finally found the exact design he used for his imbalanced pm wheel's hidden internal mechanics, was I able to actually measure those other seven angles and then finally find them in the second portrait. Let me now just quickly list them starting with the 9:00 weighted lever to be complete and moving counterclockwise around the drum to end with the weighted lever that has its pivot pin's center axis at the drum's 10:30 and then tell how these angles can be found in the second DT portrait.

They are: 9:00 weighted lever = 62.5°, 7:30 weighted lever = 31.25°, 6:00 weighted lever = 0°, 4:30 weighted lever = 0°, 3:00 weighted lever = 2°, 1:30 weighted lever = 12.5°, 12:00 weighted lever = 22.5°, and, finally, 10:30 weighted lever = 43°.

The reader should refer to the full second DT portrait given in Figure 17 as he reads the following analysis of the clues about these angles of a one-directional wheel's eight weighted levers' main arms relative their respective lever's parallel pair of radial frame pieces. This is so he can also see the German text below the portrait which contains *very* important alphanumeric clues.

The 31.25° angle for the 7:30 weighted lever can be found by extending a diagonal line segment downward and to the right along the lower edge of the lower leg of the divider which is cut off by the portrait's bottom border. However, as the end of the diagonal line segment is extended below the bottom border of the portrait, it will finally reach the dot in the letter "i" in the word "*wird*" on the first line of text below the portrait. We take the alphanumeric value of "i" as 9 since that letter is the ninth letter of the alphabet, that of "r" as 18 since it is the eighteenth letter of the alphabet, and that of "d" as 4 since it is the fourth letter of the alphabet. Adding all of these values together then gives us 9 + 18 + 4 = 31.

Next, notice the "ch" in the word "*noch*" on the second line of text below the portrait and which is located below the word "*wird*" on the first line. The "c" has an alphanumeric value of 3 and "h" a value of 8. Multiplying them gives us 3 x 8 = 24. The top of the "h" is formed into a "?" mark and it touches the "i" in "*wird*". In this case the contact between the "h" and the "i" indicates that we must add the alphanumeric value of "i" which is normally 9 to the product we got for "ch" which was 24. But, in this case we do *not* give a value of 9 to the "i", but, rather, treat it like a Roman numeral which has a value of 1. We can then write the sum of "ch" and "i" as 24 + 1 = 25. There are two dots in these three letters. There is the dot over the "i" and the dot at the top curling end of the "?" mark which is quite obvious. Two dots tell us to move the decimal point two places to the left of 25 and that then gives us 0.25. We then perform a final addition of 31 + 0.25 = 31.25 and add the degree symbol to get the angle 31.25°.

The reader may notice that this value is exactly *half* of the angle Bessler gave us for the center lines of the 9:00 weighted lever's main arms relative to the center lines of that lever's parallel pair of radial frame pieces. This interesting relationship is also suggested by the little dark knob on the top of the *divider* that points to the "L" square which represents the 9:00 weighted lever. This symbolism tells us that the angular value of 62.5° for the 9:00 lever is "divided" in two or halved to give an angular value of 31.25° for a weighted lever whose pivot pin's center axis was located at the drum's 7:30 position.

The 0° for the 6:00 weighted lever is indicated by the pointed end of the long steel file, a symbol for this lever, resting on the left visible foot of the metal lamp. This contact symbolizes the contact between the 6:00 weighted lever's main arms and their wooden radial stop piece which caused alignment of the center lines of the lever's main arms and the lever's radial frame pieces that only happens when the angle between these center lines was 0°. The angular value of 0° for the 6:00 weighted lever's parallel pair of main arms is also suggested by the little circular hole at the end of the file which resembles the numeral "0".

The 0° for the 4:30 weighted lever also is suggested by some contact in the second portrait. We get that by extending a vertical line segment downward from the rightmost vertical lead ingot which represents the 4:30 weighted lever. That line then intersects the virgule or right leaning slash mark at the end of the first line of text under the portrait. As noted earlier, that slash mark looks like an arrow that points the reader's attention down to the "*NB*" at the end of the second line of text under the portrait. One then notices that Bessler fused the two letters in "*NB*" together by putting the right vertical stroke of the "N" into *contact* with the left vertical stroke of the "B". This is the contact symbol we seek which tells us that the main arms of the 4:30 weighted lever were in contact with their radial stop piece and, hence, the angle that the center lines of those main arms made with the center lines of the lever's radial frame pieces was 0°.

The clue for the 2° angle of the 3:00 weighted lever is little more difficult to find. It will be noticed that, while carefully balanced on the carpenter's plumb bob, the 14 inch ruler is *not* perfectly horizontal. Starting from the 14 inch mark at its right end and going out toward the 0 inch mark at its left end, the ruler rises up horizontally through a distance of about 0.5 inch = 8/16 inch and the length of the ruler actually forms the hypotenuse of a very thin right triangle with the horizontal. The tangent of an angle in a right triangle is equal to the side opposite the angle divided by the adjacent side or, in this case, 0.5 inch / 14 inch = 0.0357. If one has the tangent of an angle, then he can get the angle it corresponds to by taking the arctangent of that value. The arctangent of 0.0357 is very close to 2° which is then the angle that the center lines of the main arms of the 3:00 weighted lever made with the center lines of that lever's radial frame pieces.

The 12.5° angle for the 1:30 weighted lever represented by the leftmost and smallest of the three vertically oriented lead ingots is easily found. The bottom of that ingot touches the lamp whose vertical orientation presents the two hands of a clock at either 12:00 noon or 12:00 midnight. That symbolizes the 12° that are needed. The 0.5° is provided by the inverted "V", the Roman numeral for the number 5 and the justification for moving the decimal point one place to the left of the 5 is indicated by the point of contact between the small ingot and the lamp's foot taking place on the left side of that ingot. By the supplying the degree symbols and the addition of these values, we get 12° + 0.5° = 12.5° which is the value we sought.

The 22.5° for the 12:00 weighted lever, represented by the metal lamp, and is gotten by using the upper conical top of the lamp. If one was to look down upon the top of the conical cover from above, he would notice that it appeared to be a circle divided into eight sectors by the thick ribs on the sides of the cover. The center lines between two of these eight ribs would then be 45°. By studying the conical cover in Figure 17, one will notice that he only sees four of the eight ribs because he is only seeing one side or *half* of the cover. This indicates that we are to take *half* of the angular value of 45° to get a value of 22.5°. The justification for a final angular value with a decimal point is indicated by the little metal sphere at the apex of the conical cover.

Finally, the 43° for the 10:30 weighted lever is gotten by simple measuring the angle of the conical sides of the carpenter's plumb bob which is very close to 43°. And, with this value, we see that Bessler has provided sufficient clues in the second DT portrait to tell the reverse engineer what the various angular intervals were between the center lines of the weighted levers' parallel pairs of mains arms and their respective parallel pairs of radial frame pieces for all eight of the weighted levers found in a one-directional wheel. However, he really only blatantly gives away the angular value for the 9:00 weighted lever and the reverse engineer will have next to no chance of finding the others until other portrait clues and much, much construction, either with physical or virtual computer models, has finally led him to the exact design Bessler found and used.

Now it's time to discuss what is *the* most important clue contained in the second DT portrait and perhaps *all* of the Bessler literature! It involves that odd looking little circular sector shaped piece of paper to the left side of the lamp's tripodal base that has six concentric circular arcs of various radii on it that are crossed by six straight lines at various points.

I mentioned in the last chapter that wherever an optical axis appears in the second portrait, that is a signal from Bessler that one must carefully study what that axis, represented by an extended imaginary line from the center of a lens to an object in the portrait, points to. I noted that if one extends a vertical line segment downward from Bessler's right eye, it will eventually reach the center axis of the pivot screw or bolt that holds the duckbill goniometer arms together. Therefore, that indicates that both the symbolism of the goniometer is important as well as is that of any items that its arms point to or which are included within the 45° angular sweep of those arms.

Although it is not visible in the second DT portrait because it is hidden beneath the upper arm of the divider, the pointed end of the goniometer's lower arm actually points to the origin of the circular sector paper piece containing the concentric arcs and straight lines. By this symbolism, the inventor is telling us that we must focus our attention on that origin which is the starting point from which the lengths of radii to other points in the circular sector can be measured. But, what would the lengths of those radii being measured represent?

The solution to this mystery is indicated by another item near the circular sector shaped piece of paper. That item is the "L" square near its left side whose shorter arm points to the left side straight radial edge of the circular sector.

We saw in the discussion of Figure 18 above that the "L" square represents the weighted lever of a one-directional wheel whose pivot pin's center axis was located at the 9:00 position of a clockwise rotating drum. Then, in Figure 19, it was mentioned that the longer arm of the "L" square, which points down and to the left in the second DT portrait, represents the 9:00 weighted lever's parallel pair of main or L arms. It turns out that the short arm of the "L" square that points to the circular sector paper piece represents the A arm of the weighted lever and the left side straight radial edge of the circular sector paper piece that it points to, therefore, *also* represents the A arm of a lever whose pivot pin's center axis is located at the drum's 9:00 position. From this it becomes obvious that the origin of the circular sector paper piece actually represents the center axis of the 9:00 weighted lever's pivot pin. The location of the origin of the circular sector piece is labeled "P" in Figure 20 which the reader should now locate at the end of this chapter and study.

The points on the circular sector paper piece's left side straight radial edge that the four straight lines contact then represent the center axes of the four hook attachment pins for the four types of coordinating cords attached to a weighted lever's A arm and the distances from the center axis of a lever's pivot pin to the center axes of those hook attachment pins is actually shown on the circular sector paper piece! This, obviously, is information that would be of tremendous value to any reverse engineer trying to duplicate the same precise counter balancing of weighted levers that allowed Bessler's imbalanced perpetual motion wheels to continue to maintain the center of gravity of their eight active weighted levers on the drum's descending side during clockwise drum rotation.

It will be noted in Figure 20, as previously stated, that, measuring outward from the origin at "P", each of the concentric arcs of the circular sector paper piece represents a collection of points at a precise fixed distance from the origin. I have identified all of the arcs on the circular sector paper piece and labeled all six of them at the bottom of Figure 20. Additionally, for all of the radial distances shown in the figure, I have provided two values. The shorter of the distances in inches is for the 3 feet diameter, one-directional Gera prototype wheel and it is followed by the acronym "GPW" which, obviously, stands for "Gera Prototype Wheel". The longer of the distances, again in inches, is for *both* of Bessler's 12 feet diameter, bidirectional Merseburg and Kassel wheels and is followed by the acronym "MKW".

The various center axis to center axis distances for the 12 feet diameter Merseburg and Kassel wheels are always four times the analogous distances for the 3 feet diameter Gera prototype wheel and this is to be expected since the horizontal and vertical dimensions of the two internal one-directional wheels contained within the enlarged drum of the Merseburg and Kassel wheels as seen in an axial view were four times greater than the same dimensions found inside of the 3 feet diameter Gera prototype. Thus, the widths and lengths of all of the 3 arm side pieces of the weighted levers in the Merseburg and Kassel wheels' drums were four times those in the Gera prototype wheel and the distance between the center axis of a weighted lever's pivot pin and the center axis of the axle was four time greater in those two bidirectional wheels than it was in the Gera prototype wheel.

There are even some nearby clues that tell us that we must multiply the distances indicated in the circular sector paper chart for the Gera prototype wheel by a factor of four to obtain the analogous values used in the Merseburg and Kassel wheels. To see these extra clues, the reader should briefly refer back to Figure 18.

In Figure 18 one sees that the "L" square and the lower arm of the duckbill goniometer form the numeral "4" which has been rotated clockwise through an angle of a little greater than 90° from its normal orientation. After one recognizes this number, he may also notice that the lower arm of the goniometer and the longer arm of the "L" square form the letter "X" which is the arithmetic symbol for multiplication. Thus, this clue is telling us that we will need to multiply the axes' center separation distances values indicated in the circular sector paper piece by a factor of 4. Why would Bessler want us to do that? The only reason that makes sense is so that we could convert numerical distance values provided by the circular sector paper piece into the values needed for something else which means converting the values for the Gera prototype wheel into those needed for other wheels. But which wheels?

Note that there is a pen near the upper, actually back, edge of the shop table. One end of it seems to be covered by a piece of belt fabric on Bessler's overcoat while the other end of it, possibly holding a metal nib points to and appears to touch the left side of the large rim of the lamp's conical cover (on an actual three dimensional tabletop, it would not touch the lamp cover's rim). The upper arm of the goniometer points to this end of the pen while the point of the goniometer's lower arm is hidden behind the upper arm of the divider and probably touches the upper edge of the large steel file whose pointed tip with the small hole near it rests on the lamp's left foot. Taken together, the pen, the two duckbill goniometer arms, and the metal file arm form the letter "M", but, like the numeral "4" formed by the lower goniometer arm and the "L" square, this letter has also been rotated more than 90° clockwise from its normal orientation. This letter "M" that can be made from the three items on the workshop table tells us that the quadrupled distance values we get by multiplying the center axis to center axis distances revealed in the circular sector paper piece by a factor of 4 apply to the Merseburg wheel since the first letter of the town that wheel was constructed in also begins with the letter "M".

And, what of the Kassel wheel? Is there some symbolism near the circular sector paper piece that tells us that we must multiply the center axis to center axis separation distance values in that paper piece by a factor of 4 in order to also use them for the sixteen weighted levers of the bidirectional Kassel wheel's two internal one-directional wheels? Yes, it's also there.

Notice that the "L" square and the upper arm of the duckbill goniometer form the letter "K" which is rotated clockwise through an angle of a little *less* than 90° from its normal orientation. Since the word "Kassel" begins with the letter "K", this indicates that the symbolic suggestion to multiply the circular sector paper piece's various center axes' separation distances given for the 3 feet diameter Gera prototype wheel by a factor of 4 *also* applies to the Kassel wheel.

Let us now return to Figure 20 to discuss the various coordinating cord hook attachment pin center axes' separation distances it reveals and how one uses the circular sector shaped paper piece to determine them.

As previously mentioned, the origin point of the circular sector paper piece, "P", represents the center axis of a weighted lever's pivot pin and the various concentric arcs surrounding that origin represent various distances, in inches, as measured outward from the center axis of a weighted lever's pivot pin. If one studies the concentric arcs drawn on the circular sector paper piece, he will see that there are six of them with the fifth and sixth ones very close together and the sixth arc actually forming the outermost curving edge of the paper piece. The place to start in decoding this important symbol in the second DT portrait begins with determining the meanings of these concentric arcs.

After much study, I realized that each of the concentric curving arcs on the circular sector paper piece represents a collection of points that are all a *fixed* radial distance from the center axis of the pivot pin of a weighted lever in the 3 feet diameter Gera prototype wheel which is represented by the circular sector's origin at "P". The increment of distance is 0.5 inch = 8/16 inch which is suggested by all of the symbols for the number 5 and a decimal point that surround the circular sector shaped paper piece (note that in Figure 20 I did not provide equivalent separation distance measurements to a sixteenth of an inch because there was not enough room in the figure, but one can quickly find them by simply multiplying 16 by the decimal *portion* of each center axis to center axis separation distance indicated). The number 5, thus, is represented by the various Roman numeral "V" shapes like those of the duckbill goniometer, the divider whose upper leg covers the tip of the lower arm of the goniometer, and the cross sections of the carpenter's plumb bob and the lamp's conical cover. The decimal point is symbolized by the points at the ends of the arms of the goniometer and divider arms. Indeed, the circular sector paper piece, with its origin point at "P" and "V" shape, is itself a symbol for 0.5.

Looking at the *bottom* straight radial edge of the circular sector paper piece, we notice that the distance from the origin point "P" to the first and nearest of the concentric arcs seems to be about double the distance from the first to second arcs. We then assign a radial distance of 2 x 0.5 inches = 1.0 inch to the *first* concentric arc for use with the weighted levers found inside of the 3 feet diameter Gera prototype wheel which is indicated at the bottom of Figure 20 by the "GPW" after that value and is it then quadrupled to 4.0 inches for use with the weighted levers found inside of the two internal one-directional wheels of the bidirectional Merseburg and Kassel wheels which is indicated at the bottom of Figure 20 by the "MKW" after that value since all of the distance values for use with the

Merseburg wheel's weighted levers are also used for the weighted levers inside of the Kassel wheel.

Moving on to the second concentric arc from the origin point "P" of the circular sector paper piece, the radial distance becomes 0.5 inches greater than the radial distance from the first concentric arc or becomes 1.5 inches for use with the 3 feet diameter Gera prototype wheel's weighted levers and four times as much or 6.0 inches for use with the Merseburg and Kassel wheels' weighted levers. The third concentric arc from the origin point "P" represents a radial distance of 2.0 inches from the circular sector paper piece's origin for use with the Gera prototype wheel's weighted levers and, after quadrupling, 8.0 inches from the origin for use with the Merseburg and Kassel wheels' weighted levers. The radial distance represented by the fourth concentric arc is 2.5 inches from the circular sector paper piece's origin and is for use with the Gera prototype wheel's weighted levers and, after quadrupling, 10.0 inches from the origin for use with the Merseburg and Kassel wheels' weighted levers. The sixth and outermost concentric arc has a radial distance of 3.0 inches from the origin of the circular sector paper piece for use with the Gera prototype wheel's weighted levers and, after quadrupling, 12.0 inches from the origin for use with the Merseburg and Kassel wheels' weighted levers.

But, what about the *fifth* concentric arc?

No, I did not forget it. It is a special arc whose distance is not a multiple of 0.5 inches and is 2.8 inches from the origin "P" of the circular sector paper piece for use with the Gera prototype wheel and 11.2 inches from the origin for use with the Merseburg and Kassel wheels. This arc is the only one on the circular sector paper piece that has one of the nearby outside items pointing directly to it. That outside item is the beveled or curving tip of the shorter arm of the "L" square drafting tool. This arc is also the only one with a tiny visible break in it implying that it can somehow provide us with two pieces of important information about something. But, why does Bessler bring it to our attention and why did he use the "L" square drafting tool in the second DT portrait to do so?

Here's what I think is *one* possible answer. If one takes the center axis to center axis separation distance between the center of gravity of the lead ingot end weights held between the ends of the main arms of the little wooden levers in the Gera prototype wheel and the pivot pins of those levers, that distance is exactly 3.5 inches. If one then divides that value by the distance of the fifth arc on the circular sector paper piece from its origin "P", which is 2.8 inches, he will obtain the value of 1.25. What's so special about that value and how does it relate to the "L" square drafting tool?

We saw in chapter 7 that, in order to achieve bidirectionality, Bessler had to equip each of the sixteen weighted levers in his Merseburg and Kassel wheels with two types of cat's claw gravity latches, an inner and outer latch. Each latch consisted of two parallel metal pieces that had an unusual "L" shape to them and, remarkably, looked almost exactly like the "L" square found in the second DT portrait! If the reader briefly refers to Figure 13(b), he will notice that the two important distance values given for *either* of the longer or shorter arms in each of the "L" shaped metal pieces used in *either* type of latch has a ratio of exactly 1.25! Thus, the extra, apparently unused, fifth arc in the circular sector paper piece was intended, because one of the arms of the "L" square drafting tool points to it, to *verify* the distance values used for the fabrication of the metal pieces used to make the gravity latches of his bidirectional wheels. Indeed, this is the only clue I've found that provides this verification.

Now that the reader sees how the various *possible* separation distances from the center axis of the pivot pin of a weighted lever of either the Gera prototype wheel or the Merseburg and Kassel wheels are symbolized by the concentric arcs shown in the circular sector paper piece of the second DT portrait, it's time to see how they are used to provide a reverse engineer of Bessler's imbalanced perpetual motion wheels with the *actual* separation distances from the center axis of a weighted lever's pivot pin to the center axes of the four different types of coordinating cord hook attachment pins located between the lever's pair of parallel A arms in either of these different diameter wheels.

As I mentioned above, the left straight radial edge of the circular sector paper piece shown in Figure 20 represents the parallel pair of A arms of a weighted lever and it's the 9:00 weighted lever in particular. As the reader studies that edge, he will notice that the ends of four *straight* lines make contact with it. The lines are arranged into two almost parallel pairs which cross each other.

These are four lines, which represent the four types of coordinating cords found in the 3 feet diameter Gera prototype wheel's drum are, starting in order from the one attached *closest* to the circular sector paper piece's origin "P", the spring cord whose metal end hook attaches to the A1 spring cord hook attachment pin whose center axis is 0.5 inch = 8/16 inch from the center axis of its weighted lever's pivot pin, the stop cord whose metal end hook attaches to the A2 stop cord hook attachment pin whose center axis is 0.933 inch = 14.93/16 inch from the center axis of its weighted lever's pivot pin, the long lifter cord whose metal end hook attaches to the A3 long lifter cord hook attachment pin whose center axis is 1.4 inches = 1-6.4/16 inches from the center axis of its weighted lever's pivot pin, and, finally, the main cord whose metal end hook attaches to the A4 main cord hook attachment pin whose center axis is 1.866 inches = 1-13.87/16 inches from the center axis of its weighted lever's pivot pin.

On the left side of Figure 20, all of these coordinating cord hook attachment pin center axis separation distances from the center axis of the Gera prototype wheel's 9:00 weighted levers' pivot pin, which are revealed in the circular sector paper piece of the second DT portrait, are indicated by the acronym "GPW" placed after the value which, again, stands for "Gera Prototype Wheel". These distance values, of course, also apply to the parallel pairs of A arms of all of the eight weighted levers inside of the Gera prototype and not just to the parallel pair of A arms of the weighted lever whose pivot pin's center axis is located at the 9:00 position of a clockwise rotating drum which is what is represented by the circular sector paper piece.

On the left side of the figure I have also provided the corresponding quadrupled distances for use in locating the centers axes of the A arms coordinating cord hook attachment pins in the weighted levers of the Merseburg and Kassel wheels and these values are followed by the acronym "MKW" which, again, stands for "Merseburg and Kassel Wheels".

Thus, for both the Merseburg and Kassel wheels, starting again with the straight line representing a coordinating cord that is attached to a point on the left radial edge of the circular sector paper piece which represents a weighted lever' parallel pair of A arms and is attached closest to the circular sector paper piece's origin "P" that represents the center axis of a weighted lever's pivot pin, we again have the spring cord whose metal end hook attaches to the A1 spring cord hook attachment pin whose center axis is quadrupled to 2.0 inches from the center axis of its weighted lever's pivot pin, the stop cord whose metal end hook attaches to the A2 stop cord hook attachment pin whose center axis is quadrupled to 3.733 inches = 3-11.73/16 inches from the center axis of its weighted lever's pivot pin, the long lifter cord whose metal end hook attaches to the A3 long lifter cord hook attachment pin whose center axis is quadrupled to 5.6 inches = 5-9.6/16 inches from the center axis of its weighted lever's pivot pin, and, finally, the main cord whose metal end hook attaches to the A4 main cord hook attachment pin whose center axis is quadrupled to 7.466 inches = 7-7.47/16 inches from the center axis of its weighted lever's pivot pin. In Figure 20 all of these hook attachment pin center axis separation distances from the center axis of the Merseburg and Kassel wheels' weighted levers' pivot pins are indicated by the acronym "MKW" placed after the value which, again, stands for "Merseburg and Kassel Wheels". And, again, all of these center axis to center axis separation distances apply to the parallel pairs of A arms of all of the eight weighted levers found in each of the internal one-directional wheels of these bidirectional wheels.

Above, I have mostly discussed the separation distance parameters of the A arms' four hook attachment pins to which the four types of coordinating cords are attached to the weighted levers inside either Bessler's 3 feet diameter, one-directional Gera prototype wheel or either of his 12 feet diameter, bidirectional Merseburg or Kassel wheels. All of these four types of cords also have metal hooks attached to their other ends which, in turn, are attached to coordinating cord hook attachment pins on the main or B arms of *other* weighted levers or to anchor pins attached to a wheel's drum. Now, it's time to indicate the various distance parameter values associated with these other hook attachment pins and relate them to the cylindrical sector paper piece in Figure 20.

On the right side of Figure 20, the reader will find labels that indicate the hook attachment pins to which the metal hooks at the other ends of the four types of cords originating from a weighted lever's A arms are attached.

For the spring cord of the 9:00 weighted lever in the Gera prototype wheel indicated in the circular sector paper piece of the second DT portrait, the attachment (actually of the opposite end of the extension spring whose other end is attached to the spring cord) is to the spring hook attachment anchor pin embedded between the lever's leading parallel pair of the drum's *outer* octagonal frame pieces which has its center axis located at a distance, labeled "GPW", of 16.5 inches = 16-8/16 inches from the center axis of the axle and, for the 12 feet diameter, bidirectional Merseburg and Kassel wheels, labeled "MKW", that distance is quadrupled to 66 inches from the center axis of the axle.

The hook at the other end of the stop cord, for the Gera prototype wheel, is attached to a drum stop cord hook anchor attachment pin embedded in the weighted lever's leading parallel pair of the drum's *inner* octagonal frame pieces whose center axis is located at a distance, labeled "GPW", of 12.934 inches = 12-14.94/16 inches from the center axis of the axle. For the Merseburg and Kassel wheels, this distance, labeled "MKW", is quadrupled to 51.736 inches = 51-11.78/16 inches.

Moving on to the long lifter cord of a 9:00 weighted lever in the Gera prototype wheel, we see that the hook at its other end is attached to the B1 long lifter cord hook attachment pin held between the parallel pair of B arms of a leading weighted lever, but, those B arms belong to the weighted lever whose pivot pin's center axis is at the drum's 12:00 position. The center axis of that B1 hook attachment pin is located at a distance of 2.1 inches = 2-1.6/16 inches, labeled "GPW", from the center axis of the 12:00 weighted lever's pivot pin. For both the Merseburg and Kassel wheels, this distance, labeled "MKW", is quadrupled to 8.4 inches = 8-6.4/16 inches.

Finally, there is the main cord whose other end's metal hook is attached, for the Gera prototype wheel's 9:00 weighted lever, to the L1 main cord hook attachment pin that is held between the parallel pair of main arms the 10:30 weighted lever whose center axis distance from the center axis of that lever's pivot pin, labeled "GPW", is 2.6 inches = 2-9.6/16 inches. The analogous pin center axis to pin center axis distance in the Merseburg and Kassel wheels, labeled "MKW", is quadrupled to 10.4 inches = 10-6.4/16 inches.

The distance of the center axis of an L1 hook attachment pin from the center axis of the pivot pin of a Gera prototype wheel's weighted lever of 2.6 inches = 2-9.6/16 inches was something that took me quite a bit of frustrating work to finally determine. From examining highly magnified views of the circular sector paper piece in the second DT portrait, one is quickly led to believe that the other hook of the 9:00 weighted lever's main cord should be attached to an L1 hook attachment pin whose center axis is 2.8 inches = 2-12.8/16 inches from the center axis of the 10:30 weighted lever's pivot pin and that fifth arc on the circular sector paper piece whose radial distance is 2.8 inches from the origin "P" of the circular sector certain seems to reinforce this *false* notion.

This false notion is also suggested by a mathematical pattern derived from the distances of the center axes of the 9:00 weighted lever's A arms' coordinating cord hook attachment pins from the center axis of that lever's pivot pin.

The A1 spring cord hook attachment pin distance is 0.933 inches which is approximately 2.8 inches / 3, the A2 stop cord hook attachment pin distance is 1.4 inches which is 2.8 inches / 2, and the A3 long lifter cord hook attachment pin distance is 1.866 inches which is 2.8 inches / 1.5. So, the next logical quotient, that would be for the center axis of an A5 hook attachment pin if it existed which it does not, would be 2.8 inches which is, obviously, 2.8 inches / 1. Since there is no A5 hook attachment pin held by a weighted lever's A arms, one then begins to think that Bessler must have used the value of 2.8 inches / 1 = 2.8 inches for the distance of some other hook attachment pin's center axis from its lever's pivot pin's center axis and then notices that, indeed, there is a concentric arc in the circular sector paper piece that is located at exactly 2.8 inches from the circular sector's origin. To my great surprise, however, this proved *not* to be the case.

If one very carefully inspects the main cord line symbol in Figure 20 as it stretches almost horizontally across the circular sector paper piece from the 9:00 weighted lever's A arms, represented by the paper piece's left radial edge, to the end of the straight line on the paper piece that represents the main or L arms of the 10:30 weighted lever (which is labeled in the figure), he will note that the main cord line symbol certainly appears to contact the end of the 10:30 weighted lever's main or L arms line symbol at a distance of exactly 2.8 inches from the origin of the circular sector paper piece. However, even without great magnification, one will notice that there is a small ink smear between the end of the main cord line symbol and the end of the main or L arms line symbol of the l0:30 weighted lever to which the main cord line symbol connects.

At first I thought that this ink smear was a mishap that only appeared on the printed second DT portrait of the particular copy of DT that I was studying, but then I notice that it was on *all* of the different copies that I was able to find and study. That meant that it would either have been a gross engraving mistake or *purposely* placed into the engraving of the circular sector paper piece. Only through much work with my computer models was I able to determine that it was, indeed, put into the portrait on purpose. If one follows the *left* vertical edge of that small "smear" downward a *very* short distance from the nearly horizontal line representing the main cord until that edge intersects the diagonal straight line representing the main or L1 arms of the 10:30 weighted lever, he will discover that the edge intersects the main arms at a distance of 2.6 inches from the origin of the circular sector paper piece.

We see from this example that, while Bessler misleads the reverse engineer of his imbalanced pm wheels' mechanics about where to attach the main cord hook from a weighted lever to the main or L arms of its immediate leading weighted lever, he does provide a single additional clue, if properly interpreted, which will get that person back on the right path again to making a successful analysis of the mechanics the inventor used.

There is still one straight line in the circular sector paper piece for which I have not given an interpretation. It is that prominent vertical line that extends from the origin "P" of the circular sector paper piece up to its outer top curving edge and which is not labeled in Figure 20. It makes a noticeably exact angle of 45° with the straight diagonally oriented line on the right side of the paper piece that represents the parallel pair of main arms of the 10:30 weighted lever. Thus, this vertical line actually represents the A arms of the 10:30 weighted lever. Why did Bessler include it?

I think the simple reason is indicated by the fact that, as all four of the coordinating cords are extended away from their cord hook attachment pins on the 9:00 weighted lever's A arms, they must pass through that vertical line and this was Bessler's way of indicating to a future reverse engineer of his wheels that the arrangement of the hook attachment pins on the A arms of the 10:30 weighted lever, that is, their various center axis to center axis separation distances from the pivot pin of the 10:30 weighted lever, were *identical* to the arrangement used on the 9:00 weighted lever's A arms. This can then be generalized to all of the weighted levers found in all of Bessler's wheels regardless of their size or whether they were one-directional or bidirectional.

With this, I have concluded a general analysis of the second DT portrait's very important symbol of the circular sector paper piece and, hopefully, the reader and craftsman will now more fully realize why I consider it "the" most important clue in the Bessler literature.

Many of the locations of the center axes of the various coordinating cord hook attachment pins it provides information about can be accurately determined by just making careful measurements off of high resolution enlargements of the second DT portrait and applying some simple arithmetical analysis to the arc whose separation distance from the circular sector paper piece's origin "P" is 2.8 inches = 2-12.8/16. However, even with the various coordinating cord hook attachment pin center axes locations relative to a weighted lever's pivot pin's center axis obtained from the circular sector paper piece, the reverse engineer is still far from having enough information to successfully duplicate one of Bessler's wheels. For example, he will still not know such things as the maximum separation distances between two attachment pin center axes allowed by the four types of coordinating cords, the masses and shapes of the various end weights used at the ends of the parallel pairs of main arms of the weighted levers in the different diameter wheels, or the spring constants of the various parallel pairs of extension springs attached to the weighted levers in Bessler's various wheels. Where, one wonders, is that information being hidden?

In the remainder of this chapter, I will indicate the other clues contained in the second DT portrait that provide that still missing information necessary for successful wheel duplication.

As I continued to study the symbolism of the second DT portrait's circular sector paper piece, I wondered if any of the other items near it could provide additional clues to the exact separation distances between the center axles of the Gera prototype wheel's 9:00 weighted lever's A arm's four coordinating cord hook attachment pins and the lever's pivot pin. I was hoping that any numerical values derived from such extra clues would then verify the values that I had obtained graphically from the circular sector paper piece. Eventually, my attention was drawn to the pointed end of the upper arm of the large duckbill goniometer that is located above the paper piece. Then, I thought of one of the basic intuitive axioms of Euclidean geometry which states that through any two points in space only a single straight line may be drawn. Was it possible that Bessler used this simple axiom to hide additional clues in the second DT portrait?

With this in mind, I wondered what would happen if I extended straight line segments from the pointed end of the goniometer's upper arm *through* the various intersection points on the left straight radial edge of the circular sector paper pieces where the straight line segments representing a weighted lever's four types of coordinating cords were attached to the paper piece's left radial edge that represented the A arms of the Gera prototype wheel's 9:00 weighted lever. More importantly, I wondered where the end points of those straight line segments I drew would arrive if I extended them down toward and then *past* the bottom border of the portrait.

To my utter astonishment, I found that the ends of the line segments that I extended out of the portrait area itself eventually arrived at letters in the German text under the portrait that, when *properly* alphanumerically analyzed, actually gave me the *exact* separation distant values for the center axes of the four types of coordinating cord hook attachment pins in the A arms represented by the intersection points on the left straight radial edge of the circular sector paper piece from the center axis of the pivot pin in the 9:00 weighted lever of the Gera prototype wheel! The extended line segments from the top goniometer arm's pointed tip also gave me some very important additional distant values. I realized that these were all *very* important clues that, most likely, no one other than Bessler and now I ever knew about!

In order to fully understand what I am describing here, the reader is now directed to Figure 21(a) at the end of this chapter. It shows the result of extending straight line segments downward from the pointed tip of the goniometer's upper arm, through various intersection points on the circular sector paper piece, and then farther downward to various letters below the second DT portrait.

Near the solid black circle "A" label at the bottom of the figure, we see that the line segment that passes through the point that corresponds to the A1 spring cord hook attachment pin's center axis on the circular sector paper piece's left straight radial edge eventually arrives at the letter "c" in the letter combination of "ch" in the word "*noch*" on the second line of text below the portrait. "c" has an alphanumeric value of 3 since that letter is the third letter of the alphabet. "h" has a value of 8 because it is the eight letter of the alphabet. If one studies a magnified image of the second DT portrait, he will notice that the top of the letter "c" actually touches the letter "h". This is a signal Bessler used to let the reverse engineer studying his DT portraits know that some arithmetic operation had to be performed using the alphanumeric values of the letters involved. In this case, those operations could be 3 + 8, 8 − 3, 3 x 8, 3/8, or 8/3. The one he intended us to use was 8 − 3 = 5. The "?" mark formed from top portion of the letter "h" curls to the *left* and ends with a large dot. This then tells us that we must move the decimal point one place over to the left of 5 to obtain 0.5. Adding the units of inches then gives us 0.5 inch = 8/16 inch which just happens to be the exact center axis to center axis separation distance between an A1 spring cord hook attachment pin and the pivot pin of the 9:00 weighted lever used inside of the 3 feet diameter Gera prototype wheel! This distance value will also apply to all of the other seven weighted levers inside of the drum of that wheel.

Moving on to the location of the center axis of the A2 stop cord hook attachment pin indicated on the left straight radial edge of the circular sector paper piece, we follow the extended line segment from the pointed tip of the goniometer's upper arm through that intersection point and downward until it reaches the letter "o" in the word "*noch*" on the second line of text below the second portrait. We then notice that the lower right vertical downward stroke of the letter "n" terminates with a serif that extends to the right and touches the letter "o". Again, Bessler is telling the reverse engineer by the two connected letters that he must perform an arithmetic operation involving the alphanumeric values of these two letters. The value for "n" is 14 since it is the fourteenth letter of the alphabet and 15 for "o" since it is the fifteenth letter of the alphabet. If one then divides the alphanumeric value of "n" by that of "o", he gets 14/15 = 0.933. Adding the units of inches then gives us 0.933 inch = 14.93/16 inch which is the correct distance for the center axis of the A2 stop cord hook attachment pin from the center axis of the 9:00 weighted lever's pivot pin in the Gera prototype wheel! Since the division automatically puts the decimal point in the correct location, there is no need to have a decimal point symbol near the letters. The letter combination of "no" is indicated by the solid black circle "B" label in the figure. This value will also apply to the other seven weighted levers inside of the drum of the Gera prototype wheel.

Next, we follow an extended line segment from the duckbill goniometer's pointed upper arm downward and through the intersection point on the circular sector paper piece that represents the center axis of the A3 long lifter cord hook attachment pin to which a metal hook at one end of the lever's long lifter cord is attached. That line segment, when extended below the portrait's bottom border, eventually arrives at the letter "n" in the word "*noch*" and is indicated by the solid black circle "C" label in Figure 21(a). The "n", as previously stated, has an alphanumeric value of 14 and we only use this single letter in the analysis. Immediately to the *left* of the letter "n", there is the word "*lobt*" and the pointed serif at the bottom of its last letter, "t", which points to the "n" in "*noch*" serves as a decimal point symbol that tells us to move the decimal point in 14 one position over to the left to obtain 1.4. Adding the units of inches to this then gives us a value of 1.4 inches = 1-6.4/16 inches which is the separation distance between the center axes of the A3 long lifter cord hook attachment pin and the pivot pin of the weighted levers found in the 3 feet diameter Gera prototype wheel's drum. This value will be the same for the other seven weighted levers in the drum of that wheel.

Finally for the A arms of a weighted lever in the Gera prototype wheel, we follow the extended line segment down from the pointed tip of the duckbill goniometer's upper arm, through the intersection point on the left straight radial edge of the circular sector paper piece that corresponds to the center axis location of the A4 main cord hook attachment pin and note that, as the line extends below the portrait, it eventually reaches the tiny serif in the upper left corner of the letter "n" in the word "*noch*" on the second line of text below the portrait. As I studied this clue, I noticed that serif at the upper left corner of the "n" was actually pointing downward to the serif at the bottom the "t" in the preceding word "*lobt*". I could almost imagine a line connecting the two serifs and that is exactly what Bessler also wanted an observant reverse engineer of his secret imbalanced pm wheel mechanics to do. That connection, even though only implied, suggests that one must somehow use the alphanumeric value of "t", which is 20, to alter the value of the "n" in "*noch*". In Figure 21(a), the imaginary connection between the letters "t" and "n" is indicated by the solid black circle "D" label.

The correct interpretation of this clue is gotten when one uses the left pointing top left serif on the letter "n" to justify moving the decimal point in the normal value of "t", which is 20, one place to the left to give it an altered value of 2. Because of the suggested connection between the ending letter "t" in the word "*lobt*" and the beginning letter "n" in the word "*noch*", one then multiplies the altered value of the "t" times the value of the *two* connected letters "no" in the word "*noch*", which was previously shown to be 0.933 which then gives 2 x 0.933 = 1.866. After adding the distance units of inches, this value then becomes 1.866 inches and represents the *exact* center axis to center axis distance between the A4 main cord hook attachment pin and pivot pin of the 9:00 weighted lever inside of the Gera prototype wheel's drum. Its value will also apply to the other seven weighted lever inside of that wheel's drum.

As was previously mentioned, with this analysis using the alphanumeric values of certain letters and groups of letters in the German text below the second DT portrait, it is possible to obtain the center axis locations of *all* of the coordinating cord hook attachment pins held by the parallel pairs of A arms of all of the weighted levers of the one-directional, 3 feet diameter Gera prototype wheel relative to the center axis of a lever's pivot pin. *After* these distance values are multiplied by a factor or 4, the resulting larger distance values will be valid for all of the weighted levers that were used within the bidirectional drums of the Merseburg and Kassel wheels.

We are, however, not quite done with our discussion of Figure 21(a). There is still another extended line segment from the pointed tip of the goniometer's upper arm that passes through another intersection point on the circular sector paper piece which symbolizes the location of the center axis of a coordinating cord hook attachment pin. That line segment also then continues on to reach a group of letters in the text below the portrait. That group of letters actually involves three letters on the first line and one on the second line of text below the portrait! The group of three letters on the first line is the "lle" in the word "*vollendet*" and the letter on the second line is the "d" in the word "*fchändet*" which is indicated by the solid black circle "E" label in Figure 21(a). Although correctly alphanumerically analyzing this clue might seem impossible, it really is rather easy as I will now demonstrate.

The reader may recall that I above pointed out the difficulty of determining from the straight line segments drawn on the circular sector paper piece exactly where the center axis of the L1 main cord hook attachment pin held by the A arms of the 10:30 weighted lever was located relative to the center axis of the pivot pin of that lever. It was that main cord hook attachment pin to which the little metal hook at the *other* end of the 9:00

weighted lever's main cord was attached. I mentioned that Bessler provided a clue in the paper piece in the form of a small downward ink smear at the end of the straight line segment representing that cord on the circular sector paper piece that indicated that the distance from the center axis of the 10:30 weighted lever's L1 main cord hook attachment pin to the center axis of that lever's pivot pin was exactly 2.6 inches = 2-9.6/16 inches. Now, using Bessler's additional and previously unsuspected alphanumeric clues beneath the portrait, it is possible to verify the exact value of this distance.

To do this, we simply perform several arithmetic operations on the alphanumeric values of the individual letters in the "lle" letter group because that requirement is indicated by all of these three letters being connected to each other. However, before we do that we must change the value of the *two* letter "l's" which are normally 12 each since "l" is the twelfth letter of the alphabet. We do that by adding up the individual digits in 12 to get 1 + 2 = 3. Then, using the normal value of 5 for "e", we can write (3 x 3) + 5 = 9 + 5 = 14. I used a multiplication operation here because the "x" symbol for multiplication can be found in the interlacing flourishes at the top of the two "l's" which indicates their altered values must be multiplied. Next, because the alphanumeric value of "d" is normally 4, but the pointed tip at the top of the "d" points to the *right*, we move the decimal point of the 4 one place to the right and that *temporarily* changes its value to 40. Finally, we subtract the value we obtained for the letter group "lle" on the first line of text from the altered value of the "d" on the second line of text. That then gives us 40 − 14 = 26 which, after moving the decimal point one place back again to the left gives us 2.6. Adding the units of inches then gives us 2.6 inches = 2-9.6/16 inches and this then represents the distance between the center axis of the 10:30 weighted lever's L1 main cord hook attachment pin and the center axis of that lever's pivot pin in the Gera prototype wheel. This value will also apply to the other seven weighted levers in that wheel and, after being multiplied by a factor of 4 to all of the weighted levers within either of the two internal one-directional wheels found within either the Merseburg or Kassel wheels.

There is, however, one remaining coordinating cord hook attachment pin's center axis that is not shown as an intersection point *on* the circular sector paper piece in the second DT portrait. It is actually located off of the paper piece and located on the symbol for the 12:00 weighted lever which is the metal lamp with the tripodal feet attached to its base. If one extends the right end of the straight line segment representing the long lifter cord on the circular sector paper piece to the right and off of the paper piece which was done using a dashed white line in Figure 21(a), it will reach the top part of the inside corner of the angle formed by the two visible feet of the lamp's three feet. Once the importance of that corner point is realized, one then extends a straight line segment downward from the pointed tip of the upper arm of the duckbill goniometer so that it passes through that inside corner point, then out of the bottom border of the portrait, and, finally, reaches the pointed end on the bottom right portion of the letter "e" in the word "*vollendet*" at the end of the first line of text under the portrait which is located near the black solid circle "F" label in Figure 21(a).

The alphanumeric analysis is now done using the letter group "det" at the end of the word "*vollendet*" even though only the "e" and the "t" are connected to each other. We add the normal values of the connected letters "e" and "t" to get 5 + 20 = 25. Then, again using the Roman numeral method for assigning values to numerals, we subtract the normal value of the "d", which is 4, that is to the left side of the "et" letter group from the value we got for the "et" letter group, which was 25, to get 25 − 4 = 21. The pointed lower tip, a symbol for a decimal point, of the following virgule mark which points to the left tells us we must move the unwritten decimal point of 21 one place to the left so we then write it as 2.1 and, after adding the units of inches, we get 2.1 inches = 2-1.6/16 inches. This value then represents the distance between the center axes of the 12:00 weighted lever's B1 long lifter cord hook attachment pin and that lever's pivot pin in the 3 feet diameter Gera prototype wheel's drum. The value also applies to the other seven weighted levers in that wheel's drum as well and must be multiplied by a factor of 4 for the weighted levers found inside the bidirectional drums of the Merseburg and Kassel wheels.

To summarize the results of the alphanumerical analyses done in Figure 21(a), we see that they provided us with the locations, relative to the center axis of their lever's pivot pin, of the center axes of the A1 spring cord, A2 stop cord, A3 long lifter cord, and A4 main cord hook attachment pins held between the parallel pair of A arms of the weighted lever whose pivot pin center axis was located at the 9:00 position of the Gera prototype wheel's clockwise rotating drum. These analyses further provided us with the locations of the center axes of the 10:30 weighted lever's L1 main cord hook attachment pin held between that lever's parallel pair of main or L arms and of the 12:00 weighted lever's B1 long lifter cord hook attachment pin held between that lever's parallel pair of B arms with both distance values being relative to the center axes of those levers' pivot pins. That's a total of six distance parameter values which agree with those that can be graphically determined by just making careful measurements off of the various intersection points of straight line segments that appear in the circular sector paper piece of the second DT portrait.

This now concludes the discussion of Figure 21(a) and it's time to move on to Figure 21(b) which shows how the straight lines segments representing the four types of coordinating cords attached to the 9:00 weighted lever in the Gera prototype wheel can be used, when extended in length, along with some of the other items on the top of the shop table in the second DT portrait to determine other critically important parameter values for these cords. The reader should now find Figure 21(b) and study it.

We begin with that previously encountered group of three connected letters immediately below the solid black circle "A" label in Figure 21(b). The three letters belong to the group of letters consisting of the "lle" in "*vollendet*" on the first line of text below the portrait. These are contacted by a straight line segment that extends downward from the pointed tip of the goniometer's upper arm and then passes through that tiny ink blob (not a smear) that is located just below the point there the *first* diagonal line segment from the left straight radial edge of the circular sector paper piece representing the spring cord intersects the *fifth* arc on that paper piece whose distance from the origin of the paper piece is 2.8 inches = 2-12.8/16 inches.

As the reader may notice, this line is identical to the line used in Figure 21(a) that pointed to the lacey top of the same first letter "l" in the letter group "lle". However, in Figure 21(b) this new line segment is not extended until it reaches the pointed top of the letter "d" in the word "*fchändet*" located on the second line of text and, most importantly, this new line segment will give us information about the spring cords rather than the main cords used in the Gera prototype wheel's drum. This time we will also be performing the alphanumeric analysis of the first text line's letter group "lle" a little differently to get important distance parameter value information about the spring cords which are attached, in parallel pairs, to each of the eight weighted levers inside of the drum of the Gera prototype wheel.

Interestingly, that this group of three letters does provide information about the parallel pair of spring cords, which includes their attached helical coil extension springs whose hooks at their other ends are attached to the spring hook attachment anchor pin held in a weighted lever's leading parallel pair of outer octagonal frame pieces, is indicate by the decorative looping, spring-like flourishes at the tops of the two "l's". The intermeshing of the flourishes is a further clue, in addition to the two letters being in contact through the serif at the bottom of the left "l", that we must combine their values in some way.

If we just multiply the normal alphanumeric values of the two "l's" together, we get 12 x 12 = 144 which, after adding inch units, equals 144 inches or the diameters of both the bidirectional Merseburg and Kassel wheels. However, this is not the correct arithmetic manipulation that must be used. Rather, we again add the individual digits in each "l" to get 1 + 2 = 3 as was done for an analysis using the previous Figure 12(a), but, unlike in that analysis, this time we do *not* then multiply their altered values together which would give us 3 x 3 = 9. Rather, this time for Figure 12(b), we *add* these altered values for the two "l's" together to get 3 + 3 = 6. But, we must not forget that there is also another letter in the three letter group we are working on and that is the "e" at the end of "lle" whose normal alphanumeric value is 5. The "ll" makes contact with the "e" through the serif at the bottom of the right "l" and that tells us that we must somehow arithmetically combine the two values we have. Before we do that, however, we notice that the pointed tip of the letter "d" on the text line below the "lle" points to the *left* side of the letter "e". This clue tells us that we must move the decimal point one place to the left of the value of "e", which is 5, to obtain 0.5. Finally, we add the derived value of "ll", which is 6, to the altered value of "e", which is 0.5, to get 6 + 0.5 = 6.5. Adding the inch units turns it into 6.5 inches.

What significance does this distance value of 6.5 inches have for the weighted lever whose pivot pin center axis is located at the 9:00 position of the clockwise rotating drum of the Gera prototype wheel?

It turns out that it is the exact maximum separation distance allowed by the lever's stop cord between the center axes of the lever's A arms' A1 spring cord hook attachment pin and the spring hook attachment anchor pin embedded in the lever's leading parallel pair of outer octagonal frame pieces. If we just add the derived values of the two "1's" together, we would have 3 + 3 = 6 or, after adding length units, 6 inches. This value is the *minimum* separation distance between these center axes when a Gera prototype wheel's weighted lever's pivot pin's center axis was traveling around the lower quadrant of the descending side of a clockwise rotating drum toward its 6:00 position while the lever's two parallel extension springs were unstretched and the parallel pair of main arms of the lever were in contact with their wooden radial stop piece. However, as the weighted lever's pivot pin's center axis continued to travel upward on the drum ascending side from the drum's 6:00 to 9:00 positions, the separation distance between these axes slowly increased by 0.5 inch which, obviously, is indicated by the value of the letter "e" after we move its decimal point one place to the left as required by the pointy upper tip of the "d" which points to that side of the letter "e". That added stretch distance for each of the lever's two parallel extension springs was then only 0.5 inch.

Each of the eight weighted levers inside of the drum of the 3 feet diameter Gera prototype wheel would experience this same amount of stretching for each of its two extensions springs from their normal unstretched lengths as the lever's pivot pin axis approached the drum's 9:00 position. The involvement of the letter "d" in the analysis with its alphanumeric value of 4 tells us that this stretch distance was multiplied by a factor of 4 for the extension springs used by the weighted levers inside of the drums of the bidirectional Merseburg and Kassel wheels.

After seeing all of the precise distance parameter values for the center axes of the various coordinating cord hook attachment pins inside of the Gera prototype wheel's weighted levers that emerged by extending straight line segments downward from the pointed tip of the goniometer's upper arm, through the various intersection points in the circular sector paper piece, and then farther until they reached various letters in the two lines of text below the second DT portrait, I began it wonder what would happen if I actually extended the *lengths* of the four line segments drawn on the circular sector paper piece, which themselves symbolized the four types of coordinating cords used in Bessler's one-directional wheels and which confusingly crisscross each other on the paper piece, so that they contacted various points on other items on the shop's tabletop.

I made many attempts to do this over the years that gave me no useful information until the day arrived when I began to have success with my efforts and that day was only a few weeks before the "official" rediscovery day and date of the secret imbalanced pm wheel mechanics Bessler found and used which, as mentioned in the beginning of this volume, was on Friday, April 13th, 2018! Those successful results were provided by the various straight line segments extending away from the circular sector paper piece that are shown in Figure 21(b) and which will now be discussed in detail.

I decided to start with the first straight line segment in the circular sector paper piece that represents the spring cord attached to the A1 spring cord hook attachment pin whose center axis represented by the intersection point on the paper piece's left straight radial edge that is closest to the circular sector's origin at "P" (indicated in Figure 20, but not in Figures 21(a) or 21(b)) and see where that straight line segment's extension to the upper right of the second DT portrait would lead.

As I extended it, it rose past the lower right side of the conical cover on the metal lamp and then reached the small cylindrical screw barrel microscope which stands vertically on the back right side of the shop table. I knew that this microscope contained an optical axis, so I stopped extending the line segment at the point where it would intersect that optical axis. The fact that this type of microscope usually contained a compression spring whose tension on the barrel's screw threads created enough friction to prevent the barrel from unscrewing while the instrument was being used so that its focused image of a small object would not become blurry also seemed like a positive detail since I was, after all, *looking* for information about a coordinating cord and *spring* combination, although, the springs attached to the spring cords in Bessler's wheels were the extension type and not the compression type that would have been used in the little microscope.

Since I knew Bessler liked to use optical axes to direct the attention of a viewer of the second DT portrait to items in it which contained some important information, I immediately extended the vertical optical axis of the microscope both upward and downward. This is indicated in Figure 21(b) by the vertical dashed white line on the right side of the second portrait. The top of the extended axis rose until it reached the 4:30 clock time position on the dial of the clock hanging on the side of the organ in the background. That seemed to have little to do with a spring rope attached to one of the Gera prototype wheel's weighted levers that was located at the clockwise rotating drum's 9:00 position, so I dismissed that clock time. Next, I extended the optical axis of the screw barrel microscope downward and, as soon as it passed through the bottom of the portrait, it reached the dot-like curl at the top of the letter "t" which is at the end of the word *"vollendet"* on the first line of text below the portrait. Since the "e" to the left of the "t" touched the left side of the horizontal bar that crossed the "t" that implied that the alphanumeric values of the two letters needed to be combined arithmetically. Yet, there was also that right leaning slash mark or virgule at the very end of the first line of text and to the right of the "t" which looked like an arrow pointing down toward the two fused letters in the abbreviation *"NB"* on the second line of text below the portrait. That implied that one had to somehow include the derived values gotten for the letter group "NB" with those of the "et" group! The possible ways of doing that seemed to make an accurate analysis impossible and, as a result, I made no progress with it for many weeks.

Then I took another look at the point where the diagonally ascending line segment from the circular sector paper piece representing the spring cord intersected the optical axis of the screw barrel microscope. Perhaps, I thought, Bessler wanted the reverse engineer to look at *right* angles to that optical axis?

Extending a horizontal line segment to the right of the intersection point on the microscope's optical axis only took the end of the line segment to the right side border of the portrait and beyond that there was no text with whose letters I could perform an alphanumeric analysis. But, when I extended the line to the *left* of the screw barrel microscope's optical axis, I discovered that it intersected a point on the right side of the conical cover of the metal lamp where an odd looking "knob" appeared to be attached. Under high magnification, I realized that this knob was actually some sort of spring-like handle that was *attached* to the hidden back side of the conical cover.

I had seen similar handles in the past that had, in larger versions, been attached to things like cast iron frying pans and tea pots as well as soldering irons (the kind with tips that were actually heated in a flame!) and their purpose was to dissipate the heat transferred to them from some object that they were attached to so that the temperature of the handle would not be high enough to burn the hand of a person holding the handle. I thought that, perhaps, the metal cover of the lamp, made of iron or brass, could be detached from the lamp and then used to melt small ingots of lead to cast items from them.

To do this, the cover might be inverted and placed in a pile of glowing coals to heat it up to the point at which the lead ingots it contained would melt which is 621° Fahrenheit and over a thousand degrees Fahrenheit lower than the melting point of brass which is in the range of 1652° to 1724° Fahrenheit depending upon the percentages of the various metals in this alloy. At that time, the cover would be lifted off of the coals by its spring-like, wire wound handle which would only be warm to the touch and the molten lead could then be poured into a mold. (Warning! Any craftsman intending to cast his own lead weights to use in replicas of Bessler's wheels needs to keep in mind that the dust from solid lead and the fumes from molten lead are toxic. The melting should only be done, preferably outdoors, while wearing a protective respirator mask and coveralls. When handling the finished castings, one should wear disposable gloves and limit skin contact with the metal as much as possible. One can use steel weights instead of lead, but in that case the *volumes* of any steel end weights used must be increased by a factor of 1.44 in order to compensate for the lower density of steel compared to the density of lead. To accommodate the larger volume steel weights, one might have to alter the shapes of the wooden levers used which might then increase the volumes of their various components and make them more massive than they should be. Changing over to steel end weights instead of using lead ones is certainly possible and desirable in order to minimize exposure to a toxic heavy metal, but it must be done using very careful planning.)

I also realized that, when viewed from above, the conical cover of the lamp would appear as a circle with the little ball at its pointed top looking like a much smaller circle at its center as the eight ribs on the cover seemed to radiate out from that smaller circle to a thick outer rim.

In other words, the lamp's conical cover, when viewed from above, was actually a symbol for a cross sectional axial view of the drum of one of Bessler's wheels complete with an axle and eight radial frame pieces that extended out from the axle and connected with a rim wall that formed the outermost circumference of the drum. The spring-like handle on the side of the lamp's conical cover then represented one of a weighted lever's two parallel extension springs that was attached to a point *somewhere* inside of that drum!

Now, finally, I had something whose meaning was inescapable. But, where was the end of that spring symbol located inside of the drum? In order to find out, I needed to continue extending the line segment which I had drawn, even if a change in its direction was necessary, from the end of the spring-like handle that contacted the lamp's cover to some other object in the second DT portrait.

Since the pointed tip of the duckbill goniometer's upper arm proved so useful in providing information about the locations of the center axes of the coordinating cord hook attachment pins held between the A arms of the 9:00 weighted lever of the Gera prototype wheel relative to the center axis of that lever's pivot pin, I extended the line segment in a new direction from the end of the spring-like handle touching the lamp's conical cover to the pointed tip at the end of the goniometer's upper arm. Once there, I wondered where it should be extended to next.

I looked about and tried various other points in the second DT portrait, but none of them seemed to make sense in terms of the distance value they provided about where the center axis of the spring hook attachment anchor pin should be located in the drum relative to the center axis of the Gera prototype wheel's axle. Then I made an interesting discovery.

I found that if I extended the straight line segment to the origin point of the protractor on the left side of the duckbill goniometer's pivot screw or bolt, then the line segment perfectly followed along the outermost third lower edge of the goniometer's upper arm. That seemed to me like too much of a coincidence to be one. I continued to extend the line segment a short distance farther to the left of the protractor's origin and discovered that its end arrived at a small, but noticeable dark spot on the right leg of the closed divider located at the point indicated by the nearby solid white circle "B" label in Figure 21(b). That divider's right leg pointed down and to the left and was aligned with the straight base of the protractor whose midpoint contains its origin. I then decided to again change the direction of the extending segment line and start extending another line segment in the same direction that the right leg of the divider pointed. Finally, the extending new line segment's end arrived at the top of the letter "e" in the word "*Werck*" on the first line of text below the portrait.

As I stared at the collection of four connected zigzagging line segments from the circular sector paper piece that extended its symbol for a weighted lever's spring cord all the way over to the left side of the second DT portrait during the course of the next few days, I finally realized the number value hidden at the end of its path was actually the distance, in inches, between the center axes of the drum's spring hook attachment anchor pin and the axle of the 3 feet diameter Gera prototype wheel!

To get that hidden value, one simply multiplies the number 4 represented by the combination of the "L" square and the lower arm of the duckbill goniometer by the number 5 represented by the Roman numeral "V" shape of the goniometer to get 4 x 5 = 20. Since the protractor to the left of the goniometer's pivot screw or bolt looks like a capital "d" or "D", we use its normal alphanumeric value of 4 and then *subtract* it from the other value we derived to get 20 – 4 = 16.

Next, following the line segment downward and to the left from the right side leg of the closed divider that aligns with the base of the protractor, we arrive at the "e" in the world "*Werck*" and give that letter a value of 0.5 because the pointed leg of the divider points to the left and that tells us to move the decimal point one place to the left of the 5. Finally, we simply add these values to get 16 + 0.5 = 16.5. Adding the units of inches then turns this into 16.5 inches. This then was the distance of the center axis of a spring hook attachment anchor pin from the center axis of the axle in the Gera prototype wheel and the value that actually worked in my computer models of Bessler's first working one-directional wheel. Since the spring hook attachment anchor pin's ends are embedded in the weighed lever's leading parallel pair of outer octagonal frame pieces at their midpoints, this distance was the one that would be measured from the center axis of the axle to the center axes of the holes drilled into the midpoints of the outer octagonal frame pieces into which the weighted lever's spring hook attachment anchor pin had been inserted.

With this information, I decided to revisit the screw barrel microscope's optical axis which is represented by the dashed vertical white line on the right side of the second DT portrait in Figure 21(b). I again followed that axis up to where it intercepted the 4:30 position of the dial of the clock hung on the side of the organ in the background. Then I realized what the true meaning of this clue was. The time of 4:30 pm, when counting from midnight and using a 24 hour clock (sometimes referred to as "military time"), could be written as 16:30 hours or 16 hours and 30 minutes. Since 30 minutes equal 0.5 hours, this could also be written as 16.5 hours! If we substitute the units of inches for the units of hours, then we get 16.5 inches which is the *same* value obtained by the zigzagging line that ended over on the left side of the portrait at the "e" in "*Werck*"!

Thinking it might just be a coincidence, I decided to follow the vertical optical axis of the screw barrel microscope downward to the various letters at the end of the two lines of text below the bottom right corner of the second DT portrait. I wondered if I could also derive an alphanumeric value of 16.5 from them.

The optical axis of the screw barrel microscope, after leaving the bottom border of the portrait, immediately contacted the tightly curled dot at the top of the "t" at the end of the word "*vollendet*" at the end of the first line of text under the portrait.

Next, I followed the virgule or right leaning slash mark to the right of the "et" at the end of the word "*vollendet*" downward and noticed that it pointed to the letter "N" in the letter group "*NB*" located directly below the letter group "et". The letter "N" looks, after breaking it up into two pieces, like the Roman numerals "I" for 1 and "V" for 5 placed next to each other as in "IV". "IV" has a numerical value of 4 because, when a smaller value Roman numeral is placed to the left side of a larger value Roman numeral, we must subtract the lower value from the higher value to obtain the final value of the group of Roman numerals which means we can write "IV" as 5 − 1 = 4 and assign that value of 4 to the Roman numeral "IV". I then decided to try this method of subtraction with the letters in "et" at the end of the word "*vollendet*" on the first line of text. Since "e" has an alphanumeric value of 5 and "t" a value of 20, I wrote 20 − 5 = 15. Then, because of the lower, arrowhead-like end of the virgule pointing to the abbreviation "NB", I simply placed a decimal point between the separate "I" and "V" pieces that could be made from the single letter "N" in "*NB*" which, as Roman numerals, could then be written as I.V even though Roman numerals used no decimal points! After substituting in their Arabic numerals equivalents, I.V then became 1.5. Adding this value to the value of 15 previously obtained, alphanumerically, from the "et" on the line above the "NB" then gave me 15 + 1.5 = 16.5 and, adding the units of inches, gave me 16.5 inches!

So, there is was. *Three* separate but *identical* distance parameter values for the location of the center axis of a spring hook attachment anchor pin relative to the Gera prototype's axle center axis gotten from the optical axis of the screw barrel microscope in the second DT portrait. I realized the probability of this happening by chance was next to zero and it, therefore, most highly likely had to have been deliberately and carefully placed into the second portrait by Bessler for future reverse engineers to find.

Next, we can move on to the stop cord, one of whose end hooks is attached to the stop cord hook attachment pin whose center axis is represented in the circular sector paper piece by the second farthest intersection point from its origin on the right straight radial edge of the paper piece and is also known as the A2 stop cord hook attachment pin in the nomenclature system I use to identify these pins.

The straight diagonal line that extends upward and to the right away from that intersection point exits through a small gap in the fifth arc whose radius is 2.8 inches = 2-12.8/16 inches. Again, I extended a straight line segment from the pointed tip of the upper duckbill goniometer arm downward through the point where the line segment on the paper piece representing the stop cord passed through the gap in the fifth arc, and finally extended it until it reached the letter "o" in the word "*vollendet*" on the first line of text under the second DT portrait which is indicated by the black solid circle "C" label in Figure 21(b). Analyzing this clue was actually one of the easiest of things to do.

My previous work with computer models of the Gera prototype wheel indicated that the stop cord, when taut, would only allow the center axes of the 9:00 weighted lever's A2 stop cord hook attachment pin and the stop cord hook attachment anchor pin embedded at the midpoints of the lever's leading parallel pair of inner octagonal frame pieces to reach a maximum separation distance of 6 inches from each other. The alphanumeric value of the letter "o" is 15 because it is the fifteenth letter of the alphabet. If one adds the individual digits in 15 he gets 1 + 5 = 6 and, adding the distance units of inches, this then becomes 6 inches.

But, I wondered if there was also a clue hidden in the second DT portrait that would provide me with the center axis to center axis distance between the stop cord hook attachment *anchor* pin and the axle. I already knew that this distance value was *exactly* 12.934 inches. (This distance value is gotten by using simple trigonometry. One multiplies the center axis to center axis distance between one of the Gera prototype wheel's weighted lever's pivot pins and the drum's axle which is exactly 14 inches by the cosine of 22.5° to get 14 inches x cosine 22.5° = 14 inches x 0.92388 = 12.934 inches.)

The clues whose interpretation led to the separation distance parameter value of 12.934 inches, however, proved to be some of the most difficult to find in the second portrait. But, with much patience they can be found. Here's what they are.

One must continue to extend the line segment representing the stop cord in the circular sector paper piece shown in Figure 21(b) diagonally upward and to the right until it reaches the *upper* edge of the 14 inch ruler that is carefully balanced on the carpenter's plumb bob. At that point, the line segment ends and a second one is started and extended to the left along the top end of the ruler. When that second line segment reaches the outside corner of the "L" square, it ends and third one is started that is extended diagonally downward and toward the left side of the second DT portrait. As it is extended, it must just touch the outer curving edge of the semi-circular protractor so that it is tangent to that edge. The line segment is then extended farther until it passes below the bottom border of the portrait and finally ends at a small dot that is part of the letter "Z" in word "*Zeügs*" on the first line of text below the portrait as shown in Figure 12(b). The area where this occurs is near the black solid circle "D" label in the figure.

"Z" has an alphanumeric value of 26 because it is the last letter of the 26 letter English and German alphabets. Notice that there is a letter on the line of text below the "Z" that has had a part of it crossed out by Bessler. One might think that was done to correct a grammatical mistake by changing the letter "m" into "n" so that the German word "*ihn*" would be used instead of "*ihm*". But, that's not the only reason. Bessler wants to put a noticeable "X" into the portrait text there to attract the reverse engineer's attention and, once again, uses the standard "X marks the spot" notation of treasure maps (remember he was once a treasure hunter) to indicate there is a "treasure" to be found in this portion of the text with that treasure being important information about where the center axis of the stop cord hook attachment anchor pin must be located relative to the center axis of the Gera prototype wheel's axle.

By making the change in letters, the inventor tells us to use the alphanumeric value of "n" which is 14 instead of the value of "m" which is 13. Why?

The reason is to make sure that when we perform the subtraction of the alphanumeric value of "n" from that of "Z", suggested by the removal or *subtraction* of the right part of the letter "m" with the "X" that crosses it out, we will get 26 – 14 = 12. If this had been done with the value of "m" instead, we would have gotten 26 – 13 = 13 which, while it is a number associated with God, does not have the first two digits, 12, of the value we seek which is 12.934.

At this point a reverse engineer of Bessler's imbalanced pm wheel mechanics has the first two digits he needs, but what about the remaining 0.934 that must be added to the 12?

Aside from the anomalous "X" for the letter correction on the second line of text near the black solid circle "D" label in Figure 21(b), there is something else there that is unusual. It's that swirling and looping flourish that extends all the way from the bottom of the letter "g" near the end of the word "*Zeügs*" over to the bottom of the first letter "Z" in that same word and which then makes contact with that little dot that is part of the letter "Z". Along the way, however, that elaborate flourish from the letter "g" also touches another flourish that had been added to the top of the "h" in the corrected word "*ihn*". That second flourish seems a bit inappropriate because it does not issue from the end of the diagonal serif at the top of the "h" as one might expect, but, rather, separately from the vertical stroke of the "h". How can one interpret all of this?

The contact between the odd flourish at the top of the letter "h" in the corrected word "*ihn*" and the flourish from the bottom of the letter "g" that touches the letter "Z" in "*Zeügs*" tells us that we must add some new value derived from the "h" to the value we previously obtained, which was 12, by subtracting the value of the letter "n from the value of the "Z" above it. That new value we must add to the 12, as mentioned above, has to be *exactly* 0.934. To obtain it, we note that the alphanumeric value of "h" is 8 and that of the "i" to the left of it is normally 9. However, we treat the letter "i" as though it is the Roman numeral "I" and assign it a value of 1. Then we add the Roman numeral value of the "i" to the alphanumeric value of "h" to get 1 + 8 = 9. Because of the dot in the "i" which is to the left of the letter "h", we move the decimal point one place to the left of the derived value of 9 to get 0.9.

Next, we take note of that odd flourish that issues from the vertical stroke of the letter "h" and which connected that letter to the flourish from the "g" that touches the "Z" in "Zeügs". That flourish from the "h" also loops its way along to the left under the last four letters in the word "*Werck*" on the first line of text, but, in doing so, it seems to avoid the letter "k" in the word "*Werck*". The meaning of this is obvious.

We must use the alphanumeric values of the letter group "erc" in "*Werck*", but *not* include the value of the letter "k" in that group. We use the normal value of "c" which is 3 and, by adding the individual digits in the normal value of "r" which is 18, we can convert the value of "r" into 1 + 8 = 9. By then subtracting, using the Roman numeral value determination method, the value of the normal value of "e", which is 5, from that of the derived value of "r", which is 9, we can get a derived value for the group "er" which is 9 – 5 = 4. Placing the derived value of the "er" group to the right of the normal alphanumeric value of "c" we get the number 34. (This reversal of the ordering of the values is necessary because Bessler wanted the portrait analyst to read the alphanumeric values of the "c" and "er" in the word "*Werck*" to in reverse order from right to left starting from the letter "Z" the word "Zeügs".) Note the two loops and the final third dot-like curl in the flourish under the letter group "erc" in the word "Werck". This tells us that we must move the decimal point three positions to the left of the number 34 to obtain 0.034.

Finally, we have three derived alphanumeric values which must be added together and we now do this to obtain 12 + 0.9 + 0.034 = 12.934 and, with the inch units added, this becomes 12.934 inches.

Admittedly, deriving this value, which, again, is for the center axis to center axis separation distance between the stop cord hook attachment anchor pins and the axle in the 3 feet diameter, Gera prototype wheel, is a challenge and pushes the portrait analyst's decoding abilities to their limits. Bessler purposely made it difficult because he knew how valuable the information it supplied would be to a reverse engineer hoping to successfully duplicate one of the inventor's wheels someday. Yet, he does provide many subtle hints along the way to aid in the derivation of this distance parameter. Additionally, I must again remind the reader that this distance value, as all of the others derived from the various clues in the second DT portrait (as well as in the first DT portrait to be treated in the next chapter) must be multiplied by a factor of 4 in order to use them inside the drum's of replicas of the Merseburg and Kassel wheels.

By now the reader may realize that *each* of the four types of coordinating cords found in Bessler's imbalanced pm wheels had *three* distinct distance parameter values associated with it.

The first distance parameter value, relative to the center axis of a weighted lever's pivot pin, gave the location of the center axis of an A arms' coordinating cord hook attachment pin, either an A1 spring cord, A2 stop cord, A3 long lifter cord, or A4 main cord hook attachment pin, to which one end of a particular type of coordinating cord was attached by a small metal hook.

The second distance parameter value, relative to the center axis of a *leading* weighted lever's pivot pin, gave the location of the center axis of the L1 main cord or B1 long lifter cord hook attachment pin to which the other end of a main cord or long lifter cord from a lagging weighted lever was attached by a small metal hook. The second distance parameter value, relative to the center axis of wheel's *axle*, could also give the location of the center axis of a drum's leading hook attachment *anchor* pin to which a stop cord or an extension spring was attached by a small metal hook.

And the third distance parameter value was the maximum allowed *separation* distance between the center axes of *two* adjacent weighted levers' A4 and L1 main cord hook attachment pins that were connected to each other by a stretched and taut main cord and its two metal end hooks or between the center axes of two less adjacent weighted levers' A3 and B1 long lifter cord hook attachment pins that were connected to each other by a stretched and taut long lifter cord and its two metal end hooks.

The third distance parameter value could also be the maximum allowed separation distance between the center axes of a weighted lever's stop cord hook attachment pin and the lever's leading stop cord hook attachment anchor pin permitted by a stretched and taut stop cord and its two metal end hooks or between the center axes of a weighted lever's spring cord hook attachment pin and the lever's leading spring hook attachment anchor pin that was permitted by the lever's *stop* cord so as not to overstretch an extension spring and thereby damage it.

Thus, for the four types of coordinating cords found in one of Bessler's one-directional wheels, there were actually a *dozen* different distance parameters that needed to be known by anyone attempting to duplicate its hidden imbalanced pm mechanics!

Basically, the reverse engineer needed to know what the four types of coordinating cords were and what their individual functions were. Once he knew that, then he needed to know the locations of the center axes of the two attachment pins to which the two metal hooks tied to the two ends of each cord were attached. And, finally, for each of the four types of coordinating cords, he needed to know the maximum allowed separation distance between the center axes of the two attachment pins that the cord's end hooks were attached to when the cord was taut or, in the case of the spring cord, its attached extension spring had stretched through some maximum allowed distance. The inability to determine these details, obviously, was the major reason why Bessler's secret imbalanced pm wheel mechanics were not previously rediscovered.

In the alphanumerical analyses previously done using Figure 21(a), it was possible to determine six of the dozen possible distance parameter values for the four types of coordinating cords. So far, for the analyses done using Figure 21(b), another four of the dozen possible distance parameters values for the four types of coordinating cords have been determined. That's a total of ten of the dozen possible distance parameter values for the four types of cords. The two remaining distance parameter values that need to be determined are for the maximum separation distances allowed by the long lifter and main cords located between the center axes of the weighted lever hook attachment pins that they connect when those cords are stretched and taut.

Let us begin by determining the maximum allowed separation distance between the center axes of one lever's A3 long lifter cord hook attachment pin and another leading lever's B1 long lifter cord hook attachment pin by a stretched and taut long lifter cord and its two attached metal end hooks that connected the two pins.

To obtain this separation distance parameter value for the long lifter cord represented by a black line segment drawn on the second DT portrait's circular sector paper piece, we extend both ends of *another* white line segment drawn *over* the length of that cord's black line segment as shown in Figure 21(b) (note that since the black line segment of the circular sector paper piece is covered by the white line segment in this figure, it is not visible as it was in Figure 21(a)). The left end of the white line segment extends to the left and rises slightly until its end intersects the origin of the semicircular protractor to the immediate left of the duckbill goniometer's pivot screw or bolt. The right end of the white line segment drawn on top of the paper piece's black line segment representing the long lifter cord is then extended toward the right side of the portrait and slightly downward until its end reaches the point at which the bottom left side of the smallest lead ingot just touches the right visible foot of the tripodal metal lamp. That intersection point is near the white solid circle "E" label in the figure.

At first glance, it might seem impossible to get any useful numerical information out this extended white line segment, but the analysis is actually far easier than one might suspect. The "D" shaped protractor, because it resembles the capital of the fourth letter of the alphabet, is given an alphanumeric value of 4. Next, notice that the "X" formed by the lower goniometer arm and the "L" square is also the symbol for the arithmetic operation of multiplication and that indicates we must perform a multiplication. The extended right end of the white line segment as it moved toward the smallest ingot first reached the point of the upper inner corner of the angle formed by the two visible feet of the upright metal lamp and we can take the two feet as representing an inverted "V" and treat it like the Roman numeral "V" by giving it a numerical value of 5. We then perform the suggested multiplication and get 4 x 5 = 20.

Finally, we can view the smallest vertical lead ingot whose lower end touches the right side of the right visible foot of the lamp as being a symbol for the Roman numeral "I" to which we give a numerical value of 1, but, because the *point* of contact of the ingot is on its left side, we treat the contact point like a decimal point and change the value to 0.1. Passing directly over the midpoint of the smallest lead ingot, there is the largest horizontal lead ingot that the three vertical ingots support and lift off of the tabletop. That largest horizontal ingot is a symbol for the minus sign which signals us that a subtraction must be performed. To do that, we write 20 − 0.1 = 19.9 and add the units of inches to obtain 19.9 inches. This value is the maximum allowed center axis to center axis distance between the A3 long lifter cord hook attachment pin, in either coordinating cord layer 2 or 4, held by the A arms of the 7:30 weighted lever and the B1 long lifter cord hook attachment pin held by the B arms of the 10:30 weighted lever in the Gera prototype wheel during about the first third of each 45° segment of clockwise drum rotation when the long lifter cord was taut and the counterclockwise rotation of the weighted lever about its pivot pin leaving the drum's 7:30 position was being used to rapidly rotate the weighted lever leaving the drum's 10:30 position clockwise about its pivot pin. And, as always, this distance parameter for the long lifter cords must be multiplied by a factor of 4 in order to be applied to the long lifter cords used in the bidirectional Merseburg and Kassel wheels.

There is now only the need to find the remaining allowed center axis to center axis distance for the main cords in Bessler's 3 feet diameter Gera prototype wheel.

We again draw a straight white line segment on top of the straight black line segment on the second DT portrait's circular sector paper piece that represents the main cord. We extend the left end of that white line segment until it reaches the origin of the "D" shaped protractor located to the left of the duckbill goniometer's pivot screw or bolt. The right end of the white line is next extended to the right of the portrait and appears to pass in front of the bottom of the cylindrical vertical support piece of the lamp. Finally, we stop extending the white line segment when it reaches the left side of the lower portion of the smallest lead ingot that is visible below the larger horizontal ingot that the three vertically oriented ingots together support and lift off of the surface of the table top. The region in the second portrait where this happens is indicated in Figure 21(b) by the white solid circle "F" label.

Again, this looks like an assortment of symbols which is completely random and from which no useful numerical information could be extracted. Starting from the left, there is the "D" shaped protractor, the lower arm of the duckbill goniometer resting on top of the longer arm of the "L" square, the base of the lamp, and, finally, the smallest lead ingot of the three vertically oriented ingots. By now the reader may already be anticipating how numerical values can be gotten from these items.

Here's how.

Because of its shape like a tilted capital letter "D", we give the protractor a numerical value of 4. Since the goniometer arm resting on the "L" square actually does look like the number 4, even though rotated clockwise through over 90°, we also give it a numerical value of 4. The white line segment appears to pass in front of the lamp's cylindrical vertical support piece in the perspective view of the second DT portrait, but, since the circular sector paper piece actually lies flat on the shop's table top, the line segment should really be shown extending to the right and momentarily disappearing from the reader's view as it passes *behind* the lamp's cylindrical vertical support piece and *under* the unseen *third* tripodal foot of the lamp. Because of this we can assign the unseen third foot of the lamp a numerical value of 3.

Finally, the extended right end of the white line segment touches the side of the smallest lead ingot near a point close to the larger horizontal ingot which is supported by all three of the vertically oriented ingots. This suggests that we must use *all* three of these ingots together. In the previous determination of the symbolism of the smallest ingot, it was treated like the Roman numeral "I" and given a numerical value of 1. Now, we combine all three of the vertically oriented ingots and treat them like the Roman numeral "III" which has a numerical value of 3. Because the smallest ingot touches the right foot of the lamp at a single point on the ingot's *left* side, we use this as a symbol for a decimal point and assign the numerical value of 0.3 to the three vertically oriented lead ingots.

But, how are we to manipulate these four numerical values arithmetically?

The simplest thing to do is just add the four values together and this is suggested the symbolism of the larger horizontal lead ingot that lies on top of the three smaller vertically oriented ingots. Indeed, the four ingots together actually form three "+" signs that are in a horizontal row! So, we can now write $4 + 4 + 3 + 0.3 = 11.3$. Adding the units of inches gives us 11.3 inches which was the maximum separation distance allowed by a main cord and its two attached end hooks between the center axes of a weighted lever's A4 main cord hook attachment pin and the adjacent leading lever's L1 main cord hook attachment pin when the cord was taut inside of Bessler's one-directional, 3 feet diameter Gera prototype wheel. And, yet again, we will need to multiply this value by a factor of 4 for the main cords found in the bidirectional Merseburg and Kassel wheels.

This now completes the derivation of the dozen distance parameter values associated with the four types of coordinating cords found in the inventor's Gera prototype wheel and demonstrates the great importance of the circular sector paper piece shown in the second DT portrait. While all of the dozen numerical values are shown alphanumerically, geometrically, and symbolically elsewhere in the second portrait, that seemingly unimportant paper piece serves as a sort of directional compass that helps point the reverse engineer of Bessler's secret imbalanced pm wheel mechanics toward them all.

We now leave the discussion of Figure 21(b) and move on to Figure 22(a) which the reader should locate at the end of this chapter and study.

Figure 22(a) shows the lower right corner of the second DT portrait which is rich with numerical clues about the extension springs used in Bessler's Gera prototype, Merseburg, and Kassel wheels, the masses of their weighted levers' end weights and unweighted levers, and the separation distances between the centers of gravity of the end weights and the center axes of their levers' pivot pins. Unfortunately, the congestion of these various clues makes this area of the second portrait one of the most difficult to analyze, but it is well worth the effort.

The first thing the DT portrait analyst will notice is the presence of the screw barrel microscope that dominates the items in the portrait's corner and the letters under it. Since Bessler uses the optical axes of lenses to direct the reader's attention to various parts of the second portrait, let's do as he wants us to and extend a white line segment, indicated by the white solid circle "A" label, downward along the optical axis of the microscope. The white line segment runs along the right side of the middle, vertically oriented lead ingot and, upon exiting the bottom border of the portrait, arrives at the tightly curled top of the letter "t" in the word *vollendet* at the end of the first line of text under the portrait.

Since the microscope contains a tensioning spring and the middle vertically oriented lead ingot represents a weighted lever found in the inventor's Merseburg wheel, we are getting some information about the two extension springs attached to each of those levers in that 12 feet diameter bidirectional wheel. "t" has an alphanumeric value of 20 and, by adding the spring constant dimensions of pounds per inch to it, we obtain 20 pounds per inch for the *total* spring constant of the *two* parallel extension springs attached to each of the Merseburg wheel's sixteen weighted levers. The letter "e" is next to and touches the letter "t" and there are multiple point symbols for a decimal point near it, so we can alter the value of "e" and use a value of 0.5 for the letter instead of its normal alphanumeric value of 5. Since the "e" touches the "t", we must perform an arithmetic operation. If we choose multiplication, then we can write 0.5 x 20 = 10 and, supplying the appropriate spring constant dimensions, this becomes 10 pounds per inch which is the spring constant value for a *single* extension spring attached to one of the Merseburg wheel's weighted levers.

At this time the reader may notice that I added two short white line segments to the ends of two obvious short black line segments attached to the right end of the 14 inch ruler, which is carefully balanced on top of the carpenter's plumb bob, so they can reach two of the lead ingot weights. The bottom short white line segment touches the left side of the middle lead ingot and the upper much longer white line segment touches the left side of the right lead ingot. Since the middle ingot represents a weighted lever in the Merseburg wheel, the larger one to its right must represent a weighted lever in the Kassel wheel which was the next, more massive bidirectional wheel that Bessler constructed. But, why do those two short black line segments extending away from the right end of the 14 inch ruler point to these symbols for two weighted levers with different masses? The answer is obvious. This clue was intended to tell a future reverse engineer that the two different types of weighted levers used in the two bidirectional wheels both used the *same* distance, 14 inches, between the centers of gravity of their trios of cylindrical lead end weights at the ends of their main arms and the center axes of their levers' pivot pins.

Next, we move on to another vertical white line segment that extends downward from the outer end of the spring-like handle attached to the right side of the lamp's conical cover, runs along the center of the smallest leftmost lead ingot weight where the white solid circle "B" label is located, passes through the bottom border the portrait, passes through the pointed top tip of the letter "d" in "*vollendet*" on the first line of text below the second portrait, and, finally, reaches and stops at the top dot of the two vertically arranged dots in the colon to the left of the abbreviation "*NB*" at the end of the second line of text below the portrait.

Since the descending white vertical line segment begins at the spring-like handle attached to the lamp's conical cover, we know that it will be providing us with some information about the extension springs found in the 3 feet diameter Gera prototype wheel. As the line segment descends downward from the handle on the cover, it passes through the right end of the 14 inch ruler that is visible on the right side of the vertical cylindrical support piece of the lamp. Notice that we only see 3.5 inches = 3-8/16 inches of the end of the ruler. That length is the exact separation distance between the center of gravity of a lead ingot weight at the end of the main arms of a weighted lever in the Gera prototype wheel and the center axis of the lever's pivot pin. So, we actually have double confirmation that we will be receiving information about an extension spring used in the Gera prototype wheel.

Yet another confirmation that the Gera prototype wheel is the focus of the items that this line segment passes near or through is provided by the oddly formed "n" to the left of the "d" in the word "*vollendet*" on the first line of text under the second DT portrait. This "n" appears to be made from the top half of a letter "n" joined to the bottom half of a letter "u"! This joining of the parts of both letters suggests that an addition operation should be performed on the two normal alphanumeric values of these letters which then gives us 14 + 21 = 35. Moving the decimal point one place to the left and adding distance units of inches then gives us 3.5 inches = 3-8/16 inches which, as mentioned above, is the exact separation distance between the center of gravity of one of the Gera prototype wheel's lead ingot end weights and its lever's pivot pin. Passing a right leaning diagonal line segment through the odd letter "n" can then split it into two letter "r's" that are in contact with each other. The "r" on right side, however, is inverted from its normal orientation. The two "r's" each have a normal alphanumeric value of 18 and adding them together gives us 36. Adding the units of inches to this gives us 36 inches which just so happens to be the diameter, in inches, of Bessler's one-directional, 3 feet diameter Gera prototype wheel.

There is also another interesting little "mistake" that occurs near the white solid circle "B" label that I need to point out. Notice the trapezoidal shape of the left end of the largest horizontally oriented lead ingot and how its smaller base points toward the lamp. To the immediate left of it one sees a decorative flange on the bottom of the lamp's vertical cylindrical support piece. But, there seems to be something wrong with the engraved image used for the printed image. The right side of the flange does not go completely around to the hidden side of the vertical cylindrical support piece as it does on the left side of the flange. The result is that the right side of the flange resembles a tiny hook!

This was no mistake, but yet another clue from Bessler to tell someone trying to duplicate his 3 feet diameter Gera prototype wheel that the smaller base of the lead ingots used at the ends of its weighed levers must face in toward the main coordinating cord hook attached to a lever's L1 main cord hook attachment pin that was held between the main or L arms of the lever and, thus, toward a lever's pivot pin. This orientation of the ingots was illustrated in Figures 4 and 7(a) at the end of Chapter 5 and was intended by the inventor to minimize the wear to the trailing main cord of a weighted lever whose pivot pin's center was approaching the drum's 9:00 position when the cord made momentary rubbing contact with the lead ingot end weight as the center lines of the lever's main arms swung counterclockwise around the lever's pivot pin and away from the center lines of the lever's parallel pair of radial frame pieces so as to form an angle of exactly 62.5° between the two sets of center lines.

With a free running, maximum terminal rotation rate of over 60 rotations per minute, *each* of the drum's eight main cords would be experiencing this momentary rubbing against its leading lever's lead ingot weight at a rate of over once per second! Bessler probably first tried mounting the lead ingots with their smaller bases facing out away from a weighted lever's pivot pin, but discovered that orientation caused the main cords to rub up against a raised lengthwise edge of the ingot and thereby quickly fail in use.

Below the second portrait, the white line segment with the white solid circle "B" label near it turns black and passes through the top pointed tip of the letter "d" and finally ends at the top dot of the colon on the second line of text. We can assign a value of 0.4 to the "d" because the line segment passes exactly through its pointed tip which is a symbol that tells us a decimal point is required.

On the second line of text, one will notice that Bessler made the left vertical part of the letter "N" in the abbreviation "*NB*" darker and bolder than the diagonal piece of the letter and the intention was for the reverse engineer to use that left vertical part of the letter "N" as the Roman numeral "I" and give it a value of 1. To the left of the colon, we see the letter group "et" and if we use one of the dots in the colon to justify a decimal point shift for the "t" we could give that letter a value of 2 instead of its normal alphanumeric value of 20. We could then, by eliminating the colon, write the three number values for the letter group "etI" as 521. Since Roman numerals have a tendency to reverse the expected ordering of the Arabic numerals that would be used to express numbers, we then rewrite the three values we have next to each other as 125 in order to convert to the Arabic method for ordering the individual digits within a numeral. Since we already used one of the dots in the colon to justify the shifting of the decimal point in the value of "t" to make it 2 instead of 20, let us now use the remaining dot to justify turning the reversed order, Arabic numeral 125 we obtained into a fraction and give it a value of 0.125.

We now have two number values that, because of the vertical black line segment that connects their sources, require us to perform an arithmetic operation on them. One value is the 0.4 gotten from the letter "d" at the end of the word "*vollendet*" on the first line of text below the second portrait and the other value is the 0.125 gotten from the "etI" on the second line of text. If we subtract the second value from the first as suggested by their vertical alignment, we get 0.4 − 0.125 = 0.375. Adding the units of pounds then gives us 0.375 pounds = 6 ounces and this is the mass of each of the lead ingot weights attached to the ends of the main arms of the weighted levers inside of the Gera prototype wheel!

The mass of the *un*weighted levers used in the Gera prototype wheel was 0.078125 pounds = 1.25 ounces each and, obviously, the value in pounds is too long a string of digits to use alphanumeric coding to directly express. However, this value can be obtained by simply multiplying the value derived for the letter group "etI" by a factor of 0.625 or 0.125 x 0.625 = 0.078125 which, adding the mass units, becomes 0.078125 pounds. Can we find the factor 0.625 anywhere near the letter group "etI" in the bottom right corner of the second DT portrait? Yes, it's there.

We simply take the larger letter group "*NB*" at the end of the second line of text and break the "N" into two pieces or "I" and "V". The "IV" is treated like the Roman numeral for 4 and the "B" is given its normal value of 2 since it is the second letter of the alphabet. Adding them gives 4 + 2 = 6 and we use the left pointing arrowhead at the bottom of the virgule at the end of the first line of text to justify moving the decimal point of the 6 one place to the left to obtain 0.6. Directly vertically above the "V" and "B" on the second line of text, we find the letter group "et". We give the individual letters in "et" their normal alphanumeric values and then add them to get 5 + 20 = 25. Since there is one dot at the top and two pointed serifs below and to the right of it in the letter "t" we use that to justify moving the decimal point in 25 *three* places to the left to give us 0.025. Finally, we add the two derived values for the letter groups "NB" and "et" together to get 0.6 + 0.025 = 0.625 which is the factor needed to obtain the unweighted lever mass in a Gera prototype wheel from the derived value of the letter group "etI" which was shown above to be 0.125.

Next, we must consider the line segment in Figure 22(a) that has the white solid circle "C" label near its lower end. It starts at the lowermost corner of the trapezoidal cross section of the upper end of the vertically oriented lead ingot on the right which represents a weighted lever in the Kassel wheel, extends diagonally downward and to the left from that point through the contact point that the bottom end of the smallest ingot on the left makes with the right visible foot of the lamp, and finally exits the bottom border of the portrait, turns black, and reaches and ends at the tiny serif on the left side of the middle vertical portion of the second letter "l" in the word "*vollendet*" on the first line of text under the portrait. Actually, it reaches the serif in the "l" of the group of connected letters "lle". If we add the normal alphanumeric values of these three letters, we get 12 + 12 + 5 = 29 and, adding mass units of pounds, the sum becomes 29 pounds. The reader will recognize this as the mass of a single weighted lever in the bidirectional Kassel wheel.

Note that the pointed upper tip of the "d" in the word "*fchändet*" on the second line of text, which was previously used to indicate the insertion of a decimal point, can now be considered to indicate that we must separate the letter "e" from the letters "ll" since the tip looks a little like a small knife cutting through the connection between the "ll" and the "e". When we do that, we then have two values which are 12 + 12 = 24 and 5. Adding the mass units of pounds to these then gives us 24 pounds and 5 pounds. The 24 pounds was the total mass of the three 8 pounds cylindrical lead end weights at the end of each weighted lever in the Kassel wheel and the 5 pounds was the mass of the unweighted lever which was the mass of a lever *before* the three end weights and their attachment bolts were attached to the lever.

If we add the individual digits in the normal alphanumeric value of "l", which is 12, we get 1 + 2 = 3. Using 3 as the altered alphanumeric value for "l" we can then evaluate the value of the letter group "lle" by writing (3 x 3) + 5 = 9 + 5 = 14 which, with distance units of inches added, becomes 14 inches or the center axis to center axis distance between center of gravity of the trio of 8 pound cylindrical lead end weights at the end of a Kassel wheel weighted lever (located on the center axis of the middle weight's attachment bolt) and the pivot pin of the lever. This distance of 14 inches is also symbolized by the *marked* length of the 14 inch ruler balanced on the carpenter's plumb bob in the second DT portrait (perusal of the ruler, however, will indicate that, unlike modern rulers, its total length is a fraction of an inch longer than 14 inches).

To obtain information about the two extension springs attached to each of the weighted levers inside the drum of the Kassel wheel, we again extend a white line segment down from the lowest point of the upper trapezoidal end of the right ingot that represents a weighted lever in that wheel. This time, however, the line segment is vertical and extends down along the frontmost vertical straight edge of the ingot until it passes the bottom border of the second DT portrait, turns black and reaches the top of the virgule at the end of the first line of text near the black solid circular "CC" label. Another short black diagonal line segment is then extended down and to the left along the length of the virgule until it reaches the single vertical piece shared by the right side of the "N" and the left side of the "B" in the abbreviation "*NB*" on the second line of text under the portrait. The analysis of this clue is somewhat complicated.

We treat the "N" like the Roman numeral IV and give it a value of 4. To the left of the "*NB*", we notice the letter group "et" and that the top of the "t" does not end in a tightly curled dot as it does in the "t" in "et" at the end of the word "*vollendet*" on the *first* line of text. Rather, the "t" in the "et" at the end of the word "*fchändet*" on the *second* line of text looks like a curled dot that had been unwound and then *stretched* out until its end was attached to another dot between the "e" and the "t". This is an obvious symbol for a stretched extension spring! The dot between the "e" and "t" and to the left side of the "t" is taken as meaning that we must move the decimal point of the normal value of "t", which is 20, one place to the left to convert the value to 2. Next, we multiply the connected letters in "et" to obtain 5 x 2 = 10. Note the colon or ":" symbol between the "et" and the "N". The upper dot of the colon is also a symbol for the arithmetic operation of multiplication and was introduced by Gottfried Leibniz in 1698 and, no doubt, Bessler would have been familiar with it. We can use it as a symbol to indicate we must multiply the values for "et" and "N" to get 10 x 4 = 40. Adding the spring constant units of pounds per square inch then gives us 40 pounds per inch which is the *total* spring constant for each *pair* of parallel extension springs attached to a weighted lever in the Kassel wheel.

For the spring constant of a single spring, we look at the virgule at the end of the first line of text that points diagonally downward and toward the left to the place where the "N" and the "B" in the abbreviation "*NB*" overlap. The virgule is almost identical to the mathematical symbol known as a "solidus" or division slash introduced into mathematics in 1718 by Thomas Twinning and another symbol Bessler would have been familiar with. We previously saw how a value of 40 could be derived from the letter group "et:N" that is to the left of the letter "B" which has a normal alphanumeric value of 2. We then perform the suggested division and get (etN) / B = 40 / 2 = 20. Adding the spring constant units of pounds per inch turns this into 20 pounds per inch which is the spring constant of a *single* extension spring found in the bidirectional Kassel wheel.

Finally, we need to analyze the encoded information for the weighted levers in the Merseburg wheel which are gotten by using the middle lead ingot in the second DT portrait.

Again, we extend a white line segment diagonally downward and to the left from the bottom corner point of the trapezoidal cross section at the top of the middle vertically oriented ingot. That line segment, whose bottom end is near the black solid circle "D" label, again passes through the contact point between the left side of the bottom end of the smallest lead ingot and the visible right foot of the lamp, then turns black as it extends past the portrait's bottom border to pass through the "e" in the letter group "lle" in "*vollendet*" on the first line of text under the portrait, then through the "d" under that letter group, and, finally, reaches and ends at the serif at the bottom right of the letter "n" in "*fchändet*" on the second line of text below the second portrait.

We notice that the "ä" and "n" in "*fchändet*" are connected and form the letter group "än" which means some arithmetic operation involving their two alphanumeric values will be needed. There is also an odd looking umlaut symbol placed over the "ä" to the left of the "n" that is made up of two right leaning *parallel* slashes (which, unfortunately, are partially obscured by the solid black circle "D" label in Figure 22(a) so that the reader will need to view Figure 21(b) to better see them). This is a symbol for a parallel pair of springs attached to a weighted lever in the Merseburg wheel. If we add the normal value of the "n" in the "än" group to the altered value of the "e" in the "lle" group (gotten by the pointed tip of the "d" in "*fchändet*" suggesting we move the decimal point one place to the left of the normal value, 5, of "e"), we will get 14 + 0.5 = 14.5. Adding the mass units of pounds then gives us 14.5 pounds which is the mass of a weighted lever inside of the Merseburg wheel. To get the mass of the *three* 4 pounds cylindrical lead end weights and their attachment bolts attached to the ends of the main arms in each of these weighted levers, we simply multiply the altered value of the "e" in "lle" by the sum of the normal values of the two "l's" in "lle" to get 0.5 x (12 + 12) = 0.5 x 24 = 12. Adding the mass units of pounds turns this into 12 pounds. By subtracting the mass of the three end weights from the total mass of the weighted lever we get 14.5 pounds − 12 pounds = 2.5 pounds which is the mass of an unweighted lever in the Merseburg wheel.

But what about the total spring constant of the *two* extension springs attached to each of the weighted levers in the Merseburg wheel?

To get its value, we add the normal alphanumeric values of the connected letters in the letter group "än" in the word "*fchändet*" on the second line of text to the normal value of the "e" in the letter group "lle" in the word "*vollendet*" on the first line of text below the second DT portrait. That yields 1 + 14 + 5 = 20 (note that, despite having an umlaut symbol over it, "ä" has the same alphanumeric value as the letter "a") and, adding the mass units of pounds, this becomes 20 pounds per inch for *two* extension springs or 10 pounds for *each* of the two parallel springs. As is to be expected, each of the extension springs in the Merseburg wheel had a spring constant, 10 pounds per inch, which was half of the spring constant, 20 pounds per inch, of each extension spring found in the Kassel wheel because the mass of a weighted lever inside of the drum of the Merseburg wheel, which was 14.5 pounds, was half of the mass of a weighted lever inside of the Kassel wheel which was 29 pounds.

It is interesting to note that, in Figure 22(a), the point of contact between the bottom of the smallest lead ingot and the visible right foot of the lamp functions somewhat like the pointed end of the upper duckbill goniometer arm in Figure 21(a) and the lower corners of the trapezoidal cross sections at the tops of the middle and right vertically oriented lead ingots in Figure 22(a) function somewhat like the intersection points on the left straight radial edge of the circular sector paper piece in Figure 21(a). In both cases the pairs of points involved direct a straight line segment off of the second DT portrait to letters in the lines of text under the portrait whose accurate alphanumerical analyses can provide the reverse engineer of Bessler's secret imbalanced pm wheel mechanics with important numerical values for the many parameters of distance, mass, and spring tension incorporated into the parts used by that mechanics.

Now it's time to discuss Figure 22(b) which is devoted to a more detailed analysis of the "L" square symbol in the second DT portrait. While the "L" square generally represents a weighted lever whose pivot pin's center axis has traveled clockwise around an axle's center axis from the drum's 7:30 to 9:00 positions, for this analysis it represents one of the cat's claw gravity latches that Bessler used to give his Merseburg and Kassel wheels the bidirectional ability to rotate either clockwise or counterclockwise as viewed from one of the circular sides of their 12 feet diameter drums. Let's see what parameter values for these gravity activated latches can be derived from the portion of the second portrait containing the "L" square (the reader may want to refer back to Figure 13(b) at the end of Chapter 7 to see how the parameter values obtained in Figure 22(b) relate to them).

It turns out that, contrary to expectation and perhaps to confuse reverse engineers, Bessler used the *longer* end of the "L" square drafting tool to represent the *shorter*, counterweight carrying parallel pair of arms of a cat's claw gravity latch and used the "L" square's *shorter* end to represent the *longer*, square notch containing parallel pair arms of a gravity latch. Thus, in Figure 22(b), the longer end of the "L" square with the white solid circle "A" label near it symbolizes the shorter arms of a gravity latch and the shorter end of the "L" square with the white solid circle "B" label near it symbolizes the longer arms of a gravity latch.

If, starting at the little hanging hole at the end of the longer arm of the "L" square, one extends a line segment diagonally downward and to the left, then it will exit the bottom border of the second DT portrait, turn black, pass between two curled dots on the left side of the capital "W" in the word "*Würde*" on the first line of text under the portrait, and then continue on until it reaches and ends at the "d" in the word "*weder*" on the second line of text. "W" is the 23rd letter of the alphabet with an alphanumeric value of 23. If we add its individual digits together, we get 2 + 3 = 5. We can use one of the dots near the letter to justify moving this derived value's decimal point one place to the left to obtain 0.5 and, adding mass units of pounds, turn it into 0.5 pounds which was the mass of the lead counterweight and its attachment bolt used in each of the two types of cat's claw gravity latches, inner and outer, found in the bidirectional Merseburg wheel. This mass is equivalent to 8 ounces and that value can be gotten by adding the derived value of "W", 3, to a similarly derived value for the letter "l" in the word "*lobt*" below the "W" whose upper curled dot serif touches the middle of the "W". The normal value for "l" which is 12th letter of the alphabet is 12 and its derived value gotten by adding its individual digits is 1 + 2 = 3. Adding these two derived values for the "W" and the "l", as suggested by the two letters being in contact, gives us 5 + 3 = 8. Supplying the mass units of ounces converts this into 8 ounces. The derived value of 0.5 for the "W" can also, after adding inch units, represent the width of the arms in the gravity latches used in the bidirectional Merseburg wheel or 0.5 inch = 8/16 inch.

The cutoff flat ends of the parallel pair of shorter, counterweight carrying arms of a gravity latch were located 4 inches from the center axis of their pivot holes and that distance is indicated by the value of the letter "d" at the end of the line segment with the white solid circle "A" label near it after adding the distance units of inches to it to get 4 inches.

However, the center axis of a counterweight's little attachment bolt was closer to the center axis of the lever's pivot rod hole than 4 inches. Its separation distance was only 3.2 inches = 3-3.2/16 inches. That value is gotten by extending a *vertical* line segment downward from the little hanging hole at the end of the "L" square's longer arm that then exits the bottom border of the second DT portrait, travels down between the letters "W" and "ü" in the word "Würde" on the first line of text as it touches both of the upper corner serifs in them to connect the letters, and finally ends at the serif that joins together the two connected letters in the letter group "ob" in the word "*lobt*" on the second line of text below the portrait.

Since the "W" and the "ü" were carefully positioned by Bessler so that their serifs would be connected by a vertical line segment placed between them, he intended some arithmetic operation to be performed on their alphanumeric values. If we perform a subtraction, we get 23 – 21 = 2. The black vertical line segment then extends downward to the two connected letters in the word group "ob" on the second line of text. Their connection suggests that their normal individual alphanumeric values, 15 and 2, also must be arithmetically manipulated. If we choose multiplication, then we can write 15 x 2 = 30. Finally, we add together the value we derived for the line segment connected "W" and "ü" in the word "*Würde*" on the first line of text under the portrait with the value derived for the letter group "ob" on the second line of text to get 2 + 30 = 32. We can use one of the dots at the end of a flourish issuing from the left side of the "W" to justify moving the decimal point one place to the left to obtain 3.2 and, after adding distance units of inches, finally obtain a distance parameter of 3.2 inches = 3-3.2/16 inches which was the separation distance between the center axes of a counterweight attachment bolt and a gravity latch's pivot rod hole. Also notice the left leaning umlaut symbol over the "ü" in "Würde". It is the only umlaut in either the first or second DT portrait that leans to the left and here it symbolizes the two types of cat's claw gravity latches, inner and outer, that were assigned to each weighted lever in one of Bessler's bidirectional wheels. This symbol also tells us that the distance and mass parameter values being revealed will apply to both types of latches.

There's one additional clue provided by the letters in the word "*lobt*" on the second line of text below the portrait. Note the right pointing serif at the bottom of the "l" which is a symbol for a decimal point. Also, the letter "o" in the letter group "ob" looks somewhat like a zero or 0 and the pointed serif of the "l" to its left implies we should move its decimal point one place to the left to get 0.0. Next, the "b" attached to the letter "o" has an elongated, straight, right leaning serif at its top that makes the letter look like the numeral 6 so we can take the altered value of this "b" to be 6. The last letter in the word "*lobt*" is the "t" with a normal value of 20 and its lower right corner serif points up and to the right and toward the serif at the top left corner of the letter "n" in the following word "*noch*" and we can imagine a short and invisible line segment connecting the separated letters "t" and "n" together. That connection implies that we add their values and, after converting the normal value of the "n" from 14 to 5 by adding its individual digits, we can then write 20 + 5 = 25. So, we now have three derived values that are 0.0, 6, and 25. If we then just put them in a row without adding them, we get 0.0625. After supply the units of inches, that value becomes 0.0625 inch = 1/16 inch which was the thickness of the "L" shaped steel pieces used for the construction of the two types of gravity latches, both inner and outer, that were used in *both* of the bidirectional Merseburg and Kassel wheels.

Let's now focus on the shorter arm of the "L" square drafting tool in the second DT portrait that symbolizes the longer, square notch containing parallel pair of arms in a gravity latch and is located near the white solid circular "B" label on the right side of Figure 22(b).

As was done for the longer arm of the "L" square which symbolized the shorter, counterweight holding parallel pair of arms of a gravity latch, for the shorter arm of the "L" square we again extend two white line segments away from the little hanging hole near its pointed end. One line segment extends diagonally to the right and downward, passes over the visible left foot of the lamp, touches the bottommost corner of the visible right foot of the lamp, exits the bottom of the portrait, briefly turns black, and immediately reaches and ends at the curled dot serif at the top of the letter "t" in the two letter group "et" at the end of the word "*vollendet*" at the end of the first line of text under the second portrait. Let us begin by analyzing this line segment and the shop table items it touches first.

The white portion of the line segment first passes over the left visible foot of the metal lamp that forms an inverted "V" with the other visible right foot. Since "V" is the Roman numeral for 5, we simply add units of inches to convert it to 5 inches which was the horizontal distance, in inches, between the center axis of the pivot rod hole of a gravity latch of either type, inner or outer, and the pointed tips at the beveled ends of its longer arms. The white line segment, before exiting the portrait, passes over the bottom pointed corner of the visible right foot of the lamp and this indicates we can move the decimal point one place to the left of the numerical value for "V" (note that inverting the "Λ" formed by the lamp's two visible feet moves the corner that the line segment passes over to the upper *left* corner of the resulting "V" and, thus, justifies moving the decimal point the corner represents one place to the left). This gives us a value of 0.5 and, after adding the appropriate distance units, the distance parameter of 0.5 inches. This parameter is the width of a gravity latch's two parallel longer arms and also of its two parallel shorter arms.

As the white line segment indicated by the white solid circle "B" label exits the bottom border of the second DT portrait, it turns black and almost immediately reaches and ends at the curled dot serif at the top of the "t" at the end of the word "*vollendet*" on the first line text under the portrait. If we add the normal values of the two connected letters in the letter group "et" together, we obtain 5 + 20 = 25. The slightly left leaning dot at the top of the "t" and the left pointing arrowhead-like bottom end of the virgule at the end of the first line of text justify moving the unwritten decimal point of the value 25 *two* places to the left to obtain 0.25. Adding distance units of inches then converts this to 0.25 inch = 4/16 inch. What does this value have to do with the cat's claw gravity latches Bessler used? It corresponds to the length of *one* inside edge of the square notches used at the ends of the longer pairs of parallel arms of either type of latch, inner or outer, in the Merseburg wheel. How can we be assured that this is the correct interpretation?

Notice the nearly 45° slope of the right leaning portion of the serif at the top of the letter "t" at the end of the word "*vollendet*". It suggests that one should extend another line segment diagonally downward and to the left of the "t". I did this and the result was the extra black line segment on the right side of Figure 22(b). It just touches the top of the "e" in the letter group "et" in the world "*vollendet*", passes through the center of the "d" to the left of the letter group "et", touches the serif in the lower right corner of the odd "n" to the left of the "d", and finally reaches an unusual point in the looping flourish at the top of the "t" in the letter group "et" in the word "*fchändet*" on the second line of text under the portrait where the flourish crosses over itself.

That point in the middle of the looping flourish at the top of the second line's "t" represents the point at the center of an "x" shaped multiplication symbol. It tells us that we are can multiply the derived values for the two "et" letter groups, 0.25, by each other to get 0.25 x 0.25 which, by adding distance units, becomes 0.25 inch x 0.25 inch = 4/16 inch x 4/16 inch and this was the area, 0.0625 square inches, of one of the square notches at the ends of the parallel pair of longer arms of either type of gravity latch whether it was the inner or outer type.

Next, we need to discuss the *vertical* white line segment that extends away from the little hanging hole at the end of the shorter arm of the "L" square. That line segment passes through the end of the lower arm of the duckbill goniometer, the two legs of the divider and the steel file that they straddle, exits the bottom border the second DT portrait, turns black, passes through the letter "w" in the word "*wird*" on the first line of text under the portrait and finally reaches the top of the "c" in the connected letter group "ch" in the word "*noch*" on the second line of text. The top part of the "h" in "*noch*" is shaped into a very obvious "?" mark and was obviously intended to make the reader of DT wonder about the meaning of that symbol.

There are multiple interpretations that can be given to the letters that this black portion of the line segment passes through and contacts. For example, we can subtract, using the method of evaluating the value of Roman numerals, the normal values of the letters in "ch" to get 8 − 3 = 5. Then, adding inch units, we get 5 inches which is the horizontal distance, in inches, between the pointed tips of the beveled ends of the parallel pair of longer arms of either type of gravity latch, inner or outer, and the center axis of their pivot rod holes. The question mark, "?", at the top of the letter "h" in the word "*noch*" on the second line of text touches the bottom of the "i" in the word "*wird*" on the first line of text. If we use the "i" as a Roman numeral, we can give it a value of 1 and then subtract it from the value we derived for "ch" that was 5 to get 5 − 1 = 4 and which, after adding inch units, becomes 4 inches. This distance parameter corresponds to the distance, in inches, between the center axis of the pivot rod holes of a parallel pair of longer gravity latch arms of either type, inner or outer, and the center axis of the weighted lever's L1 main cord hook attachment pin that would be secured within the square notches at the ends of the latch's longer arms when the weighted lever's main arms were locked down into contact with their lever's wooden radial stop piece.

That these distance parameters applied to the two bidirectional Merseburg and Kassel wheel's gravity latches is indicated by the two connected letters in the letter group "wi" in the word "*wird*" on the first line of text. The "w's" vertical parts have been squeezed together and make the letter look more like an "m" instead of a "w". If we consider it to be an "m", then its normal alphanumeric value would be 13. Since it is connected to the letter "i" to its right which looks like the Roman numeral whose value is 1, we can then add these two values together to get 13 + 1 = 14 which, with inch units added, becomes 14 inches. This, of course, is the exact distance along the center lines of the main arms, in inches between the center of gravity of a trio of cylindrical lead end weights and the center axis of the lever's pivot pin in the weighted levers used in *both* of the bidirectional Merseburg and Kassel wheels.

However, all of the mass and distant parameters derived from the symbols and letters in Figure 22(b) for the two types of cat's claw gravity latches, inner and outer, only pertain to the latches found in the Merseburg wheel. For the Kassel wheel, the masses and, therefore, thicknesses of the disc-like lead counter weights attached to the ends of the shorter arms of a gravity latch, the widths of these steel arms, and the inside edge lengths of the square notches at the ends of the longer arms had to all be doubled in value. However, the thickness of each arm (0.0625 inch = 1/16 inch) and the two length parameters for each of the arms, longer and shorter (4 inches and 5 inches for the longer arms and 4 inches and 3.2 inches = 3-3.2/16 inches for the shorter arms), for both types of latches, inner and outer, remain the same when discussing the cat's claw latches used in the Kassel wheel.

As a final comment on Figure 22(b) before moving on, I'd like to direct the reader's attention to another possible bit of whimsy on the part of Bessler when he was arranging the various clues involving the "L" square drafting tool in the second DT portrait which pertain to the gravity activated latches used in his bidirectional wheels. I have been referring to them as "cat's claw gravity latches" because the actions of their pairs of notch containing longer arms on a lever's L1 main cord hook attachment pin reminded me of an actual cat's claw when the animal is hunting prey. During a bidirectional drum's rotation, the internal one-directional wheel that is undergoing retrograde rotation will have each of its eight weighed levers' parallel pair of main arms momentarily locked down against their wooden radial stop piece as their lever's pivot pin's center axis passes through the drum's 12:00 position simultaneously by both of the two types of gravity latches, inner and outer, which means, for a brief time, the lever's L1 main cord hook attachment pin is held by the square notches at the ends of *four* of the latches' longer arms. When a cat "plays" with its prey, usually a mouse, before consuming it, it may hold the crippled animal in place with one of its forepaws using either four or five of the claws in that paw. The similarity to this feline behavior by the actions of the notched ends of the gravity latches' longer arms is so great that it would be difficult to think that Bessler did not also notice it. If so, did he leave us a clue in the second DT portrait that he did? Yes, that clue actually exists!

On the right side of the section of the second DT portrait shown in Figure 22(b) the reader will find an enlargement of the middle portion of the capital letter "W" at the beginning of the word "*Würde*" on the first line of text under the portrait. The middle portion of the "W" looks somewhat like the silhouette of a cartoon mouse that has spun around to his right side and is raising his right paw to fend off something, like a cat, that is coming toward the mouse from the left. We can see the mouse's two large ears at the top which are above a snout that points to the left. At the bottom there appears to be two legs and a tail trailing out to the left that something, perhaps a cat's paw, is stepping down on in order to immobilize the mouse. An out of place looking flourish from the left side of the letter "W" swoops over to the right and, like a cat's claw, comes down upon and appears to be grabbing the head of the mouse. It is interesting to note that the left side of the capital "W" almost looks like it is running toward the mouse in order to catch him. Is this middle portion of the "W" in "*Würde*" really supposed to be a symbol for a mouse which tells us the inventor is imparting parameter values for the cat's claw gravity latches used in his bidirectional wheels with the alphanumeric values of the letters near the "W" or is this just the result of a ornate calligraphic embellishment and the psychological phenomenon of "pareidolia" which causes the human mind to "see" familiar objects in random shapes? I lean toward believing the former because the middle portion of the capital "W" in the earlier word "*Werck*" on the same first line of text under the second DT portrait (refer to Figure 21(b) to see that capital "W") looks nothing like a mouse!

Now we must move on to Figure 23 which gives us some important information about the various layers of coordinating cords Bessler used in his imbalance perpetual motion wheels based on items on the left side of the second DT portrait. However, before getting to this chapter's Figure 23, the reader should first briefly revisit the last chapter's Figure 17 because items are contained in it that will now be mentioned which could not be fitted into Figure 23 without reducing the sizes of the items that are shown in that figure.

The focus now is on the left side of the second DT portrait and in particular the items, as seen in Figure 17, that are in the background behind and above Bessler's right hand which holds a pair of armless eyeglasses.

Between the thermometer that is partially hidden by the left border of the second portrait and the hanging saw, which we learned in the last chapter is a symbol for the Arabic numeral 9 and the 9:00 position of a clockwise rotating drum, there are two carpenter's planes that hang by their handles from hooks attached to the back wall of Bessler's work shop.

On the sides of the two vertically arranged carpenter planes, we see two large steel chisels with wooden handles. Chisels are used to remove material from pieces of metal, stone, or wood. When a large amount of material is to be removed, the chisel is usually hammered at one end to force its other sharpened end into the material, but, for the removal of smaller amounts of material, the chisel is smaller and, in the case of softer materials like wood, can be forced into the material by merely pushing on an attached wooden handle. If the sharpened end is not flat, but, rather, concave, the chisel is referred to as a "gouge" and these are almost exclusively used in woodworking to cut grooves into the surfaces of pieces of wood that align with the grain of the wood.

If one extends a vertical line segment downward from the sharpened ends of both of the large chisels, they will each reach a small tool rack attached to the background wall that holds five small tools. The small lower tool rack nearest the globe (which is behind its left side that ascends if the globe is turning clockwise) holds five chisels or gouges of varying lengths. The other small upper tool rack is located between the upper of the two eyeglass lenses and the trigger guard of the shotgun. This second upper rack also holds five tools, four are small chisels or gouges, but the one in the middle with a large ball shaped handle is not a chisel or gouge, but, rather, a tool known as an "awl". (Note that a closer examination of high resolution images of this upper tool rack, however, reveals that there are an additional two tools located *behind* the leftmost and rightmost tools in the rack.). An awl is used to punch small holes into material like leather or into pieces of wood to serve as pilot holes into which larger diameter nails or screws may be more easily inserted. The oversized ball-shaped handle of an awl allows one to apply great puncturing force with its pointed steel tip while spreading the reactional force the user's hand feels over a larger area of skin and underlying muscles and bones so as not to cause injury to the hand especially if the tool is used to punch a large number of holes into some item such as a leather belt.

Chisels and gouges can be used to make distinct grooves or channels in wood. In the case of the second DT portrait, their presence in the small tool racks is symbolic of the five coordinating cord layers that subdivided the volume of space between the sixteen inward facing surfaces of the eight parallel pairs of 3 arm side pieces of a one-directional wheel's eight weighted levers.

Leaving Figure 17, I will now return to making references to the second DT portrait's left side items which are shown in Figure 23. Because there was not enough space to show an enlargement of the entirety of the left side of the second portrait, a significant portion of it had to be removed. However, I have preserved the vertical alignment of the remaining items which is what is really important to understanding what their symbolism can tell us above the layered system of coordinating cords Bessler used inside his wheels.

If we just consider the lower tool rack near the left border of the second DT portrait and extend a vertical line segment, indicated by the black solid circle "A" label, downward from the tip of its leftmost tool to the first line of text below the portrait as is done in Figure 23, then that line segment will reach and end at the pointed tip of the right curling top of the letter "d" in the German word "*des*" on the first line of text below the second portrait. The "d" represents the village of Draschwitz, whose name also begins with a capital "d", in which the last of Bessler's one-directional wheels was constructed and publicly exhibited. At the end of the word "*des*" we see the letter "s" that, if one closed up the upper part of the letter, would look like the letter "g" and would then stand for the town of Gera where the first two of Bessler's working one-directional wheels, his 3 feet diameter prototype and his second 4.5 feet diameter one, were constructed with only the second one being publicly exhibited there.

If we consider the word "*des*", we will notice that it has the letter "e" in the middle of it. "e", being the fifth letter of the alphabet in both English and German, has an alphanumeric value of 5 and here it tells us that the Gera and Draschwitz wheels, being one-directional wheels, each contained five coordinating cord layers.

But, we are not done yet with the information that can be extracted from the word "*des*" (or "*deg*") at the beginning of the first line of text under the second DT portrait.

We take the letter "d", but, instead of using its normal value of 4, assign it an altered value of 40 after *temporarily* moving its decimal point one place to the right which we are directed to do because of the point at the top of the letter which points to the right, but also looks like it is under tension and, like the end of stretched bow, will eventually snap back into a more vertical orientation. We then subtract the value of the "e" which is equal to 5 from that of the altered value for "d" to get 40 – 5 = 35. Once we have this value, we allow its unwritten decimal point to return one place to the left again to get a value of 3.5. Adding distance units of inches converts this to 3.5 inches = 3-8/16 inches. By now the reader should immediately recognize that this value is the separation distance between the center of gravity of a lead ingot end weight attached to the ends of the main arms of a weighted lever in the 3 feet diameter Gera prototype wheel and the center axis of the lever's pivot pin.

That distance was tripled to 10.5 inches = 10-8/16 inches for the Draschwitz wheel and that value can also be found in the word "*des*". To obtain it, we must take the normal alphanumeric value of "s" which is 19 and then subtract the *sum* of the normal values of the "d" and "e" to the left of that letter to get 19 − (4 + 5) = 19 − 9 = 10. We are now almost to the distance value of the 10.5 inches = 10-8/16 inches that is needed, but where do we obtain the 0.5 to add to this result?

Where to obtain the value of 0.5 becomes obvious from studying a greatly magnified image of the "s" at the end of "*des*". The top part of the "s" is almost in contact with the "W" in the next word "*Werck*" on the first line of text under the second portrait ("*Werck*" in English means "Invention" and refers, of course, to the inventor's wheels). Bessler even drew a tiny *eyeball* into the flourish work on the top left side of the letter "W" that looks toward the "s" to tell us that we must study the connection between the "s" and the "W" very carefully. (Indeed, the entire left side of the letter "W" looks like a face that is turned toward the word "*des*" with the upper right part of the letter "s" in "*des*" forming its nose. There's even an open mouth in the face that seems to be expressing surprise at what one will discover from the word "des" by studying it! I've included an enlargement of that face, near the black solid circle "C" label, at the bottom of Figure 23.) There is also another swirling flourish that emerges from the bottom left side of the "W" and ends right below the "s" with a noticeable curled dot that is also a symbol the inventor used for a decimal point. That flourish loops around and crosses itself to form a sort of distorted "+" sign. All of these subtle clues tell us that we must somehow *add* a value derived from the letter "W" in the word "*Werck*" to the value of 10 we derived for the letters in the word "*des*".

The solution now becomes rather easy to find. "W" has an alphanumeric value of 23 since it is the twenty-third letter of the alphabet. We simple add its two digits together to get 2 + 3 = 5 and then move its decimal point one place to the left to obtain 0.5. This value is then added to the 10 to get 10 + 0.5 = 10.5 which, after adding the distance units of inches, becomes 10.5 inches = 10-8/16 inches or the distance between the center of gravity of a lead ingot end weight attached between the ends of a parallel pair of main arms in a weighted lever inside of the 9 feet diameter, one-directional Draschwitz wheel and the center axis of the lever's pivot pin.

But, we still have another wheel we can get some information about from the word "*des*" and the letter "W". That wheel is the *second* Gera wheel, having a drum diameter of 4.5 feet, which was the first one-directional wheel Bessler publicly exhibited. Can one find the separation distance between the center of gravity of one of its lead ingot end weights and the center axis of its weighted lever's pivot pin in this one-directional wheel? With the understanding that the reader now has of the alphanumeric encryption method for parameter values that Bessler used, that distance can also easily be found.

Since the second Gera wheel's drum was 4.5 feet in diameter compared to the prototype's drum diameter of 3 feet, all of the separation distances between center axes in the second Gera wheel's weighted levers, as seen in an axial view of the drum, would have to be multiplied by a factor of (4.5 / 3) = 1.5. Because the lead ingot center of gravity to its weighted lever pivot pin's center axis separation distance in the Gera prototype wheel was 3.5 inches = 3-8/16 inches, this means that the analogous distance in the second Gera wheel was 3.5 inches x 1.5 = 5.25 inches = 5-4/16 inches.

To obtain this value, we take the "d" in "*des*" as again having a *temporary* value of 40, the "e" as having its normal value of 5, and use the letter "s" as a "g" with a value of 7. Adding these three values added together give us 40 + 5 + 7 = 52. If undoing the temporary decimal point move in the letter "d" also moves the decimal points in the normal values of the "e" and the "g" one place to the left as well, then the sum can be written as 4 + 0.5 + 0.7 = 5.2. We then use the 5 gotten from the "W" in "*Werck*" by adding the values of the individual digits in its normal value of 23 and move its decimal point *two* places to the left to obtain 0.05. This double place movement of the decimal point is justified by the two noticeable dots that appear in the left moving flourish issuing from the bottom of the letter "W" in the word "*Werck*" on the first line of text under the second portrait. Adding that altered value for the letter "W" to the sum obtained from the altered values of the three letters in "*des*" then gives us 5.2 + 0.05 = 5.25 and, adding the distance units of inches, this becomes 5.25 inches = 5-4/16 inches which is the distance parameter we sought for the second, 4.5 feet diameter one-directional Gera wheel.

The reader may find this alphanumeric coding system that Bessler used simple and fairly easy to understand once it is explained in detail. However, discovering *how* he was using it was not something that I managed to do in a mere matter of weeks or months. Rather, it took me *years* of consistent effort backed up by a tremendous number of computer models in order to finally locate and accurately interpret particular encoded part parameter values in the second DT portrait that he describes with his simple alphanumeric coding method. Often, it would take me weeks to months to determine whether a *single* parameter value that I thought that I had "discovered" was actually even a parameter value at all!

Now the reader needs to notice an interesting alignment that takes place on the left side of the second DT portrait in Figure 23. In involves the centers of the two circular armless eyeglass lenses and the large sphere shaped wooden handle of the awl located in the middle of the *upper* rack of tools. These three centers are all collinear which means that they all lie on the same line segment drawn between the center of the lower lens and the center of the awl's spherical wooden handle. Could that alignment of centers and the awl's handle having the same diameter of the eyeglasses' lenses be just a coincidence? Not likely. If we extend a white vertical line segment downward from the center of the lower lens, it will touch the wooden handle of the *center* tool located in the *lower* tool rack. Then, if as is done in Figure 23, ones extends a vertical line segment downward from the end of the tool on the right side of the lower tool rack, it will pass directly through the left side of the globe, through the largest section of the horizontally oriented telescope, and finally extend past the portrait's bottom border and reach and end at the letter "W" in the German word "*Werck*" on the first line of text below the second DT portrait. From this we see that the lower lens of the armless pair of eyeglasses is attached to the lower tool rack and that rack and its five tools represent the five coordinating cord layers found in the inventor's first three one-directional wheels.

One can also extend another line segment, indicated by the white solid circle "B" label in the figure, downward from the center of the ball shaped handle of the awl located in the middle of the upper tool rack which will pass through the right side of the globe and the first section of the horizontal telescope to, eventually, reach the letter "r" in the middle of the word "*Werck*" on the first line of text and the letter "n" in the word "*man*" on the second line of text under the second DT portrait.

In this symbolism, the lower tool rack is, through the optical axis of the armless eyeglasses and the handle of the awl, connected to the upper tool rack and both racks *together* represent the *two* internal one-directional wheels found inside of the enlarged drums of the bidirectional Merseburg and Kassel wheels. The vertical line segment from the center of the upper tool rack's awl eventually passes through the middle of the word "*Werck*" on the first line of text below the second portrait. The letter "W" at the beginning of "*Werck*" stands for Weissenstein Castle and the "k" at the end of that word stands for the city of Kassel, Saxony in which that castle was located and in which the bidirectional Kassel wheel was constructed. The letter "W" in "*Werck*" can also be inverted to obtain the letter "M" which, obviously, stands for the town of Merseburg, Saxony in which the bidirectional Merseburg wheel was constructed. Bessler is telling a future reverse engineer of his wheels with these clues that his one-directional Gera wheels and his one-directional Draschwitz wheel contained five coordinating cord layers, but his bidirectional wheels actually contained *two* one-directional wheels that *each* contained five layers so that a total of ten coordinating cord layers were found in his bidirectional wheels.

Since each weighted lever inside of a one-directional wheel had five coordinating cords (there were actually only four different types of cords, but the spring cord was used twice for each weighted lever in order to attach a parallel pair of extension springs to it) in five separate layers attached to the coordinating cord hook attachment pins held by the lever's parallel pair of A arms, that meant that each one-directional wheel Bessler constructed had to contain forty cords to precisely coordinate the motion of its eight weighted levers during drum rotation in order to keep their center of gravity on the axle's descending side. Thus, for a bidirectional wheel, even though one of its two internal one-directional wheels was always inactive during drum rotation in either direction, a total of eighty coordinating cords in ten layers were required! This may, indeed, sound like an enormous number of cords to put into an imbalanced pm wheel, but it is actually the *minimum* needed to achieve bidirectionality and the price Bessler had to pay for this capability which he hoped would convince skeptics that he had found a genuine, working perpetual motion. By spreading most of the cords around the large outer circumference of a bidirectional wheel's enlarged drum and using color coded coordinating cords, the difficulty of having to use such a large number of cords could be kept manageable.

The line segment with the "B" label, however, does not stop with the letter "r" in "*Werck*" on the first line of text under the second portrait and can be further extended downward until it reaches and ends at the letter "n" at the end of the word "*man*" on the second line of text below the portrait. The alphanumeric value of "n" is 14 and, adding units of inches, we get 14 inches which was the separation distance between the center of gravity of the trio of cylindrical lead weights at the ends of a weighted lever's main arms and the center axis of the lever's pivot pin in *both* of the Merseburg and Kassel wheels.

If we add the alphanumeric values of the three letters "erc" in the connected letter group of "erck" in the word "*Werck*" with the value of the "n" in "*man*", we get the sum of 5 + 18 + 3 + 14 = 40. The normal value of the "W" at the beginning of the word "*Werck*" is 23, but, by adding its individual digits, we get 2 + 3 = 5. We can then use the curled dot at the end of the flourish that issues from the bottom of the letter "W" and moves to the left to justify moving the decimal point for 5 one place to the left to get the altered value for "W" of 0.5. Next, we multiply this altered value of "W" by the previous sum of 40 to get 0.5 x 40 = 20 and, after adding the units for a spring constant, this becomes 20 pounds per inch which was the *combined* spring constant for the parallel *pair* of the extension springs attached to a weighted lever in the Merseburg wheel. Dividing this value by 2 gives us the spring constant for each *individual* extension spring in this wheel or 10 pounds per inch.

But, what about the combined and individual spring constants in the Kassel wheel which were 40 and 20 pounds per inch? How do we get them? Their values simply require multiplying the combined and individual spring constant values for the Merseburg wheel's extension springs by a factor of 2. This multiplication is suggested by the altered value for the letter "k" at the end of the word "*Werck*". "k", being the 11th letter of the alphabet, has an alphanumeric value of 11, but, by adding its individual digits together, we get 1 + 1 = 2 which provides the required factor needed to convert the Merseburg wheel extension spring constants, combined and individual, into those that were used in the Kassel wheel.

With the above numerical analysis completed, it now becomes possible to understand why Bessler included an extra two tools that are located *behind* the two *end* tools in the upper tool rack. The leftmost and rightmost *front* tools in the upper rack represent the parallel pair of extension springs attached to each of the weighted levers in the bidirectional Merseburg wheel with an individual spring having a spring constant of 10 pounds per inch and the parallel pair of springs having a combined spring constant of 20 pounds per inch. The two additional tools located *behind* these leftmost and rightmost front tools tell us that, for the weighted levers in the bidirectional Kassel wheel, these values were doubled to a spring constant of 20 pounds per inch for an individual spring and 40 pounds per inch for the combined spring constant of a parallel pair of springs.

Now, it is time to conclude this chapter with a discussion of Figure 24 which gives the reverse engineer important information about the shape of the 3 arm levers Bessler used in all of his wheels. "Shape" here refers to the angular relationships between the center lines of an unweighted lever's three parallel pairs of arms (again, note that a "center line" of an arm is an imaginary line that extended through the center axis of the lever's pivot pin and the center axes of any coordinating cord hook attachment pins whose ends are embedded in the wood of the lever arm).

I previously mentioned that, in the second DT portrait, Bessler would use the optical axis of a lens system to tell the reader to carefully study any items to which the optical axis passed through as it extended away from the lens or lenses involved. Those optical axes also applied to the ones from the lenses in his eyes. If one extends a vertical line segment downward from *his* right eye in the second portrait, whose bottom end is indicated by the white line segment with the arrowhead on it near the white solid circle "A" label, then that line segment will exactly reach and end at the center axis of the screw or bolt that holds the two arms of the duckbill goniometer together. That is a clue that the reverse engineer should study the various items on the shop table's top that the upper and lower arms of the goniometer point to as well as any items contained within the 45° sweep of the two arms.

The first clue in Figure 24 about the shape of the 3 arm weighted levers Bessler used in all of his wheels comes from the inclusion of the shorter arm of the "L" square within the 45° sweep of the spread goniometer arms. The shorter arm of the "L" square points to the left straight radial edge of the circular sector paper piece and, like that radial edge, represents the parallel pair of A arms of one of the weighted levers found in Bessler's wheels. Since that shorter arm of the "L" square is included within the 45° sweep of the goniometer's two arms, that tells us that the center lines of the A arms of a weighted lever were arranged at an angle of 45° to the center lines of some other parallel pair of arms in the lever. But, which other parallel pair of arms?

That other pair of arms is the parallel pair of main arms of a weighted lever and it is symbolized by the longer arm of the "L" square to which its shorter arm is attached. We know that the longer arm of the "L" square in this analysis, near the white solid circle "B" label, represents the main arms of a weighted lever because it points to the smaller section of the telescope's tube whose sections get progressively larger in diameter and length as one moves along the length of the telescope from its eyepiece to its other end that is not seen because it extends beyond the left border of the second DT portrait. The various size telescope sections represent the end weights of progressively larger sizes and masses that Bessler attached to the ends of the main arms of his wheels' weighted levers beginning with his 3 feet diameter, one-direction Gera prototype wheel and ending with his 12 feet diameter, bidirectional Kassel wheel.

At this point, the reverse engineer has the angular relationship between the center lines of two of the parallel pairs of arms in a weighted lever, the main and A arms, which is 45°, but there is still the third parallel pair of arms which must be accounted for.

The third parallel pair of arms in a weighted lever, the B arms, is represented in the second portrait by the top straight edge of the 14 inch ruler that is carefully balanced on the carpenter's plumb bob near the white solid circle "C" label. If one measures the angle between that top edge of the ruler and the top edge of the "L" square's shorter arm, he will find that the angle is exactly 22.5° as indicated in the figure. This suggests that the angle between the center lines of the B arms in a weighted lever and its A arms was exactly 22.5°. But, that would then mean that the center lines of the B arms were either located exactly *between* the center lines of the main and A arms or that the center lines of the B arms were outside of the angle between the center lines of the main and A arms and were then either 22.5° away from the center lines of the A arms *and* 45° + 22.5° = 67.5° away from the center lines of the main arms or that the center lines of the B arms were 22.5° away from the center lines of the main arms and 45° + 22.5° = 67.5° away from the center lines of the A arms. This would then imply that the center lines of all three parallel pairs of arms where contained *within* an angular sweep of only either 45° or 67.5°. If this was the case, then one would have a rather lopsided 3 arm lever and it would be impossible to connect any coordinating cords from the "inside" parallel pair of arms to any of the other weighed levers in a wheel without those cords being blocked by the two "outside" pairs of parallel arms. This is clearly not the correct angle between the center lines of a weighted lever's B arms and its A arms. So, if this is not correct, then what is?

The correct angular relationship between the center lines of a weighted lever's B and A arms is gotten by extending a line segment to the left of the second DT portrait along the top edge of the 14 inch ruler as shown in Figure 24. That line segment will pass through the center axis of the screw or bolt that holds the two arms of the goniometer together and, shortly after, intersect the right outer circular perimeter of the protractor at a most interesting point. That point is located exactly on the dividing line between the 21st and 22nd angular increments on the protractor if the increments are counted starting at the bottom left side of the protractor. Each of the 24 angular increments on the 180° protractor's outer circular perimeter represents 180°/24 = 7.5°.

Thus, since the extended line segment from the top edge of the balanced 14 inch ruler intersects the dividing line *between* the 21st and 22nd angular increments measured from the bottom left part of the protractor, we see that Bessler is here providing us with an angle equal to 21 x 7.5° = 157.5°. If, however, the number of angular increments is counted starting from the upper right corner of the protractor, then the dividing line between the angular increments that the line segment extended from the top edge of the 14 inch ruler intersects would equal an angle of 3 x 7.5° = 22.5°. We rejected an angle of 22.5° or 67.5° between the center lines of the B and A arms of a weighted lever for the reason given in the previous paragraph, so that means that the correct angular relationship between the center lines of the B and A arms within a weighted lever had to have been 157.5°.

If there are only three parallel pairs of arms in a weighted lever and the angle between the center lines of the lever's main arms and A arms is 45° and the angle between the center lines of its B arms and A arms is 157.5°, then that means that the smallest angle between the center lines of the B arms and the main arms of the lever must *also* be 157.5°. It is this angular symmetry which gives a weighted lever at the 12:00 position of a drum the appearance of the letter "Y". The upper left branch of the letter represents the end weight carrying main arms of a weighted lever, the upper right branch of the letter represents the four coordinating cord hook attachment pins holding A arms of the lever, and the bottom vertical half of the letter represents the B arms of the lever which only hold a single coordinating cord hook attachment pin which is the B1 long lifter cord hook attachment pin. If we add up the three angles between the center lines of the three arms of a weighted lever, we get 45° + 157.5° + 157.5° = 360° which shows that all of the angles are correct and that we have the correct axial geometric shape for the weighted levers that the inventor used.

In the second DT portrait, Bessler hints that a weighted lever whose pivot pin's center axis is located at the drum's 12:00 position will have a shape resembling the letter "Y" by having the 14 inch ruler's right end extend to the right and pass behind the base of the metal lamp as it does so. The lamp, as previously mentioned, is a symbol for a weighted lever at the drum's 12:00 position and its visible shape in the second portrait does resemble the letter "Y". However, to make the clue less obvious, the "Y" symbol he shows us is inverted. Perhaps his intention was that the reverse engineer would, as we have now just done, see the connection between the metal lamp and the 180° protractor that the line segment extended along the top edge of the ruler would reach and he would then get the idea that the inverted "Y" represented by the lamp needed to be rotated vertically through an angle of 180° so that it would then represent the letter in its normal orientation and the *approximate* shape of a weighted lever whose pivot pin's center axis had reached the 12:00 position of a clockwise rotating drum. (I wrote "approximate" shape because, obviously, the angle between the center lines of the lamp's two visible tripodal feet that represent a weighted lever's main and A arms would be 120° and not the actual angle of 45° that was used in Bessler's weighted levers.)

This chapter on the clues contained in the second DT portrait now concludes and, again, I want to emphasize that there are probably more clues in this portrait which I have not found. They await the efforts of future "DT portraitologists".

One marvels at the precision of these clues. They give the reverse engineer all of the basic information about Bessler's secret imbalanced perpetual motion wheel mechanics that are necessary in order to duplicate all of his wheels. I can only imagine that Bessler must have spent weeks, if not months, coming up with the special geometric arrangements of items and their alignments with the text wording under the portrait in order to preserve this information. Then he would have made a precise sketch of the second portrait, as well as the first portrait, and handed them over to a professional artist to engrave onto a printing plate for him. Of course, that artist, Christian Fritzsch, would have had no idea of the location and meanings of the various clues. He would only have been told that he had to make one of the most precise engravings ever made and would be amply compensated for his effort. He did his work well because, today, these portraits and the clues they contain will, hopefully soon, lead to the first successful duplication of the inventor's wheels in the last three centuries.

In the next chapter I will, as I did in this chapter, deal with the many clues found in what is usually referred to as the "first DT portrait" (even though it was really the *second* one that the eyes of a reader of DT would see only after he had first lifted away and folded over the paper flap covering it which had what is usually referred to as the "second DT portrait" printed on it). It will be seen that practically all of the parameter values encoded into the second portrait and discussed in this chapter will be found in the first portrait that is treated in the next chapter, but Bessler's methods of encoding them there will be different.

Figure 18 - Second DT Portrait Table Items

Each of a one-directional wheel's eight weighted levers are represented by one of the items. The item also represents the weighted lever after a 45 degree rotation of the wheel has taken place.

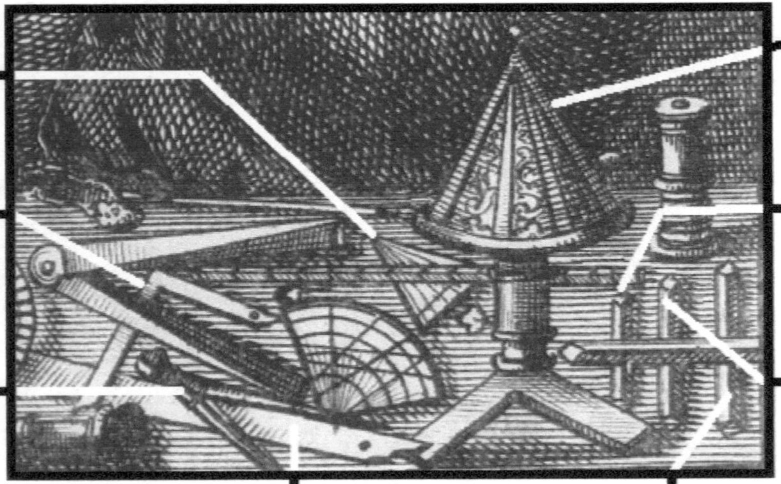

- Carpenter's plumb bob is weighted lever at 10:30 after wheel moves it from 9:00 to 10:30

- L square is weighted lever at 9:00 after wheel moves it from 7:30 to 9:00

- Divider is weighted lever at 7:30 after wheel moves it from 6:00 to 7:30

- Metal file is weighted lever at 6:00 after wheel moves it from 4:30 to 6:00

- Right ingot is weighted lever at 4:30 after wheel moves it from 3:00 to 4:30

- Middle ingot is weighted lever at 3:00 after wheel moves it from 1:30 to 3:00

- Left ingot is weighted lever at 1:30 after wheel moves it from 12:00 to 1:30

- Lamp is weighted lever at 12:00 after wheel moves it from 10:30 to 12:00

Figure 19 - Second DT Portrait Clues Give Angle of the 9:00 Weighted Lever

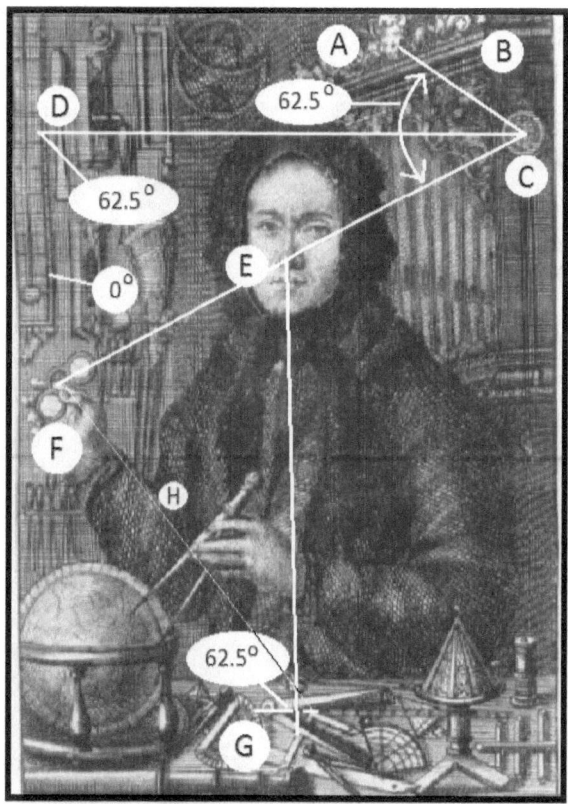

A = Middle Cherub on Top of Organ
B = Upper Right Corner Cherub of Organ
C = Center of Clock Dial
D = Temperature on Thermometer (62.5 Degrees)
E = Tip of Bessler's Nose
F = Nose Space under Eyeglasses' Bridge
G = Point where Longer "L" Square Arm crosses under Lower Duckbill Goniometer Arm
H = Line that connects Eyeglasses' Nose Bridge with corner of "L" Square

Figure 20 - Second DT Portrait Circular Sector Paper Piece Shows Cord Attachment Points

The left edge of the circular sector paper piece gives separation distances, in inches, between the center axes of the various coordinating cord hook attachment pins on a weighted lever's A arms and the lever's pivot pin. The separation distances between the center axes of the other hook attachment pins and their pivot pins or the axle to which the cords are attached are given on the right side of the figure. Values for the 3 feet diameter Gera prototype wheel are labeled "GPW" and those for the 12 feet diameter Merseburg and Kassel wheels are labeled "MKW".

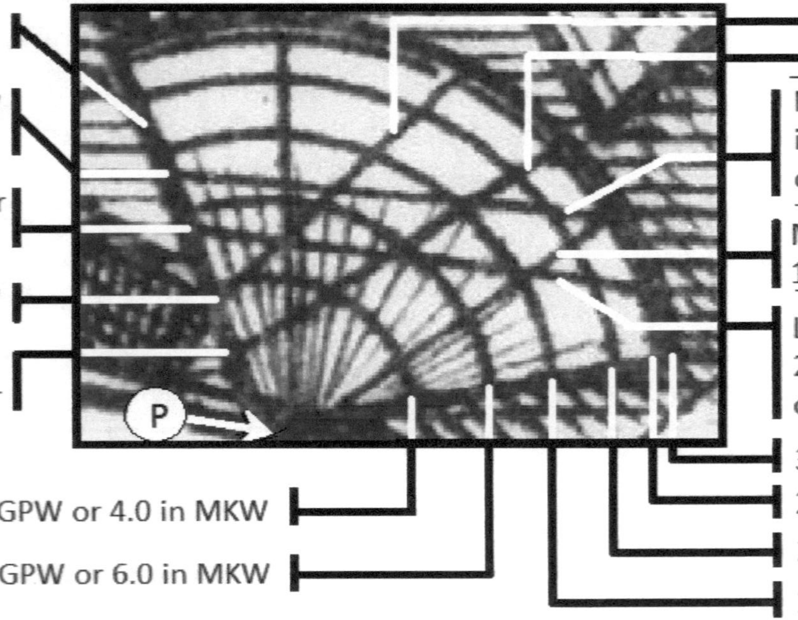

A Arms of 9:00 Lever

A4 is at 1.866 in GPW or 7.466 in MKW

A3 is at 1.4 in GPW or 5.6 in MKW

A2 is at 0.933 in GPW or 3.733 in MKW

A1 is at 0.5 in GPW or 2.0 in MKW

1.0 in GPW or 4.0 in MKW

1.5 in GPW or 6.0 in MKW

2.0 in GPW or 8.0 in MKW

2.5 in GPW or 10.0 in MKW

2.8 in GPW or 11.2 in MKW

3.0 in GPW or 12.0 in MKW

Long Lifter Cord to B1 at 2.1 in GPW or 8.4 in MKW on B Arms of 12:00 Lever

Main or L Arms of 10:30 Lever

Main Cord to L1 at 2.6 in GPW or 10.4 in MKW on L Arms of 10:30 Lever

Spring Cord to 16.5 in GPW or 66.0 in MKW from center axis of axle

Stop Cord to 12.934 in GPW or 51.736 in MKW from center axis of axle

Figure 21(a) - Second DT Portrait Clues That Verify The Circular Sector Paper Piece's Cord Attachment Points

Figure 21(b) - Some Additional Cord Clues

Figure 22(a) - Second DT Portrait Clues for Spring Constants, Weight and Lever Masses, and Main Arm Lengths

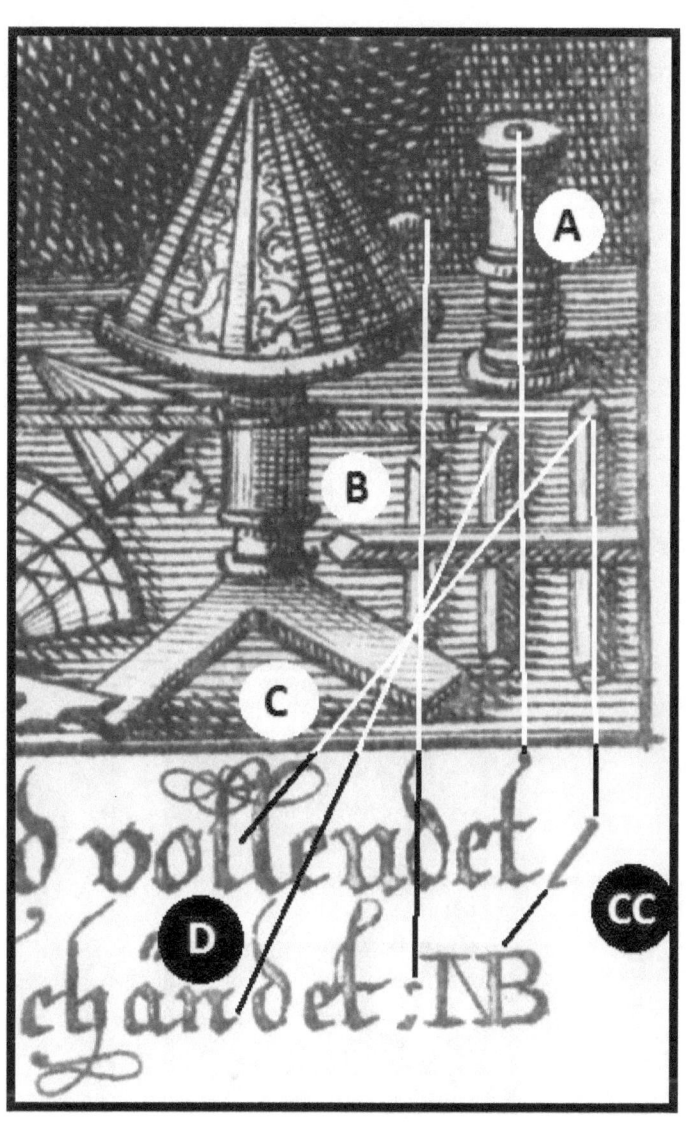

Figure 22(b) - Second DT Portrait "L" Square Gives Gravity Latch Parameters

Symbol of mouse in middle of capital "W" in "Würde"

Figure 23 - Second DT Portrait Clues about Cord Layers

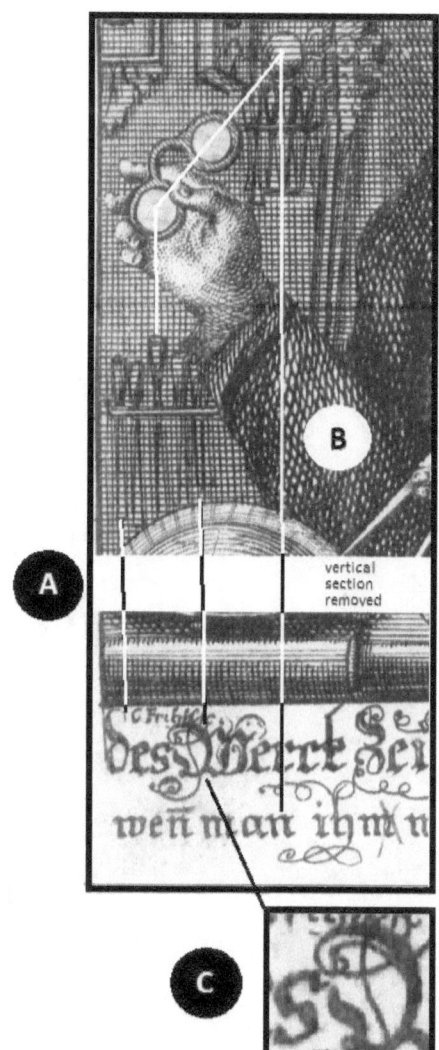

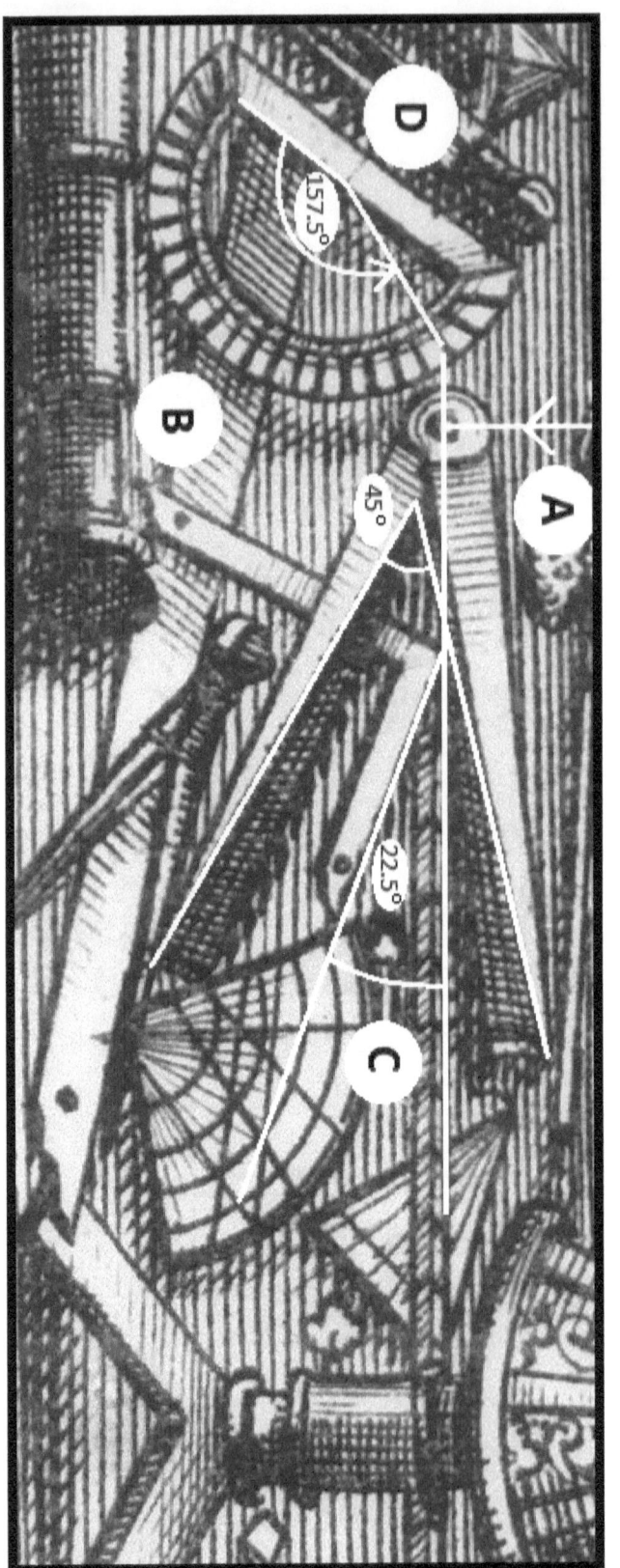

Figure 24 - Second DT Portrait Clues Give Lever Shape

CHAPTER 10

Clues Found in the First DT Portrait

In this chapter, as the title implies, I will present the locations and interpretations of the various alphanumerical and geometrical clues concerning Bessler's imbalanced pm wheel mechanics that are found in the first DT portrait which, as mentioned in previous chapters, was really the second portrait that the inventor intended a reader of DT to view. In this portrait we see Bessler surrounded by the various symbols of a man of culture and great learning. There are books, jars filled with medicinals, and he is richly dressed and wearing an expensive wig that was the symbol of the educated aristocrat in his day. Since no chair is visible in the portrait and the table top before him seems below his waist level, I've always assumed that, rather than sitting, he is actually standing behind the table despite the blanket that is draped over his left side. He is relaxed, confident, and is directing the reader's attention to the book on the table in front of him with his right hand's extended index finger. Behind him is a curtain which is partially drawn aside on the left side of the portrait. Many a reader of an original copy of DT has viewed this portrait over the centuries and wondered about the significances of the various items found it in.

The symbolism of the background curtain and the book on the table that he points to is rather obvious. The curtain is hiding whatever is behind it and it represents the cloth coverings on the sides of Bessler's Merseburg and Kassel wheels that concealed their internal mechanics. But, the curtain appears to be *partially* drawn toward the right side of the portrait. This indicates that the reader will find *some* of the information about the inventor's imbalanced pm wheel mechanics in this first DT portrait and that knowledge is symbolized by the book on the table which would certainly be a repository of information if its subject was mechanics.

Besides these obvious symbolisms, however, the book serves as a means of locating many of the component parameter values used in Bessler's wheels because most of them are actually located near the book. To begin locating and analyzing them, the reader should now find Figure 25 at the end of this chapter and refer to it as he reads the following material.

We saw in the last chapter that covered the symbols represented by the various items on Bessler's shop table in the second DT portrait that all eight of a one-directional wheel's weighted levers were present. Not too surprisingly, the first portrait also has symbols for a one-directional wheel's weighted levers which show them when their pivot pins' center axes are located at a drum's 7:30, 9:00, 10:30, 12:00, 1:30, 3:00, and 4:30 positions *after* a 45° clockwise rotation of a rotating drum has carried their pivot pins' center axes to those locations. Thus, as in the second DT portrait, the weighted lever symbols in the first DT portrait can represent a lever whose pivot pin's center axis is *already* at a particular drum location or a lever whose pivot pin's center axis has just completed the process of moving clockwise through an arc of 45° around the center axis of the drum's axle to a particular clock dial time drum position. It's important that the student of the two DT portraits keep this dual meaning in mind for the weighted lever symbols in both portraits.

However, there is one difference with the weighted lever symbols shown in the first DT portrait. There is no symbol present that I can find for the weighted lever whose pivot pin's center axis was located at the drum's 6:00 position after that axis has traveled there from the drum's 4:30 position. Why? My best guess is that this was Bessler's way of telling the reverse engineer of his wheels that this particular weighted lever did not play an active role, during any 45° segment of clockwise drum rotation, in helping to rapidly lift the end weight(s) carrying main arms of the 9:00 weighted lever closer (but not enough to touch) to their radial stop piece as that lever's pivot pin's center axis approached the 10:30 position of the drum. Since this particular weighed lever, that is, the one whose pivot pin axis traveled from the drum's 4:30 to 6:00 positions, did not actively participate, he chose not to represent it with a portrait symbol. The reason that lever did not participate, as the reader learned in earlier chapters, was because, during any 45° segment of clockwise drum rotation, that lever's main arms were resting on their wooden radial stop piece and would do so until after their lever's pivot pin's center axis passed the drum's 6:00 position.

Let us now proceed to locate the various symbols for the seven weighted levers of a one-directional wheel which are shown in the first DT portrait and which I labeled with letters inside of solid circles, either black or white, in Figure 25.

We can begin with the item labeled "A" which is the leaning vase on the table top. This vase is symbolic of a weighed lever that travels from the drum's 6:00 to 7:30 positions. If one draws a straight line segment (indicated in Figure 25 by the white line segment that has two arrowheads at its ends) through the central axis of the vase and extends it downward, it will reach the letter "F" in the name "C. Fritzsch" written under the bottom left corner of the portrait who was the engraver of the two DT portraits. "F" is the 6th letter of the alphabet and here it tells the reverse engineer that the vase symbol represents a weighted lever that starts a 45° segment of clockwise rotation of its pivot pin's center axis around the center axis of a wheel's axle at the drum's 6:00 position.

To find how the vase symbol tells us the drum location at which the 45° segment of clockwise rotation of the drum finally ends, we must extend the white line segment along the vase's central axis up through the open mouth of the vase. When this is done, the arrowhead at the upper end of the line segment will touch the front right side edge of the third book that sits on the table top.

This book, however, is actually the nineteenth item on the left side of the portrait. To get this cardinal number of the book, one must begin by counting from the first medicinal jar on the left side of the *top* shelf in the background. Only half of that jar is visible in the first portrait. There are four shelves and each appears to be able to hold a maximum of four jars. That means that there are a total of sixteen jars on the four shelves even though some of are hidden by the left side of the curtain. On the table top there are three books and if, starting with the leftmost of these, we add them to the quantity of sixteen items on the shelves, we find that the rightmost of the three books is the nineteenth item.

The extended diagonal line segment from the mouth of the tilted vase touches the right *side* of the 3rd book or the nineteenth item on the left side of the portrait. If there was a fourth book on the table top to the right of the third book, then it would be the twentieth item. So, we can consider the right side of the third standing book on the table to be a region that is *halfway* between the nineteenth and twentieth items. That makes the right *side* of the third book correspond to item #19.5. What is the significance of this? If we now rewrite this number in terms of the hours on the dial of a clock, it would correspond to the time 19:30 or 19 hours plus 30 minutes which are 0.5 of a 60 minute hour. But, what position of a drum does 19 hours and 30 minutes correspond to? If one begins counting clockwise around the dial of a clock starting from 12:00 midnight, he will discover that 19 hours and 30 minutes later, he will arrive at the 7:30 position of the clock's dial! Thus, this very difficult to locate and interpret clue gives us the ending drum position of a weighted lever's pivot pin's center axis as it travels through a 45° segment of clockwise drum rotation starting from the drum's 6:00 position.

Now that we have located the symbol for the 7:30 weighted lever of a one-directional wheel as its pivot pin's center axis makes its way from the drum's 6:00 to 7:30 positions, the next logical lever symbol to look for is the one that represents the 9:00 weighted lever's pivot pin's center axis as it makes its way from the drum's 7:30 position to its 9:00 position. This symbol, surprisingly, is located on the right side of the first DT portrait and is Bessler's left hand whose thumb and index finger appear to be holding the bottom of the book on the table top so that its top edge does not slide off of the skull which is supporting it. But, to confirm that the Bessler's left hand is, in fact, a symbol for a weighted lever traveling between the drum's 7:30 and 9:00 positions, we must be able to use the hand to find these clock times in the portrait or the letters under the portrait. Fortunately, unlike the vase symbol, doing this for the hand symbol is very easy. Here's how to do it.

One notes that the thumb and index finger of Bessler's left hand form the number 7. There are also three other fingers: his middle, ring, and pinky fingers. By multiplying the number 3 by 10 we get 3 x 10 = 30 which is then placed next to the 7, but separated from it by an added colon symbol or ":". This then gives us 7:30 which is the clock dial time that corresponds to the starting drum position of the center axis of the pivot pin of the weighted symbolized by Bessler's left hand. After 45° of clockwise drum rotation, that lever's pivot pin's center arrives at the drum's 9:00 position. That time is also obtainable by studying the hand symbol.

In Figure 25, four of the fingers of Bessler's left hand point downward and to the word "*ORFFYREUS*" which on the first line of text under the first DT portrait. The number of letters in the word "*ORFFYREUS*" is 9 and this represents the finally drum position of the center axis of the pivot pin of a weighted lever after the drum has rotated clockwise through an angle of 45° and carried it away from the drum's 7:30 position. So, with this interpretation of the word to which his fingers point, we have decoded the symbolism of Bessler's left hand in the first portrait as far as which of a one-directional wheel's eight weighted levers it represents.

Before proceeding further, however, I must again point out here, as became obvious in the last chapter's analysis of the second DT portrait's many symbols, that many of Bessler's portrait symbols can actually have *multiple* meanings which will become apparent as the analysis of the first DT portrait symbols continues in this chapter. Why would he do something like that? The obvious answer is to make it as difficult as possible for a future reverse engineer to determine the exact parameter values of the imbalanced pm wheel mechanics he had spent a decade and about one hundred attempts to find. I must also again mention that, while I did manage to find what I believe are the most important clues hidden in the first DT portrait, no doubt future research into them by others will find additional ones. With these matters noted, let us now continue the identification of the various weighted lever symbols found in the first portrait which can then be added to the two found so far that represent the 7:30 and 9:00 weighted levers after their pivot pin center axes had completed a 45° clockwise rotation around the center axis of the wheel's axle.

The inclined book on the table top in Figure 25, labeled "C", is a symbol for the 10:30 weighted lever whose pivot pin's center axis has traveled from the drum's 9:00 position to its 10:30 position during a 45° clockwise rotation of the wheel's drum. The 9:00 position is indicated by the lowest corner of the book that rests on the table top. It forms a downward pointing arrowhead that points toward the letter "Y" in the word "*ORFFYREUS*" on the first line of text below the first DT portrait.

As was mentioned above, there are nine letters in the word "*ORFFYREUS*" and, thus, the word represents a weighted lever whose pivot pin's center axis is at the drum's 9:00 position. Also notice that the letter "Y" in the word "*ORFFYREUS*" looks like the letter "I" placed next to the letter "J". Again, this was no error made by Bessler.

We can treat the letter "I" like the Roman numeral and give it a numerical value of 1. We then use the normal alphanumeric value for "Y" of 25 and multiply its individual digits to get 2 x 5 = 10. Finally, using the Roman numeral method to determine the value of two numerals placed next to each other, we subtract the value of the letter "I" from that of the letter "J" to get 10 − 1 = 9 which again tells us that the starting position of the weighted lever whose pivot pin's center axis was at the drum's 10:30 position was the drum's 9:00 position.

To analyze how the inclined book is a symbol for a weighted lever whose pivot pin's center axis was at the 10:30 position of the drum, we note that the orientation of the spine of the book itself is insufficient because it only forms an angle of 33° with the horizontal surface of the table top and not the 45° angle that would be formed with the horizontal by a radial line segment drawn from the center axis of a drum's axle to the drum's 10:30 position which symbolizes a clock dial time of 10:30. Thus, one might conclude that the inclined book cannot be used to obtain the drum's 10:30 position. That, however, would be a premature conclusion to reach.

If one extends a straight line segment along the left front edge of the book's cover upward and to the left as shown in Figure 25 by a white line segment with a small arrowhead on it, then that line segment will reach an exposed section of the hem of the blanket wrapped around the inventor's right side which is labeled with a lower case black "d" inside of a white solid circle and is enclosed by a small white rectangle. Attached to the lower left corner of the main portion of Figure 25, the reader will find an enlargement of that section of the blanket hem that is labeled with a lower case white "d" inside of a black solid circle.

Study of the embroidered pattern on that small section of the blanket's visible hem shows an "X" which is the Roman numeral with a value 10. Just above the "X" is a odd little pattern that looks like two objects being split apart by a third object with a rounded end that is rising between them. We can interpret this pattern as suggesting that something must be divided into two equal parts. If one divides an hour of 60 minutes into two equal parts, then each part will be equal to 30 minutes. Now, if we take the 10 represented by the Roman numeral "X" and let it represent the time of 10:00 and then add 30 minutes to it, we get the clock dial time of 10:30. So, this portion of the blanket hem, pointed to by the left front edge of the inclined book on the table top, allows us to conclude that the book actually represents a weighted lever whose pivot pin's center axis has traveled 45° clockwise from the drum's 9:00 position to its 10:30 position.

The blanket wrapped partially around Bessler's upper body, labeled "D" in Figure 25, is a symbol for a weighted lever whose pivot pin's center axis has, after a 45° clockwise rotation of the drum, been carried from the drum's 10:30 to 12:00 positions.

As was mentioned previously, the odd embroidered pattern on the visible portion of the blanket's hem at "d" represents the drum's 10:30 position so this can also be used to represent the starting position of a weighted lever represented by the blanket. However, finding the symbol that represents the drum's 12:00 position is a bit more difficult.

The drum's 12:00 position is actually represented by the twelfth visible button in the first DT portrait that is labeled with the lower case "e" in the white solid circle in the main portion of Figure 25 which is also shown in the enlarged view attached near the upper left corner of the main figure and labeled with a white "e" inside of a black solid circle. This twelfth button's ordinal number is gotten by counting the buttons starting from the farthest one to the left on the right sleeve of Bessler's jacket which is the 1st, then continuing upward along the center of the jacket to its collar, and, finally, downward along his shirt which is worn under the jacket. As can be seen from the enlarged view of the twelfth button labeled "e", we only see a small portion of the right side of the twelfth button. That indicates that there's something hidden represented by that button which the reverse engineer of Bessler's wheels must find.

If one very carefully locates certain key points on the blanket formed by the intersection of two or more crease lines or the kinks in a single crease line, then one will find that he can connect them with straight line segments so that all of the line segments will converge exactly on the twelfth button of Bessler's shirt! This then allows us to associate the drum's position of 12:00 with the blanket he wears and conclude that the blanket represents a weighted lever whose pivot pin's center axis has traveled clockwise from a turning drum's 10:30 to 12:00 positions.

The weighted lever whose pivot pin's center axis moves clockwise through an arc of 45° from the drum's 12:00 to 1:30 positions is represented by Bessler's raised right hand which is labeled "E" in Figure 25. A short horizontal line segment extended from the hand's thumb to the right points, as indicated by the small arrowhead at the right end of the line segment, to the twelfth mostly hidden button on his shirt which is labeled "e" and inside of a white solid circle. Thus, we can associate the 12:00 drum position with the right hand symbol for a weighted lever.

The extended index finger of Bessler's right hand in the first portrait is a symbol for the number 1 because this finger is the one a person is most likely to raise when indicating this number to someone else. Note that his middle finger, however, is bent toward his abdomen so that only about *half* of its length from the knuckle to the finger's tip is visible and projects out from behind his extended index finger and points to the right side of the portrait. Since we only see about half of Bessler's middle finger, we can interpret this as standing for the number 0.5. Half or 0.5 of an hour is 30 minutes. We now interpret the number 1 we obtained from his extended index finger as the clock dial time of 1:00 and add 30 minutes to that to obtain 1:30. So, this now attaches the drum position of 1:30 to the right hand symbol to verify that, along with the interpretation of the partial twelfth shirt button representing the drum's 12:00 position, his right hand represents a weighted lever that has reached the drum's 1:30 position after its pivot pin's center axis was carried there from the drum's 12:00 position by 45° of rotation of a clockwise turning drum.

At "F" in Figure 25 is located the skull that has no jawbone and which supports the top of the inclined book and keeps it from falling down onto the surface of the table top (and, as noted in a past chapter, the skull is also supported by the book so that it does not roll over to the right and land on its upper teeth). The skull here is a symbol that represents a weighted lever whose pivot pin's center axis has traveled from the 1:30 to 3:00 positions of a clockwise turning drum.

Associating these drum positions with the skull is actually quite easy. One need only to extend a vertical line segment upward from the skull's empty left eye socket and note that it passes through Bessler's right hand

and finally a small arrowhead placed on the line segment's end reaches the third button on his sleeve. That hand, as previously interpreted, represents the drum's 1:30 position and the third button on his sleeve, obviously, stands for the drum's 3:00 position. With these interpretations, we see that the jawless skull becomes a symbol for a weighted lever whose pivot pin's center axis has reached the drum's 3:00 position after traveling through an arc of 45° from a clockwise rotating drum's 1:30 position.

Finally, to complete our discussion of Figure 25, we must locate the symbol that represents a weighted lever whose pivot pin's center axis is located at the drum's 4:30 position after traveling there from the drum's 3:00 position. This weighted lever is actually symbolized, quite surprisingly, by the top of the table itself!

To get the drum position of 3:00 we note that only three of the fingers on Bessler left hand, this middle, ring, and pinky fingers, actually seem to touch the top of the table. To get the drum position of 4:30 we must examine the lower right hand corner of the first DT portrait which is labeled "G" in Figure 25. Note that the right side of the table makes a perfect 45° angle with the horizontal lower edge of the portrait. We can now take the number 45 from this angle and insert a decimal point and units of hours to obtain 4.5 hours which, in clock dial time, becomes 4:30.

As previously mentioned, I could find no obvious symbol in the first portrait which represents the weighted lever whose pivot pin's center axis travels between the drum's 4:30 and 6:00 positions during each 45° segment of clockwise drum rotation. During this time, a weighted lever's parallel pair of main arms was in physical contact with the lever's wooden radial stop piece and the center of gravity of the lever's lead end weight(s) was as far away from the center axis of the wheel's axle as possible. However, from studying the first DT portrait in order to identify the locations of the symbols for the various weighted levers for Figure 25, it occurred to me that, perhaps, the bottom horizontal border of the portrait itself represents that inactive lever. The right side of the bottom horizontal border ends at the right corner where we previously saw there was a symbol for the drum's 4:30 position that was labeled "G". The left side of the bottom horizontal border of the portrait ends near the bottom of the tilted vase, labeled "A", whose center axis, when extended by a white line segment, points diagonally downward to the letter "F" in the name of the portraits' engraver, "C. Fritzsch". The letter "F" is the sixth letter of the alphabet and would therefore represent the drum's 6:00 position.

Choosing the bottom border of the first portrait to represent a weighted lever certainly seems rather strange. But, in a way, it also makes some sense. The weighted lever moving between the drum's 4:30 and 6:00 positions had a pair of parallel main arms that, because their leading surfaces are in contact with the lever's wooden radial stop piece, had center lines that are actually located inside of the lever's parallel pair of radial frame pieces. In this orientation and by using our imaginary x-ray vision, all four arms would appear overlapped in an axial view and seem to form a single arm with a single center line. Perhaps this alignment of four center lines to form a single line segment reminded Bessler of a simple straight line segment and that is why he chose to represent it in the first DT portrait by the portrait's horizontal bottom border. Of course, this is only speculation and, perhaps, some future DT portrait analyst will find another symbol in the first portrait for a weighted lever whose pivot pin's center axis has traveled from a clockwise turning drum's 4:30 position to its 6:00 position.

Now we move on to Figure 26 where another important hidden clue in the first DT portrait is located and interpreted that, again as was done in the previous chapter for the second DT portrait, gives the reverse engineer the precise angle of 62.5° that the center lines of the main arms of the 9:00 weighted lever must make with the center lines of the parallel pair of radial frame pieces of that lever when one has all eight of the weighted levers of a one-directional wheel's clockwise rotating drum in the precise configuration they must have at the *beginning* the *first* 45° segment of clockwise rotation that follows the release of the self-starting wheel's drum.

At the end of that first segment of drum rotation, however, this angle was reduced slightly by the centrifugal forces acting on the levers with the reduction slowly increasing as the drum approached its maximum terminal rotation rate. And, of course, this resulted in the horizontal displacement of the center of gravity of all eight weighted levers moving slightly closer to a vertical line passed through the center axis of the axle which, in turn, resulted in a slight decrease in the torque being applied to the axle by the eight weighted levers. When a wheel was rotating at its maximum free running rotation rate, its eight weighted levers' center of gravity would be located almost directly under the center axis of the axle, but not quite at the location of its *punctum quietus*.

It is possible to graphically derive this critical angle from Bessler's left hand in the first portrait. Here's how to find it.

One begins by drawing a horizontal line segment that is tangent to the bottom curve of his thumb as shown in Figure 26. The next straight line segment is extended diagonally downward along a somewhat flat skin surface between the first joint of his thumb and the first joint of his index finger. If that second line segment is extended below the bottom border of the first portrait, it will eventually reach the left almost vertical part of the letter "Y" in the word "*ORFYRREUS*" on the first line of text below the portrait. Careful measurement of the interior angle formed by these two line segments will show it is exactly 62.5° as indicated in the figure.

There is also an easy to find alphanumeric clue in the first line of text that gives this angle. To get it we note that the second line segment extended down along the first joint of Bessler's index finger will reach the letter "Y" which is the twenty-fifth letter of the alphabet and, consequently, has an alphanumeric value of 25. We then move the unwritten decimal point of this value one place to the left to obtain the altered value of 2.5. Next, we notice that the letter immediately to the left of the "Y" is the letter "F". "F" is the sixth letter of the alphabet and has a normal alphanumeric value of 6, but, to compensate for our previous decimal point movement one place to left, we now move the decimal point of the value of "F" one place to the right to obtain the altered value of 60. Finally, we just add the two altered values together and attach the units of degrees to obtain 2.5 + 60 = 62.5 = 62.5°.

Next, we must tackle what are the most important clues in the first DT portrait which, unfortunately, are also somewhat difficult ones to decode. They involve analyzing the orientations of the fingers of Bessler's left hand in Figure 27 and showing how one can obtain the separation distances of the center axes of the four A arm coordinating cord hook attachment pins from the center axis of a weighted lever's pivot pin in the Gera prototype wheel. These critically important distance parameter values were obtained in the second DT portrait from an analysis of the circular sector paper piece that directed the reverse engineer to other portions of that portrait that contained alphanumerically encoded information. Bessler repeats these parameter values in the first DT portrait, but uses a different method, mostly alphanumeric, to conceal them.

One might wonder why the inventor would bother repeating the same parameter values in the first DT portrait when he had already provided them in the second portrait. The answer to this mystery is simply that he was afraid that, if the invention ever sold, some rival inventor could claim that the Bessler whose face was depicted in the first portrait was not the same man in the second portrait who actually invented the imbalanced pm wheel mechanics that some buyer was paying a huge sum of money for. It was this nagging thought which originally motivated Bessler to cut out oval holes in the printed image of the original face in the second portrait showing him in his workshop so that, when it was placed over the first portrait, the *same* face would appear in both portraits. Aside from that, however, Bessler also needed to be able to prove that the man shown in the first portrait had not just attached his face to the second portrait in an effort to make it look like he was also the inventor of the working pm wheel when he actually was not. To do that, the man shown in the first portrait also had to prove that he had *detailed* knowledge of the actual mechanics used in the wheel as would be expected of the man who actually discovered those mechanics and whose various parameter values were hidden in the second DT portrait. Providing identical parameter values in both portraits was Bessler's way of being able to prove that, indeed, the man in the first portrait knew the exact same details of the imbalanced pm wheel mechanics as did the man in the second portrait and, therefore, was the same man.

Let us now located and properly interpret the first portrait clues that provide these duplicated separation distance parameter values.

We begin by noting the unusual arrangement of the fingers of Bessler's left hand. As previously mentioned, his thumb and index finger obviously form the Arabic numeral 7 which was used as part of a previous clue that was evaluated. The relationship of the tips of his middle and ring fingers, however, is a bit more difficult to discern. They can be viewed as either touching each other or crossing without touching each other. This ambiguity is, as I eventually learned, not coincidental and its proper interpretation gives some important information about the coordinating cords in the inventor's wheels. Finally, there is the pinky finger that seems to be unnaturally displaced from the hand's ring finger.

After much study of the first portrait, I eventually realized that it's possible to draw straight line segments along those fingers such that, when extended, they reach particular letters in the first and second lines of text under the portrait. The evaluation of the alphanumeric values of these letters then gives one the separation distances of the various A arm coordinating cord hook attachment pin center axes from the center axis of a weighted lever's pivot pin in the 3 feet diameter Gera prototype wheel. The distance of the B arm long lifter cord hook attachment pin's center axis from its lever's pivot pin's center axis is also provided as is the distance of the center axis of the spring hook attachment anchor pin from the center axis of the Gera prototype wheel's axle.

(Note that in the remainder of this volume I will dispense with expressing distance parameters in both decimal form and fractional form to the sixteenth of an inch. That was done in earlier chapters for the convenience of the craftsman wishing to replicate one of Bessler's wheels. The study of the portrait clues really does not require it and the presence of the fractional distance values would only tend to confuse the alphanumerical analyses to follow. One can, as mentioned in an earlier chapter, simply multiply the decimal portion of any distance parameter value by sixteen to obtain the number of sixteenths of an inch it is equivalent to for the purpose of constructing models of Bessler's various wheels.)

We begin with the finger labeled "A" in Figure 27 which is the pinky of Bessler's left hand. This pinky represents the A1 spring cord hook attachment pin which is closest to a weighted lever's pivot pin. As seen in the section of the first portrait shown in the Figure 27, the pinky is bent so as to move it away from the ring finger and this was done to indicate to the reverse engineer that there was something quite different about the A1 spring cord hook attachment pin as compared to the other three coordinating cord hook attachment pins held by a weighed lever's parallel pair of A arms. And, indeed, that is because, unlike the other coordinating cord hook attachment pins that, via two tied on metal end hooks, only connected a *single* length of cord to some other lever's coordinating cord hook attachment pin or to a drum anchor pin such as the stop cord hook attachment anchor pin held between a lever's leading parallel pair of *inner* octagonal frame pieces, the A1 spring cord hook attachment pin was attached to a hook which was then attached to a short length of cord which, again via a metal hook, was attached to one end of a helical extension spring. The other end of that spring was then equipped with a third identical hook that was then finally fastened to its lever's spring hook anchor pin which was embedded in the lever's leading parallel pair of *outer* octagonal frame pieces. The A1 spring cord hook attachment pin was also different from the other three cord hook attachment pins held by a weighted lever's parallel pair of A arms because it was the only cord hook attachment pin which had *two* spring cord hooks attached to it.

A straight line segment drawn diagonally downward to the left from the pinky's last knuckle and through its nail will reach the letter "E" in the word "*ORFFYREUS*" on the first line of text under the portrait. "E", being the 5th letter of the alphabet, has an alphanumeric value of 5. Moving the decimal point one place to the left as suggested by the tip of the pinky bent in that direction and adding units of inches then turns this into 0.5 inch which was the separation distance between the center axes of the A1 spring cord hook attachment pin and the pivot pin of a weighted lever used in the Gera prototype wheel.

The straight line segment extended diagonally downward and to the right from the pinky's largest knuckle and through its next smaller knuckle exits the bottom border of the first DT portrait, passes through the curled dot at the bottom left corner of the semicolon, ";", at the end of the first line of text under the portrait, reaches the pointed tip at the bottom swirl of the last letter "c" in the shortened word "*Comerc*" on the second line of text below the portrait, then immediately passes through the bottom dot of the colon, ":", between the words "*Comerc*" and "*Rath*" on the second line of text, and, finally, reaches the letter "M" in the word "*Mobél*" on the third line of text under the first portrait. (Note that Bessler adds a colon, ":", after some of the words under the first DT portrait to indicate that he has shortened a word by eliminating some of its last letters.)

If one looks at the semicolon at the end of the first line of text under the portrait, it will be noticed that it bears a similarity, depending upon the angle of the missing piece that must be added, to either the Arabic numeral 5 or the numeral 3. For this analysis, we take it as the numeral 5 and use the curled dot at its bottom left side to justify a decimal point move one place to the left to obtain 0.5. Next, we see that the lower end of the extended line segment touches both the letters "c" and "M" and we can then add together the normal alphanumeric values of these letters with the derived numerical value for the semicolon to get 3 + 13 + 0.5 = 16.5. This addition operation is justified by the fact that the semicolon and the two letters are connected by the extended line segment from Bessler's pinky. Adding distance units of inches converts this sum to 16.5 inches which is the exact distance between the center axes of a drum's spring hook attachment anchor pin and the axle in the 3 feet diameter Gera prototype wheel. It's also important to note here that this distance parameter value as well as all others gotten from analysis of the symbols and letters in the first DT portrait will also apply to the analogous distance parameters in the 12 feet diameter, bidirectional Merseburg and Kassel wheels *after* the values for the Gera prototype wheel have been multiplied by a factor of 4.

But, what about the *minimum* separation distance between the center axes of the A1 spring cord hook attachment pin and the spring hook attachment anchor pin held between a weighted lever's leading parallel pair of outer octagonal frame pieces when the lever's main arms were in contact with their wooden radial stop piece and the lever's two extension springs were unstretched? Is there any information about that provided by the pinky? Yes, it's also there.

To obtain that distance, which was 6 inches, we simply use the semicolon at the end of the first line of text under the portrait as the numeral 3 instead of the numeral 5. The former value is then added to the alphanumeric value of the letter "c" on the second line of text to obtain 3 + 3 = 6. After adding the distance units of inches, that value becomes 6 inches.

So, we see from the above analysis of the symbolism of Bessler's left hand pinky in the first DT portrait that it can be used to obtain three important distance parameters used in the construction of Bessler's 3 feet diameter Gera prototype wheel. The pinky represents the A1 spring cord hook attachment pin held between the A arms of a weighted lever in the wheel and the bend in the finger then directs two line segments to various alphanumeric clues in the lines of text under the portrait from which the center axes separation distance values of the A1 spring cord hook attachment pin relative to its lever's pivot pin or 0.5 inch and the lever's leading spring hook attachment anchor pin in the drum relative to the drum's axle or 16.5 inches can be found. The minimum separation distance between the center axes of the A1 hook attachment pin and axle or 6 inches was also provided.

Moving up along knuckles of the inventor's left hand, it is now time to interpret the meaning of Bessler's ring finger, labeled "B" in Figure 27, and see what distance parameter values it can provide us with for the Gera prototype wheel. As in the case of his pinky finger, we will see that we can use his ring finger to direct two straight line segments downward to letters below the first DT portrait whose alphanumeric analysis will then give us some important separation distance values.

As the reader may have surmised by now, Bessler's ring finger represents the A2 stop cord hook attachment pin whose center axis was located at a distance of 0.933 inches from the center axis of its lever's pivot pin. The coordinating cord whose little metal end hook was attached to that A2 hook attachment pin is referred to as the "stop cord" and its other end's hook was attached to the stop cord hook attachment anchor pin whose ends were embedded in the midpoints of its weighted lever's leading parallel pair of *inner* octagonal frame pieces. The center axis of the stop cord hook attachment anchor pin was located exactly 12.934 inches away from the center axis of drum of the 3 feet diameter Gera prototype wheel and the maximum separation distance allowed between the center axes of these two hook attachment pins by the stop cord was exactly 6 inches. Now let's see how we can use the ring finger in the first portrait to obtain these important distance parameter values.

We can start by extending a straight line segment diagonally downward and to the left from the next smallest knuckle down from the largest knuckle of the ring finger and through the tip of that finger until it reaches the letter "Y" in the word "*ORFYRREUS*" on the first line of text below the first portrait. "Y" has a normal alphanumeric value of 25 because "Y" is the twenty-fifth letter of the alphabet. However, when one studies the "Y", it actually looks like it was engraved by placing the letter "I" next to the letter "J"! So, Bessler is telling the reverse engineer of his wheels that he must treat the letter "Y" as though it was a letter *group* consisting of the connected letters "I" and "J". That means that we must work with the normal alphanumeric values of both letters which are 9 and 10 respectively.

It will also be noted that the long lower tail of the letter "J" swings down and to the left until its end reaches and touches the umlaut symbol over the letter "u" in the word "*Für*" on the second line of text under the first DT portrait (note that in German, the umlaut marks placed over a vowel only change the way that the letter is pronounced, but a vowel with this symbol is *not* considered and additional letter to the 26 letter in the German alphabet). The normal alphanumeric value of "u" is 21 because "u" is the twenty-first letter in the alphabet. However, in this case we assign it an altered value of 3 by adding its individual digits together to get 2 + 1 = 3. Since the letter "J" contacts the umlaut marks over the "u", this means we must perform some arithmetic operation involving the two letters "J" and "u". Because the right part of the umlaut mark that the lower end of the "J" touches looks like a miniature solidus or division sign, we divide the normal value of "J" by the altered value of "u" to get 10 / 3 = 3.3 (the actual value is 3.333..., but we do not need this degree of accuracy) and use the pointed tip of the bottom of the "J" that points to the left and the umlaut's two short slashes to justify a decimal point move two places to the left to change this value to 0.033. Next, we return to the normal value of the "I" that was to the left of the "J" inside of the "Y" on the first line of text under the first DT portrait. "I" has a normal value of 9, but we again use the bottom tip of the "J" or "Y" to justify moving its decimal point one place to the left to obtain 0.9. Finally, we add these two derived values for the "I" and "J" that make up the letter "Y" together to obtain 0.9 + 0.033 = 0.933. Adding the distance units of inches then turns this into 0.933 inch which is the separation distance between the center axes of the A2 stop cord hook attachment pin and its weighted lever's pivot pin in the Gera prototype wheel.

Next, we must find the separation distance between the center axes of a weighted lever's stop cord hook attachment anchor pin and the drum's axle. This distance, for the Gera prototype wheel was *exactly* 12.934 inches. Here's how to find it.

We again use the two knuckles in Bessler's ring finger that are the largest and next largest for that finger. A line segment extended downward through the right sides of those knuckles in the ring finger will pass through the bottom border of the first DT portrait and reach the top of the letter "E" in the word "*ORFFYREUS*" on the first line of text below the portrait. Note that the second letter "R" and the letter "E" in

"*ORFFYREUS*" are in contact with each other and that implies that some arithmetic operation must be done with them. We can subtract the normal alphanumeric value of "E" from that of "R" to get 18 − 5 = 13.

If we next extend a line segment downward from the right tip of the serif at the bottom of the left vertical part of the second capital letter "R" in "*ORFFYREUS*" on the first line of text under the portrait and through the right sides of the top and bottom serifs of the left vertical part of the capital letter "H" in the shortened word "*Hess*" on the second line of text below the first portrait, the line segment will finally reach the tiny gap in the middle of the elongated swirl that forms the letter "d" in the word "*des*" on the third line of text below the portrait. This alignment of four intersection points on the line segment was done on purpose by Bessler to show us that the derived values for these letters must be arithmetically combined in some way.

We can use the two dots in the colon to the left of the capital "H" to justify moving the unwritten decimal point of its normal alphanumeric value two places to the left to give this letter the altered value of 0.08. Now look at the "d" in the word "*des*" on the third line of text below the first portrait. Its top becomes a long hook-like flourish that extends to the left, touches the top end of the letter "t" at the end of the shortened word "*Invent*" and appears to be drawing the letter "t" toward itself. The contact between the two letters also tells us that an arithmetic operation must be performed and we choose multiplication of the normal values of "t" and "d" to get 20 x 4 = 80, but, because of the two dots in the colon to the left of the "d", the bottom one of which is touched by the bottom serif of the "t", we move the unwritten decimal point of 80 two places to the left to alter its value to 0.8. If we then multiply the altered values of the "H" and the letter group "td", we get 0.08 x 0.8 = 0.064.

Another "mistake" in the engraving of the letters below the first DT portrait gives the final altered value that we need. Notice the two letters "e" and "s" in the word "*des*" on the third line of text below the portrait. These letters are connected by the long curving bottom of the flourish-like "s" at the end of the shortened word "*Hess*" on the second line of text which implies that some arithmetic operation must be done with the two letters. However, notice the engraving "mistake" that made the "s" in "*des*" look like the number 3! Again, not a real mistake, but Bessler's way of saying that we need to use an altered value for "s" of 3 instead of its normal value of 19. We do that and choose the arithmetic operation of subtraction for the two letters by subtracting the Bessler suggested altered value of "s" from the normal alphanumeric value of "e" to get 5 – 3 = 2. There are *three* point symbols to the left side of their letters near the "s" that looks like the number 3 and they are the dot at the bottom of the capital letter "P" at the beginning of the next word to the right of "*des*", the dot at the bottom of the "s" that looks like the number 3, and, finally, the pointed tip of the end of the bottom of the letter "s" at the end of the shortened word "*Hess*" on the second line of text that connects the two letters, "e" and "s", in the word "des" on the third line of text under the first portrait. We can use these three decimal point symbols to justify moving the unwritten decimal point three places to the left of the value derived for the letter group "es" in "*des*" to obtain 0.002.

We finally have three derived values to use in our analysis: 13, 0.064, and 0.002. If we subtract the sum of the last two derived values from the first derived value, we get 13 – (0.064 + 0.002) = 13 – 0.066 = 12.934. Adding units of inches gives us 12.934 inches which was the exact distance from the center axis a stop cord hook attachment anchor pin to the center axis of the axle in the Gera prototype wheel!

Admittedly, the derivation of this distance parameter is a difficult one and probably the most difficult in the first DT portrait. Happily the derivation of the maximum allowed separation distance between the center axes of the A2 stop cord hook attachment pin and the stop cord hook attachment anchor pin in the drum, which was 6 inches, is much easier to obtain. We need only go to the third line of text below the first portrait and notice the letter "t" to the left of the "d" in "*des*" that the looping top of the letter "d" appears to be embracing. It will be noticed that the "t" is actually part of the two letter group "nt" at the end of the shortened word "*Invent*". The connection between the two letters tells us to perform an arithmetic operation on them and, using a Roman numeral type subtraction of their normal alphanumeric values, we can write 20 – 14 = 6. Adding the distance units of inches then converts this value to 6 inches.

Now we move on to Bessler's middle finger, near the "C" label in Figure 27, which represents the A3 long lifter cord hook attachment pin held by the parallel pair of A arms of one of the 3 foot diameter Gera prototype wheel's weighted levers. As we previously saw, the distance of the center axis of the A3 long lifter cord hook attachment pin from the center axis of its weighted lever's pivot pin was exactly 1.4 inches. The small metal hook that fastens to this attachment pin had a long lifter coordinating cord tied to it and the hook tied to the other end of that cord was then attached to a *leading* weighted lever's B1 long lifter cord hook attachment pin. That B1 long lifter cord hook attachment pin was located between the B arms of a leading weighted lever whose pivot pin's center axis was located 90° clockwise around the wheel's axle away from the center axis of the pivot pin of the weighted lever from whose A3 long lifter cord hook attachment pin the long lifter cord originated. That B1 long lifter cord hook attachment pin's center axis, in turn, is then located at a distance of exactly 2.1 inches from *its* weighted lever's pivot pin's center axis in the 3 foot diameter Gera prototype wheel. Let's now see how the middle finger of Bessler's left hand in the first DT portrait can be used to find these two important distance parameter values of 1.4 inches and 2.1 inches.

We begin by drawing a straight line segment along the straight left side of Bessler's middle finger. That line segment is extended downward until it reaches the letter "E" in word "*ORFFYREUS*" on the first line of text below the first DT portrait and we notice that the "E" and the "R" to its left are just barely connected to form a two letter group. The connection tells us Bessler wants us to perform an arithmetic operation on the values we obtain for the two letters. We assign the normal alphanumeric value of 5 to "E" because it is the 5th letter of the alphabet. We also use the letter "R", but do not use the normal alphanumeric value of "R" which is 18 since "R" is the 18th letter of the alphabet. Rather, we add together the individual digits in "R's" value of 18 to obtain 1 + 8 = 9 and then assign the altered value of 9 to the letter "R". Adding the normal value of "E" and the altered value of "R" together then gives us 5 + 9 = 14. We use the somewhat pinched connection between the "R" and the "E" to justify inserting a decimal point between the individual digits in the sum we obtained to convert it to 1.4. Supplying units of inches then turns this into 1.4 inches which is the distance of the center axis of an A3 long lifter cord hook attachment pin from the center axis of its weighted lever's pivot pin in the 3 foot diameter Gera prototype wheel.

The long lifter cord attached by a little metal hook to the A3 long lifter cord hook attachment pin of the 9:00 weighted lever has its other end attached by an identical little metal hook to the B1 long lifter cord hook attachment pin of the 12:00 weighted lever. The distance of the center axis of that B1 long lifter cord hook attachment pin from the center axis of its lever's pivot pin is 2.1 inches. To obtain this value, we note that the bottom right serif of the letter "E" in the letter group "RE" in the word "*ORFFYREUS*" has a wedge shape whose pointed end points to the letter "U" to the immediate right of the letter group. We use the normal alphanumeric value of "U" which is 21 and use the pointed tip of the "E's" wedge shaped serif to the left of the "U" to justify moving the "U" value's unwritten decimal point one place to the left. That then gives us 2.1 and, adding units of inches, this becomes 2.1 inches which is the distance of the center axis of the 12:00 weighted lever's B1 long lifter hook attachment pin from the center axis of the 12:00 lever's pivot pin.

Is it possible, using Bessler's middle finger in Figure 27, to find the maximum separation distance allowed by the long lifter cord and its two attached end hooks between the center axes of the 7:30 weighted lever's A3 long lifter cord hook attachment pin and the 10:30 weighted lever's B1 long lifter cord hook attachment pin? Again, the answer is "yes". That distance parameter had a value of 19.9 inches and is achieved when a long lifter cord in either coordinating cord layer 2 or 4 is stretched taut between the 7:30 and 10:30 weighted levers at the beginning of each 45° segment of drum rotation. By the end of the segment, however, the distance is slightly shorter and the long lifter cord slackens. Here's is how to find this important value.

We use the letter "Y" to the left of the letter group "RE" in the word "*ORFFYREUS*" on the first line of text under the portrait. We again treat the "Y" as though it is the two letters "I" and "J" placed next to each other. The normal alphanumeric value of "I" is 9 and the value of "J" is 10. Since the two letters are connected, Bessler wants us to perform an arithmetic operation with them. We add them to obtain 9 + 10 = 19. To the right of the "Y" we see the "R" in the letter group "RE". Using just the "R" which has a normal alphanumeric value of 18, we add its individual digits to obtain 1 + 8 = 9. Since the pointed bottom tip of the "Y" is a symbol for a decimal point and it points to the left, we move the decimal point one place to the left of 9 to obtain 0.9. Finally, we add this altered value for "R" to the sum of the normal values for the individual letters "I" and "J" that the letter "Y" can be considered to be made from to obtain 19 + 0.9 = 19.9. Adding units of inches then converts this to 19.9 inches which is the maximum separation distance allowed between the center axes of the 7:30 weighted lever's A3 long lifter cord hook attachment pin and the 10:30 lever's B1 long lifter cord hook attachment pin by the long lifter cord and its two attached end hooks during about the first 15° of each 45° segment of clockwise drum rotation.

Finally, we must analyze the index finger on Bessler's left hand in the first DT portrait which is near the "D" label in Figure 27 and see what numerical parameter values it can provide.

The index finger represents the A4 main cord hook attachment pin held at the ends of the A arms of a weighted lever and, in the case of the Gera prototype wheel being treated, its center axis was exactly 1.866 inches from the center axis of a lever's pivot pin. The small metal hook that was fastened to this pin had a main cord tied to it the other end of which was tied to an identical hook which was fastened to the L1 main cord hook attachment pin held by the main arms of the next adjacent leading weighted lever which was the 10:30 weighted lever if the main cord originated from the 9:00 weighted lever. The center axis of the L1 main cord hook attachment pin on that leading 10:30 weighted lever was located exactly 2.6 inches from the center axis of that lever's pivot pin.

The maximum separation distance allowed between the center axes of one lever's A4 main cord hook attachment pin and its leading lever's L1 main cord attachment pin was exactly 11.3 inches, but this separation distance was only achieved between a lever leaving the drum's 9:00 position and its leading lever leaving the drum's 10:30 position *after* the initial 12° of each 45° segment of clockwise drum rotation had taken place. Now let us see how these three distance parameter values can be derived from the two line segments extended downward from the inventor's index finger as shown in Figure 27.

We start by drawing two line segments down along the left and right sides of the index finger. One line segment, labeled "D" in Figure 27, is extended from the largest knuckle downward along the left side of the finger to the next knuckle, passes through the oddly pointed tip of the finger, exits the bottom of the first DT portrait, and reaches the top of the letter "R" in the word "*ORFFYREUS*" on the first line of text below the portrait. A second line segment, not labeled in Figure 27, starts along the flat *right* side of the index finger just below the second knuckle from the top largest knuckle and is also extended downward and through that odd pointed tip of the index finger. It travels diagonally downward and to the left, exits the bottom of the first DT portrait, and finally reaches the second letter "F" in the word "*ORFFYREUS*". We will begin the analysis with this second line segment.

The second line segment reaches the second "F" in the word "*ORFFYREUS*" which has a normal alphanumeric value of 6. However, notice the wedge shaped serifs on the top right and middle part of the letter. These two serifs have a *total* of *three* little triangular corners on them that point vertically away from the horizontal pieces of the letter to which they are attached. The two opposed triangular corners of the middle piece actually form a little arrowhead that points to the left! We use this to justify moving its unwritten decimal point three places to the left to convert the value of the second letter "F" from 6 to 0.006. Next, we move to the left and consider the *first* letter "F" in the word "*ORFFYREUS*". It also has wedge shaped serifs on its top right and middle parts, but the middle part one only looks like it has a single triangular corner pointing down on it and a little toward the second "F". This is because the top triangular corner on the first "F's" middle part was not as prominently engraved as was the one on the second letter "F". Again, this would have been done purposely by Bessler and is an important clue for the DT portraitologist to find. Thus, while the second "F" has a total of three triangular corners, the first "F" only has a total of two. We can use this to justify moving the unwritten decimal point two places to the left of the normal value of the first "F" to convert its normal value to 0.06.

Finally, we turn our attention to the letter "R" which is to the left of the two "F's" in the word "*ORFFYREUS*" on the first line of text under the first DT portrait. It has a normal alphanumeric value of 18. However, there's something a bit unusual about this "R" as compared to the second "R" in "*ORFFYREUS*". It will be noted that, unlike in the second "R", *all* of lower end of the upper right loop of the first "R" actually swings around to the left to make a point contact with the middle of the vertical part of the letter (which is actually slanted a bit to the right). This was purposely done to tell the portraitologist that he must shift the unwritten decimal point of the normal value of "R", which is 18, one place to the left to obtain 1.8.

Since the letter "R" and the letter "F" to its right side are connected so as to form the letter group "RF", that tells us that we must apply an arithmetic operation to them. Because there are actually two "F's" to right of the "R", this also suggests that we must apply that operation to the second "F" as well. The simplest arithmetic operation to apply to all of the altered alphanumerical values of these three letters is addition and doing that gives us $1.8 + 0.06 + 0.006 = 1.866$. Adding units of inches then turns this into 1.866 inches which is the distance between the center axis of the A4 main cord hook attachment pin of a weighted lever in the Gera prototype wheel and the center axis of that levers pivot pin.

Next, we look at the first line segment mentioned earlier, labeled "D" in Figure 27, that was extended downward until it exited the bottom border of the first DT portrait and reached the second letter "R" in the word "*ORFFYREUS*" on the first line of text under the portrait. The distance parameter value information it provides us with is easy to determine. We can use the normal alphanumeric value of that second "R" which is 18.

If the line segment is extended farther down, it will touch the right tip of the serif across the top of the left vertical part of the letter "H" in the shortened word "*Hess*" on the second line of text below the first portrait. "H" has a normal alphanumeric value of 8. Because the line segment connects the two letters and, if extended to the bottom of the "H" would form a "+" sign with the horizontal middle part of that letter, we then add the two normal letter values of the "R" and "H" together to get $18 + 8 = 26$. Using the top dot of the colon to the left of the "H" which is near its middle "+" sign as justification, we then move the unwritten decimal point one place to the left of the sum we obtained to get 2.6. Adding units of inches then converts this to 2.6 inches which is the center axis to center axis distance between the main arms' L1 main cord hook attachment pin and the pivot pin in one of the weighted levers found in the 3 feet diameter Gera prototype wheel.

So far with the analysis of Bessler's left hand in the first DT portrait as depicted in Figure 27, we have obtained eleven of the dozen critically important distance parameter values for the Gera prototype wheel associated with its four types of coordinating cords and the hook attachment pins to which their metal end hooks were attached. We are still missing the maximum separation distance between the center axes of a weighted lever's A4 main cord hook attachment pin and the L1 main cord hook attachment pin of the next leading weighted lever that is allowed by a main cord and its two attached metal end hooks. That distance value is 11.3 inches and this value is fairly easily found in the first portrait.

To obtain it, we again look at the letter "Y" in the middle of the word "*ORFFYREUS*" on the first line of text below the first DT portrait which is located between the "F" and "R" letters which are reached by the two extended line segments drawn from Bessler's index finger. Again, we consider the letter "Y" to be formed from two separate letters, an "I" and a "J", which are in contact with each other suggesting that some arithmetic operation must be performed using their values. This time, however, we treat the "I" like a Roman numeral and give it a value of 1 instead of its normal value of 9 while the "J" is given its normal value of 10. Adding these two values together gives us 1 + 10 = 11. Next, we follow the large curving bottom of the "J" down and to the left and note that its pointed end touches the left dash of the umlaut mark over the "u" in the word "*Für*" on the second line of text under the first portrait. "u" has a normal alphanumeric value of 21, but we add its individual digits together to get 2 + 1 = 3. We then use the pointed end of the bottom curve of the "J" that points to the left of the portrait to justify moving a decimal point one place to the left so that we can turn the altered value of "u" into 0.3. Adding this to the values we previously got for the letters "I" and "J" gives us 11 + 0.3 = 11.3. Supplying units of inches converts this to 11.3 inches which is the exact maximum allowed distance between the center axes of a weighted lever's A4 main cord hook attachment pin and the L1 main cord hook attachment in the leading lever inside of the 3 feet diameter, Gera prototype wheel.

It is obvious from the analysis of the dozen distance parameter values that can be obtained from the inventor's single left hand shown in the first DT portrait that Bessler's skill in using alphanumeric coding greatly improved after his use of it in the second DT portrait which I maintain was actually the first one he completed and intended to be the first one that a reader of DT would see.

One might assume from the analyses so far presented that Bessler first planned and drew the arrangement and orientations of the various items in the two DT portraits and then added the text below the portraits as a final step. However, I think the opposite was actually the case. In other words, he first carefully arranged the words and lettering in the text lines and, after he was satisfied with the encoded alphanumeric parameter values and their positions in each line of text, he later completed the items in the portraits so that straight line segments extended from them would reach certain letters below the portraits. It must have taken him weeks to months to get the many parameter values for the 3 feet diameter Gera prototype wheel encoded into *both* portraits so they would provide the same values in different ways.

The two DT portraits are a true testimony to the man's tenacity and perfectionism which are the personal qualities one would expect from someone who had successfully constructed the world's only known *working* imbalanced pm wheel, the 3 feet diameter Gera prototype wheel, and who even went on to construct up to a *half dozen* copies of it of various larger sizes and capabilities. Unfortunately, while such qualities allowed him to achieve success where countless others had failed, they also served as an impediment to him achieving business success with his invention. As a result, the world has been, until now, denied the amazing mechanics he had found including what advanced versions that might have evolved from it during the past three centuries.

Let us now move on to Figure 28 in which some important clues concerning the number of extension springs used in Bessler's wheels are located and interpreted. The reader should locate and study this figure as he reads through the next section of this chapter.

In Figure 28 the section of the first DT portrait showing Bessler's head is the focus and one immediately notices the large wig he is wearing. It is dark, shoulder length, and covered with many curls.

Such elaborate head dressings, known as "perukes", were the style throughout Europe for about two centuries following an outbreak of syphilis that reached its peak in Europe in 1580. They had no antibiotics to treat this venereal disease back then and one of its many distressing symptoms was the formation of patchy bald spots on a victim's head. Males who were naturally going bald and did not actually have the disease certainly did not want to even be suspected of being syphilitic because of the moral stigma associated with the infection. So they began to wear wigs made of goat, horse, and human hair to assure others they were free of the infection. Even males who were not naturally balding or actually losing their hair due to syphilis would shave their heads so that the wigs would fit properly.

The wigs were usually powdered and scented with lavender or orange which helped mask body odors. Head lice were also widespread and, when people started wearing wigs, these pesky parasites simply started infecting the wigs. Upon discovering that one's wig was harboring lice, the wig would be sent off to a wigmaker who would then place it into boiling water to delouse it.

The perukes were expensive to make and the larger and more elaborate the wig, the more expensive it was. Because of this, large elaborate wigs quickly became a status symbol for the wealthier upper classes. It is for this reason that paintings of kings from the 16th and 17th centuries show them wearing wigs whose curls flow almost halfway down their backs.

In the first DT portrait section shown in Figure 28, Bessler's shoulder length peruke is intended to signal the reader that he is a man of refinement and some wealth and to be considered on a par with a member of the upper classes. The wig he is shown wearing in the engraved portrait is also probably one that he actually owned and would have worn when interacting with aristocrats or the agents of wealthy potential buyers for his inventions. This, of course, means that when he was not wearing the wig or some other head covering, he *might* have been naturally bald or, if not, regularly shaved his head!

In studying the wig Bessler wears in the first DT portrait, I almost immediately noticed that one can actually count the open ends of is various curls. Doing so, I found that there are sixteen curls on *each* side of the central part of the wig. The meaning of this was obvious. The half of the wig on each side of its part (indicated by the nearly vertical white line segment in Figure 28) represents one of the internal one-directional wheels contained within the enlarged drum of a bidirectional wheel. I've placed white circles on each of the open ends of the curls each one of which corresponds to a single extension spring attached, via its spring cord and its end hook, to a weighted lever's A1 spring cord hook attachment pin. Since there were eight weighted levers in each of the internal one-directional wheels contained within a bidirectional wheel's enlarged drum and each lever was attached to two extension springs, this means that each one-directional wheel, represented by the portion of Bessler's wig on each side of its part, had 8 x 2 = 16 extension springs in it. Thus, from the wig shown in the first DT portrait, one can conclude that the bidirectional Merseburg and Kassel wheels each contained a total of thirty-two extension springs.

Now one must look at the lower rightmost portion of Bessler's wig in Figure 28 that covers part of *his* left shoulder in the first portrait. There is a single curl there, labeled "A", that is noticeably farther from the other curls and purposely placed there to draw the portrait analyst's attention to that area. If one extends a vertical line segment downward from the center of that curl, through the bottom of the first DT portrait, and then through the tip of Bessler's pinky finger which, as we learned above represents the A1 spring cord hook attachment pin, the line segment will reach the top of the letter "U" at the end of the word "*ORFFYREUS*" on the first line of text below the portrait (which I again had to remove a vertical section from in order to fit enlarged images of the relevant portions onto a single page). Interpreting the meaning of these particular letters associated with the inventor's wig curls is now fairly easy to do.

The normal alphanumeric value of letter "U" is 21 and the letter "S" to the right of the "U" has a normal value of 19. There is a short horizontal serif at the end of the bottom part of the "S" which points to the "U" and we can use this as justification to add the normal values of the two numbers to get 21 + 19 = 40. Adding spring constant units of pounds per inch to this turns it into 40 pounds per inch which is the *combined* spring constant of the two parallel extension springs attached to each weighted lever inside of the Kassel wheel's drum. Again, note the semicolon to the right of the "S" at the end of the word "*ORFFYREUS*" on the first line of text under the first DT portrait. If we supply the missing piece to it to make it into the number 5 and use the dot at its bottom left side to justify a decimal point movement one place to the left, then we can assign a numerical value of 0.5 to this semicolon. Next, we multiply the combined spring tension of the two parallel extension springs attached to each of the Kassel wheels weighted levers by 0.5 to obtain 40 pounds per inch x 0.5 = 20 pounds per inch for the spring constant of a *single* spring in that 12 feet diameter, bidirectional wheel.

Can we obtain any information about the spring constants of the extension springs used in the Merseburg wheel from this?

To get those spring constant parameter values, we treat the "U" as though it was made up of two Roman numeral "I's" placed next to each other that form the numeral "II" which has a numerical value of 2. The "S" to the right of the "U" in "*ORFFYREUS*" is then altered by adding up the individual digits in the letter's normal alphanumeric value of 19 to get 1 + 9 = 10. This value is then multiplied by the altered value of the "U" to get 2 x 10 = 20 and, adding the spring constant units of pounds per inch, this then becomes 20 pounds per inch which is the combined spring constant of a *pair* of parallel extension springs that were attached to each of the sixteen weighted levers found in the bidirectional Merseburg wheel. To get the spring constant of a single spring, one need only multiply this by the value of 0.5 previously assigned to the semicolon at the end of the first line of text under the first portrait to get 0.5 x 20 pounds per inch = 10 pounds per inch.

To complete the analyses of the spring constants using the wig curls and the left hand pinky finger of Bessler shown in the first DT portrait, we need to consider the bottommost curl on the *other* side of the inventor's wig. The open end of that curl appears near the top button of his jacket and has the "B" label next it.

If we extend a line segment diagonally downward and to the right of the portrait at the correct angle, it will also pass through the pointed tip of the inventor's pinky finger, exit the bottom of the first DT portrait, pass through the letter "S" in "*ORFFYREUS*", go on to pass through the bottom right pointed tip of the letter "c" in the shortened word "*Comerc*" on the second line of text, and, finally, reach and end at the pointed tip of the serif at the bottom right of the letter "M" in the word "*Mobél*" on the third line of text under the portrait. Adding the normal alphanumeric values of the "c" and "M" together gives us 3 + 13 = 16. We use this value as a divisor of the combined spring constant we got for a parallel pair of extension springs in the Kassel wheel to get 40 pounds per inch / 16 = 2.5 pounds per inch which is the combined spring constant for the parallel pair of extension springs attached to each one of the weighted levers found in the one-directional, 3 feet diameter Gera prototype wheel. To obtain the spring constant of a single extension spring from that wheel, we multiply the combined spring constant value by the assigned numerical value for the semicolon at the end of the first line of text under the portrait to give us 2.5 pounds per inch x 0.5 = 1.25 pounds per inch.

How do we know that these spring constant values apply to the Gera prototype wheel?

Again, Bessler provides an obvious clue in the first DT portrait. We simply add up the normal alphanumeric values of the three letters touched by the second line segment that we extended diagonally downward from the lowest curl in Bessler's wig, labeled "B", which passed through the letters, "S", "c", and "M" on the three lines of text under the first portrait. That addition operation then gives us 19 + 3 + 13 = 35. We then use the left pointing horizontal serif at the bottom of the "S" to justify moving the decimal point one place to the left to obtain 3.5. Finally, we supply distance units of inches and this becomes 3.5 inches which is the exact separation distance between the center of gravity of a lead ingot end weight and the center axis of a pivot pin in a weighted lever inside of the Gera prototype wheel.

Now we must move on to the treatment of the section of the first DT portrait shown in Figure 29 which gives important parameter values for the masses of the end weights and the unweighted levers to which they were attached. It also provides the separation distance between an end weight's center of gravity and its lever's pivot pin's center axis for several of Bessler's wheels.

We can begin by extending a straight line segment, labeled "A" in the figure, along the bottom edge of the visible *front* cover of the inclined book, whose unseen back cover's upper end is supported off of the surface of the table by the skull, diagonally downward and toward the lower left corner of the first DT portrait. As the line segment exits the bottom border of the portrait, it passes through the swirling and highly decorative letters that are the initials, "JEEB", in Bessler's full legal name of "Johann Ernst Elias Bessler" and which take up the beginnings of the first and second lines of text under the portrait. The capital letter "O" in the next word "*ORFFYREUS*" is also intermeshed with the final "B" in "JEEB" to show the reader of DT that the inventor also used the name "*Orffyreus*" which he had derived from the word "Bessler" via a simple letter substitution method described back in Chapter 1. The letter "O" in "*ORFFYREUS*" is also an obvious symbol for the empty drum of one of the inventor's wheels. Note that the "O" is connected to the letter "B" to its left which, in turn, is connected to the letter "E" to its left. The two letters "E" and "B" also appear like *two* letter "B's" which have been placed *back to back*.

This symbolism using the letters "E" and "B" in the initials "JEEB" is a clue that the enlarged drums of Bessler's bidirectional Merseburg and Kassel wheels actually contained *two* internal one-directional wheels which were mounted on the axle in a back to back fashion. Thus, when the front one-directional wheel was turning in its normal self-starting clockwise direction, the back one-directional wheel would be forced to rotate in a counterclockwise direction (as viewed from the back side of the drum) which was the opposite of its normal self-starting clockwise direction of rotation. On the other hand, when the back one-directional wheel was rotating in its normal self-starting clockwise direction, the front one-directional wheel would be forced to rotate in a counterclockwise direction (as viewed from the front side of the drum) which was the opposite of its normal self-starting clockwise direction of rotation. Additionally, the presence of two internal one-directional wheels is reinforced by the contact of the letter "B", which has an alphanumeric value of 2, with the letter "O" which a symbol for a wheel's drum.

As the line segment labeled "A" in Figure 29 is extended farther, it eventually passes through the letter "P" and finally reaches the bottom right side of the letter "M" in the abbreviation "*M:P*", which stands for the words "Medical Practitioner", that is located on the third line of text under the first portrait in its lower left corner. If we add the normal alphanumeric values of "M" and "P" we get 13 + 16 = 29. Adding the mass units of pounds converts this to 29 pounds which is the mass of each of the sixteen weighted levers used in the bidirectional Kassel wheel.

Next, we move on to the line segment labeled "B" which starts at the bottommost part of the thumb of Bessler's left hand, and, when extended diagonally downward and to the left of the first DT portrait, exits the bottom border of the portrait, reaches the rightmost point of the top right part of the first letter "F" in the word "*ORFFYREUS*" on the first line of text, and continues to the second line of text under the portrait where it touches both of the top parts of the two letters "c" and "h" in the word

"*Hoch*" before finally reaching and ending at the pointed, lower right end of the letter "c".

The analysis of these clues is fairly simple. We note that the first letter under the portrait that the line segment "B" touches is the first "F" in the *connected* letter group "RF" in the word "*ORFFYREUS*". Their connection by lower serifs tells us we must perform an arithmetic operation on the two letters. If we simply add together their normal alphanumeric values, we get 18 + 6 = 24 and, after adding mass units of pounds, this becomes 24 pounds. This was the total mass of the three cylindrical lead end weights attached to each unweighted lever in the Kassel wheel. We can then follow the line segment down to the second line of text and note that it touches both of the letters "c" and "h" and connects them which implies we must apply an arithmetic operation to these letters. If we choose multiplication of their normal alphanumeric values, we get 3 x 8 = 24 which, with mass units of pounds, becomes 24 pounds that, again, is the total mass of the three cylindrical weights attached to each of the Kassel wheel's unweighted levers.

However, one can also choose to apply the arithmetic operation of subtraction to the two letters of "c" and "h". If the subtraction is done as happens when determining the value of a group of Roman numerals based on their positions with respect to each other where the smaller value of a numeral to the left of a larger value numeral is subtracted from the larger value numeral (for example "IV" = "V" – "I" = 5 – 1 = 4), then we could subtract the normal alphanumeric value of the letter "c" from that of the "h" and write 8 – 3 = 5. Adding mass units of pounds to this value converts it to 5 pounds which was the mass of an *unweighted* lever inside of the Kassel wheel. The total mass of a weighted lever in the Kassel wheel was then just the sum of the mass of its trio of cylindrical lead end weights (each weight having a mass of *almost* 8 pounds), their three steel attachment bolts (each one of whose mass adds to the "almost" 8 pound mass of the cylindrical lead end weight so that sum of the two masses is very close to 8 pounds), and their unweighted lever. Adding the mass of the three cylindrical lead end weights *and* their three attachment bolts used in the Kassel wheel to the mass of its unweighted lever gives us 24 pounds + 5 pounds = 29 pounds which was the total mass of a single Kassel wheel weighted lever as was determined previously using the letters on the third line of text under the first DT portrait touched by the line segment labeled "A".

We're not quite done yet with the information that can be gotten by alphanumerically evaluating the letters that the line segment "B" passes through and touches in Figure 29.

If we add together the values of the individual digits in the letter "R", which has a normal alphanumeric value of 18, of the letter group "RF" in "*ORFFYREUS*", we get 1+ 8 = 9 and can use 9 as the altered value for "R". Since the "R" and "F" letters are in contact, we again add them together to get 9 + 6 = 14. Adding distance units of inches converts this to 14 inches which was the distance between the center of gravity of the trio of cylindrical lead ends weights held between the ends of a parallel pair of mains arms and the center axis of a pivot pin in one of the weighted levers in *either* the Merseburg or Kassel wheel. Next, we can take the value we got by subtracting, Roman numeral style, the adjacent letters "c" and "h" which was 8 – 3 = 5. Then, using a reversed direction for the pointed lower end of the "c" (again inspired by the Roman numeral style of reversing the direction of subtraction) to justify a decimal point movement one place to the left, we can write the 5 as 0.5. We then add this to the value for "RF" we got above to write 14 + 0.5 = 14.5. Adding mass units of pounds then converts this to 14.5 pounds which was the total mass of a single weighted lever found inside of the bidirectional Merseburg wheel.

But, what about the mass of an *unweighted* lever found inside of the Merseburg wheel? It was 2.5 pounds and it can be found by extending the line segment "B" farther downward until it reaches and ends at the capital letter "I" in the shortened word "*Invent*" on the third line of text under the first DT portrait. "I" has an alphanumeric value of 9 and we then add to it the altered difference value we previously got for the two letters "c" and "h" on the second line of text which was 0.5 to get 9 + 0.5 = 9.5. Next, we note the two capital "F's" in the word "*ORFFYREUS*" on the first line of text and add together their normal alphanumeric values to obtain 6 + 6 = 12. Finally, we subtract the first sum from the second sum to get 12 – 9.5 = 2.5. Adding mass units of pounds turns this into 2.5 pounds or the mass of an unweighted lever used inside of the Merseburg wheel.

We now move on to the line segment labeled "C" in Figure 29 which provides a very important distance parameter used in both the Merseburg and Kassel wheels. It starts at the point of contact between the bottom of the book's spine and the table's top and descends directly vertically downward. After it exits the bottom border of the first DT portrait, it passes through the "Y" in "*ORFFYREUS*", through the capital "M" on the second line, and, finally, stops on the third line of text under the first portrait at the top of the letter "t" at the end of the word "*Invent*" where the "t" is touched by the swirling upper part of the "d" in the following word "*des*".

Evaluation of the parameter value in this clue is fairly simple. We start by breaking up the letter "Y" into two separate, but connected letters, "I" and "J", and then adding together the normal alphanumeric values of these two letters to get 9 + 10 = 19. Then we use the normal value of "M" which is 13. Finally, on the third line of text, we use the normal value of the letter "t" at the end of the word "*Invent*" which is 20 and the normal value of the letter "d" in "*des*" which is 4. We then note that the pointed end of the extended top of the letter "d" points to the center of the letter "t" which looks like a "+" sign and indicates that we need to do an addition. Doing that with the normal values of the "t" and "d" gives us 20 + 4 = 24.

So, after all of this is done we have the following three values to work with: 19, 13, and 24. Adding them together then gives us 19 + 13 + 24 = 56. Adding distance units of inches turns this into 56 inches which is the exact distance between the center axes of the weighted lever pivot pins and the axle in both the Merseburg and Kassel wheels!

The next line segment we can consider would be the one that is extended diagonally downward and to the right along the visible *back* edge of the book's spine and, after exiting the bottom border of the first DT portrait, reaches the capital letter "E" in the word "*ORFFYREUS*" on the first line of text below the portrait near the "D" label. The letter "E" is part of the two letter group "RE" because it's lower left serif touches the letter "R" to the left of it. This contact implies that an arithmetic operation be done on the two letters and we chose multiplication to get 18 x 5 = 90. Next, we notice the letter "U" to the right of the "E" and "U" has a normal alphanumeric value of 21. Finally, below the letter "U" on the second line of text under the first portrait is the capital letter "C" in the shortened word "*Comerc*" which very faintly touches the bottom of the "U" above it indicating that the two letters need to be arithmetically combined in some way. Since the normal value of "C" is 3, we add it to the normal value of "U" to get 21 + 3 = 24.

We now have two values to work with: 90 and 24. Subtracting the smaller value from the larger one gives us 90 − 24 = 66 and, after adding distance units of inches, this becomes 66 inches. What is this distance's significance? It happens to be the exact distance between the center axes of a spring hook attachment anchor pin and the axle in both the Merseburg and Kassel wheels! As the reader may recall, the separation distance between the center axes of the spring hook attachment anchor pin and the axle in the 3 feet diameter Gera prototype wheel was 16.5 inches. Thus, we expect the analogous parameter value in both the 12 feet diameter, bidirectional Merseburg and Kassel wheels to be four times this value and, indeed, it is because 4 x 16.5 inches = 66 inches.

After finding this value alphanumerically encoded into the first DT portrait, one naturally wonders if the other important distance parameter for the center axis of the stop cord hook attachment anchor pin is there as well. The separation distance between the center axis of this anchor pin and the center axis of the axle in either of the 12 feet diameter, bidirectional wheels was exactly 51.736 inches. Amazingly, it is also there, but is far more difficult to find.

We begin by noting that 51.736 = 55 − 3.264 so if we can find these two values in the first portrait, we need only subtract them to obtain the value of the distance parameter we seek. Obtaining the 55 is easy because to get it, we only have to multiply the normal value of the "E", which is 5, in "*ORFFYREUS*" on the first line of text by an altered value for the "U" to its right, gotten by treating it as though it is the Arabic numeral 11 which it resembles, to get 5 x 11 = 55. For the value of 3.264 that must be subtracted from the 55 to get the value we seek, we note the letter "C" on the second line of text whose top part barely touches the "U" indicating some arithmetic operation must be performed using the two letters. Since "C" has a normal value of 3, we need to use this letter in the analysis.

The top of the "C" curls down to a point on its right side indicating a decimal point belongs there which would be the normal position for it. But, notice how the two pointed ends of the "C" seem to be coming together on the left side of the letter "o" in order to pinch it. This implies that another value must be derived from the rest of the letters in "*Comerc*" and added to the value of the letter "C" which is 3. That additional value must be 0.264.

To get that value, we must first get a value of 264 which will then later have its unwritten decimal point moved *three* places to the left to give us 0.264. We start by multiplying the normal value of the "o" in "*Comerc*", which is 15, by the altered value of the "m" to the right of it which is gotten by adding its individual digits in its normal alphanumeric value together to get 1 + 3 = 4 and that then give us 15 x 4 = 60. But, since the "m" has a bar over it meaning that the letter "m" is actually repeated again in the *unshortened* version of the word "*Comerc*", we use the altered value of "m" as a factor *twice* and write 15 x 4 x 4 = 240. This value then becomes the derived value for the letter group "om". We then use the two letters, "e" and "r" to the right of the letter "m" in "*Comerc*" and add the normal value of the "e", which is 5, to the altered value of "r", gotten by adding its individual digits together to get 1 + 8 = 9, and obtain 5 + 9 = 14. This value is then added to the value we got previously for the two letter group "om" to give us 240 + 14 = 254. Unfortunately, this number is still 10 *less* than the value of 264 that we need.

I almost gave up on finding the extra 10 needed to obtain the desired value until something on the third line of text under the first DT portrait caught my eye. Notice the letter "t" at the end of the shortened word "*Perpet*" on the third line of text. The top of the "t" is bent diagonally down and to the left to form an arrowhead or hook that points to the "e" in the word "*Comerc*" and the suggestion is that we must add the normal alphanumeric value of the letter "t" or 20 into the value we obtained for the "omer" portion of the word "*Comerc*" on the second line of text which we saw above was equal to 254. Doing that gives us 254 + 20 = 274 which is now 10 *more* than the value of 264 that we seek! All is not lost, however.

To the left side of the "t" we see the letter "e" with a normal alphanumeric value of 5. But, we use the point of the left pointing serif at the top of the "t" to justify moving the decimal point of the "e's" value one place to the left. That then gives the "e" the altered value of 0.5. We then multiply this altered value times the normal value of the "t" to get 0.5 x 20 = 10 and adding this value to the 254 previously obtained, finally, gives us 254 + 10 = 264 which is the exact value we need. Using the two pointed ends of the capital letter "C" to the left of the letters "omer" in "*Comerc*" and the left pointing tip of the serif at the top of the letter "t" in the shortened word *"Perpet"* as justification, we move the unwritten decimal point of 264 three places to the left to turn it into 0.264. This value is then added to the normal alphanumeric value of "C" to get 3 + 0.264 = 3.264. Finally, this value is *subtracted* from the 55 we got earlier for the product of the "E" and "U" in *"ORFFYREUS"* on the first line of text under the first DT portrait to obtain 55 − 3.264 = 51.736. Adding distance units of inches turns this into 51.736 inches which is the exact distance between the center axes of the stop cord hook attachment anchor pins and the axle in both the Merseburg and Kassel wheels!

To conclude the discussion of Figure 29, I must now provide an alphanumeric analysis for the letters touched by the line segment extended diagonally downward and to the right along the front edge of the book's spine which is attached to the front left side edge of the book's cover. It is labeled "E" in the figure. It will be noted that the line segment passes directly through the place where Bessler's ring fingertip is partially hidden behind his middle fingertip. The implication of this symbolism is that we are going to be obtain some hidden information about the two internal one-directional wheels found in the inventor's bidirectional Merseburg and Kassel wheels which were arranged so that each one's back internal wheel would appear immediately behind its front internal wheel when an uncovered drum was viewed from the front end of the enlarged drum's axle.

We start by taking the normal alphanumeric value of the letter "S" at the end of the word *"ORFFYREUS"* which is the first letter below the first DT portrait that the line segment reaches. That value is 19 and, after assigning a value of 3 to the semicolon to the right of the "S" by adding a short diagonal line segment between its separated upper and lower parts, we can subtract the smaller value from the larger to get 19 − 3 = 16. This value represents the total number of weighted levers found in both the Merseburg and Kassel wheels. If we consider the "U" to the left of the "S" to be the Roman number "II", then we can use its value of 2 as a factor to multiply the 16 by which then gives us 2 x 16 = 32. This number represents the total number of *both* the cat's claw gravity latches *and* the extension springs found in both the Merseburg and Kassel wheels. Thus, there were sixteen pairs of gravity latches, each consisting of an inner and outer cat's claw gravity latch, used by these wheels' sixteen weighted levers and sixteen parallel pairs of extension spring attached to these wheels' sixteen weighted levers. The parallel arrangement of the individual "I" letters in the Roman numeral "II" also suggests that *each* of the sixteen weighted levers in both the Merseburg and Kassel wheels had both a side by side pair of gravity latches capable of latching and holding onto its main arms' steel L1 main cord hook attachment pin as well as a parallel pair of extensions springs attached to its wooden A arms' steel A1 spring cord hook attachment pin.

If we take the normal alphanumeric value of the "U" in *"ORFFYREUS"* which is 21, then we can multiply it by the normal value of the capital letter in the word *"Comerc"* on the line below it to get 3 x 21 = 63. Next, we find the altered value of the letter "S" to the right of the "U", which is gotten by adding its individual digits together to get 1 + 9 = 10. Adding these two derived values together then gives us 63 + 10 = 73. Finally, we again assign a value of 3 to the semicolon at the end of the first line of text under the first DT portrait and subtract this value from the previous one to get 73 – 3 = 70. Adding distance units of inches turns this into 70 inches which is the distance between the center of gravity of the three cylindrical lead end weights and the center axis of the drum, in either the Merseburg or Kassel wheels, when a weighted lever was locked down so that the center lines of its main arms were parallel to the center lines of the lever's parallel pair of wooden radial support pieces.

This alignment only occurred for a weighted lever when its particular internal one-directional wheel was forced to rotate counter to its normal self-starting clockwise direction and, thus, each of its eight weighted levers had their main arms in contact with their radial stop pieces. Such a symmetrical alignment of the disabled internal one-directional wheel's eight weighted levers inside of a bidirectional wheel's drum was only possible when one or both of the two types of gravity latches, inner or outer, was grasping the sides of each lever's L1 main cord hook attachment pin with the square notches at the ends of its longer arms.

This distance parameter value is important because we can subtract it from the *radius* of a 12 feet diameter, bidirectional wheel to get the separation distance between the center of gravity of its weighted lever's three cylindrical lead end weights and the *outside* surface of its enlarged drum's rim wall. The radius of the 12 feet drums Bessler used for his bidirectional wheels was 6 feet which equals 6 feet x 12 inches per foot =

72 inches. The distance between the center of gravity of a weighted lever's *middle* of three cylindrical lead end weights and the rim wall's *outer* surface was therefore 72 inches − 70 inches = 2 inches. Because the outer rim wall was 0.25 inches thick and the wooden bridge piece it was attached to was 0.5 inch thick, that means that the distance between the middle end weight's center of gravity and the *inner* surface of a wooden bridge piece was only 2 inches − (0.25 inch + 0.5 inch) = 2 inch − 0.75 inch = 1.25 inches for both the Merseburg and Kassel wheels.

Since the cylindrical lead end weights used in the Merseburg wheel were 1.78 inches in diameter, they therefore had a radius of 0.89 inch and the gap between the outer surface of the *middle* of a trio of these end weights and the inner surface of the bridge piece was 1.25 inches − 0.89 inch = 0.36 inch. The cylindrical lead end weights used in the Kassel wheel were 2.04 inches in diameter and therefore had a radius of 1.02 inches. The gap between the outer surface of the middle of a trio of these end weights and the inner surface of the bridge piece was 1.25 inches − 1.02 inches = 0.23 inch. These are rather tight tolerances, but my computer models indicate that they are sufficient to allow the mains arms of the weighted levers in either the Merseburg or Kassel wheels to swing into alignment with their radial frame pieces and be locked down against their wooden radial stop pieces without any contact taking place between any of a lever's three cylindrical lead end weights and the inner surface of the bridge piece attached to the ends of the lever's parallel pair of radial frame pieces.

This now completes the analysis of the various clues hidden in the section of the first DT portrait shown in Figure 29 where we saw that extended line segments from the edges of the inclined book on the table top direct the DT portraitologist to important parameter values related to the various weighted lever masses and center axes separation distances found in the bidirectional Merseburg and Kassel wheels. Now it is time that we moved on to the final figure in this chapter which is Figure 30 that the reader will find at the end of this chapter and should study for a while before proceeding.

This last figure provides some very important clues concerning the geometric arrangements of the parallel pairs of 3 arm side pieces used to make the weighted levers in Bessler's wheels and some clues concerning the unique system of layers that Bessler divided the interior volume of each set of eight weighted levers into in order to prevent mutual coordinating cord contact and rubbing while a wheel was in a state of rotation.

After finding an abundance of information in the first DT portrait concerning a weighted lever's end weights and coordinating cord hook attachment pins, I began to wonder if there was any information about the actual shape of a weighted lever in the portrait. The reader may recall that in the last chapter there was an analysis done in Figure 24 that provided some important clues concerning the actual shape of the weighted levers and, using geometric logic, it became apparent that they had a shape that resembled the letter "Y". Since Bessler needed to repeat the same parametric values in both DT portraits, it seemed to me that he should also have provided information about lever shape in the first DT portrait as he had in the second portrait. But, where was it?

It took much effort to find where the clues concerning lever shape were finally hidden in the first portrait and, like most of the inventor's hidden information in the two portraits, it only finally becomes obvious *after* one knows exactly what the numerical values are of the information he is looking for in the portraits. This is certainly a very frustrating way to encode mechanical information into a drawing, but it is the way Bessler chose to do it because, while he wanted the details of his secret imbalanced pm wheel mechanics to be preserved, he did not want to make deciphering them an easy task. That it has taken over three centuries to do so attests to that fact!

Thus, the challenge in using the two DT portraits' many concealed construction clues is to first find the information one needs *independently* of the portraits and then, through much testing of real or virtual model imbalanced wheels, confirm that what one has found *may* actually be valid useable information. After that, the DT portrait clues pertaining to that information, *if* correctly located and interpreted, serve to verify that the information is actually genuine. But still one must do much hunting around for the verifying clues in the two portraits and do much arithmetical and geometrical analysis in the process. What I have revealed in this and the previous chapter may to some seem childishly simple, yet it took me many months to locate and properly interpret the clues even after I was very sure of the parameter values my computer modeling had revealed and which I was then looking to confirm with the portrait clues.

By analogy, Bessler, when he placed his wheels' mechanical details into the two portraits, was like a locksmith who had constructed an elaborate combination lock which requires dialing in about two dozen numbers to open it! He, of course, knows the combination to the lock and can dial it in and unlock the device in a minute or so. The DT portraitologist studying the two portraits' parameter value clues, however, is like someone who finds that closed lock after the locksmith lost it and decides to try to find the combination that will open it. He must carefully listen to the lock's internal mechanism as he patiently dials in one number after another and listens to hear if there was any subtle change coming from the lock's internal mechanism that might indicate the number he just dialed in is working. He is never sure of exactly how many numbers he must find and dial in. But, when he finally does determine the correct sequence of all two dozen numbers and the last number is dialed in, only then will the lock suddenly open for him. While the locksmith may have been able to open the lock in less than a minute, it takes the other person possibly months or even years of daily effort to be able to do so. Of course, once the finder of the lost lock manages to finally open it the first time, thereafter, like the locksmith, he will also be able to open it in a matter of a minute or so whenever he wants.

As I studied the first DT portrait's central inclined book on the table top and Bessler's left hand fingers, I noticed two anomalies that drew my attention.

The first one was that on a book which has a total of four sharply defined *outermost* corners on its front *and* back covers which do *not* meet with the book's spine, I was only seeing *one* of them. With the orientation of the inclined book on the table top, if Bessler's left hand was *not* present, I would expect to see three of the sharp cover corners which would be the front top right side corner, the front bottom right side corner, and the back bottom right side corner. However, with Bessler's left hand present, my view of the front bottom right side corner and the back bottom right corner was blocked. Thus, I was restricted to only viewing the top right side corner of the book's front cover.

I eventually realized this was done on purpose because the inventor wanted someone seriously studying the first DT portrait to use the *point* at the top right side corner of the inclined book's front cover in the portrait as an origin point which would serve in a manner similar to the pointed tip of the upper arm of the duckbill goniometer found in the second DT portrait; that is, this origin point, which I have labeled "P" in the Figure 30, would serve as the starting point from which to extend line segments of varying lengths to other points in the first DT portrait (I also used the letter "P" for this origin point, as I did in Figure 20, to suggest the word "pivot" since the line segments extending away from it could be swung around it in different directions as their other ends searched for a second point in the first DT portrait to make contact with).

The second anomaly I noticed concerned the middle and ring fingers of Bessler's left hand in the first portrait. Aside from them being the only two fingers that touch each other on his left hand, I noticed that the shadow for the two fingers was cast in the *wrong* direction! Their shadow is cast out toward the front of the table nearest the viewer, but the shadow of the nearby bent pinky finger is diagonally up and to the right side of the portrait which should be impossible if the hand was being illuminated by a single light source such as a sunny window or a brightly burning oil lamp. Again, I can only emphasize that this was not an engraving mistake, but was deliberately done so as the draw a DT portraitologist's attention to it.

I decided, as shown in Figure 30, to extend a line segment diagonally downward and to the right from the top right corner of the inclined book's front cover, labeled "P", and see what would happen if I let it travel through the point where the ring finger of Bessler's left hand was touching or actually tucked slightly behind the middle finger of that hand. The line segment then exited the bottom border of the first portrait and reached the top right side of the letter "U" in the word "*ORFFYREUS*" on the first line of text under the first DT portrait. I labeled that point of contact "M".

So, then I had my first line segment, "MP", but what was its significance?

If we consider the letter "U" to the left of the letter "S" at the end of the word "*ORFFYREUS*" to be the Roman numeral "II" which has a value of 2, we can add it to the normal alphanumeric value of the "S" to get 2 + 19 = 21. By then using the short horizontal serif at the bottom end of the "S" that points to the left to justify moving this sum's unwritten decimal point one place to the left, we can alter its value to become 2.1. Next, we add the missing piece to the semicolon at the end of the first line of text under the first portrait to turn it into the number 5 and, using the curled dot at the bottom left of the number to justify moving its decimal point one place to the left, write it as 0.5. Adding this to our previous sum gives us 2.1 + 0.5 = 2.6. Finally, adding distance units of inches, we convert this value to 2.6 inches.

The reader should recognize this distance value by now. It's the exact separation distance between the center axis of the L1 main cord hook attachment pin held by the main arms of a weighted lever and the center axis of that lever's pivot pin in the 3 feet diameter Gera prototype wheel! This means that the line segment "MP" that Bessler coaxed us into drawing in the first DT portrait is actually a symbol for the center lines of the main arms in the weighted levers of the Gera prototype wheel. The next logical question that should occur to the analyst of DT portrait symbols is whether or not there is another line segment that can be extended away from the pointed upper right corner of the inclined book's front cover, labeled "P", to another point in the text under the first DT portrait that will represent the center lines of the A arms found in the weighted levers of the Gera prototype wheel. There is and finding it is now made much easier by the discovery of the significance of line segment "MP" that was just made.

As the reader may recall from the analysis done using the second DT portrait's work shop table items shown in Figure 24, there was an angular sweep of 45° between the center lines of the A and main arms in the three arm weighted levers Bessler used in his imbalanced pm wheels. That means that now, in Figure 30, we should be able to draw the line segment symbolizing a weighted lever's A arms' center lines in the Gera prototype wheel by simply again starting the line segment at point "P" at the upper right corner of the inclined book's front cover and extending the segment downward until its lower end reaches some point, labeled "A" on a letter in the text under the portrait so that the line segment "AP" forms an angle of 45° with the line segment "MP" which we already know is a symbol for the center lines of the main arms of the weighted levers inside of the Gera prototype wheel.

When that line segment "AP" is extended, as shown in Figure 30, it will reach the first letter "R" in the word "*ORFFYREUS*" on the first line of text under the first DT portrait. We use the normal value of the letter which is 18, but use the unusual left pointing contact made with the middle horizontal piece of the letter by its inward curving upper right loop to justify a decimal point movement one place to the left that turns the "R's" value of 18 into 1.8. Since the letter "F" to the right of the "R" has two downward pointing spike-like serifs on the right side of its top piece and on its middle piece which looks a little like an arrowhead pointing to the left, we use them to justify moving the unwritten decimal point of the normal alphanumeric value of "F", which is 6, two places to the left to alter it to 0.06. We then move onto the second "F" to the right of the "R" and use its *three* pointed right side serifs to justify a three place decimal point move to the left, also suggested by the arrowhead-like middle projection of the letter that points to the left, that converts the normal alphanumeric value of that letter "F" into the altered value of 0.006. Adding the values for the three letters, "RFF", together then gives us 1.8 + 0.06 + 0.006 = 1.866. Finally, providing distance units of inches converts this to 1.866 inches. Again, the reader should recognize this distance parameter value which is the separation distance between the center axes of the A4 main cord hook attachment pin held between the ends of a weighted lever's A arms and the lever's pivot pin which was used in the lever's found in the Gera prototype wheel.

Again, looking at Figure 30, we see that we have the "MP" line segment that represents the center lines of the main arms of a Gera prototype wheel's weighted lever and the line segment "AP" which represents the center lines of the A arms of that lever. Their discovery then leads one to finally ask if a line segment can be extended from the point "P" to some point "B" in the text below the first portrait which will then represent the center lines of the B arms of a weighted lever in the Gera prototype wheel. Such a line segment can be so extended and here's how it's done.

From studying the inclined book's front cover, one will notice that there is a somewhat distinct diagonal dividing line between the lighter upper left portion and the darker lower right portion of the cover. If one extends a line segment from the upper right corner of the book's front cover, labeled "P", along that dividing line, it will eventually exit the bottom border of the portrait and finally reach the top left part of the letter "Y" in the word "*ORFFYREUS*" on the first line of text below the first DT portrait.

"Y" has a normal alphanumeric value of 25 since it is the twenty-fifth letter of the alphabet.

Next, one notices that the long lower "tail" of the "Y" sweeps down and to the left and exits the left side of the cursive font letter "M" on the second line of text under the first DT portrait. That implies an arithmetic operation be performed on the two letters. First, however, we need to get an altered value for "M" by adding the individual digits of it normal alphanumeric value, which is 13, together as suggested by the addition sign, "+", formed by the bottom tail of the "Y" with the leftmost vertical portion of the "M". The altered value for the "M" then becomes 1 + 3 = 4. Subtracting that altered value from the normal alphanumeric value of the letter "Y" gives us 25 − 4 = 21. We then use the left pointing end of the bottom tail of the "Y" to justify a move of the unwritten decimal point of this value one place to the left to convert it into 2.1. Adding distance units of inches to this result then gives us 2.1 inches. The reader should recognize that this distance corresponds to the exact distance between the center axes of the B1 long lifter cord hook attachment pin in one of the Gera wheel's weighted levers and the pivot pin of that lever!

So, at this point a DT portraitologist can rightly conclude that Bessler had provided him with the critical axes separation distances for all three of the parallel pairs of arms found in each of the 3 feet diameter, one-directional Gera prototype wheel's weighted levers. He should also rightly conclude that the line segment "MP" symbolized the center lines of the main arms in a Gera prototype wheel's lever, the line segment "AP" symbolized the center lines of the A arms, and, now, the newly found line segment "BP" symbolizes the center lines of the B arms. If one measures the two angular sweeps between the two line segments "AP" and "BP" and between the two line segments "BP" and "MP", he will discover that they are both equal to 22.5°. Thus, we see that the line segment "BP" actually bisects the 45° angle formed by the two line segments "AP" and "MP". At first glance, this, of course, suggests that a weighted lever's B arms were located *between* the lever's main and B arms and all three arms were, therefore, located on *one* side of the lever's pivot pin.

However, as was noted in the last chapter dealing with the clues in the second DT portrait where the shape of the weighted levers Bessler used was discussed using Figure 24, a lever shape that placed all three of its parallel pairs of arms near each other and *within* an angular sweep of only 45° was one that would result in a wheel's long lifter coordinating cords attached between the A and B arms of the levers they connected becoming hopelessly entangled with the three types of arms and would therefore make coordination of a one-directional drum's eight weighted levers physically impossible during drum rotation. And this would, of course, then make keeping the center of gravity of all eight weighted levers on the descending side of the axle during drum rotation impossible.

So, one of the three parallel pairs of arms that made up a weighed lever had to be located on the opposite side of the lever's pivot pin from the other two parallel pairs of arms. This detail is confirmed by the shape of the letter that the extended line segment "BP" reaches in the first line of text below the first DT portrait in Figure 30. It is the letter "Y" in the word "*ORFFYREUS*". If one looks at this letter, he will notice that it has two "arms" in its top half and only one, although curving to the left, in its lower half. Thus, this clue is Bessler's way of telling us that it was the parallel pair of B arms, whose center lines were symbolized by the line segment "BP", that was on the opposite side of the pivot pin of a weighted lever from the side where the parallel pair of main arms and the parallel pair of A arms were located.

Bessler also provides another interesting clue as to how the arms of his weighted levers were oriented with respect to the drum when it was viewed from its front side.

If we extend the line segment "BP" past the point "P" *upward* and toward the left, it will reach that 12th button on Bessler's shirt near the point labeled "B' " and we can then talk about a longer line segment "B' " that has its lower end touching the letter "Y" on the first line of text under the first DT portrait and the its higher end touching the 12th button.

That button, being half hidden behind his outer jacket suggests that properly analyzing its meaning can provide us with some hidden information in the first DT portrait of importance. It does, indeed! The reader saw earlier in this chapter that this button was a symbol for the clock time of either 12:00 noon or 12:00 midnight as well as the 12:00 position of an imbalanced pm wheel's drum. Now we see that, as the longer line segment passes through the point "P" at the upper right corner of the inclined book's front cover, we can associate the clock time of 12:00 and the drum position of 12:00 with the letter "Y" or with *something* that is shaped like the letter "Y". Bessler is telling us here that there was something located at the 12:00 position of his wheels' drums that was shaped like the letter "Y" and this can only refer to a 3 arm weighted lever. Since the partially exposed 12th button is also a symbol for the clock time of 12:00, it also refers to the *two* hands, which are actually more like extended arms, of a clock's dial at that time which *both* point upward. Thus, the inventor is telling us that a weighted lever at the 12:00 position of a drum had *two* of its arms also pointing upward away from the drum's axle and toward its rim wall.

The arms of a weighted lever are its main, A, and B arms and their center lines, as are shown in Figure 30, are symbolized by the line segments "MP", "AP", and "BP", but the upward extension of the end of the line segment "BP" at the origin P to reach the symbol of the button can also be thought of as causing *only* the line segment "BP" to rotate 180° so as to form the new line segment "B' P". When that is done, we now have the weighted lever whose three parallel pairs of arms are arranged into a shape that resembles the letter "Y" although, in the section of the first DT portrait shown in Figure 30, we see that the "Y" shape of a weighted lever suggested by the three line segments "MP", "AP", and "B' P", is approximately inverted compared to what the actual orientation of a lever looks like at a drum's 12:00 position.

At the bottom left of Figure 30, I show an axial view of what the actual weighted lever of the Gera prototype wheel looked like when its pivot pin's center axis was located at the drum's 12:00 position and a small wood clamp was used to hold its main arms in contact with their wooden radial stop piece (in the actual ready to run wheel with the clamp removed, the orientation of the weighted lever shown in the figure would be rotated 22.5° counterclockwise so that its parallel pair of B arms would be vertically oriented).

As can be seen, all of the angles that can be drawn between the various line segments I added to the section of the first DT portrait shown in the *upper* part of Figure 30 are also present in the actual Gera prototype wheel's weighted lever shown in the *lower* left part of Figure 30.

There is a 45° angle located between the center lines of the main and A arms of the weighted lever. There are the two 22.5° angles formed between the center lines of the lever's B arms and the two extensions made through the center axis of the lever's pivot pin of the lever's A arms and main arms which, in the section of the first DT portrait shown in the top part of Figure 30, appear as thin white line segments extended upward from the common end point "P" of the two line segments "AP" and "MP". There are also the *two* 157.5° angles that the center lines of the B arms of the actual weighted lever made with the center lines of the lever's main arms and A arms and which are also shown in the section of the first DT portrait shown in the top of Figure 30 between the line segment "B' P" and the line segments "MP" and "AP". As the reader may recall, the angle of 157.5° was cleverly provided by the "D"-shaped protractor in the second DT portrait.

Thus, from a detailed study and analysis of the various line segments that can be drawn between the upper right corner of the inclined book's front cover, labeled "P", and other points in the portrait and in the first line of text below the first DT portrait, I have absolutely no doubt that a weighted lever whose pivot pin's center axis was located at the 12:00 position of the Gera prototype wheel's drum, when the lever's main arms were vertically oriented, would have the *exact* arrangement and sizes of its three parallel pairs of arms as illustrated in the lower left part of Figure 30.

At the lower right of Figure 30 I have provided an enlargement of the letter "Y" in the word "*ORFFYREUS*" on the first line of text below the first DT portrait in order to show how much the actual weighted lever at the 12:00 position of the Gera prototype wheel's drum resembled that letter. I've also added the arm names to that enlarged letter "Y" to show how each of its parts corresponds to a parallel pair of arms in the actual weighted lever. Now imagine the weighted lever on the bottom left of the figure being rotated 22.5° counterclockwise into the actual position it had in the prototype when its main arms were not locked down against their wooden radial stop piece. In that configuration the center lines of the lever's parallel pair of B arms would be vertically oriented and the similarity to the letter "Y" would become even greater.

Now we need to begin concluding this chapter by taking a closer look at the spine of the inclined book shown in the *continuation* of Figure 30 and see what hidden clues it contains about the mechanics of Bessler's imbalanced pm wheel mechanics. The reader should now find that continuation of Figure 30 at the end of this chapter and briefly study it before continuing.

We can begin with the top of the book's spine near the "A" label through which the line segment "AP" is extended (this line segment should *not* be confused with the similarly lettered segment used in the previous figure because it has nothing to do with the center lines of a weighted lever's parallel pair of A arms). One finds that there are two very closely spaced ribs there that appear to be only a fraction of an inch apart. Careful study of these two parallel ribs shows that they look quite different from the others spanning the width of lower parts of the spine. These two ribs actually appear to run across the *entire* width of the book's front cover and all the way over to its upper right corner labeled "P"!

The meaning of their symbolism becomes obvious after one studies Figures 12(c) and 12(e) in Chapter 6. The topmost rib represents a weighted lever's radial frame piece and the next lower rib represents *one* of a lever's two parallel 3 arm side pieces from which its parallel pair of main arms was made. In the Gera prototype wheel, the gap between the inside surface of a radial frame piece and the outside surface of a weighted lever's main arm was only 0.125 inch. In the Merseburg and Kassel wheels, that gap was increased to 0.25 inch. Those narrow gaps in Bessler's wheels are symbolized by the narrow gap between the two ribs at the very top of the inclined book's spine.

When one views the opposite, bottommost two closely spaced ribs on the book's spine, however, there are some noticeable differences between them and the two topmost ribs of the spine. The two bottommost ribs do not extend across the entire width of the book's front cover and we notice that the book must have been accidentally dropped on the bottom part of its spine which then caused the bottommost rib to be pushed even closer to the nearby rib above it as an unsightly inwardly folded crease formed above the bottommost rib. This was Bessler's way of telling future reverse engineers and craftsmen trying to duplicate his wheels that they needed to use as much precision as possible in their constructions in order to prevent the inner surface of a wheel's weighted lever's radial frame piece from making rubbing contact with the outer surface of the lever's main arm. Such repeated contacts will do nothing except generate a lot of energy wasting heat and noise with every rotation of the wheel's drum.

Moving down to the next much larger gap below the two closely spaced ribs at the top of the book's spine we can safely conclude that this larger gap represents *one* of a weighted lever's two spring cord layers, either layer 1 or 5, which contained one of the two extension springs that was attached to the A1 spring cord hook attachment pin held by a lever's parallel pair of A arms. This symbolism for this larger gap is confirmed by the decorative pattern that spans its width. That pattern is a zigzag and an obvious symbol for a *single* extension spring. Bessler is telling the reverse engineer of his wheels here that there is some important information about a wheel's extension springs that can be obtained from this larger gap. But, what could that information be?

So far in the analysis of the various parameters for the levers and the coordinating cords of the Gera prototype wheel hidden in the first DT portrait, I have shown how the separation distances between the center axes of the various A arms' coordinating cord hook attachment pins and their lever's pivot pin could be found. But, there is one set of distance parameters which have not yet been revealed. They have to do with the maximum allowed separation distances between the center axes a weighted lever's A arms' hook attachment pins and the main and B arms' hook attachment pins of other leading weighted levers and also between a lever's A arms' hook attachment pins and the lever's leading drum hook attachment anchor pins for the *four* types of coordinating cords and their attached metal end hooks found in the Gera prototype wheel. Since that information was given in the second DT portrait, one can be assured that it must, somewhere, also be located in the first DT portrait as well. Here's how to find the values of these four critically important separation distances in the first DT portrait.

Again we start at the upper right corner of the inclined book's front cover, labeled "P", and extend a line segment to the left and through the first large gap at the top of the spine until it reaches a pointed corner at the top of the skull's empty right eye socket labeled "A". Another line segment is then started at "A" and extended vertically downward until it reaches the letter "h" in the shortened word "*Mathem*" on the third line of text below the first DT portrait at the point labeled "A'". Note that the line segment "AP" first passes through an usually dark spot on one of the upper corners of the single zigzagging pattern (indicated by a small white arrow in the enlarged view labeled "F" at the bottom of the continuation of Figure 30) in the first large gap near the top of the book's spine which was previously seen to represent a single extension spring in a spring cord layer and that the vertical line segment "AA'" then passes through the coil-like cursive capital letter "E" on the first line of text under the portrait. Thus, Bessler is telling the DT portraitologist here that he can find some information having to do with the *springs* used in the Gera prototype wheel.

To alphanumerically analyze this clue, we use the left pointing serif at the top of the letter "t" to the left of the letter "h" in the word "*Mathem*" to move the unwritten decimal point of the normal value of "t" one place to the left which turns its value of 20 into 2. Then, we simply use the Roman method of determining the value of several numerals next to each other and subtract the altered value of the "t" from the normal value of "h", which is 8, to get 8 – 2 = 6. We can also add an altered value of the letter "e" to the right of the letter "h" to this derived value to get 6 + 0.5 = 6.5. Adding the units of inches to these two values turns them into 6 inches and 6.5 inches.

These two values, 6 inches and 6.5 inches, represent the minimum and maximum allowed separation distances between the center axes of a Gera prototype wheel's weighted lever's A1 spring cord hook attachment pin and that lever's leading spring hook attachment anchor pin located at the midpoints of its leading parallel pair of outer octagonal frame pieces. When the main arms of a lever were in physical contact with their wooden radial stop piece, the minimum separation distance was 6 inches. This separation distance only existed for a weighted lever as its pivot pin's center axis traveled clockwise from a position of about 12° clockwise past the drum's 3:00 position until the center axis reached the drum's 6:00 position.

As the pivot pin axis of a weighted lever was approaching the drum's 9:00 position, however, the center lines of the lever's main arms would have swung counterclockwise around the lever's pivot pin and away from alignment with the center lines of the lever's parallel pair of radial frame pieces so as to form an angle of 62.5° with them. By the time that happened, the two extension springs attached to the weighted lever would have *each* been allowed to stretch through an additional distance of exactly 0.5 inch from their normal unstretched length of 6 inches and the separation distance between the center axes of the lever's A1 spring cord hook attachment pin and the lever's leading spring hook attachment anchor pin would have reached its maximum allowed distance of 6.5 inches (note that, ultimately, this allowed stretch distance was achieved when the stop cord attached to the lever's A2 stop cord hook attachment pin was finally stretched taut).

To get the analogous minimum and maximum separation distances for the bidirectional Merseburg and Kassel wheels, these distances must be multiplied by a factor of 4. In fact, the minimum value for these two bidirectional wheels, 24 inches, can be gotten from the first DT portrait by simply adding the normal alphanumeric value of the letter "a" in "*Mathem*" to the altered value of the "t" in that same word and then multiply their sum by the normal value of the "h" to the right of the "t" to get (1 + 2) x 8 = 3 x 8 = 24. Adding distance units of inches then turns this into 24 inches. To get the maximum separation distance value for these two 12 feet diameter, bidirectional wheels, which was 26 inches, we simply add together the normal values of the two letters at the beginning and end of the word "*Mathem*" to get 13 + 13 = 26. Adding units of inches then converts this to 26 inches.

These distance parameter values tell us that the extension springs in both the Merseburg and Kassel wheels were only allowed to stretch a maximum distance of 26 inches − 24 inches = 2 inches. Since, in order to prevent damage to the extension springs, their stretch distance could not exceed 50% of their *unstretched* lengths, we can conclude that the extension springs used inside of the two bidirectional wheels' drums could not have had an unstretched length of *less* than 4 inches. With a minimum separation distance of 24 inches, we see that, *if* an unstretched extension spring and its single attached end hook (the one that was attached to its spring hook attachment anchor pin) was *about* 7 inches in length, then the spring cord and its *two* end hooks (one of which was attached to the side of the spring *not* having its own hook while the other one was attached to a weighted lever's A1 arms' spring cord hook attachment pin) would have been *about* 17 inches in length.

As we move down the spine of the inclined book, we reach the next large space located between two ribs and notice it has no discernible pattern in it. We ignore it and continue down to the next lower and middle space of the spine which has the most decorative patterns on it of the four spaces which have such decorations. In the center of that middle space there is a noticeable item which is hard to precisely make out. My best guess is that it is a key of some sort and its inclusion was intended to be a symbol to tell the student of the first DT portrait that using that little key in conjunction with the upper right corner point, "P", of the book's front cover would reveal more information about the maximum separation distances allowed between center axes in the Gera prototype wheel. My guess proved to again be correct. In particular, I found that the clues provided by the middle space of the book's spine pertained to the coordinating cords, the stop and main cords, found in the drum's middle or #3 coordinating cord layer.

If one extends a straight line segment downward and to the left from point "P" and through the top rounded part of the key (indicated by a white arrow in the enlarged view which is shown at the bottom of the continuation of Figure 30 and labeled "G"), it will arrive at the top of the capital letter "O" in "*ORFFYREUS*" at the point where the top right side of the letter forms a hook that catches onto a loop on the top left side of the letter which is labeled "B" in the continuation of Figure 30. That is an obvious symbol for a cord's end hook, but which cord? The letter involved is "O" which has a normal alphanumeric value of 15. Adding this number's individual digits together gives us 1 + 5 = 6 and, after adding distance units of inches, that sum becomes 6 inches. This separation distance was the maximum allowed between a weighted lever's A2 stop cord hook attachment pin and the lever's leading stop cord hook attachment anchor pin in the Gera prototype wheel so this clue in the first DT portrait relates to the *stop* cords and their two attached end hooks in that wheel when the cords were taut. This maximum separation distance was attained when a weighted lever whose pivot pin's center axis was moving clockwise around the wheel's drum had reached a location about 6° short of the drum's 9:00 position and would persist until the lever's pivot pin center axis had passed the drum's 9:00 position and the angle of 62.5° between the center lines of the lever's main arms and the lever's radial frame pieces began to decrease because the main arms of the lever were being lifted closer to their wooden radial stop piece attached to the lever's nearest parallel pair of radial frame pieces. This separation distance parameter was quadrupled in the cases of the Merseburg and Kassel wheels to 24 inches. That value can also be gotten from the word "*ORFFYREUS*" on the first line of text under the first DT portrait by adding the altered value of the "O", which was 6, to the normal alphanumeric value of the letter "R" to the right of it, that is 18, to get 6 + 18 = 24 which, after adding inch units, becomes 24 inches.

But, the middle coordinating cord layer in the drums of Bessler's one-directional wheels, cord layer #3, contained *two* types of cords. There were the eight stop cords and also the eight *main* cords. Information about the maximum separation distance allowed by each main cord and its two end hooks can also be gotten using the "key" symbol in the large center space of the inclined book's spine.

To do this, we again start at point "P" at the top right corner of the book's front cover and extend a line segment downward and a little to the left. However, this time the line segment passes through the tip at the *bottom* end of the key symbol and is extended until it reaches the top right part of the capital "H" in the word "*Hoch*" on the second line of text under the first DT portrait. This point is labeled "C" in the continuation of Figure 30. My computer models of the 3 feet diameter Gera prototype wheel and the clues hidden in the second DT portrait showed conclusively that the maximum allowed center axes' separation distance between a weighted lever's A4 main cord hook attachment pin and the L1 main cord hook attachment pin of the next leading lever was exactly 11.3 inches when a main cord was taut. This separation distance is easily gotten using the letter "H" and some of the letters surrounding it.

It will be noticed that the two letters "H" and "O" overlap and are, thus, in contact with each other and so too are the letters "R" and "F" in the word "*ORFFYREUS*" on the first line of text under the portrait. Both pairs of touching letters are then joined together by the line segment "CP" that passes between them. This tells us we must alphanumerically evaluate each pair of letters separately and then combine the results in some way. To evaluate the value of the "H" and "O" pair, we simply subtract, justified by the horizontal middle piece of the "H" which looks like a minus sign, the alphanumeric value of the "H" from that of the "O" to get 15 − 8 = 7 and then use the pointed curl inside of the top of the letter "O" to justify a decimal point move one place to the left. That turns the 7 into 0.7. Next we subtract the normal alphanumeric value of the "F" from that of the "R" to the left of it in the letter group "RF" to get 18 − 6 = 12. Finally, we subtract the first derived value from the second to obtain 12 − 0.7 = 11.3. Supplying distance units of inches turns this into 11.3 inches or the maximum separation distance allowed by a weighted lever's main cord *and* its two end hooks between the center axes of a weighted lever's A4 main cord hook attachment pin and the L1 main cord hook attachment pin of its leading lever in the Gera prototype wheel when the cord was taut. This separation distance parameter must then be multiplied by a factor of 4 for use in both the Merseburg and Kassel wheels which then gives us a value of 4 x 11.3 inches = 45.2 inches. Is that value found in the first DT portrait? Again the answer is "yes".

To get that value, we simply add up the normal alphanumeric values in the first four letters of the word "*ORFFYREUS*" or the letters "ORFF" to get 15 + 18 + 6 + 6 = 45. To obtain the 0.2 to add to this, we again notice that the "Y" in "*ORFFYREUS*" can be written as an "I" and a "J" next to each other. However, the "I" is not formed from a single almost vertical piece, but rather two closely spaced pieces that make it look like an elongated "II" which is the Roman numeral with a value of 2. Using the left pointing tip of the bottom of the attached "J" as a justification for a decimal point shift one place to the left, we change the 2 into 0.2. That is then added to the previous sum gotten for the four letters "ORFF" to give us 45 + 0.2 = 45.2. Adding distance units of inches turns this into the maximum allowed separation distance value for the adjacent center axes of the A4 and L1 main cord hook attachment pins found in Bessler's two 12 feet diameter, bidirectional wheels or into a value of 45.2 inches when their enlarged drums' main cords were taut.

Having obtained the maximum allowed center axes separation distances associated with the stop and main coordinating cords found in the middle coordinating cord layer or layer #3 of the Gera prototype wheel and the two 12 feet diameter bidirectional wheels using the middle large gap of the inclined book's spine, it's time to move onto the next large gap below it.

That gap represents either coordinating cord layer #2 or #4 which *each* only contained four long lifter cords. Again, my computer models as well as the symbols in the second DT portrait conclusively indicated that the center axes' maximum separation distance allowed by the Gera prototype wheel's long lifter cords and their two attached end hooks, when taut, between a weighted lever's A3 long lifter cord hook attachment pin and the B1 long lifter cord hook attachment pin of the lever whose pivot pin's axis led that originating lever's pivot pin axis by 90° was exactly 19.9

inches. For the Merseburg and Kassel wheels, this distance parameter increased to four times as much or 79.6 inches. Here's how to find these parameter values hidden in the section of the first DT portrait shown in the continuation of Figure 30.

Although it is obscured by the *two* joined white line segments in the continuation of Figure 30, there is an odd cusped line at the top corner of the design in the large space below the middle large space of the inclined book's spine (it is indicated by a white arrow in an enlarged view at the bottom of the continuation of Figure 30 which is labeled "H"). After the cusp, the direction of the line in the design suddenly bends downward and this was the inventor's way of symbolically telling the DT portraitologist that he must extend a *second* line segment vertically downward from the point of the cusp and through the bottom border of the first DT portrait until its end reaches the letters of the text below the first DT portrait. I've indicated the location on the book's spine of that cusp point with the smallest diameter solid white circle "D" label and show a line segment, "DP", drawn from the inclined book's front cover's upper right corner, labeled "P", to the point "D" located on cusp. Then, to indicate the deflection of the direction of extension of the first line segment "DP", a *second* line segment, labeled "DD'", is started and extended vertically downward until it touches the letter "v" in the shortened word *"Invent"* on the third line of text under the first DT portrait. However, as the line segment "DD'" was extended to the letter "v", it *initially* passed through the first letter "F" in the word *"ORFFYREUS"* on the first line of text below the portrait and then touched the lower left part of the letter "F" in the word *"Für"* on the second line of text.

We can evaluate this clue by using the normal alphanumeric values of the two letter "F's" that the line segment "DD'" passes through which are both 6. Then we determine the altered values of the letter "v" in the shortened word *"Invent"* on the third line and the letter "u" in the word *"Für"* on the second line of text gotten by adding together their individual digits which for the "v" gives us 2 + 2 = 4 and for the "u" gives us 2 + 1 = 3. Adding together the values for these four letters gives us 6 + 6 + 4 + 3 = 19. Finally, we add in the value of the letter "r" at the end of the word "Für", however, we do not use its normal value of 18. We must, as was done with the "v" and the "u", alter its value of 18 by adding together its individual digits which gives us 1 + 8 = 9. But, we're still not done altering its value. Notice the left pointing serif at the top left of the letter "r". It appears sharper and more prominent than it does on the other "r's" in the text under the first DT portrait. This was not an accident, but, rather, an instruction from Bessler to move the unwritten decimal point of the altered value for the "r" we got one place to the left to turn it into 0.9. Adding this to the previous result we got for the sum of the letters "F", "v", and "u" then gives us 19 + 0.9 = 19.9. Adding distant units of inches makes this 19.9 inches which was the maximum allowed separation distance between the center axes of a weighted lever's A3 long lifter cord hook attachment pin and the B1 long lifter cord attachment pin of the lever whose pivot pin's center axis led the pivot pin center axis of the cord's originating lever by 90° in the Gera prototype wheel's drum.

Can we also find the corresponding separation distance in the Merseburg and Kassel wheels which was four times greater or 4 x 19.9 inches = 79.6 inches? It is there, but well hidden.

We must first *multiply* the normal alphanumeric value of the first letter "F" in the word "*ORFFYREUS*" on the first line of text under the first DT portrait by the normal value of the capital letter "F" in the word "*Für*" on the second line of text. That gives us 6 x 6 = 36. Next, we add the *normal* values of the letter "v" in the shortened word "*Invent*" on the third line of text and the normal value of the "u" in the word "Für" on the second line of text to the previously obtained product to get 36 + 22 + 21 = 79. If we can find a value of 0.6 nearby, then that can be added to this sum to obtain the correct value.

The reader will notice from studying the continuation of Figure 30 that the right part of the umlaut symbol over the "u" in "*Für*" is touched by the letter "Y" in the word "*ORFFYREUS*" on the first line of text under the portrait. The bottom tail of the "Y" also passes through the letter "M" on the second line of text. All of these contacts indicate that one must perform some arithmetic operations on all of the letters involved.

If we again break the "Y" up into the two letters "I" and "J" that are in contact with each other, then we can just use the normal alphanumeric value of "J" which is 10. The normal value of the letter "M", through whose left arch the tail of the "J" passes, is 13. We then derive an altered value for the "M" by adding together the individual digits of its normal value to get 1 + 3 = 4. Subtracting this altered value from the normal value of the "J" then gives us 10 − 4 = 6. We're still not done. We then use the left pointing lower tip of the "J" that touches the umlaut over the "u" in "*Für*" to justify moving the unwritten decimal point one place to the left to turn this derived result for the letter "J" into 0.6. Finally, we add this last derived value for the letter "J" to the result that we previously got when we added the product of the two letter "F's" to the normal values of the letters "v" and "u" to finally get 79 + 0.6 = 79.6. Adding units of inches then makes this 79.6 inches which was the maximum separation distance allowed between hook attachment pin center axes by a taut long lifter cord and its two metal end hooks in the Merseberg and Kassel wheels.

Finally, we have to consider the lowest and last large gap at the bottom of the inclined book's spine. This space can, like the topmost larger space on the spine, also represent *either* coordinating cord layer #1 or #5 each of which contained eight spring cords and their attached end hooks as well as the eight extension springs each of which, by one of its end hooks, was attached to a weighted lever's leading spring hook attachment anchor pin held between a pair of parallel outer octagonal frame pieces. By carefully studying the pattern inside of this bottommost space, it can be seen that it actually contains *two* side by side, zigzagging patterns as compared to the single zigzagging pattern found in the topmost larger space on the book's spine. The symbolism is obvious. This was Bessler's way of showing that the spring constant of each extension spring used in the Kassel wheel was double that of a spring used in the Merseburg wheel or 20 pounds per inch as compared to 10 pounds per inch.

However, despite these differences we should expect the maximum and minimum separation distances allowed by the spring cord and its attached extension spring and hooks between the center axes of a weighted lever's A1 spring cord hook attachment pin and the lever's leading spring hook attachment anchor pin to be identical with the values obtained using the single zigzagging spring symbol found earlier in the topmost larger space of the book's spine. Those separation distance values for the 3 feet diameter, one-directional Gera prototype wheel were a minimum value of 6 inches and a maximum value of 6.5 inches. For *both* the bidirectional Merseburg and Kassel wheels, these separation distance parameter values were increased by a factor of four to a minimum value of 4 x 6 inches = 24 inches and a maximum value of 4 x 6.5 inches = 26 inches which indicates that, when an extension spring in one of these wheels had achieved its maximum stretched length, it was 2 inches longer than its normal unstretched length. As will be seen, amazingly, all of these important parameter values can be found by using only four letters under the first DT portrait!

To begin the alphanumeric analysis, we can extend a line segment from the upper right corner of the book's front cover, labeled "P", downward and to the right to a particular point, labeled "E", located in the bottommost larger space on the book's spine. That point, is, unfortunately obscured in the section of the first DT portrait shown in the continuation of Figure 30 by the width of the white line segment, but it is a noticeably darker spot on one of the corners of one the zigzagging spring symbols near the topmost corner of the larger space at the bottom of the spine (it is, however, indicated by a white arrow in the enlarged view, labeled "I" at the bottom of the continuation of Figure 30). Again, this spot was made a bit darker than the surrounding ones on purpose by Bessler in order to draw the attention of a future DT portraitologist to it.

From that darker spot, labeled "E", we then extend a second line segment, "EE'", vertically downward and past the bottom border of the portrait, through the second letter "F" in the word "*ORFFYREUS*" on the first line of text under the first DT portrait, through the end of the tail of the "Y" in that word that sweeps down into the second line of text, through the letter "r" in the word "*Für*" on the second line of text, and finally stop extending the line segment when its end reaches the letter "n" in the shortened word "*Invent*" on the third line of text below the portrait. For the analysis to follow, we can actually ignore the letters that the line segment "EE'" passed through on the first and second lines of text under the portrait.

The minimum and allowed maximum separation distances between the center axes of an A1 spring cord hook attachment pin and the associated drum's spring hook attachment anchor pin that were connected together by a spring cord, its attached extension spring, and their three associated metal hooks in the Gera prototype wheel of 6 and 6.5 inches can be easily found by just using the connected letters in the group "nt" in the shortened word "Invent" on the third line of text under the first DT portrait.

To get the minimum center axes' separation distance, we subtract, using the Roman numeral method for determining values, the smaller normal alphanumeric value of the letter "n" from the larger value of the "t" to get 20 − 14 = 6 which, after adding the distance units of inches, becomes 6 inches. To get the maximum allowed center axes' separation distance of 6.5 inches which was attained after an extension spring had stretched 0.5 inch as its weighted lever's pivot pin approached the 9:00 position of the clockwise turning drum, we add in an altered value for the letter "e" to the left of the connected letters in the group "nt" which is obtained *after* we have moved the unwritten decimal point of the "e's" alphanumeric normal value of 5 one place to the left. This decimal point move is justified by the left pointing flourish that swings over the top of the "e" from the letter "M" on the second line of text, to give us an altered value of 0.5 for the letter "e". We then just add distance units of inches to this value to turn it into 0.5 inches. Finally, performing the necessary addition gives us 6 inches + 0.5 inches = 6.5 inches for the maximum separation distance between the center axes of an A1 spring cord hook attachment pin and its assigned spring hook attachment anchor pin in the drum of the 3 feet diameter, one-directional Gera prototype wheel.

For the corresponding values in the bidirectional Merseburg and Kassel wheels, these values must both be multiplied by a factor of 4. That factor is provided by the letter "d" in the word "*des*" to the right of the shortened word "*Invent*" on the third line of text under the first DT portrait. Notice how the looping top of the "d" actually touches the top of the "t" in "*Invent*" and seems to be trying to pull the letter group "nt" toward itself. Again, this was an intentional clue inserted into the first DT portrait by Bessler to show us that we must use an arithmetic operation, in this case multiplication, with the separation distance values we obtained for the Gera prototype wheel using the letters in the word "*Invent*". In order to get the analogous minimum separation distance between the center axes of a weighted lever's A1 spring cord hook attachment pin and the lever's leading spring hook attachment anchor pin in the drum of *either* of the two 12 feet diameter, bidirectional wheels, we write 6 inches x 4 = 24 inches. For the analogous maximum allowed separation distance between the center axes of the pins involved, we write 6.5 inches x 4 = 26 inches.

Again, the difference in the values of these two distance parameters demonstrates that the maximum allowed stretch distance for each of the extension springs in either of the Merseburg and Kassel wheels was limited to only 2 inches regardless of the unstretched lengths of their extension springs which could not be *less* than 4 inches in order to prevent permanent damage to the springs as they were alternately stretched and relaxed once during each drum rotation in whichever of a bidirectional wheel's two internal one-directional wheels was the active one at the time.

Some readers may question the ultimate value of these many clues encoded, both alphanumerically and geometrically, into the two DT portraits. They will reason that Bessler was able to find a design for an imbalanced pm wheel that worked without having such clues available to him.

That is, of course, true. All that Bessler really had available to him was an excellent education considering the social class he came from, skillful crafting abilities honed by years of practicing many trades, incredible tenacity, an unshakable faith that perpetual motion was possible and that he, with God's divine help, would find it, an usually invigorating dream, and phenomenal luck. He had no hidden clues to guide him to eventual success. If he had been supplied with them by someone else, then his successful construction of a working imbalanced pm wheel would not have been his invention, but merely a duplicate of someone else's invention. Any craftsman attempting to duplicate one of Bessler's self-moving wheels using the information in this volume should keep this in mind. It is not his invention that he is building, but only a duplicate of one Bessler constructed about three centuries ago. However, any improvements to Bessler's basic design that a present day craftsman makes will be the product of his creativity and his invention which he can legitimately patent in his name.

However, Bessler also knew that any future reverse engineer who might try to duplicate one of his wheels and their secret mechanics would, most likely, not have all of these factors working in his favor. He would need a little help in the form of the hidden clues in the two DT portraits. Bessler also wanted to make sure that those clues would only be recognizable to a craftsman who was actually building wheels and constantly modifying them in an attempt to come closer to a working design. Thus, the DT portrait clues serve as sign posts on the road to success that actually subtly steer the craftsman in the direction his designs should take. Indeed, if it had not been for the thousands of hours I personally invested in the construction of virtual computer model wheels in my search for Bessler's design, these clues would have meant nothing to me as they have to thousands of others who saw them over the past three centuries and then moved on to other matters. Now, however, with the publication of this volume, these clues and their meanings can be available to all craftsmen to use in their duplications of Bessler's various wheels.

As this chapter closes, the reader should now have a fairly exact knowledge of the previously unknown mechanics that Bessler used in his imbalanced pm wheels. In the next and final chapter, I will be summarizing, in table form, the various parameter values for the wheels that are the subject of the clues in the two DT portraits which means the parameter values for the 3 feet diameter, one-directional Gera prototype wheel and the two 12 feet diameter, bidirectional Merseburg and Kassel wheels. However, there were several *other* wheels that Bessler constructed or planned to construct whose parameter values I can not find in the hidden clues of the two DT portraits. That does not mean that they are not there, just that I was not able to find them. To determine the parameter values for those other wheels requires that the reverse engineer use his knowledge of the details of the construction of the wheels whose parameter values are now known to extrapolate the values for those other wheels.

In the next chapter I will cover the fairly simple arithmetic method that allows one to determine the parameter values for the second Gera, Draschwitz, Merseburg, Kassel, super, and Karlshafen wheels using only the information Bessler provided in the two DT portraits for the parameters of his 3 feet diameter Gera prototype wheel. The parameter values used in all of Bessler's wheels will then be neatly summarized in two large tables that the craftsman can refer to in order to quickly obtain the critically important parameter values he needs to begin constructing a replica of any one of Bessler's wheels. These tables should then allow him to decide which wheel size is most appropriate for him to attempt duplicating.

From studying the arithmetic methods used to extrapolate the various parameter values of the Gera prototype wheel to other of Bessler's wheels, the craftsman will find that he will eventually be able to design a custom wheel having almost any desired drum diameter, drum thickness, and start up power output. Ideally, such a custom wheel should have the most startup power output for the smallest drum diameter and thickness possible. However, even when using modern materials in its construction, there will always be practical limits as to how much start up power one can obtain from a wheel whose drum has a given diameter and thickness. To push the wheel's start up power output beyond that limit will require the craftsman to turn inventor and try to find modifications to Bessler's basic imbalanced pm wheel mechanics that will improve its performance. Such efforts will, no doubt, be the subject of much discussion in the future after Bessler's original wheels are successfully replicated using the information in this volume.

Again, as I've mentioned several times before, I *strongly* recommend that any craftsman, regardless of his skill level or available equipment and facilities, begin exactly where Bessler did with the construction of his smallest 3 feet diameter, one-directional Gera prototype "toy" wheel. Its construction should be considered as necessary "basic training" for acquiring the skills one will need in order to successfully pursue larger and far more powerful Bessler wheel replicas.

Figure 25 - First DT Portrait Table Items

Figure 26 - First DT Portrait Clues Give Angle of 9:00 Lever

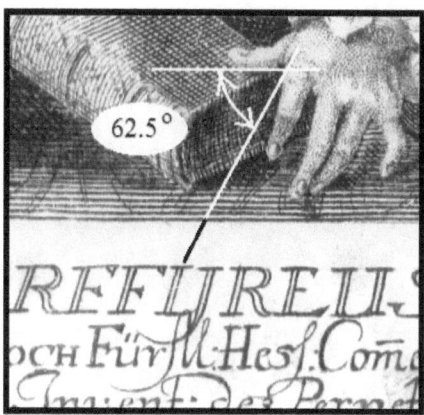

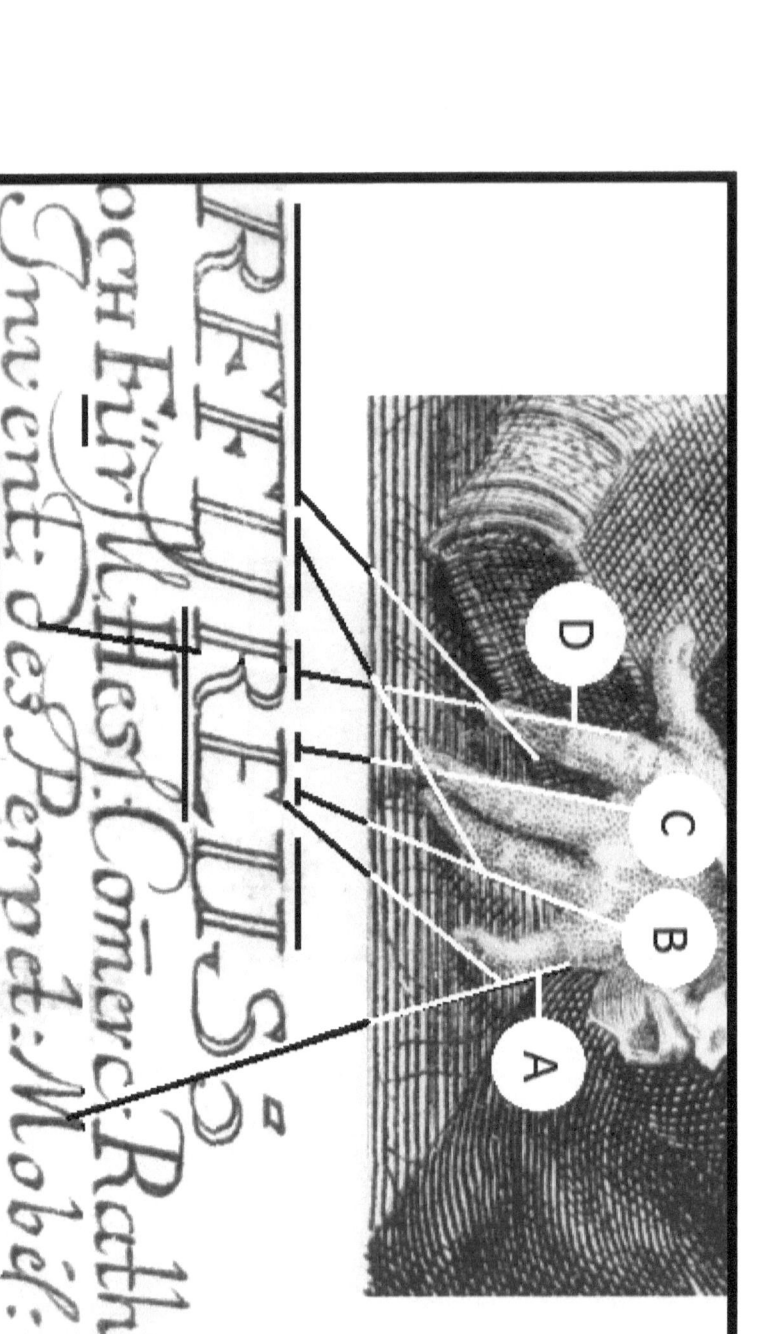

Figure 27 - First DT Portrait Hand Gives Cord Attachments

Figure 28 - First DT Portrait Wig Gives Spring Clues

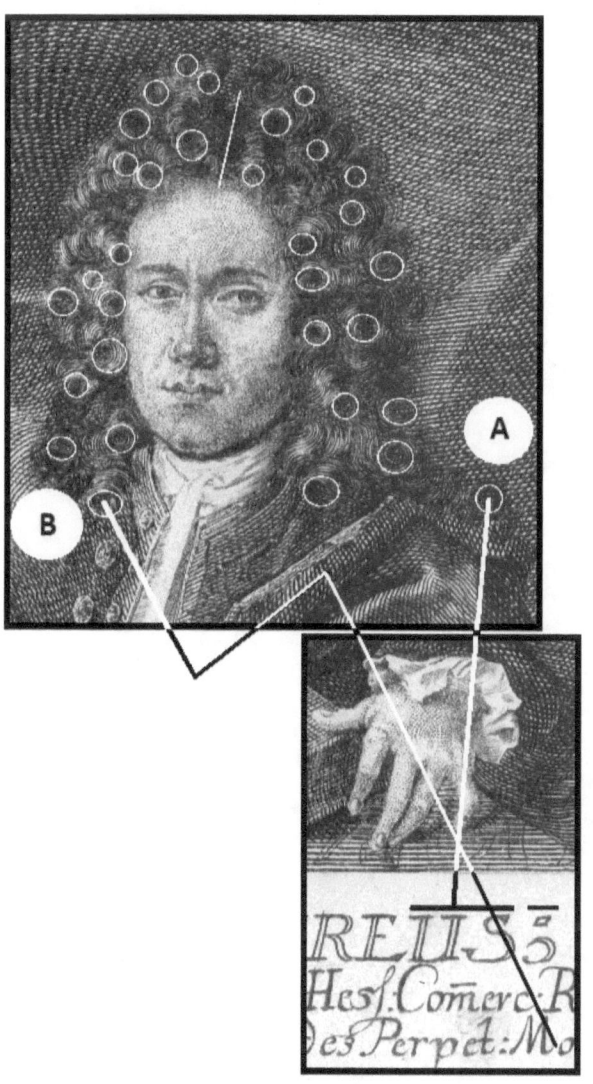

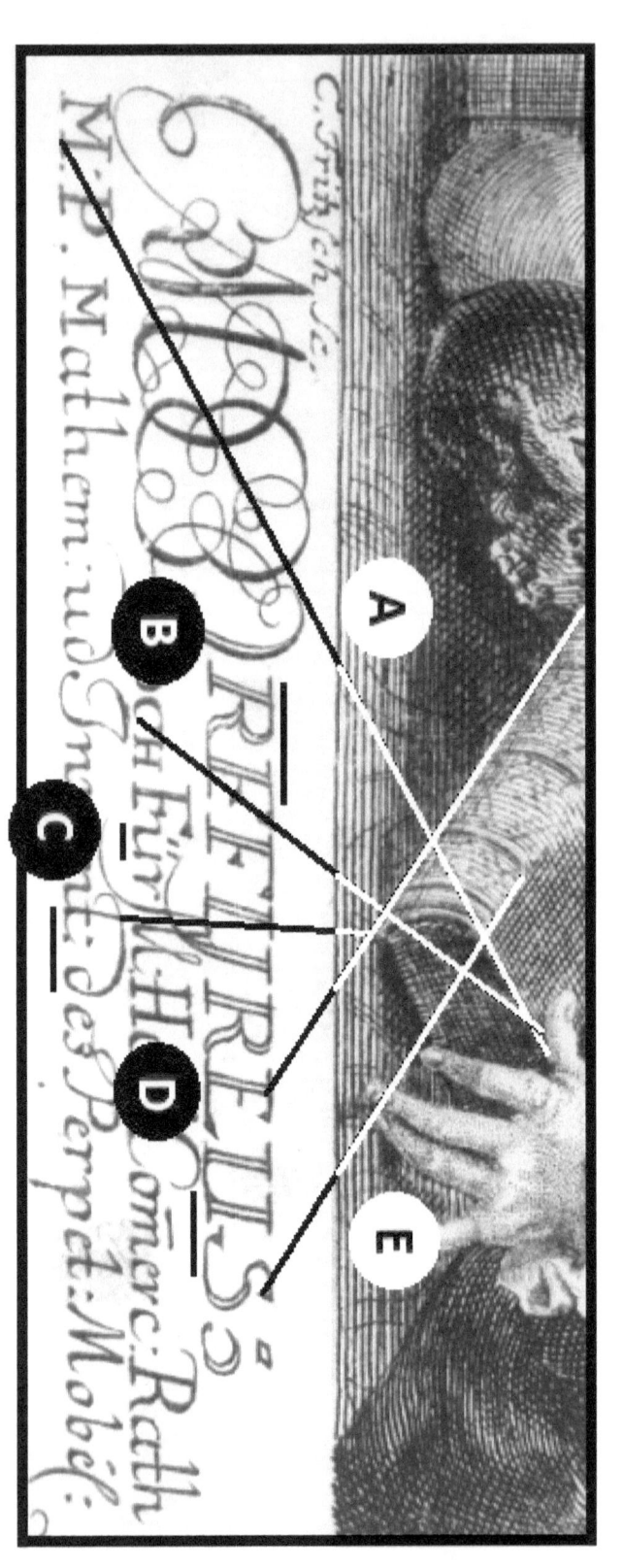

Figure 29 - First DT Portrait Clues about Weight and Lever Masses, Axle Center to Lever Pivot Center Distances, and Lever Lengths

Figure 30 - First DT Portrait Clues about Lever Shape and Cord Layers

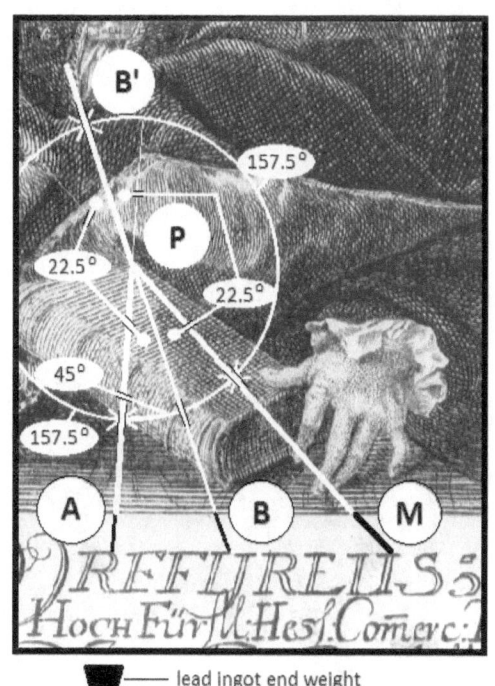

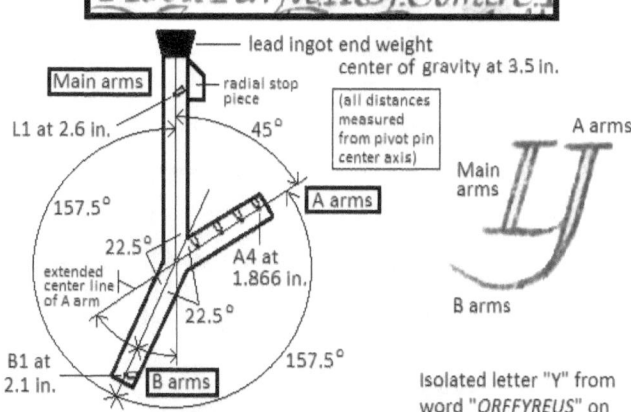

Figure 30 - continued

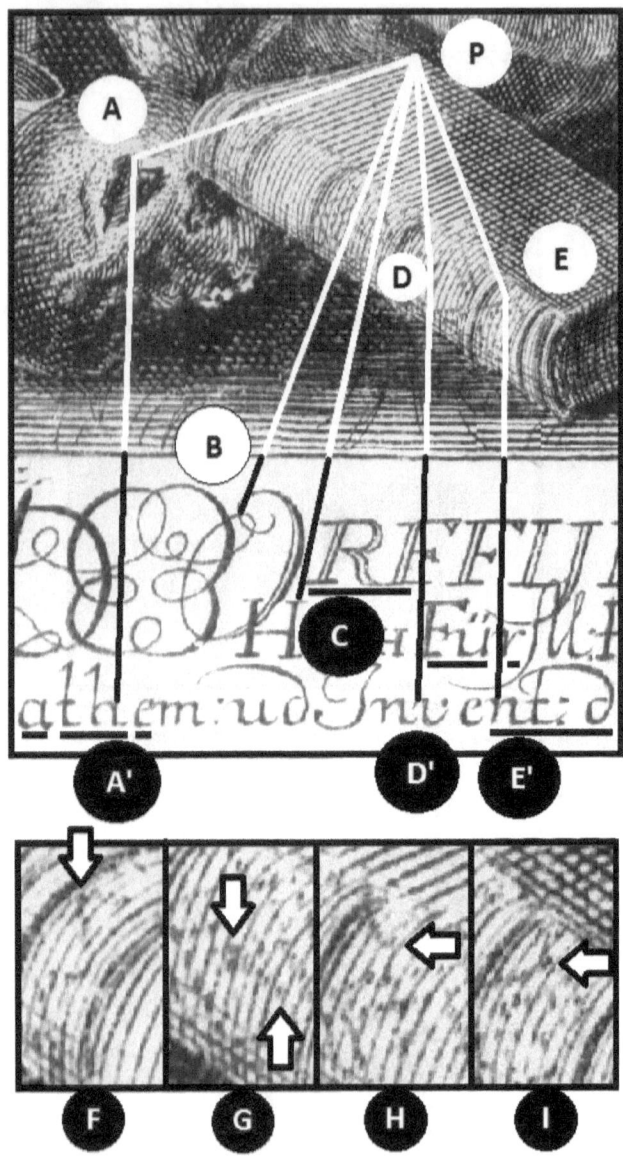

CHAPTER 11

WHAT ABOUT THE SECOND GERA, DRASCHWITZ, AND KARLSHAFEN WHEELS?

Study of the various imbalanced pm wheel mechanics parameter clues given in the two DT portraits only provides specific information for Bessler's 3 feet diameter, one-directional Gera prototype "toy" wheel and his two 12 feet diameter, bidirectional Merseburg and Kassel wheels. How then is the reverse engineer to determine the parameters of the 4.5 feet diameter, one-directional *second* Gera wheel that was publicly exhibited in that town, the next constructed 9 feet diameter, one-directional Draschwitz wheel, the contemplated, but never constructed, 40 feet diameter, one-directional "super wheel" or "*großes rad*" (pronounced like "grosses rad") that Karl envisioned perpetually powering his cherished cascade, or the two 6 feet diameter, one-directional wheels constructed at Bessler's new home in Karlshafen in a desperate attempt to finally sell his invention after the destruction of his Kassel wheel at Weissenstein Castle?

The answer to this mystery is that Bessler expected the successful reverse engineer of his wheels' secret mechanics to learn enough from the details of his 3 feet diameter Gera prototype wheel and two push start required bidirectional wheels given in the two DT portraits to then be able to build a wheel of *any* diameter and start up power output. This knowledge could then, of course, be applied to determine the various component parameter values for all of his imbalanced pm wheels that were not specifically covered by the clues in the two DT portraits. Before I provide the reader with a simple method for doing this, let us first review some of the parameters of the inventor's 3 feet diameter Gera prototype and bidirectional Merseburg and Kassel wheels.

Earlier in this volume we saw that there were a variety of critical parameters that the parts used in the 3 feet diameter Gera prototype wheel *had* to have in order for it to function like a genuine imbalanced perpetual motion wheel.

The mass of one of its little lead ingot end weights was 6 ounces or 0.375 pound and the mass of its *un*weighted wooden lever to which the ingot was attached by two small steel screws was 1.25 ounces or 0.078125 pound. The total mass of a weighted lever in this prototype wheel was, therefore, 6 ounces + 1.25 ounces = 7.25 ounces or 0.375 pound + 0.078125 pound = 0.453125 pound.

Each of the Gera prototype wheel's eight weighed levers consisted of two parallel wooden 3 arm side pieces that were held rigidly together by the steel coordinating cord hook attachment pins that penetrated them and the wooden coordinating cord hook keepers that were glued between the side pieces. Thus, each lever consisted of three distinct parallel pairs of arms that were connected to each other where the lever's pivot pin holes were located. These two pivot pin holes were tightly fitted with two thin metal sleeves which were lubricated so that the entire lever could easily swing around a steel pivot pin whose ends were embedded in a parallel pair of wooden radial frame pieces which helped form the wheel's drum.

A lever's parallel pair of main or L arms held the lever's 6 ounce lead ingot end weight at its ends and also the L1 main cord hook attachment pin and its wooden hook keeper which were located closer to the lever's pivot pin than the attachment screws that secured the lead ingot end weight to the arms.

The main or L arms' L1 main cord hook attachment pin *received* a main cord that originated from another lever which immediately *lagged* any lever being considered. The parallel pair of A arms, oriented at a 45° angle to the main arms, held the lever's A1 spring cord, A2 stop cord, A3 long lifter cord, and A4 main cord hook attachment pins and their respective wooden hook keepers from which the four different types of coordinating cords originated from the lever's A arms. And, finally, there was the parallel pair of B arms, oriented at an angle of 157.5° from *both* the lever's main arms and A arms, that held the B1 long lifter cord hook attachment pin and its wooden hook keeper. The B1 long lifter cord hook attachment pin *received* a long lifter cord which originated from the A3 long lifter cord hook attachment pin of another lever whose pivot pin's center axis lagged the center axis of any lever being considered by 90°.

One learned that each of the center axes of the 3 feet diameter Gera prototype wheel's weighted lever's two lead ingot end weight attachment screws and of its six coordinating cord hook attachment pins held by its three parallel pairs of arms had to each be at a precise distances from the center axis of the lever's pivot pin holes and, of course, when mounted on the lever's pivot pin, from the center axis of the pivot pin about which they would all rotate during drum rotation (actually, they were all very carefully restricted by the lever's wooden radial stop piece and its taut stop cord to only swinging about the lever's pivot pin *within* an arc of 62.5°).

The center axes of the attachment screws that held the center of gravity of the 6 ounces lead ingot weight at the end of the parallel pair of main or L arms was 3.5 inches from the center axis of a lever's pivot pin while the center axis of the L1 main cord hook attachment pin was 2.6 inches from the center axis of the pivot pin.

For a weighted lever's parallel pair of A arms, the A1 spring cord hook attachment pin's center axis was 0.5 inch from the center axis of the pivot pin, the A2 stop cord hook attachment pin's center axis was 0.933 inch from the center axis of the pivot pin, the A3 long lifter cord hook attachment pin's center axis was 1.4 inches from the center axis of the pivot pin, and, finally, the A4 main cord hook attachment pin's center axis was 1.866 inches from the center axis of the lever's pivot pin.

For the remaining third parallel pair of B arms, the B1 long lifter cord hook attachment pin's center axis was 2.1 inches from the lever's pivot pin center axis. (Note that each lever had two main cord hook attachment pins, one held by the main or L arms and the other by the A arms, and two long lifter cord hook attachment pins, one held by the A arms and the other by the B arms, because each lever was *both* the origin of a main cord and a long lifter cord that went out to two leading levers and the recipient of an incoming main cord and a long lifter cord from two lagging levers. This is why, even though there were only four different types of coordinating cords used in Bessler's wheels, each weighted lever had to have six cord hook attachment pins.)

In order to perfectly balance the five weighted levers of the Gera prototype wheel from the drum's 9:00 position to its 3:00 position required that each lever be attached to two helical coil extension springs *each* having a spring constant of exactly 20 ounces per inch or 1.25 pounds per inch. The combined spring constant for the two parallel extension springs attached to each lever was 40 ounces per inch or 2.5 pounds per inch.

We also saw that the distance from the center axis of the Gera prototype wheel's small wooden axle to the center axes of its eight weighted levers' pivot pins was exactly 14 inches and that the distance from the center axis of each weighted lever's pivot pin to the center of gravity of its 6 ounce lead ingot weight (which was located on the coinciding center axes of the ingot's attachment screws) was 3.5 inches. The distance from the center axis of the axle to the outside surface of the outer rim wall of this "toy" prototype wheel was exactly 18 inches.

These distance parameters are interesting because they show that the distance between the center axis of the wheel's axle and the center axis of a weighted lever's pivot pin was exactly four times the distance between the lever's pivot pin's center axis and the center axis of its lead ingot end weight's attachment screws, on which the center of gravity of the ingot weight was located which can be expressed as 14 inches = 4 x 3.5 inches.

The distance from the center axis of the 3 feet diameter Gera prototype wheel's axle to the center axis of one of its weighted lever's pivot pins divided by the distance from the center axis of the axle to the outer surface of the drum's attached outer rim wall was (14 inches / 18 inches) = 0.777... which is a never ending decimal string of 7's that would have had special significance for a numerologist like Bessler. 7 is a number considered to be lucky in numerology and to also have some divine connotations to it since, before the invention of the telescope, there were only seven heavenly bodies in the sky that were visible to the unaided eyes of humans. I suspect that Bessler purposely incorporated this ratio into his wheel's mechanics for these reasons. It was his way of acknowledging God's role in helping him find the secret of a working imbalanced pm wheel and he also hoped it would bring him the luck he needed in order to sell the invention for the large lump sum of money he demanded.

Once one obtains these parameter values for the 3 feet diameter Gera prototype wheel from the many clues hidden in the two DT portraits, he can then immediately determine what the analogous values were in his 12 feet diameter Merseburg and Kassel wheels. How? By simply following one of the most prominent and easily interpreted clues in the second DT portrait. I'm referring to the lower arm of the duckbill goniometer which sits on top of the larger arm of the "L" square drafting tool. Together, the two arms form the letter "x" which is the symbol used in arithmetic for multiplication. So, this clue is Bessler's way of telling us that we need to perform a multiplication. As one further studies the goniometer's lower arm and both arms of the "L" square drafting tool, he becomes aware that they form the numeral 4 after it has been rotated a little over 90° clockwise from its normal orientation. The interpretation of this clue is obvious. Bessler is telling us that we must multiply some numerical values by a factor of 4. And, indeed, that is exactly what the reverse engineer must do as he "scales up" from Bessler's 3 feet diameter, one directional Gera prototype wheel to the inventor's 12 feet diameter, bidirectional Merseburg and Kassel wheels. The first multiplication can immediately be made with the 3 feet diameter of the Gera prototype since 4 x 3 feet = 12 feet. But, the multiplication by 4 does not stop there.

Next, we multiply both the axle center axis to weighted lever pivot pin center axis distance in the Gera prototype wheel by 4 and determine that this distance in the inventor's 12 feet bidirectional wheels was 4 x 14 inches = 56 inches. The analogous distance from a weighed lever's pivot pin's center axis to the center axis of the attachment bolt of the *middle* one of three cylindrical lead end weights held between the ends of the main arms of a 12 feet diameter, bidirectional wheel becomes 4 times the distance in the 3 feet diameter Gera prototype wheel or 4 x 3.5 inches = 14 inches for the 12 feet diameter wheels. This distance parameter for the lever's in the 12 feet diameter, bidirectional wheels is symbolized by the 14 inch ruler balanced on the carpenter's plumb bob in the second DT portrait.

In the second DT portrait, the short arm of the duckbill goniometer points to the left straight radial side of the circular sector paper piece. In Chapter 9 it was noted that this portion of the chart represents the parallel pair of A arms of the 9:00 weighted lever of the 3 feet diameter Gera prototype wheel. It was also shown in Figure 20 that the intersection points of four of the straight lines drawn on the paper piece with the left straight radial side of the paper piece were located at distances from the circular sector's origin point, labeled "P", that corresponded to the distances of the various A arms' coordinating cord hook attachment pins' center axes from the center axis of a weighted lever's pivot pin in the 3 feet diameter Gera prototype wheel. And, again the symbolism of the goniometer's lower arm crossing over the "L" square's longer arm to form the numeral 4 as well as a multiplication sign tells us that we must multiply these distance parameter values by a factor of 4.

Doing this then converts the distance between the A1 spring cord hook attachment pin's center axis and its lever's pivot pin's center axis in the 3 feet diameter Gera prototype wheel to 4 x 0.5 inch = 2.0 inches for the analogous distance in the 12 feet diameter bidirectional wheels. The distance between the A2 stop cord hook attachment pin's center axis and its lever's pivot pin's center axis in the Gera prototype wheel was 0.933 inches and in the 12 feet diameter bidirectional wheels that distance parameter was 4 x 0.933 inch = 3.73 inches. The distance between the A3 long lifter cord hook attachment pin's center axis and its lever's pivot pin's center axis in the Gera prototype wheel was 1.4 inches and in the 12 feet diameter bidirectional wheels that distance parameter was 4 x 1.4 inches = 5.6 inches. And, finally, the distance between the A4 main cord hook attachment pin's center axis and its lever's pivot pin's center axis in the prototype was 1.866 inches and in the 12 feet diameter bidirectional wheels that distance parameter was 4 x 1.866 inches = 7.47 inches.

Bessler's suggestion that the reverse engineer multiply the A arms' four coordinating cord hook attachment pin center axes distance parameters by a factor of 4 also applies to the L1 main cord hook attachment pin and B1 long lifter cord hook attachment pin held, respectively, between the main arms and the B arms of the Gera prototype's weighted levers. Thus, the distance between the L1 hook attachment pin's center axis and its lever's pivot pin's center axis was 2.6 inches and in 12 feet diameter bidirectional wheels this was 4 x 2.6 inches = 10.4 inches. The distance between the B1 long lifter cord hook attachment pin's center axis and its lever's pivot pin's center axis was 2.1 inches in the Gera prototype wheel and this was 4 x 2.1 inches = 8.4 inches in the 12 feet diameter bidirectional wheels.

Finally, we must deal with the method used to scale up the masses of the lead end weights and the spring constants of the extension springs used in the 3 feet diameter, one-directional Gera prototype wheel to obtain the corresponding masses and spring constants used in the 12 feet diameter, bidirectional Merseburg and Kassel wheels. This, however, is a bit more complicated than just multiplying the parameter values for these parts found in the 3 feet diameter Gera prototype wheel by a factor of 4.

Assuming that the density of the materials remain the same, which is the case for the lead used in the end weights of the weighted levers of a wheel regardless of its diameter, then, as one scales up from the 3 feet diameter Gera prototype wheel to a 12 feet diameter bidirectional wheel, the increase in the masses of the end weights will only be proportional to the increase in their *volumes*. The lead weights in the 3 foot diameter Gera prototype wheel had a mass of exactly 6 ounces = 0.375 pounds and were shaped like ingots which had isosceles trapezoidal and rectangular cross sections.

If we imagine all three of the dimensions of a Gera prototype lever's single lead ingot end weight being increased by a factor of 4 in order to make a single large ingot to be used in a 12 feet diameter one-directional wheel, then the ingot's volume *and* mass would increase by a factor of 4 x 4 x 4 = 64. If we then multiply the mass of the Gera prototype wheel's single lead ingot end weight by a factor of 64, then it will give us the mass of the larger single ingot weight needed for a 12 feet diameter one-directional wheel. Doing that gives us 64 x 0.375 pounds = 24 pounds.

Of course, Bessler did not use a single large lead ingot weight at the ends of the main arms of the weighted levers in this two bidirectional wheels, but, instead, opted to use three cylindrical lead end weights whose center axes were parallel to each other. The mass of 24 pounds derived here was made by attaching *three* closely placed cylindrical lead end weights *and* their steel attachment bolts, with each weight/ bolt combination having a total mass of 8 pounds, to the ends of the main arms of each of the Kassel wheel's levers. In the case of his earlier Merseburg wheel, the total mass of the three cylindrical lead end weights and their three attachment bolts was only 12 pounds because *each* of a weighted lever's three end weights *including* its attachment bolt only weighed 4 pounds.

(The witnesses at the official testing of the Merseburg wheel on October 31st, 1715, during the translocation of the drum and axle to another set of vertical support planks when the drum's forty-eight weights and their forty-eight steel attachment bolts had to be removed in order to lighten the wheel, were allowed to handle at least one of these cylindrical lead end weights which the inventor had wrapped in a handkerchief and forbade them from touching either of its circular ends. It's possible that, after removal from the ends of a lever's parallel pair of main arms, Bessler just reinserted the bolts into the shafts drilled through the center axes of the weights so as to minimize the chance of losing a bolt as the weights were being temporarily put aside until it was time to reattach them to their levers. That could explain why he did not want any of the witnesses touching the ends of the weight(s) he removed from the Merseburg wheel. If they had done so, then they may have felt the slotted head of an attachment bolt at one end of the cylindrical lead weight and the threaded section protruding from the other end and that would have immediately told them how the weights were attached to the ends of the levers. And, if the weight(s) he passed around did not include the attachment bolt(s), then the witnesses might still have surmised how the weights were attached to the ends of the levers by just feeling the end openings of the shafts drilled through the center axis of a weight.)

One can also determine the mass of an *un*weighted levers used in the 12 feet diameter, bidirectional Kassel wheel by simply multiplying the mass of an unweighted lever used in the 3 feet diameter Gera prototype wheel by a factor of 64 or 64 x 0.078125 pound = 5 pounds. This mass must again be halved for the mass of the unweighted levers used in the Merseburg wheel or 0.5 x 5 pounds = 2.5 pounds. (The reader should keep in mind that these masses for the unweighted levers in both bidirectional wheels also included the masses of their seven attached metal cord hooks, half of the masses of the seven coordinating cords attached to those seven metal hooks, and half the masses of the two extension springs if their ends, via hooks, were free to rotated about their common spring hook attachment anchor pin embedded in the outer octagonal frame pieces of the drum. If, however, the ends of the springs were fixed to the drum by some means so that the springs were supported, then none of the mass of the two springs need be included in determining the mass of an unweighted lever in either of the two 12 feet diameter, bidirectional wheels.)

So, we see from this that, while one can simply multiply the various distance parameters of a weighted lever in the 3 feet diameter Gera prototype wheel by a factor of 4 to obtain the analogous distance parameters for a weighted lever in the 12 feet diameter, bidirectional Kassel wheel, we must multiply the mass of a Gera prototype wheel's lead ingot weight by a factor of 4 x 4 x 4 = 64 in order to obtain the analogous total mass of a trio of cylindrical lead end weights *and* their attachment bolts attached to the main arms of an unweighted lever in the Kassel wheel. In the case of the Merseburg wheel, however, this multiplication factor was reduced to 32. These two mass scaling factors, 64 and 32, could also be multiplied by the mass of an unweighted lever in the Gera prototype wheel, which was 0.078125 pound, to obtain the masses of the unweighted levers used in the Kassel and Merseburg wheels which were seen to be 5 and 2.5 pounds, respectively.

Next, we must determine how the spring constants of the extension springs used in the 3 feet diameter Gera prototype wheel were scaled up so that they could be used for the larger and more massive weighted levers in a 12 foot diameter wheel.

One might immediately think that all that need be done to determine the spring constant of a single extension spring used in the Kassel wheel would be to simply multiply the spring constant of an extension spring used in the 3 feet diameter Gera prototype wheel, which was 20 ounces per inch or 1.25 pounds per inch, by a factor of 64 to obtain a value of 1280 ounces per inch or 80 pounds per inch since the mass of a weighted lever in the Kassel wheel was 64 times the mass of a weighted lever in the 3 feet diameter Gera prototype wheel. However, if extension springs each having a spring constant of 80 pounds per inch had actually been used in the Kassel wheel, then they would have prevented the precise balance among the six weighted levers from the drum's 7:30 to 3:00 positions that was *critically* necessary in order to produce the pm effect.

The reason is because, just as the various distance parameters of a weighted lever's coordinating cord hook attachment pins' center axes relative to their lever's pivot pin's center axis in the Kassel wheel were four times the corresponding distance parameters for the lever's in the 3 feet diameter Gera prototype wheel, the amount of *stretch* from their relaxed unstretched lengths of the extension springs attached to the weighted levers in the Kassel wheel were also four times greater than their corresponding stretched distances in the 3 feet diameter Gera prototype wheel. Thus, while an extension spring attached to the 9:00 weighted lever in the

Gera prototype wheel was stretched 0.5 inches, the corresponding stretch distance for an extension spring attached to the 9:00 weighted lever in the Kassel wheel was four times as great or 4 x 0.5 inch = 2 inches. This stretch distance of 2 inches also applied to the extension springs attached to the half as massive weighted levers in the Merseburg wheel.

If one simply multiplies the value of the spring constant of a Gera prototype wheel's extension spring by a factor of 64 for use in the Kassel wheel and then stretched that spring through a distance four times greater than its smaller version was stretched in the Gera prototype wheel, then that spring would supply a force to the particular weighted lever that it was attached to that was four times greater than it needed to be in order to provide the precise balance among the six levers from the drum's 7:30 to 3:00 positions. That excessive force would actually prevent the weighted levers from "settling" into the particular configuration they needed to have in order to keep the center of gravity of the Kassel wheel's eight active weighted levers on the drum's descending side so that a chronic state of imbalance would be maintained despite drum rotation.

So, in determining the increased spring constant parameter for an extension spring used in the Kassel wheel, this fourfold increase in the maximum allowed stretch distance of an extension spring, when its weighted lever's pivot pin's center axis approached the drum's 9:00 position, must be taken into account and compensated for. To do this, we need to first multiply the spring constant of a Gera prototype wheel's extension spring by a factor of 64 to adjust for the increased mass of the three cylindrical lead end weights attached to each weighted levers in the Kassel wheel and then we must *divide* that product by 4 in order to compensate for the fourfold increase in the maximum allowed stretch distance of an extension spring in the Kassel wheel. Thus, by doing this we are compensating for the increased stretch distance of a spring by decreasing the amount of contractive force that stretch creates.

Since we must multiply the Gera prototype's spring constant by a factor of 64 and then divide the resulting product by 4, that is equivalent to just multiplying the Gera prototype wheel's spring constant by a factor of 16 in order to obtain the spring constant value of an extension spring in the Kassel wheel. Doing so give us (64/4) x 1.25 pounds per inch = 16 x 1.25 pounds per inch = 20 pounds per inch. Since the extension springs attached to the weighted levers in the Merseburg wheel were, like the springs in the Kassel wheel, also stretched by a maximum allowed distance of 2 inches as the pivot pin center axes of the weighted levers they were attached to approached the drum's 9:00 position, but the mass of a weighted lever in the Merseburg wheel, 14.5 pounds, was only half of the mass of a lever in the Kassel wheel, 29 pounds, we must half this spring constant value of 20 pounds per inch used for the Kassel wheel springs to give us 10 pounds per inch for the spring constant of a spring used in the Merseburg wheel.

At this point I can now very briefly summarize the general method that must be used to predict what parameters must be used for wheels with diameters greater than Bessler's 3 feet diameter, one-directional Gera prototype wheel.

The method begins by determining what factor is needed to convert the diameter of the drum of Bessler's 3 feet diameter Gera prototype wheel into the diameter of a wheel with a larger size diameter.

For example, if one wishes to construct a wheel with a drum diameter of 6 feet, then he must start, obviously, by multiplying the diameter of the Gera prototype by a factor of 2.

The craftsman must then also, when designing the weighted levers for the larger diameter wheel and determining the locations of their pivot pins, multiply all of the distances between the center axes of the various pins in the 3 feet diameter Gera prototype wheel, whether located between the parallel pairs of arms of a lever or between the octagonal frame pieces of the wheel's drum, by a factor of 2 in order to scale them up to determine the larger corresponding values that will need to be used in the 6 feet diameter one-directional wheel that he might want to construct.

Thus, the lever length or distance of 3.5 inches in the Gera prototype wheel measured from the center axis of a lever's pivot pin to the center of gravity of its lead ingot end weight, which was located on its two attachment screws' center axes, would become 2 x 3.5 inches = 7.0 inches for the levers to be used in a one-directional wheel which was 6 feet in diameter. For the distance from the center axis of the axle to the center axis of a weighted lever's pivot pin in the 6 feet diameter wheel, one simply multiplies the corresponding distance in the 3 feet diameter Gera prototype wheel, 14 inches, by a factor of 2 to obtain 2 x 14 inches = 28 inches.

The distance values for all of the various center axes of the four types of coordinating cord hook attachment pins of a weighted lever relative to the center axis of the lever's pivot pin in the Gera prototype wheel are then also multiplied by a factor of 2 to obtain the correct corresponding distance values for use in the weighted levers of the 6 feet diameter wheel.

Thus, in the 6 feet diameter wheel, the distance value for the L1 main cord hook attachment pin's center axis is 2 x 2.6 inches = 5.2 inches from the center axis of the lever's pivot pin, the A1 spring cord hook attachment pin's center axis is 2 x 0.5 inches = 1.0 inches from the center axis of the lever's pivot pin, the A2 stop cord hook attachment pin's center axis is 2 x 0.933 inches = 1.87 inches from the center axis of the lever's pivot pin, the A3 long lifter cord hook attachment pin's center axis is 2 x 1.4 inches = 2.8 inches from the center axis of the lever's pivot pin, the A4 main cord hook attachment pin's center axis is 2 x 1.866 inches = 3.73 inches from the center axis of the lever's pivot pin, and, finally, the B1 main cord hook attachment pin's center axis is 2 x 2.1 inches = 4.2 inches from the center axis of the lever's pivot pin.

Determining the mass of a weighted lever and the spring constants of each of the two extension springs attached to a weighted lever in the 6 feet diameter wheel that the craftsman wants to build, however, is a bit more complicated and can be done using one of two approaches. One can begin by deciding what the total mass of his wheel's weighted levers will be and, after that it done, then calculate the spring constants of each of the two extension springs he will need to precisely balance that lever mass after the weighted lever is installed in the drum and the center axis of the lever's pivot pin is located at the 9:00 position of a clockwise rotating drum. Or, he can begin by selecting the value of the spring constant for each of the two extension springs that he will use and, after that is done, calculate the mass of the weighted lever which will be precisely balanced by two of the extension springs when the center axis of a weighted lever's pivot pin is located at the 9:00 position of a clockwise rotating drum.

Unless one is actually designing and manufacturing his own custom springs, which Bessler most likely did not do, he will have to rely upon a commercial source for the springs to use in his Bessler type imbalanced pm wheel. While such sources may stock a wide variety of extension springs with various sizes and spring constants, they may not have any with some unusual values if a craftsman needs such after first starting the planning of his construction project by selecting some nice whole number value for the mass of a weighted lever to which two of the springs will be attached. For this reason it is better to start one's planning of a wheel by first selecting the spring constants of the springs he will be using and then determining the mass of a weighted lever appropriate for two springs with those constants. Let's see how to apply this to our example of a 6 feet diameter wheel which was, most likely, the diameter of the two one-directional Karlshafen wheels Bessler constructed after the destruction of his last 12 feet diameter, bidirectional Kassel wheel in Weissenstein Castle in the city of Kassel, Saxony on August 17th, 1721.

If one starts by selecting extension springs, sixteen of which will be needed for the eight weighted levers of his one-directional wheel, with a spring constant of 5 pounds per inch each, he will have selected a spring constant that is exactly four times that of a spring used in Bessler's 3 feet diameter Gera prototype wheel. *If* his 6 feet diameter wheel used weighted levers that were the *same* physical size, as seen in an axial view, as those found in the Gera prototype wheel so that the spring cord hooks were also attached to A1 spring cord hook attachment pins whose center axes were 0.5 inch away from the center axis of a lever's pivot pin, then the weighted levers would have to have four times the mass of the weighted levers used in the prototype wheel or 4 x 0.453125 pounds = 1.8125 pounds in order to also stretch the extension springs attached to them through a distance of 0.5 inch when a weighted lever's pivot pin's center axis was at the drum's 9:00 position. However, because the craftsman plans to double the drum's diameter as well as all of the cord hook attachment pin distance parameters for the weighted levers in his 6 feet diameter wheel, that means that the A1 spring cord hook attachment pin center axes are now located at a distance of 1 inch from the center axes of the levers' pivot pins. This then requires that the mass of a weighted lever be able to stretch an attached extension spring through a distance that is doubled to 1 inch. The only way this can happen is if the mass of the weighted lever in the planned 6 feet diameter wheel is further doubled to 2 x 1.8125 pounds = 3.625 pounds.

However, the craftsman's planning is not yet done.

The mass of a weighted lever in one of Bessler's wheels was the sum of the mass of its end weight(s) including their attachment screws or bolts and the mass of the unweighted lever that they were attached to which included the full mass of the seven metal hooks attached to the lever's six coordinating cord hook attachment pins and half the masses of the seven cords attached to the hooks and the two extension springs which were free to rotate about their spring hook attachment anchor pins embedded in the drum's radial frame pieces. One cannot just select any end weight and unweighted lever mass so that they add up to the mass of the weighted lever he needs for a particular spring constant of the extension springs he has decided to use. There is a specific *ratio* between these two masses that must be used. That ratio *must* be identical to that found in the Gera prototype wheel that had a lead ingot end weight with a mass of 6 ounces and an unweighted lever with a mass of 1.25 ounces which, when joined, resulted in a total weighted lever mass of 7.25 ounces.

From these two values, we can derive multipliers that will convert the mass of a weighted lever that is determined to be necessary for a given selected spring constant of an extension spring into the portion of that mass that will be contained in the lever's end weight(s) and its (their) attachment screws or bolt(s) and the portion of the mass that will be in the unweighted lever. To find the mass of the end weight(s) and their attachment screws or bolt(s), we multiply the mass of the weighted lever needed by the factor 6/7.25 = 0.827586. To find the mass of the unweighted lever, we multiply the mass of the weighted lever needed by the factor 1.25/7.25 = 0.172414.

It was shown above that the selection of an extension spring with a spring constant of 5 pounds per inch then obliged the craftsman to use weighted levers with a total mass of 3.625 pounds in his 6 feet diameter Bessler wheel replica. Using the just derived multipliers, we see that this weighted lever mass must be divided up between end weight(s) and their attachment screws or bolt(s) with a mass of 0.827586 x 3.625 pounds = 3 pounds and an unweighted lever with a mass of 0.172414 x 3.625 pounds = 0.625 pounds.

Should the craftsman decided to double the start up power output of his 6 feet diameter replica wheel, then he need only double the spring constants of a weighted lever's two extension springs to 10 pounds per inch each while also doubling the mass of a weighted lever to 7.25 pounds which means that the mass of the end weight(s) and their attachment screws or bolt(s) will be doubled to 6 pounds and the mass of the unweighted lever to which the end weight(s) will be attached will be doubled to 1.25 pounds.

So, at this point the craftsman is finally ready to begin designing the drum and weighted levers that will be used in his 6 feet diameter, one-directional wheel. I purposely chose this example because I think that the 6 feet diameter wheel is exactly what Bessler began constructing after the destruction of his Kassel wheel in an attempt to finally find a buyer for his invention. The design is convenient because its drum's diameter can easily fit into a room whose floor to ceiling height is only 8 feet as may have been the case for the rooms of his home in Karlshafen. The weighted levers have L1 center axis to pivot pin center axis distances of 7 inches which is a convenient size for cutting their 3 arm side pieces from a single sheet of wood. To obtain a single cylindrical lead end weight with a mass of 3 pounds, Bessler would only have had to cut off one quarter of the length of a 4 pound cylindrical lead end weight that he had salvaged after he destroyed the Merseburg wheel.

So let me now summarize the various steps that a craftsman needs to take if he wishes to construct a Bessler type imbalanced pm wheel of *any* diameter using extensions springs having *any* readily commercially available spring constant. He can do it rather easily by using a series of simple equations I've derived which can be referred to as the "Bessler imbalanced pm wheel design equations".

Firstly, the craftsman must calculate a "scaling factor", S, for his planned wheel which is:

$$S = (\text{planned wheel diameter in feet}) / 3 \text{ feet}$$

S is simply the outer rim wall diameter measured between the opposite *outside* surfaces of the drum of the wheel he wishes to construct divided by the outer rim wall diameter of the Gera prototype wheel which was 3 feet. The outer rim wall is not absolutely critical to the functioning of a Bessler type imbalanced pm wheel and may be eliminated from a wheel's construction if so desired. However, its presence helps reduce energy wasting aerodynamic drag that the frame pieces of a completely exposed drum will experience during drum rotation.

Secondly, after the scaling factor, S, is determined, all of the various separation distances in the Gera prototype wheel between the center axes of the cord hook attachment pins (as well as the center axes of the screws needed to affix the center of gravity of a single end weight to the ends of the main arms or the center axis of the attachment bolts of the middle one of a trio of cylindrical lead weights to the ends of the main arms) in the three parallel pairs of arms and their lever's pivot pin, between a lever's pivot pin and the drum's axle, between the spring hook attachment anchor pin and the drum's axle, between the stop cord hook attachment anchor pin and the drum's axle, and, finally, between the cord hook attachment pins of *different* weighted levers which are connected together by the four types of coordinating cords *and* their attached end hooks (which are the maximum distances when the cords are taut) must be multiplied by the scaling factor S. Thus, the scaling factor, S, must be applied to a total of fourteen distance parameters found in an axial view of the 3 feet diameter Gera prototype wheel!

Thirdly, the craftsman must select the spring constant of a *single* one of the two extension springs that he will attach, via a spring cord and its two metal end hooks, to the A1 spring cord hook attachment pin of each of the weighted levers in his wheel. Once that is done, he must determine the mass of the weighted lever he will need in order for the two extension springs he has selected for attachment to each of his weighted levers to result in the six levers from the drum's 7:30 to 3:00 positions in his wheel being perfectly balanced by the various coordinating cords connecting them together. The mass of the required weighted lever can be gotten by using this equation:

Mass of weighted lever in pounds = (0.453125 pounds) x S x (spring constant selected in pounds per inch / 1.25 pounds per inch)

Fourthly, after the craftsman has found the mass of each of the weighted levers necessary to make his wheel display the pm effect using the particular spring constant for *each* of the two extension springs he has selected to attach to each lever's A1 spring cord hook attachment pin via a spring cord, he must determine how the mass of one of its weighted levers is divided up between its end weight(s) and the unweighted lever to which the weight(s) are attached. For this he must use the following two equations:

Mass of end weight(s) in pounds = 0.827586 x Mass of weighted lever in pounds

Mass of unweighted lever in pounds = 0.172414 x Mass of weighted lever in pounds

As a check of one's calculations, he should add together the values of the masses of the end weight(s) and their attachment screws or bolt(s) and the unweighted lever to which they will be attached to make sure that their sum equals the required mass of the weighted level.

Fifthly and finally, the craftsman will probably want to have some idea of what the start up power output of his Bessler type imbalanced pm wheel might be when it is finished. This can be done by using the start up power output of the Kassel wheel as a reference. As I mentioned in Chapter 1, this wheel had enough technical information provided about it in DT to estimate that its start up power output was about 50 watts. Of course, like all of the inventor's wheels, that initial power output would slowly decline as the wheel's rotation rate increased toward its maximum terminal value when the wheel was allowed to run freely with no outside machines being operated by its axle.

The start up power output of an imbalanced pm wheel is proportional to the product of the total mass of the wheel's eight active weighted levers and the horizontal displacement distance of the center of gravity of that total mass from a vertical line passing through the center axis of the wheel's axle. However, we can also find the start up power output of *any* planned wheel by just *dividing* the *product* of the mass of the end weight(s) and their attachment screws or bolt(s) of a *single* one of the planned wheel's weighted levers and the planned wheel's drum diameter by the *product* of the mass of the three cylindrical lead end weights and their attachment bolts of a single Kassel wheel's weighted lever and the diameter of that wheel's 12 feet diameter drum. Once that quotient is obtained it is multiplied by the Kassel wheel's maximum estimated start up power output of 50 watts to obtain the start up power output of the wheel that the craftsman plans to construct. To put this description into the form of an equation, one can write:

Start up power of planned wheel = 50 watts x [(mass of end weight(s) *and* attachment screws or bolts of planned wheel in pounds x planned wheel diameter in feet) / (24 pounds x 12 feet)]

From inspection of this equation, it becomes obvious that one could construct a 6 feet diameter, one-directional wheel that was as powerful at start up as the 12 feet diameter, bidirectional Kassel wheel was, but, in order to do so, the 6 feet diameter wheel would require weighted levers each of whose parallel pair of main arms holds three cylindrical lead end weights and their three attachment bolts with a total mass of 48 pounds. This would result in weighted lever with a mass of 58 pounds because each unweighted lever to which the three cylindrical lead end weights were attached to would have a mass of 10 pounds. This also means that each of the 6 feet diameter wheel's three cylindrical lead end weights would, with its attachment bolt, have to have a mass of 16 pounds which would be double the mass of a cylindrical lead end weight, including its attachment bolt, inside of the Kassel wheel. The 6 feet diameter drum of such a high power yet compact one-directional wheel would probably also require a drum 18 inches thick as did the Kassel wheel.

In order to better assist the craftsmen who may wish to attempt to duplicate one of Bessler's imbalanced pm wheels using the details of their previously secret mechanics which are amply verified by the two DT portraits' many hidden parameter value clues without him having to apply the "Bessler imbalanced pm wheel design equations" and thereby possibly making a math error in the process, I have included two tables at the end of this chapter that conveniently summarize all of the important parameter values for each of the wheels Bessler actually constructed or planned to construct. The reader, however, should keep in mind that only the parameter values for the Gera prototype, Merseburg, and Kassel wheels are directly obtainable from the two DT portraits (future DT portraitologists may, however, find others). The parameter values for the second Gera, Draschwitz, Karlshafen, and super wheels, however, are based upon extrapolations, using the "Bessler imbalanced pm wheel design equations", which were made by using the Gera prototype wheel's parameter values (with the single exception of the wheels' start up power outputs which were made using the estimated power output of the Kassel wheel). I consider these extrapolated parameter values to be the most probable ones that Bessler would have used in these other wheels and remain confident that they are accurate.

Table 1 shows the various *external* parameter values for all of Bessler's wheels. These wheels, eight in number, are shown in the first column of the table labeled "Where built" at its top and are listed downward in the order of their construction. Near the top of the column are the two wheels he constructed in the town of Gera, followed by the much larger diameter Draschwitz, Merseburg, and Kassel wheels. Then there are the two wheels constructed at Bessler's home in Karlshafen, one after the destruction of the Kassel wheel and the second much later just before his death. The last wheel at the bottom of the column presented a bit of a problem as to where it should be placed in the column. It was the inventor's planned, but not constructed "super wheel". Since it was probably conceived before or during the construction of his Kassel wheel in order to impress Count Karl, I was tempted to place it between the Merseburg and Kassel wheels and, thus, higher up in the column. However, because it was only planned and not constructed and would also have had the largest diameter of all of Bessler's wheels, I decided to put it at the very bottom of the column and, furthermore, to emphasize its differences by placing a thicker black horizontal border across the width of the table to separate its row of parameter values from the rows of values for his other wheels which were actually constructed.

The next five vertical columns of Table 1, collectively labeled "Drum, axle, and mass values", provide the various external parameter values for the eight wheels listed in the first column. Here one finds the values for the physical dimensions of the various wheels' drums and axles, in feet and inches, as well as estimates for their masses in pounds. Many of these values will be familiar to the student of Bessler's inventions from the available translations of his written works.

The next three vertical columns of Table 1, collectively labeled "Wheel performance values", show the maximum terminal rotation rates or speeds when no outside machines were attached to their axles, estimated start up power outputs, and directional capabilities of the eight wheels. These three columns comprise the maximum performance values for Bessler's wheels.

The last two vertical columns of Table 1, collectively labeled "Important years", provide the year each wheel was constructed as well as the year it was destroyed. The fact that Bessler's 40 foot diameter super wheel was never constructed and, consequently, could not be destroyed is indicated at the bottom of the last rightmost column of Table 1 by the words "Not Applicable".

Next, we come to Table 2 which contained too much important information to fit onto a single page! It also lists all eight of Bessler's wheels in its leftmost columns followed by twenty-one columns that provide the twenty-two critical *internal* parameter values for the weighted levers and their drum that allowed each of those eight wheels to remain in a state of persistent imbalance as they rotated or, in the case of the never constructed super wheel, would have rotated. These parameter values were derived from the careful construction of about two thousand computer model wheels and my years of analysis of the two DT portraits to locate their many carefully concealed parameter value clues that verified that the values obtained from my computer modeling were, in fact, correct. But, again, I must point out that, despite much effort, I have only been able to find the hidden verifying parameter value clues pertaining to Bessler's 3 feet diameter, one-directional Gera prototype wheel and his two 12 feet diameter, bidirectional Merseburg and Kassel wheels in the two DT portraits. All of the other values were obtained by extrapolation of the parameter values given for the Gera prototype wheel in the two portraits. And, as in the case of Table 1, I have placed the parameter values for his planned, but never constructed 40 feet diameter, one-directional super wheel into the second table's bottom row and separated them with a thick horizontal line from the single row above that provides the internal parameter values for the *two* identical Karlshafen wheels.

Starting from the left side of Table 2, there are four vertical columns, collectively labeled "Important pin and post location values within a drum", that contain important distance parameter values, in inches, for the center axes of various pins and posts within a wheel's drum relative to the center axis of the drum. In the case of the super wheel on the bottom row of the table, I have also provided the distance parameter values in feet as well as inches because of the longer distances involved.

The first column of the group of four, labeled "Axle to Pivot Pin (inches)", gives the distance, in inches, between the center axis of a wheel's axle and the center axis of one of its weighted levers' pivot pins.

Moving to the right, the next and second column of the group of four, labeled "Axle to OOAP/IOAP (inches)", gives the distances, in inches, between the center axis of a wheel's axle and the center axes of *two* of the drum's hook attachment anchor pins. The first value is the distance between the axle's center axis and center axis of a drum's spring hook attachment anchor pin whose ends were embedded at the midpoints of a lever's leading parallel pair of *outer* octagonal frame pieces (abbreviated as "OOAP" standing for "Outer Octagonal Anchor Pin" at the top of the column). The second distance indicated is the distance between the axle's center axis and the center axis of the stop cord hook attachment anchor pin whose ends were embedded at the midpoints of a lever's leading parallel pair of *inner* octagonal frame pieces (abbreviated as "IOAP" standing for "Inner Octagonal Anchor Pin" at the top of the column).

The third column of the group of four, labeled "Axle to Outer Cord Guide Post (inches)", gives the distance, in inches, between the center axis of a wheel's axle and the center axis of one of its drum's *outer* long lifter cord guide posts which was held between a parallel pair of radial frame pieces. The next and forth column of the group of four, labeled "Axle to Inner Cord Guide Post (inches)", gives the distance, in inches, between the center axis of a wheel's axle and the center axis of one of its drum's *inner* long lifter cord guide posts.

The difference in the distances between the center axes of the two types of long lifter cord guide posts and the axle's center axis assured that, when a long lifter cord was taut between two weighted levers whose pivot pins' center axes were separated by an angle of 90° on the drum's ascending side, the long lifter cord would not make rubbing contact with either of the two types of guide posts. As mentioned in an earlier chapter, the outer surfaces of these wooden posts would have been sanded smooth and then varnished so that, when dry, they would be as smooth and slippery as possible in order to allow the long lifter cords to easily side over them when the cords did make contact with them in other of a rotating drum's fixed clock time positions.

The sixth vertical column from the left side of Table 2, labeled both "Lever length" and "Pivot Pin to End Wt CoG at L2 (inches)", provides the distance values for what I have in earlier chapters referred to as the "lever length" of the weighted levers used in Bessler's wheels. This was just the distance, in inches, between the center of gravity of a lever's end weight(s), which was located *on* the center axis of its two attachment screws or bolt, and the center axis of a weighted lever's pivot pin. The actual flat ends of the wooden main arms would, however, extended a bit farther than this distance as indicated in Figure 11(b) at the end of Chapter 6.

The remaining six right side columns of the first part of Table 2, collectively labeled "Coordinating cord hook attachment pin location values within a weighted lever", provide some of the most important distance parameter values for the weighted levers used in Bessler's wheels. The values shown represent the distances, in inches, from the center axis of a weighted lever's pivot pin to the center axes of its three arms' coordinating cord hook attachment pins. Since each weighted lever's parallel pair of 3 arm side pieces held, *not* counting the end weight attachment screws or bolt(s), a total of six coordinating cord hook attachment pins, six columns of values are required to list the values for each of Bessler's eight wheels.

The first column of the group of six columns, labeled "Pivot Pin to L1 Pin (inches)", provides the distance, in inches, between the center axis of a weighted lever's pivot pin and the center axis of the L1 main cord hook attachment pin held by the main arms for all of Bessler's seven constructed wheels and his never constructed super wheel. The next four columns to the right of that column provide the distances, in inches, from the center axis of a weighted lever's pivot pin to the center axes of the A1 spring cord, A2 stop cord, A3 long lifter cord, and A4 main cord hook attachment pins held by a lever's parallel pair of A arms for all of Bessler's eight wheels. These four distance parameter columns are, moving left to the right, labeled "Pivot Pin to A1 Pin (inches)", "Pivot Pin to A2 Pin (inches)", "Pivot Pin to A3 Pin (inches)", and "Pivot Pin to A4 Pin (inches)". The remaining and last column on the right side of the first part of Table 2 labeled "Pivot Pin to B1 Pin (inches)" provides the distances, again in inches, from the center axis of a weighted lever's pivot pin to the center axis of the B1 hook attachment pin held by a weighted lever's parallel pair of B arms for all of Bessler's seven constructed wheels and his never constructed super wheel.

This completes the treatment of the twelve center axes' separation distance parameter values listed in the first part of Table 2. Now it is time to turn our attention to the remaining ten values provided in the continuation of Table 2 and the reader should now locate and study that extension of Table 2.

The second through fifth columns of the continuation of Table 2, collectively labeled "Coordinating cord values", provide the various distance parameters for the four different types of coordinating cords used in Bessler's wheels. These distances, however, are not the lengths, in inches, of the coordinating cords themselves which will always be a little shorter that what is listed. Rather, they are the *maximum* allowed separation distances between the center axes of the two different coordinating cord hook attachment pins which are connected by a particular type of coordinating cord *and* its two attached metal end hooks.

The first column of the four, labeled "main cord A4 to L1 (inches)", shows the maximum allowed distances between the center axes of the A4 and L1 main cord hook attachment pins when a main cord is stretched tight. This maximum separation distance was only achieved by the three *taut* main cords between the four weighed levers whose pivot pins' center axes were located at the 10:30, 12:00, 1:30, and 3:00 positions of the drum of a one-directional wheel designed to spontaneously start rotating in a clockwise direction.

The second column of the four, labeled "long lifter cord A3 to B1 (inches)", shows the maximum allowed distance between the center axes of the A3 and B1 long lifter cord hook attachment pins when the long lifter cord was stretched tight by the two weighted levers it connected together. This distance was only achieved by the long lifter cord that connected the B1 long lifter cord hook attachment pin of a weighted lever whose pivot pin's center axis had just passed the drum's 9:00 position with the A3 long lifter cord hook attachment pin whose weighted lever's pivot pin's center axis had just passed the drum's 6:00 position. The long lifter cord, when tight, played a critically important role in causing the main arms of a weighted lever whose pivot pin's center axis was approaching the 10:30

position of a clockwise rotating drum to rapidly rotate clockwise around their lever's pivot pin so as to bring the leading surfaces of the parallel pair of main arms closer to their wooden radial stop piece, but not yet close enough for physical contact to occur.

The third column of this group of four columns, labeled "stop cord A2 to IOAP (inches)", shows the maximum allowed distance between the center axis of a weighted lever's A2 stop cord hook attachment pin and the center axis of the lever's stop cord hook attachment anchor pin held between the midpoints of a lever's leading parallel pair of inner octagonal frame pieces when the stop cord was stretched tight. This distance was briefly achieved when the pivot pin center axis of a weighted lever was about 6° away from a clockwise rotating drum's 9:00 position and persisted until shortly after the pivot pin's center axis passed the drum's 9:00 position and the perfectly extension spring balanced weighted lever began to rapidly rotate clockwise about its pivot pin due to the torque applied to its B arms by the long lifter cord connecting the lever's B1 long lifter cord hook attachment pin to the A3 long lifter cord hook attachment pin of the then counterclockwise rotating weighted lever whose pivot pin's center axis had passed the drum's 6:00 position.

Finally, the fourth and last column in the group of four columns, labeled "spring + cord A1 to OOAP (inches)", shows the *minimum* allowed distance between the center axes of a weighted lever's A1 spring cord hook attachment pin and the lever's spring hook attachment anchor pin held between the midpoints of a leading parallel pair of outer octagonal frame pieces when the lever's two extension springs were *unstretched*. This minimum distance was achieved by a weighted lever whose pivot pin's center axis had moved about 12° clockwise past the drum's 3:00 position, but had not yet passed the drum's 6:00 position. Between these two drum positions, the two parallel extension springs attached by spring cords to a lever's A1 spring cord hook attachment pin were unstretched and the short spring cord connecting one end of each extension spring to the weighted lever was only very slightly taut due to the force applied to it by the weight of the extension spring and the centrifugal force acting on it if the one end of the spring was free to rotate about its spring hook attachment anchor pin. At all other locations inside of the drum, the extension springs would be stretched from their normal, fully contracted lengths and, of course, their attached spring cords would be taut due to the spring tension applied to them.

We are now ready to proceed to the next group of three vertical columns of parameter values in the Table 2 continuation which is labeled "Lever mass values".

The first column in this group of three columns, labeled "end weight mass (pounds)", gives the mass, in pounds (and also ounces for the Gera prototype wheel), of a single end weight attached to the end of a lever's parallel pair of main arms by either two screws tightened into holes formed into the soft lead of a single ingot's flattened ends or by a single long steel bolt that passed through a shaft drilled along the center axis of a single cylindrical lead end weight. Where there was more than one weight attached to the ends of a lever's main arms, the mass shown is the total for all of the weights. In the cases of the Merseburg and Kassel wheels there were three cylindrical lead end weights attached to each weighted lever's parallel pair of main arms. For the super wheel, nine or more large lead disc weights would have been used that, like washers, would have been slid, one by one, onto a large steel bolt at the ends of a lever's parallel pair of main arms. The mass of the lead end weights shown in the column also *includes* the mass of the steel screws or attachment bolts which secured the weights to the end of their lever's main arms.

The second column in the group of three columns, labeled "unweighted lever mass* (pounds)", provides the masses, in pounds (and also ounces for the Gera prototype wheel), for the *unweighted* levers used in Bessler's wheels. An "unweighted" lever is one which does not have any end weight(s) and their attachment screws or bolt(s) affixed to the end of its parallel pair of main arms. The mass of an unweighted lever is the sum of the masses of its two wooden 3 arm side pieces including the two thin metal sleeves that line their pivot holes, its six steel coordinating cord hook attachment pins, its six wooden coordinating cord hook keepers, the seven metal hooks attached to the six steel coordinating cord hook attachment pins, half the masses of the seven coordinating cords attached to the seven metal hooks, and half the masses of the two extension springs attached to a weighted lever if their spring hooks are free to rotate about their common spring hook attachment anchor pin (if they are not free to so rotate because the ends of the two springs are physically attached to the drum and part of it, then the masses of the two extension springs are not included in the mass of an unweighted lever). Interestingly, one can always obtain the required mass of an unweighted lever by simply dividing the total mass of the lever's end weight(s) by 4.8.

The third and last column of the group of three columns, labeled "weighted lever mass (pounds)" provides the mass, in pounds (and also ounces for the Gera prototype wheel), of a *complete* weighted lever in each of Bessler wheels. The masses shown represent the *sum* of the mass of a lever's end weight(s), which includes the mass of their attachment screws or bolt(s), and the mass of the unweighted lever to which the end weight(s) and their attachment screws or bolt(s) were attached. Interestingly, one can always obtain the mass of the complete weighted lever by multiplying the mass of its unweighted lever by 5.8.

The final group of three vertical columns on the right side of the Table 2 continuation, collectively labeled "Spring values", provides important parameter values for *each* of the two extension springs attached to each weighted lever within one of Bessler's wheels.

The first of the three columns is labeled "minimum spring length (inches)" and, as the label implies, it lists the minimum length, in inches, that an *unstretched* extension spring must have in order to *not* experience permanent damage because it has been stretched beyond its elastic limit when the pivot pin center axis of the weighted lever it is attached to reaches the 9:00 position of a one-directional wheel's clockwise rotating drum and the spring reaches it maximum stretched length.

At the bottom of the column, I provide a minimum unstretched length of 14 inches for a single spring or a single group of parallel springs that would have to be attached to the large iron lever's A1 spring cord hook attachment pin in either coordinating cord layer 1 or layer 5 of Bessler's proposed 40 foot diameter, one-directional super wheel. However, as mentioned in Chapter 1, the super wheel, if it had ever been constructed, probably would have used two flat, metal bar type steel springs instead of the two helical coil extension springs found in Bessler's other wheels. Such a metal bar type spring will also have an elastic limit which limits how far an exposed unfixed end of the bar can flex before the bar is damaged and that limit will depend upon the physical dimensions of the particular bar used and the metal alloy from which it is made. The rarest of craftsmen who may wish to construct the ultimate wheel that Bessler envisioned but never built, will have to do much experimentation with such metal bar type springs to determine their elastic flexing limits in order to make sure that they are not exceeded as the pivot pin's center axis of a super wheel's weighted lever passes through its drum's 9:00 position and the two metal bar type springs attached to the 1160 pound weighted lever are flexed to their maximum extent.

The second of the three columns in the group, labeled "maximum stretch at 9:00 (inches)", gives the maximum allowed distance which an unstretched extension spring will be stretched through as the pivot pin center axis of the weighted lever that the spring is attached to reaches the 9:00 position of the drum of a clockwise rotating, one-directional wheel. As can be seen, Bessler designed his 3 feet diameter Gera prototype wheel so that each of its unstretched minimum 1 inch long extension springs would stretch through a maximum distance of exactly 0.5 inch as its weighed lever's pivot pin's center axis reached the 9:00 drum position of a clockwise rotating, one-directional wheel. These maximum stretch distances then increase in proportion to the increase in diameter of a wheel so that, as one proceeds from the 3 feet diameter Gera prototype to the four times larger 12 feet diameter, bidirectional Merseburg and Kassel wheels, the maximum stretch distance of an extension spring increases from 0.5 inch to 2 inches since 4 x 0.5 inch = 2 inches.

The bottommost value of the second column provides the maximum stretching distance of a single extension spring or single group of such springs or the flexing distance of single flat bar type spring of the two springs that would be used on each of the super wheel's massive weighted levers. Since the super wheel's 40 feet diameter drum was 13.333 times that of the 3 feet diameter Gera prototype wheel, the free end of the super wheel's flat, bar type spring that was attached to a chain that connected it to a weighted lever's A1 spring cord (now chain!) hook attachment pin would flex or bend through a maximum distance equal to 6.67 inches when the weighted lever's pivot pin's center axis was located at the drum's 9:00 position. This distance is gotten by simply multiply the maximum stretch distance of an extension spring in the Gera prototype wheel by 13.333 to get 13.333 x 0.5 inch = 6.67 inches.

This now finally brings us to the third column of the group of three columns collectively labeled "Spring values" in the continuation of Table 2, which is labeled "spring constant (pounds/inch)". This last column provides the spring constant, in pounds per inch of stretch (and ounces per inch of stretch for the Gera prototype wheel), for a single one of the two extension springs that were attached, via two spring cords, to the A1 spring cord hook attachment pins of the weighted levers inside of all of Bessler's wheels.

The 3 feet diameter Gera prototype wheel's extension springs were the smallest and weakest with each having a spring constant value of only 1.25 pounds per inch = 20 ounces per inch. Then, for the next two wheels Bessler constructed, the second Gera and Draschwitz wheels, he used springs whose spring constants were, respectively, 2 and 8 times greater than the value used in the Gera prototype wheel. When the inventor constructed his most powerful 12 feet diameter, bidirectional Merseburg and Kassel wheels, he used extension springs whose spring constants were, respectively, 8 and 16 times greater than those of the springs used in the Gera prototype wheel. For the two "last chance for a sale" 6 feet diameter, one-directional wheels constructed at his home in Karlshafen after the destruction of the Kassel wheel, Bessler was forced to use springs with a spring constant value of only 5 pounds per inch. For his 40 foot diameter super wheel, if it had ever been constructed and used only two extension springs on each of its weighted levers, then each spring would have required a spring constant of 240 pounds per inch. If, however, each weighted lever was perfectly balanced when its pivot pin's center axis was located at the drum's 9:00 position by two flat, bar type springs, then this stretch distance parameter would become a flex distance parameter for the unfastened end of each bar type spring to which a spring cord (or chain!) was attached by a large metal hook.

This now concludes this volume's treatment of the subject of Bessler's wheels and their previously secret mechanics. With the various columns containing thirty-two external and internal structural parameter values provided in Tables 1 and 2 for all of Bessler's constructed and conceived wheels, the serious craftsman should be well prepared to contemplate undertaking the duplication of one of the inventor's wheels. Once again, however, I would strongly advise anyone interested in undertaking such a duplication to first attempt the construction of his 3 feet diameter, one-directional Gera prototype "toy" wheel. Its construction can serve to teach one the basics of the mechanics employed in his larger and more powerful wheels and will aid in the later replication of one should that be undertaken. I must also remind the craftsman that, while I give values in the columns to three decimal places, such precision is not required to produce a functioning Bessler wheel. However, the closer one can get to the parameter values listed, the better his wheel will function.

With the long awaited disclosure of the internal details of the imbalanced perpetual motion wheel mechanics Bessler found and used, it should not be long before someone manages to produce a working physical duplicate of one of them. That will be big news in the world of science and should help vindicate Bessler and gain him his rightful place in the history of science and technology even though such recognition will be about three centuries belated. He will be viewed as an honest individual who really did invent working perpetual motion machines even though, in the absolute sense of the term, they were not really capable of running eternally. And, hopefully, the interest produced by successful duplication will help to stimulate further and more intense research into the subject of self-motive machinery.

Table 1 - External Parameter Values for All of Bessler's Wheels

Where built	Drum, axle, and mass values					Wheel performance values			Important years	
	Drum Diameter (feet)	Drum Thickness (inches)	Axle Diameter (inches)	Axle Length (feet)	Drum & Axle Mass (pounds)	Maximum Speed (rpm)	Maximum Power (watts)*	Wheel Direction Capability	Year Built	Year Destroyed
Gera (prototype)	3	4	1.5	1	6.1	V 60	0.2	one	1711	1727
Gera (second)	4.5	4	4	2.5	30	60	0.9	one	1712	1713
Drasch-witz	9	6	6	2.5	120	V 50 ^ 60	14	one	1713	1714
Merse-burg	12	12	6	6	550	40	25	two	1715	1716
Kassel	12	18	8	6	1100	26	50	two	1717	1721
Karlshafen (2 built)	6	5	5	2.5	75	50	3.1	one	1720's and 1740's	1720's and 1740's
Kassel (not built)	40	IV 40	28	10	15,000	^ 15	6,667	one	planned 1717 not built	Not Applicable

*All wheel maximum start up power values are extrapolated values based on the Kassel wheel having an estimated maximum start up power of 50 watts.

Table 2 - Internal Parameter Values for All of Bessler's Wheels

Where built	Important pin and post location values within a drum				Lever length	Coordinating cord hook attachment pin location values within a weighted lever					
	Axle to Pivot Pin (inches)	Axle to OOAP/IOAP (inches)	Axle to Outer Cord Guide Post (inches)	Axle to Inner Cord Guide Post (inches)	Pivot Pin to End Wt CoG at L2 (inches)	Pivot Pin to L1 Pin (inches)	Pivot Pin to A1 Pin (inches)	Pivot Pin to A2 Pin (inches)	Pivot Pin to A3 Pin (inches)	Pivot Pin to A4 Pin (inches)	Pivot Pin to B1 Pin (inches)
Gera (prototype)	14	16.5 / 12.934	10	9	3.5	2.6	0.5	0.933	1.4	1.867	2.1
Gera (second)	21	24.75 / 19.401	15	13.5	5.25	3.9	0.75	1.4	2.1	2.8	3.15
Draschwitz	42	49.5 / 38.802	30	27	10.5	7.8	1.5	2.8	4.2	5.6	6.3
Merseburg	56	66 / 51.736	40	36	14	10.4	2	3.73	5.6	7.47	8.4
Kassel	56	66 / 51.736	40	36	14	10.4	2	3.73	5.6	7.47	8.4
Karlshafen (2 built)	28	33 / 25.868	20	18	7	5.2	1	1.87	2.8	3.73	4.2
Kassel (not built)	186.67 (15.56 ft)	220 / 172.453 (18.33 ft / 14.368 ft)	133.33 (11.11 ft)	120 (10 ft)	46.67 (3.89 ft)	34.67 (2.89 ft)	6.67 (0.56 ft)	12.44 (1.04 ft)	18.67 (1.56 ft)	24.89 (2.07 ft)	28 (2.33 ft)

Table 2 - continued

Where built	Coordinating cord values				Lever mass values			Spring values	
	main cord A4 to L1 (inches)	long lifter cord A3 to B1 (inches)	stop cord A2 to I0AP (inches)	spring + cord A1 to O0AP (inches)	end weight mass (pounds)	unweighted lever mass* (pounds)	weighted lever mass (pounds)	minimum spring length (inches)	maximum stretch at 9:00 (inches)	spring constant (pounds / inch)
Gera (prototype)	11.3	19.9	6	6	0.375 (6 oz)	0.078125 (1.25 oz)	0.453125 (7.25 oz)	1	0.5	1.25 (20 oz/in)
Gera (second)	16.95	29.85	9	9	1.125 one ingot	0.234	1.359	1.5	0.75	2.5
Drasch-witz	33.9	59.7	18	18	9 one ingot	1.875	10.875	3	1.5	10
Merse-burg	45.2	79.6	24	24	12 3 cylinders	2.5	14.5	4	2	10
Kassel	45.2	79.6	24	24	24 3 cylinders	5	29	4	2	20
Karlshafen (2 built)	22.6	39.8	12	12	3 one cylinder	0.625	3.625	2	1	5
Kassel (not built)	150.67 (12.55 ft)	265.33 (22.11 ft)	80 (6.67 ft)	80 (6.67 ft)	960 nine+ discs	200	1160	14	6.67	240

* The unweighted lever mass should include the full masses of any coordinating cord metal attachment hooks and half the masses of any coordinating cords and unfixed springs attached to the spring cord hooks.

EPILOGUE

The Future of Perpetual Motion Devices

By the conclusion of this volume, the reader, like the author, will hopefully have reached the conclusion that an early 18th century inventor, one Johann Ernst Elias Bessler, actually did manage to construct several working imbalanced perpetual motion wheels which, due to a combination of paranoia and bad luck, were never sold, but whose previously secret mechanics have now been rediscovered. Of course, the wheels he constructed were not truly "perpetual" in the absolute sense of the word. That's because, being powered by the energy content of the masses of their active components, which were just carefully coordinated weighted levers, they could only output mechanical energy nonstop for a *finite* amount of time because their weighted levers only contained a finite amount of mass. However, depending upon whether a wheel was outputting energy at its maximum rate during wheel start up (about 50 watts for the 12 feet diameter, bidirectional Kassel wheel) or at the minimum rate required to overcome only its own various aerodynamic drags and bearing frictions when the wheel was running freely with no outside machinery attached to its axle, that energy content of the weighted levers was enough to allow the wheel to continuously output mechanical energy from tens of millions to tens of *billions* of years!

Certainly, even at its maximum rate of energy output after start up, such an enormous performance duration would be far more than sufficient for an average person to declare the inventor's wheels as being "perpetual" as far as their functioning was concerned. However, these durations would not be enough for a strict purist to accept them as being perpetual. For the purist, however, a wheel must, in both principle *and* practice, be able to continuously output energy for all of eternity. In order to work, therefore, the purist's perpetual motion wheel would actually have to somehow manufacture energy out of nothing and, as far as current 21st century physics is concerned, that is a physical impossibility and, most likely, will remain so for the foreseeable future.

One of the unfortunate aspects of Bessler's wheels was that, as energy sources, they all had a very low "power to mass ratio" which is the power output of a source divided by its mass. A modern spacecraft grade photovoltaic panel might have a power to mass ratio of about 35 watts per pound. By comparison, the Kassel wheel outputted about 50 watts, but weighted about 1100 pounds which gave it a power to mass ratio of about (50 watts / 1100 pounds) = 0.045 watts per pound. So, we see that Bessler's Kassel wheel, at its maximum start up power output, only had a power to weight ratio that was 0.0013 times or 0.13% that of a photovoltaic cell used on such things as satellites and space probes.

That power output, however, was only produced by the active one of the bidirectional wheel's two internal one-directional wheels. If we calculate the power to mass ratio using only that one wheel, we can subtract the 232 pounds of mass of the inactive internal one-directional wheel (gotten from the mass of eight inactive weighted levers weighing 29 pounds each) and write (50 watts / (1100 pounds − 232 pounds)) = (50 watts / 868 pounds) = 0.058 watts per pound. That improves the situation somewhat and now the increased power to mass ratio is 0.0017 times or 0.17% that of a photovoltaic cell. But, even so, the power to mass ratio of the Kassel wheel is less than 1% that of the photovoltaic cell and thus miniscule compared to it.

From this, it would seem doubtful that Bessler's wheels could be a useful source of energy for the modern world even though, neglecting construction and maintenance costs, the energy obtained from them was "free". Some, of course, may object and say that perhaps, once his wheels are duplicated and again turning and being studied minutely by physicists and engineers, this situation could change as their power to mass ratios were dramatically increased so as to compete with a photovoltaic panel. Again, while I cannot absolutely exclude that possibility, I suspect, from my own research with his imbalanced pm wheel mechanics, that this will be a most difficult task although, of course, I wish those attempting to modify his mechanics for far greater power output the best of luck with their efforts.

For example, in the Kassel wheel, its maximum start up power output of 50 watts was all due to the center of gravity of its active, internal one-directional wheel's eight 29 pound weighted levers with a total mass of 232 pounds being horizontally offset from the center axis of the axle by only about 0.25 inch! If some advanced redesign of the shape of this wheel's weighted levers and the coordinating cords that interconnected them could move the levers' center of gravity out to a horizontal displacement of 10 inches away from the wheel's axle, then the start up power output of the wheel would be made (10 inches / 0.25 inch) = 40 times greater and would then climb to 40 x 50 watts = 2000 watts. If both of the two internal, one-directional wheels inside of the Kassel wheel's enlarged drum were then arranged so that their spontaneous start up directions of rotation were the *same* and they both were outputting mechanical energy as the drum rotated, the power output of the redesigned Kassel wheel would be doubled to 4,000 watts.

That outputted mechanical energy, of course, would be used to drive a small 120 or 240 VAC electrical generator that would convert it into electrical power. This level of power output begins to approach what would be needed to power a small home. Any outputted power not being used by the home could be stored in lithium ion batteries for times when the power demand of the home exceeded the wheel's maximum power output of 4,000 watts. When the batteries were fully charged, any excess power generated could then be sent out into a nation's electrical power distribution system or "grid" and thereby sold to a homeowner's local electrical utility company. The money earned would be used for the maintenance of the improved, now one-directional Kassel wheel and the electrical generator to which it was connected.

I've often wondered where the world would be right now if Peter the Great had not died on his way to visit Bessler in 1725 and had the opportunity to view the first 6 feet diameter, one-directional "last chance for a sale" wheel the inventor had constructed at his home in Karlshafen, Saxony after the destruction of the bidirectional Kassel wheel at Weissenstein Castle in Kassel, Saxony on August 17th, 1721. The czar certainly had the money to purchase the wheel and, most likely, would have done so after seeing Bessler do a few impressive demonstrations with it and then being told by the inventor that it was just the thing Peter needed to prove to Europe that he was serious about modernizing Russia and bringing it educationally, scientifically, and technologically up to par with his European neighbors whom he envied.

Once the 6 feet diameter, one-directional wheel was back in Russia, Peter's court engineers would have immediately disassembled it and, using its parts as guides, began manufacturing working duplicates of it. Next, they would have increased the size, mass, and start up power outputs of their replicas. That would have been followed by decades of effort by their best mathematicians to improve the design and move the center of gravity of its active weighted levers as *far away*, horizontally, from an imaginary *vertical* line passing through the center axis of a wheel's axle as possible while also trying to move that center of gravity as *close*, vertically, to an imaginary *horizontal* line passing through the center axis of a wheel's axle as possible. The more those goals could be achieved, the greater the start up torque and power output of a wheel would be.

All of this would have begun in the early 18th century and today, nearly three centuries later, we might have the greatly improved versions of Bessler's original wheels in widespread use around our planet. Perhaps the Industrial Revolution would even have completely bypassed the use of fossil fuels and we would, today, have a much cleaner planet without a trace of Climate Change to worry about!

However, as nice as these intriguing scenarios sound, they are only hypothetical in nature. Right now, the reality is that the first duplicates of Bessler's imbalanced pm wheels will certainly amaze everyone, but that amazement may wear off rather quickly when they realize one would need a mass reduced, about 868 pound version of the Kassel wheel that was 12 feet diameter just to illuminate a *single* 50 watt incandescent light bulb! People, in general, will quite rightly question the practical usefulness of such a wheel just as they did with all of Bessler's constructed wheels during his demonstrations of them.

So, what is the real importance of Bessler's wheels?

The importance is that they graphically indicate that it is physically possible to construct a machine that will continuously output the energy associated with the masses of its active or moving parts until a purposeful effort is made to prevent them from doing so. Such a machine does not rely upon an external energy supply or upon a conventional internal one. It can, in principle, also run continuously for many millions of years depending upon the rate at which it outputs energy.

But, as we saw, the power output of Bessler's wheels was very small considering how massive they were. This, of course, was due to them relying upon the Earth's gravity field to serve as a catalyst for the extraction of their parts' energy content. Gravity is known to be the weakest of the four forces of nature which are the gravitational, electromagnetic, and strong and weak nuclear forces. One then wonders if any of the other forces other than gravity could be employed to drain the energy from a wheel's moving internal components.

I can see no way to employ the strong and weak forces that bind together the various particles or nucleons found within the nuclei of atoms. There may, indeed, come a time in the future when this is possible and we will be able to, on demand, directly tap the nuclear energy of even stable atomic nuclei in devices that are compact and light weight and do so with none of the radiological hazards associated with the kinds of nuclear reactions that take place inside either weapons or power plants. That time, however, is not the present.

That then leaves the remaining force of nature to try using in our "perpetual" motion devices: electromagnetism. This essentially means that it might be possible to construct a continuously running rotary device that would then use either electric or magnetic field to enable the extraction of the energy associated with the mass of that device's active parts. Of these, the magnetic field, particularly that of permanent magnets, would seem to be the most convenient one to use. Modern rare earth alloy magnets can be made with far stronger magnetic fields than found in the common steel or Alnico (an acronym derived from its component elements of aluminum, nickel, and cobalt) magnets of a few decades ago. As a consequence we can expect the torques and power outputs of devices using these more powerful magnets to be far greater than those of Bessler's wheels. While the Kassel wheel had to be 12 feet in diameter and use eight active weighted levers with a total mass of 232 pounds in order to output a mere 50 watts, a permanent magnet motor, by comparison, might be able to output 1000's of watts and do so with a mass of less than 100 pounds.

What might such a permanent magnet motor look like?

I have provided a design for a relatively simple and easy to construct permanent magnet motor in Figure 31 at the end of this epilog that might be able to run continuously once released. Obviously, it does not use the same system of levers and coordinating cords as did Bessler's wheels, but it does have the same inherent asymmetry of torques constantly present that also allowed Bessler's imbalanced pm wheels to do the "impossible". The reader should locate and study Figure 31 as he reads the following description of the device and its method of operation.

In this figure we see a central rotor disc that is attached to an axle that would be bearing mounted and free to rotate in either a clockwise or counterclockwise direction (note that the bearings are not shown in the figure).

The rotor material is nonmagnetic and made of some plastic material like polystyrene or nylon and, for a quickly fabricated prototype, even wood may be used. The axle is steel.

Securely attached to the outer rim of the rotor disc with a strong adhesive are six disc magnets which can be either Alnico or ceramic. However, for maximum effect, one might consider using the far more powerful neodymium or rare earth alloy type permanent magnets although they will have to be attached to the rim of the rotor by nonmagnetic brass screws passing through holes in their centers or by having the entire body of the disc magnet tightly pressed into holes bored into the rim wall that act as nests for them. All of the disc magnets attached to the rotor disc are oriented so that their north magnetic poles (indicated by the solid black half of their rectangular side views shown in Figure 31) face outwards away from the rotor's attached axle and their south magnetic poles face inwards toward the rotor axle.

The rotor with its six rim disc magnets is situated so that its axle's center axis is located along the center axis of a cylindrical hole in a block of material which serves as a stator body and remains stationary while the rotor and its axle can rotate inside of it. As in the case of the rotor, the stator material itself is nonmagnetic and made of some plastic material or even of wood. Around the inside curving surface of this stator's cylindrical hole are located eight disc magnets. Again, depending upon how powerful the eight stator magnets are, they can be either glued to or pressed into holes formed in the curving cylindrical wall of the stator. Unlike the rotor magnets, all of the north magnetic poles of the stator magnets point inward toward the rotor's steel axle.

Because the north poles of all eight stator magnets are close to the north poles of the six rotor magnets, there will be strong repulsive magnetic forces acting between the stator and the rotor. If there were only the fourteen disc magnets shown in the design of Figure 31, then the magnetic repulsions acting between them would quickly cause the rotor to find a position in which the torques tending to make it rotate clockwise were exactly balanced by the torques tending to make it rotate counterclockwise and the rotor would then "fall" into this orientation and not be able to leave it unless an outside torque was applied to either the rotor or its attached axle in order to help it rotate out of this lowest magnetic potential energy configuration and then fall over into the next one that was available. When the rotor did finally fall over into that next lowest magnetic potential energy configuration, it would oscillate rapidly until it finally dissipated all of the extra magnetic potential energy given to it by an outside source, like the inventor's hand, by causing heating of the surrounding air and the axle bearings.

Inventors working with such a design would say that there were multiple "sticking points" produced by the stator magnets that were causing the rotor magnets to be forced into certain orientations about the rotor axle. However, the design shown in Figure 31 has an additional component which, hopefully, will prevent this from happening and which may actually allow such an arrangement of stator and rotor magnets to produce a perpetually rotating permanent magnet motor that could continuously output mechanical energy.

The reader will notice that there is something that has been added to the side of each of the rotor and stator magnets. Each of the six rotor disc magnets and eight stator magnets has a small circle on one side of it that represents a steel ball bearing that has been glued to the side of the magnet and the nearest curving surface of the rotor or stator. For disc magnets that are embedded in nests in the rotor's rim or the stator's inner cylindrical wall, each would have a ball bearing embedded next to it in its own separate nest.

The ball bearings serve the purpose of distorting or warping the shapes of the magnetic fields as they emerge from the north and south pole faces of a rotor or stator disc magnet. Although I have not drawn in the individual magnetic fields (represented by the convention of curves with arrowheads on them that indicate the "direction" of a magnetic field line at the location of the arrowhead) that surround the rotor and stator magnets since that would make Figure 31 excessively cluttered, the reader can imagine them as a bundle of curving lines that emerge from the north pole faces of each disc shaped magnet, loop around on all sides of it, and then reenter the south pole face of each magnet. *Inside* of each magnet, the magnetic field lines can be considered as continuing from the south pole face back to the north pole face again.

Ordinarily, the shape of each disc magnet's external magnetic field is perfectly cylindrically symmetrical around the circumference of the disc. It is this symmetry which is actually responsible for the "sticking points" mentioned above that cause the rotor of the device to fall into a low magnetic potential energy "well" provided by the stator magnets from which the rotor will not emerge unless provided with external energy. Adding a steel ball bearing to the side of each of the permanent magnets, however, severely distorts the shapes of these symmetrical magnetic fields and replaces them with asymmetric or "lopsided" magnetic fields.

Once the steel ball bearings are added to the fourteen disc magnets of the rotor and stator, the individual magnetic field of each permanent magnet will tend to bunch up on the side of the stator where the ball bearing is located. This happens because it is easier for the magnetic field lines to pass though the steel of a ball bearing than through the air surrounding the rest of the circumference of the disc magnet. In fact, we can imagine that most of a disc magnet's external magnetic field will actually be leaving the magnet's north pole face and curving over toward and then entering the nearest surface of the magnet's nearby spherical ball bearing. It will travel through the diameter of the ball bearing, emerge from its opposite side, and then curve about and reenter the south pole of the disc magnet whereupon it will continue through the thickness of the magnet to again reemerge from its north pole.

Now consider the effect that the distortions of the external magnetic fields of a rotor and stator magnet will have as the rotor and its attached axle rotates so as to reduce the distance between a rotor and stator magnet.

If there were no ball bearings present, then, as a clockwise rotating rotor's rim magnet approached one of the stator magnets, the repulsive force between the two magnets would rise to a maximum at their closest approach. As the rotor continued to turn and the two magnets began to separate, the repulsive force acting between them would steadily decrease. If one was to graph the strength of the repulsive magnetic force versus the position of the rotor magnet, he would find that the curving graph line would form a perfectly symmetrical peak about the position of the rotor that placed the magnet as close to the stator magnet as possible.

When, however, the steel ball bearings are present on the *opposite* sides of approaching rotor and stator magnets, the situation changes dramatically. Because the ball bearings are distorting the three dimensional shapes of the magnetic fields of the rotor and stator magnets, the shape of the graph of the repulsive force acting between the magnets as they come closest together and then separate versus the position of the rotor will not be symmetrical, but, rather, asymmetrical or lopsided just as is the shape of the magnetic fields of each magnet. This asymmetry is indicated in the graph drawn below the permanent magnet motor design shown in Figure 31 which graphs the magnitude of the repulsive force acting between a single pair of rotor and stator magnets as they approach each other and then separate during a 60° segment of rotor rotation.

As can be seen in that graph, as a rotor and stator magnet approach each other during the *first* 30° of a 60° segment of clockwise rotor rotation, the repulsive force, indicated in the figure above by the two arrowheads at the ends of the *thinnest* line segment which are pointing away from each other, begins to increase exponentially *after* 7.5° of rotor rotation and that portion of the graph is labeled "A". At "B" the rotor has turned exactly 30° clockwise through the 60° segment of rotor rotation and the repulsive force between the north poles of the two magnets is at a maximum, indicated by the two arrowheads at the ends of the *thickest* line segment which are pointing away from each other. This is because this position of the rotor causes the a stator and rotor magnet to be as physically close to each other as the design will allow them to be. During the next 22.5° of clockwise rotor rotation, the repulsive force acting between the two magnets, indicated by the two arrowheads at the ends of the *medium* thick line segment which are pointing away from each other, steadily decreases and this portion of the curve is labeled "C". It will be noted that, while the increase of repulsive magnetic force during the "A" portion of rotor rotation is exponential, the decrease in this force is *logarithmic* during the "C" portion of the rotor's rotation.

In essence, whenever a rotor and stator magnet approach each other, their lopsided magnetic fields will cause the weaker portions of their external magnetic fields to repel each other. However, when the magnets separate again, it is the *stronger* parts of their lopsided magnetic fields which repel each other. When the two magnets are closest to each other, the repulsive force acting between them is strong, but it is directed toward the center axis of the rotor and, therefore, produces no torque on the rotor. It is only when the two magnets are approaching or parting that the magnetic repulsive forces between them can produce a torque on the rotor.

Because of this difference in the intensity of the repulsive force acting between a rotor and stator magnet that are approaching each other as compared to the repulsive force acting between them as they are separating from each other, this means that *less* energy is needed to bring the two magnets together than is released by them when they separate! Thus, there is a small net output of mechanical energy from the two magnets as the rotor turns through 60°. Consequently, for the clockwise motion of the rotor shown in Figure 31, there will be a net clockwise torque produced by the interactions of all of the permanent disc magnets shown which will cause the rotor to accelerate or run at a constant speed while performing outside work.

Where does this outputted energy come from?

The answer is that, just as in the case of the weighted levers found in Bessler's imbalanced perpetual motion wheels, the energy being outputted by this permanent magnetic motor would be derived from the masses of the permanent magnets themselves and, as the device is allowed to run and continuously outputs energy, the mass of its fourteen disc shaped magnets will constantly decrease. As in the case of Bessler's wheels, the magnetic rotor shown might be able, with regular maintenance, to run for millions of years before all of the mass energy of its magnets was completely exhausted and the magnets then became completely massless.

In Figure 31 it will be noted that, at any moment of rotor rotation, two of the rotor magnets are on the "A" portion of their repulsive force versus rotor position curves and experiencing repulsive forces from their nearest stator magnets that are trying to turn the rotor in a counterclockwise direction, two of the rotor magnets are very near or at the "B" peak of their curves and experiencing a maximum repulsive force on them from their nearest stator magnets that is trying to push them toward the rotor's axle and, thus, apply little or no torque to the rotor, and two of the rotor magnets are on the "C" portion of their curves and experiencing repulsive forces that are trying to turn the rotor they are attached to in a clockwise direction. When all of the magnitudes of all of these clockwise and counterclockwise torques acting on the rotor and its attached axle are added up, however, it will be seen that there is a net clockwise torque acting on them. Thus, the rotor should always be experiencing a clockwise torque as it rotates although there will be some small changes in its magnitude during rotation.

To even out any rapid periodic variations in the clockwise axle torque of this permanent magnet motor, anyone experimenting with this design can consider mounting several separate rotors, each with six disc magnets attached to its outer rim, on a common axle. These extra rotors, however, must not have all of their magnets in perfect alignment with each other, but, rather, be mounted on their shared axle so that the magnets of the different rotors are in a staggered arrangement with respect to each other.

Thus, the second rotor behind the front one might have its six disc magnets all 10° out of phase with the first rotor's six magnets. The third rotor would be mounted so that its six magnets were all 20° out of phase with the first rotor's six magnets. The fourth rotor would then have its six magnets all 30° out of phase with the first rotor's six magnets. The fifth rotor would have its six magnets all 40° out of phase with the first rotor's six magnets. And, finally, the last and sixth rotor's six magnets would all be 50° out of phase with the first rotor's six magnets. This staggered pattern can be repeated for as many rotors as the builder wishes to add to the axle in order to both increase and smooth out the torque acting on it. This staggered pattern, however, is *not* used for the extra stator magnets that will need to be attached to the inside wall of the cylindrical hole in the block of nonmagnetic material that forms the body of the stator. For each extra rotor added to the motor's axle, another set of eight disc magnets must be attached to the curving inner wall of the stator's cylindrical hole so that the plane defined by the centers of those eight stator magnets will also contain the centers of the added rotor's six magnets. If this is not done, then the rotor axle will not experience maximum torque acting on it.

One should also bear in mind that as more and more magnets are added to the design, its gross weight will increase. This is an unavoidable consequence of using permanent magnets made from metallic elements. The use of ceramic magnets can keep weight down, but at the expense of less axle torque being produced by a given number of rotors.

By using powerful neodymium permanent magnets for the rotor and stator magnets and constructing a device with several dozen rotors, it may be possible to make a motor small enough to fit into the engine compartment of an ultra-light automobile (that's one with a gross weight of 500 pounds or less without its passengers!) and which would have sufficient power output to give the vehicle a robust acceleration. For this, I estimate that, at a minimum, a permanent magnet motor outputting about 50 horsepower would be required. This is equivalent to a power output of about 37,285 watts. I believe that a permanent magnet motor of the type I describe above could produce this level of power output and it could be made small enough so that its total weight was less than 100 pounds or less than 20% of the gross weight of an ultra-light automobile.

The production of such vehicles could help in our generation's effort to combat the ever worsening effects of Climate Change and the potential damage they can do to our planet's biosphere. Aside from drastically reducing the use of fossil fuels and the carbon dioxide green house gases they emit, such vehicles would only need to occasionally visit a service station for periodic maintenance or necessary repairs. Using some sort of magnetic braking system could eliminate the need for hydraulic brake part replacements. There would be no starting, ignition, lubrication, cooling, or emission control systems to eventually fail. No fans, radiators, hoses, timing chains, fuel injectors, ignition wiring, spark plugs, fuel tanks, mufflers, or exhaust pipes in need of periodic and expensive replacement.

While the magnetic motor was directly mechanically used to drive the car's rear wheels, only a single drive belt from it to an alternator would be needed to power the entire car's electrical system with a previously charged lithium battery providing backup power when the alternator's output was not sufficient to meet the electrical power demands of the vehicle. All exterior and interior lights would be high efficiency LED lamps that might never need replacing. With a tubular frame aluminum or magnesium chassis and a carbon fiber reinforced plastic body, such a car would be virtually immune to corrosion. Its tires would be solid and made from some high tech elastic material that would be deflation proof and would carry a vehicle many times more miles than today's tires can before they were replaced and their worn away remains recycled to make fresh tires for other drivers' cars.

Like a conventional internal combustion engine, the rotary permanent magnet motor that powers this hypothetical free energy car of the future must be such that its power output can be quickly varied from none to maximum. This control could be easily achieved by using eight long nonmagnetic plastic control rods that would be able to slide along the length of the motor's stator structure and parallel to its rotor's axle. Each of these control rods would have a number of steel ball bearings embedded in it along its length whose spacing would correspond to the spacing between the magnet carrying rotors attached to the device's axle.

A simple cable connection between an accelerator pedal in the passenger compartment and a plate attached to the ends of all of the rods on one side of the motor would allow the driver, by applying foot pressure to the accelerator pedal, to move all of the control rods at once so that all of their ball bearings were slid from inactive positions between the rotor discs to positions that then placed them next to their assigned stator magnets so that they could distort their external magnetic fields. As this alignment was slowly attained, the torque produced in the rotor's axle would increase and, at some point, the car would begin to move and pick up speed. Continuing to depress the accelerator would rapidly accelerate the ultra-light vehicle up to its top speed which could be limited to 60 miles per hour so as not to exceed the crash survivability protection provided by its various automatically deployed airbags and the lightweight tubular chassis and body of the vehicle.

By providing a simple reverse gear manual transmission, it would be possible to reverse the torque of the permanent magnet motor by simply moving a lever from its "drive" to "reverse" positions. With a direct drive propulsion system like this, there is no need to attach the motor to a heavy automatic transmission before using it to power the car's rear wheels. When the car's magnetic braking system was used, it would not only rapidly slow and then stop the wheels from turning, but it would also do the same to the vehicle's entire drive train including the permanent magnet motor's rotor. This would do no damage to the motor and, upon releasing the brakes, it would, depending upon the pressure applied to the accelerator pedal which then adjusted the positions of all of its stator magnets' magnetic field distorting ball bearings, immediately begin accelerating the car again at the desire rate and in either a forward or reverse direction. Thus, such a vehicle would require no manual clutch or fluid torque converter as found in automatic transmissions.

Despite the pessimism of the today's scientific orthodoxy, based on my research into the mystery of Bessler's imbalanced wheel mechanics, I am confident that the future of self-motive machinery is very bright, indeed. I have no doubt that, in the coming decades, we will finally see commercially viable free energy devices come into widespread use in our homes and vehicles. Their use will be one of the many approaches used to undo the damage to our planet's atmosphere and biosphere caused by an over reliance on fossil fuels in the past and still, unfortunately, in the present.

Figure 31 - A Permanent Magnet Motor Design Using Distorted Magnetic Fields

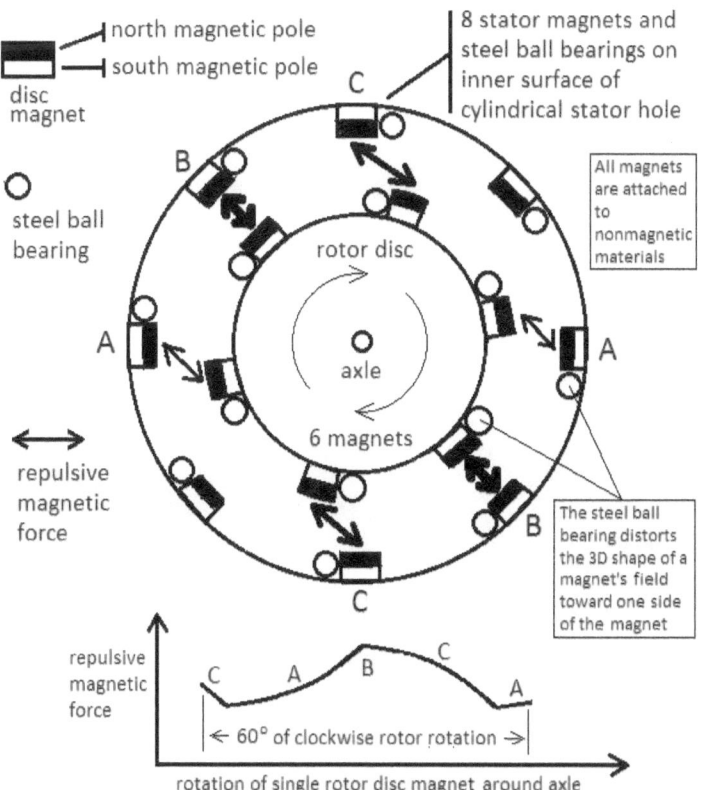

During each 60° of rotor rotation, two of the rotor magnets are approaching an A stator magnet and experiencing increasing repulsion that produces counterclockwise torque, two rotor magnets are nearest B stator magnets and experiencing maximum repulsion toward the rotor axle that produces no rotor torque, and two rotor magnets are separating from a C stator magnet and experiencing decreasing repulsion that produces clockwise torque. The clockwise torques due to magnetic repulsion acting on the rotor are always stronger than the counterclockwise torques to produce a net clockwise torque that drives the rotor and its attached axle.

ABOUT THE AUTHOR

Kenneth W. Behrendt has been a lifelong student of phenomena in the fields of ufology and the paranormal. Although professionally trained as a chemist, he has been investigating and writing about the UFO phenomenon since the early 1980s. He has had several personal sightings of UFOs during his lifetime that have convinced him in the reality of these objects and considers their study to be of great importance to humanity. He has also maintained a lifetime interest in the history of so-called "perpetual motion machines" and, in particular, the fabulous self-moving wheels of Johann Bessler. The author currently resides in suburban New Jersey, where he continues his researches in the areas of ufology, paranormal phenomena, and free energy physics.

www.ingramcontent.com/pod-product-compliance
Lightning Source LLC
Chambersburg PA
CBHW020717180526
45163CB00001B/7